THE SCIENCE OF
ENTOMOLOGY

A personal library is a lifelong source of enrichment and distinction. Consider this book an investment in your future and add it to your personal library.

3rd Edition

THE SCIENCE OF ENTOMOLOGY

William S. Romoser
Ohio University

John G. Stoffolano, Jr.
University of Massachusetts

Wm. C. Brown Publishers
Dubuque, Iowa • Melbourne, Australia • Oxford, England

Book Team

Editor *Kevin Kane*
Developmental Editor *Jim Dagget*
Production Editor *Michelle M. Campbell*
Designer *Kristyn A. Kalnes*
Art Editor *Kathleen Huinker*
Photo Editor *Shirley Lanners*
Permissions Editor *Vicki Krug*

Wm. C. Brown Publishers
A Division of Wm. C. Brown Communications, Inc.

Vice President and General Manager *Beverly Kolz*
Director of Sales and Marketing *John W. Calhoun*
Marketing Manager *Carol J. Mills*
Advertising Manager *Amy Schmitz*
Director of Production *Colleen A. Yonda*
Manager of Visuals and Design *Faye M. Schilling*
Design Manager *Jac Tilton*
Art Manager *Janice Roerig*
Publishing Services Manager *Karen J. Slaght*
Permissions/Records Manager *Connie Allendorf*

Wm. C. Brown Communications, Inc.

President and Chief Executive Officer *G. Franklin Lewis*
Corporate Vice President, President of WCB Manufacturing *Roger Meyer*
Vice President and Chief Financial Officer *Robert Chesterman*

Cover photographs: top front and center back © John Gerlach/Tom Stack & Associates; bottom front © Milton Rand/Tom Stack & Associates; background © Dee Wilder/Tom Stack & Associates.

Copyedited by Brown Editorial Service

The credits section for this book begins on page 509 and is considered an extension of the copyright page.

Library of Congress Catalog Card Number: 92–81634

ISBN 0–697–03349–X

Printed in the United States of America by Wm. C. Brown Communications, Inc., 2460 Kerper Boulevard, Dubuque, IA 52001

10 9 8 7 6 5 4 3 2 1

EXPANDED

Preface

The magnitude of the role played by insects in the scheme of life is undisputed. In adaptive diversity and number of species, they are among the most successful of all organisms. The relatively few that cause problems have taxed our ingenuity to its fullest throughout history, and the battle will probably never be over. Thus the science of entomology is a vital applied science as well as one of the major areas of basic biology. In keeping with this fact, we have treated entomology from both basic and applied points of view.

Our objective in the third edition of this text has been to continue to provide a broad, balanced introduction to the topic of entomology for use in a one-quarter or one-semester general course. At the same time, we hope professional entomologists will find it useful as an up-to-date review and source of literature references.

The discussion of the literature of entomology has been retained in the introductory chapter from earlier editions with the hope that the student will be encouraged to make full use of the vast amount of information available. The remainder of the text is developed around four topics: structure and function spanning the cellular to organismal levels of biological organization (Part One); insects in an environmental context (Part Two); unity and diversity as reflected in insect systematics and as the result of organic evolution (Part Three); and finally, applied entomology (Part Four).

Substantial changes have been made throughout the book. All chapters have been updated and/or expanded; a new chapter on insects and plants has been added; a new chapter on integrated pest management in agroecosystems has been added; the coverage of the orders of Insecta has been increased; the glossary has been expanded to include new terms; new illustrations have been added, and a few old ones omitted; and the number of literature citations has been increased.

Part One, "Structure and Function," begins with a discussion of the diverse roles played by the integumentary system and subsequently elaborates on the structure and function of the insect skeleton. This is followed by discussions of the nervous, glandular, and muscular systems, and then the alimentary and remaining systems. The placement of the control and effector systems first facilitates discussions of regulation of the alimentary and remaining systems. Because the chapter on reproduction and morphogenesis stresses anatomy and physiology, it follows the chapters on anatomy and physiology. The chapters dealing with sensory mechanisms, locomotion, behavior, follow in a continuous and logical sequence.

Part Two, "Insects and Their Environment," begins with a discussion of insect populations and how these populations are influenced by, and react to, the physical, chemical, and biological factors in their environment. A separate chapter is devoted to the relationships between insects and plants.

Part Three, "Unity and Diversity," begins with a discussion of basic systematics, a topic that is all too often neglected in this day and age of molecular biology, and yet a topic that provides a foundation for the rest of biological science. The remainder of the first chapter in this section of the book centers around the origin and evolution of insects. The following two chapters provide an overview of the orders and selected major families in the class Insecta. We have tried to show how the various groups relate to one another, as well as provide information regarding their biology and medical, economic, and ecological significance.

Three chapters are included in Part Four, "Applied Entomology." In chapter 14, the many ways insects are beneficial and are harmful are discussed. In chapter 15, the various methods available to control insects are considered. In chapter 16, we use the topic of integrated pest management in agroecosystems to illustrate approaches to insect control that are designed to be compatible with the need to minimize environmental damage.

We have arranged the topics in the sequence we think most appropriate for dealing with the various aspects of entomology. However, each chapter can be read and understood with minimal reference to other chapters. Thus, this text should be amenable to any organizational framework a given instructor may choose to follow. A reference list for each chapter, consisting mainly of major review papers, monographs, and specialized textbooks is given at the end of the book.

In addition to those persons who contributed to the first and second editions, we wish to express our sincere appreciation to the following individuals who have played important roles in the development of this edition. "Guest authors" Lance A. Durden (Institute of Arthropodology and Parasitology; Georgia Southern University), Bruce A. McPheron (Pennsylvania State University), Dave Ferro (University of Massachusetts), Patrice Morrow (University of Minnesota), and Margaret Lowman (Williams College), have greatly enriched the text by providing insightful revisions of chapters from the last edition or by writing new chapters. Although, unfortunately, in most cases their identities to us remain pleasant voices over the phone, the senior editorial staff at Wm. C. Brown—Kevin Kane, Carol Mills, and Michelle Campbell—have been delightful to work with and extraordinary in their patience and guidance in developing this text from the untidy first draft to the finished product. We also wish to thank the following Wm. C. Brown editorial staff members for their contributions to our project: Shirley Lanners, Connie Gibbs, Kathy Huinker, Kristyn Kalnes, and Vicki Krug. Copy editor Lynn Brown and her staff at Brown Editorial Service have done a diligent job of policing our grammatical blunders, run-on sentences, and various other unintentional assaults on the English language.

We also wish to thank our professional colleagues whose constructive comments have proven very helpful in producing an accurate, up-to-date text. These colleagues are as follows:

Chih-Ming Yin
University of Massachusetts, Amherst

Donald G. Cochran
Virginia Polytechnic Institute and State University

Robin Leech
Northern Alberta Institute of Technology/University of Alberta

Hal C. Reed
Oral Roberts University

A. G. Scarbrough
Towson State University

We offer a special thank you to Roger Meola, Texas A & M University, who has provided encouragement and advice throughout the life of this text.

As with past editions, the development of this third edition has been a challenging, demanding, at times frustrating, but ultimately very rewarding, experience. We are enthusiastic about our field; we are pleased to have an opportunity to share it with you; and we wish you well in your entomological quest.

W. S. R.
J. G. S.

CHAPTER 1

Introduction

Insects are arthropods, the largest group in the Animal Kingdom. Arthropods are characterized by a segmented body that bears a varied number of paired and segmented appendages; bilateral symmetry; an exoskeleton that contains the nitrogenous polysaccharide, chitin; and various internal features, such as an open circulatory system, Malpighian tubules (generally), and in most a system of ventilatory tubules (the tracheae and tracheoles).

Insects can be differentiated from the vast majority of other arthropods by several rather distinct traits. Among these are three well-defined body regions: a head, a thorax, and an abdomen; three pairs of legs in the adult stage; commonly one or two pairs of wings; a single pair of segmented antennae on the head; and several less obvious but equally distinctive characteristics. The name *Hexapoda* (six legs) is commonly applied to insects. However, the name *Insecta* is preferable, because there is some question as to whether all arthropods with six legs in the adult stage actually belong in the same class (Sharov 1966). Insecta literally means "in-cut," which describes the segmented appearance of the members of this class. The arthropods will be discussed in more detail when we consider the evolution of insects.

Significance of Insects

Insects as a group are highly successful organisms. Their significance can be looked upon from two standpoints: their tremendous success relative to organisms other than human beings and their extreme importance from the human point of view.

One useful measure of the success of insects is the number of extant (as opposed to extinct) species. Insects probably outnumber all the other species of animals and all the species of plants combined. Estimates based on the current rates of description of new species of insects run from one to several million, and several large groups have hardly been studied. Arnett (1985) estimates the number of described insect species in the world to be 751,012 and points out the sad fact that we may never know the actual number due to the rapid destruction of habitat worldwide. Studies of tropical insect fauna lead to widely divergent estimates of the total world insect fauna (described plus undescribed) ranging from somewhere between 5 and 10 million (Gaston 1991) to 30 million or more (Erwin 1982, 1988).

Other important criteria for success include the span of geologic time traversed by a group of organisms and their adaptability to various environmental situations. Insects are thought to have arisen in the Devonian era, approximately 400 million years ago. Mammals as a group are approximately 230 million years old; modern humans arose perhaps 1 million years ago. In this sense, insects have not invaded the human world; we have invaded theirs! The adaptability of the basic insectan plan has been phenomenal. Insects can be found in nearly every conceivable terrestrial habitat. As you proceed with your study of insects, you will come to realize the seemingly unlimited adaptability of insects and gain insight as to how they have reached their position of success.

From earliest times people have seen certain insect species as arch enemies. Although the pest species make up a very small proportion of the total number of insect species, members of this group are chronic troublemakers, destroying annually millions of dollars worth of agricultural crops, fruits, shade trees and ornamental plants, stored products of various sorts, household items, and other valuable material goods (Davidson and Lyon 1987). Pest species act as vectors of the causative agents of human and domestic animal and wildlife diseases, and their direct attacks cause irritation, blood loss, and sometimes death (Harwood and James 1979, Kettle 1984).

However, there are two sides to the picture. Insects provide many highly valued goods and services. Such insect products as honey and beeswax, silk, shellac, and cochineal are used for a variety of applications, ranging from sweetening biscuits to constituting one of the basic components of many cosmetics. In addition, there are many indirect benefits of insect activities, such as plant pollination and involvement in nutrient cycling.

Although there is much that can be said both for and against insects as they relate to humans, the vast majority of insects are neutral, neither bestowing benefit nor causing harm.

The Science of Entomology

Entomology is a specialized field within the biological sciences, because it is concerned with "living" systems. The biological sciences can be divided into basic divisions, such as morphology, physiology, genetics, and ecology, and into taxonomic divisions, such as ornithology, mycology, and bacteriology. Entomology, as the study of a very specific group of organisms, is a taxonomic division. Therefore, we can approach the science of entomology by considering the "basic" divisions as they apply to insects: insect morphology, physiology, ecology, and so on.

The study of insects has played and continues to play a major role in the development of biology. This is evident upon examination of a current general biology text. Among the entomological topics, one might find the following:

> Redi's experiments with maggots and spontaneous generation; coevolution of plants and pollinating insects; the mechanisms of sex determination in insects; mutations in *Drosophila;* linkage groups, sex-linkage, chromosomal mapping, and induced mutations in *Drosophila;* chromosomal puffs in *Drosophila* and their induction by ecdysone; pheromones; behavioral genetics of cricket singing; spatial orientation of the digger wasp; migration of the monarch butterfly; circadian rhythms and adult emergence in *Drosophila pseudoobscura;* several other behavioral examples using insects (including behavior of the honey bee); termite–protozoan mutualism; temperature control in termite mounds and beehives; bees and orchids; ants and acacias; fig wasps and figs; industrial melanism; specialization in the *Drosophila willistoni* complex; pesticide problems. . . .

Further evidence for the major role played by entomology in the development of biology is that Nobel Prizes have been awarded to several scientists who studied insects, e.g., Karl von Frisch, Niko Tinbergen, and Konrad Lorenz.

For an excellent introduction to entomology as a science, see Wigglesworth (1976).

Entomological Information

A serious study of entomology requires a knowledge of resources available that help entomologists to acquire information about insects. The oldest and still one of the most important means is by word of mouth. This is the method employed in the classroom, at scientific meetings, conferences, and so on. Perhaps its most valuable aspect is the opportunity for a two-way exchange of information. Ideally, in this situation the lines of communication are wide open and a minimal amount of ambiguity should be the result. Another, equally important—in fact, essential—means of acquiring scientific information is consultation of the literature. The advantage of this method is that one can go back in time as far as one wishes, but the two-way exchange of information is impossible if the author of a piece of recorded work is no longer living. Both these means of acquiring information should be considered to be prerequisites for a third means— personal investigation. This method is, of course, the source of new information. All three means are essential to the existence of science, and certainly no one method could take precedence over the other two. However, let us delve further into the use of entomological literature.

The literature of entomology consists of a wide variety of publications ranging from rather popularized accounts intended for the layperson to highly technical treatises on very specific aspects of the science. The information available on insects is vast and is growing so rapidly that no one person can stay abreast of and learn more than comparatively small portions of it. There is also a great deal of specialization, the concentration of effort on a single topic or a group of closely related topics. Thus, among entomologists there are, for example, insect physiologists, ecologists, morphologists, systematists, toxicologists, and economic entomologists. Generally, the specialization goes even further; someone in one of the preceding groups may concentrate on a particular insect species or group of insects, or on a specific topic, or both. For example, there are specialists in the systematics of a particular family of beetles or in mosquito physiology.

The tremendous number of entomological publications makes a comprehensive review inappropriate within the context of this book. However, a brief discussion of some of the different types of entomological literature may be helpful. More extensive treatments of the literature of entomology and zoology in general can be found in Smith, Reid, and Luchsinger (1980), Chamberlin (1952), and Blackwelder (1967). The recent compendium by Gilbert and Hamilton (1990) deserves special mention.

Publications pertinent to entomology (for that matter, the entire field of zoology) can be divided into two basic groups. The first group includes all those publications, irrespective of type, that contain the actual information about animals (insects). This group presents the products of zoological (entomological) research. The second group includes all the publications that attempt to coordinate the vast information contained in the first group, making it more readily available to investigators. Examples of publications from each of these groups are presented. This method of classification, as with most, is certainly not without exception, as will be shown.

Publications That Contain the Actual Information about Entomology

Textbooks
A textbook is generally designed to give the reader an understanding of the basic principles involved in a given subject. A textbook may be quite general in scope, presenting a survey or overview of an entire field—for example, a general entomology textbook such as the one you are reading or any of several others, including Borror, Triplehorn, and Johnson (1989), C.S.I.R.O. (1991), Richards and Davies (1977, 1978) and Evans (1984). Other textbooks deal with a particular area within the field. For example, there are texts devoted to insect physiology—Chapman (1982), Blum (1985), and Wigglesworth (1972); insect ecology—Price (1984); insect behavior—Matthews and Matthews (1978), Atkins (1980); medical entomology—Harwood and James (1979), Kettle (1984); or economic entomology—Davidson and Lyon (1987). Books of this type treat a subject in more depth than the general text and are more likely to be used in advanced courses.

Monographs
A monograph is limited in scope, dealing with only a very small area within a science. However, monographs are usually comprehensive in their coverage of pertinent literature and handling of subject matter. Excellent examples are Dethier's superb *The Hungry Fly* (1976) and the 1990 Pulitzer Prize-winning book, *The Ants* by Hölldobler and Wilson.

Symposia
Symposia proceedings are published collections of the presentations of several specialists in a given area who have met to consider a specific aspect of science. A recent example, which deals in great depth with a particular group of mosquito-borne viruses, is *California Serogroup Viruses* (Calisher and Thompson 1983) which is the published proceedings of an international symposium.

Lectures
A lecture is an oral discourse presented to an audience or class, particularly for instructional purposes. Good examples of entomological lectures are the talks presented by outstanding scientists to the general sessions usually held several times during each annual meeting of the various entomological societies throughout the world. These talks are commonly published in bulletins issued periodically by these societies.

Essays
Essays are analytical or interpretative expositions that usually deal with a given topic from a rather personal or limited point of view. A well-known example is the book *Silent Spring* by Rachel Carson.

Reference Works
Reference books generally attempt to offer comprehensive coverage of a specific area and are designed to be consulted as needed rather than read and digested in their entirety. This category may seem rather arbitrary, since any piece of literature that is consulted can be classified as a reference, and rigorously documented and highly technical entomological texts, such as Richards and Davies (1977, 1978) and *The Insects of Australia* (C.S.I.R.O. 1991), can equally well be viewed as reference works. An especially valuable recent reference work is the thirteen-volume set entitled *Comprehensive Insect Physiology, Biochemistry, and Pharmacology* (Kerkut and Gilbert 1985).

Pamphlets
Pamphlets are usually rather brief writings on a specific topic, commonly geared to the persons who put many of the findings of entomological research into practice: farmers, exterminators, and so on. Examples of pamphlets are the *Farmer's Bulletins* published by the U.S. Department of Agriculture, which pertain to the biology and control of economically important species of insects.

Reports
From time to time, groups of experts are called together to investigate or to discuss an issue or problem of particular significance. For example, the World Health Organization (WHO) periodically sponsors meetings of expert committees on various problems pertaining to international health matters, including topics such as malaria and similar diseases with which insects are involved. These committees usually submit reports describing the problems discussed and the conclusions reached in the course of the meeting.

Series
Serial publications are issued periodically (sometimes at irregular intervals) and have a certain unity of subject matter; that is, a particular series deals, volume after volume, with more or less the same general subject matter. Series are published by most professional societies and many private and governmental concerns in the form of journals, bulletins, miscellaneous publications, yearbooks, and so on. Hammack (1970) and more recently

Gilbert and Hamilton (1990) provide useful and comprehensive descriptions of the serial literature pertinent to entomology. Some examples of entomological journals are *Annals of the Entomological Society of America, Journal of Economic Entomology, Environmental Entomology, Journal of Insect Physiology, Systematic Entomology, Ecological Entomology, Physiological Entomology, Psyche, Journal of Insect Behavior, Boletín de La Asociacion Española de la Entomologia, Canadian Entomologist, Deutsche Entomologische Zeitschrift (Berliner Entomologische Zeitschrift), Bulletin de La Société Entomologique de La France.*

Publications That Attempt to Coordinate the Literature

From the time scientists realized the fantastic rate of growth of the literature of science, many very useful attempts to coordinate and integrate the works in various areas have been made. In this section we want to discuss briefly some of the more common publications in this category. Most of these publications are quite expensive and are seldom purchased by an individual.

Bibliographies

A bibliography is to the literature of science as an index is to a book, the basic difference being that a bibliography lists only publications of various sorts on a single topic, instead of the subjects, authors, and so on, that are found in the index of a single book. Bibliographies appear in different forms. For example, one form is a list of pertinent references at the end of a chapter of a book or a scientific paper. Another type of bibliography is issued periodically and lists current publications in a particular field. Important examples of this type are *Zoological Record, Bioresearch Index, Bibliography of Agriculture, Cumulated Index Medicus,* and *Current Contents.* The *Zoological Record* has a very broad coverage, both foreign and domestic, and is arranged according to taxonomic groups and by subjects. The section on insects is quite extensive and contains references to papers of interest to most entomologists, not just taxonomists. *Bioresearch Index* is a monthly publication that furnishes bibliographies from various journals and contains citations of research papers. The *Bibliography of Agriculture,* a monthly publication, generally contains a large number of references to entomological papers. The last issue of each year is a cumulative subject index. *Cumulated Index Medicus* is issued four times yearly, covers much of the foreign and domestic medical literature, and may contain references of interest, particularly to medical entomologists. *Current Contents* reprints the tables of contents of many journals, several entomologically oriented ones included. It is issued weekly and is probably one of the best ways to keep abreast of the most current literature. This is especially important when one is working in a very active area in which a number of researchers are publishing extensively and often. The recent availability of current contents on computer diskettes has greatly enhanced the process of using this publication.

Abstracting Journals

Abstracting journals contain brief descriptions or abstracts of the results reported in the journals that fall within their scope. Abstracts are extremely useful because they give an investigator a better idea of the content of a given reference than does a mere title listing, although they do serve also as bibliographies. This kind of information helps one decide whether or not to consult the actual references and compensates somewhat for the fact that some publications are extremely difficult to obtain or translate from a foreign language. Abstracting journals are usually extensively cross-indexed, which makes them efficient to use. One of the most significant examples of this type of publication is *Biological Abstracts.* This bimonthly publication is comprehensive in its coverage of the literature of both theoretical and applied biology. In addition to the volumes containing the abstracts, a semimonthly publication, *B.A.S.I.C.,* provides an elaborate computerized subject index to the issues of *Biological Abstracts.* A cumulative subject index based on all the issues of *B.A.S.I.C.* is published semiannually. Since January 1970 the abstracts and citations of research papers pertaining to insects and arachnids in *Biological Abstracts* and *Bioresearch Index* have been compiled into a separate publication, *Abstracts of Entomology.* One issue of this publication corresponds to two issues of *Biological Abstracts* and one issue of *Bioresearch Index.*

Other abstracting journals that specialize in covering entomological literature are *Entomology Abstracts* and *Review of Applied Entomology. Entomology Abstracts* is published monthly and covers a wide variety of entomological topics. *Review of Applied Entomology* is published in two series: Series A, Agricultural, and Series B, Medical and Veterinary. It is well indexed, both by subject and author, and contains abstracts covering a wide variety of entomological topics. Other periodical publications that are at least partly abstracting journals are *Biologisches Zentralblatt, Physiological Abstracts, Tropical Diseases Bulletin, Apicultural Abstracts,* and several others from various countries.

Another very useful monthly publication is *Dissertation Abstracts.* This contains abstracts of all dissertations (i.e., the printed results of doctoral student research) by contributing institutions in the United States and Canada. It is arranged by the type of subject matter. These abstracts are useful since there is commonly a significant period of time between the writing of a dissertation and the publication of a research paper or papers based on it. If one decides on the basis of a given abstract that more information is necessary, he or she may readily obtain, for a fee, a microfilm or printed copy of an entire dissertation. In a similar vein, *Dissertation Abstracts* publishes abstracts from contributing worldwide institutions.

Review Journals

Review journals contain papers that discuss the literature on a rather specific topic in a given field. Review papers not only bring together information from the pertinent literature on a given topic but also commonly contain useful syntheses of information that may not occur in any other type of publication. In this sense they may be classed in either of the two rather arbitrary categories we have used to discuss entomological literature. Two very important review journals in the field of entomology are the *Annual Review of Entomology* and *Advances in Insect Physiology*. Other review journals that may contain reviews of entomological interest are the *Annual Review of Ecology and Systematics, Annual Review of Physiology, Annual Review of Phytopathology,* and the *Annual Review of Medicine*. In addition, review papers may appear in journals, bulletins, and so on, which contain other types of articles.

Taxonomic Indexes and Catalogs

Taxonomic indexes include literature references to such items as the original description of a given genus or species and revisions of genera. These publications are quite useful for tracing the taxonomic literature pertinent to a given group and for determining the systematic position of a given genus or species. Especially important indexes are *Nomenclator Zoologicus* edited by A. A. Neave, *Zoological Record,* and *Biological Abstracts. Nomenclator Zoologicus* lists the names of genera and subgenera of all zoological groups from 1758, the year of the publication of the 10th edition of Carl Linne's (Linnaeus) *Systema Naturae,* to 1950. *Zoological Record* contains the names of all new genera described each year and pertinent literature references from 1864. *Biological Abstracts,* in the section "Systematic Zoology," provides references to the original descriptions of genera and subgenera of animals since 1935. Smith (1958) points out that "Neave's *Nomenclator Zoologicus* and *Biological Abstracts* serve admirably as a complete generic index from 1758 to the present." He further suggests that ". . . for names published since the most recent issues of *Biological Abstracts,* journals in which new genera of the various groups might be expected to occur must be consulted."

For species indexes and catalogs similar to the generic ones just described, one must refer to one or more of several currently available. Sherborn's *Index Animalium* is the only general species index available and covers all the specific names proposed for animals from 1758 through 1800. Smith, Reid, and Luchsinger (1980),

Chamberlin (1952) and Blackwelder (1967) each contain lists of catalogs for various insectan and other groups.

Science Citation Index

The Science Citation Index is published by the organization that publishes *Current Contents* and is composed of two sets of indexes, a citation index and a source index, both of which are cumulative. Its objective is to list and index the current and past research papers that cite a given reference. It enables an investigator to begin with a given reference and find other references that have cited the "starting reference." Because both the "starting reference" and "citing reference" are likely to pertain to the same or very closely related topics, one is able to proceed forward or backward in time, using "citing references" as "starting references" in a cyclical manner and by doing so accumulate references on a given topic.

Union List of Serials and New Serial Titles

Most libraries do not have complete sets of all journals useful in entomology or any other science. However, they generally have agreements with other libraries, whereby volumes can be borrowed or copies of particular papers can be obtained, that is, interlibrary loans. *The Union List of Serials in Libraries of the U.S. and Canada* and *New Serial Titles* are listings of all journals and of the major libraries that house these journals. Thus, by consulting these lists, one may determine which libraries have the journal he or she is seeking. Both lists are indexed by journal name and by subject.

Books In and Out of Print

Books in Print is an annual listing of books currently available on the commercial market. This list is composed of four volumes: Volumes 1 and 2, titles; and Volumes 3 and 4, authors plus author and title indexes. Out-of-print books may be found by consulting the *A. B. Bookman's Weekly,* various companies that specialize in such books, and major libraries. Facsimiles of out-of-print books are available from University Microfilms International in Ann Arbor, Michigan.

Databases

In recent years the computer has come to play an important role in the management of scientific literature. Several major databases, such as computer-accessible indexes and bibliographies, are maintained and can be tapped for literature searches. Among the databases useful in entomology are AGRICOLA which includes, among others, citations from the *Bibliography of Agriculture,* Bioscience Information Service (BIOSIS), which includes citations from *Biological Abstracts* and *Bioresearch Index,* and Medical Literature Analysis and Retrieval System (MEDLARS), which includes citations from *Cumulated Index Medicus.* These databases are accessible through information centers such as Biographic Retrieval Service (BRS) in Schenectady, New York, which provides the databases as well as others. Users communicate with such information centers through computer terminals via the telephone. Computer searching of databases offers a means to search extensive bodies of literature very rapidly. In some cases, abstracts of research papers can be obtained in addition to citations.

The Language of Entomology

There are a number of publications that are very helpful in learning the technical terminology that necessarily exists in entomology. Borror's *Dictionary of Word Roots and Combining Forms* (1960) enables one to determine roughly the word roots of scientific terms and is thus useful in learning and understanding these terms. Leftwich (1976) and Nichols (1989) are also valuable in dealing with complicated and often confusing entomological terms. A preliminary thesaurus of entomology (Foote 1977) is available, the role of which is explained as follows.

> A carefully built thesaurus is, then, the vocabulary of a specific field of activity. It is also the guide required if more than one person is to become involved meaningfully in storing and retrieving information. Without specifically defining every term, the thesaurus specifies those terms which are to be used for specific "things" and concepts and interrelates them in a way that makes the application clear.

The History of Entomology

Persons interested in the history of entomology will find useful information in the following references: Cloudsley-Thompson (1976); Cushing (1957); Essig (1931); Howard (1930); Mallis (1971); Osborn (1937); Richards and Davies (1978); Smith, Mittler, and Smith (1973); Southwood (1977), and Wigglesworth (1976). In addition, the *Annual Review of Entomology* often contains biographical information about entomologists.

PART ONE   # Structure and Function

In the following seven chapters, we take a broad view of structure and function, moving from an examination of each of the systems composing an insect to a discussion of all the systems acting in concert, producing insect behavior. We begin in chapter 2 with the integument as the boundary between the "outside" and the "inside" of an insect, and as such, a major functioning system, including determination of insect skeletal form.

In chapter 3, we examine the insect nervous, muscular, and glandular systems. An understanding of the systems involved in coordinating all insect activities sets the stage for learning how the other insect systems are regulated.

We then proceed to the alimentary, circulatory, ventilatory, and excretory systems in chapter 4. These systems directly serve the needs of insect cells and are involved in the procurement and transport of nutrients as raw materials for biosynthesis and sources of energy for cellular work; procurement and transport of oxygen; and the removal of cellular wastes, including carbon dioxide, from the insect.

In chapter 5, we consider insect reproduction and development from a structural/functional standpoint.

Chapters 6 through 8 deal with the collection of information from the environment (via the various sensory mechanisms), as well as light and sound production; structural and functional aspects of movement through the environment (locomotion); and finally, the responses of insects to changes in both the internal and external environment, that is, insect behavior.

CHAPTER 2

The Integumentary System

The general body covering, or *integument,* is a complex organ of diverse structure and function. Its most obvious role is that of forming a supportive shell, the *skeleton.* The insect skeleton differs from the internal skeleton, or endoskeleton, of vertebrates in being located on the outside of the body; that is, it is an *exoskeleton.* Even the internal integumental processes are continuous with the external shell. A modified form of the cuticular part of the integument lines portions of the alimentary canal, the tracheal system, genital ducts, and the ducts of the various dermal glands.

The insect skeleton affords almost unlimited area for muscle attachment. In many cases the elasticity of parts of the skeleton opposes muscular contraction in a manner similar to the mutual opposition of antagonistic muscles in vertebrates. For example, muscle contraction causes retraction of the pretarsus (terminal segment of tarsus or "foot"), and extension is due to skeletal elasticity. Some elastic regions are involved in the storage and release of energy during jumping and flying. Elasticity may, in some cases, aid in the inspiration of gases during ventilatory movements. The insect skeleton is hardened, or *sclerotized,* to varying degrees and thus serves as a protective armor for the organs encased by it. Other properties of the integument also play a role in the protection of the insect. For example, the integument prevents or retards the entry of many pathogens and insecticides. The integument may be relatively impermeable to water, thereby helping to minimize water loss. External portions of the sensory and exocrine structures are intimately associated with the exoskeleton, which lies between the external environment and the remainder of the insect. Characteristics of the integument usually determine body coloration.

Although some regions of the exoskeleton are more flexible than others, there are definite limits to expansion. Growth is facilitated by the periodic shedding and renewal of part of the integument, the process of molting.

Extensive information on the insect integument may be found in Hepburn (1976, 1985), King & Akai (1982), Neville (1975), Richards (1951, 1978), and Binnington and Retnakaran (1991).

Histology of the Integument

Basic Components

The insect integument (figure 2.1*a*) can be divided into three basic parts: one cellular layer, the *epidermis;* and two noncellular layers, the *basement membrane* (*basal lamina*) which is entad of the epidermis and the *cuticle* which is ectad of the epidermis.

The epidermis is typically one cell thick; adjacent cells are joined by septate desmosomes. The epidermis secretes the cuticle and may secrete the basal lamina. Because the epidermis secretes the cuticle, it is most active during molting. Interspersed among the epidermal cells are *dermal glands,* some of which play a part in secreting a portion of the cuticle. As you will see in chapter 3, other types of single-celled or multicellular dermal (exocrine) glands carry out a variety of functions such as secretion of defensive substances, silk, and pheromones.

The basal lamina is generally 0.5 micrometer or less in thickness and in electron micrographs appears as a continuous, amorphous granular layer. It is apparently composed of a mucopolysaccharide.

The cuticle comprises up to one half the dry weight in certain species. The properties of cuticle vary both spatially and temporally according to the demands of function and growth. The chemical and physical properties of cuticle in a given region at a given time reflect the expression of specific genes in the epidermal cells. Many techniques, as outlined in Miller (1980), are useful in studies of cuticle structure and function.

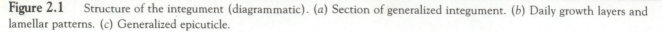

Figure 2.1 Structure of the integument (diagrammatic). (*a*) Section of generalized integument. (*b*) Daily growth layers and lamellar patterns. (*c*) Generalized epicuticle.

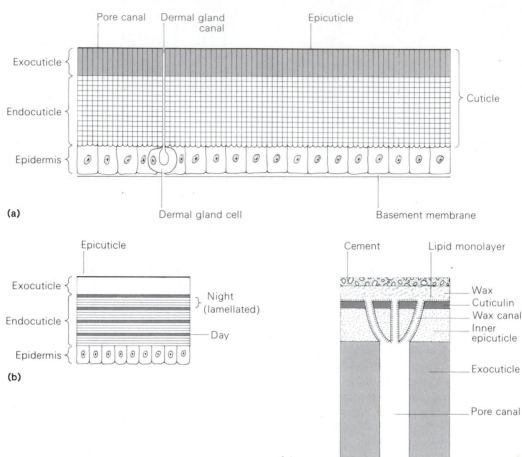

With the techniques of light microscopy and various stains, the insect cuticle can be seen to consist of a number of different layers. There are two major layers: the very thin, outer *epicuticle* (0.03–4.0 micrometers thick) and the much thicker, inner *procuticle* (up to 200 micrometers thick). In hardened regions, the procuticle is typically divisible into at least two layers of variable thickness. The outer layer is called *exocuticle* and the inner layer is called *endocuticle*. The endocuticle may or may not be pigmented, the exocuticle is often pigmented, and the epicuticle is unpigmented. Compared with hardened regions, softer areas of cuticle are typically comprised of only endocuticle and epicuticle.

Under the light microscope, with polarizing filters, the endocuticle can be seen to be composed of successive light and dark layers (figure 2.1*b*), which in all insects studied thus far, except the beetles and Apterygota, correspond to daily growth layers (Neville, 1970). The dark layers are those deposited during the day and the lighter layers are those deposited during the night. The light layers are further subdivided into alternating layers of light and dark lamellae. Also, typically, striations running vertically through the cuticle can be seen.

The use of the electron microscope has resulted in the development of models to explain the lamellae and striations. In the helicoidal model, the lamellar patterns are considered to be the result of the orientation of layers of microfibrils of chitin and probably protein embedded in a protein matrix. The microfibrils are laid down in layered sheets (figure 2.2*a*). Within each sheet, the microfibrils are parallel to one another, but in successive sheets they are aligned at regularly changing angles. This arrangement would account for the parabolic patterning of endocuticle when cut transversely and observed with the electron microscope (figure 2.2*b*). In the dark growth layers deposited during the day, the microfibrils would be oriented in successive sheets in a "preferred" direction and hence would not have the appearance of the lighter nighttime layers. The vertical striations have been shown to be tiny tubes, the *pore canals* (figures 2.1*a, c* and 2.2*b*), which extend from the epidermal layer nearly to the external surface of the epicuticle. The pore canals are usually 1 micrometer or less in diameter, are ribbonlike in appearance, and apparently serve as connecting tubes between the cellular and the cuticular layers. The pore canals run a spiral course through the helicoidally

Figure 2.2   The "helicoidal" model of endocuticular structure. (*a*) Helicoidal and preferred structure of layers of endocuticle. The parallel lines in each layer represent microfibrils. Note the parabolic effect in transverse section. (*b*) Transverse section of endocuticle showing parabolic effect and appearance of pore canals as they pass through helicoidal layers. *Redrawn with modifications from Neville, 1970.*

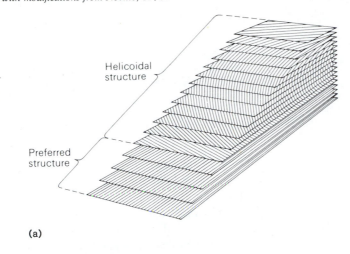

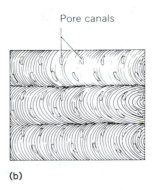

(a)
(b)

arranged layers of endocuticle and may number from several thousand to well over a million per square millimeter of integument.

In contrast to the endocuticle, which is digested by molting fluid, the exocuticle is a highly stabilized layer of the procuticle that is resistant to the action of molting fluid and is shed as the *exuvia* during *ecdysis*. Exocuticle is typically thin in soft-bodied insects and thicker in hard-bodied insects.

In addition to the exocuticle and endocuticle, two other procuticular layers have been identified: the *mesocuticle* and a granular *subcuticle*. Mesocuticle lies between exocuticle and endocuticle and is differentiated on the basis of staining reaction. Subcuticle is probably newly secreted endocuticle that has not yet reached its final structural configuration.

The epicuticle, despite its comparative thinness, is complex and possesses characteristics that make it an important layer of the cuticle. Although it appears as a thin, sometimes indiscernible, line under the light microscope, the electron microscope has revealed a multilayered structure that is penetrated by wax canals 60–130 Ångstroms in diameter containing wax filaments (Locke 1974). At least four layers, of varying thickness depending on the insect species and life stage, have been described in the epicuticle (figure 2.1*c*):

1. The outer *cement layer,* sometimes called tectocuticle ("roof" cuticle), less than 0.1 micrometer thick.
2. The *wax layer.*
3. The *cuticulin layer.*
4. The *inner epicuticle.*

The cement layer is secreted by dermal glands and may be similar to shellac, a substance secreted by a particular group of insects in the suborder Homoptera. This

layer, outermost when present, probably determines the surface properties of the cuticle—that is, whether the cuticle will be *hydrophobic* (water-repellent) or *hydrophilic* (water-attractant). The cement layer also probably serves as a protective barrier for the more vulnerable layers beneath and may actually contain antifungal components. The epicuticle is apparently absent in many adult insects that possess cuticular scales. In some insects—cockroaches, for example—it may form a spongy meshwork in which wax molecules can move about. In other insects, such as many Homoptera, a waxy bloom appears on the surface of the cement layer (Locke 1974). The wax layer is thought to consist of an ordered monolayer of lipid directly associated with the cuticulin layer and a more ectad, less well-ordered lipid layer. Experimental evidence indicates that this layer is responsible for many of the permeability characteristics of the cuticle. The cuticulin layer has been found in every insect in which the epicuticle has been investigated and covers the entire integumental surface, including the tracheoles and gland ducts. The cuticulin layer is thought to be composed of lipoprotein and is 100–200 Ångstroms thick. The innermost layer, the proteinous inner epicuticle, is about 1.0 micrometer thick and appears as a homogeneous, dense, refractile layer. It may be the layer that limits the amount of expansion of cuticle possible between molts. The pore canals terminate beneath this inner layer of epicuticle; the dermal gland canals communicate with the surface of the epicuticle.

Tiny *wax canals* approximately 100 Ångstroms in diameter are evidently continuous with pore canals, penetrating the inner epicuticle and cuticulin. Because they are continuous with pore canals, these wax canals are evidently involved in transporting wax molecules to the epicuticle (Locke 1974, Wigglesworth 1976).

Figure 2.3 Structural formula of chitin.

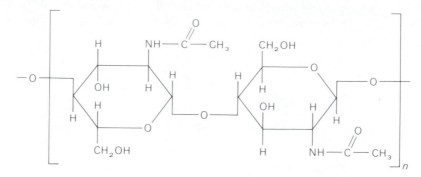

Chemical Composition of the Cuticle

Quantitatively, the polysaccharide *chitin* and various structural proteins are the major cuticular constituents. Chitin (figure 2.3) is a high-molecular-weight polymer of *N*-acetyl-D-glucosamine with the empirical formula $(C_8H_{13}O_5N)_n$. Among the three forms of known chitin, α-, β, and γ, α-chitin is the overwhelmingly dominant form. It may make up from one fourth to more than one half of the dry weight of the exocuticle and endocuticle. Chitin has not been found in the epicuticle. Chitin probably does not exist in a pure state naturally, but is combined with a protein as a glycoprotein. Apparently chitin chains are attached to one another by hydrogen bonds, forming elongate microfibrils. It seems likely that chitin microfibrils and protein chains are in intimate combination with one another and that this complex is impregnated with loosely bound protein (Hackman 1976).

Cuticular proteins usually make up more than 50% of the dry weight of cuticle. Included in this category are *arthropodins,* a group of soluble proteins; *resilin,* a protein that forms a rubberlike framework and is found sometimes in pure form in skeletal articulations; and *sclerotins,* stabilized proteins that are responsible for the hard, horny character of cuticle.

Other constituents of the cuticle include polyhydric phenols and quinones, which play a role in the *sclerotization* (hardening) and *melanization* (darkening) processes; lipids of various sorts associated with the epicuticle; enzymes (nonstructural proteins), which catalyze the many complex biochemical reactions involved in molting and subsequent processes; and very small amounts of inorganic compounds.

Barrett (1991), Richards (1978), Anderson (1979), Cohen (1987,1991), Hopkins and Kramer (1991), and de Renobales et al. (1991) discuss the chemistry of cuticle and cite several useful references.

Sclerotization

Chitin is not the agent responsible for the hardness of the cuticle, although it undoubtedly lends strength to it. In fact, highly sclerotized skeletal regions may contain less chitin than softer membranous areas.

The hardening of insect cuticle is due to the tanning of protein, which follows *eclosion* (emergence of an immature stage from the egg) or ecdysis (shedding of the cuticle) and subsequent expansion of an insect. Proteins are rendered hard, dark, and insoluble by the linkage of adjacent polypeptide chains and the blocking of reactive groups by the tanning substances (phenols and quinones). These tanned proteins are the sclerotins mentioned above. The degree of hardness resulting from the tanning process is highly variable. Segmental plates of caterpillars (Lepidoptera larvae) are nearly unsclerotized, whereas the mandibles of certain beetles are capable of biting through metals such as lead, tin, and copper.

In a very few species, for example, in puparial cases of the flies *Musca autumnalis* and *M. fergusoni,* calcium carbonate (lime) is responsible for the hardness of the external shell.

For information on the chemistry of sclerotization, see Hackman (1974), Anderson (1976, 1979, 1991), Richards (1978), and Sugumaran (1991).

Physical Properties of Cuticle

One need only consider the form and various activities of an insect to realize the great importance of the physical characteristics of cuticle. Cuticle in the same insect must be rigid, elastic, impermeable, permeable, flexible, and so on, as appropriate to a given structure and/or function. Physical properties are, of course, a function of the kinds and patterns of organization of molecules that compose cuticle.

The insect exoskeleton is a shell in the engineering sense, that is, a supporting structure on the surface that is thin relative to total size. The exoskeleton is essentially made up of a series of tubular, curved, and spherical panels. Physical calculations based on these shapes show that shells are appropriate only for relatively small animals like insects; that is, as an organism increases in size, the necessary thickness and weight of a shell increase at a greater rate than body mass, and eventually the shell becomes prohibitively bulky. Insect cuticle serves nicely as a shell and weight for weight compares favorably with

Figure 2.4   Scanning electron micrographs of iridescent scales on a wing of the Buckeye butterfly (Family Satyridae) showing the fine structure responsible for the light-breaking effect. 1, scale; 2, socket. (Scale lines: (*a*) 20 μm; (*b*) 2 μm; (*c*) 1 μm.) *Courtesy of Walter J. Humphreys.*

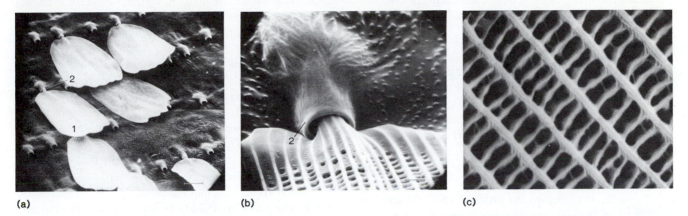

(a)   (b)   (c)

metals. For example, the tensile strength of steel is 100 kg/mm² while that of cuticle is 10 kg/mm², but steel is eight times as heavy as cuticle (Locke 1973).

The physical characteristics of cuticle in a given region may change periodically. For example, when the bloodsucking bug *Rhodnius prolixus* takes a blood meal, a "plasticizing factor" is secreted, under control of the central nervous system, that changes the pH of cuticle and as a result changes the degree of bonding between cuticular proteins (Bennet-Clark 1962). Thus the cuticle becomes more flexible at a time when the abdomen will be drastically expanded.

See Hepburn and Joffe (1976) and Locke (1974) for useful discussions of the physical properties of cuticle. Bennet-Clark (1976) discusses physical properties of cuticle in jumping insects.

Coloration

Insect color may be due to various pigments present in the cuticle, scales on the cuticle, epidermal cells, or fat body; to physical characteristics of the cuticle and/or scales; or to a combination of pigments and physical characteristics.

Most insect pigments appear to be metabolic by-products derived from various pigments in food. Coloration is commonly produced by complex mixtures of different pigments, and the same color may be achieved in different ways. Examples of insect pigments are the very common *melanins* (yellow, brown, black), the *carotenoids* (red and yellow) derived from plant tissues, the widely distributed *pterins* (red, yellow, white), and the *ommochromes* (red, yellow, brown). Pterins and ommochromes are commonly found as eye pigments. Other pigments include *anthraquinones* (red, orange), which are common in scale insects (Homoptera); *aphins* (red, orange, yellow), confined to aphids; *chlorophyll deriva-*

tives (greenish); *hemoglobin derivatives* (reddish); *anthoxanthins* (whitish, yellow); *insectoverdins* (green); and *flavines* (greenish yellow).

For discussions of pigments and pigment metabolism, see Fuzeau-Braesch (1972), Needham (1978), and Wigglesworth (1972).

Physical coloration is due to the effects of various structural configurations on light (Hinton 1976). Colors result from the reflection of light in all directions (scattering) by irregular surfaces or granules beneath the surface; from optical interference between light rays reflected from successive thin layers; and from the light-breaking effect of minute, regular, cuticular striations. Whites in several butterflies (and probably most insects) are produced by the light-scattering effect of the irregular surfaces of scales. In several dragonflies, very small, randomly dispersed granules in the epidermal cells produce a blue color, *Tyndall blue,* by scattering light. (The scattering of light by colloidal-sized particles is called the Tyndall effect.) Most iridescent colors of insects are produced by interference (figure 2.4), as in the brilliant metallic coloration of the *Morpho* spp. butterflies and of several beetles. It has been shown that some pierid butterflies produce interference colors in the ultraviolet part of the spectrum (Hinton 1976).

Some insects are able to change color reversibly. This is accomplished in epidermal cells or in the cuticle (Hinton 1976). Some insects (certain Odonata and Orthoptera among others) change color by movements of pigment granules in epidermal cells. Some beetles in the genus *Dynastes* and several tortoise beetles can change cuticular color. The cuticle of Hercules beetles (*Dynastes hercules*) can appear greenish yellow or black depending on atmospheric moisture. The epicuticle of the elytra is transparent and covers a spongy yellow layer. Beneath the yellow layer, the cuticle is black. Under dry conditions (e.g., daytime), the spongy yellow layer contains air and yellow is reflected back through the epicuticle. However, when humidity increases (e.g., nighttime), the spongy layer becomes fluid-filled and allows light to pass through, revealing the black layer. These color changes evidently help protect the beetle from predators.

Permeability Characteristics

Insects, being essentially terrestrial animals, are continually faced with the problem of losing water, especially in extremely arid habitats. The small size of insects makes this problem particularly acute, because transpiration (water loss) rate varies inversely with the ratio of surface area to volume. Thus the smaller an organism, the greater the amount of surface area per unit volume and hence the greater the tendency to lose water.

Experiments, such as the application of abrasives or dessicants or various organic solvents that dissolve a portion of the epicuticle, have demonstrated that the abrasion, adsorption, or dissolution of the epicuticular layers causes an appreciable increase in the rate of transpiration, often resulting in the death of the insect. On the basis of these and other kinds of experiments, it is felt that at least part of the barrier to the exit of water from an insect lies in the epicuticle, particularly the wax layer. Further, the wax layer is absent in many insects and other arthropods found in very moist terrestrial or aquatic habitats.

Transpiration rate in insects varies directly with temperature (figure 2.5). However, this relationship does not produce a linear curve throughout its length. Above a certain temperature—depending on the species of insect—within an approximate range between 28°C and 60°C, transpiration rate increases abruptly. The temperature at which this occurs has been called the critical temperature, or *transition temperature*. This abrupt change may be brought about by disruption of the oriented lipid monolayer in the wax layer of the epicuticle (figure 2.1c) at the critical temperature, which is somewhat below the melting point of the wax. Permeability of the insect cuticle to substances other than water may be a function of other layers of the epicuticle and of the exocuticle, mesocuticle, and endocuticle (Richards 1978).

The permeability of the integument to water may vary at different times (Locke 1965). This variability may be related to phase changes in polar molecules—hydrophilic (water-soluble) at one end and hydrophobic (non-water-soluble) at the other end—composing the wax

filaments. These filaments could be liquid crystals composed of lipid and water, which change from the middle phase (figure 2.6), in which they impart impermeability, to the reversed middle phase or complex hexagonal phase, in which they allow at least some passage of water. Phase changes may occur in response to changes in temperature and humidity.

The epidermis may also play a vital role in integumental permeability (Ebeling 1976). Injection of most insecticides adversely affects an insect's ability to regulate water loss. This action is presumably due to interference with normal functioning of the epidermal cells. Dead firebrats, *Thermobia domestica*, lose water more rapidly than live ones, suggesting again that an active epidermis is involved in control of water loss.

For further information on cuticular permeability, see Ebeling (1974, 1976), Noble-Nesbitt (1991), and Wharton and Richards (1978).

Insects conserve water in a variety of ways in addition to the integumental mechanisms mentioned above. These will be mentioned in appropriate sections of other chapters.

Molting

The periodic shedding of cuticle followed by formation of a new cuticle is a mechanism facilitating growth despite a more or less inflexible integument. Molting has been the subject of numerous investigations, and as a result a general description is possible (figure 2.7). At the onset of molting, the epidermal cells show much activity. They generally increase in size and often increase in number. During the molting process the epidermal cells separate from the old cuticle and begin to secrete the new. The initial separation of old cuticle and epidermis is called *apolysis* (Jenkin and Hinton 1966). The epidermal cells then secrete the *molting* fluid, or *exuvial fluid*. It contains the chitinase and protease that are capable of digesting endocuticle, which may make up to 80–90% of the old cuticle. The products of this digestion are then absorbed by the epidermal cells. Thus it may be said that a significant portion of the cuticle never entirely leaves the metabolic pool of the insect.

At apolysis, a thin, homogeneous, transparent *exuvial membrane* may appear between the epidermis and the old cuticle. Its origin and function are not established, but it is resistant to molting-fluid enzymes, persists throughout molting, and is shed with the undigested old cuticle.

Figure 2.5   Rate of transpiration in relation to temperature. (*a*) From dead insects. (*b*) From a cockroach nymph at constant saturation deficiency (arrow indicates critical temperature). +, air temperature; •, cuticle temperature.
(a) redrawn with slight modifications from Wigglesworth, 1945; (b) redrawn with slight modifications from Beament, 1958.

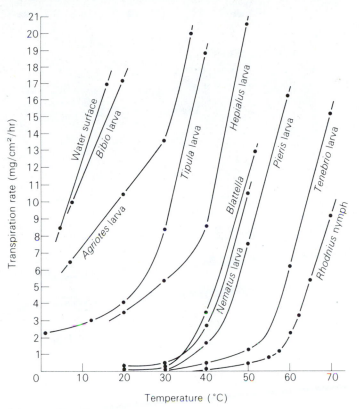

(a)

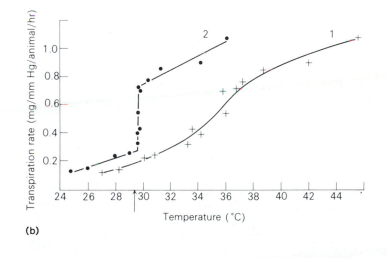

(b)

Figure 2.6 Diagram of wax molecule. (*b*) Lipid–water liquid crystalline phases in wax canals.
Redrawn from Locke, 1965.

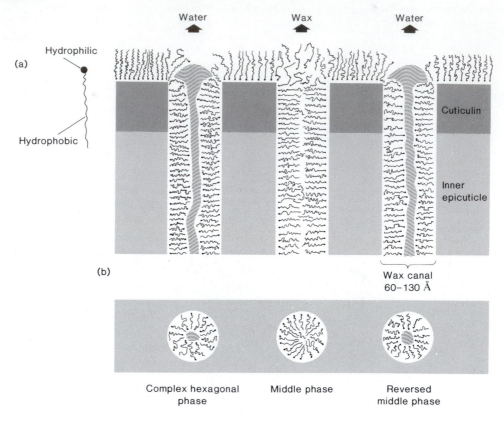

Figure 2.7 Molting process. (*a*) Section of integument prior to molting. (*b*) After apolysis, digestion of old endocuticle in progress and new cuticle being deposited. (*c*) Just prior to ecdysis.
Redrawn with slight modifications from Wigglesworth, 1959.

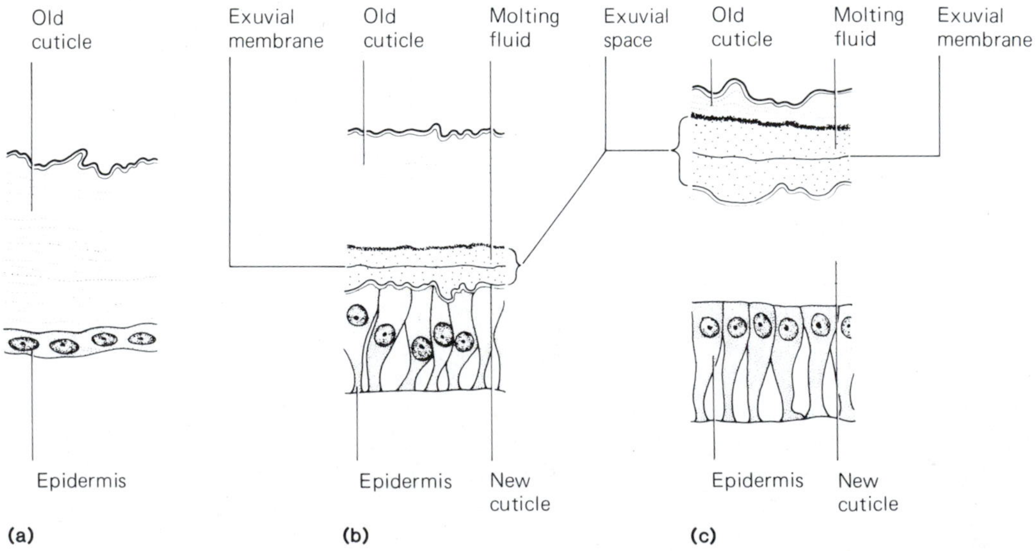

Figure 2.8 Ecdysis (diagrammatic). (*a*) Section of integument prior to molting. (*b*) Section following digestion of endocuticle with line of weakness where endocuticle was previously in contact with epicuticle. (*c*) Splitting of head capsule along preformed ecdysial line.

Redrawn from Snodgrass, 1960.

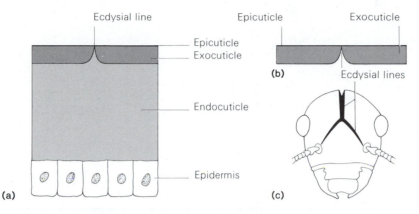

Exocuticle and epicuticle are resistant to the action of the molting fluid and make up the portion of the integument that is shed at ecdysis. As the old endocuticle is digested, forming the *exuvial space,* new cuticle is deposited, cuticulin layer first, then the inner epicuticle, and finally exocuticle and endocuticle, the raw materials of which have come at least in part from digestion of the old endocuticle. The new cuticle is typically wrinkled beneath the old, indicative of the greater surface area to be occupied in the "expanded" insect after the remaining old cuticle is shed. The inner epicuticle probably ultimately determines the maximum to which cuticle can be expanded.

The wax layers are laid down shortly before ecdysis, assuring the waterproofing of the newly emerged insect. The cement layer is the last secreted, forming soon after ecdysis. It is secreted by dermal glands, the canals of which perforate the wax layers and hence allow the cement layer to be formed over the wax layers. Pore canals apparently serve as routes for secretion of the wax layer.

When the secretion of the new cuticle is complete, the insect emerges (the act of *ecdysis*), leaving behind what remains of the old cuticle and the tracheal and gland duct linings, that is, the *exuvia.* This process is facilitated by *ecdysial cleavage lines* beneath which only epicuticle and endocuticle are present (figure 2.8). Because the endocuticle is digested during the molting process, a line of weakness develops. When ready to emerge, the insect may gulp air or water and increase the hydrostatic pressure of the blood by contracting various body muscles. These actions exert an internal force on the ecdysial lines, and the old cuticle subsequently splits wherever they are located. These lines of weakness are usually located on the dorsum of the head and thorax with an anterior–posterior

orientation. Following ecdysis an insect may eat the exuvia and hence reclaim nearly all nutrients that may have been lost during molting. Sclerotization and melanization occur after ecdysis.

Zacharuk (1976) reviews the structural changes and Wolbert and Schafer (1991) review the macromolecular changes that occur during molting.

External Integumentary Processes

The integument of various insects bears a great number of different external processes, and these can be classed (figure 2.9) as noncellular or cellular. Noncellular processes are composed entirely of cuticle and may take any of several forms, such as spines, ridges, or nodules. Some noncellular processes are in the form of *microtrichia* (minute fixed hairs) that lack the basal articulation characteristic of *macrotrichia* (*setae*), which are cellular processes.

Cellular processes may be further broken down into multicellular processes and unicellular processes. Multicellular processes are hollow outgrowths of the integument and are lined with epidermal cells. They generally take the form of spines and are found, for example, on the hind tibiae of certain Orthoptera. Unicellular processes are all referred to as setae (macrotrichia), although they show an extensive diversity of form. They are commonly hairlike, but may be flattened into scales (figure 2.4), may bear branches and appear plumose, or may take on other shapes. The setal shaft is formed by a protoplasmic outgrowth of a specialized hair-forming, or *trichogen cell.* This projection is surrounded by a setal membrane and lies within a socket. The membrane and the socket are formed by a second cell, the *tormogen cell,*

Figure 2.9   External integumentary processes.

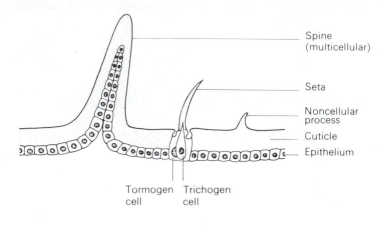

Spine
(multicellular)

Seta

Noncellular
process

Cuticle

Epithelium

Tormogen
cell

Trichogen
cell

Figure 2.10   Sclerites (denoted by ✕), sutures, and internal integumentary processes (diagrammatic).

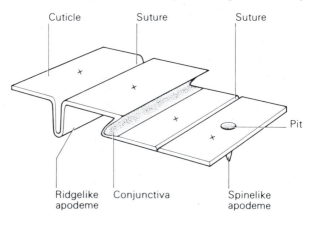

Cuticle   Suture   Suture

Pit

Ridgelike
apodeme

Conjunctiva

Spinelike
apodeme

or socket-forming cell. Setae have various functions. For example, many are connected to a sensory nerve cell, some are associated with a specialized gland that may secrete a toxic substance, and the scale form often contains the pigment or possesses the physical characteristics responsible for much of the coloration and patterning.

The Insect Skeleton

The insect skeleton is composed of a series of plates, or *sclerites,* of varying degrees of hardness (figure 2.10). These sclerites are separated either by very soft membranous areas, called *conjunctivae,* or by external grooves that indicate internal inflections or very narrow lines of thin, sometimes flexible integument. The external grooves are referred to as *sutures* or *sulci* (sing., *sulcus*). The combination of hardened plates joined by conjunctivae allows body movement and articulation of the appendicular joints. The internal inflections of the integument denoted externally by sutures or pits are *apodemes* and may be multicellular or unicellular. These inflections may be spinelike or ridgelike and serve to strengthen the exo-

skeleton or as points for muscle attachment. A spinelike inflection is called an *apophysis.*

Segmentation

The insect body, like that of other arthropods, is divided into a series of segments. Examination of an insect will reveal that many of these segments are rather different from one another. For example, some bear appendages, others do not; the appendages of one segment may be unlike those of another; some segments may be quite movable relative to the others, whereas some adjacent segments are fused or partially fused. One aspect becomes immediately apparent—the obvious functional specialization of the different segments. These specialized functions will be discussed later, but first we want to consider briefly the hypothetical explanation of the origin of these segments.

The primitive arthropod is thought to have been composed of a series of essentially identical segments called *somites,* or *metameres.* Most of these somites apparently carried a pair of appendages. The anterior *acron,*

Figure 2.11   Types of segmentation (tergites only). (*a*) Primary. (*b*) and (*c*) Secondary (adjacent secondary segments overlapping in *c*).
Redrawn with modifications from Snodgrass, 1935.

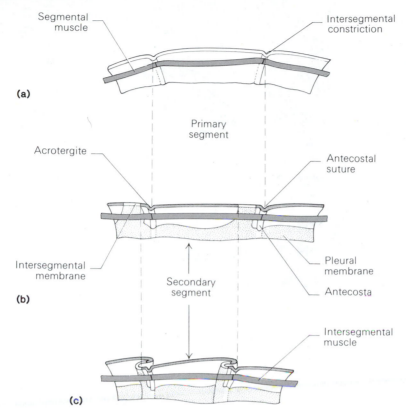

or *prostomium,* and the posterior *telson,* or *periproct,* in which the posterior opening of the alimentary canal, the *anus,* was located, did not bear appendages. The mouth, the anterior opening to the alimentary canal, opened between the acron and the first postoral segment. During the course of evolution these segments fused in different ways, forming the variously divided body regions of modern arthropods. The appendages took on specialized roles appropriate to the body region in which they were located, or disappeared altogether.

The segments in the primitive arthropod were marked externally by constrictions of the integument (figure 2.11*a*). Internally, these constricted regions formed folds where the principal longitudinal muscle bands, the *segmental muscles,* were attached. This arrangement of body units has been called *primary segmentation.* This type of segmentation is found today among the soft-bodied wormlike larvae of several insects—for example, the larvae of Lepidoptera—and in all arthropod embryos.

Another type of segmentation (figure 2.11*b, c*), *secondary segmentation,* is found in adult and many larval insects. In this type of segmentation, the membranous areas between adjacent segments do not coincide with the points of attachment of longitudinal muscles, but are slightly anterior to them. Secondary segmentation is con-

sidered to be a derivative of primary segmentation; the infolded region of a primary segment has become sclerotized, and a membranous area has developed just anterior to it. Secondary segmentation is well illustrated by a pregenital abdominal segment of a generalized insect.

The typical abdominal segment consists of a dorsal sclerotized region, the *tergum* (figure 2.11*b, c*) (see also figure 2.24), and a ventral sclerotized region, the *sternum.* Individual sclerites that form subdivisions of the tergum or sternum are called *tergites* or *sternites,* respectively. The tergum and sternum are separated by a lateral membranous area, the *pleural membrane.* The external groove that corresponds to the infolded portion forms an internal ridge that, in some cases, may be quite pronounced. This internal ridge or apodeme is called the *antecosta.* A small sclerite is demarcated by the antecostal suture and the secondary intersegmental membrane, both in the tergite and in the sternite. The sclerite associated with the tergum is the *acrotergite;* that with the sternum, the *acrosternite.* There is an advantage to overlapping hardened plates, for these plates can then have a protective function and also become involved in locomotion. Because the arrangement of longitudinal muscles in secondarily segmented insects does not coincide with the definitive segments, these muscles are referred to as *intersegmental muscles.*

Figure 2.12   Generalized pterygote insect.

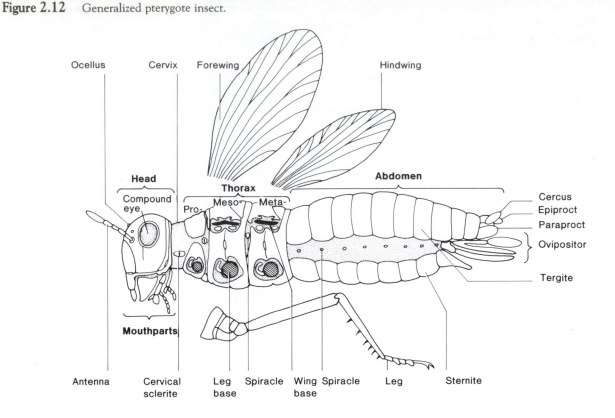

The General Insect Plan

Despite the many features that insects have in common with one another, they display a tremendous diversity in form. This being the case, it is useful to have a hypothetical generalized insect form that subsequently will serve as a conceptual basis for interpreting variations in insect structure, or morphology. For good accounts of general insect morphology and access to literature, see Chapman (1982), Richards and Davies (1977), and Snodgrass (1935, 1963). Matsuda (1976) outlines the major principles of morphology. Richards (1973) discusses the history of insect morphology.

Tagmata

The segments of the insect body are divided into three well-defined regions, or *tagmata:* a head, a thorax, and an abdomen (figure 2.12). The head bears the organs of ingestion, or *mouthparts; compound eyes;* simple eyes, or *dorsal ocelli,* which may be lacking; and a pair of appendages called *antennae.* The thorax is composed of three basic segments named by their relative positions from anterior to posterior as follows: *prothorax, mesothorax,* and *metathorax.* Each of the three thoracic segments bears a pair of legs in most immature and nearly all adult insects. Wings, if present, are found only on the mesothorax and metathorax. The class Insecta is usually divided into two groups (see chapter 12) on the basis of the presence or absence of wings: *Apterygota,* primitively

wingless insects, and *Pterygota,* winged or secondarily wingless insects. The abdomen consists of a varying number of legless segments, the primitive number being considered on the basis of embryological studies to be 11 plus a terminal segment, the *periproct,* or *telson,* which bears the anus. The external genitalia are borne on one or more of the posterior segments.

Head

The insect head (figure 2.13) is a composite structure that has evolved from the fusion of the prostomium with a number of postoral segments and modifications of the appendages of these segments into the organs of ingestion, the mouthparts. There is no general agreement as to the number of primitive segments that contributed to the evolution of the head. Opinions vary from four to seven segments plus the acron (Matsuda 1965, Rempel 1975). (See chapter 11.) Basically, the head is composed of a hardened capsule, the *cranium,* which bears the antennae, eyes, and mouthparts. There is little evidence of segmentation in the mature insect head, and some of the lines or sutures that give it a segmented appearance are probably of functional significance rather than relating to any primitive segmentation. However, one of the sutures, the *postoccipital suture* of the cranium, is considered to have persisted, separating the maxillary and labial segments. The head is attached to the thorax by means of a membranous region, the neck, or *cervix.* The cervical membrane is flexible and allows movement of the head.

Figure 2.13   Generalized insect head. (*a*) Anterior view. (*b*) Posterior view. (*c*) Lateral view. (*d*) Top of head capsule cut away to show tentorium.

Redrawn with slight modifications from Snodgrass, 1935.

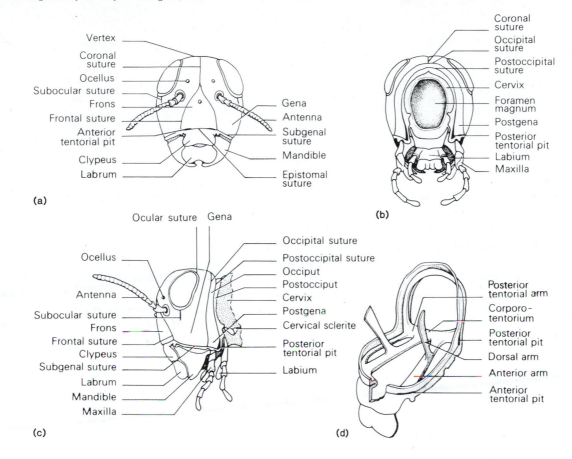

The neck typically bears two pairs of lateral plates, the *cervical sclerites*. Muscles from both the head and the thorax attach to these sclerites and are involved with certain movements of the head. The cervix may be derived in part from the prothorax and in part from the labial segment of the head.

Cranial Structure

The cranium is divided into various regions by a series of sutures. These sutures may be quite apparent in some insects and completely lacking in others. Therefore, this description will pertain to a generalized insect cranium and will mention sutures and regions that may or may not be readily discernable in a given insect. The *epicranial suture* is shaped like an inverted Y, the stem forming the dorsal midline of the cranium and the arms diverging ventrally across the anterior portion of the head. The stem of the Y is called the *coronal suture;* the arms form the *frontal sutures*. The frontal sutures are commonly obscure, especially in adults, and often lacking altogether.

The region delimited by the frontal sutures is called the *frons,* and the dorsal portion of the cranium bisected by the coronal suture is the *vertex*. These three sutures

are lines along which the shed cuticle of the cranium splits during ecdysis (figure 2.8*c*). For this reason it is preferable to call them ecdysial cleavage lines (Snodgrass 1963). Because the ecdysial cleavage lines are poorly developed in many insects, a better definition of frons is the frontal region of the cranium, which bears the *median ocellus,* if present, and internally bears the origins of the muscles of the anterior mouthpart, the *labrum*. Using this definition, the ecdysial cleavage lines would be located on the frons instead of defining it.

The *occipital suture* forms a line from the posterior termination of the coronal suture to just above the mandibles on either side of the cranium. The postoccipital suture lies posterior to, and in the same plane as, the occipital suture. This suture surrounds the posterior opening of the head capsule, the *foramen magnum,* through which the internal organs communicate between the head and thorax. The postoccipital suture forms an internal ridge to which are attached the muscles that move the head. On either side of the cranium immediately above the bases of the mandibles and maxillae are the *subgenal sutures*. Internally, these sutures form ridges that add strength to that portion of the cranium.

Figure 2.14 Generalized insect antenna.

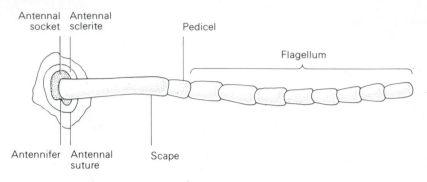

Above each subgenal suture and anterior to the occipital suture is a "cheek," or *gena*. A *postgena* lies adjacent to the gena, but posterior to the occipital suture. The postgenae are delimited posteriorly by the postoccipital suture. Dorsally, the region between the occipital and postoccipital sutures is called the *occiput*. The plate posterior to the postoccipital suture, which surrounds the better part of the foramen magnum, is the *postocciput*. This structure generally bears a bilateral pair of processes upon which the cervical sclerites articulate. The subgenal sutures may be connected across the front of the cranium, just beneath the frontal suture, by the *epistomal suture,* or *frontoclypeal suture.* In a fashion similar to the postoccipital and subgenal sutures, the epistomal suture forms an inflected ridge.

A lobelike structure, the *clypeus,* lies immediately ventral to the epistomal suture. In many insects, the clypeus is hinged with the foremost mouthpart, the labrum. The compound eyes are commonly surrounded by the *ocular sutures.* Similarly, an *antennal suture* surrounds the base of each antenna. In addition, *subocular sutures* may be present, running vertically beneath the compound eyes.

Tentorium

At each of four points on the cranium, the cuticle forms an armlike inflection. Internally these inflections join, forming a framework. This internal framework, the *tentorium* (figure 2.13*d*), affords many points for muscle attachment, contributes appreciably to the rigidity of the head capsule, and provides support for the brain and the part of the foregut that passes through the head. Externally the points of inflection form the anterior and posterior *tentorial pits.* The anterior tentorial pits are generally contained in the epistomal suture, although in some insects they may be more closely associated with the subgenal sutures. The posterior tentorial pits open bilaterally in the postoccipital suture immediately adjacent to the subgenal sutures. Internally the tentorium is composed of a pair of anterior arms, which arise from the anterior pits and fuse with the posterior arms, which correspondingly arise from the posterior pits. At the point

of fusion, a transverse bar, the *tentorial bridge,* or *corporotentorium* is formed. In addition, a third pair of arms may arise dorsally from the anterior arms.

Compound Eyes and Ocelli

On each side of the head is a compound eye, so called because each eye is composed of many individual units. Each eye unit is an *ommatidium.* The surface of each eye is divided into a large number of usually hexagonal facets. These facets are the *corneal lenses,* and they are the cuticular parts of the ommatidia. The dorsal ocelli, or simple eyes, are commonly three in number and are located on the anterior portion of the cranium, one on either side of the coronal suture and the third between the frontal sutures.

Antennae

The antennae are paired appendages that articulate with the head capsule and are located on the anterior portion near the compound eyes. Some evidence suggests that they have been derived from appendages associated with a primitive segment (Rempel 1975). Antennae are usually strictly sensory appendages bearing large numbers of sensilla (see chapter 6). Typically, they are composed of a series of segments and the generalized form (figure 2.14) is filament-like in appearance. There are three basic parts of an antenna in most insects:

1. A basal *scape.*
2. A *pedicel.*
3. A distally located *flagellum,* which is usually long and composed of a number of subsegments.

The pedicel in most insects (except Collembola and Diplura) contains the Johnston's organ, a special sensory structure that is discussed in chapter 6. The scape articulates in an antennal socket, the integument between the antenna and head capsule being membranous and flexible. The rim of the antennal socket associated with the cranium characteristically contains an articular point, the *antennifer.* Antennae are moved about by muscles inserted on the base of the scape. Schneider (1964) provides a review of the research literature on the structure and function of antennae.

Figure 2.15   Generalized mandibulate mouthparts as illustrated by a cockroach.
Redrawn from James and Harwood, 1969.

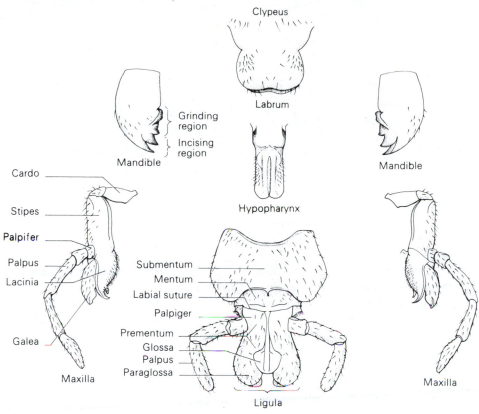

Mouthparts

The mouthparts of the generalized mandibulate type (figure 2.15) are considered to be the primitive form. They typically consist of the following:

1. An anterior "upper lip," or labrum.
2. The hypopharynx.
3. A pair of mandibles.
4. A pair of maxillae.
5. A posterior "lower lip," or *labium*.

There is general agreement that the mandibles, maxillae, and labium arose from the appendages of three primitive segments. The labrum may also represent primitive segmental appendages (Rempel 1975, see figure 11.13).

The labrum is suspended from and articulates with the clypeus by a narrow membrane, which allows considerable movement. The integument of the labrum forms a lining that runs dorsally across the inside of the clypeus and terminates at the true mouth. This epipharyngeal wall forms the dorsal lining of the *preoral cavity* formed by the mouthparts and is continuous with the lining of the pharynx. Commonly the epipharyngeal wall bears a lobe referred to as the *epipharynx*. The *hypopharynx* lies in the preoral cavity somewhat in the way a tongue does

(figure 2.16). The portion of the preoral cavity between the hypopharynx and labrum is the *cibarium*. The portion of the preoral cavity between the hypopharynx and labium forms the *salivarium*.

The *mandibles* (figure 2.15) are a pair of highly sclerotized unsegmented jaws, each of which forms two articulations with the cranium, a posterior articulation with the postgena and an anterior secondary articulation near the tentorial pit. Each mandible has a proximal molar or grinding region and a distal incisor or cutting region.

The paired *maxillae* serve as accessory jaws, aiding in holding and chewing food. They are somewhat more complex than the mandibles. Each is composed of the proximal *cardo,* which bears the *stipes,* which in turn bears two distal lobes, the *galea* and *lacinia*. The galea, a comparatively unsclerotized lobe, is located lateral to the mesal lacinia, which is more sclerotized and contains teeth on its inner edge. The galea commonly covers the lacinia when the mouthparts are "closed." In addition, the stipes also bears a lateral sclerite, the *palpifer,* to which is attached a five-segmented *maxillary palpus*.

The labium is a composite structure formed from the fusion of two primitive segmental appendages, and its

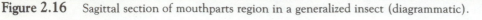

Figure 2.16   Sagittal section of mouthparts region in a generalized insect (diagrammatic).

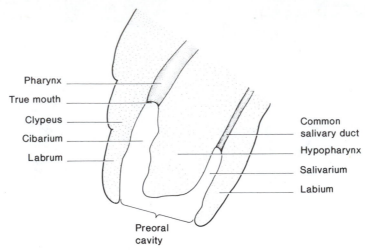

parts can be seen to correspond to those of the maxillae. It consists of a basal *postmentum* attached to the cervix ventral to the foramen magnum and is closely associated with the postoccipital region near the posterior tentorial pits. The postmentum is commonly divided transversely into two portions, a proximal *submentum* and a distal *mentum*. The apical portion of the labium is the *prementum*, which is hinged to the postmentum by the *labial suture*. The *prementum* bears laterally a pair of segmented *labial palpi* and distally four lobes, two inner lobes, the *glossae*, located between two outer lobes, the *paraglossae*. The labial palpi are attached to lateral sclerites on the prelabium, the *palpigers*.

The muscles responsible for the movement of the mouthparts are attached at various points on the head capsule and tentorium. One needs only to observe a grasshopper or cockroach eating to appreciate the intricate, highly coordinated movements of which the mouthparts are capable.

Thorax

As mentioned previously, the insect thorax is composed of three segments: an anterior prothorax, a mesothorax, and a posterior metathorax. Functionally, the thorax is the locomotive tagma because it bears the legs and, if present, the wings. Wings are borne on either or both the mesothoracic and metathoracic segments. These two segments are thus often referred to collectively as the *pterothorax* (ptero = wing). In most winged insects the prothorax is usually separate from, and somewhat less developed than, the remaining segments. In many insects at least part of the first abdominal segment has become intimately associated with the thorax, and in many of the Hymenoptera it has literally become a part of the thorax, being separated from the rest of the abdomen by a constriction.

Each thoracic segment typically can be divided into four distinct regions (figure 2.17a): a dorsal tergum, or *notum;* a pair of bilateral *pleura* (sing., *pleuron*); and a ventral *sternum.* Each of these regions is commonly subdivided into two or more sclerites. The legs arise on the pleura; the wings articulate between the notal and pleural regions. *Spiracles* (figure 2.12), the external openings of the ventilatory system, are usually found one in each of the pleural regions between the prothorax and mesothorax (mesothoracic spiracles) and one in each of the pleural regions between the mesothorax and metathorax (metathoracic spiracles). Prothoracic spiracles are atypical.

Matsuda (1970, 1979, 1981) and Manton (1972) should be consulted for rigorous treatments of the morphology of the insect thorax. Additional information on wing musculature and the functional morphology of the pterothorax may be found in chapters 3, 7, and 11 respectively.

Thoracic Terga

The thoracic terga (nota) of apterygote and some immature pterygote insects are rather simple when compared to the more modified terga of the adult winged forms. Each is merely a single plate associated with the remaining two plates by secondary segmentation. The dorsal longitudinal muscles of the thorax are attached to the antecostae as described in the discussion of secondary segmentation in an abdominal segment. The terga of most pterygote insects are more complex, being divided into smaller sclerites by various sutures. These divisions of the terga, as well as the other modifications of the generalized thoracic segment, arose as a result of the evolution of wings and flight. Typically, the tergum of a wing-bearing segment is composed of two parts: an *alinotum* and a *postnotum* (figures 2.18a and 2.19). The alinotum bears the wings; the postnotum bears the internally inflected *phragma,* which may be viewed as the modified

Figure 2.17 (*a*) Cross section of a generalized thoracic segment. (*b*) Sternal apophyses in the form of a furca (diagrammatic).

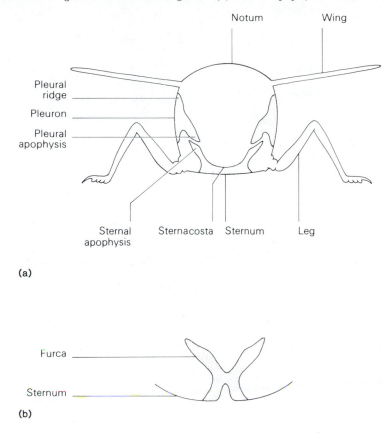

(a)

(b)

antecosta of the next segment posterior. In the case of the metathorax this would be the antecosta of the first abdominal segment. The phragmata are more platelike than ridgelike and afford a comparatively large surface area for the attachment of the dorsal longitudinal wing muscles, which are extremely important in the flight of insects.

According to Snodgrass (1935), the intersegmental membrane present in the more primitive apterygote insects is reduced or lacking and the acrotergite of the next segment posterior has become the postnotum. Correlated with the presence of wings is a strengthening of the alinotum afforded by various internally inflected ridges. Although there are different sutures associated with these internal ridges in various insects, one, the *scutoscutellar suture* (figure 2.18*a*), is present in nearly all winged forms. It lies in the posterior portion of the alinotum and it is shaped somewhat like a V, with the bottom part directed anteriorly. This suture divides the alinotum into an anterior *scutum* and a posterior *scutellum*. Many insects also have a transverse or *prescutal suture*, which lies on the anterior part of the alinotum dividing off a small anterior plate, the *prescutum*. Processes that serve as articular points for the wings have developed on the lateral margins of the scutum. These are the *anterior notal process* and the *posterior notal process*.

Thoracic Pleura

In pterygote insects an internal inflection, denoted externally by the pleural suture, divides the pleuron of each thoracic segment into two parts: an anterior *episternum* and a posterior *epimeron* (figure 2.18*b*). Internally this inflection forms the pleural ridge (figure 2.17*a*), which gives additional strength to the pleuron and which bears, in pterothoracic segments, the *pleural apophysis*, an armlike projection that is directed ventrally and is usually associated with a sternal apophysis (figure 2.17*a*). The prefixes pro-, meso-, and meta- are commonly used in combination with epimeron and episternum; for example, the epimeron of the mesothorax is the mesepimeron. The postnotum of a wing-bearing segment is usually united with the epimeron, forming the *postalar bridge*. The pleuron is usually supported ventrally by the *precoxal bridges* and *postcoxal bridges,* which are fused with the sternum. A small sclerite, the *trochantin* (figure 2.18*b*), is present in many of the more generalized wing-bearing insects. When present it usually bears one of the points of articulation of the coxa. A second articular process is generally located at the ventral extremity of the pleural suture. Dorsally, the *pleural wing process* serves as a ventral articulation point for the wing. Two small sclerites, one anterior (*basalar*) and one posterior (*subalar*) to the pleural wing process, are important in wing movements (more in chapter 7).

Figure 2.18 Generalized wing-bearing thoracic segment. (*a*) Notum. (*b*) Pleuron. (*c*) Sternum. (*a–b*) *redrawn from Snodgrass, 1935, (a) with modifications.*

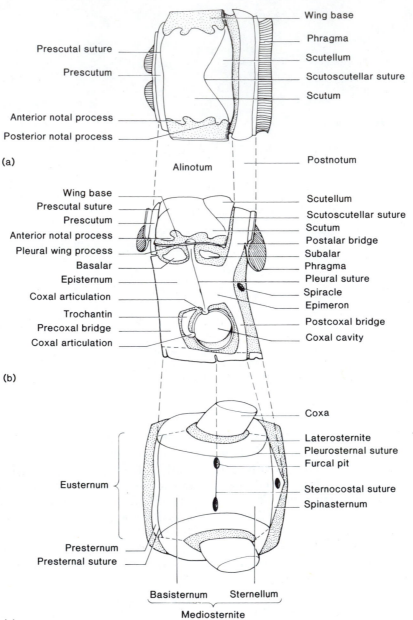

Figure 2.19   Sagittal section of dorsum of generalized pterygote insect thorax (diagrammatic).
Redrawn with modifications from Snodgrass, 1935.

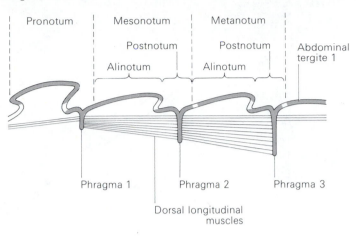

Figure 2.20   Hypothetical primitive arrangement of subcoxal sclerites.
Redrawn from Snodgrass, 1935.

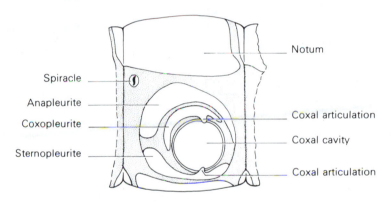

The pleural region in the apterygotes is composed of a group of subcoxal sclerites associated with the *coxa,* the basal segment of the leg. On the basis of comparative morphological studies it has been suggested that the primitive subcoxal sclerites were three in number (figure 2.20), a dorsally located *anapleurite,* forming a rather crude semicircle around the coxa; a smaller *coxopleurite,* concentric with and ventral to the anapleurite; and a ventral *sternopleurite.* Presumably the coxopleurite and the sternopleurite each carried a process with which the coxa articulated. With the advent of wings, the anapleurite and coxopleurite apparently fused, forming a sclerotic pleuron, and the ventral sternopleurite joined with the sternum, resulting in the formation of a continuous sclerotic body beneath the notum and the wings (Snodgrass, 1935). Other interpretations, however, disagree with this "subcoxal" theory of origin of the pleuron (Manton, 1972; see also the discussion of insect phylogeny in chapter 11).

Thoracic Sterna

In many generalized pterygote forms, the sternum (figure 2.18c) of a thoracic segment is composed of a segmental plate, or *eusternum,* and a smaller, posterior plate, apparently derived from an intersegmental plate, the *spinasternum.* The eusternum is commonly divided into three parts: the small, anterior *presternum,* the *basisternum,* and the *sternellum.* The presternum is separated from the rest of the sternum by the *presternal suture.* A pair of internal projections, the *sternal apophyses* (figure 2.17a) arise by invagination of the sternum and are indicated externally by the *furcal pits* (figure 2.18c). The furcal pits are often connected by the *sternacostal suture* which is associated with an apodeme, the *sternacosta* (figure 2.17a), which connects the sternal apophyses. The furcal pits and sternacostal suture form the division between the basisternum and sternellum (figure 2.18c). Lateral plates,

Figure 2.21   Generalized insect leg.

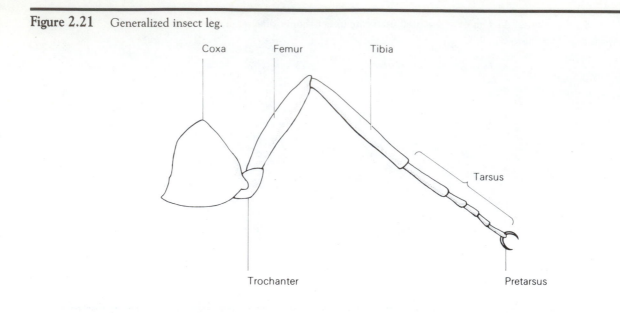

Figure 2.22   Cross section of wing of dragonfly nymph.
Redrawn from Comstock, 1918.

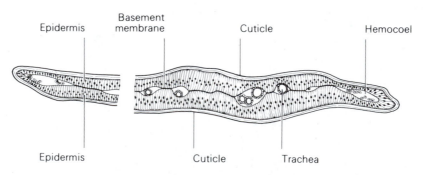

or *laterosternites,* are sometimes divided from the sternum by *pleurosternal sutures.* The sternal apophyses are often in the form of a V diverging from the sternacosta (figure 2.17*b*). In this case they are termed *furcae* (sing., *furca*). Furcalike structures also arise from the sternites of some apterygotes. The spinasternum bears an internally inflected median process, the *spina,* that forms an externally apparent *spinal pit* (figure 2.18*c*). The various internal inflections of the sternum provide strength and areas for muscle attachment in the thorax.

Legs

The generalized insect leg consists of six segments (figure 2.21):

1. A basal coxa, which articulates with the thorax in the pleural region (figure 2.18*b*).
2. A small trochanter.
3. A femur.
4. A tibia.
5. A segmented tarsus.
6. A pretarsus.

The coxa is often divided into two parts, the posterior and (usually the larger part) being called the *meron*. The *trochanter* articulates with the coxa, but usually forms an immovable attachment with the femur. The *femur* and *tibia* are typically the longest leg segments. The *tarsus,* which is derived from a single segment, is usually divided into individual *tarsomeres*. The *pretarsus* may consist of a single claw, but it is usually composed of a pair of moveable claws and one or more pads or bristles. Legs are usually looked upon as the principal organs of terrestrial locomotion, although, as will be seen, they have undergone many modifications and have been adapted to a wide variety of functions, including swimming, prey capture, and digging.

Wings

The wings arise as outgrowths of the integument between the tergal and pleural sclerites (figure 2.17*a*). They are thus composed of two layers of integument (figure 2.22). A series of tracheae grow between these integumentary layers, and when a wing is fully developed, they run within

Figure 2.23 Hypothetical primitive pattern of wing venation. Longitudinal veins: A = anal; C = costa; Cu = cubitus; M = media; R = radius; Rs = radial sector; Sc = subcosta. Cross veins: h = humeral; m = medial; m-cu = mediocubital; r = radial; r-m = radiomedial; s = sectorial. Costal, apical, and anal margins are indicated. *Redrawn from Comstock, 1918.*

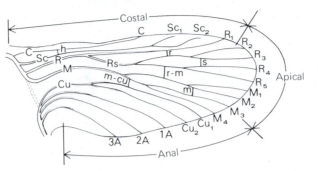

the longitudinal and transverse supportive framework, the *wing veins.* The cuticle is often thicker in the region of these veins, lending further rigidity. Because the wings are outgrowths of the integument, the space between the epidermal layers is continuous with the body cavity, or *hemocoel,* of the insect. This space is usually evident only around the veins, because in the other parts of the wing, in the *cells,* for example, the two layers of integument become closely appressed to one another. In many instances, blood cells, or *hemocytes,* can be seen circulating in the wing immediately on either side of a vein. In many insects complete circulation of hemolymph in the wing is made possible by the presence of numerous cross veins. The wings are, of course, the organs of aerial locomotion in most cases, but, like the legs, they have undergone extensive adaptive modification.

Wings articulate dorsally with the anterior and posterior notal processes (figure 2.18*a, b*) and ventrally with the pleural wing processes (figure 2.18*b*). Refer to figures 7.9 and 7.10 for further details on the articulation of wings with the thorax.

Comparative studies of wing venation in many species have led to the development of various hypothetical generalized patterns. One such pattern is given in figure 2.23. The patterns of wing venation in different insects, interpreted on the basis of the generalized pattern developed by Comstock (1918) and others, are extremely useful in the identification of many insects.

Major *longitudinal veins* (figure 2.23) have been identified by their location relative to one another, their form, their association with basal sclerites (figure 7.10), and the presence of tracheae. The study of fossil forms has also helped identify the major longitudinal veins. Longitudinal veins are connected by *cross veins* (figure 2.23). The combination of longitudinal veins and cross veins, or longitudinal veins reaching the wing margin, divides a wing into various cells. In the more primitive, generalized insect orders, such as Orthoptera, the wings fold in a pleated fashion so that the longitudinal veins lie on top of crests (convex veins) or within troughs (concave

veins). Concave veins are always concave, and convex veins are always convex; for example, the main stem of the radius is always convex. Comstock (1918) and Hamilton (1972) deal in detail with wing venation.

The edges, or *margins,* of wings are named as follows (figure 2.23): the anterior margin, or *costal margin;* the posterior margin, or *anal margin;* and the outer margin, or *apical margin.*

Abdomen

The segmentation of the insect abdomen has already been discussed. As previously mentioned, the primitive number of abdominal segments is considered to have been 11 true metameres plus a terminal segment, the periproct or telson, that contained the anus. The tendency in insectan evolution has been toward a reduction in the number of segments, and in the generalized insect abdomen (figure 2.24) there are 11 segments, the eleventh being reduced and divided into lobes that surround the anus. This terminal segment may bear a pair of appendages, the *cerci.* These are considered to be serially homologous with the legs and mouthparts. The plates of the eleventh segment are generally three in number: the *epiproct* dorsal to the anus, and the two *paraprocts* on either side of the anus. The abdominal segments are usually numbered from anterior to posterior, number one being immediately posterior to the metathorax. Spiracles, the external openings of the ventilatory system, are typically found one on either side of the first eight abdominal segments.

In the generalized female pterygote insect, modified appendages of the eighth and ninth abdominal segments form the *ovipositor,* or egg-laying apparatus, which is composed of two pairs of basal *valvifers.* The valvifers in turn bear the *valvulae,* one pair on the eighth segmental appendage and two pairs on the ninth segmental appendage. The female *gonopore* (reproductive opening) is usually on or posterior to the eighth or ninth segment. The male external copulatory apparatus, *penis,* or *aedeagus,* is usually borne on the ninth abdominal segment.

Figure 2.24 Generalized insect abdomen.
Redrawn from DuPorte, 1961.

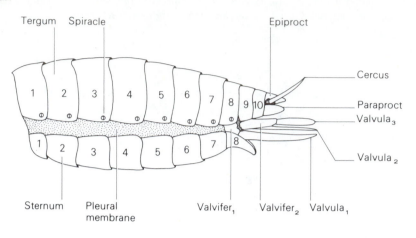

Variations on the General Insect Plan

Matsuda (1976) deals specifically with the morphology of the insect abdomen. Snodgrass (1957), Smith (1969), Tuxen (1970), and Scudder (1971) deal specifically with genitalia.

One of the major reasons for the tremendous success of insects has undoubtedly been the seemingly endless potential of the basic plan to undergo evolutionary modification. It appears as though no structure has escaped some modification. Insects thus are excellent organisms with which to demonstrate the phenomenon of adaptation. Having presented the external structure of a hypothetical, generalized insect as a basis for interpreting modifications, we turn now to some of the more obvious variations in the insect skeleton.

Matsuda (1965, 1970, and 1976), Borror, Triplehorn, and Johnson (1989), C.S.I.R.O. (1991), and Richards and Davies (1977) provide additional information on the external anatomy of members of the various insect orders and families.

Patterns of External Integumentary Processes

Setae are commonly arranged in constant patterns that have often been useful in systematics, for example, among members of the suborder Cyclorrhapha, in the order Diptera (true flies). These arrangements and associated nomenclature are referred to as *chaetotaxy.*

Modifications of the Head

Antennae

Although insect antennae vary greatly in length, overall size, size of the individual segments, segmentation, setation, and other aspects, they can usually be described as being of a particular type or combination of types (e.g., capitate–lamellate). In some insects antennae are weakly developed, as in some larval Hymenoptera and some larval Diptera, or absent, as in the Protura. Antennal type is commonly of value in the assignment of a given insect to a family and may, in certain instances, serve to differentiate the sexes, as, for example, in mosquitoes. Male mosquitoes bear distinctly plumose antennae; the female's antennae are less feathery in appearance and bear comparatively few whorls of hairs. This is an example of *sexual dimorphism,* a structural difference between the two sexes. Most modifications of the antennae occur in the flagellum. Antennal structure is closely related to function. For example, the plumose antennae of many moths provide extensive surface area, and the large numbers of olfactory sensilla that cover them respond to chemical cues associated with mate location.

Commonly occurring antennal types with examples of insects possessing them are presented in figure 2.25.

In most insects antennae serve exclusively as sensory structures. However, they are involved in grasping prey in larval *Chaoborus* spp. (Diptera, Chaoboridae), and they serve as claspers by which males of several kinds of insects hold females during copulation.

Figure 2.25 Antennal types. (*a*) Filiform, grasshopper. (*b*) Moniliform, wrinkled bark beetle. (*c*) Capitate, skin beetle. (*d*) Clavate, carrion beetle. (*e*) Setaceous, dragonfly. (*f*) Serrate, click beetle. (*g*) Pectinate, fire-colored beetle. (*h*) Plumose, male mosquito. (*i*) Aristate, flesh fly. (*j*) Stylate, horse fly. (*k*) Lamellate, scarab beetle. (*l*) Flabellate, cedar beetle. (*m*) Geniculate, honey bee.

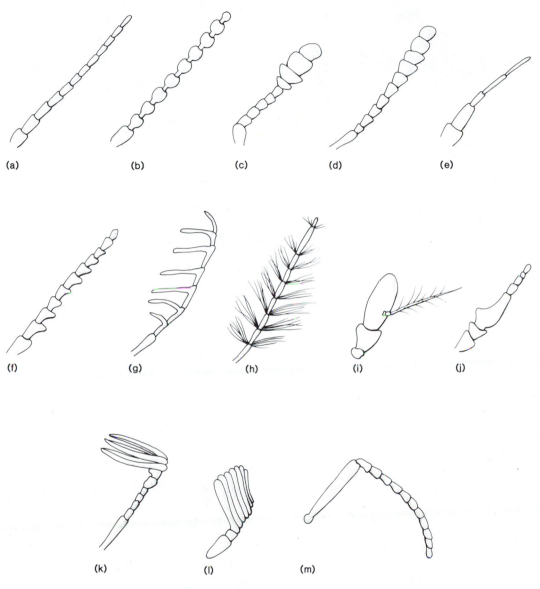

Compound Eyes and Ocelli

Compound eyes are characteristic of most adult insects and many nymphs, but are absent in a majority of the larvae of insects with complete metamorphosis. In these larvae, structures called stemmata are the light receptors. Compound eyes are also absent in the soldier castes of some termite species, some species of fleas, and certain species of springtails and other primitive, wingless insects. When present (figure 2.26), compound eyes occur in many diverse—sometimes bizarre—forms. Some are exceedingly large and contain many facets, as in adult dragonflies (figure 2.26*a*); others are quite small and have few facets, as in many thrips (figure 2.26*b*). In some insects, the compound eyes are actually divided, appearing as two pairs. Examples are found among the beetles in the families Gyrinidae (figure 2.26*c*) and Cerambycidae, and among mayflies in the genus *Cloeon* (figure 2.26*d*). In the latter the anterior division is borne upon a stalk-like outgrowth of the head capsule. The compound eyes of males of the flies in the genera *Bibio* (figure 2.26*e*) and

Figure 2.26 Variations in compound eyes. (*a*) Dragonfly. (*b*) Thrips. (*c*) Whirligig beetle. (*d*) Mayfly, *Cloeon* sp. (*e*) March fly, *Bibio* sp. (*f*) Blow fly, *Phormia* sp., male. (*g*) Blow fly, female.

(*a*) redrawn from Snodgrass, 1954; (*b*) redrawn from Essig, 1958; (*c*) redrawn from Boreal Labs key card; (*d*) redrawn from Comstock, 1940; (*e*) redrawn from Richards and Davies, 1977; (*f*) and (*g*) redrawn from Folsom and Wardle, 1934.

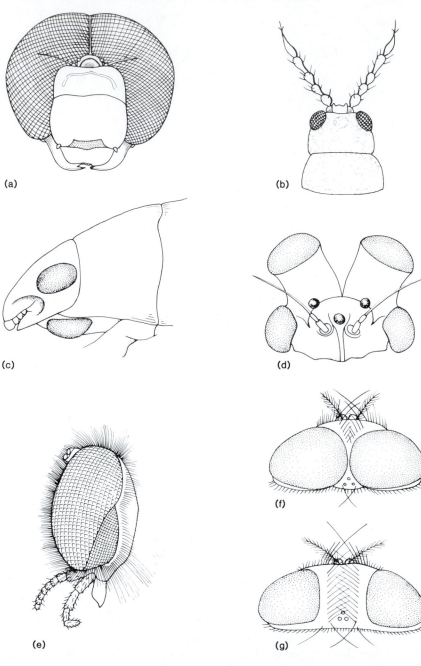

Simulium have two rather distinct areas of different-sized facets, giving the appearance of a compound compound eye. The compound eyes, like the antennae, may be involved in sexual dimorphism, as is the case in the fly *Phormia regina.* In the male fly the compound eyes are large and meet at the dorsal midline of the head; in the female they are smaller and do not meet at the dorsal midline of the head (figure 2.26*f, g*).

Many larval and adult insects have *stemmata,* or *lateral ocelli* (figure 2.27*a*) instead of compound eyes.

Stemmata are the only eyes in most larvae of insects with complete metamorphosis. In other insects (e.g., springtails, silverfish, fleas), stemmata are the only "eyes" of the imaginal stage; compound eyes never appear. The number of stemmata is quite variable, some insects having as few as one on each side of the head (e.g., many fleas) whereas others have many (e.g., 12 in *Lepisma* and 50 in some adult Strepsiptera).

Figure 2.27 Ocelli. (*a*) Stemmata (lateral ocelli) of a caterpillar. (*b*) Two dorsal ocelli, stink bug. (*c*) Three dorsal ocelli, cicada.

(a) redrawn from Snodgrass, 1961; (b) redrawn with modifications from Snodgrass, 1935; (c) redrawn from Boreal Labs key card.

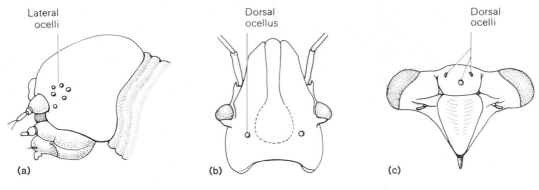

Lateral ocelli

Dorsal ocellus

Dorsal ocelli

(a)   (b)   (c)

Figure 2.28 Scanning electron micrographs of mandibulate and haustellate mouthparts. (*a*) Head of a termite, mandibulate type. (*b*) Head of the spider bug, *Stenolemus* sp., haustellate type. 1, Labrum; 2, mandible; 3, maxillary palp; 4, labial palp; 5, antenna; 6, compound eye; 7, beak (contains stylets). Highly magnified.

Courtesy of Walter J. Humphreys.

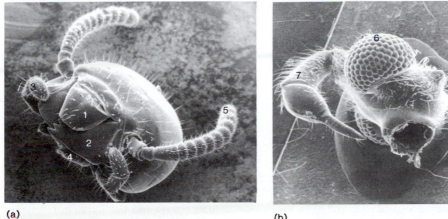

(a)   (b)

Many insects possess simple eyes, the dorsal ocelli, in addition to compound eyes. When present (figure 2.27*b, c*), these ocelli vary in number from one to three: one is very uncommon; two are found in most Hemiptera and in several other insects; and three occur in many members of the order Homoptera and others. Ocelli may also vary in position. For example, the most common position for three ocelli is one in each of the angles formed by the ecdysial cleavage lines. In members of the genus *Perla* in the order Plecoptera all three lie within the angle of the arms of the Y formed by the coronal and frontal sutures.

Mouthpart Structure

Upon examination of the mouthparts of various insects, it seems that there are nearly as many variations in mouthpart structure and function as there are different feeding situations. The mouthparts discussed previously are considered to represent the primitive condition, for they are found in the more generalized insectan forms and in most cases show greater resemblance to primitive locomotor appendages from which, based on comparative morphological and ontogenetic data, they were derived. Specialized mouthparts can, in most instances, be homologized with the generalized type. In a number of insects, mayfly adults (Ephemeroptera) among others, the mouthparts are reduced and nonfunctional.

Mouthparts can be very broadly classified into two groups: biting, or *mandibulate,* and sucking, or *haustellate* (figure 2.28). The mouthparts already discussed represent the mandibulate type. An outstanding characteristic of this type is the presence of a pair of well-developed, usually highly sclerotized, mandibles that articulate at two points with the head capsule and are capable of lateral movement (figure 2.28*a*). Mandibulate mouthparts are generally adapted to chewing activities, the mandibles acting as cutting and grinding structures. However, there are many exceptions (figure 2.29). For

Figure 2.29 Variations in form of mandibles. (*a*) Ground beetle, *Calosoma* sp. with left maxilla and labial palp removed. (*b*) Male dobsonfly, *Corydalus cornutus*. (*c*) Antlion larva, *Myrmeleon* sp.
(*a*) *redrawn from Essig, 1958;* (*b*) *redrawn from Packard, 1898;* (*c*) *redrawn from Peterson, 1951.*

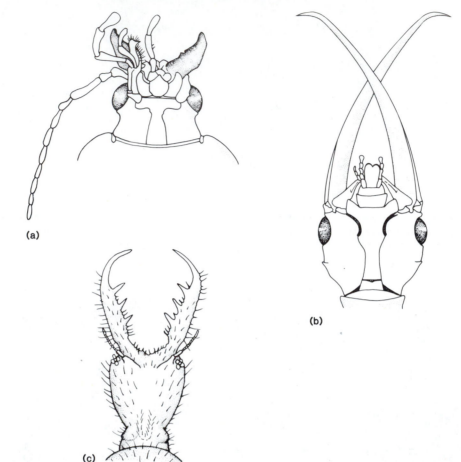

(a)

(b)

(c)

example, in many predaceous beetles and ants, the mandibles are elongate, grasping structures, well adapted for catching and holding prey (figure 2.29*a*). Similarly developed mandibles in the male dobsonfly hold the female during copulatory activities (figure 2.29*b*). The female dobsonfly does not have such impressively developed mandibles, and dobsonfly mandibles provide another example of sexual dimorphism. In some insects, such as antlion larvae (figure 2.29*c*), the maxillae and mandibles are elongate and grasping and together form a food channel through which the body fluids of prey are sucked. Although these particular mouthparts are functionally sucking, they are obvious modifications of the chewing mandibulate type. In pollen-feeding and dung-feeding beetles, the mandibles are more or less flattened and serve to mold dung or pollen into small pellets or balls.

Haustellate mouthparts (figure 2.28*b*) are generally adapted for sucking activities of various sorts. Many are characterized by the presence of *stylets* (figure 2.30), which are swordlike or needlelike modifications of one or more of the generalized mouthpart structures. Stylets may be formed from a combination of one or more of the

mouthparts and the hypopharynx. Stylets enable the insects that possess them to pierce or at least abrade plant or animal tissues and subsequently feed on the fluids that exude or are pumped from the host.

However, not all haustellate mouthparts have piercing stylets. For example, the mouthparts found in most butterflies and moths, in the nonbiting muscoid flies, and in many higher hymenopterous insects lack stylets. Because the mouthparts of each of these groups lack stylets, they are incapable of penetrating tissues. These insects are thus obliged to feed on exposed fluids or soluble solids of various sorts, such as sugar. In the vast majority of butterflies and moths, an elongate sucking tube or proboscis is formed from the galeae of the maxillae (figure 2.31*a*). The remaining mouthparts are reduced or absent. These insects feed mainly on flower nectar. When inactive, the proboscis is coiled beneath the head. This type of mouthpart structure and method of feeding is commonly referred to as *siphoning*. The nonbiting muscoid flies have a rather peculiar method of feeding, often referred to as *sponging* (figure 2.31*b*). A basal segment, the *rostrum,* which is made up of a part of the clypeus and

Figure 2.30   Examples of stylate haustellate mouthparts. (*a*) Sagittal section of head of a sucking louse. (*b–d*) Mouthparts spread out to show details of structure: (*b*) mosquito; (*c*) cicada; (*d*) flea.
(a) after Vogel, 1921; (b) redrawn from Snodgrass, 1959; (c) redrawn from Snodgrass, 1935; (d) redrawn with slight modifications from James and Harwood, 1969.

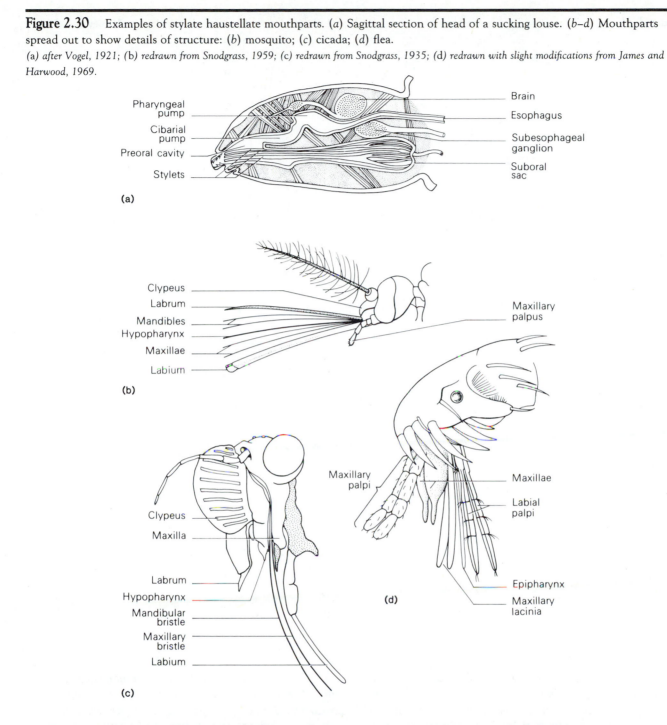

basal portions of the maxillae, bears distally a fleshy retractile proboscis that represents the labium. The apical portion of this proboscis bears the spongelike *labella* in which are located many tiny channels that ultimately converge into the food channel formed by the labrum–epipharynx and hypopharynx. The labella are capable of taking up exposed liquids. These nonbiting muscoid flies egest salivary secretions onto solid foods, so they are quite able to feed on such materials as solid sugar (e.g., dried honeydew).

The more advanced hymenopterous insects have an altogether different "sucking" arrangement (figure 2.31*c*). The labrum and mandibles usually resemble those found in typically chewing insects. For this reason these mouthparts could as easily be included with the mandibulate group as with the haustellate group. However, the maxillae and labium are quite different from the labrum and mandibles, having become united, as in the honey bee, to form a sort of lapping structure. The elongate glossae of the labium form the tubular part of this structure, which is well adapted for thrusting into flower nectaries. This type of mouthpart structure has been functionally classified as chewing-lapping.

Figure 2.31 Examples of nonstylate, haustellate mouthparts. (*a*) Moth. (*b*) House fly, *Musca domestica*. (*c*) Honey bee, *Apis mellifera*.

(a) *redrawn from Snodgrass, 1961; (b) redrawn from James and Harwood, 1969; (c) redrawn from Herms and James, 1961.*

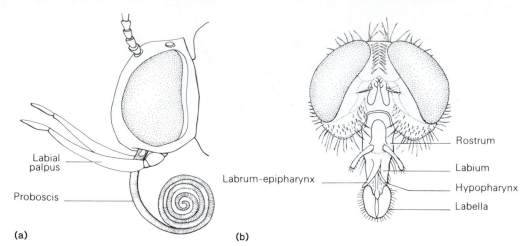

(a)

(b)

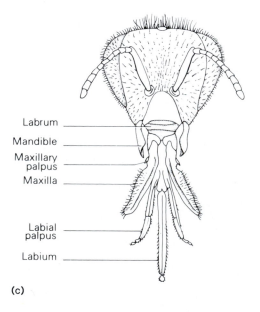

(c)

In most sucking insects, the cibarial region has become modified and functions as a pump. The pharyngeal region also commonly forms a pump. An example of both types of food pumps is found in adult mosquitoes (see figure 4.2).

Mouthparts of Immature Forms

In many instances the mouthparts of immature forms are essentially identical to those of the adult. However, in some, the differences are rather profound. For example, the nymphs (aquatic immatures) of dragonflies and damselflies (figure 2.32*a, b*) present an unusual modification

of the labium. This structure is quite enlarged, and an "elbow" is found at the junction of the prementum and postmentum. The apical portion of the prementum bears a pair of grasping jaws. When the nymph is at rest or stalking prey, the elbow of the labium is flexed and part of the prementum with its jaws forms a "mask" over the other mouthparts. When prey is within striking range, the labium is thrust out extremely fast, the jaws grasping the prey until it can be brought within reach of the maxillae and mandibles (see figure 6.15). In the lower Diptera, the mouthparts, although fundamentally chewing, are adapted for filter feeding, as in larval mosquitoes

Figure 2.32   Examples of mouthparts from immature forms. (*a*) Dragonfly nymph. (*b*) Damselfly nymph. (*c*) Mosquito larva. (*d*) Larva of a true fly (maggot). (*e*) Moth larva (caterpillar).

(a) redrawn from Snodgrass, 1954; (b) and (e) redrawn from Peterson, 1948; (c) redrawn from Snodgrass, 1959; (d) redrawn from Metcalf, Flint, and Metcalf, 1962.

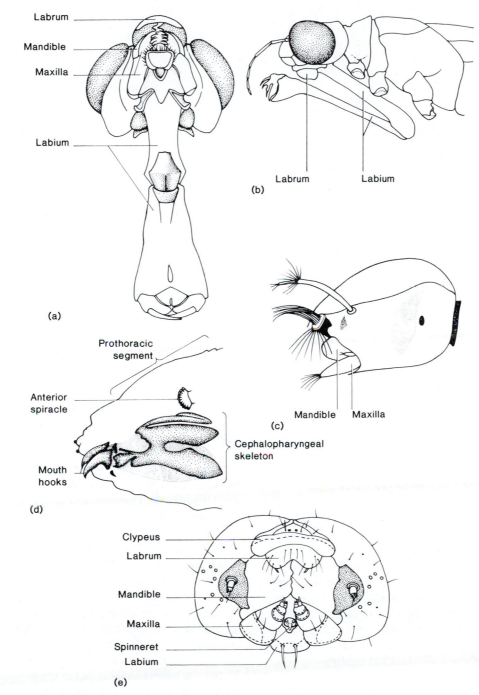

(figure 2.32*c*). In the larvae of higher Diptera, the mouthparts are extremely reduced and have been invaginated into the head, forming the *cephalopharyngeal skeleton,* which anteriorly bears the vertically moving *mouth hooks* (figure 2.32*d*). Lepidopterous larvae (figure 2.32*e*) and others, such as larval trichopterans and hymenopterans, have a rather elaborate apparatus for spin-

ning silk into cocoons. This spinning apparatus, the *spinneret,* is composed of the maxillae, hypopharynx, and labium. Neuropteran larvae, such as antlions and aphidlions, have a grasping-sucking modification of the mandibles and maxillae similar to that in the predaceous diving beetle described earlier.

Figure 2.33 Positions of mouthparts relative to the head capsule. (*a*) Hypognathous. (*b*) Prognathous. (*c*) Opisthognathous. *Redrawn from Snodgrass, 1960.*

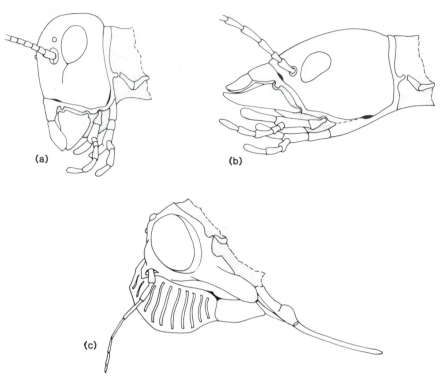

(a)

(b)

(c)

Position of the Mouthparts

There are basically three positions of the mouthparts relative to the head capsule: *hypognathous, prognathous,* and *opisthognathous* (figure 2.33*a–c*). In the hypognathous condition, the mouthparts hang ventrally from the head capsule. This is considered to be the most primitive condition of the three since the mouthparts are apparently modified locomotor appendages and have retained a similar position relative to the insect body. The prognathous condition is characterized by the anteriorly directed position of the mouthparts. Correlated with this modification is an elongation of the genal and postgenal regions and some flattening of the head capsule. Opisthognathous insects are those in which the mouthparts are directed ventroposteriorly relative to the head capsule. This condition is found mainly among the Hemiptera and enables them to place the sucking beak between the legs and out of the way when not feeding.

Head Capsule

The overall shape and structure of the head vary with the position of the mouthparts relative to the head capsule and the necessity for rigidity and muscle attachment associated with the mouthparts. The various sutures and cleavage lines described earlier for the generalized insect may or may not be present and, if present, may be highly modified or reduced. As a result, the cranial regions, which are largely defined by the sutures, also show a great

amount of variation. The internal framework, the tentorium, also varies considerably, again correlated with the necessity for muscle attachment and strengthening support.

In many insects with the hypognathous condition, the occipital foramen (foramen magnum) is closed ventrally, separating the postmentum of the labium from the cervical membrane. This closure is due to the fusion of lobes of the posterior or hypostomal region of the subgenae, and the resulting structure is referred to as a *hypostomal bridge* (figure 2.34*a, b*). In other insects, lobes of the postgenae converge mesially on the hypostomal bridge and, in some cases, may fuse to form a *postgenal bridge* (figure 2.34*c, d*).

In some prognathous insects, the parts on the ventral side of the head are essentially equivalent to the posterior parts in the hypognathous forms. However, in some—for example, several species of beetles and neuropterous insects—the hypostomal regions of the postgenae have fused with one another in a fashion similar to the formation of the hypostomal bridge in hypognathous insects and formed a structure called the *gula* (figure 2.34*e*). In many instances, the gula fuses with the submentum, forming a composite structure, the *gulamentum.*

Modifications of the frontoclypeal region of the head are often correlated with the size of the cibarial region and the pharynx or with the development of the cibarial pump. The clypeus may be quite pronounced, as in many

Figure 2.34 Modifications of the head capsule. See text for explanation. (*a*), (*b*), and (*e*) Diagrammatic. (*c*) Honey bee. (*d*) Vespid wasp. (*f*) Cicada.

(a–d) redrawn from Snodgrass, 1960; (e) redrawn from Snodgrass, 1959; (f) redrawn from Snodgrass, 1935.

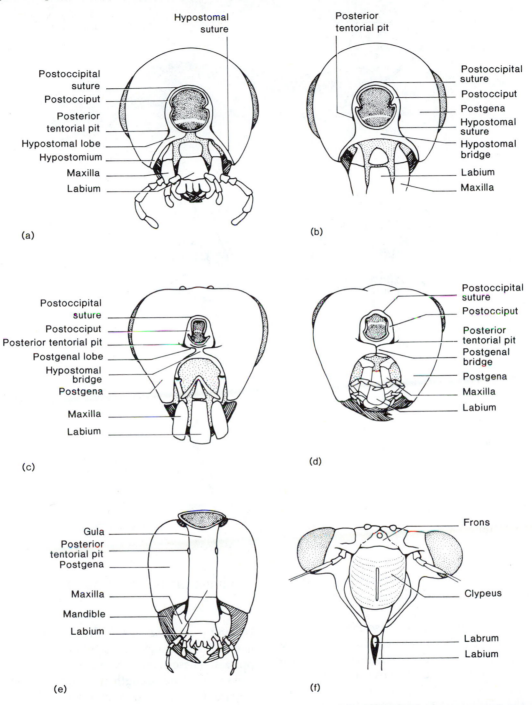

Figure 2.35 Modifications of the pronotum. (*a*) Grasshopper, *Melanoplus bilituratus*. (*b*) German cockroach, *Blattella germanica*. (*c*) Treehopper, *Thelia bimaculata*. (*d*) Hercules beetle, *Dynastes hercules*. (*a*) *redrawn from Ross, 1962; (b) after U.S. Public Health Service; (c) redrawn from Borror and DeLong, 1964; (d) redrawn from Essig, 1942.*

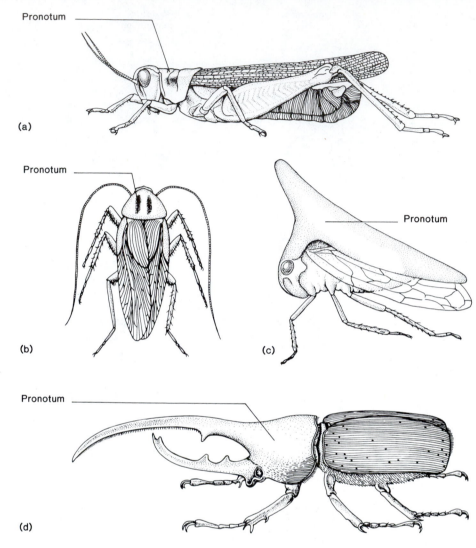

Pronotum

(a)

Pronotum

(b)

Pronotum

(c)

Pronotum

(d)

Hemiptera (figure 2.34*f*), where it presents a broad surface area for the attachment of the muscles of the sucking pump.

The tentorial structure already described is that of a generalized pterygote insect and would be found, for example, in members of the order Orthoptera. In the entognathous apterygote insects (those in which the mouthparts are not externally apparent), the tentorium is absent. In the ectognathous apterygotes (those in which the mouthparts are externally evident) anterior arms arise from the anterior tentorial pits. However, these arms and the posterior bridge do not connect. In some of the higher orders there has been a tendency toward a reduction of the tentorium, and in some the anterior and posterior arms do not connect.

Modifications of the Thorax

Nota

In wing-bearing segments, the terga or nota vary in the number, nature, and location of internally inflected ridges present. These ridges are marked externally by sutures and serve to lend strength and rigidity to the tergum. The terga of the prothorax of pterygote insects are quite different from those of the wing-bearing segments because they are not directly associated with the wing mechanism. Consequently, any sutures associated with internal ridges on the pronotum cannot be considered equivalent to those in the pterothorax. The pronotum is commonly quite small and undeveloped compared to the nota of the wing-bearing segments. However, in many insects it is quite pronounced, forming a pronotal shield (figure 2.35*a, b*). In some insect species it has taken on rather bizarre shapes, often mimicking an environmental characteristic such as a thorn (figure 2.35c) or an elongate projection over the head (figure 2.35d).

Pleura

A pleuron of a winged segment and that of the prothorax are generally similar. However, a propleuron is usually less developed, and there may be secondary modification in a wing-bearing pleuron associated with the wing mechanism. For example, the episternum or epimeron may be subdivided into dorsal and ventral plates, or an anterior plate may be separated from the episternum. The precoxal and postcoxal areas of the pleuron may be separated from the episternum and epimeron. The trochantin is fairly well developed in more generalized pterygotes, somewhat less so in higher orders, and absent in some orders. In certain dipteran species, the meron of the coxa has actually become quite pronounced, forming a part of the pleuron.

Sterna

In the more generalized pterygote insects the major sternal sclerites are usually present, although they may be reduced, leaving rather large membranous areas between them. However, in the higher pterygote orders, there is considerable modification, such as fusion of various plates and formation of secondary sclerites. It is thus often very difficult to identify homologies between these highly modified sterna and those of the more generalized insects.

Modifications of the Legs

Insect legs, although typically ambulatory in function, have been modified extensively in several directions. Both the immature and adult stages of most insects have thoracic legs. However, there are many examples of apodous (lacking legs) larvae (e.g., fly maggots) and even of apodous adults (e.g., female scale insects). Typically developed insect legs are *cursorial;* that is, they are adapted for walking and running. The cockroach is a good example of an insect with cursorial legs (figure 2.36a). In some insects, such as the mole cricket and the nymphs of the periodical cicada, the forelegs are highly modified, bearing heavily sclerotized digging claws (figure 2.36b). *Fossorial* is the term commonly used to denote the adaptation for digging. The forelegs of certain other insects (e.g., the praying mantis) (figure 2.36c) are *raptorial* or modified for grabbing and holding prey. The forelegs have not been the only ones to undergo modification. For example, the femora of the hindlegs of grasshoppers and katydids are enlarged, accommodating the muscles used in jumping (figure 2.36d). Legs adapted for this kind of activity are commonly referred to as *saltatorial*.

The legs of several aquatic insects are modified in such a way that they facilitate swimming, as is the case with adult dytiscid beetles (figure 2.36e), which bear two rows of "swimming hairs" on the edges of the flattened tibiae and tarsi of the middle and hindlegs. These hairs are attached to the legs by movable joints. When the legs are thrust posteriorly during the act of swimming, the distal ends of these hairs move out from the legs, greatly expanding the surface area being applied against the water in the paddling action. As the legs are brought anteriorly in the recovery stroke, the hairs become pressed very close to the legs, reducing the surface area applied against the water, much like the feathering of a paddle when paddling a boat in the wind. The term *natatorial* applies to swimming legs. Although similar in their gross morphology to the middle and hindlegs, the forelegs of the Protura are carried in an elevated position anterior to the body. It has been said that these are principally sensory in function and that they suggest a step in the evolution of antennae.

The legs of many insects bear various specialized structures. For example, the legs of honey bees bear structures that are used during their pollen-collecting activities. One of these structures is the *corbiculum* (figure 2.37a), or *pollen basket,* composed of two rows of hairs on the outer surface of the hind tibia, where the pollen collected by a foraging worker is stored for transport back to the hive. The forelegs of males of some species of diving beetles bear large suction discs on the tarsi (figure 2.37b). They are used to hold the female during copulation.

The hind femora of certain species of short-horned grasshoppers have short peglike structures with which they rub the forewings and produce a sound. The legs of different insects may bear sensory structures of various types (see chapter 6). Several species of flies (e.g., blow flies and house flies) "taste" by means of chemosensilla on the tarsi of their forelegs. The long-horned grasshoppers and crickets possess oval auditory organs, or *tympana,* at the base of each front tibia (figure 2.37c; see also figures 6.8 and 12.15b).

The tarsal and pretarsal segments are also variously modified. Padlike *pulvilli* may be found on the lower surface of each tarsal segment, as in several members of the order Orthoptera, or in association with each *ungue,* or pretarsal claw, as in the flies (figure 2.38a). A bulbous lobelike structure, the *arolium* (figure 2.38b), may be present between the claws. Some insects have a spinelike or lobelike structure, the *empodium* (figure 2.38a), which arises from the distal part of the *unguitractor plate* and also is located between the claws.

Figure 2.36 Modifications of insect legs. (*a*) Cursorial (running) foreleg of a cockroach. (*b*) Fossorial (digging) foreleg of a mole cricket. (*c*) Raptorial (grasping) foreleg of a praying mantis. (*d*) Saltatorial (jumping) hindleg of a grasshopper. (*e*) Natatorial (swimming) leg of a water beetle.

(a–d) redrawn from Boreal Labs key card; (e) redrawn from Folsom and Wardle, 1934.

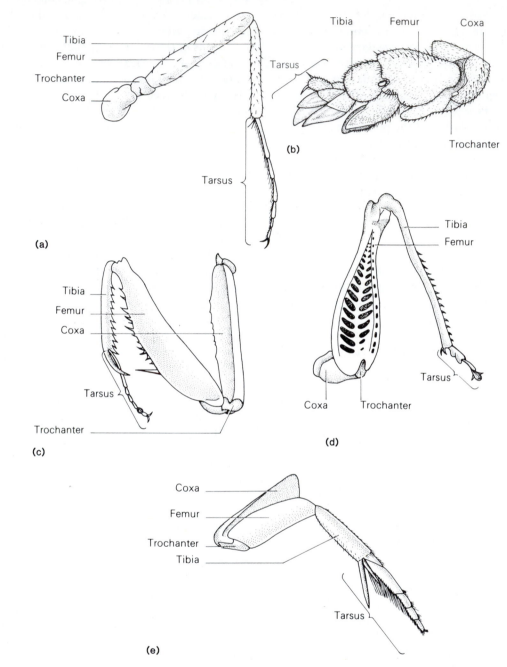

Figure 2.37 Specialized structures borne on the legs. (*a*) Corbiculum or pollen basket on the hind tibia of the honey bee. (*b*) Suction discs on the fore tarsus of a male diving beetle. (*c*) Tympanic organs on the fore tibia of a long-horned grasshopper. (*a*) *redrawn from Snodgrass, 1956; (b) redrawn from Folsom and Wardle, 1934; (c) redrawn from Packard, 1898.*

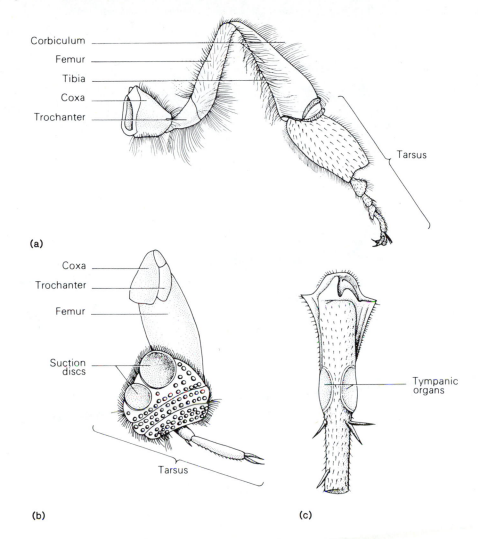

Figure 2.38 Tarsal structures. (*a*) Robber fly tarsus. (*b*) Grasshopper tarsus. *Redrawn from Snodgrass, 1935.*

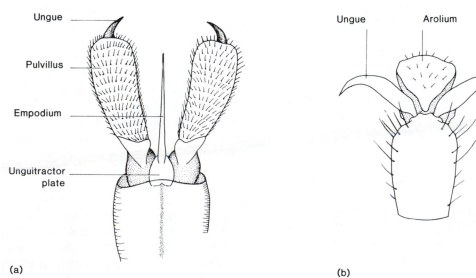

Figure 2.39 Variations in wing structure. (*a*) Wings of a wasp with reduced venation. (*b*) Wings of a dragonfly with an elaborate network of veins. (*c*) Ladybird beetle with left elytron (forewing) and hindwing extended. (*d*) Hemelytron (forewing) of a true bug. (*e*) Lateral view of the thorax of a true fly.

(a), (c), and (d) redrawn from Essig, 1958; (b) redrawn from Boreal Labs key card; (e) redrawn from Smart, 1959.

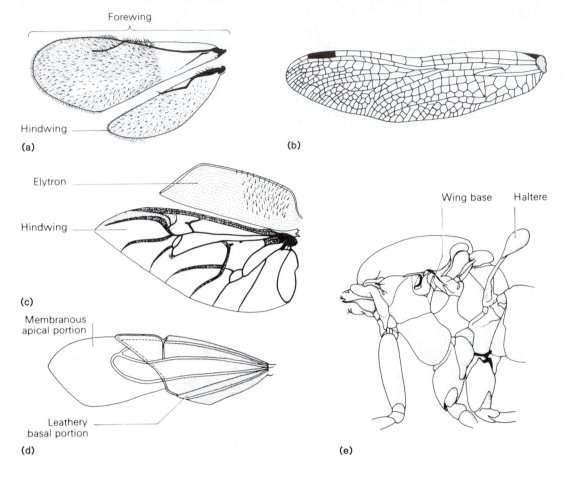

Modifications of the Wings

Insects may bear a single pair of wings, two pairs, or none at all. Many wingless insects are grouped with the winged forms (Pterygota; see chapter 12) on the basis of developmental and morphological similarities. Among these insects, the apterous condition is considered to be secondary, having developed from a winged ancestor. On the other hand, some insects (Apterygota; see chapter 12) and their ancestors never had wings.

There is considerable variation in the structure of insect wings. Examples of some of these variations are presented in the following paragraphs.

Size

The wings may be quite large, as in many of the larger butterflies and moths, or extremely small, as in many of the wasps and flies. The Atlas moth of Australia has been said to have a wingspan of 14 inches. In many insects, there is a tendency for the hindwings to be smaller than the forewings (figure 2.39*a*).

Venation

Because of the tremendous variation in wing venation, it has been used as a source of taxonomic characters. Venation ranges from the extensively reduced and simplified system found in many of the wasps (figure 2.39*a*) to the highly complex network in the wings of dragonflies and damselflies (figure 2.39*b*). The regions between principal veins in many of the more primitive insect groups, for example, Odonata (figure 2.39*b*), contain irregular networks of veins. These networks probably represent what remains of the *archedictyon* described from insect fossils. Veins also vary in thickness; for example, those of the periodical cicada are quite thick, whereas those in the scorpionfly are very thin and delicate.

Function and Texture

The most obvious function of wings is, of course, flight. However, in several instances the wings have been modified or are at least used for different purposes. In the beetles, the hindwings are membranous and fold beneath the forewings, which are usually quite hard and form a

Figure 2.40   Wing-coupling mechanisms. (*a*) Wings of the honey bee. (*b*) Wings of a frenate moth (Psychidae). (*c*) Wings of a jugate moth (Hepialidae).
Redrawn from Comstock, 1918.

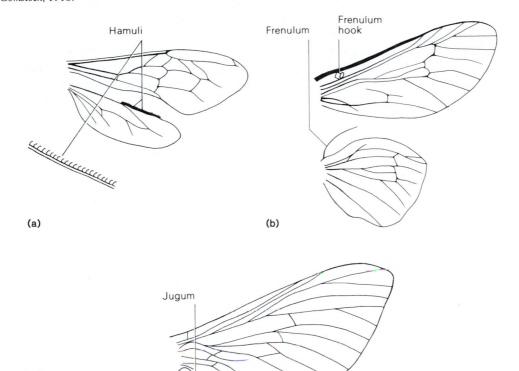

protective armor for the membranous hindwings when not in use. These modified forewings (figure 2.39*c*) are called *elytra* (sing., *elytron*). Elytra may be sculptured in various ways. Among the Hemiptera, the forewings are only partly hardened, the distal portions remaining membranous and containing veins. These structures (figure 2.39*d*) are appropriately named *hemelytra* (sing., *hemelytron*), or "half"-elytra. The forewings of orthopterans are parchmentlike and probably afford similar protection to the hindwings. A parchmentlike wing is called a *tegmen* (pl., *tegmina*). In some insects the wings are used for the production of sound. Several examples of insects that produce sound with their wings are to be found among the Orthoptera. The field cricket is well known for this activity. The true flies possess a single pair of well-developed forewings and a pair of highly modified hindwings, the *halteres* (figure 2.39*e*). These club-shaped structures are important in the stability of flight of these insects. In the very hot days of midsummer, honey bees fan their wings in a community effort and thereby reduce the temperature within the hive.

Relationship to One Another

Among the insects with two pairs of wings, the wings may work separately as in the dragonflies, damselflies, mayflies, and Neuroptera. However, in many of the higher pterygote insects, the forewings and hindwings are coupled to one another in various ways, resulting in each pair of wings acting together as a unit. Examples of wing-coupling mechanisms include tiny hooks, or *hamuli,* found among the Hymenoptera (figure 2.40*a*) and the spinelike *frenulum* (figure 2.40*b*) or lobelike *jugum* (figure 2.40*c*) found in the Lepidoptera. In insects that have wing-coupling mechanisms, the hindwings are usually somewhat smaller than the forewings. The tendency toward a reduction of the hindwings has, of course, reached its maximum in the true flies, which have lost the hindwings as such, the remnants being the previously mentioned halteres.

Resting Position

When not being utilized for flight, the wings are held in various positions relative to the body. Members of the primitive orders Ephemeroptera and Odonata are unable to flex the wings over the abdomen and hence when at rest hold them vertically over the dorsum (mayflies and damselflies) or horizontally (dragonflies). Other insects (bees, wasps, etc.) are able to flex the wings over the abdomen at rest. Many Homoptera and others (e.g., Neuroptera) hold the wings rooflike over the abdomen.

Coloration

Although many insects have *hyaline* (clear) or opaque unpigmented wings, there are groups in which wing coloration is especially well developed. The coloration may be due to pigmentation within the integument itself or to a covering of minute pigmented scales, or the covering scales may physically resemble thin layers or diffraction gratings in their effects upon impinging light. The bright and decorative colorations on the wings of some dragonflies are a good example of the first kind of coloration. The butterflies and moths as a group possess the second type of coloration, colored scales. Variety and complexity of coloration abounds in the Lepidoptera, and many behavioral patterns are intimately linked with coloration. In several instances the coloration matches the environmental background and in this way affords a degree of protection from predators. The protective value of coloration is discussed in chapter 9.

Presence of Hairs and Scales

The scales of butterflies and moths have already been mentioned. These are considered to be modified setae. The wings of the caddisflies are covered with tiny hairs, which are also modified setae. Wings of other insects bear various types of macrotrichia and microtrichia. Although microtrichia are randomly scattered over a wing surface, macrotrichia (true setae) are not. Macrotrichia tend to be concentrated along the major veins and branches. Thus distinct rows of macrotrichia, in the absence of a vein, are assumed to indicate the location of a vein that has been lost. As such, these hairs have been useful in understanding evolutionary changes in wing veins.

Modifications of the Abdomen

In adult pterygote insects the abdominal segments (except the first) anterior to those that bear the external genitalia are usually quite simple and uniform. Each of these segments consists of a tergum and a sternum separated by a pleural membrane and never bears appendages. As explained earlier, the first abdominal segment in pterygote insects is associated more with the thorax than the abdomen, because the antecostal portion of the tergum furnishes the third phragma, to which the dorsal longitudinal wing muscles are attached. In many of the Hymenoptera the first abdominal segment, the *propodeum* (figure 2.41*a*), is completely associated with the thorax and is separated from the remaining abdominal segments (collectively, the *gaster*) by a constriction, the *petiole*.

The abdomen as a whole varies considerably both in size and in the number of segments. As pointed out earlier, the primitive number of segments is considered to have been 12. This condition occurs, for example, among adult Protura and in the embryos of some higher insects. The telson is probably represented in most insects by the membrane surrounding the anus, but in Odonata the anus is surrounded by three small plates that may represent the telson. The tendency has been toward a reduction in the number of abdominal segments. The springtails have 6 segments in their abdomen, the more generalized pterygotes have 11, and the higher pterygotes usually have 10 or fewer.

The size of the abdomen relative to the remainder of the body ranges from the tiny abdomen characteristic of several parasitic wasps (figure 2.41*b*) to the extremely large abdomen of a gravid termite queen (figure 2.41*c*).

Nongenital Abdominal Appendages

Unlike the pregenital and genital segments in adult pterygotes, the pregenital and genital segments of many larval pterygotes and many apterygote insects do bear appendages of various sorts. For example, the first three abdominal segments of adult proturans bear rather simple, bilateral appendages, the *styli*. Similarly, styli are usually borne on several of the abdominal segments of adult thysanurans (figure 2.42*a*).

Members of the order Collembola are in many ways quite different from other insects. One of the outstanding differences is the presence of three rather unique abdominal structures (figure 2.42*b*). The most anterior structure, the *collophore*, is located on the venter of the first abdominal segment. This structure is roughly cylindrical in appearance when it is protruded by the hydrostatic pressure of the hemolymph. *Collophore* means literally "that which bears glue," based on an early suggestion that

Figure 2.41 Modifications of the abdomen. (*a*) Fire ant. (*b*) Parasitic wasp, *Eusemion* sp. (*c*) Gravid termite queen. *(a) after U.S. Public Health Service; (b) redrawn from Essig, 1958; (c) redrawn from Skaife, 1961.*

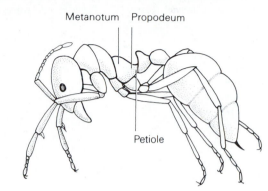

(a)

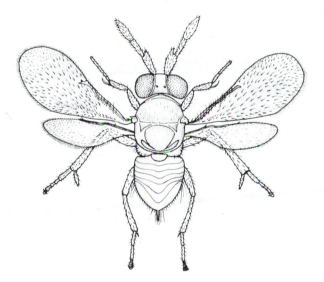

(b)

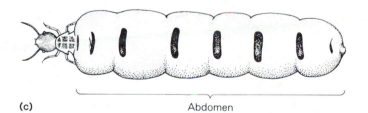

(c)

Figure 2.42 Nongenital abdominal appendages. (*a*) Venter of posterior portion of the abdomen of a silverfish. (*b*) Springtail. (*c*) Caterpillar. (*d*) Mayfly nymph. (*e*) Aphid.
(a) redrawn with modifications from Essig, 1942, after Oudemans; (b) after U.S. Public Health Service; (c) redrawn from Snodgrass, 1961; (d) and (e) redrawn from Boreal Labs key card.

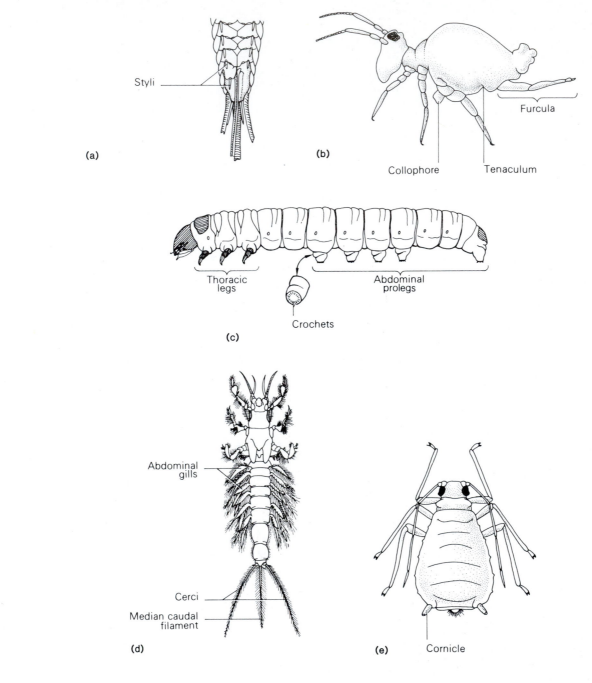

Figure 2.43 Modifications of the terminal abdominal segments. (*a*) Posterior portion of earwig abdomen with forcepslike cerci. (*b*) Stonefly nymph with feelerlike cerci. (*c*) Posterior portion of damselfly abdomen with gill-bearing epi- and paraprocts. (*a*) *redrawn from Hebard, 1934; (b) redrawn from Ross, 1962; (c) redrawn from Snodgrass, 1954.*

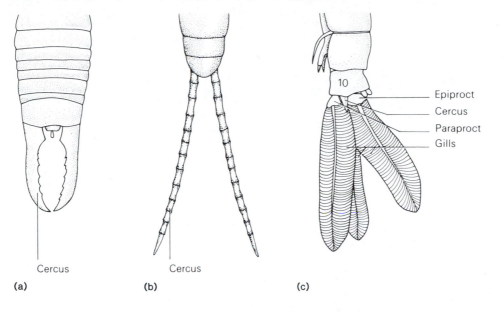

Cercus   Cercus

(a)   (b)   (c)

it might serve as an organ of adhesion (see chapter 12). The two posterior structures are the *tenaculum* on the venter of the third segment and the *furcula* on the venter of the fifth segment. The furcula is capable of being moved anteriorly, engaged by the tenaculum, and subsequently released, exerting a force against the substrate and propelling the insect through the air. Hence the name "springtail."

Abdominal appendages of pterygote larvae usually serve either a walking or a ventilatory function. An example of the first is found among the larvae in the order Lepidoptera (figure 2.42*c*), which generally bear bilateral appendages on the first four or five, and commonly the tenth, abdominal segments. These are the *prolegs* and complement the three pairs of thoracic appendages in the locomotion of the insect. Each proleg bears a series of minute hooks or *crochets*. Bilateral abdominal appendages in mayfly nymphs serve as gills (figure 2.42*d*), facilitating the absorption of oxygen from and release of carbon dioxide into the surrounding water. A pair of lobelike projections, *cornicles,* on the posterior dorsum of the abdomen is characteristic of Aphids; figure 2.42*e* (order Homoptera).

The structure of the terminal abdominal segments (e.g., the cerci and the epiprocts and paraprocts) is variously modified. For example, the cerci may be forcepslike or clasperlike (figure 2.43*a*), feelerlike (figure 2.43*b*), reduced, or absent. Cerci are characteristic of the more primitive orders, such as Ephemeroptera and Orthoptera, but are absent in the hemipteroid orders and most of the higher (holometabolous) orders, except the Mecoptera and possibly some Hymenoptera. Like the cerci, epiprocts and paraprocts may be long and filiform, may bear anal gills (figure 2.43*c*), or may be reduced or inconspicuous.

External Genitalia

In insects that have an ovipositor, this structure may show considerable variation, depending upon the situation into which the eggs must be placed. For example, cicadas deposit their eggs beneath the bark on tree twigs. Their ovipositor (figure 2.44*a*) is well adapted for this function, being composed of three rather sharp and rigid blades. Other ovipositors, such as those of the katydids (figure 2.44*b*), are constructed for depositing eggs beneath the

Figure 2.44 Modifications of the ovipositor. (*a*) Cicada, *Magicicada septendecim.* (*b*) Katydid, *Scudderia* sp. (*c*) Ichneumon wasp, *Megarhyssa lunator.* (*d*) Firebrat, *Thermobia domestica.* (*e*) Telescoping terminal abdominal segments serving as an ovipositor in the house fly, *Musca domestica.* Numerals indicate abdominal segments.
(*a*), (*b*), and (*d*) *redrawn from Snodgrass, 1935; (c) redrawn from Riley, 1888; (e) redrawn from West, 1951.*

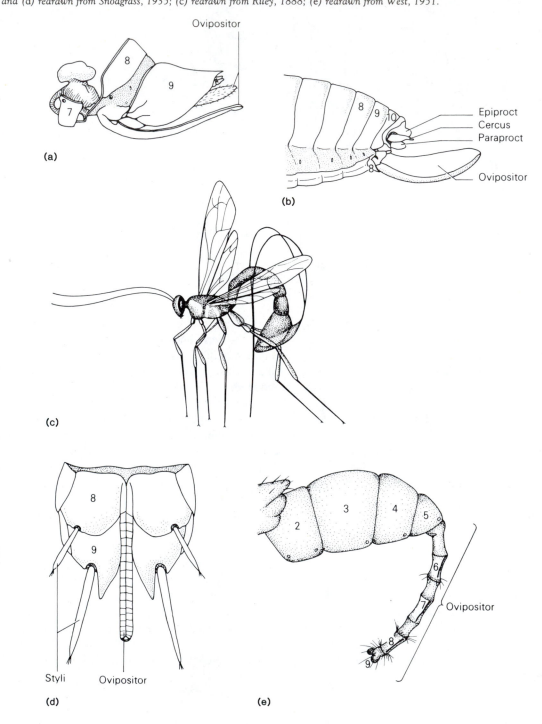

surface of the soil or in plant tissue. Some parasitic ichneumon wasps have extremely long ovipositors (figure 2.44c), which enable them in some instances to penetrate the bark of a tree and deposit an egg in a wood-boring larva.

The adult female members of the apterygote order Thysanura possess a very primitive ovipositor composed of paired appendages borne on the venter of the eighth and ninth abdominal segments (figure 2.44d). Each appendage or *gonopod* is borne on a basal *coxopodite,* which may or may not bear a stylus. Many similarities are apparent between this primitive ovipositor and that found among the pterygotes. Many insects lack an ovipositor altogether and have devised other means of egg deposition. For example, certain members of the orders Thysanoptera, Mecoptera, Lepidoptera, Coleoptera, and Diptera use the abdomen itself as an ovipositor, or *oviscapt* (figure 2.44e). Some are capable of telescoping the abdomen to greath lengths.

Mayflies (Ephemeroptera) and earwigs (Dermaptera) are exceptional in that the female genital opening (gonopore) is just behind the seventh abdominal segment.

The external genitalia of male insects are commonly extremely complex structures and probably show more inherent variation than any other insectan structure. It is for this reason that taxonomists have made extensive use of these structures in their work. Since taxonomists usually specialize in a single group, there has been a proliferation of terminologies applied to male genitalia. This obviously creates a problem for the entomology student and he or she is best referred to Tuxen (1970). Attempts to homologize the male genitalia in the diverse insectan groups include Snodgrass (1957) and Smith (1969).

Two exceptional situations deserve mention. Although dragonflies and damselflies (Odonata) are like most other insects in having the gonopore on the ninth abdominal segment, the copulatory organ in these flies is located on the venter of the second and third abdominal segments. Male mayflies have paired gonopores and penes.

Modifications of the General Body Form

The shape of many insects is an obvious adaptation to a given environmental situation. An outstanding example of this is found in the fleas. These insects are bilaterally flattened, a characteristic that enables them to move easily between the feathers or hairs of their hosts. One does not realize the efficiency of this adaptation until he or she attempts to remove one of these insects from a pet cat or dog. An example of an insect flattened dorsoventrally is the bed bug. This body shape enables the bed bug to hide in tiny cracks and crevasses between feedings. Many insects possess rather bizarre shapes and in many instances actually mimic an environmental characteristic, as in the case of protective coloration, which often provides protection from potential predators (see chapters 8 and 9). Some insects (e.g., walkingsticks, Phasmida) are elongate and tubular in shape. In the sedentary stage, the bodies of scale insects are so modified that they are hardly recognizable as insects.

CHAPTER 3

The Nervous, Glandular, and Muscular Systems

The Nervous System

The many, diverse activities of the various systems of an insect are coordinated in large part by the nervous system. This system is composed of elongated cells, or *neurons,* which carry information in the form of electrical impulses from external and internal sensilla (sensory cells) to appropriate effectors, and special cells called *glial cells,* which protect (i.e., provide a nearly constant extracellular bathing fluid), support, and provide nutrition for the neurons. The nature and location of the stimulated sensilla determine the nature of the response. The basic effectors are muscles and glands. However, there are other effectors (e.g., the light-producing organ of the firefly). The nervous system and effector organs enable an insect to adjust continually and often very quickly to changes (*stimuli*) in both the internal and external environment and to respond in a manner favorable to the maintenance of life.

Huber (1974) and Treherne (1974) provide good reviews of the structure and function of the insect nervous system. A classic, comprehensive coverage of the invertebrate nervous system is that of Bullock and Horridge (1965), whereas Lane (1985) reviews the structure of the insect nervous system. In the same volume, 5, several authors provide new information on the more detailed aspects of structure and motor function of the nervous system. Volume 9 (Behavior) of Kerkut and Gilbert (1985) contains chapters dealing with the functional aspects of the nervous system.

Structure and Function of the Nervous System

The Neuron

The basic functional unit of the nervous system is the nerve cell, or neuron (figure 3.1*a*). A neuron may be described as a thin-walled tube that varies from much less than 1.0 mm to more than 1 meter (in larger animals) in length and has a diameter between 1.0 and 500 micrometers. Insect neurons are usually comparatively small in diameter, on the order of 45–50 micrometers for the largest (much less for the smallest). Typically, a neuron consists of a cell body, the *perikaryon,* or *soma,* and one or more long, very thin fibers, or *axons* (sometimes called neurites). A branch of an axon is called a *collateral.* Associated with the cell body or near it are tiny branching processes, the *dendrites.* Similar branching processes, the *terminal arborizations,* are found at the end of an axon.

Structurally speaking, neurons may be *unipolar, bipolar,* or *multipolar* (figure 3.1*a–c*). In unipolar neurons, a single stalk from the cell body connects with the axon and a collateral. In bipolar neurons the cell body bears an axon and a single, branched or unbranched dendrite. Multipolar neurons have an axon and several branched dendrites.

The individual neurons are not connected directly to one another but communicate either electrically (electrically coupled) or via special molecules called *neurotransmitters.* The finely branching terminal arborizations of an axon come into extremely close association with the dendrites or axon of another neuron or they may end near a muscle (i.e., a neuromuscular junction). A very small but measurable distance, however, always lies between them. The region of close association between terminal arborizations and dendrites is called a *synapse,* and the space between the arborizations and dendrites is called the *synaptic cleft.* Usually the neuron from which the signal is coming is termed the presynaptic neuron and the next neuron in line is called the postsynaptic neuron. These terms will be useful when nerve to nerve chemical communication is discussed. In insects, synapses form only

Figure 3.1 Structural types of neurons. (*a*) Unipolar. (*b*) Bipolar. (*c*) Multipolar. (*d*) Relationship among sensory, motor, and interneurons.
Redrawn from DuPorte, 1961.

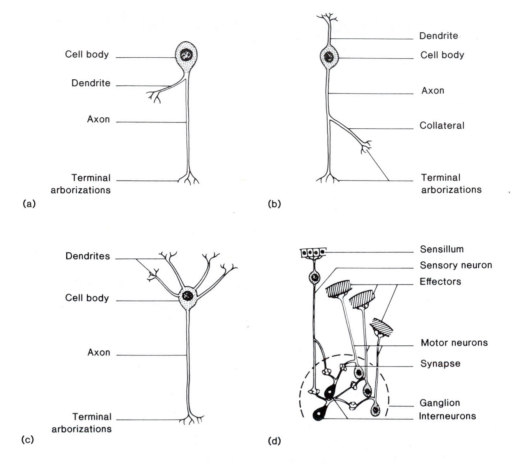

along the dendrites or axons and do not occur on the perikaryon. Readers interested in knowing more about electrical synapses or junctions are referred to Lane (1985).

Most synapses occur within the neuropile (regions of interconnected axons and dendrites) located in aggregations of neurons called *ganglia* (sing., *ganglion*). The term *ganglion* means a "swelling." Several histological components of a ganglion can be identified (figure 3.2).

1. An outer connective tissue layer, the sheath, or *neural lamella.*
2. The *perineurium,* a layer of cells beneath the neural lamella (the perineurium probably secretes the material that composes the neural lamella).
3. A region containing the neuron cell bodies with associated glial cells.
4. A central region consisting of intermingling, synapsing axons encapsulated by processes of glial cells, the *neuropile.*

Distinct *fiber tracts,* groups of axons running parallel to one another, are usually evident in the neuropile. Within a ganglion, extracellular spaces are present between the glial cells surrounding axons. The fluid in these spaces contains higher concentrations of sodium and potassium ions and a lower concentration of chloride ions than the hemolymph. Maintenance of the proper ionic concentration of this fluid is critical to neural function. It is provided by the perineurium, which forms the "blood-brain" barrier. The neural lamella, perineurium, and glial cells are apparently involved in maintaining the composition of this fluid as well as providing physical support and transporting and storing nutrients used by neurons. In addition, glial cell processes serve to insulate single axons or groups of axons from one another. Unlike vertebrates, insects and other invertebrates do not have the glial membranes compressed to form a myelin sheath. Even though they greatly outnumber the neuron cells, our knowledge of the functional significance of glial cells in insects lags considerably behind that of the neurons. Autoradiographic studies on insects and other invertebrates recently showed that the glial cells are able to "clean up" various substances such as exogenous molecules, ions, or transmitters from the extracellular spaces surrounding the neurons. Obviously, more research effort is needed on this particular aspect of the insect's nervous system. Carlson and Saint Marie (1990) provide the most recent review of the structure and function of glial cells.

Figure 3.2 Cross section of part of the caudal ganglion of the cockroach *Periplaneta* sp. (diagrammatic). Darkly shaded areas indicate extracellular spaces.
Redrawn from Smith and Treherne, 1963.

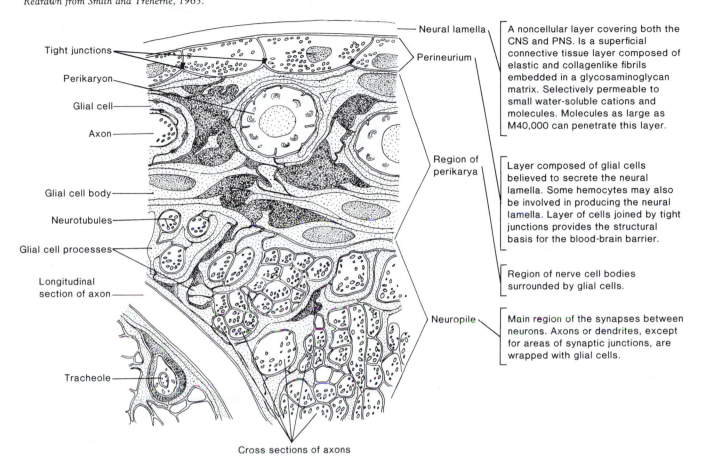

Cross sections of axons

Nerves are bundles of axons that are invested in a noncellular neural lamella, which is secreted by hemocytes and/or the underlying glial cells making up the perineurium. Individual neurons are also surrounded by a different type of glial cell than those composing the perineurium. Nerves provide connection among ganglia and between ganglia and other parts of the nervous system.

Types of Neurons

Based on structure, the insect nervous system is composed of three types of neurons (figure 3.1*a–c*). In addition, neurons are classified in ways that relate more to function.

  I. Relationship to type of input or output (figure 3.1*d*)
   A. Sensory or afferent
   B. Motor or efferent
   C. Interneuron, association, or internuncial
 II. Type of action on postsynaptic neurons
   A. Excitatory
   B. Inhibitory

III. Chemical taxonomy using immunocytochemical techniques with antibodies developed against various neurotransmitters released by the neuron
   A. Cholinergic
   B. Glutaminergic
   C. Aminergic
   D. Peptidergic

Sensory neurons, or *afferent neurons,* are usually bipolar, and cell bodies are located peripherally in the insect. The distal process, or dendrite, is associated with a sensory structure of some type (see chapter 6); the proximal process usually connects directly, without any intervening interneurons, with a motor neuron or more commonly with one or more interneurons. Some sensory neurons may be multipolar. Their distal processes may branch, sometimes elaborately, over the inner surface of the integumental wall, through perforations in the integumental wall, or over the alimentary canal, while their axons enter the ganglia of the central nervous system. Some multipolar neurons are involved with stretch receptors (see chapter 6).

Figure 3.3   Generalized insect central nervous system.

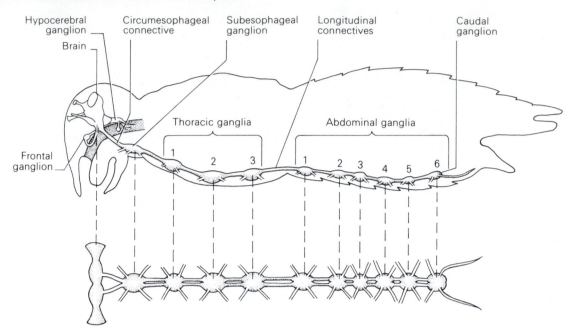

Motor neurons, or *efferent neurons,* are unipolar. The cell body lacks dendrites and is located in the periphery of a ganglion. The bundles of axons from the cell bodies form the motor nerves that activate muscles; the collateral of each neuron enters the neuropile, and the terminal arborizations connect with those of association neurons, or sensory neurons.

Interneurons, and *association neurons* (or *internuncial neurons*) also have their cell bodies located in the periphery of a ganglion and may synapse with one or more other interneurons, sensory neurons, or motor neurons. Some interneurons are quite large and are connected to "giant axons" that have very large diameters (approximately 45 micrometers) and may run the entire length of the ventral nerve cord (e.g., in *Periplaneta americana,* the American cockroach). These axons have been thought to serve as a rapid conduction system for alarm reactions and are associated with a variety of sensory-motor combinations in different insects. For example, giant axons may be involved in the sudden "jet propulsion" (i.e., the forcible expulsion of water from the rectum) escape response of dragonfly larvae (see chapter 12). It was previously thought that the escape response elicited by a puff of air across the cerci of the American cockroach (*Periplaneta americana*) was mediated via the giant axons. However, stimulation of these axons fails to induce the escape response, and, further, the escape response can be induced in individuals in which the cell bodies of the giant axons have been severed (Parnas and Dagan 1971, and others cited in Huber 1974).

The functional significance of excitatory and inhibitory neurons, plus neurons based on the type of neurotransmitter released, will be discussed under the section on nervous coordination and integration.

Based on anatomy alone, the insect nervous system is divided into three major parts which are all interconnected:

1. The central nervous system
2. The visceral, or sympathetic, nervous system
3. The peripheral nervous system

Central Nervous System (CNS)

The insect's *central nervous system* (CNS) (figure 3.3) is composed of a double chain of ganglia joined by lateral and longitudinal connectives. The anterior ganglion, the *brain,* is very complex and is located dorsal to the foregut in the head. It is usually connected by *circumesophageal connectives* to a ganglion ventral to the foregut. This is the *subesophageal ganglion,* which is also highly complex, being composed of three fused ganglia representing the mandibular, maxillary, and labial segments. The subesophageal ganglion innervates sense organs and muscles associated with the mouthparts, salivary glands, and the neck region. In addition, the subesophageal ganglion in many insects has an excitatory or inhibitory influence on the motor activity of the whole insect. Posterior to the subesophageal ganglion are, typically, three segmental *thoracic ganglia,* each containing the sensory and motor center for its respective segment. Two pairs of major nerves, one pair supplying the legs and the other the remaining musculature of each segment, arise from each ganglion. In winged insects, the mesothoracic and metathoracic ganglia each give rise to a third pair of nerves associated with the wing musculature. In a dissection of the 6th instar larva of the gypsy moth (figure 3.4) one can see the brain, the frontal commissures connecting the brain to the frontal ganglion, the

Figure 3.4 Whole mount dissection of part of the central nervous system and endocrine glands of the gypsy moth larva. FG = frontal ganglion; Br = brain; SOG = subesophageal ganglion; T-1 = first thoracic ganglion; CC = corpus cardiacum; CA = corpus allatum; NCC = nervi corpora cardiaca; CeC = circumesophageal connectives; Fc = frontal commissure. *Courtesy of Chih-Ming Yin.*

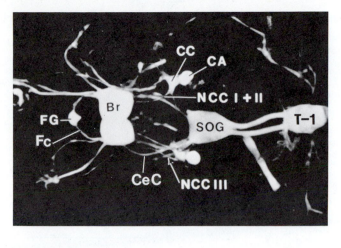

Figure 3.5 Variation in the concentration of ventral chain ganglia in two species of beetles. *Redrawn from Packard, 1898.*

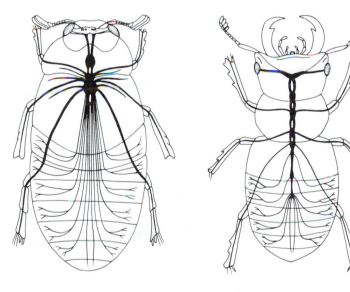

circumesophageal connectives linking the brain to the subesophageal ganglion, the connectives going from the subesophageal ganglion to the first thoracic ganglion, and the nerves going to the corpora cardiaca and corpora allata. In some insects—e.g., adult Hymenoptera, Diptera, and some Coleoptera—(figure 3.5), the thoracic ganglia may be more or less fused longitudinally, forming what appears to be a single neural mass in the thorax.

In the more primitive insects each of the first several abdominal segments contains a ganglion, the first 8 segments in apterygote insects, 7 in dragonfly and damselfly nymphs, and 5 or 6 in grasshoppers and their relatives. However, there has been a tendency toward reduction in the number of abdominal ganglia; for example, several adult flies have only one, which is partially fused with the large single thoracic ganglion. Patterns of fusion of ganglia sometimes differ between the larval and adult stages of the same species.

In those insects that possess abdominal ganglia, the most posterior, the *caudal ganglion,* is always compound and furnishes the sensory and motor nerves for the genitalia. This ganglion is therefore intimately involved in the control of copulation and oviposition. The other abdominal ganglia typically give rise to a pair of nerves to the segmental muscles.

Although ganglia are associated with specific body segments, it should not be assumed that a given ganglion provides innervation only for its own segment, because muscles of one segment may receive nerves from a ganglion associated with a different segment.

Figure 3.6 Brain and stomatogastric nervous system of the grasshopper *Dissosteira carolina;* (*a*) Anterior view. (*b*) Lateral view.

Redrawn with slight modifications from Snodgrass, 1935.

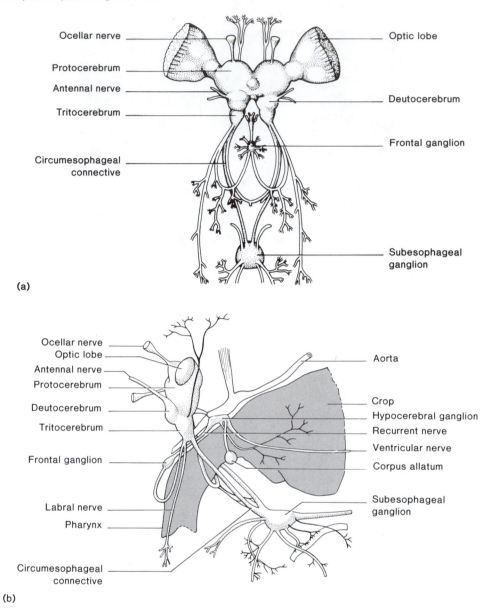

(a)

(b)

The insect brain (figures 3.6 and 3.7) is a very complex structure, apparently formed from the fusion of three separate, paired segmental ganglia of primitive lobopod ancestral types and composed of at least three lobes from dorsal to ventral, the protocerebrum, deutocerebrum, and tritocerebrum.

The *protocerebrum* is the most complex part of the brain. Several distinct cell masses and regions of neuropile have been identified.

1. Optic lobes (associated with the compound eyes).
2. Ocellar centers (associated with the dorsal ocelli).
3. Central body.
4. Protocerebral bridge.
5. Pars intercerebralis.
6. Corpora pedunculata ("mushroom bodies").
7. Antennal lobes.

The *optic lobes* receive sensory input from the compound eyes and are composed of three neuropiles and associated perikarya. The *ocellar centers* are associated with the bases of the nerves from the ocelli. A neuropile mass, the *central body,* connects the two lobes of the protocerebrum and is located in the center of the protocerebrum dorsal to the esophagus. It receives axons from various parts of the brain and may be the source of premotor outflow from the brain. The *protocerebral bridge* is a mass of neuropile located medially. It is associated with axons

Figure 3.7 Diagram showing major neuropilar regions (shaded) of the brain and some connections between these regions. Black dots indicate location of perikarya.

Redrawn from Chapman, 1971.

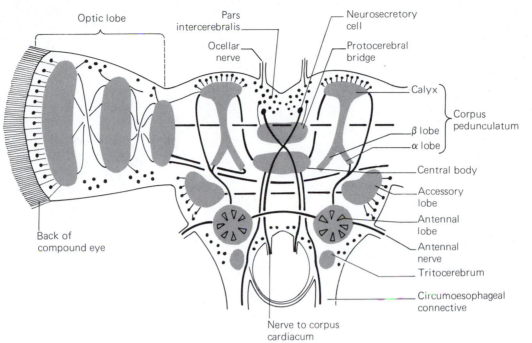

from many parts of the brain, except the corpora pedunculata. The *pars intercerebralis* is located in the dorsal median region above the protocerebral bridge and central body. It contains two groups of *neurosecretory cells* that transport neurosecretory material to the *corpus cardiacum* (to be discussed under glandular system). Each corpus pedunculatum is composed of a central stalk that splits ventrally into α and β lobes and is capped dorsally by a mass of neuropile and associated perikarya, the *calyx*. The *corpora pedunculata* contain interneurons, which do not extend outside of these bodies, and terminal portions of axons that enter from perikarya located in other parts of the brain. Axons that connect with the calyx and α lobe apparently are mainly sensory; those connecting with the β lobe are premotor axons, which in turn synapse with motor fibers.

In view of the complexity of the protocerebrum, it is not surprising that this part of the brain is considered to be the location of the "higher centers" in the central nervous system, which control the most complex insect behavior. The fact that the corpora pedunculata are comparatively large in the social Hymenoptera (ants, bees, and wasps) and small in less behaviorally sophisticated insects (true bugs, flies, etc.) supports this idea.

The *deutocerebrum* contains the *antennal* or *olfactory* ("sense of smell") *lobes*, which receive both sensory and motor axons from the antennae. The antennal

lobes (figures 3.7 and 3.8) are important because they are the centers for receiving and processing information concerning host selection, locating mates using species-specific volatile chemical signals termed *pheromones*, finding food, locating oviposition or egg-laying sites, and many more activities relating to the insect's perception (i.e., either attraction or avoidance) of the diverse volatile cues of its environment. Within each antennal lobe are located spheroidal collections of neuropile called *glomeruli*. These regions have been structurally, physiologically, and biochemically analyzed with respect to one specific, but extremely meaningful, volatile substance (i.e., sex pheromone) for some insect species. Failure of the sexes to come together for mating may result if the volatile substance is not present. The antennal lobes are connected to one another by a central fiber tract, or commissure. Tracts of olfactory fibers connect the antennal lobes and corpora pedunculata of the protocerebrum.

The *tritocerebrum* connects the brain to the stomatogastric nervous system via the *frontal ganglion* and to the ventral chain of ganglia (beginning with the subesophageal ganglion) via the circumesophageal connectives. The tritocerebrum also receives nerves from the labrum and possibly sensory fibers from the head capsule.

See Strausfeld (1976) for a well-illustrated treatise on the fly brain. Howse (1975) reviews the literature pertinent to brain structure and behavior. Mobbs (1985)

Figure 3.8 (*a*) Diagram of the electroantennogram technique for recording from *Manduca sexta* adult. (*b*) An enlargement showing how one can fill an individual neuron with a dye (hexamine cobaltic chloride) and also record from the identified neuron using a glass microelectrode (GME). AN = antennal nerves; OL = optic lobe; SOG = subesophageal ganglion; AL = antennal lobe. Notice the filled neuron at the tip of the microelectrode. (*c*) Whole mount preparation of the terminal abdominal ganglion of the cricket, *Acheta domesticus*, showing a neuron stained intracellularly with a microelectrode and cobaltic chloride.

(b) taken from Christensen and Hildebrand, 1987; (c) courtesy of R. K. Murphey.

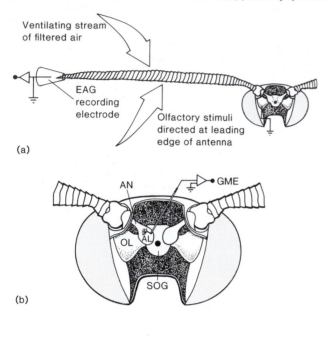

provides an excellent review of insect brain structure and Gupta (1987) covers the evolution, development, structure, and functions of the arthropod brain. See Homberg, Christensen, and Hildebrand (1987) for information on the deutocerebrum.

In the past, information concerning the insect nervous system relied heavily on traditional studies (injecting the vital dye methylene blue into the insect for obtaining whole mounts and using silver nitrate for histological sections) of the various neurons, with some attempt to study how the system is integrated. The integrative aspect of the nervous system, however, is difficult to study without first knowing how the system is interconnected. In order to understand how the system is integrated or "wired," it is critical to know which receptors and neurons are linked with and/or communicate with the various parts of the central nervous system, with one another, or with specific glands and/or muscle sets. In order to obtain "maps" of the "wiring" for various parts of the central nervous system, techniques for labeling and tracing individual neurons to visualize their connections have been developed. One major and initial advantage of

using the insect systems as models in neurobiological studies is the peripheral location of the perikarya within the ganglion (figure 3.2). Using a microscope, the neurobiologists are able to insert very fine glass microelectrodes into a specific cell body (figure 3.8) of a ganglion or brain lobe. While they are recording from this cell body (using electrophysiological techniques), neurobiologists can simultaneously inject different dyes (markers). Through a process called iontophoresis, the dye is transported throughout the neuron, which then can be visualized in various ways depending on the dye injected. At the same time, nerves can be cut and their ends inserted into a solution of cobalt chloride, which if left for a period of time, will, by passive diffusion, completely label the neuron (figure 3.8*b* and *c*). The numerous techniques used by neurobiologists to label "lines" are explained in greater detail in Kater and Nicholson (1973), Strausfeld and Miller (1980), and Nässel (1987). Recently, either monoclonal or polyclonal antibodies have been produced against various antigens associated with neurons and then used in immunocytochemical techniques for marking or labeling neurons. These techniques, along with those

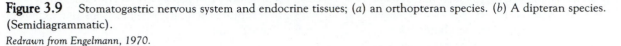

Figure 3.9   Stomatogastric nervous system and endocrine tissues; (*a*) an orthopteran species. (*b*) A dipteran species. (Semidiagrammatic).

Redrawn from Engelmann, 1970.

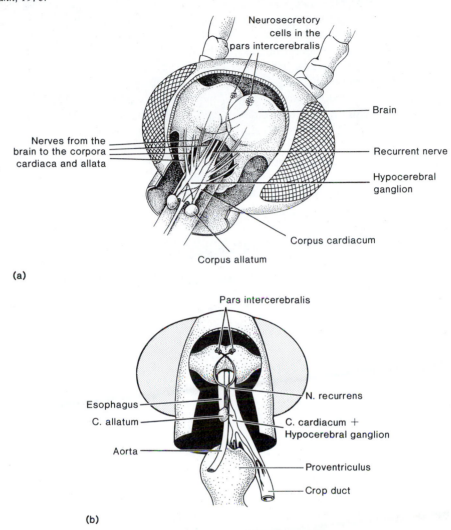

(a)

(b)

Neurosecretory cells in the pars intercerebralis

Brain

Nerves from the brain to the corpora cardiaca and allata

Recurrent nerve

Hypocerebral ganglion

Corpus cardiacum

Corpus allatum

Pars intercerebralis

N. recurrens

Esophagus

C. allatum

C. cardiacum + Hypocerebral ganglion

Aorta

Proventriculus

Crop duct

developed by electrophysiologists, are making insect models extremely powerful systems for studies in neurobiology. At the same time, if the animal rights groups foster more legislation making research with vertebrate systems more difficult, insects will probably be used even more for basic neurobiological research.

Visceral Nervous System (VNS)

The *visceral nervous system* is often referred to as the sympathetic nervous system of insects; however, we will follow the suggestion of Penzlin (1985), who in his discussion of the stomatogastric nervous system of insects states that since structures in the insect's nervous system are not truly homologous with those of the vertebrates, the term "sympathetic" system in insects, as well as other arthropods, should not be used. The VNS is made up of three separate subsystems:

1. Stomatogastric (stomodael), associated with the brain, salivary glands, and foregut.

2. Ventral visceral, associated with the ventral nerve cord.

3. Caudal visceral, associated with the posterior segments of the abdomen.

The *stomatogastric system* (figure 3.9) arises during embryogenesis from the dorsal or the dorsal and lateral walls of the stomodaeum and eventually becomes connected to the brain. Its various components typically include a *frontal ganglion*, which lies on the dorsal midline of the foregut just anterior to the brain. The frontal ganglion connects with the brain by bilateral nerves. The *recurrent nerve* arises medially from the frontal ganglion and extends beneath and posterior to the brain. Beyond this point there is considerable variation in the stomodael system among different kinds of insects. Therefore, the description that follows will pertain only to one of the more common arrangements. The recurrent nerve ends posteriorly in a *hypocerebral ganglion,* which may give rise to one or two *gastric nerves,* or *ventricular nerves,* which continue posteriorly and eventually terminate with a *ventricular ganglion*. Also associated with the hypocerebral ganglion are two pairs of endocrine glands, the *corpora cardiaca* (sing., *corpus cardiacum*) and the *corpora allata* (sing., *corpus allatum*). The stomatogastric system apparently exerts some control over the movements of the gut and possibly, in certain insect species, labral muscles, mandibular muscles, and the salivary glands. The corpora cardiaca and corpora allata are involved with hormone secretion and are thus considered as parts of the endocrine system. For more information on the stomatogastric nervous system, see Miller (1975) and Penzlin (1985).

The *ventral visceral system* is associated with the ganglia of the ventral nerve cord. From each segmental ganglion a single median nerve arises and divides into two lateral nerves, which innervate the muscles regulating the closing and the opening of the spiracles of the segment in which they are located. Depending on the species, these median and lateral nerves may be lacking altogether.

The nerves of the *caudal visceral system* arise from the caudal ganglion of the ventral chain and supply the posterior portions of the hindgut and the internal reproductive organs.

Peripheral Nervous System (PNS)

All the nerves emanating from the ganglia of the central and visceral nervous systems compose the *peripheral nervous system*. The distal processes, dendrites, of sensory neurons within these nerves are associated with a sensory structure (see chapter 6), whereas the axon usually synapses with neurons within a ganglion of the central nervous system. Nerves also typically contain motor fibers, the cell bodies of which are located in ganglia of the central nervous system and fibers of which end in muscles, glands, and other effector structures. The components of the PNS serve as the "windows" of the insect. Through peripherally located *sensilla* (sing., *sensillum*) and sense organs, information about the environment traverses the normally impenetrable cuticle, thus providing the insect's central nervous system with an essential contact with its surrounding environment whether it be in water, on land, in the soil, inside a plant, or elsewhere. These sensilla are located all over the insect's body but are generally concentrated on specific and prominent structures such as the antenna, tarsi, palps, labellum, ovipositor, and cerci. Sense organs such as the eyes and tympana also provide information. A classical book on insect senses is that of Dethier (1963), and an updated account of this important system is provided in volume 6 of Kerkut and Gilbert (1985).

Insects possess numerous and various sensilla. These neurons convert environmental stimuli into meaningful signals in the form of nerve impulses. Three important concepts of sensory physiology are worth mentioning here. They are receptor specificity, receptor sensitivity, and receptor transduction.

The external environment contains a variety of stimuli (modalities) (e.g., light, temperature, gravity) which are constantly bombarding peripheral sensilla and sense organs. Because of receptor specificity, however, a given type of receptor or sensillum is generally far more sensitive to one type of stimulus than another. Thus, a chemosensillum located on the tip of a fly's tarsus is specifically "tuned" to various chemicals, which are generally in the liquid phase, but it is not "tuned" to sound or light stimuli. Also, as we will see later, each tarsal receptor of a fly generally has five sensilla within the tarsal hair (figure 3.10). One of these responds only to mechanical stimulation, one is called the salt-best sensillum because it was first shown to respond to salt, one is the water-best sensillum because it was shown to respond to water, another is the sugar-best sensillum, and the specificity of the fifth cell remains unknown.

The spectrum of a specific environmental stimulus (e.g., varying concentrations of the sugar fructose, which is one of the major sugars in plant nectar) is generally much broader than that which a specific sensillum type (e.g., chemosensillum on a fly's tarsus) is "tuned" to. Thus, most stimuli cover a broad range, from a concentrated signal to a dilute one, and each sensillum only responds to a specific "window" or segment of that range. For example, the sugar-best sensillum has an electrophysiological, receptor threshold of 0.001 M. Below that concentration the sensillum does not respond to the very dilute concentration of fructose. In nature such a situation could exist following a rain. Similar examples could be given for other sensilla and stimuli.

Numerous factors have been shown to influence the sensitivity of a receptor threshold. Some of these include past experience, genetics, age, sex, circadian rhythms, hormones, and biogenic amines. The mechanisms whereby some of these factors influence receptor sensitivity will be discussed in chapters 6 and 8.

Receptor transduction is the process whereby the receptor acts as a transducer, which is a device that changes energy from one form (e.g., sugar molecule in solution) into another form that is meaningful to the insect (e.g., electrical energy of the nerve). We know very little about the exact mechanisms of transduction in most insect sense organs, or sensilla. Some evidence and theories have been provided for the olfactory sensilla (Mayer and Mankin 1985) and the mechanosensilla (French 1988). A most stimulating reference concerning the molecular biology of odorant-binding proteins, olfactory reception and transduction, and the molecular biophysics and membrane function of vertebrates is that of Margolis and Getchell (1988).

Figure 3.10   A taste hair on either the tarsus or labellum of the blow fly, *Phormia regina*, showing the mechanosensillum and four chemosensilla associated with the hair. Diagrams 1–3 show the on-off of the stimulus, the change in receptor potential, and action potentials following stimulation.
Modified and adapted from Dethier (1976).

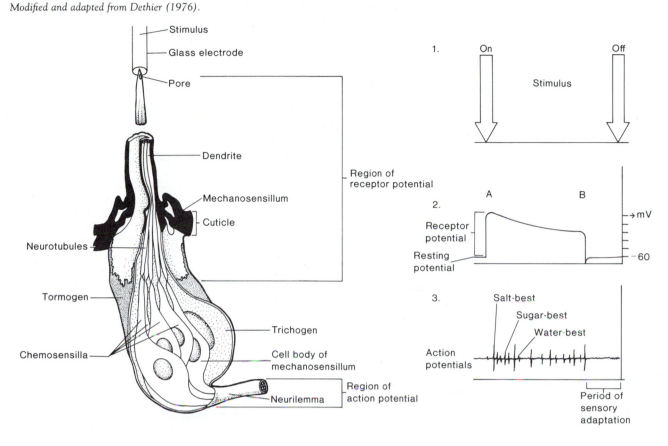

Nervous Integration

Information about the external and internal environment is continuously conveyed from sensilla to the central nervous system where it is integrated in a way that appropriate behavioral and regulatory changes are made. Figure 3.11 provides a simplified model of the interrelationships among the various parts of the nervous system. In this section we want to take a closer look at the actual operation of the nervous system (additional information may be found in chapter 8). Although many aspects of insect neurobiology have been studied, only beginnings have been made in understanding the detailed links between neural function and behavior. Biophysically, nervous conduction in insects has not been found to differ significantly from that in other invertebrates.

Transformation of Stimuli in Receptors or Sensilla

Input (nervous impulses that are really action potentials and often called "spikes") from a single sensillum to the central nervous system usually consists of the following events:

1. Stimulus (external or internal).
2. Transduction of stimulus to receptor potential.

3. Receptor potential produced via depolarization of dendrite or cell body.
4. Action potential produced via depolarization in the axon of the sensory cell.
5. Release of chemical neurotransmitter at the presynaptic membrane.
6. Numerous biochemical events at the postsynaptic membrane.
7. Receptor potential in the next neuron (postsynaptic) in line.
8. Action potential, and so on.

Stimuli range from electromagnetic radiation—which includes different wave lengths representing ultraviolet, infrared radiation, and light as we know it—to changes in pH (see chapter 6). Exactly how the energy of a given stimulus is transformed (changed) into the receptor potential is not completely understood, but a change in membrane permeability of the dendrite is involved.

The blow fly, *Phormia regina* (order Diptera, family Calliphoridae), tarsal chemosensillum will be used here as a case study to describe how stimuli lead to action potentials, which represent the input to the central nervous

Figure 3.11   A model of the major interrelationships of the insect nervous system.
Modified from Cornwell, 1968.

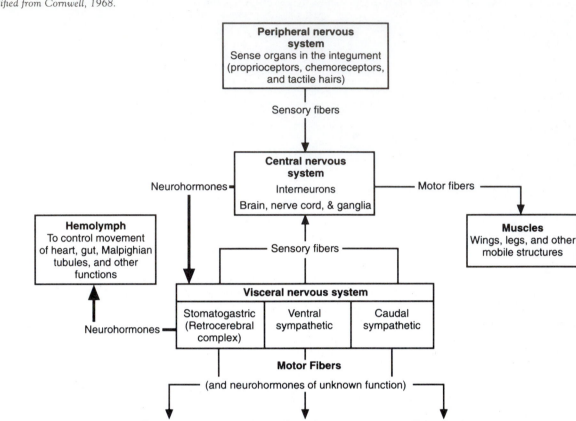

system for integration into overt behavior. Before presenting the case study, however, one must understand what is taking place at the level of the cell membrane of the neuron. The cell membrane of a resting axon—that is, an axon that is not responding with an impulse—is actively maintained, with respect to the extracellular fluid surrounding the nerve, in a polarized state called the *membrane potential*. The inside of the axon is negatively charged (approximately 70–80 mV depending on the cell-type) relative to the outside. When at a resting state and not transmitting an impulse, the membrane potential is referred to as the *resting potential*. The resting potential exists solely because of the barrier (cell membrane) between the inside of the cell and the extracellular fluid. Also, this polarity is maintained by the active pumping of Na^+ out of the cell by the sodium pump of the cell membrane. These events will now be explained using our blow fly model.

As described previously under receptor specificity, the tarsal hairs of the blow fly contain 5 bipolar neurons (4 chemosensilla and 1 mechanosensillum) (figure 3.10). When a fly lands on a punctured apple, the fly contacts the juice via its tarsal receptors. Such a stimulus contacts the dendritic receptive region (after passing through a sensillar fluid) of the hair via an open pore (figure 3.10). Once contacted (figure 3.10), the stimulus is referred to as being "on." If the stimulus (i.e., the concentration of sugar in the apple juice) is above the receptor threshold, the cell membrane, which is at the resting potential (figure 3.10) (i.e., for the fly about 60 mV), becomes depolarized and converts this chemical signal into an electrical receptor potential. Receptor potentials are graded; that is, their strengths vary with the strength of the stimulus. This receptor potential (shown as the rise at "A") gradually declines (shown as "B") and then returns to the resting potential even though the stimulus is still "on." In the fly, the receptor potential is produced somewhere between the region of stimulus contact to the region in which the neurotubules and possibly the cell body are located. At the base of the cell body, the receptor potential leads to action potentials (depolarizations of the axon), which travel directly to the central nervous system without a synapse. Thus, it is possible to stimulate one hair and produce a behavioral response (such as proboscis extension). Whereas receptor potentials are graded, the action potential is not. It is of constant amplitude and is all or none in that if the receptor potential is above threshold, a full-strength action potential occurs. The action potential or

Table 3.1 Neurotransmitter and Acetylcholine Content in the Central Nervous System[a]

Neurotransmitter content of *Locusta migratoria* central nervous system

	Acetylcholine	5-Hydroxytryptamine	Dopamine	Noradrenaline
μg/g	111	2.3	1.3	0.2
nmol/g	762	10.6	8.6	1.4

Acetylcholine content in the central nervous system of various species

Species	nmol/mg protein	μg/g weight
Fly (Musca)	9.0	150
Cockroach	8.5	136
Locust	7.7	111
Guinea pig	0.3	4.8
Rat	0.2	3.4

[a]From Breer (1987) Table 18.1, p. 416.

spikes for different sensilla generally have different shapes and magnitudes (i.e., height of spike in mV), which can be used to identify which chemosensillum is responding. Thus, as seen in figure 3.10, the salt-best sensillum produces the largest spike, the sugar-best the next in amplitude, and the water-best produces the smallest spike. Notice that once the receptor potential returns to the resting potential, action potentials are no longer produced, even though the stimulus is still contacting the hair. This results from *sensory adaptation* and is a characteristic of almost all sensilla. Later, we will use this tarsal chemosensillum model to explain various aspects of how feeding behavior is regulated.

Synaptic Transmission

One of the most active areas of current research in insect biology is neurochemistry. This basically translates into identifying the different types of neurotransmitters and neuromodulators involved in synaptic transmission, studying the diverse biochemical reactions taking place at the terminal arborizations—those occurring at the synaptic cleft (space between presynaptic neuron and postsynaptic neuron, 5–25 nm)—and, finally, studying events at the postsynaptic membrane.

The arrival of an action potential at terminal arborizations induces a calcium-dependent release of a chemical transmitter (stored in vesicles or synaptosomes, which are 30–100 nm in diameter and located inside the neuron at the synaptic region) via exocytosis of the membrane into the synaptic cleft—the transmitter crosses the cleft where it temporarily binds to a receptor protein on the postsynaptic membrane. This binding of the transmitter sets into motion a series of events that ultimately leads to depolarization of the postsynaptic neuron and so on to the next neuron in the circuit.

Acetylcholine has been identified as one among several possible transmitter substances in insects (table 3.1).

The importance of acetylcholine in the central nervous system of insects, compared with vertebrates, is shown in table 3.1. It is now established that sensory and interneurons of insects are cholinergic while efferent neurons are not. The reverse is true for vertebrates. As soon as acetylcholine is released into the synaptic cleft, the enzyme acetylcholinesterase begins to break it down. Without the normal breakdown of neurotransmitters in the synaptic cleft region, normal neuronal functioning is greatly impaired or stopped. In fact, a strong interest in cholinergic synapses exists because many insecticides (i.e., carbamates and organophosphates) act on the cholinergic system. These insecticides act as inhibitors of acetylcholinesterase, the enzyme that removes the acetylcholine from the synaptic region, thus interfering with normal neurotransmission. More will be said about the functional role of neurotransmitters and neuromodulators in chapter 8. In some cases there is a tight junction between the terminal arborization of one neuron and the dendrites of the next, and the transmission is electrical rather than chemical.

The outcome of a nervous impulse depends on a number of factors. For example, an impulse arriving at a given synapse may induce a subthreshold generator potential and have no effect. However, if a whole series of impulses arrive at a synapse in rapid succession, the resulting generator potentials may reinforce one another sufficiently to exceed threshold and stimulate an action potential (*temporal summation*). When several axons synapse with a single neuron, impulses arriving simultaneously may also summate and induce an action potential (*spatial summation*).

At any given moment, the ganglia, including the brain, receive input from a variety of sensory axons. They integrate the diverse impulses and exert regulatory influences both by stimulation and by inhibition. For example, stimulation of certain parts of the cricket (*Acheta* sp.,

Figure 3.12 Examples of exocrine glands. (*a*) Simple unicellular gland. (*b*) Unicellular gland with ductule cell. (*c*) Simple aggregation of gland cells. (*d*) Aggregation of gland cells with common duct. (*e*) Complex aggregation of gland cells with common duct and reservoir.

Modified from various sources.

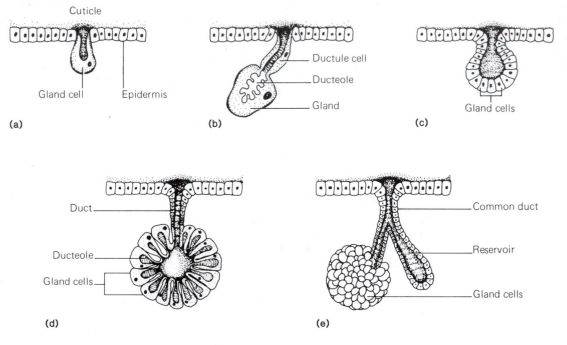

(a)

(b)

(c)

(d)

(e)

Gryllidae, Orthoptera) brain stimulates stridulation ("singing"), whereas stimulation of other parts exerts an inhibitory effect.

Many axons fire spontaneously without input from a sensory neuron. Such activity is considered to keep the nervous system, or at least a part of it, in a highly sensitive state. Thus, a stimulus that would otherwise be subthreshold might induce an action potential by summating with a spontaneous discharge.

Pichon and Manaranche (1985) provide an extensive review of the biochemistry of the nervous system. Information on insect neuropeptides can be found in Holman, Nachman, and Wright (1990).

Exocrine and Endocrine Glands

Here we consider a diverse group of cells and tissues that secrete a wide variety of substances with an equally wide variety of functions. Secretion of a specific product or products as a major function is what unites this diverse group. All cells secrete to some extent, but secretion is the main function of the cells and tissues discussed in this section. There are basically two types of glands: *exocrine* and *endocrine*. Exocrine glands discharge their products via apertures or ducts into the external world or into lumens of various viscera, for example, the reproductive tract or the alimentary canal. Endocrine glands are typically ductless, and their secretions are usually released directly into the *hemolymph.*

Exocrine Glands

Exocrine glands may be single cells or small aggregates of secretory cells (figure 3.12). A single secretory cell, for example, one that secretes a toxic substance, may contain an *intracellular ducteole* and may also be associated with another cell, a *ductule cell,* that forms a duct for transfer of secretions from the gland cell to the outside. More complex glands may be formed as invaginated masses composed of large numbers of cells that secrete their products into a common lumen with a single aperture that may in turn open into a common duct (e.g., salivary glands; see figure 4.1). Further, complex glands may be associated with a separate, but attached, reservoir in which large amounts of secretion can accumulate (e.g., salivary reservoirs in many insects). Externally, fine hairs may be associated with a gland opening or gland duct opening. Such hairs may aid in rapid dispersal of the secretion product(s), such as one involved in chemical communication between individuals.

Exocrine glands are generally of ectodermal origin and are found widely scattered over the insect body. The specific location of a given gland is often correlated with its function. For example, salivary glands are typically derived from the labial segment of the insect head during embryogenesis, although in the fully formed insect, they are usually located in the thorax on either side of the foregut. Glands associated with the reproductive system are closely connected with the internal genital organs (see chapter 5). Glands that secrete their products only into the external world vary a great deal in location. Strictly speaking, the epidermal cells that secrete the cuticle, as well as the secretory cells in the midgut and Malpighian tubules (see chapter 4), can be viewed as glandular epithelium. From this point of view, the epidermis emerges as the most extensively developed "gland" in an insect.

Functions of the Exocrine Glands

Functionally (i.e., based on secretion), exocrine glands can be placed into four major categories with the specific types of glands fitting into one or more of these groups.

Defense

The production of repugnatorial substances is widespread among the insects (see chapter 8). Examples include the scent ("stink") glands, located on the metepisternum or on the dorsum of the abdomen among the true bugs, that secrete a number of different *hydrocarbon derivatives*. Nasute soldiers found in certain termite species (Isoptera) secrete a sticky defensive material from the enlarged frontal gland in the head. Some insects have eversible repugnatorial glands, for example, the *osmeteria* found on the dorsum of the prothorax of swallowtail butterfly larvae and similar structures at various locations among other species of Lepidoptera. If you collect a swallowtail larva, try touching it around the head to see if you can make the osmeteria evert. If not, gently grab the larva between two fingers and gently squeeze. When the orange colored osmeteria are everted, put your nose near it to smell the odor of the defensive secretion. *Pygidial glands* at the posterior end of certain beetles also secrete a repugnatorial substance. Glands and associated structures produce and deliver the venoms and alarm pheromones found in the stinging Hymenoptera. Poison glands are also associated with urticating (stinging) setae or spines on the bodies of certain Lepidopterous larvae. These are easily detached from the insect's body, are very sharp, and can cause considerable discomfort if they come in contact with human skin.

Many insects such as aphids and bedbugs form aggregations and may be vulnerable to predators. Both, however, have alarm pheromones that cause the other members of the aggregation to disperse. When aphids are grabbed by either a syrphid larva or golden-eyed lacewing larva, they produce the alarm pheromone and release it from the cornicles. This causes the aphids in the vicinity to drop off the plant. Alarm pheromones are reviewed in volume 9 of Kerkut and Gilbert (1985), and Staddon (1979) reviews the literature on scent glands in the Hemiptera.

Intraspecific Communication

Glands in many insects produce chemicals that serve as signals of various sorts for other members of the same or different species. The various functions of these substances, called pheromones, are dealt with in chapter 8. A few examples will suffice here. In Lepidoptera a variety of glands produce sex pheromones that attract or excite a mate. These glands include the *androconia,* specialized scales and associated glands scattered or clustered among the wing scales, and eversible *hair-pencils*. Both androconia and hair-pencils provide much surface area, facilitating the dispersal of an aphrodisiac secretion. If you collect a male monarch butterfly adult, notice the dark black scent pouches on the hindwings. Also, by gently squeezing the abdomen you can force out the hairpencils. This does not hurt or injure the butterfly. Birch et al. (1990) review the scents and eversible scent structures of male moths while several chapters on pheromones are covered in volume 9 of Kerkut and Gilbert (1985). Glands in the mandibular segment of the queen honey bee produce the "queen substance," which inhibits the workers from constructing queen cells and stimulates various other behaviors in workers.

The production of nonterpenoid hydrocarbons (cuticular hydrocarbons) by the exocrine glands of some insects (i.e., termites and flies) has been shown to serve the insect in various aspects of intraspecific communication. Reviews are provided by Howard and Blomquist (1982), volume 3 of Kerkut and Gilbert (1985), and Blum (1987). Analysis of cuticular hydrocarbons is a growing area of research and has been shown to provide the systematist with a new tool of classifying species (see chapter 11).

Building Structural Materials

Examples of structural materials include various waxes, lac, and silk. The wax secreted by epidermal glands located ventrally between the overlapping sternal plates of the fourth through the seventh abdominal segments of honey bees is used to construct honeycomb. Members of the order Homoptera—scale insects (superfamily Coccoidea), white flies (Aleyrodidae), and so on—secrete waxes in various forms (powdery, filamentlike, etc.) onto the body surface. Certain scale insects, the Lacciferidae and especially *Laccifer lacca,* secrete a resinous material called lac from which commercial shellac is produced (see chapter 12). Insects in many groups produce silk, which is usually composed of fibrous proteins (Rudall and Kenchington 1971). Silk is typically produced in the form of fine threads that are then woven into various structures, including the protective pupal cocoons of many Lepidoptera, Hymenoptera, and Siphonaptera. The tunnels of Embioptera (webspinners) are constructed with silk produced in the enlarged foretarsi. Members of the Trichoptera (caddisflies) are aquatic and utilize silk to construct both larval and pupal shelters, often incorporating materials from the environment among the silk threads. Certain predaceous Trichoptera construct silken nets and snares to trap prey. The biochemistry of silk production is reviewed in volume 10 of Kerkut and Gilbert (1985).

Transport of Materials in the Liquid State

The functional significance of the salivary glands will be discussed in chapter 4 and the spermathecal and accessory reproductive glands will be discussed in chapter 5. Except for a few cases of the apparent presence of neurosecretory axons, few dermal glands have been found to be innervated. Other exocrine glands, such as the salivary glands, may (cockroaches) or may not (*Calliphora*, Calliphoridae, Diptera) be innervated (Noirot and Quennedey 1974).

Endocrine Glands

Endocrine glands commonly function in close coordination with one another, in contrast to exocrine glands, which often function independently. The endocrine system is involved with the secretion of substances called hormones (which means "to excite" in Greek). Hormones generally secrete into the hemolymph from a rather well-defined tissue, circulate throughout the insect body until the *critical titer* is reached (the concentration in the hemolymph necessary to elicit a response on the target organ or target tissue), and "excite," or stimulate, the target. Like the nervous system, the endocrine system helps the insect to adjust to external and internal environmental changes, but physiological and behavioral responses mediated by hormones occur more slowly than those mediated directly by the nervous system. It might be said that hormones are involved with long-term adjustments (i.e., they are the "calendars" of the insect), whereas nerves provide short-term adjustments (they are the "alarm clocks" of the insect) to changes in the external and/or internal environment.

The field of insect endocrinology has passed through three major generations or periods of change that have helped shape its current status. Students are urged to read the two page preface by Professor Bollenbacher in volume 7 (endocrinology) of Kerkut and Gilbert (1985) for a historical perspective of this aspect of insect biology. What follows is a brief synopsis of this preface. Technological advances are meaningless without the ingenuity of a clever investigator to use the appropriate technology to answer the meaningful questions in light of currently available techniques. The time periods presented for the three major periods of development in insect endocrinology are arbitrary, and one must be aware that they often overlap. Also, endocrinologists will always use a combination of all the available techniques developed to date.

The first generation of insect endocrinologists (1920s to 1950s) used very simple techniques. These consisted of *ligation* (separating different regions of an insect's body by tying it off with string or hair, thus separating and isolating the blood supply), *parabiosis* (connecting two different insects either directly with sealing wax or with glass tubing so that their hemolymph components are shared), *transplantation* (removing a suspected endo-

crine gland from one insect and putting it into another), and, finally, *extirpation* or *cauterization* (physically removing a gland, or using an electric current to destroy a particular region or tissue). Using these tools, scientists established the major endocrine centers or glands (the brain, corpora allata, corpora cardiaca, and prothoracic glands).

The second generation (1950s and 1960s) of endocrinologists had available many new chemical and biochemical techniques such as gas and liquid chromatography, mass spectrometry, NMR, HPLC, and fluormetry. With these techniques, many of the hormones were identified as to structure, and their biosynthetic pathways were determined.

The third generation (1970s and 1980s) of endocrinologists had available many new technological developments in the area of immunology (i.e., ELISA, immunocytochemistry, RIA, etc.) and molecular biology (gene cloning). Using the ELISA and RIA techniques, scientists were able to determine the hormone titers throughout development. By using cloning techniques and other molecular biology techniques, scientists were able to determine the molecular sequence of various peptide hormones.

You, the reader, will be part of the future generation (1990s and 2000s) and will observe the benefits of research in this field of insect biology. The major impact will come from the area of molecular biology. Endocrinologists will genetically engineer insect pests or beneficial insects with genes, which when expressed will either produce or not produce specific hormones or neuropeptides; thus having either a detrimental or beneficial effect, respectively, on the gene recipient. In addition, baculoviruses (insect viruses capable of expressing a particular gene responsible for the production of large quantities of proteins) are already being used as expression vectors of foreign genes (Maeda 1989). This process will enable the endocrinologist to isolate the gene for a particular neuropeptide he or she is interested in studying. Because neuropeptides are produced in such small quantities in insects, larger amounts are needed to study either their biochemistry or physiological effects on the insect. By putting the gene into an expression vector, scientists can use this system to produce sufficient quantities to do the studies they are currently unable to conduct.

The best identified endocrine structures whose functions are at least partially understood are the corpora cardiaca, the corpora allata, and the thoracic glands (see figure 5.27). It is a well established fact (see chapter 8) that many insects are responsive to changes in the length of day (e.g., short days causing the insect to enter diapause and long days preventing diapause). Since diapause is generally controlled by hormones, it follows that the length of day somehow has influenced the production of hormones. This is not an entirely new concept inasmuch as poultry farmers leave the lights on in the hen-

Figure 3.13 Neurosecretory cells in the brain of the mosquito *Aedes taeniorhynchus*. Cells exposed by retraction of posterior part of head capsule. 1, Neurosecretory cells; 2, brain; 3, pharyngeal muscles; 4, retracted portion of head capsule; 5, compound eye.

Courtesy of Arden O. Lea.

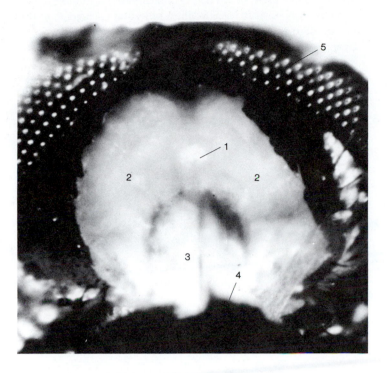

house during the winter months to keep egg production at normal levels. One can, however, ask the question how is the light stimulus (i.e., the length of day) converted into hormone signals. As previously stated, one form of energy must somehow be transformed into another form in order for this phenomenon to work. Insects, as well as other invertebrates and vertebrates, have special cells (*neurosecretory cells*) that are modified neurons that secrete neurohormones (*neurosecretory products*) and act as the link between the nervous and endocrine systems. In fact, some textbooks combine the endocrine glands and neurosecretory cells into one system called the neuroendocrine system. Neurosecretory cells are located in the protocerebrum (figure 3.13) and can be identified because of the blue color they produce from reflected light. The original of figure 3.13 is a color print showing the neurosecretory cells as blue structures. This physically produced blue color is due to the Tyndall effect which also makes the sky blue. Neurosecretory cells are also found in the other parts of the brain, in the subesophageal and other ventral chain ganglia, and in the corpus cardiacum. They are usually identified by the use of specific stains; for example, they stain deeply with paraldehyde fuchsin (figure 3.14*a–b*). Recent work has shown that there are two classes of neurosecretory cells (type-A, which stain with paraldehyde fuchsin, and type-B,

which do not). Neurosecretory cells are also characterized by the presence of electron-dense granules (figure 3.14*c*).

The locations of the corpora cardiaca and corpora allata have already been described, and specific functions will be discussed in later chapters. In addition to containing intrinsic secretory cells, the corpora cardiaca receive axons from the neurosecretory cells in the brain and serve as storage and release sites for their secretions. This is seen in figure 3.15*a*, which was obtained by photographing a preparation using the dye injection technique with cobalt. Notice the cobalt has filled the median neurosecretory cell and traveled down its axon to the corpus cardiacum. Structures with a storage and release capacity are called *neurohaemal organs*. The *thoracic glands,* found only in immature insects with the exception of the Apterygotes, are also called *prothoracic glands* or *ecdysial glands,* are irregular masses of tissue of ectodermal origin that are usually intimately associated with tracheae (figure 3.15*b*). They may or may not be innervated. They have been found to secrete the hormone that initiates the molting process, consequently the name ecdysial glands. As will be discussed in more detail in chapter 5, the secretory activity of the thoracic gland is stimulated by a hormone (*ecdysiotropin*-PTTH or prothoracicotropic hormone) from neurosecretory cells in the brain.

Figure 3.14 (*a*) Photomicrograph of neurosecretory cells exposed by staining with paraldehyde fuchsin (PF). Sagittal section of the brain of a mosquito, showing neurosecretory cells and axons running anteriorly and then posteriorly. The axons eventually reach the corpora cardiaca (see figure 3.9). 1, Neurosecretory cells; 2, axons of neurosecretory cells; 3, cuticle of pharyngeal pump; 4, recurrent nerve; 5, pharyngeal pump muscle; 6, cell bodies of neurons; 7, neuropile; 8, neural lamella. (Scale line = 50 μm.) (*b*) Photomicrograph of PAF stained MNSC of *P. regina* adult. (*c*) Electron micrograph of the same cells showing the neurosecretory granules (NSG) that are electron dense.

(a) courtesy of S. M. Meola. (b) courtesy of J. G. Stoffolano and Chih-Ming Yin.

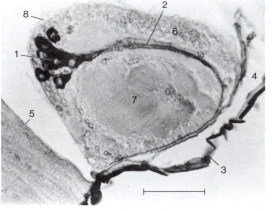

(a)

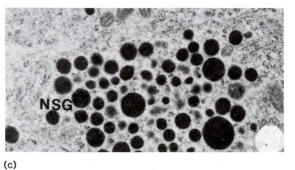

(b)

(c)

Figure 3.15 Endocrine glands. (*a*) Cobalt backfilling of 6th instar larva of the gypsy moth, *Lymantria dispar*, showing the median neurosecretory cell (MNC), the lateral neurosecretory cell (LNC), the corpus cardiacum (CC) and the nervi corpora cardiaci (NCC I & II). Note the passage of the dye between the neurosecretory cells and the storage organ, the corpus cardiacum. (*b*) Whole mount dissection of southwestern corn borer, *Diatraea grandiosella*, showing the prothoracic gland (PG), and the associated tracheal trunk (T).

Courtesy of Chih-Ming Yin.

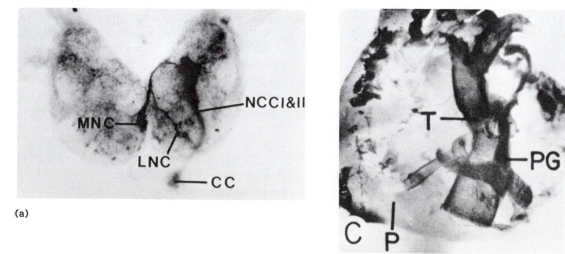

(a)

(b)

Figure 3.16 Major physiological functions regulated by neurohormones in insects. Hormonal distribution is accomplished chiefly by release into the hemolymph, but localized secretion from nerves does occur. CC = corpora cardiaca; CA = corpora allata; PG = prothoracic gland.

Modified from Cook and Holman (1985).

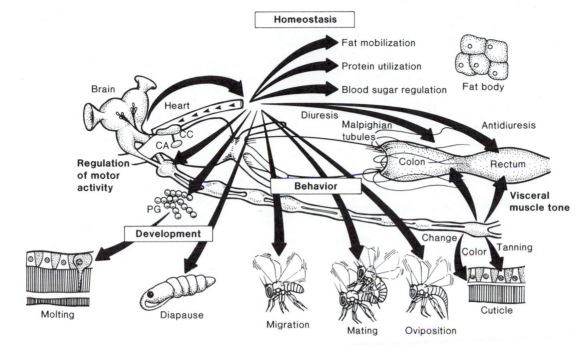

In larvae (maggots) of the higher Diptera (true flies, suborder Cyclorrhapha), there is a small ring of tissue, supported by tracheae, called the *ring gland,* or *Weisman's ring.* The different cells that compose it are considered to be homologous with the corpora allata, corpora cardiaca, and the thoracic glands.

Functions of the Endocrine Glands

Several endocrine functions have been identified in insects (see volumes 7 and 8 of Kerkut and Gilbert 1985, for an extensive treatment of this topic). Gupta (1983), Raabe (1982, 1983, 1989) and Scharrer (1987) explore the various functions of insect neurohormones.

Insect hormones and neurohormones have been studied with respect to their involvement in the following general areas of insect biology and will be discussed in greater detail in the chapters or references cited:

1. Regulation of molting (chapter 5).
2. Determination of form at metamorphosis (chapter 5).
3. Polymorphism (chapter 5).
4. Regulation of diapause (chapter 8).
5. Involvement in reproduction (chapter 5).
6. Regulation of metabolic activities and general body functions (chapters 4–6, 8, and Steele, 1976).

A diagram summarizing the major functions of insect hormones and neurohormones is presented in figure 3.16.

Specifically, hormones and neurohormones influence color changes; regulation of the excretion of water (osmoregulation); control of heartbeat and amplitude; regulation of metabolic activities, such as the maintenance of carbohydrate (trehalose) levels in the hemolymph, synthesis of proteins, and metabolism of lipids; control of sclerotization and melanization of the cuticle; control of growth and metamorphosis; (possibly) control of circadian rhythms; determination of sexual receptivity; regulation of dormancy; pheromone production and release; and regulation of migratory behavior.

The development of radioimmunological and immunocytochemical techniques was extremely important to endocrinologists. These techniques permitted vertebrate endocrinologists to survey tissues for the presence of extremely minute quantities of hormones. Another major breakthrough occurred when biologists realized that these techniques could be applied to the insect system. Since then, the number of vertebrate hormones that have been identified in insects has risen dramatically (Kramer, 1985). Even though the vertebrate hormones insulin and the pain killer neurotransmitter peptides (enkephalin and endorphin) have been found in insects, the physiological and functional roles in most cases remain obscure. This area of research should be exceptionally appealing to those individuals having the "pioneering" instinct. Kramer (1985) provides an excellent review of this exciting area.

In summary, table 3.2 provides a brief review of a selection of neurohormones and the major hormones that

T able 3.2   **A Selected List of Insect Nonneural Hormones and Neural and Peptide Hormones**

Active Principle	Origin	Target	Functional Role
I. Nonneural hormones			
A. *Immature insects*			
Ecdysone (molting hormone)	ecdysial gland	epidermis	initiates molt
Juvenile hormone	corpora allata	epidermis	controls or directs fate of metamorphosis at molt
B. *Adult insects*			
Ovarian hormone (=ecdysone)	ovarian tissue and probably follicle cells	fat body	initiates and regulates production of vitellogenin
Juvenile hormone	corpora allata	fat body	primes the fat body to become competent to produce vitellogenin
Juvenile hormone	corpora allata	accessory reproductive glands	affects development and production of glandular secretion
Juvenile hormone	corpora allata	follicle cells	activates patency and uptake of vitellogenin by follicle cells
II. Neural hormones and peptide hormones			
Ecdysiotropin (=prothoracicotropic hormone)	brain (protocerebrum)	ecdysial glands	developmental—stimulates and regulates production and release of ecdysone
Bursicon	MNSC and thoracico-abdominal ganglion	epidermis	developmental—stimulates sclerotization and melanization of cuticle
Eclosion hormone	brain of pre-ecdysis moths	abdominal ganglia	behavioral—synchronization of eclosion with photoperiod
Allatotropin	brain	corpora allata	developmental/behavioral/ and homeostasis—stimulates JH production and release homeostasis
Diuretic hormone	thoracic ganglia	Malpighian tubules and rectum	controls diuresis or fluid secretion
Mating inhibition hormone	accessory reproductive glands of male	brain	behavioral—prevents remating
Oviposition initiation hormone	accessory reproductive glands of male	oviduct?	behavioral—initiates egg-laying
Cardioaccelerator hormone	brain/corpora cardiaca	myocardium	homeostasis—increase in frequency and amplitude of muscle contraction
Proctolin	corpora cardiaca	hindgut and possibly visceral muscle in general (heart and oviduct)	homeostasis—muscle contraction, defecation, egg-laying, and heartbeat

This table was extracted from different references (vol. 11 of Kerkut and Gilbert, 1985; Menn and Börkovec, 1989; Lunt and Olsen, 1988; and Raabe, 1982). It should be noted that this is a selected list and is far from complete because more and more peptides and neural hormones are being discovered daily. Selection of a particular hormone was made because it provides information on the examples used elsewhere in this book. The role of ecdysone and juvenile hormone in behavior will be discussed in chapter 8.

Figure 3.17   Photomicrograph of a sagittal section of a mosquito showing the indirect flight muscles in the thorax and intersegmental muscles in the abdomen. 1, Longitudinal flight muscles; 2, tergosternal flight muscles; 3, intersegmental muscle. (Scale line = 0.5 millimeter.)
Tissue prepared by S. M. Meola.

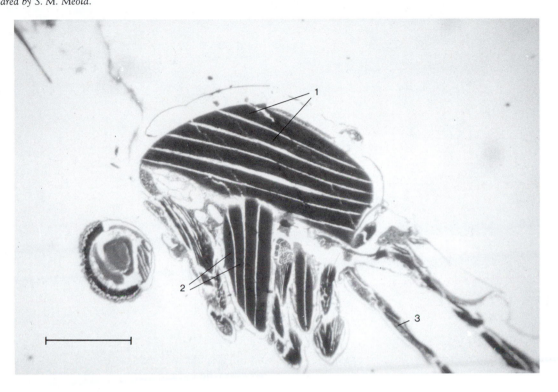

are documented as to their active principle; origin or site of production; target organ or tissue; and development, or homeostatic or behavioral role.

The Muscular System

The conversion of chemically stored energy into mechanical energy occurs within muscle tissue. This conversion is associated with the muscle shortening in length, or *contraction*. Muscle contraction in turn produces a variety of results, depending upon the location, attachments, and degree of stimulation of the contracting muscles. Muscle contraction is responsible for most forms of locomotion, for the maintenance of posture, and for movements of the viscera, such as the peristaltic propulsion of food along the alimentary canal. Quantitatively speaking, muscle is probably the most abundant tissue in the higher animals.

Muscles are commonly classified as striated, those containing definite transverse lines or striations, and smooth, those lacking these striations. Insect muscles are all of the striated variety, although in some types of muscle, the striations are extremely difficult to discern. Insect muscles are typically colorless or grayish in appearance, but may be tinged with yellow, orange, or brown, as is wing muscle in many insects. In view of the small size of insects, one might assume that they have comparatively few muscles. Actually, the opposite is true;

larger insects possess perhaps two or three times as many individual muscles as, for example, humans do.

Insect muscles can be divided into two groups, based on location: *skeletal* and *visceral*. Skeletal muscles (figure 3.17) are attached at both ends to regions of the integument and are those associated with the maintenance of posture and the various movements of the skeleton. Visceral muscles are involved with movements of various internal organs, such as the alimentary canal, the ovaries, and the Malpighian tubules. Visceral muscles commonly attach to other visceral muscles.

Skeletal Muscles

The "evolutionary decision" of arthropods to exploit the exoskeleton rather than the endoskeleton as their major system for support, protection, and movement had profound effects on specific aspects of their method of growth, physiology, and muscle attachment. Unlike vertebrates where muscle is attached to an internal skeleton via structures called tendons (myotendonous connection), insects and other arthropods evolved a different muscle-skeletal attachment arrangement termed myocuticular connection. Skeletal muscle in most cases has a somewhat stationary attachment area, the *origin,* and a movable attachment area, the *insertion.* Areas of attachment

Figure 3.18 Skeletal muscle attachments. (*a*) Muscle attached directly to epidermal cells. (*b–d*) Muscles attached to cuticle by means of tonofibrillae. (*e*) Muscle attached to apodeme.
Redrawn from Richards, 1951.

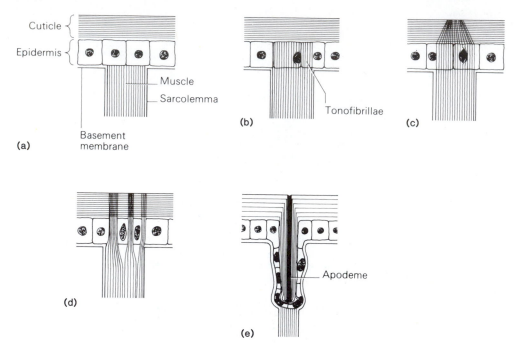

may be either directly on the body wall (figure 3.18*a–d*) or on the inner surface of an apodeme (figure 3.18*e*). Three major apodemes serve as a functional endoskeleton by providing tissue support and surface area for muscle attachment: the furca (thorax), tentorium (head), and phragma (thorax). A number of different means exist by which muscle is attached to integument, but in all cases where it is present, the outer membrane of a muscle fiber, the *sarcolemma*, is continuous with the basement membrane of the epidermis. Typically, fibrillar structures called *tonofibrillae* aid in attaching muscles to the cuticle. These tonofibrillae are really microtubules produced by specialized epidermal cells called *tendinous epidermal cells*. They provide the connection between the myofibrils of the underlying muscles and the vertically oriented muscle attachment fibers of the overlying cuticle (figure 3.18*b–d*). Ultrastructural studies have revealed that tonofibrillae are bundles of microtubules that extend from desmosomes, which link them to muscle cells, to hemidesmosomes at the junction between the epidermis and the cuticle (figure 3.19). Additional fibers run from the hemidesmosomes through pore canals to the epicuticle (Caveney 1969). Desmosomes are local points of attachment between cells. Hemidesmosomes ("half" desmosomes) and the additional fibers apparently serve as local points of attachment between epidermal cells and cuticle. In some cases, muscle fibers may attach directly to unmodified epidermal cells (figure 3.19*a*). The attachments between the cuticle and epidermal cells are resistant to molting fluid, and hence muscles remain functionally attached to the cuticle between apolysis and ecdysis.

Groups of Skeletal Muscles
The skeletal muscles of the head effect three movements: of the entire head, of mouthparts, and of antennae. Muscles that originate on the anterior part of the prothorax and insert on the tentorium and various parts of the head capsule are responsible for head movement. Mouthparts and antennae are moved both by *extrinsic muscles,* which originate on the inner surface of the head capsule or tentorium and insert within a given appendage, and by *intrinsic muscles,* whose origins and insertions are entirely within a given appendage (figure 3.20*a*).

Thoracic skeletal muscles are involved mainly with the legs and wings. Legs (figure 3.20*b*), like mouthparts and antennae, are operated by both extrinsic and intrinsic muscles; the extrinsic muscles are responsible for leg movement and the intrinsic muscles for movement of the individual segments. The origins of the various extrinsic leg muscles are on the terga, pleura, and sterna of each thoracic segment.

In the typical wing-bearing segment, well-developed longitudinal muscles run between the phragmata of the pterothorax, and dorsoventral muscles (tergosternal)

Figure 3.19 Diagram showing ultrastructural details of muscle attachment to cuticle (in Apterygota). (*b–e*) Represent specific regions in (*a*). The parts of a desmosome are identified in (*e*).
Redrawn from Caveney, 1969.

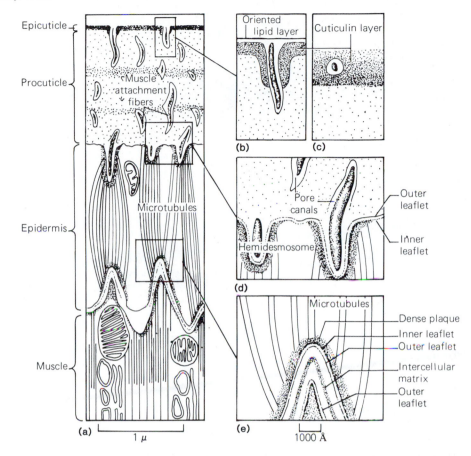

attach to the tergum and sternum of each segment (figures 3.17 and 3.21; see also figure 7.9). Contraction of the longitudinal muscles causes arching of the tergum, which, owing to the construction of the thorax, causes the depression of the wings. The action of the dorsoventral muscles is antagonistic to that of the longitudinals, contraction resulting in the depression of the tergum and consequent elevation of the wings. Because the action of these two sets of muscles on the wings is indirect, they are referred to as *indirect wing muscles*. In addition to indirect wing muscles, there are muscles that attach directly to the bases of the wings. *Direct wing muscles* are responsible for various wing movements during flight and for the flexing of the wings over the abdomen in those insects that possess this ability. In some insects direct muscles provide the main force of propulsion. Their action will be considered in more detail in chapter 7.

In addition to the muscles associated directly or indirectly with leg and wing movements, there are muscles that run between the pleura and terga or sterna and lateral intersegmental muscles. Also, because there are spiracles located in the thorax, muscles associated with their closure mechanisms are present.

The skeletal musculature of the abdomen is somewhat simpler than that of the head and thorax. The most prominent abdominal muscles are the longitudinal intersegmental or segmental ones (figure 3.22). These run between successive antecostae of both the terga and the sterna. Also present are lateral abdominal muscles, which may be oblique in orientation, the *oblique sternals,* but typically are dorsoventral in orientation, the *tergosternals.* These may be inter- or intrasegmental. Spiracular muscles are also present. In addition, special muscles are involved in the various movements of the copulatory structures, ovipositor, and cerci.

For more detailed information on the musculature of insects, consult Chapman (1971), Richards and Davies (1977), and Snodgrass (1935, 1952).

Figure 3.20   Extrinsic and intrinsic skeletal muscles. (*a*) Labium of a cricket. (*b*) Metathoracic leg of a grasshopper. *Redrawn from Snodgrass, 1935.*

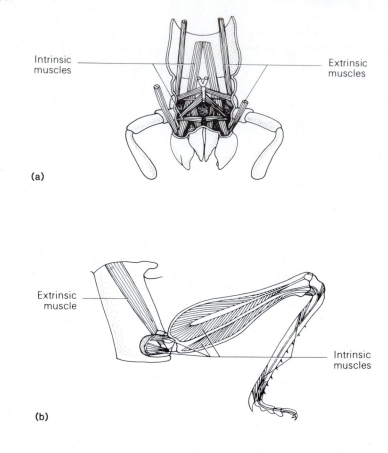

(a)

Intrinsic muscles

Extrinsic muscles

Extrinsic muscle

Intrinsic muscles

(b)

Figure 3.21   Cross section of the mesothorax showing the action of the indirect flight muscles. (*a*) Notum depressed, wings elevated. (*b*) Notum arched and elevated, wings depressed. Arrows indicate direction of movement. *Redrawn from Snodgrass, 1963.*

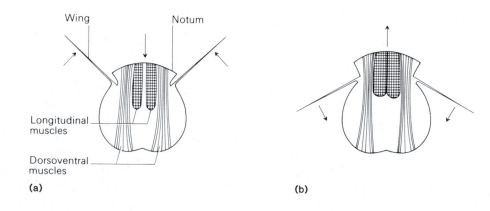

Wing

Notum

Longitudinal muscles

Dorsoventral muscles

(a)

(b)

Figure 3.22   Skeletal musculature of the ventral portion of the abdomen (diagrammatic). (*a*) Dorsal view. (*b*) Lateral view.

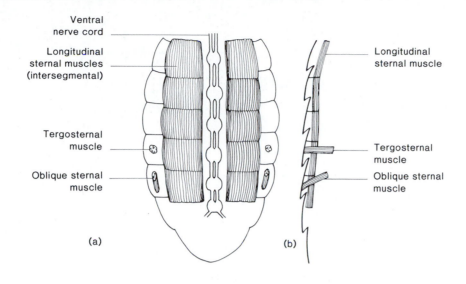

Ventral
nerve cord

Longitudinal
sternal muscles
(intersegmental)

Tergosternal
muscle

Oblique sternal
muscle

(a)

Longitudinal
sternal muscle

Tergosternal
muscle

Oblique sternal
muscle

(b)

Structure of Skeletal Muscles

Insect muscles represent the "engines" or "motors" of the organism, and both a source of oxygen and fuel are essential for the conversion of stored energy into mechanical energy for movement and locomotion. In addition, because of their size (i.e., thickness) and shape (i.e., elongated and tubular), flight muscle fibers initially faced two major constraints to functioning efficiently. Both constraints were dealt with early in insect evolution and had to do with the relative thickness of the muscle fiber and how oxygen and the depolarization effect of the sarcolemma were able to reach deeper into the fiber, thus reaching all myofibrils.

The basic structural unit of a muscle is the *muscle fiber* (figure 3.23). Typically, it is composed of an outer membranous layer, the sarcolemma (cell membrane), and the inner *sarcoplasm* (cytoplasm). Within the sarcoplasm is a bundle of tiny fibers, the *myofibrils,* each of which has a cross-sectional diameter of about 1 micrometer. Numerous cell nuclei and mitochondria and sarcoplasmic (endoplasmic) reticula are also located within the sarcoplasm. The two constraints previously mentioned for flight muscle were dealt with by evolving a T-system and tracheolar system that penetrated deeper into the muscle fiber. The T-system (transverse tubular system) is produced by deep invaginations of the cell membrane or sarcolemma into the muscle fiber. This permits the depolarization reaction, which initially takes place on the surface of the fiber, to penetrate deeper into the muscle fiber. The deep penetration of the tracheolar system permits gaseous exchange at sites which, without this system, would not have rapid gaseous exchange. The myofibril is the contractile element. By weight a muscle

fiber is approximately 20% protein; the remaining 80% is primarily water, plus small quantities of salts and metabolic substances (Huxley 1965).

Ultrastructural studies of insect muscle fibers reveal regular arrays of thick and thin filaments organized in the same fashion as vertebrate muscles (figures 3.23 and 3.24). As in vertebrate muscle, the thick filaments are considered to be composed of the protein myosin, the thin filaments of the protein actin. The actin filaments are more numerous and in cross section can be seen to be arranged in regular arrays around myosin filaments. For example, in flight muscle each myosin filament is surrounded by 6 actin filaments; in leg muscle, arrays of 9–12 actin filaments have been observed. In longitudinal section actin filaments are seen to be anchored in a region called the Z-line. A sarcomere is a longitudinal division of a fibril from one Z-line to the next. Sarcomere length in insects ranges from about 2.5 to more than 9.0 micrometers (Hoyle 1974). The myosin filaments are located centrally and do not reach the Z-line; the actin filaments, on the other hand, are attached at the Z-lines, but do not reach the center of a sarcomere. Regularly occurring branches extend from the myosin filaments to the actin filaments, forming cross-bridges (figure 3.23). These cross-bridges are currently thought to be the globular "heads" of the myosin molecule and are the structures that interact with the actin filaments during muscle contraction. The organization of the thick and thin filaments is responsible for the transverse striations mentioned earlier. There are only actin filaments in the I-bands and only myosin filaments in the H-band. Hence these bands do not stain as deeply as do the A-bands, the regions where both actin and myosin filaments overlap. Patterns of

Figure 3.23 Structure of a generalized muscle fiber (diagrammatic).
Modified from Herried, 1977.

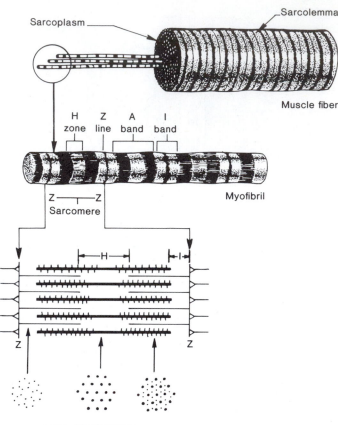

x-sec. Myofilaments

transverse bands are also evident when fresh unstained muscle fibers are viewed under an ordinary light microscope with the condenser stopped down or under a polarizing light microscope. The banding pattern changes when a muscle is induced to contract. Smith (1968, 1984) should be consulted for a survey of insect muscle ultrastructure.

Individual insect skeletal muscle fibers are organized into discrete morphological entities, the *muscle units* (figure 3.25). The fibers, typically 10–20, composing each unit are enveloped by a tracheolated membrane. Muscle units, in turn, compose the various skeletal muscles. Larger muscles are composed of a few to several muscle units, whereas a very small muscle may consist of a single unit (Hoyle 1983). This is in distinct contrast to vertebrate muscles, which are typically composed of very large numbers of muscle units.

Smith (1984) notes that no comparative survey of tracheation in insect muscles has been made. In general, the tracheoles pass close to the cell surface for small muscles, which include most visceral and intersegmental fibers. In contrast, however, the tracheoles penetrate close

to, but still remain outside, the sarcolemma for both asynchronous and synchronous flight muscles (Smith, 1984). More will be said about tracheoles in chapter 4.

Muscle Contraction

As in the vertebrate muscle fibers, contraction of insect muscle fibers is thought to involve the sliding of the myosin and actin filaments past one another with the consequent shortening of the sarcomeres. Current evidence suggests that this is accomplished by the cross-bridges, which are the heads of the myosin molecules acting as "independent force generators," engaging with the actin during contraction, thus pulling the actin filaments from both sides towards the center, and disengaging during relaxation.

Contraction of a fiber is induced by the arrival of a nerve impulse from the motoneuron, which leads to release of a transmitter substance [believed to be L-glutamate (Aidley 1985)]. The membrane of a resting muscle fiber is maintained in a polarized state (i.e., the inside is on the average -40 mV to -60 mV relative to the outside). Arrival of the transmitter substance causes

Figure 3.24 Ultrastructure of skeletal muscle in the mosquito *Aedes aegypti*. (*a*) Cross section (9600×): 1, center of a single myofibril; 2, mitochondria. (*b*) Longitudinal section (25,400×): 1–2 and 3–4, I-bands; 5, mitochondrion; the dark, thick (myosin) and thin (actin) parallel, horizontal lines between 2 and 3 myofilaments. (*c*) Portions of cross sections of myofibrils showing arrays of thick (myosin) and thin (actin) myofilaments (73,000×): 1, parts of T-system.
Courtesy of M. Catherine Walker.

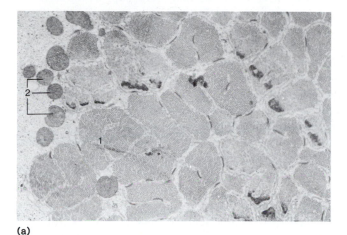

(a)

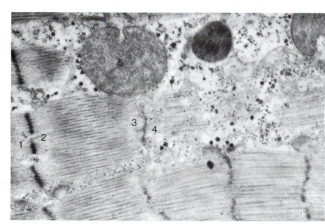

(b)

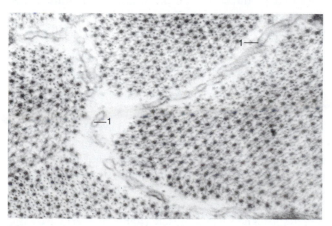

(c)

Figure 3.25 Innervation of a typical muscle unit showing fast and slow axons and multiterminal endings.
Redrawn from Hoyle, 1965.

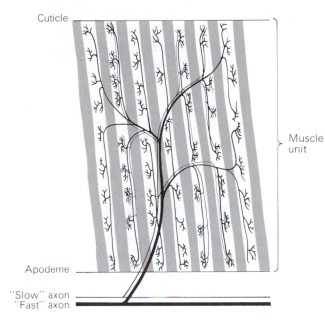

a change in permeability by activating specific ion channels to open resulting in an influx of sodium ions and consequent depolarization. The potential difference across the muscle membrane returns to its original level with the movement of potassium ions out of the cell. The change in membrane potential induced by the nervous impulse is called the postsynaptic potential. Inhibitory motoneurons can synapse near the regions of the excitatory neuromuscular junctions by releasing a transmitter (believed to be gamma-aminobutyric acid) that results in inhibition by causing hyperpolarization of the sarcolemma. Without inhibition, the postsynaptic potential spreads from the neuromuscular junctions; it decreases rapidly, and hence its effect is localized. Now we see the importance of the T-system. The T-system, already discussed, helps carry the postsynaptic potential deep into muscle cells. Here the electrical signal causes the sarcoplasmic reticula to release large quantities of calcium ions, which in turn initiates the cascade of events involving the myosin heads to attach to and pull the actin filaments towards the center of the sarcomere. The energy for muscle contraction comes from the hydrolysis of ATP, which is bound to the outside of the heads of the myosin filaments, to ADP and the enzyme (ATPase) that catalyzes the breakdown of ATP and is thought to be localized in the heads of the myosin. The action of calcium ions on the whole cascade appears to depend on two other accessory proteins (tropomyosin and troponin). In the absence of calcium ions these two accessory proteins inhibit the myosin heads from interacting with the actin filaments, whereas the presence of calcium ions facilitates the engagement between the two filaments, thus facilitating contraction. Muscle relaxation, then, appears to involve the sequestration of calcium ions by the sarcoplasmic reticula and consequent termination of ATPase activity. For a more detailed description of muscle contraction see Aidley (1985) and Maruyama (1985).

Energy for Muscle Contraction

The biochemical processes that result in the production of adenosine triphosphate (ATP) occur in the cytosol and mitochondria of muscle. The phosphate that replenishes the ATP utilized in muscle contraction is furnished by the phosphogen, arginine phosphate, which has been identified in a number of insects (in vertebrate muscle the phosphogen is creatine phosphate). Carbohydrates are commonly the primary fuel for muscle contraction, although many insects utilize fat and occasionally protein or amino acids. For example, the amino acid proline serves as the primary flight fuel in the tsetse fly. For further information on this topic see the reviews by Bailey (1975), Crabtree and Newsholme (1975), Gilmour (1965), Rees (1977), Downer (1981), and Goldsworthy (1983).

Innervation and Contraction of Skeletal Muscle

In contrast with the single end-plate characteristic of vertebrate neuromuscular junctions, insect motor axons branch out, forming terminals at several regularly spaced points along a fiber (figure 3.25). This is called *polyneuronal innervation*. Because the postsynaptic potential decays rapidly and influences only a portion of a fiber, the occurrence of many terminals promotes stimulation of a whole fiber.

In the majority of cases the number of axons that innervate a given muscle unit is two. However, a third has been identified in the jumping hindlegs of *Romalea* and other locusts. In a typical muscle unit (figure 3.25) one axon is fast and the other slow. An individual fiber may receive branches from one or both kinds of axons. The designations "fast" and "slow" refer to the kind of contraction produced. Stimulation of the fast axon characteristically produces a postsynaptic potential of constant size and a rapid, powerful twitch. A rapid train of impulses along a fast axon produces a sustained strong contraction (*tetanus*). Stimulation of the slow axon has a different effect. A single impulse produces a single, weak postsynaptic potential and a single, weak twitch. A train of closely spaced impulses induces successively greater postsynaptic potentials and contractions, that is, a graded response. This graded responsiveness of muscles to impulses arriving along the slow axon affords precision in control of muscle contraction. This method of control of contraction is associated with the small size of insect muscles as compared with those of vertebrates. Vertebrates effect slow, precise movements by varying the number of different muscle units stimulated within a given muscle.

Fast axons are involved in activities like jumping that require sudden strong muscle contraction. Slow axons are involved with muscle tonus and mediate actions that require slow, smooth, and precisely controlled contractions. The term *tonus* refers to the continuous tension applied by a muscle as opposed to alternate contraction and relaxation. Muscles in the tonic state support an insect in a given stance. In soft-bodied larvae, such as caterpillars, muscles in the tonic state maintain the rigidity of the exoskeleton. The third motor axon mentioned above, when present, may function in an inhibitory fashion. Although fast and slow axons often function independently, they also commonly function in such a way that the action of one reinforces the action of the other.

Chapman (1971), Usherwood (1975), Hoyle (1974), Maruyama (1985), and Aidley (1985) all contain information on innervation and contraction of insect muscles. Pichon and Ashcroft (1985), however, provide new information on resting potentials of nerve and muscle cells, excitatory and inhibitory synapses, and regulation of the ionic currents in muscle fibers. The topic of neurotransmission and neuromodulation of skeletal muscles is reviewed by Piek (1985).

Physiological Types of Insect Muscle

Insect muscles can be classified as *resonating* (*asynchronous*) and *nonresonating* (*synchronous*). Resonating muscle in insects is the fibrillar type. In appropriate mechanical situations it is capable of undergoing several successive contractions when stimulated with a single nervous impulse. This type of muscle is almost exclusively associated with flight, but it is also associated with the halteres of the true flies and with the sound-production mechanism of the cicada. Nonresonating muscles include the other types of muscle fibers described earlier and are involved with movement other than that produced by fibrillar muscle. The differences between the resonating and nonresonating types of muscle will become more clear when we discuss the flight of insects.

Muscle Power

Great feats of strength have been ascribed to insects. For example, it has been said that if a flea were the size of a human being, it would be able to leap over tall buildings. This is a common fallacy. It is true that some insects can leap comparatively great distances and that some are capable of lifting or moving objects many times heavier than themselves, but it is *not* true that if these insects were to increase in size, their strength would increase correspondingly. The force a muscle is able to exert varies directly with its cross-sectional area, whereas the volume or mass of a body varies cubically. As an insect body would increase in size, its volume and mass would increase at a greater rate than a cross-sectional area of muscle, and hence the muscles would become relatively weaker. Thus we can say that insect muscles are relatively stronger than those of larger animals only because of the comparatively small size of insects. Actually, when the absolute muscle strength (maximum force a muscle can apply per square centimeter) of vertebrates and insects is compared, it is found that there are no great differences (Wigglesworth 1972).

Although the forces exerted by insect and vertebrate muscles do not differ much, the work per unit time (i.e., power output) capability of insect flight muscles is comparatively large. In fact, active insect flight muscles display the highest metabolic rates of any known tissue (Chapman 1971).

Visceral Muscle

As mentioned earlier, visceral muscles are associated with the various internal organs, including the muscles of the dorsal vessel, accessory pulsatile organs, alary muscles, dorsal and ventral diaphragm, the alimentary canal, Malpighian tubules, reproductive organs, pheromone glands, and labial or salivary glands. Visceral muscles may be extrinsic or intrinsic. Extrinsic visceral muscles originate on the integument and insert on organs; for example, the muscles responsible for the dilation of the cibarial or pharyngeal pumps (see chapter 4). Intrinsic visceral muscles occur in regular meshes of circular and longitudinal strands or in irregular networks surrounding a given organ.

Some visceral muscles, like the cibarial and pharyngeal dilators, are indistinguishable histologically from skeletal muscle. However, most visceral muscles are different from skeletal muscles. The contractile material in visceral muscles is not grouped into distinct myofibrils, but simply fills all or part of a fiber. However, the contractile material is composed of thick and thin filaments, apparently myosin and actin. Whereas insect flight muscle typically has six thin filaments arrayed around a single thick filament, visceral muscle tends to have a higher thin filament/thick filament ratio: 10 or 11 or 12 thin to one thick filament. The T-system may be regular or irregular, or it may be only weakly developed. Visceral fibers tend to have longer sarcomere lengths than do flight muscles.

The insect visceral nervous system (i.e., both ventral and caudal) innervates the visceral muscles by both classical neurons and neurons termed neurosecretory cells. Unlike some other neurosecretory cells, the neurosecretory cells innervating the visceral muscles directly release their secretion onto the muscles and do not rely on a hemolymph transport system. Their secretions are released at regions termed neurosecretomotor junctions.

Huddart (1985) provides an excellent review of visceral muscle neurotransmitters. Basically he categorizes them into three groups: the monoamines (e.g., serotonin or 5-Hydroxytryptamine = 5-HT), the amino acids (e.g., L-glutamate and α-aminobutyric acid = GABA), and peptides and hormones (e.g., proctolin, cardioaccelerating hormone, refer to table 3.2 for others). More information can be obtained on these transmitters and their role in the excitation-contraction coupling in visceral muscle by reading Huddart's (1985) review. The specific functions of visceral muscles are dealt with, as appropriate, in other chapters. The best reviews of insect visceral muscles are those of Huddart (1985) and Miller (1975).

Muscle Development and Maintenance

During development from the immature stages to the adult, the type of transformation that occurs in the muscles of hemimetabolous insects differs greatly from holometabolous insects. In the former, the transition is gradual and muscle sets of the adult are already present in the larval forms, whereas in holometabolous insects most larval muscle sets degenerate with new ones being formed in the adult. At the same time, insects may possess specific muscle sets that are used only for a brief period in their development (e.g., moths escaping from the pupal cuticle and cocoon). The fate of these muscle sets (i.e., degeneration or not) and the factors that regulate this process are reviewed by Finlayson (1975).

The giant silkmoth, *Antheraea polyphemus*, is an excellent model system to investigate the factors regulating muscle degeneration. The developing adult, housed within both the pupal cuticle and silken cocoon, has a muscular system composed mainly of intersegmental muscles located in abdominal segments 4–6. These muscles are similar to those shown in figure 3.22 and function mainly in assisting the adults' escape from the pupal case and cocoon. Once the adult moth has emerged (adult eclosion) from the pupal case and cocoon, these muscle sets are no longer "necessary" and undergo rapid degeneration or programmed cell death (Finlayson 1975). Ligation experiments in the 1960s showed that if the abdomen was tied off and separated from the head and thorax prior to adult emergence, these muscles did not exhibit rapid degeneration, whereas muscles in the abdomens that were ligated and isolated after adult eclosion degenerated within 30 hours. These experiments suggested to the investigators that some "factor" from the anterior region, either head and/or thorax, was responsible for triggering muscle degeneration. This research was continued by Truman and his colleagues who identified the "anterior factors" responsible for muscle degeneration (Schwartz and Truman 1984).

Truman's laboratory showed that the two types of muscle degeneration that occur following adult eclosion (i.e., slow degeneration taking about 6 days versus rapid degeneration occurring within 30 hours following eclosion) were both under hormonal control. Ecdysone (20-hydroxyecdysone) is a steroid hormone produced in the ecdysial glands and is important in activating the molt (see table 3.2). The process of slow muscle degeneration is activated by ecdysone, whereas the rapid, programmed cell-death depends on both ecdysone and a peptide eclosion hormone, which is produced in the brain (more will be said about this peptide in chapter 8). The eclosion hormone acts directly on the muscles but has no effect until these muscles have been made competent to respond to the eclosion hormone by previously being exposed to a decline in the ecdysone titer. In fact, this decline occurs just prior to adult eclosion and is a good cue to set the stage for the action of the eclosion hormone. This regulatory control scheme applies only to *A. polyphemus* and may differ in other moths and insects. In addition to endocrine control of muscle degeneration, neural control has also been demonstrated (Finlayson 1975).

Adult insects differ little from most vertebrates in that they often have a time in their lives when the muscular system is more extensively developed, more finely tuned, and biochemically more efficient than at other times. This is best exemplified in the insect flight system and, in most insects, peak performance usually correlates with specific events in the life history that are most demanding on the flight system (e.g., finding a mate, migrating, or locating an egg-laying site). More will be said about this in chapters 7 and 9. Johnson (1969) reviews this topic of muscle development and efficiency and also provides information on the effect of diapause (see chapters 5 and 8) on muscle development.

CHAPTER 4

Alimentary, Circulatory, Ventilatory, and Excretory Systems

Every cell in the insect body, regardless of its function, requires a source of energy, a source of oxygen, and the raw materials with which to carry out its own maintenance and synthesizing activities. As a result of carrying out these processes, a cell produces carbon dioxide and other waste products. Those systems directly involved in the transport of nutrients and oxygen to the individual cells and the removal of accumulated waste materials and carbon dioxide are discussed in this chapter.

The Alimentary System

The alimentary system is involved in the initial steps in the transport of nutrients to individual cells. Ingestion, trituration (chewing), digestion, absorption into the hemolymph, and egestion are all associated with this system. Insects possess a tube (often coiled), the alimentary canal, which extends from the anterior oral opening, the mouth, to the posterior anus. The gut is formed by a one-cell-thick layer of epithelial cells, and a non-cellular basement membrane (basal lamina) is present on the hemocoel side of this layer of cells.

It must be emphasized here that the generalized insect used by most textbooks is based on the orthopteroid plan (e.g., grasshoppers, crickets, and locusts). Thus, figures 3.3, 3.6, 3.20, and those to follow, use this plan as the basic model to help you understand how the insect systems are organized, how they function, and how they interact. The decision to use the orthopteroid plan is based partly on the phylogenetic position of this group. The orthopteroid orders exemplify the basic, primitive design from which other insect groups evolved modifications and specializations. In addition, their size and economy and ease of collection make them excellent specimens for in which to study anatomy, histology, etc. The following references cover the anatomy and mor-

phology of some members of this group: Jones (1981), Cornwell (1968), Guthrie and Tindall (1968), and Chapman and Joern (1990).

One of the major reasons for the biological success of insects is their ability to eat, digest, and utilize an enormous diversity of foods. This ability allows one to better appreciate the extreme diversity observed in the modifications and specializations of the alimentary system of insects. The structural and biochemical modifications of the alimentary system of a particular species reflect the type of food eaten. Because of the differences in dietary requirements between the different life stages and the differences between the sexes, one finds both structural and functional differences in the way foods are obtained, stored, processed, and absorbed (e.g., butterfly larvae chew up plant material, whereas adults may suck up only floral nectar; female mosquitoes bite and suck up a vertebrate blood meal, whereas males do not). This chapter will impart a sense of this diversity by providing appropriate examples. Terra (1990) takes an evolutionary approach to the study of the alimentary system while Goodchild (1966) reviews the evolution of the alimentary canal in Hemiptera.

The insect alimentary canal (figure 4.1) is divided into three distinct regions, the anterior *foregut,* or *stomodaeum,* the *midgut,* or *mesenteron,* and the posterior *hindgut,* or *proctodaeum.* The foregut and hindgut are lined with a chitinous *intima,* which is continuous with the cuticle of the integument. The hindgut and foregut are unsclerotized, contain endocuticle and a two-layered epicuticle, and are flexible. At the molt, both foregut and hindgut and their contents are shed. Because termites house in their hindguts the symbiotic organisms necessary for the digestion of wood, each molt could leave the termite extremely vulnerable to starvation, because they

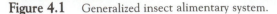

Figure 4.1 Generalized insect alimentary system.

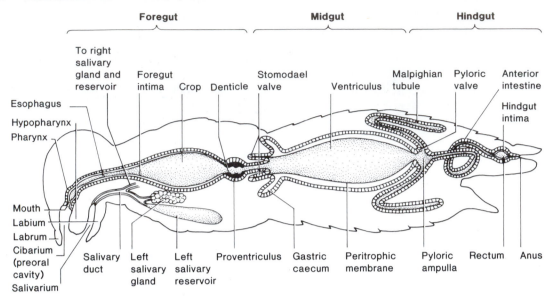

would be unable to digest wood (i.e., cellulose). This dilemma and the way the termites solved it, that is by anal trophallaxis (the passage of symbionts from one generation to the next by the exchange of alimentary liquids between individuals), is considered one of the major evolutionary driving forces leading to sociality in this group of insects. The foregut and hindgut arise as invaginations of the ectoderm, which also gives rise to the integument, and the gut epithelial cells are continuous with the epidermal cells. The midgut is generally believed to be of endodermal origin. Longitudinal and circular (intrinsic) muscles are usually associated with each of the three regions and, by means of rhythmical *peristaltic contractions*, move food along the alimentary canal. Some muscles originate on the integument and attach to certain parts of the alimentary canal (i.e., extrinsic muscles); for example, the muscles that dilate the pharyngeal pump. The anterior alimentary canal muscles are innervated by the stomatogastric system, and the posterior muscles are innervated by nerves from the posterior ganglion of the ventral chain (see chapter 3).

In addition to extrinsic muscles, tracheae and tracheoles (discussed later in this chapter) provide support for the gut.

The alimentary canal tends to be shorter in species that exist on high-protein diets and longer in those with high-carbohydrate diets, but there are many exceptions.

As discussed in chapter 2, the alimentary canal plays a major role in the process of molting. By taking air or water into the alimentary tract and with the aid of specialized muscle sets, the insect increases its internal hemolymph pressure, which is important in rupturing the cuticle at the lines of weakness (i.e., ecdysial suture) during molting.

Foregut

Although there is considerable variation in the foregut, several morphological regions can usually be recognized. The *true mouth* lies at the base of hypopharynx within the *cibarium* (preoral cavity) formed by the mouthparts. The true mouth communicates directly with the *pharynx,* a structure that varies greatly among different insects. The cibarium or pharynx or both may be highly modified, forming pumps with well-developed extrinsic visceral musculature. These pumps, cibarial and/or pharyngeal, are most highly developed in sucking insects, such as Lepidoptera, Hemiptera, and many Diptera (figure 4.2). Next to the pharynx is the *esophagus,* which is commonly enlarged posteriorly to form the *crop* (figure 4.1). In some insects the posterior enlargement may be in the form of one or more blind sacs, or *diverticula* (figure 4.2). Immediately posterior to the crop is the *proventriculus* (figure 4.1). The luminal side of this structure often bears sclerotized denticles (teeth) or spines. The proventriculus typically communicates with the midgut by means of an intussusception, which consists of both foregut and midgut tissue. This structure will be called the *stomodael valve.* Midgut tissue (*cardial epithelium*) surrounds the foregut portion of the stomodael valve. The true flies (Diptera) lack a proventriculus in the sense described. In these insects, the stomodael valve lies between the esophagus—or diverticulum(a), if present—and the midgut posterior to the cardia and is commonly referred to by investigators who study Diptera as the proventriculus.

The foregut with its various morphological divisions serves mainly as a conducting tube, carrying food from the cibarial cavity to the midgut. The enlarged crop

Figure 4.2   Alimentary canal of a mosquito.

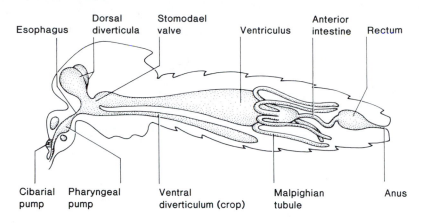

Midgut

functions, at least in part, as a site of temporary food storage and partial digestion in some cases.

The proventriculus, when armed with denticles, may serve as a grinding structure in addition to the mouthparts. Spines in the proventricular region may act together as a food sieve or filter. In honey bees, for example, proventricular spines allow the movement of pollen into the midgut without admitting ingested flower nectar. The stomodael valve is developed to varying degrees in different insects. Whether this structure actually acts as a valve in most insects is not known. The proventriculus is often called the gizzard because of its similar function to the gizzard of birds. Birds also have a crop for storing food. Many birds swallow stones which are passed to the highly muscular gizzard. Here the muscular action and presence of stones act like a grinding mill to physically break down the food. Chapman (1985) discusses phylogenetic differences in the proventriculus of different insect groups as structure relates to type of food eaten. Hemiptera are fluid-feeders, and they lack a proventriculus. Again, structure complements function. It might help to think like an insect architect. Ask questions about design as it relates to function. Observe the insect in its natural environment and see how body design relates to function.

In addition to the saclike diverticula of some insects (e.g., crop of flies), some species that feed as larvae on resinous plants may have diverticula present in the esophagus, which store resins. With the aid of powerful circular muscles around these diverticula, the insect is able to defend itself by ejecting the resins onto a predator. The resins act as deterrents and are extremely effective in protecting the larva (Eisner et al. 1974).

Although the foregut is not the major digestive region of the alimentary canal, some digestion may occur in the crop by the action of salivary enzymes and enzymes regurgitated from the midgut—for example, in the Orthoptera. Except for the possible passage of small amounts of lipid in certain insects, the foregut intima is impermeable. Thus the foregut probably plays no major role in the absorption of materials into the hemolymph.

The midgut (figure 4.1) typically begins with the cardial epithelium (*cardia*) associated with the stomodael valve. Immediately posterior to the cardia there is commonly a group of diverticula, the *gastric caeca*. The number of these caeca varies in different species, and similar pouches may be present in other sections of the midgut. The remainder of the midgut, the *ventriculus,* is usually a somewhat enlarged sac and serves as the insect's stomach. In some insects, the midgut is divided into distinct regions; for example, two, three, and four regions have been identified among various true bugs (Hemiptera).

The midgut epithelium of most insects is composed of three basic cell types: columnar digestive cells, regenerative cells, and endocrine cells. The major cell type is the digestive cell, which is typically columnar with a striated border formed by *microvilli* on the luminal side (figure 4.3). On the hemocoelic side, the basal plasma membrane is characteristically infolded, and mitochondria are associated with these folds. These cells usually contain extensive rough endoplasmic reticulum, much of which is probably involved in the synthesis of digestive enzymes. The microvilli and folded basal plasma membrane provide the extensive surface area one would expect to find in actively absorbing and secreting cells. Recently, considerable attention has been given to the ultrastructure of the midgut because it serves as the major site for enzyme induction processes against toxic substances that are either in the natural diet or applied by humans as insecticides (Terriere 1984), it is the major site for many pathogens and/or parasites to gain entrance to either the insect host or insect vector, and it may prove (as shown in the vertebrates, see DelValle and Yamada 1990) to be the largest endocrine organ of the insect. Chapman (1985) provides a list of references that focus on the ultrastructure of the midgut, whereas an excellent review chapter on the ultrastructure of the digestive tract is provided by Martoja and Ballan-Dufrançais (1984). Billingsley (1990) discusses the midgut ultrastructure of blood feeding insects.

Figure 4.3   Generalized midgut epithelial cell.

Redrawn from Berridge, 1970.

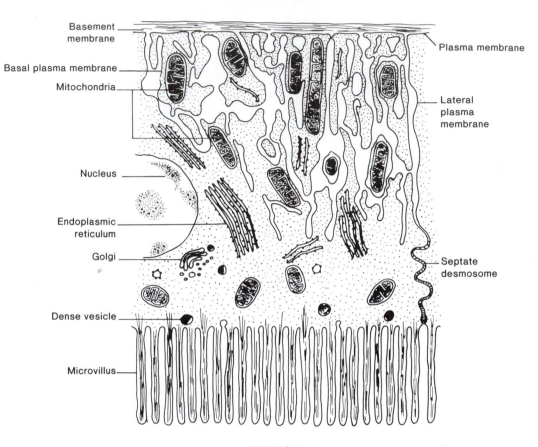

At the bases of the midgut epithelial cells are small *regenerative cells,* or *replacement cells.* These cells replace the actively functioning gut cells that die or that degenerate as a result of *holocrine secretion.* In holocrine secretion a cell breaks apart completely when it releases its products. Regenerative cells may be dispersed individually among the epithelial cells (e.g., in caterpillars and true flies) or may be concentrated in discrete groups as *nidi* (e.g., in grasshoppers and relatives, dragonflies, and damselflies) or *crypts* (e.g., in many beetles).

The midgut is the principal site for the production and secretion of digestive enzymes, digestion, and absorption. Depending on the species and dietary intake, the insect midgut may differ considerably with respect to regions of secretion and absorption. Functionally, the midgut can be divided into different regions, which absorb different components of the food materials (Chapman 1985, Dow 1986, Terra 1990).

When present, the gastric caeca (figure 4.1) function to increase the surface area for either secretion of digestive enzymes or absorption of water, ions, glucose, cholesterol, amino acids, and possibly other nutrients not yet tested. Digestion in certain insects also takes place in the

gastric caeca (Terra 1990). However, according to Terra (1990), the evolutionary trend in the holometabolous insects is towards the loss of the caeca. More will be said about these structures when countercurrents in the midgut are discussed.

A peritrophic membrane (figures 4.1 and 4.4), which is composed of chitin fibrils in a protein-carbohydrate matrix, is secreted by either the cardial epithelium or the general ventricular epithelium, or by both. This membrane (peri = around; trophic = food) surrounds the food bolus and may protect the epithelial cells from possible abrasion by food particles. That the peritrophic membrane acts in other ways is suggested by its presence in insects that do not ingest solid, potentially abrasive food particles (i.e., fluid-feeders of various types) and its absence in some insects that do. The Hemiptera, which are fluid-feeders, lack a peritrophic membrane. In some beetle larvae, just before pupation, the peritrophic membrane continues to be produced after feeding has ceased and the gut emptied of undigested material. This membrane is extruded from the anus and collapses and dries, forming a thread that is used like silk to form a cocoon (Kenchington 1976).

Figure 4.4 Generalized models of the insect gut showing the compartmentalization. (*a*) Created by the peritrophic membrane. (*b*) The countercurrent fluxes of water that are responsible for the movement of solutes.
Taken and modified from Terra, 1990.

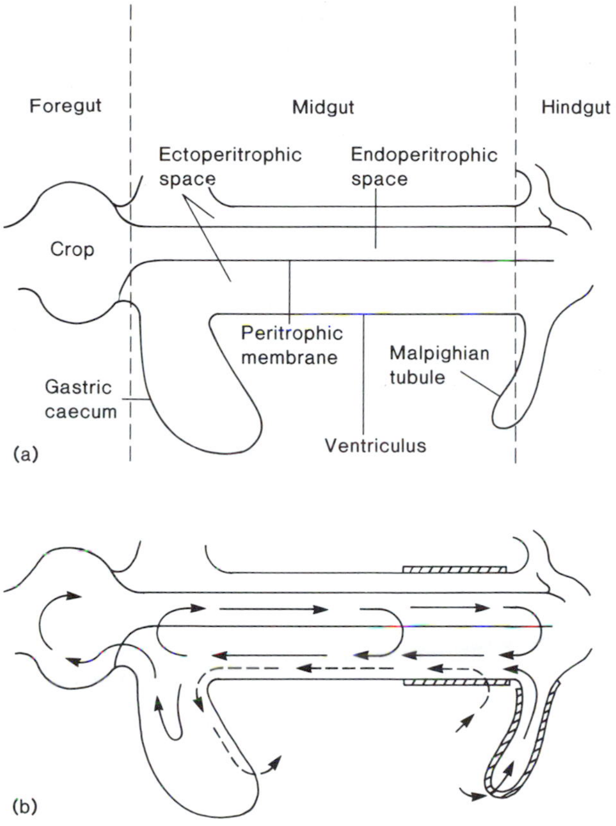

Richards and Richards (1977) review the literature on peritrophic membranes in general, and Peters (1976) reviews the literature dealing specifically with the Diptera, Chapman (1985), Peters (1992), Spence (1991), Terra (1990), and Billingsley (1990) discuss the functions of the peritrophic membrane.

Hindgut

The hindgut (figure 4.1) is composed of cuboidal epithelial cells. It commences with the *pylorus,* which is associated with a variable number of typically slender, elongate excretory structures, the *Malpighian tubules,* and which usually contains a valvular structure, the *pyloric valve.* The Malpighian tubules can be used as landmarks indicating the beginning of the hindgut. The hindgut is divisible into a tubular *anterior intestine,* just posterior to the Malpighian tubules, and a highly muscularized, enlarged *rectum,* which terminates with the anus. The anterior intestine may be differentiated into an anterior *ileum* and posterior *colon.* The rectum usually contains a number of pads, or *papillae* (usually six), that project into the lumen. These structures receive an ex-

tensive supply of tracheae and are metabolically very active. They play an especially important role in the excretory system.

The hindgut is the major region of the insect involved in recycling. Here needed materials are "reclaimed" while excess or waste materials are "trashed." Functions of the hindgut include the following:

1. Water absorption from urine and feces.
2. Ion absorption from urine and feces.
3. Cryptonephridial system for water conservation (the Malpighian tubules come into close contact with the hindgut, thus facilitating water and ion regulation of the hemolymph).
4. Pheromone production (e.g., male fruit flies of *Dacus tryoni* have rectal pouches that produce a sex attractant while some scolytid beetles produce an aggregation pheromone that passes out in the feces).
5. Respiration in larval dragonflies (suborder Anisoptera).
6. Modifications in structure for housing symbiotic microorganisms.

Interested individuals can read more about these functions by consulting Chapman (1985).

Generally, little or no digestion occurs in the hindgut. It serves primarily to carry undigested food material away from the midgut, ultimately egesting it from the insect. In the case of insects housing symbionts in the hindgut, however, digested products produced by the symbionts such as short-chain fatty acids, acetate, and butyrate are absorbed by the hindgut. Unlike the foregut intima, the hindgut intima is permeable and allows the passage of at least relatively small molecules. Before egestion, the hindgut absorbs, to varying degrees, water, salts, and amino acids previously removed from the hemolymph by the Malpighian tubules. Thus it plays a major role in the water and salt balance of an insect.

Digestion

Although the initial stages of chemical (i.e., enzymatic) breakdown of ingested food may occur in the foregut, most digestion occurs in the midgut. Even though the insects that eat particulate foods do not have teeth as we know them, and they cannot masticate their food over and over as we do, they do have structures, mainly in the proventriculus, that physically break down the large food particles into smaller particles. This not only facilitates passage through the alimentary canal but increases the surface area for enzymatic action. Digestion occurs by a series of progressive enzymatically catalyzed steps, each producing a simpler substance until molecules of absorbable size or nature are produced. For example, polysaccharides are broken down into small chains, disaccharides, and finally into simple, absorbable monosaccharides (e.g., glucose); proteins are broken down into peptones, small polypeptides, dipeptides, and finally into amino acids, which are absorbable.

The current thinking concerning solute (i.e., enzymes and digested molecules) movements in the midgut (sometimes including the foregut and hindgut) has generated new ideas about digestion and absorption in insects and has given new significance to the peritrophic membrane. Terra (1990) proposes a model for one dipterous larva where three phases of digestion occur in three separate areas of the midgut. The importance of the peritrophic membrane in creating this compartmentalization of the midgut is shown in figure 4.4. Thus, there exists an endoperitrophic space and an ectoperitrophic space. At least for one dipteran larva, the first phase of digestion occurs inside the endoperitrophic space where the larger molecules are unable to pass through the peritrophic membrane filter. Once reduced by enzymatic action to molecules capable of passing through the membrane filter, they enter the ectoperitrophic fluid where another group of enzymes causes further digestion (second phase). The final stage of digestion (third phase) occurs mainly among the microvilli of the midgut caecal cells. How does the ectoperitrophic fluid find its way forward to the caeca, which are located in the opposite direction of the normal posterior movement of the food in the endoperitrophic space? This is accomplished by the countercurrent water fluxes that are produced by differential water absorption from the caecal lumen into the hemolymph and secretion from the hemolymph into the ectoperitrophic space at either the Malpighian tubules or the posterior midgut (figure 4.4; Terra 1990). The countercurrent movement of water in the midguts of other insects may be different depending on the phylogenetic position of the group (Terra 1990). Evidence is accumulating that shows that some digestive enzymes remain trapped in the glycocalyx matrix covering most midgut microvilli and that final steps in digestion of specific food molecules may occur here prior to entrance into the hemolymph of the final digested products.

There is some correlation between the kinds of food material eaten by and the kinds of enzymes present in a given insect. Thus, one would expect an insect, such as a cockroach, that eats many different kinds of food to have more enzymes than an insect, such as a tsetse fly, that feeds primarily on blood. In addition, different enzymes may be secreted by different parts of the midgut epithelium. It is also likely (i.e., especially in insects that feed on seeds or are predators) that enzymes that act in the midgut actually originate in the salivary glands. In addition, digestion in many of these insects is extra-intestinal. Once the powerful enzymes have acted, the digested material is either sucked up or eaten. For more information about seed-feeders and those insects involved in extra-intestinal digestion, see Applebaum (1985), Dow (1986), and Slansky and Panizzi (1987).

Because over one half of the insect orders contain insects that feed on plants and over one half of all known species are phytophagous (figure 4.5), a few words are in order concerning problems faced by plant feeders. Also, because phytophagous insects are the number one competitor with human beings for food and fiber products, we may be able to learn some lessons from studying their basic biology, which in turn may be applied to new and novel control strategies that are not based on chemical pesticides.

One novel way to confer insect resistance in plants is to modify the plant genetically by introducing foreign genes that will express the desired form of resistance (Meeusen and Warren 1989). Protease inhibitors are widely present in plants and appear to confer pest resistance by preventing digestion of the plant protein by the herbivore (in this case, the insect). English scientists, using this information, introduced the trypsin inhibitor gene (encodes for a serine protease inhibitor), which imparts naturally-occurring resistance to the cowpea plant (*Vigna unguiculata*) into the tobacco plant. These tobacco plants were then resistant to the tobacco budworm, *Heliothis virescens*. Failure of the insect to grow normally is attributed to the interference of digestion by the protease inhibitor preventing normal protein digestion from occurring. Because insects and plants are constantly engaged in this "arms war," one would expect the insect to have evolved strategies to counter these biochemical toxins (Rosenthal and Janzen 1979, Brattsten and Ahmad 1986, Barbosa and Letourneau 1988, Rosenthal and Berenbaum 1992). Terra (1990) discusses one case in which hemipterans feeding on seeds rich in trypsin inhibitors evolved cathepsin-like proteinases, instead of using trypsinase, to circumvent the problem of proteinase inhibitors from the seeds interfering with digestion. This area of research is in its infancy and is awaiting inquisitive investigators to take up the call and to exploit its potential both from the basic and the applied angles. Interested readers are referred to the following references for more information: Applebaum (1985) and Ryan (1979, 1989).

Phytophagous insects are food limited, not because of the quantity of food available but because of the poor quality of food. In response, unique morphological and biochemical modifications of the alimentary system to obtain adequate nutrients have evolved in insects. An example is the filter mechanism of the midgut of certain sap feeders (e.g., aphids) to concentrate the extremely low level of nutrients in sap. What do human beings do to tree sap to obtain maple syrup?

Bruce Ames reported that, by weight, humans ingest about 10,000 times more natural pesticides (found in plants) than synthetic ones (insecticides) (Abelson 1989). Brattsten (1979) also alludes to this problem when referring to insects as eating from "the poisoned platter." These natural pesticides (toxins) are the plant's defensive mechanism against herbivores (e.g., ruminants like sheep,

Figure 4.5 Table of major insect feeding habits as a function of theoretical phylogeny. *Taken from Dow, 1986.*

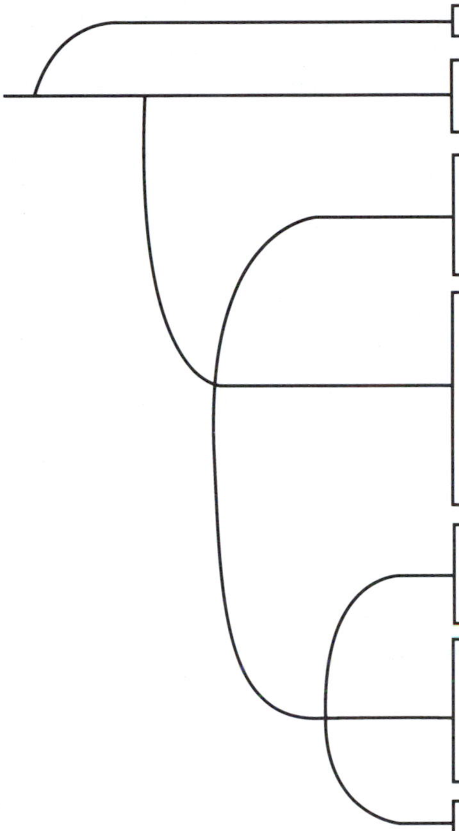

Order	Scavenger	Carnivore	Moss fern fungi	Phytophagy	Number of species (approx.)
Apterygota	yes	yes	yes		3000
Paleoptera					
Ephemeroptera	yes			yes	2000
Odonata		yes			5000
Orthopteroid					
Orthoptera	yes	yes	yes	yes	20000
Dermaptera	yes		yes	yes	1200
Isoptera	yes		yes	yes	2000
Embiidina	yes	yes			200
Hemipteroid					
Plecoptera	yes	yes	yes	yes	1200
Zoraptera			yes		20
Psocoptera			yes		1700
Mallophaga		yes			2600
Anoplura					540
Thysanoptera		yes	yes	yes	4500
Hemiptera		yes		yes	50000
Homoptera				yes	32000
Neuropteroid					
Neuroptera		yes			4000
Coleoptera	yes	yes	yes	yes	300000
Strepsiptera		yes		yes	300
Panorpoid					
Mecoptera	yes	yes			400
Trichoptera	yes	yes			7000
Lepidoptera				yes	112000
Diptera	yes	yes	yes	yes	100000
Siphonaptera		yes			1300
Hymenopteroid					
Hymenoptera		yes		yes	100000

goats, etc., and insects) and may constitute from 5%–10% of the plant's dry weight (Abelson 1989). Rosenthal and Janzen (1979), Ishaaya (1986), Barbosa and Letourneau (1988), and Brattsten and Ahmad (1986) and Rosenthal and Berenbaum (1992) provide more information about this exciting area of insect-plant interactions. What is more pertinent here, however, is to understand how this relates to the alimentary system and how the insects deal with these phytotoxins. The gut is the first part of the insect that encounters food that is ingested. As aptly stated by Brattsten (1979), "Herbivores expose themselves to the hazard of being poisoned by every meal." In those situations in which insects or other herbivores feed successfully on toxic plants, these insects have evolved various strategies to deal with the problem. One of these strategies is to degrade or detoxify the toxins rapidly using enzymes (referred to as microsomal mixed function oxidases or MFOs) located in and produced either by midgut or by fat body tissue. Another mechanism is to rapidly excrete the toxins before they can exert their effect. Brattsten (1979) provides an excellent review of plant toxins and the role of MFOs in detoxifying them.

Several proteases, lipases, and carbohydrases have been found in insects, some of which are rather unusual. For example, cellulase has been found in several wood-boring insects. In some, microbes may furnish this enzyme, but others such as *Ctenolepisma lineata,* a thysanuran, are able to digest cellulose in the absence of intestinal microflora. Other enzymes in the unusual category include chitinase, lichenase, lignocellulase, hemicellulase, and hyaluronidase. Certain insects are able to digest ordinarily stable substances. For example, chewing lice and a few other insects are able to break down keratin, a protein that occurs in wood, hair, and feathers. Larvae of the wax moth, *Galleria mellonella,* are able to digest beeswax.

Midgut pH (typically pH 6–8; Dadd 1970), buffering capacity, and oxidation-reduction potential are important factors in the digestive process. These factors vary from species to species and may also vary from one region of the midgut to another within the same insect.

Absorption

Three factors affecting absorption are: the presence of microvilli, which increase the surface area for both absorption and digestion; the functional differences in membrane permeability of various regions of the digestive tract; and the presence of a countercurrent. Turunen (1985) noted that even in insects lacking a peritrophic membrane the anterior midgut appears to be the major site of absorption, thus some type of countercurrent system must exist even for these species. An extensive list showing the types of chemicals absorbed, the species of insect involved, and the region of absorption is provided by Turunen (1985).

It appears that simple sugars (monosaccharides) and lipids diffuse down concentration gradients between the midgut lumen and hemolymph. The diffusion of simple sugars like glucose and fructose is enhanced by the rapid conversion of these sugars to trehalose in the fat body, a process (called *facilitated diffusion*) that maintains a concentration gradient across the gut epithelium. Although it has been generally assumed that carbohydrates and proteins must be broken down to monosaccharides and amino acids in order to be absorbed, there is some evidence that disaccharides and peptide fragments might be absorbed intact. The midgut caeca appear to be particularly active in lipid absorption.

Insects are unique in that they maintain rather high levels of free amino acid stores in the hemolymph, thus many amino acids have to be actively absorbed against a concentration gradient. An interesting evolutionary adaptation of the insect's gut involves handling of the amino acid L-glutamate. Because L-glutamate is believed to be an important neurotransmitter in the somatic neuromuscular sites in insects and is commonly found in the insect's food, the ability of the gut to keep it out of circulation in the hemolymph should be important. In fact, glutamate is poorly absorbed by the midgut, and the gastric caeca metabolize L-glutamate before it reaches the hemolymph. Turunen (1985) notes that this is an excellent example of how the alimentary system is involved in the insect's defense against noxious substances in the diet.

Terra (1990) reports that digestion is not the problem for insects feeding on plant sap, because the nutrients present in the food can readily be absorbed as taken in, however, the main problem is with absorption. Plant phloem fluid has a great amount of sucrose but very low amounts of amino acids, and xylem fluid is also poor in amino acids. In order to obtain sufficient amounts of amino acids and other nutrients, sap-suckers possess various mechanisms that are extremely effective in concentrating the necessary nutrients (i.e., amino acids) from a dilute food source by eliminating water. A similar mechanism operates in blood feeders and will be discussed later. The filter chamber, present in the Cicadoidea and Cercopidae (order Homoptera), is a modification of the anterior midgut, which in combination with the Malpighian tubules facilitates water removal and concentration of the desired nutrients prior to absorption. Goodchild (1966) and Chapman (1985) provide more details about the filter chamber.

One area of alimentary physiology that needs more research attention is the mechanisms by which many naturally-occurring plant xenobiotics are absorbed and sequestered by phytophagous insects. Many of these chemicals have been shown to provide the insect with some sort of protection from predators. A good example is the role of the cardenolides, which are extracted from milkweed plants, in conferring protection on the monarch butterfly, *Danaus plexippus,* from avian predators (Brower et al. 1988). Other examples can be found in Brattsten and Ahmad (1986) and Barbosa and Letourneau (1988).

The active transport of sodium may play a key role in the passive absorption (diffusion) of other molecules (Berridge 1970). The plasma membranes on the hemocoel side of the midgut epithelium are considered to be less permeable than those on the luminal side. Thus when sodium molecules are pumped (active transport) from the midgut cells across the plasma membrane and into the hemocoel, they are replaced by sodium molecules diffusing into the midgut cells from the lumen. The movement of sodium molecules across the cells tends to produce a water gradient between the lumen and the cells, that is, to concentrate water in the lumen. Hence, water would diffuse into the cells. Movement of water from the lumen would, in turn, tend to concentrate other molecules that would then diffuse down gradients into the cells. In short, the work necessary to produce the gradients for diffusion (a passive process) of water and other absorbable molecules would be the active transport of sodium ions. Turunen (1985) provides an update of active transport mechanisms involved in absorption. Absorption in the hindgut will be discussed later in this chapter.

Regulation of the Alimentary System

Regulation of the alimentary system in insects involves control of food movement, control of enzyme secretion, and control of absorption. The alimentary canal is regulated in part through the action of the stomatogastric nervous system (see chapter 3). The control of feeding is dealt with in chapter 8. Dethier (1976) deals extensively with the regulation of the alimentary system in the blow fly, *Phormia regina.* Gelperin (1971), Bernays and Simpson (1982), and Bernays (1985) review the literature on regulation of feeding, including gut emptying, and Chapman (1985) covers coordination of digestion. Food is ingested by the actions of the mouthparts, cibarium, and pharynx, and is typically stored in the crop. It is then released gradually, via the stomodael valve, into the midgut, where digestion and absorption occur. In most insects that have been studied, stretch receptors associated with the crop provide information to the brain (via the frontal ganglion) regarding crop distension and help prevent overfilling of this organ. In some insects, stretch receptors in the abdominal wall have a similar role.

The destination of ingested food may vary with the kind of food. For example, in female mosquitoes sugar meals (flower nectar) are directed to the diverticula and blood meals are directed to the midgut. Sensilla in the roof of the cibarial pump, acting via the frontal ganglion, are thought to be involved in this so-called "switch mechanism." In other blood-feeding insects, such as tsetse flies, ingested blood goes to the crop first.

Control of passage of food from the crop to the midgut (rate of crop emptying) has been studied mainly in the cockroach, *Periplaneta americana* (Davey and Treherne 1963), and in the blow fly, *Phormia regina* (Gelperin 1966). Passage of food from the cockroach crop is inversely related to the osmotic pressure of the food—that is, the higher the concentration of food, the slower the passage. In the blow fly, crop emptying appears to be controlled by the osmotic pressure of the hemolymph—the higher the osmotic pressure of the hemolymph, the slower the rate of crop emptying. Osmotic receptors have been identified in the wall of the cockroach pharynx, but the receptors in the blow fly have not been found.

Applebaum (1985) reports that two mechanisms for the control of enzyme secretion in the insect gut have been suggested:

1. *Secretogogue:* a substance in the ingested material may stimulate enzyme secretion.
2. *Hormonal:* stimuli inducing enzyme secretion are hormones that act via the hemolymph. There is increasing evidence for this mechanism (House 1974b).

Applebaum (1985) also notes that there is no reason to conclude that these two mechanisms operate independently of one another, and recent evidence suggests that secretogogue control may be an immediate response to food, whereas hormonal control is more related to developmental and environmental effects. Nervous control is highly unlikely because the midgut is sparsely innervated or not at all (Dadd 1970, Huddart 1985).

Absorption appears to be controlled by the availability of absorbable molecules, release of food material from the crop being so regulated that digestion and subsequent absorption occur at an optimal rate for a given circumstance.

Many insects ingest foods with a very high water content. Some of these insects (e.g., butterflies and many true flies) store the dilute food in the impermeable crop and pass it gradually to the midgut. In others (e.g., many blood-feeding insects) food may go to the midgut where excess water is rapidly absorbed in the hemolymph and then excreted via the Malpighian tubules. Both mecha-

nisms probably prevent extensive dilution of the hemolymph, and removal of water concentrates solid food, increasing the efficiency of digestion. As previously stated, in spittlebugs and cicadas, the more posterior regions of the alimentary canal (posterior midgut or anterior hindgut) are closely associated with the more anterior regions (posterior foregut or anterior midgut). Such associations are called *filter chambers*. Although there is much variation in structural detail, these filter chambers apparently all facilitate the passage of water directly across the closely associated epithelial layers, bypassing most of the midgut. This action concentrates solid food in the esophagus or anterior midgut from which the food then passes into the region of digestion and absorption.

Movements of the alimentary canal (mainly foregut and hindgut) that complement the actions of the digestive enzymes and aid absorption have been reported to be under neural or neurosecretory control in some insects (Miller 1975, Huddart 1985, Cook and Holman 1985). In others, no neural connections can be found, and gut movements in these cases are assumed to be myogenic. Hormonal stimuli may also have a great deal to do with the rate of gut movements. In cockroaches, a neurohormone (proctolin—see table 3.2; O'Shea and Adams 1986; and volume 11 of Kerkut and Gilbert 1985) is released that stimulates peristalsis of the hindgut but not the foregut or midgut, and also increases the heartbeat rate.

Salivary Glands

Although there may be glands associated with the mandibles and maxillae, salivary glands are typically associated with the labial segment. The *salivary glands* or *labial glands* (figure 4.1), two in number, lie ventral to the foregut in the head and thorax and occasionally extend posteriorly into the abdomen. Depending on the type of food eaten and the insect species involved, salivary glands vary in size, shape, and the type of secretion produced. Two basic types of salivary glands exist: *acinar* (i.e., resembling a cluster of grapes) and *tubular*. Odonata and Orthoptera have the acinar type while Diptera, Lepidoptera, and Hymenoptera have the tubular type. In the acinar type, each "grape," or *acinus*, bears a tiny duct that communicates with other similar ducts, eventually forming a *lateral salivary duct*. Lateral salivary ducts run anteriorly and merge as the *common salivary duct*, which empties between the base of the hypopharynx and the base of the labium. This region is called the *salivarium* and in some sucking insects forms a salivary syringe that "injects" saliva into whatever is being pierced. The lateral salivary ducts may communicate with salivary reservoirs, as in the cockroaches. Depending on stage and species, salivary glands may be a single cluster of cells or may be composed of two or more lobes. In some species, salivary glands are absent (e.g., most Coleoptera).

The secretory products of the salivary glands are generally clear fluids that serve a variety of functions in different insects.

1. They moisten the mouthparts and serve as a lubricant (possibly original function).
2. They act as a food solvent.
3. They serve as a medium for digestive enzymes and various anticoagulins and agglutinins.
4. They secrete silk in larval Lepidoptera (caterpillars) and Hymenoptera (bees, wasps, and relatives).
5. They are used to "glue" puparial cases to the substrate in certain flies.
6. They serve for the production of toxins.
7. They secrete antimicrobial factors (e.g., in certain blow fly larvae).

The most common enzymes found in insect saliva are amylase and invertase, although lipase and protease also have been found. Aphids secrete a pectinase that aids their mouthparts in the penetration of plant tissues. The spreading factor, hyaluronidase, which attacks a constituent of the intercellular matrix of many animals, has been found in the assassin bug.

In the predatory bugs (e.g., assassin bugs, family Reduviidae) and the robber flies (Asilidae) the salivary glands produce not only powerful digestive enzymes permitting extraintestinal digestion but also produce various toxins that help immobilize the prey (Zlotkin 1985, Schmidt 1982, Piek 1985, Tu 1984).

As succinctly stated by Ribeiro (1987), "Hematophagy is polyphyletic in origin, but convergent evolution has equipped many blood-feeding arthropods similarly for locating" and ingesting blood. In fact, the same can be said for blood-feeding animals. One of the major problems for blood feeders is to overcome blood clotting and cessation of bleeding (= hemostasis). It appears that the saliva of the vampire bat, hookworms, leeches, ticks, and blood-sucking (*hematophagous*) insects all contain various antihemostatic (anticoagulant) agents. Current evidence, at least for mosquitoes, is that these various salivary components mainly increase the chances of the female locating a blood vessel. The evidence comes from experiments showing that when the salivary ducts are cut, operated females probe the host more and for longer periods of time than the "sham-treated" females, but operated females are still able to take a full-blood meal when a vessel is finally located. Many hematophagous insects contain a salivary apyrase (i.e., an enzyme that hydrolyzes ATP, which is released from damaged platelets, and ADP, which is released from activated platelets and injured cells, to AMP). When secreted during vessel searching (probing), the enzyme prevents hemostasis by removing ADP, which is the major chemical signal for platelet aggregation, thus preventing hemostasis. More

will be said in chapters 6 and 8 about the importance of blood ATP as a phagostimulant for hematophagous insects.

Do parasites utilize antihemostasis mechanisms to favor their survival? An interesting example of coevolution between the bird malaria parasite (*Plasmodium gallinaceum* sporozoites) and the mosquito vector (*Anopheles* spp.) exists, which favors parasite transmission. The parasites, present in the mosquito's salivary glands, reduce the production of apyrase. This creates a problem for the infected female in locating a blood vessel; consequently, the female probes more, thus favoring increased parasite transmission to the vertebrate hosts (Ribeiro 1987).

Research into the importance of insect saliva will continue to focus on the importance of allergic responses of humans to bites, the role of salivary secretions on plant gall formation, and the use of salivation by the vector as the mechanism of transmission for various plant and animal parasites and pathogens.

What regulates the production and secretion of saliva in insects? This question provides the opportunity to think about and question the validity of using information that has been shown to explain a specific phenomenon in one species to explain the same phenomenon in another species. Production and secretion of saliva in the Odonata and Orthoptera are regulated by nervous innervation from both the stomatogastric nervous system and the subesophageal ganglion, whereas in the Diptera (e.g., the adult blow fly) there is no innervation of the salivary glands and these glands are controlled by an unidentified neurohormone (believed to be 5-hydroxytryptamine, 5-HT). Salivation has been shown to be controlled by phagostimulation of external chemoreceptors on the mouthparts. This same stimulus probably also activates the salivary pump. House (1980) and House and Ginsborg (1985) review the literature on the salivary glands of insects and invertebrates. Miles (1972) reviews the literature on hemipteran saliva.

Microbiota and Digestion

Under normal circumstances, all insects possess intestinal symbionts that may or may not contribute to their host's well-being. Among the intestinal microflora and microfauna are found bacteria, protozoa, and fungi of various species. In some cases, as will be discussed later, these microbes may provide substances of nutritive value (e.g., vitamins or nitrogen) for their hosts. In other cases these symbionts may synthesize an enzyme that enables the insect to digest substances it would otherwise be unable to digest. Leafhoppers (order Homoptera, family Cicadellidae) harbor yeasts capable of digesting starch and sucrose. Bacteria that ferment cellulose are probably always present in the alimentary tracts of wood-feeding insects. In some insects, such as the lamellicorn beetles (stag, bess, and scarab beetles), a specialized region of the gut houses these cellulose-fermenting bacteria. In the lamellicorn beetles a portion of the hindgut is enlarged, forming a "fermentation chamber." Wood-feeding termites similarly possess an enlarged sac in the hindgut, but in this case the sac harbors mainly cellulase-secreting flagellate protozoans required for the survival of the termites on a cellulose diet. Certain microbes may aid significantly in the processing of the food; for example, in the larvae of the blow fly, *Lucilia* sp., microbes produce an alkaline state that aids in the liquefaction of ingested animal tissues. Fungi may aid certain wood-feeding insects in the breakdown of cellulose. The review by Campbell (1989) discusses the role of microbial symbiotes in herbivorous insects while Cruden and Markovetz (1987) discuss microbes in the cockroach gut.

Insect Nutrition

Given a source of chemical energy and the basic raw materials, insects, like other heterotrophic animals, are able to synthesize many, if not most, of the more complex molecules necessary for the maintenance of life. However, certain molecules they cannot synthesize, but which are necessary in one way or another for survival and reproduction, also are part of the nutritional requirements of insects. Not only are specific molecules required, but the quantities of each are of great importance. Too little of one may result in the impairment of some vital function of the insect. Too much of a given molecule in the diet results in excretion of the amount in excess, and, because excretion is an energy-utilizing process, this is metabolically inefficient. High-energy-yielding molecules such as carbohydrates and lipids are needed in comparatively large amounts (i.e., grams per kilogram of body weight per day). In adult insects, amino acids, purines, and certain lipids are required in somewhat smaller, but still fairly large, amounts (milligrams per kilogram of body weight per day). Immature insects probably require larger amounts of amino acids, because they are more actively synthesizing structural protein. Other molecules, in particular vitamins and minerals, are required in comparatively minute amounts (micrograms per kilogram of body weight per day).

Dadd (1973, 1985) and House (1974a) provide reviews of the literature on insect nutrition.

Amino Acids

Amino acids are the building blocks of protein. Different insects have different requirements, depending upon which amino acids they are capable of synthesizing. Those which they cannot synthesize fall into the required category. Required amino acids may be present in the food material in a free state or more commonly in the form of protein. The qualitative and quantitative amino acid composition of a given protein determines its nutritive value.

Carbohydrates

Carbohydrates are not considered to be essential nutritive substances for most insects, but they are probably the most common source of chemical energy utilized by insects. However, many insects (e.g., many Lepidoptera) do, in fact, need them if growth and development are to occur normally.

Lipids

Lipids or fats, like carbohydrates, are good sources of chemical energy and are also important in the formation of membranes and synthesis of hormones. Most insects can probably synthesize from carbohydrate and protein most of the necessary fatty acids that make up the larger lipid molecules. However, some insect species do require certain fatty acids and other lipids in their diets. For example, certain Lepidoptera require linoleic acid for normal larval development. All insects apparently require a dietary sterol such as cholesterol, phytosterols, or ergosterol for growth and development.

Vitamins

Vitamins include a diverse group of compounds that are required in very small amounts for the normal functioning of any animal. They are not used for energy, and they do not form part of the structural framework of the insect tissues. Vitamin A is required for the normal functioning of the compound eye of the mosquito *Aedes aegypti* (Brammer and White 1969). Vitamin A is a fat-soluble vitamin. However, vitamins required by insects are principally of the water-soluble type (e.g., B complex vitamins and ascorbic acid). These compounds are coenzymes, working in conjunction with enzymes in specific metabolic reactions.

Minerals

Various minerals, like vitamins, are required by insects, in very small amounts, for normal growth and development. Minerals required by insects include potassium, phosphorus, magnesium, sodium, calcium, manganese, copper, and zinc. Some insects (e.g., the aquatic larvae of mosquitoes, which possess very thin-walled anal papillae) are able to absorb mineral ions from the water through the cuticle of these structures.

Purines and Pyrimidines

The nucleic acids, DNA and RNA, are the molecules that carry and mediate the genetic code. Although insects are able to synthesize nucleic acids, dietary nucleic acids have been shown to have an influence on growth. For example, RNA exerts a positive effect on the growth of certain fly larvae.

Water

All insects require water, whether it be from food, by drinking, from absorption through the cuticle, or from a by-product of metabolism. Insects vary greatly with respect to amounts of water needed. Some, like the mealworm, *Tenebrio molitor,* can survive and reproduce on essentially dry food. Others, for example, honey bees and muscid flies, require large amounts of water for survival. The excrement of the mealworm is hard and dry, with almost all the water having been reclaimed by the insect, while the excrement of bees and muscid flies contains large amounts of water.

Nutritional Ecology

The study of nutritional ecology brings together the studies of physiology, behavior, ecology, and evolution, and it does this within the context of nutrition (i.e., food consumption, utilization, and allocation) (Slansky and Rodriguez 1987). It is the selective pressures on how an insect gets its food, how it uses it, and how it allocates it that has led to the evolution of different life-styles, morphologies, and behaviors that permit the insect to deal most effectively with its varied food sources or lack of them. For example, the caterpillar, *Nemoria arizonania,* feeds on several oak species in Arizona. There are two broods of caterpillars each year, and the two broods differ significantly in morphology and color. The spring brood feeds on the oak catkins, and its body shape and color develop so that the larvae mimic the catkins. The summer brood, however, feeds on the leaves and develops into twig mimics. This morphologically adaptive character is apparently related to the high tannin content in the leaves (Greene 1989). When caterpillars are fed a catkin diet (which is normally low in tannins) containing an artificially elevated level of tannins, the caterpillars developed into the twig mimics. Presumably, this adaptive strategy evolved against bird predation. To further explore this exciting area, interested readers are referred to Slansky and Rodriguez (1987).

Insect Nutrition and Mass Rearing Programs

With increasing public and political pressure, plus the actions taken by the Environmental Protection Agency (EPA) and the Food and Drug Administration (FDA), more and more chemically based pesticides are being removed from use. The trend is to rely more and more on control strategies that are not based solely on chemical control. Thus as more control programs have become interested in mass rearing both pest insects for sterile male release (see chapter 15) and beneficials (i.e., parasites and predators) for biological control, the research area of artificial diets has taken on a new importance. Suitable and cost-effective diets, based on nutritional studies, must be developed so that mass reared and released insects are as effective and competitive as their natural counterparts (Rodriguez 1972, Reinecke 1985, Thompson 1986, Singh and Moore 1985).

Microbiota and Nutrition

Experiments involving the treatment of eggs to destroy microbes (for example, those in insect's food) have shown that some insects require the presence of certain of these microbes for normal growth and development and, in some instances, for survival. In many cases necessary microbes are passed from generation to generation. In some insects in which this occurs, the microbes are housed in specialized cells, *mycetocytes,* and the tissues composed of these cells, *mycetomes,* are associated with the gut, fat body, or, appropriately, the gonads. The latter would ensure the infection of any eggs produced, thus furnishing one of the possible means for bridging the gap between generations. Whether "hereditary" or not, microbes are commonly found in the alimentary canals of insects, often in the various diverticula of the midgut. The kinds of microbes reported to be likely contributors to their host's nutrition include yeasts, fungi, bacteria, and protozoans. These microbes probably benefit the insect in various ways. Suggestions include the possible fixation of atmospheric nitrogen; synthesis of protein from nitrogenous waste materials; and the provision of vitamins (particularly those of the B group), sterols, and amino acids.

For more information on microbiota and nutrition (and digestion) see Brooks (1963), Buchner (1965), Koch (1967), Cruden and Markovetz (1987), and Campbell (1989).

Controversy and Hope

This text is not intended to deceive the student into thinking that either everything is known about the biology of insects or that there are no controversies. Without controversies and people to accept the challenges these controversies present, good science fails to exist and to advance.

Are specialized cells found in the midgut correctly labeled as endocrine cells? Chapman (1985) stated that their ". . . role has not been proved." However, Brown et al. (1986) and Billingsley (1990) claim they are. Even though the vertebrate and invertebrate antibodies used by Billingsley and Brown et al. to test the midgut cells for neuropeptides reacted positively with the secretory products within the putative endocrine cells, several investigators do not feel that this is enough evidence to claim they are, in fact, endocrine cells. What additional evidence is needed to positively, without any doubt, claim they are endocrine cells? Also, are there problems with using these antibody tests? More research is needed to resolve this controversy.

Hope comes in many forms. By better understanding the structure and function of the midgut of hematophagous insects, we will be in a better position to manipulate infections of potential vectors by parasites and/or pathogens. Such a strategy could have a significant impact on important diseases of both humans and animals. The midgut is the main site where parasites and pathogens enter the insect vector's hemolymph. Any barrier (and some believe the peritrophic membrane is a major barrier) to parasite or pathogen penetration into the hemolymph of the host will prevent disease transmission. Many questions need to be answered, however, before we can hope to manipulate this aspect of the life cycle of disease agents. For example: How do the parasites or pathogens "know" they are in the midgut? By what mechanism(s) do they enter the hemocoel and cross the gut barrier?

The Circulatory System

The circulatory system of insects is different from those of vertebrates and many other invertebrates in that the major portion of the blood or hemolymph is not found within the confines of a closed system of conducting vessels. Instead, insects have an *open circulatory system,* so called because the blood bathes the internal organs directly in the body cavity or *hemocoel.* This "blood cavity" is not a true *coelom;* that is, it is not lined entirely with mesodermal tissue. Although coelomic sacs do occur in the embryo, almost the entire hemocoel is formed from the epineural sinus and lined with ectodermal and endodermal tissue (see chapter 5). The only conducting tube is the *dorsal vessel,* which is a pulsatile structure that generally extends the length of the insect from the posterior part of the abdomen to just beneath the brain in the head. The blood of insects is commonly called *hemolymph,* a term which implies that it carries out the functions of both blood and lymph, which are distinctly different fluids in vertebrates. The epithelia of all organs are separated from the hemocoel by a basement membrane. Thus, hemolymph probably does not bathe individual cells directly, the immediate environment of a cell being determined by the nature of the basement membrane. Insects, unlike many other animals, do not rely on the circulatory system for oxygen transportation. Instead, they have devoted an entire system to oxygen and other gas transport (i.e., ventilatory system). The vast majority of insects do not even contain hemoglobin.

Jones (1977) and Miller (1985b) provide useful reviews of the circulatory system.

The Dorsal Vessel and Accessory Pumping Structures

The dorsal vessel (figure 4.6) is the principal organ responsible for blood circulation. It lies along the dorsal midline of the insect body, extending from the posterior region of the abdomen to just behind or beneath the brain in the head. It may be a simple straight tube or have bulbar thickenings along its length or be a complex, branching structure. It is largely composed of circular muscle, but it may also have semicircular, oblique, helical, or longitudinal fibers and may have a connective tissue sheath around the outside. Fine elastic fibers, which arise from the dorsal integument, alimentary canal, somatic muscles, and other structures, serve as suspensors for the dorsal vessel. The dorsal vessel may or may not be extensively tracheated.

It is commonly possible to identify two major divisions of the dorsal vessel (figures 4.6 and 4.7): a posterior *heart* and anterior *aorta.* The heart is usually closed at its posterior end and bears a number of valvular openings, or *ostia,* which allow hemolymph to enter (incurrent ostia) or exit from (excurrent ostia) the heart. Ostia usually occur in laterally oriented pairs but may be dorsally or ventrally located. The heart may be constricted between successive ostia, giving it a chambered appearance, but its lumen is in nearly every instance continuous throughout. On either side of the heart, in a segmental arrangement, are the "wing" or *alary muscles,* so named because when viewed from either a dorsal or ventral aspect, they resemble wings. These fibromuscular structures are attached laterally to the body wall and vary in number from 1 to 13 pairs, depending on the species of insect. Lateral segmental vessels associated with the heart have been identified in several insects, for example, many cockroach species. The heart may extend into the thorax but is generally confined to the abdomen. The aorta extends anteriorly from the heart and opens behind or beneath the brain. It usually lacks any valvular openings and in some insects may be variously thrown into one or two vertical loops or arches, lateral kinks, or coils. Also it may have dilated portions along its length.

In addition to the pumping activities of the heart, various accessory pulsatile structures (figure 4.7) aid the movement of hemolymph. They have been found at the bases of the antennae, at the bases of and within the legs and wings, and within the mesothorax and metathorax.

Sinuses and Diaphragms

The hemocoel, or body cavity, particularly in the abdomen, is usually separated into two and sometimes three cavities, or *sinuses.* This compartmentation is produced by the presence of one or two fibromuscular septa. The one that is generally present is the *dorsal diaphragm,* or *pericardial septum.* It typically consists of two layers, which enclose the alary muscles associated with the heart. It may be imperforate, but it is usually not, containing lateral *fenestrae,* or "windows," through which hemolymph can readily pass. This diaphragm divides the dorsal *pericardial* ("around the heart") *sinus* from the *perivisceral* ("around the gut") *sinus.* In many insects, a second

Figure 4.6 Generalized insect circulatory system. x—x's in top two diagrams indicate cross-sectional plane of bottom diagram.

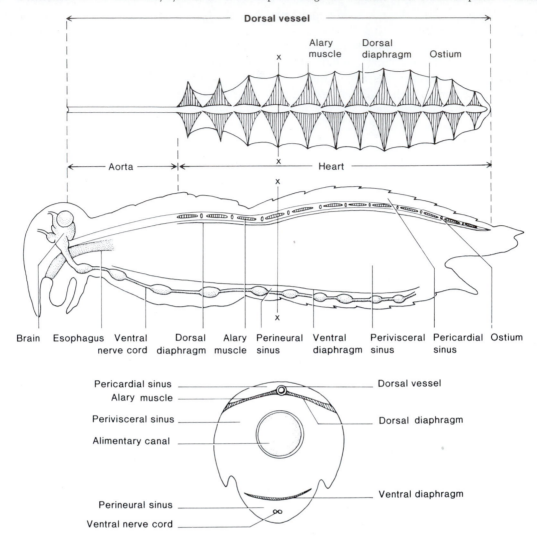

septum, the *ventral diaphragm,* is present. It also contains muscle fibers and is located ventral to the alimentary tract but dorsal to the ventral nerve cord. Like the dorsal septum, it is usually fenestrated around its periphery. The ventral diaphragm separates the perivisceral from the *perineural* ("around the nerve cord") *sinus.* Undulation of both the dorsal and the ventral diaphragm may aid appreciably in the circulation of blood.

Vertically oriented septa, attached dorsally and ventrally, have been identified in a few insects.

Cardiac Regulation

The area of research concerning innervation of the dorsal vessel has produced conflicting results. One must carefully study the reports to make sure that the nerve being investigated actually terminates and synapses in the structure under study (in this case, the dorsal vessel) and that the nerve is indeed a nerve cell and not a neurosecretory cell. In fact, many reports suggesting direct neuronal input to the heart are based solely on morphological evidence at the light microscope level. When carefully examined with the transmission electron microscope, the neuron was a neurosecretory cell, and it terminated in the dorsal diaphragm or alary muscle and not in the dorsal vessel. Quoting from Miller's (1985a) review: "The nervous basis for heartbeat control in insects is thus completely undescribed as yet. Heartbeat 'accelerator' nerves have not been found." The only documented case of cardiac neurons is that of the cockroach, *Periplaneta americana* (Miller 1985a).

Convincing evidence, both ultrastructurally and using various bioassays, exists for neurohormonal control of heartbeat. Both a cardioaccelerator neuropeptide and proctolin (the first insect neuropeptide to be structurally characterized) are myotropins that act on the heart (see table 3.2; Holman et al. 1990, Cook and Holman 1985). Do we have any behavioral or physiological evidence that supports the existence of cardioaccelerators?

Figure 4.7 Patterns of circulation. (*a*) Insect with a fully developed circulatory system. (*b*) Pattern of circulation in a house fly wing.
(*b*) *redrawn from West, 1951.*

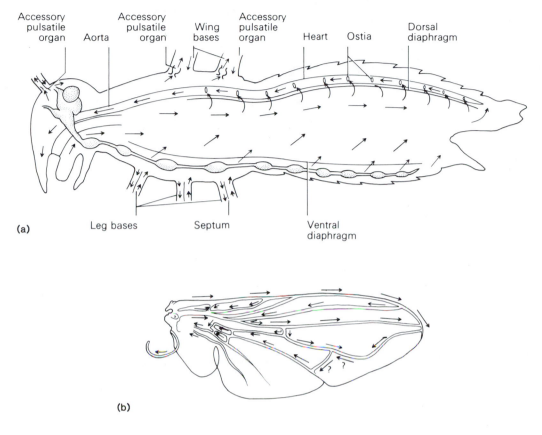

Few events in the life of a butterfly or of a moth are as important as escaping from the pupal cuticle and/or the surrounding cocoon. The precise regulation of this event by the eclosion hormone influences not only behavior (chapter 8) but activates various physiological events that assure escape from the insects' coverings and assure wing unfolding. Both events (expansion and wing unfolding) involve increase in hemolymph pressure and increase in hemolymph flow to the hemocoelic space of the wings. The importance of specific muscle sets for escaping from the pupal case and cocoon and the impact of eclosion hormone on muscle degeneration was discussed in chapter 3. These muscles, along with taking air into the digestive tract, and/or ventilatory system, all contribute to an increase in hemolymph pressure. At the same time cardiac acceleration would be important, to force hemolymph into the hemocoel of the expanding wings, as well as that of the whole insect body. As will be discussed later, the drastic changes that take place during an insect's life (e.g., increase in size at molting) are partly accommodated by a buffering system of hemolymph pressure changes versus internal air space changes that ultimately account for maintenance of volume or size.

Wasserthal's research on the nymphalid butterfly (*Caligo brasiliensis*) showed that, during normal periods, the heartbeat alternated between forward peristalsis (anterograde) and backward peristalsis (retrograde); but during adult eclosion the heartbeat rate increased significantly, and its directionality was only forward. This forward movement would certainly facilitate hemolymph movement into the wing hemocoel and aid expansion and wing unfolding. In the next section we will discuss Wasserthal's new idea concerning the possible significance of the reverse heartbeat in butterflies as it relates to air movement within the wings. See Miller (1985a) for a broader coverage of neurohormonal control of eclosion heartbeat.

With respect to function, the discovery of the neurohormone (i.e., eclosion hormone) in Truman's laboratory, which controlled heartbeat at eclosion in the tobacco hornworm, *Manduca sexta,* shows how much influence on total physiology a small molecule can have (Holman et al. 1990). At the time of adult eclosion from the pupal case and cocoon, the eclosion hormone influences many events. One of these is cardioacceleration, and its effect has been shown to last for only the 75 minutes during the eclosion period. This is the first demonstration and characterization of a cardioaccelerator in insects (Miller 1985a).

Reflex cardiac responses to external stimuli are also probably important, for example, the report by Angioy (1988) demonstrated by simultaneously recording from the heart and the tarsal chemosensillum (also, the one electrode provides the stimulus input; see figure 3.11) of the blow fly, *Calliphora vomitoria,* that when either a tarsal or labellar sensillum was stimulated a rapid change in cardiac response ensued. More experiments using this dual recording approach must be conducted, while at the same time, the biological meaning of the change in the cardiac response has to be explained.

Circulation

The general pattern of circulation in insects (figure 4.7) can be described as follows. Blood enters the heart through the ostia. It is then directed anteriorly by a wave of peristaltic contraction that passes along the dorsal vessel in the direction of the head. The direction of propagation of the wave of peristaltic contraction has been observed to reverse on occasion. Blood usually returns to general circulation in the head. With the aid of undulatory movements of the diaphragms, action of the accessory pulsatile structures, and visceral and body movements, blood circulates throughout the general body cavity and appendages. Blood returns to the pericardial sinus through the openings in the dorsal diaphragm and enters the heart. It should be noted that insect hemocytes may circulate or may remain sessile on tissues until needed, and this depends on the insect species. No reports exist that show that hemocytes enter the dorsal vessel.

The alternate phases in the heartbeat cycle, contraction (*systole*) and relaxation (*diastole*) can be measured mechanically at a single point along the heart (figure 4.8). Local decrease in heart volume (movement of the curve upward) results from contraction, and increase in heart volume (movement of the curve downward) results from relaxation. A period of rest or *diastasis* occurs between successive beats. Toward the end of this rest period, a small but definite additional expansion may occur. This is represented by the presystolic notch and may be the result of contraction of the alary muscles or blood being forced from behind into the region of measurement.

Experiments with denervated dorsal vessels or fragments of vessels have demonstrated that rhythmic contraction usually continues despite the absence of exogenous stimuli (i.e., these dorsal vessels are myogenic). Because the waves of contraction are commonly from posterior to anterior and because of older literature, people thought that a *pacemaker* (i.e., a region from which waves of contraction propagate) must exist. Miller (1985a) reviews this literature and remains firm that until proof is provided, the insect heart is myogenic and lacks a pacemaker.

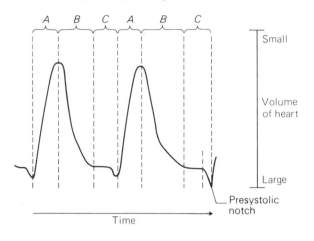

Figure 4.8   Mechanical recording of two heartbeats (cardiac cycles) in an isolated cockroach, *Periplaneta americana*, heart. (*a*) Systolic phase; (*b*) diastolic phase; (*c*) relaxation phase with presystolic notch.
Redrawn with modifications from Yeager, 1938.

The rate and amplitude of the heartbeat are affected by a variety of factors, including intensity of activity; ambient temperature; metabolic rate; developmental stage; and the presence of biologically active chemicals, insecticides, drugs, and so on. The accessory pulsatile structures are independent of the dorsal vessel, but are influenced by the same factors.

Hemolymph pressure is a function of the amount of fluid in the hemocoel and the forces exerted by the body muscles. In hard-bodied insects, average pressure is generally quite low, although by contraction of regions of the body wall, considerable hydraulic pressure can be brought to bear locally. In soft-bodied insects (e.g., caterpillars), tonic contractions of the body musculature acting in conjunction with the hemolymph maintain body shape hydrostatically ("hydrostatic skeleton").

For a review of the electrophysiology of the insect heart see Miller (1974). For information on recording techniques, cardioaccelerators, control of hemolymph pressure, and the diaphragms (topics beyond the scope of this text), see Miller (1985a, 1985b).

General Characteristics of the Hemolymph

Hemolymph is a clear fluid that is usually colorless or, because of certain pigments, slightly green or yellow. Outstanding exceptions are the red hemolymphs of some midge (Diptera, Chironomidae) larvae, certain species of backswimmers (Hemiptera, Notonectidae), and the horse bot fly (Diptera, Gasterophilidae), all of which contain the pigment hemoglobin. The hemolymph makes up approximately 5%–40% of the total body weight of an insect, depending on the species. Blood pH is usually slightly acid (between pH 6 and pH 7) and may vary slightly within a species. In a few insects the pH may be slightly alkaline, pH 7–pH 7.5.

Insect hemolymph is slightly more dense than water. The specific gravity of hemolymph typically lies between 1.015 and 1.060 and is subject to increase during periods of molting (Patton 1963). The total molecular concentration in the hemolymph is fairly high, a fact that accounts for the osmotic pressure being somewhat higher than that of mammalian blood. In the more advanced insect orders, free amino acids, organic acids, and other organic molecules play significant roles as osmolar effectors, whereas inorganic anions and cations are largely responsible for the osmotic pressure of vertebrate blood. As with vertebrates, insect blood can be separated into a fluid portion, or *plasma,* and a cellular portion, the hemocytes. In addition to the two basic fractions, several nonhemocytic elements may be found, including muscle fragments, free fat body cells, oenocytes, free crystals, spermatozoa, various parasitic organisms (e.g., bacteria, protozoans, and nematodes), tumor cells, and so on (Jones 1977).

Chemical Composition of the Hemolymph

The chemical composition of the hemolymph shows considerable variation, both qualitatively and quantitatively, among the various insectan groups and is subject to variation in the same species depending upon the physiological state, age, sex, food, and so on, of the organism. Florkin and Jeuniaux (1974) provide extensive information on this topic. Mullins (1985) presents the concept that the hemolymph is not as static a solution as many may think but that it is a dynamic fluid that changes with diet, environmental changes, and different life stages.

A brief listing of some of the major constituents that have been identified follows.

Water
As with other organisms, water is the major component of the internal body fluid of insects. The typical water content of hemolymph, 84%–92% (Buck 1953), is comparable to that found in human blood plasma. In certain insects, water content may be much lower, perhaps less than 50%.

Inorganic Constituents
Sodium, potassium, calcium, sulfur, magnesium, chloride, phosphorus, and carbonate comprise the major inorganic materials that have been identified in insect hemolymph. Chloride content is comparatively low relative to that in mammals, whereas phosphate, calcium, and magnesium commonly occur in somewhat higher concentration in many insects than in mammals.

In phytophagous insects like Lepidoptera and certain Hymenoptera and Coleoptera, magnesium and/or potassium replace sodium as major cations. This is moderately correlated with diet in that plants contain relatively high concentrations of these elements. Thus zoophagous (carnivorous) insects, which commonly take in large amounts of dietary sodium, do not show high levels of magnesium and potassium. Copper, iron, aluminum, zinc, manganese, and other metallic elements have been found in very small amounts in insect hemolymph.

Nitrogenous Waste Materials
The breakdown of proteins and amino acids produces nitrogenous waste materials. In insects, these wastes are usually in the form of uric acid, which occurs in a very high concentration in the hemolymph. It is produced mainly in the fat body and is usually excreted by the Malpighian tubules. Other nitrogenous waste materials produced by insects include urea, allantoin, allantoic acid, and ammonia, the last being formed mainly in aquatic insects. Cockroaches normally do not excrete urates but store them in special cells termed urocytes or in special male glands termed uricose glands. Both adults can pass these urates onto their progeny, which may use them for a source of nitrogen. More will be said about this topic in the discussion of excretion.

Organic Acids
Succinate, malate, fumarate, citrate, lactate, pyruvate, α-ketoglutarate, and several organic phosphate compounds are the major organic acids found in insect blood. The first seven are formed during the Krebs cycle. It has been suggested that these acids are important in balancing the cations in the blood.

Carbohydrates
This group of organic compounds is extremely important to the insect in the following ways: as the major energy source, as "identification tags" in cellular recognition and in protein translocation, and in cold stress metabolism. The functions of carbohydrates in the life processes of insects is discussed by Chippendale (1978) and their metabolism is discussed in Friedman (1985). A rather surprising discovery was that trehalose, a disaccharide composed of two glucose molecules (thereby containing twice as much energy as a single glucose molecule), is the major blood sugar in most insects. However, trehalose is not the major blood sugar in every insect; glucose, fructose, ribose, and others have also been found. Insects are believed to maintain homeostasis of their blood sugar levels through two neurohormones: the hyperglycemic decapeptide that causes elevation of hemolymph carbohydrates by mobilizing fat body carbohydrate stores (mainly glycogen) and the hypoglycemic neurohormone that lowers hemolymph carbohydrate levels by enhancing carbohydrate uptake by tissues and cells. Only very small amounts of glycogen have been found in the hemolymph. The role of carbohydrates in energy metabolism are discussed in greater detail by Downer (1981) and Friedman (1985). Much of the carbohydrate found in insect blood is in combination with proteins forming

glycoproteins. One important group of hemolymph glycoproteins is the agglutinins (lectins), which are found in some insects' plasma and serve as "scout" molecules that continually survey the hemolymph for "non-self" molecules, dead tissues, and/or foreign organisms (e.g., pathogens and parasites). In order to survive in cold climates, insects have evolved natural antifreeze or cryoprotectants. Glycerol and sorbitol, two polyhydric alcohols, are found in the hemolymph and aid in lowering the freezing point of the hemolymph. When winter is over, these chemicals can be converted back into trehalose. For further reading, consult Heinrich (1981).

Lipids

It is generally known that oil and water do not mix, yet lipids are usually found in insect hemolymph. How are they present, and what are their functions? Most insect lipids are found in combination with proteins, thus their name lipoproteins. This combining with proteins is what keeps lipids in solution. Insects, unlike mammals, which have several types of well-known transport lipoproteins (e.g., low-density and high-density, with both involved in blood cholesterol levels in humans), appear to have a single transport type called lipophorin (Greek *lipos,* "fat" and *phoros,* "bearing"). Not all insect lipoproteins are lipophorins. Lipophorins are "reusable shuttle" molecules that transport molecules from one site to another. These molecules are believed to transport digested fats (i.e., fatty acids) from the gut to various tissues, cholesterol, carotenoids and possibly xenobiotics (any foreign chemicals not normally found within an organism, such as pesticides). In addition, lipophorins may be involved in the insect defense system. Vitellogenin and vitellin are both very-high-density lipoproteins. It is suggested that the vitellogenin, a glycolipoprotein, somehow transports lipids into the eggs for use by the developing embryo. Lipophorins also transport lipids, which are the major fuel source in the locust and Lepidoptera, to the flight muscles. Lipophorins also transport hydrocarbons from their sites of synthesis to the cuticle. Oenocytes are also involved in hydrocarbon biosynthesis and release into the hemolymph, whereas a lipophorin is involved in its transport from the site of synthesis to the cuticle. By determining the cuticular hydrocarbon profile taxonomists have a new tool for use in the identification of species (see chapter 11). More information about this exciting group of chemicals is available in Chino (1985) and Shapiro et al. (1988).

Amino Acids

The presence of free amino acids in the hemolymph in concentrations higher than that reported in any other animal group is one of the outstanding characteristics of most groups of insects (especially Lepidoptera, Hymenoptera, and many Coleoptera). Insects have 100 to 300 times the free amino acids in their blood as compared to humans. These free amino acids may be either dietary or synthesized by the insect. They may represent an excess, derived from the diet, which is stored in the hemolymph prior to excretion; or they may serve as a reservoir of raw materials for the synthesis of protein needed in the construction of new cells during periods of growth and metamorphosis. Tyrosine, for example, plays an important part in the sclerotization of the cuticle. Proline, not lipids or trehalose, serves as the major flight energy source in some insects. Whatever their function(s), amino acids are major osmotic effectors (e.g., in aquatic larvae), replacing in large part inorganic ions. Chen (1985) and Mullins (1985) provide more information on amino acids as metabolic pools or "stores."

Proteins

Proteins in the hemolymph function as storage molecules, enzymes, antibactericidal proteins, carrier or binding proteins, and antifreeze nucleators. A good example of a storage protein is calliphorin, which functions as a storage depot for nutrients during larval development in the fly *Calliphora vicina.* At pupation, however, these stored nutrients are utilized for adult protein synthesis. In addition, larval storage proteins may just be reserves to provide the insect with nutrients when the larva stops feeding (e.g., pupal period) and until the adult begins to feed. Some of the enzymes found in the hemolymph are trehalase, juvenile hormone esterases, and phenoloxidases. The role of both lipoproteins and glycoproteins as carrier or binding proteins has already been discussed. Hemolymph carrier proteins for juvenile hormone and ecdysone have been identified. When insects are injected with either viable nonpathogenic bacteria or sublethal doses of pathogenic bacteria, an acquired humoral "immunity" has been demonstrated. This acquired state has been linked to the synthesis of several hemolymph proteins. One such protein is the enzyme lysozyme, which aids in hydrolyzing bacterial cell walls. In addition to the importance of glycerol and sorbitol as antifreeze agents, several hemolymph proteins have recently been found to serve as antifreeze nucleating agents and are very similar to those found in Arctic fish [see Dunn (1986), Chen (1985), and Levenbook (1985)].

Pigments

Several pigments have been identified in insect blood. Among them are hemoglobin (a conjugated protein), already mentioned; kathemoglobin, derived from the blood meals taken by the blood-sucking bug, *Rhodnius;* carotene and xanthophyll in plant-eating (phytophagous) insects; riboflavin and fluoroscyanine. In insects with thin transparent cuticle, blood pigments may determine body coloration. A review on insect pigments is presented in Kayser (1985).

Gases

Both oxygen and carbon dioxide may occur in the hemolymph, usually in very low concentrations, because the major route of gaseous exchange is via the ventilatory or tracheal system. Exceptions to this are found, of course, in those species that contain the oxygen-carrying pigment hemoglobin in their blood.

In summary, the composition of the insect hemolymph is in some respects similar to that of the internal body fluids in other animals. The role of insect hemolymph proteins as carrier or binding proteins and antibactericidal and antifreeze nucleators needs more research, and many lessons can be learned from research already conducted on vertebrate systems. However, there are many blood characteristics that are rather distinctive in insects. Among the more significant of these are the high aminoacidemia (amino acids in the hemolymph), the functioning of amino acids and other organic acids as important osmotic effectors, the presence of the unique blood sugar trehalose, the fact that oxygen is usually not carried over great distances by the blood, and the relatively high concentration of uric acid in the blood.

Functions of Hemolymph

The hemolymph is a complex mixture of a variety of materials and different cell types. This complexity is reflected in the many diverse functions of the components of the hemolymph. A brief description of some of the major functions of hemolymph follows. See Arnold (1974), Crossley (1975), Jones (1977), and Mullins (1985) for further information.

Lubricant

Hemolymph serves as a lubricant, allowing easy movement of the internal structures relative to one another.

Hydraulic Medium

Like the hydraulic fluid in an automobile braking system, insect blood is incompressible. Thus, forces that tend to reduce the blood volume in one portion of the insect's body (e.g., compression of the abdomen) are transferred hydrostatically ("hemostatically") via the hemolymph to other parts of the body and may effect a change there. A good example of this is found among the Diptera (suborder Cyclorrhapha). These flies, when ready to emerge, literally pop the preformed lid from the end of the puparium (i.e., the hardened case in which they pass the pupal stage). This is accomplished by means of the hydrostatic extrusion of the bladderlike *ptilinum* from the anterior portion of the head. In addition, the osmeterium of the swallowtail butterfly larvae is extruded by using hydrostatic pressure. Other eversible glands, such as pheromone glands of adult butterflies or certain fruit flies and hair-pencils of some male butterflies, may rely on hemostatic pressure alone or in combination with special muscles for eversion, whereas muscular action alone appears to be the mechanism for retraction.

As mentioned in chapter 2, hydrostatic pressure may oppose muscular contraction in the extension of an appendage. In addition, ecdysis involves hydrostatic pressure, and newly emerged adult insects expand the wings hydrostatically.

Transport and Storage

Like the body fluids of other animals, insect hemolymph transports various substances from one tissue or organ to another. Substances transported include nutrients (amino acids, sugars, fats, etc.) absorbed through the gut wall or released from cells that store such materials; metabolic wastes (nitrogenous materials) and foreign materials to be excreted or taken in and stored by certain cells; possibly hormones; and oxygen and carbon dioxide (usually over very short distances). Any of these may be carried in the plasma and, in some instances, may be carried by certain hemocytes. *Pinocytosis,* the intake of fluids or tiny particles (e.g., colloids) by minute infoldings of the cell wall, has been observed in insect hemocytes (Jones 1977).

In addition to transport, the hemolymph serves as an important storage pool for the raw materials used in the construction of new cells.

Heat Transfer

In the sphinx moth, *Manduca sexta,* and probably most insects, the hemolymph is involved in the transfer of heat from one body region to another (see chapter 8). See Heinrich (1981) for more on this topic.

Protection

Insects possess diverse physiological defensive mechanisms against foreign invaders or against physical damage to the cuticle. These are summarized in table 4.1. Insects do not have the classical antigen-antibody system of vertebrates. The term insect immune system is used here to refer to those mechanisms that permit the insect to resist infection by foreign organisms. The humeral (hemolymph-borne) immune factors found in the hemolymph consist of two basic types: those in which the factors do not require *de novo* synthesis and those that are inducible and require *de novo* synthesis of RNA and proteins. Inducible factors in the hemolymph include antibacterial proteins (e.g., cecropin) and lysozymes, whereas the noninducible (also referred to as constitutive) factors are the lectins (hemagglutinins) and the phenyloxidases. The phenyloxidase system, once activated, produces a cascade of chemical events that follows various biochemical pathways. One pathway leads to melanin (a dark brown pigment), which often surrounds invasive foreign organisms. This particular melanin response is noncellular and unlike the one produced by blood

Table 4.1 Physiological Insect Defense Mechanisms

External
  Deterrents and antimicrobial agents
  Cuticular barrier
Internal-external
  Reflex bleeding
  Cuticular encystment
Internal
  Coagulation
  Humoral
    Inducible factors
      Antibacterial proteins
      Lysozymes
    Constitutive factors
      Lectins (agglutinins)
      Phenyloxidases
  Cellular
    Phagocytosis
    Nodule formation
    Encapsulation

Taken from Vinson (1990).

cell involvement, yet somehow it contributes to the death of the invader. Several reviews on the humeral defense system are available (Götz and Boman 1985, Brehélin 1986, Dunn 1986, Gupta 1986, Vinson 1990). Boman and Hultmark (1987) provide an excellent review on antibacterial molecules.

General Characteristics of the Hemocytes

All insects and all life stages of an insect contain hemocytes which are of mesodermal origin, do not contain hemoglobin, and do not appear to enter the dorsal blood vessel or heart. Compared to vertebrate hematology, what we know about insect hematology is limited. This has resulted from the inability of researchers to develop isolated cultures of insect hemocytes plus years of controversy as to nomenclature. This last point has often hampered the development of many areas of science and, in fact, has even promoted erroneous conclusions, polarized camps, and fostered bitter feelings. In addition, certain areas of science have failed to progress or have become stagnant only because of the lack of the appropriate technology to answer specific questions, which at the time cannot be answered. Advancements in the areas of insect immunity, and hematology have remained slow until the recent wave of new technologies in biochemistry, immunology, and molecular biology. With these technologies, strides are being made to understand how the hemocytes recognize self from damaged self/nonself. This area of insect physiology is one of the most exciting and challenging frontiers in insect biology. Answers pro-

vided about the ways the immune system functions will aid biological control workers in preventing beneficial symbiotic organisms from being destroyed by the host's immune system while at the same time knowledge about how the insect immune system functions will give medical entomologists a tool to genetically manipulate field populations of arthropod vectors so that the host vector's immune response will destroy invading disease-producing organisms of human beings and their domestic and wild animals. Lackie (1988) provides a well-written and concise overview of where insect immunity research is headed and of some of the problems that it faces and that it still needs to address.

Origin of Hemocytes

Embryologically, all the cells in the circulatory system are of mesodermal origin. Several insects have *hemocytopoietic organs* (blood-cell-producing organs), similar to the vertebrate hemocytopoietic tissues, in which hemocytes multiply and/or differentiate. Many still consider the prohemocytes to be the basic stem cells from which others are derived (Jones 1964, Arnold 1974). The exopterygota appear to possess hemocytopoietic organs throughout their lives, whereas these organs or glands are absent from adult endopterygota (Gupta 1979a, Lackie 1988). The ability of circulating hemocytes to respond by increasing cell numbers during an "attack" by an invading organism or depletion of blood cells by wounding is still not clear. It is possible, especially in the case of adult endopterygotes, that the major supply of blood cells arises from circulating hemocytes. More information on the importance of hemocytic cell division as a contributing factor in replenishing or increasing cell numbers is drastically needed. Factors known to affect the absolute number of hemocytes in insects are developmental status, metamorphosis, stress (starvation or overcrowding), wounding, and infection by foreign organisms.

Number of Hemocytes

Counts of the total number of hemocytes per unit volume of insect hemolymph have proved to be quite variable, depending upon species, developmental stage, various physiological states, the technique applied to obtain the count, and so on.

The number of hemocytes tends to increase during the larval instars, decline in the pupal stage, and increase initially and then decline in the adult stage. Jones (1977) provides a table listing hemocyte counts in more than 100 species of insects. Among the species listed, the number of cells per microliter ranges from 10 to 167,000 and averages about 20,000.

There is evidence that hormones may be involved in the regulation of the amount of hemolymph and also in cyclic changes in the number of circulating cells.

Functions of Hemocytes

A number of morphologically distinct cells have been identified in the hemolymph of insects. Some of these hemocytes are found in all groups of insects; others are less common, and some are quite rare, found only in a few or even a single species. Much controversy has centered on hemocyte nomenclature (Gupta 1979, 1986) and exactly how many different types exist. In this case, there is more interest in the functions of the hemocytes, and no effort will be made to classify individual types. Vinson (1990) recognized the following (the first three are found in all insects): prohemocytes, plasmatocytes, granulocytes (= coagulocytes and phagocytes), oenocytoids, spherule cells, and thrombocytoids.

Hemocytes, either acting alone or in conjunction with the hemolymph, have been found to provide protection against invading parasites, inanimate particles, pathogens, and cell fragments in the following ways:

Hemostasis, Coagulation, and Plasma Precipitation

Hemocytes are involved with inhibiting hemolymph loss at a wound site by a mechanical plugging action and/or promotion of coagulation or plasma precipitation. Mechanical plugging may be a nonspecific function of hemocytes that are normally circulating and that settle out at a wound site. On the other hand, coagulation and plasma precipitation are usually functions of particular hemocytes, the granulocytes (also called coagulocytes), and have been observed ejecting material that induces the rapid formation of a fine granular precipitate or of veil-like or threadlike networks that enmesh the cells around them.

Wound Healing

Hemocytes of various types (e.g., plasmatocytes and possibly spherule cells) tend to accumulate at sites of injury where they may be phagocytic, promote coagulation by granulocytes, and form protective sheets (e.g., plasmatocytes), all of which aid healing. Plasmatocytes have been observed undergoing mitotic division in association with a wound.

Detoxification

Some hemocytes are capable of rendering toxic metabolites and certain insecticidal materials nontoxic.

Phagocytosis

Certain hemocytes, in particular plasmatocytes and granulocytes, actively ingest foreign particles of various sorts, bacteria, and cellular debris. Whether the humeral factors, in conjunction with hemocytes, are involved in the destruction of invading protozoa, viruses, bacteria, and so on, remains unclear. Phagocytic hemocytes are also important during periods of molting and metamorphosis, when many tissues are in a state of disintegration (histolysis) and fragments of cells freely invade the hemolymph.

Nodule Formation

This response by the hemocytes is usually elicited by factors (e.g., bacteria, protozoans) that are not attacked by the humeral response and appears to act as a "clearance" mechanism to remove materials from the hemolymph. Once removed, the nodules usually become attached to various tissues or organs. Nodules are aggregates of entrapped material, whether abiotic or biotic, that are usually contained within a coagulum, which results from degranulation of granulocytes. This material is then surrounded by plasmatocytes that become flattened. Nodules may or may not be melanized. Small numbers of bacteria that are injected can be destroyed by phagocytosis; however, higher numbers are usually "cleared" from the hemolymph by nodule formation.

Encapsulation

When foreign organisms are too large for either phagocytosis or nodule formation, this immune response of insects is usually successful in destroying the intruder. Encapsulation mainly involves hemocytes, which flatten out and surround or encapsulate the object in several layers of hemocytes (figure 4.9). Melanization may or may not accompany successful encapsulation. Because the topic of encapsulation is too large in scope to be treated here, only a few key issues will be focused on. One of the first steps in this hemocytic response is recognition of damaged-self or nonself. The precise mechanisms involved in recognizing damaged-self (i.e., wounding) have not been worked out. Regardless, a wound healing response occurs and is usually a combined effect of coagulation and hemocytic involvement in the repair of the wound. Several investigators have isolated a wound or injury factor that may provide the recognition signal or cue to hemocytes to initiate repair of the wound.

What does the insect consider self? Current research suggests that anything completely covered by a basement membrane is self. The cuticle of insects is produced by the epidermal cells, which are separated from the hemolymph by a basement membrane. Except for hemocytes, a basement membrane of greater thickness than that under the epidermal cells also surrounds every tissue or organ. Thus, self to the insect's immune system appears to be the basement membrane. This was demonstrated using a melanotic "tumor" forming, temperature sensitive *Drosophila* mutant. The melanotic "tumors" develop when the larva are reared at 26° C, but not at 18° C. The term tumor here is not correct since it does not represent a proliferation of tissue but is an encapsulation of its own fat body. It turns out that this temperature sensitive mutant fails to produce a normal basement

Figure 4.9 Hemocytic encapsulation, within the hemolymph of a *Drosophila* larva, around the egg of a parasitoid wasp *Leptopilina heterotoma*. Notice the large flattened hemocytes, which are called lamellocytes but considered as plasmatocytes by some.

Photograph courtesy of A. J. Nappi.

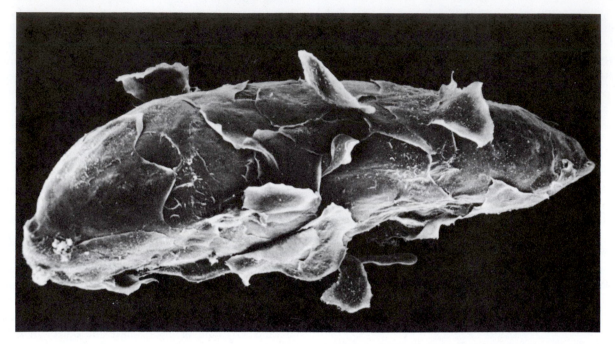

membrane around the fat body. Thus, lack of a competent basement membrane surrounding something is viewed as nonself by the hemocytes, and encapsulation takes place. Wound healing may also result from the immune system recognizing the break in the basement membrane. Apparently, the hemocytes monitor the internal hemocoel for surface "cues" that are not self. For more information on the immune response of *Drosophila*, plus its genetic aspects, consult Rizki and Rizki (1984) and Nappi and Carton (1986), respectively.

This is an opportune time to mention the importance of selecting the right model system for researching a specific problem. If one wishes to study the genetic aspects of physiological, biochemical, or behavioral mechanisms, few model organisms are better than *Drosophila melanogaster*. It is unfortunate that we do not have good genetic models for major pests. We need to develop other insect genetic models since *Drosophila's* size imposes certain restrictions on some research.

We know very little about the structure, chemical composition, and surface properties of insect basement membrane. It presumably is similar to vertebrates, which contains a nonfibrillar collagen to which proteoglycans and glycoproteins are attached. The carbohydrate moieties associated with these molecules which impart specific properties to the basement membranes are not well known for insects. See Lackie (1988) for a review of the importance of insect basement membrane in the recog-

Table 4.2   **Steps Involved in Encapsulation**

1. Foreign body (FB) randomly contacted by a granulocyte (Gr)
2. Gr recognizes the foreign body
3. Gr degranulates and material sticks to foreign body, which is followed by additional Gr attaching to foreign body
4. Gr lyses releases a hemocytic recognition factor (HRF)
5. Recognition factor attracts and recruits plasmatocytes
6. Plasmatocytes attach to FB and HRF complex
7. Plasmatocytes flatten and spread over the FB surface (figure 4.9)
8. Plasmatocytes increase the numbers of layers around the FB until it is no longer recognized as foreign

nition process. Whether it is the physicochemical properties (i.e., electrical charge or hydrophobicity), the composition and concentration of surface carbohydrates or a combination of these that makes this membrane appear as self to the insect immune factors must be elucidated. Table 4.2 provides a theoretical overview of the steps involved in the encapsulation process.

For additional readings, see Ratcliffe and Rowley (1981), Götz and Boman (1985), Boman and Hultmark (1987), Brehelin (1986), Gupta (1986), Dunn (1986), and Lackie (1986, 1988).

Formation of Other Tissues

To date, no concrete evidence exists that hemocytes contribute to the formation of other tissues. However, evidence does exist that hemocytes may be involved in the production of the basement membrane underlying the epidermal cells and possibly the sheath material surrounding locust muscles (Lackie 1988). Dean et al. (1985) argue that they find it difficult to imagine that hemocytes contribute to the basement membrane surrounding fat body. In this case the fat body cells possibly produce the basement membrane. Involvement in basement membrane formation comes from experiments using a DNA probe that is specific for *Drosophila* basement membrane collagen. Using this probe, researchers showed that it hybridizes with mRNA in hemocytes of the hemopoietic glands (referred to as lymph glands in dipterous larva).

Other Tissues Associated with the Circulatory System

Nephrocytes

These are stationary cells that occur either singly or in groups and are usually suspended in the hemolymph by strands of connective tissue. Their main function is to help maintain hemolymph homeostasis by taking in, via endocytosis, nonparticulate colloids and releasing other molecules into the hemolymph by exocytosis (figure 4.10). Nephrocytes share this clearance function of the hemolymph with some types of hemocytes, fat body, and other excretory organs such as Malpighian tubules. Their name is derived from the Greek roots (i.e., *nephros,* "kidney" and *kytos,* "container") and describes their function as being kidneylike. In fact, these cells (based on their electronegative colloids) are considered to be analogous to the vertebrate reticuloendothelial system. Notice in figure 4.10*a* that the first encounter hemolymph-borne molecules have is with the basement membrane. The significance of the basement membrane has not been fully realized in insect biology; however, as already discussed, it appears to be extremely important in recognition of self by the immune system.

The basement membrane covering the nephrocytes (and all other cells and tissues except the blood cells) serves as a major permeability barrier, which possesses both physical and electrical charge selectivity, thus only permitting the entrance of molecules of a specific size and charge. The Singer and Nicholson fluid-mosaic model of plasma membrane structure is now generally accepted and provides the basis for our understanding how cells recognize specific molecules and also how this membrane maintains its characteristic of selective permeability. For the nephrocyte, identifying which molecules to take up selectively becomes a problem. Biologists have developed an array of techniques that can be used to identify the carbohydrate moieties (i.e., the molecules that provide for recognition) protruding above the basement membrane by using lectins to locate specific sugars, by using antibodies to recognize specific molecules, and, to provide some idea of surface charge, by using ionized macromolecules such as ferritin, which is a metalloprotein. Some of these probes or tracers, such as colloidal gold and ferritin (figure 4.10*b*), can be used either *in vitro* or *in vivo* and are directly visible in the electron microscope, whereas others, such as lectins and antibodies need to be conjugated with various ultrastructural markers (i.e., colloidal gold or peroxidase). Figure 4.10*b* is an electron micrograph showing the binding of the cationized ferritin probe to anionic sites on the basement membrane covering the follicle cells of a fly. After the basement membrane, the next level of selective uptake by the nephrocytes is accomplished by receptor-mediated endocytosis. This process will be discussed in greater detail in the next chapter on reproduction. In addition, nephrocytes conduct protein synthesis and release these products into the hemolymph via exocytosis. *Pericardial cells* are nephrocytes that are located around the heart of certain insects (figure 4.11*a*). Nephrocytes are not hemocytes and are easily distinguished from them because nephrocytes are surrounded by a basement membrane. Another difference is that nephrocytes are pinocytotic, taking in only nonparticulate colloidal materials via endocytosis, whereas certain hemocytes take in particulate material by phagocytosis. Finally, nephrocytes contribute to the synthesis of hemolymph solutes such as lysozymes and possibly hemolymph pigments. Crossley (1985) provides an excellent treatment of these cells.

Fat Body

It has been said that no other metazoan cell type can compare with the insect fat body cells when it comes to the diverse functional roles they are called upon to perform. The fat body (figure 4.11*b*) is a loose meshwork of lobes composed of individual cells (*adipocytes*, sometimes, trophocytes) and is invested in connective tissue strands. Fat body is variously distributed in the insect hemocoel and its distribution depends on the species and stage of development. Usually there are two major types of fat body that differ not only in location but may also differ in function. The perivisceral fat body surrounds the

Figure 4.10 (*a*) The transport routes and selective mechanisms for entrance or exit from the cell. Notice the small molecules capable of passing the first layer of selectivity (the pores in the basement membrane), the negatively charged molecules that are attracted to the anionic sites of the nephrocytes, and the larger molecules that are unable to pass through the basement membrane. The next level of selectivity is achieved by receptor-mediated endocytosis. (*b*) and (*c*) Electron micrographs of follicle cells covering the oocyte of a fly. Notice the binding of cationized ferritin molecules to the basement membrane. Because the ferritin is evenly spaced, it is believed that the anionic sites are also evenly spaced.

(*a*) *Taken and adapted slightly from Crossley (1985); (b) and (c) photo courtesy of Giorgi, Chih-Ming Yin, and Stoffolano.*

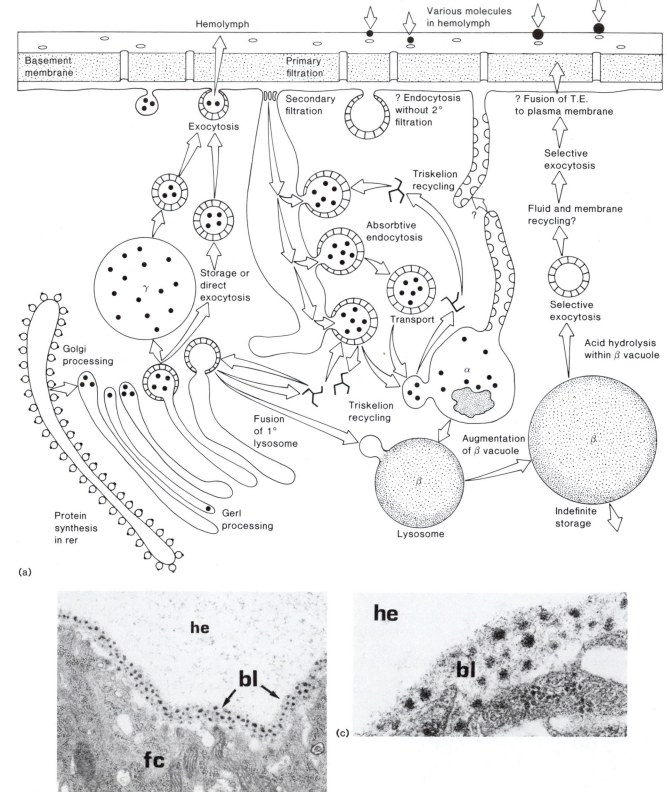

Figure 4.11 Photomicrographs of tissues associated with the circulatory system in adult mosquito, *Aedes triseriatus*. (*a*) Pericardial cell. (*b*) Fat body. (*c*) Oenocytes. 1, Pericardial cell; 2, heart; 3, midgut epithelium; 4, cuticle; 5, muscle; 6, fat body; 7, oenocytes. (Scale lines = 50 μm.)

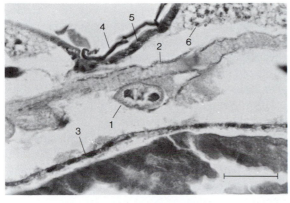

(a)

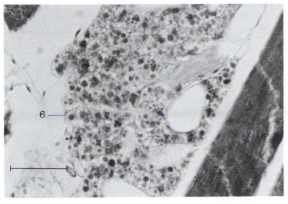

(b)

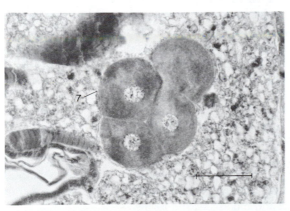

(c)

Table 4.3 Different Types of Molecules Synthesized, Secreted, and Stored by Insect Fat Body Cells

I. Synthesis and secretion.
   A. Larval-specific storage proteins (e.g., calliphorin)
   B. Vitellogenin-female specific hemolymph protein
   C. Lipoproteins and lipophorins
   D. Juvenile hormone carrier proteins
   E. Juvenile hormone esterases
   F. Hemoglobin in *Chironomus* larva—molecule is considered a storage protein
   G. Diapause proteins
II. Mobilization and storage of reserves to be used as precursors for metabolism in other tissues
   A. Lipids
   B. Glycogen
   C. Proteins

digestive tract, whereas the peripheral fat body is located under the cuticle. Both types are tissues whose diffuse structure and suspension in the hemolymph readily facilitates exchange of molecules with both the hemolymph and other tissues. Fat body cells, depending on the species, may be associated with three other cell types: urocytes, mycetocytes (found in Collembola, Thysanura and Dictyoptera), and oenocytes (to be discussed). Fat body is an unfortunate term since these cells synthesize and store proteins and carbohydrates as well as lipids. It is the major metabolic-storage tissue in insects and is the principal site for intermediary metabolism. Its role in synthesis, secretion, and mobilization of reserves for storage vary and the specific molecules involved are listed in table 4.3.

The metabolic pathways taken by the fat body cells and the types of materials stored or secreted are influenced by the following factors: molting, stress, nutrition, state of reproduction, and diapause. The major role of the immature fat body is the synthesis and storage of reserves for growth and molting for the pupal period and sometimes even for reproduction in the adult insect, whereas in the adult the fat body functions mainly in the synthesis of materials used for reproduction and flight. The overall controlling influence of various neuroendocrine secretions, juvenile hormone, and ecdysone on the metabolic state of the fat body cells and its cellular organelles is diverse and depends on the species. In general, juvenile hormone appears to be important in influencing general protein synthesis, whereas ecdysone (i.e., mainly from the ovaries), either alone or in combination with juvenile hormone, influences vitellogenin production. Price (1972) provides a review of protein and nucleic acid metabolism in the fat body, the structure of fat body is covered by Locke (1984) and Dean et al. (1985), and the physiology and biochemistry of fat body are treated by Keeley (1985).

Oenocytes

These are highly specialized cells of ectodermal origin found in nearly all insects except certain Thysanura (figure 4.11c). They may be arranged segmentally or be dispersed randomly. The function of the oenocytes has not been completely elucidated, although there is evidence that they secrete the lipoprotein that forms the cuticulin layer of the epicuticle (Wigglesworth 1976). As previously stated, oenocytes are often in close association with fat body cells. The functional significance of this association is not known. Recently, researchers have demonstrated that *in vitro* culturing of oenocytes has shown that they can produce ecdysteroids. This is an important find, in as much as several workers have shown that when the ovaries are surgically removed, some adult insects can still produce low levels of ecdysteroids. Whether these come from oenocytes must be demonstrated.

The Ventilatory System

The ventilatory system (tracheal system) involved in gaseous exchange with the environment, above all the other insect systems, remains unique in the animal kingdom. Unlike the gaseous exchange systems of many animals, it does not rely on an oxygen transport pigment (e.g., hemoglobin) and does not use the circulatory system as the vehicle for gaseous exchange. Instead, it uses a series of branching tubes called tracheae (figure 4.12). Apparently, the tracheal system evolved independently in the Arachnida, Onychophora, Insecta, and Myriapoda (Boudreaux 1979). Use of the term *ventilation* to denote this gaseous exchange seems preferable to *respiration,* which in the strict sense is an intracellular process involving the oxidative breakdown of carbohydrate to carbon dioxide and water. Miller (1974a) reviews several aspects of the ventilatory system, including ultrastructure and electrophysiological aspects. Ventilation by insect eggs will be discussed in chapter 5. Whitten (1972) discusses the comparative anatomy of the tracheal system. Wigglesworth (1983) reviews the literature on the physiology of tracheoles, and Noirot and Noirot-Timothée (1982) and Mill (1985) treat the structure and physiology of this system.

A tracheal system is present in all insects with the exception of Protura and some Collembola. These small insects live in moist habitats and are able to carry on gaseous exchange directly with the environment via the cuticle.

Structure of the Ventilatory System

Tracheae

In the vast majority of insects, gaseous exchange is accomplished by a system of branching tubules or *tracheae* (figure 4.12). In most terrestrial and many aquatic insects these structures communicate with the outside by means of comparatively small openings, the spiracles. Internally the tracheae divide and subdivide, their diameters becoming successively smaller and smaller as they probe deeply into the tissues of the insect.

Tracheae are of ectodermal origin and are continuous with the integument. In the embryo they originate in each segment independently of one another. However, with the exception of a few apterygote insects, in postembryonic immature and adult stages they do not retain this early developmental arrangement, but become joined by longitudinal trunks so that all parts of the tracheal system are in communication.

From each original developing trachea, branches are given off dorsally to the body wall, heart and aorta, various muscles, and dorsally located tissues; mesially to the alimentary tract and associated structures; and ventrally to the nerve cord, body wall, various muscles, and other tissues. Tracheae are also found inside the veins of winged insects and are involved in gaseous exchange with living cells of the wings. The tracheae are circular or somewhat elliptical in cross section. They are histologically (figure 4.13a,b) similar to the integument, being composed of a layer of epithelial cells that secrete a cuticular layer, the intima, on the luminal side (i.e., the side opposite from the hemocoel). The intima is shed along with the old integument during each molt. Near the spiracles, the cuticle (which is continuous with the epicuticle of the integument) is composed of a cuticulin layer with a chitoproteinous layer beneath and possibly a wax layer on the luminal side. Chitin is lacking in the smaller tracheal branches. The cuticular lining of the tracheae is thrown into a series of folds that usually run spirally around the lumen. These folds are called *taenidia* and lend a measure of support to the tracheae, protecting against collapse with changes in pressure. In the smaller tracheae the folds of the intima may be annular. Although tracheae are resistant to compression in a transverse direction, they can be stretched longitudinally to some extent. These properties are important in insects in which the abdomen becomes greatly distended with food, for example, in several blood-sucking forms.

Tracheoles

The tracheoles are the smallest branches of the tracheal system, ranging in size from 0.2 micrometer to 1.0 micrometer in diameter, and are the sites where gaseous exchange takes place. When examined with the electron microscope, they are seen to have very tiny taenidia.

Figure 4.12 Cross section of insect thorax showing the major tracheal branches (diagrammatic). *Redrawn from Essig, 1942.*

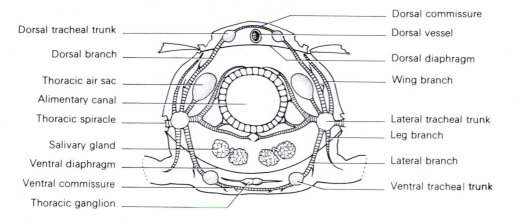

Dorsal tracheal trunk
Dorsal branch
Thoracic air sac
Alimentary canal
Thoracic spiracle
Salivary gland
Ventral diaphragm
Ventral commissure
Thoracic ganglion

Dorsal commissure
Dorsal vessel
Dorsal diaphragm
Wing branch
Lateral tracheal trunk
Leg branch
Lateral branch
Ventral tracheal trunk

Figure 4.13 Tracheal structure. (*a*) Portion of trachea. (*b*) Fine structure of trachea. (*c*) Tracheoles in close contact with muscle. (*d*) Air sacs in the honey bee.
(a) and (d) redrawn from Snodgrass, 1963; (b) redrawn from Miller, 1964; (c) redrawn from Wigglesworth, 1930.

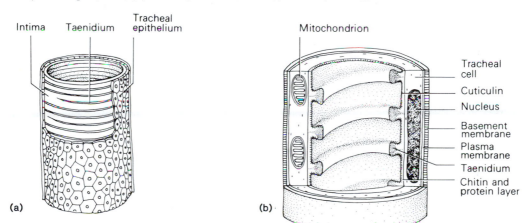

Intima Taenidium Tracheal epithelium

(a)

Mitochondrion

Tracheal cell
Cuticulin
Nucleus
Basement membrane
Plasma membrane
Taenidium
Chitin and protein layer

(b)

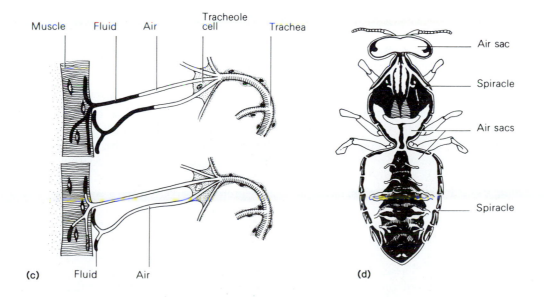

Muscle Fluid Air Tracheole cell Trachea

(c) Fluid Air

Air sac
Spiracle
Air sacs
Spiracle

(d)

Whether the lining of the tracheoles, unlike that of the tracheae, is shed or not at the molt appears to depend on the species and stage of the insect. A trachea typically ends with a tracheal end cell (also termed tracheoblast), which gives rise to several tracheoles that are all a part of this cell (figure 4.13c). Some tracheae merge gradually into a tracheole, and in other cases tracheoles arise directly and abruptly from the side of a trachea. The tracheoles have a greater permeability to gases. Because tracheoles are the major sites for gaseous exchange, ultrastructural studies show them associated with tissues or organs that have a high metabolic rate and high oxygen demand. These tissues or organs include the flight muscles, ovaries, fat body, gut epithelium, Malpighian tubules, and rectal papillae. Most people realize the importance of an oxygenated central nervous system, but few researchers working with insects consider the significance of the oxygen supply to the central nervous system or to dissected organ preparations or tissue cultures. About 6% of the house fly brain is tracheal tissue. Truman (1978) showed that failure to keep the tracheal system, the abdominal ganglion and muscle preparation intact, plus failure to adequately aerate the tracheal branches, resulted in a ventral nerve cord preparation that did not function. More about this preparation will be discussed in chapter 8. In general, tracheoles and tracheae are separated from other tissues by a basement membrane, which is believed to be produced by the tracheal epithelial cells. Tracheoles may be involved with other tissues in a variety of ways. For example, they may intertwine and form a network around certain organs, such as ovaries and testes, or they may merely lie on the surface of some tissues, such as those of the alimentary tract. Tracheoles may also be interstitial—that is, located between cells—and may actually penetrate some muscle cells. These smallest branches of the tracheal system nearly always end blindly at a diameter of about 0.2 micrometer.

Collectively the tracheoles provide a tremendous surface area for gaseous exchange. Consider that there are 1.5 million tracheoles in a fifth instar silkworm larva!

Air Sacs

Air sacs (figure 4.13d) are tracheal dilations of varying size, number, and distribution found mainly in flying insects. They are typically oval in cross section, their cuticular layers lack chitin, and the walls of the cuticular lining lack oriented taenidia. Air sacs are quite distensible and collapsible. They are especially pronounced in those insects that are the most active ventilators. Thus one may observe the alternate distension and collapse as air is taken in and released by the insect.

Several functions of air sacs have been discovered; the major function is probably to increase the volume of the tidal air (the air that is inspired and expired). The presence of large air sacs in the body cavity of an insect appreciably lowers the specific gravity, thereby aiding in flight. Air sacs may also provide room for growth of internal organs; for example, in certain female flies the air sacs provide room for the growth of the ovaries. Other observed functions include aiding in heat conservation in those large insects that must generate high temperatures for flight, assisting in hemolymph circulation, reducing the mechanical damping of flight muscles by the hemolymph, forming the tympanic cavity of the hearing organs of various insects, and forming a flexible cavity behind the sound producing mechanism (i.e., tymbal) of the cicada, which reduces the vibrational damping effect hemolymph would produce.

Spiracles

Spiracles are the external openings of tracheae. They occur in two basic types: simple and atriate. The simple type of spiracle (figure 4.14a) is merely an opening to the tracheal system. The atriate type is formed as a result of the entad migration of the primitive (simple) spiracular opening. Thus, in the fully developed atriate spiracle (figure 4.14b,c), the opening to the trachea, the tracheal orifice, lies at the bottom of a spiracular chamber, or *atrium*. In this type the opening to the external environmental is referred to as the *atrial aperture,* or *orifice.*

In addition to being permeable to oxygen and carbon dioxide, the tracheal system also readily allows the passage of water. This presents a problem for insects, because they would naturally tend to lose water very rapidly through the integument or tracheae by evaporation, a phenomenon called *transpiration*. The development of various types of spiracular closing mechanisms has been one of the evolutionary solutions to this problem. These closing mechanisms are found in most atriate spiracles. Two principal types of closing mechanisms may be found (Snodgrass 1935): the lip type (figure 4.14b), in which folds of the integument form opposing lips that can be pulled together and effect closure of the atrial aperture, and the valvular type (figure 4.14c), which lies at the inner end of the atrium and regulates the size of the tracheal orifice.

Whatever the type of mechanism, closure is ultimately accomplished by contraction of the associated muscles. Atriate spiracles with the second type of closing mechanism may be lined with tiny hairs, which form a *felt chamber*, or may bear other structures, such as *sieve plates,* which are porous covers over the atrial aperture that probably serve to retard water loss and to prevent airborne particles from entering the tracheae. In addition to these closing mechanisms and accessory structures, glandular tissue is commonly associated with spiracles. The secretions of these glands may serve as lubricants for the movable parts of the spiracular closure mechanism, may prevent water from entering the tracheae, and may

Figure 4.14 Types of spiracles. (*a*) Simple, nonatriate type. (*b*) Atriate spiracle with lip closure mechanism. (*c*) Atriate spiracle with filter apparatus and valve closure mechanism.
Redrawn from Snodgrass, 1935.

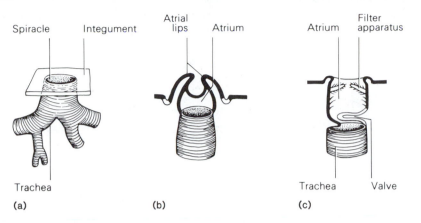

improve the seal of the closure mechanism. In some insects spiracular glands may secrete a repugnant substance. The shape and general design of the caudal spiracles are major key structures for taxonomic purposes for many dipterous larvae. Most adult insects have two pairs of thoracic spiracles; one pair for the mesothoracic segments and one pair for the metathoracic segments. Eight pairs of thoracic spiracles are present on the abdomen.

Types of Ventilatory Systems

Spiracles, tracheae, air sacs, and tracheoles compose the ventilatory systems of most insects. A few—such as most Collembola, many Protura, and certain endoparasitic wasp larvae—lack tracheae, and all gaseous exchange occurs via the integument. In the remainder of insects, tracheae and tracheoles are always present, but air sacs may be absent and spiracles nonfunctional or absent.

The organization of tracheae may be comparatively simple, as in some springtails (Collembola, Sminthuridae) and some Protura, in which tracheae arise from each spiracle, but do not connect to any other tracheae. However, tracheal organization in most insects is more complex (figures 4.12 and 4.15). There is typically a pair of *lateral longitudinal trunks* into which the spiracles open, a similar pair of *dorsal longitudinal trunks,* and often a pair of *ventral longitudinal trunks.* The dorsal, lateral, and ventral trunks are connected by more or less dorsoventrally oriented tracheae, and the longitudinal trunks on either side are connected by *transverse tracheal commissures.*

Although the basic pattern of tracheation is genetically determined, new tracheae and tracheoles can be induced to develop if an insect is reared in an atmosphere with a very low oxygen content. New tracheae and tracheoles do not develop between successive molts, but

changes in the distribution can occur at the time of molting if there is a demand. For example, if the tracheae and tracheoles in one portion of an insect are destroyed, this region will receive tracheation from an adjacent portion in the next developmental stage.

Major tracheal branching patterns are constant within a given species and are often very similar among members of a given family or order. For this reason, tracheal patterns are sometimes of value in assessing phylogenetic relationships (Whitten 1972).

Based on the presence or absence and functional or nonfunctional nature of spiracles, there are principally two types of ventilatory systems, open and closed, with a variety of modifications within each type.

The open ventilatory system (figure 4.15*a,b*) is found among most terrestrial insects and many aquatic forms and is characterized by the presence of one or more pairs of functional spiracles. In the more generalized insects, there are two lateral rows of ten spiracles each; the first two spiracles on each side are located on the mesothorax and metathorax and the following eight on the first eight abdominal segments. This is the *holopneustic* arrangement. During the evolution of the various groups of insects there has been a tendency toward reduction in the number of spiracles (*hemipneustic* arrangements), in some to the point where there is only a single pair; for example, those located at the posterior extremity of the abdomen of mosquito larvae (figure 4.15*b*).

The closed type of ventilatory system, referred to as *apneustic* (figure 4.15*c*), is found in many aquatic and endoparasitic insect larvae. There are no functional spiracles in this type of system; gaseous exchange between the atmosphere and the tracheal system occurs directly through the integument.

Figure 4.15 Representative types of tracheal systems. (*a*) Dorsal and lateral views of the open type, grasshopper. (*b*) Open type with two posterior spiracles, mosquito larva. (*c*) Closed type with no functional spiracles and cuticular ventilation, mayfly nymph (only posterior part of thorax and anterior part of abdomen shown).

(a) redrawn from Essig, 1942; (b) redrawn from Snodgrass, 1935; (c) redrawn from Packard, 1898.

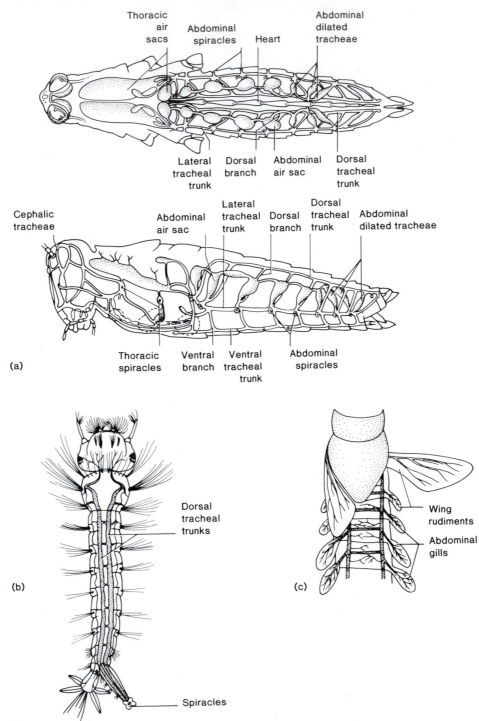

The Ventilatory Process

Passive Ventilation

Calculations based on measurements of tracheae and on the physical properties of oxygen have shown that simple diffusion from the outside of smaller insects and from well-ventilated air sacs in larger insects can supply sufficient oxygen to the body tissues to maintain life. Simple diffusion is a passive form of ventilation, that is, one in which no pumping or other movements aid the passage of gases in the tracheae and tracheoles. Use of the phrase *passive ventilation,* however, does not imply that insects ventilating exclusively in this manner do not control diffusion. Diffusion control is accomplished by the opening and closing of the spiracles. Thus, in addition to their function in the prevention of excessive water loss from the tissues, the spiracles are able to regulate to some degree the entrance and the exit of gases. The spiracles respond to decreased oxygen in the air by remaining open for longer periods of time. Increase in the carbon dioxide content of air produces a similar effect. Spiracular muscles receive nerves from the segmental ganglia of the ventral cord, and there is evidence that opening and closing are under both nervous and hormonal control. Humidity and water balance may also be involved in spiracular control.

Bulk Flow and Active Ventilation

As with hemolymph circulation, general movements of the body and viscera no doubt aid incidentally in the movement of gases (bulk flow) in the tracheal system. In larger insects passive ventilation and general body movements may be inadequate to supply the needs for oxygen when demand is high—for example, during flight. In these insects, air sacs, if present, and larger tracheae are often ventilated by rhythmical pumping movements of the body, *active ventilation.* Movements that are known to cause the inspiration and expiration of gases include peristaltic waves over the abdomen, telescoping or dorsoventral flattening of the abdomen, and, in some, movements in the thorax or even protraction and retraction of the head. The elasticity of the cuticle is also thought to play a part, especially in expiration. In addition, muscular movements other than those already mentioned, such as heartbeat and movements of the gut, may assist in ventilation by pressing against adjacent tracheae. Whatever type they may be, these pumping movements acting via the incompressible hemolymph renew the air in the tracheae and air sacs. Tracheae that are circular in cross section do not respond to compressive forces but, as mentioned, are quite extensible. Both tracheae that are oval in cross section and air sacs are collapsible and hence can serve to increase the volume of tidal air.

In certain insects, ventilatory movements and opening and closing of the spiracular valves are coordinated, sometimes producing a unidirectional flow of gases

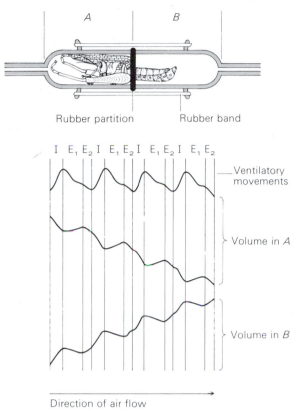

Figure 4.16   Apparatus used to demonstrate directed ventilation in a grasshopper. A rubber partition was used to isolate the thoracic (A) from the abdominal (B) spiracles. I, inspiration, thoracic spiracles open, abdominal spiracles closed; E[1], initial part of expiration, all spiracles closed; E[2], final part of expiration, abdominal spiracles open. *Redrawn from Fraenkel, 1932; apparatus drawn from photograph.*

through the body—for example, into the thoracic spiracles and out the abdominal ones (figure 4.16). This coordination mechanism has been shown to be under neural control. As in simple spiracular opening and closing, oxygen deficiency and carbon dioxide excess can both serve as regulatory stimuli of pumping movements.

As was discussed earlier, cardiac reversal occurs in various lepidopterans, in which the heart normally reverses direction between an anterograde and retrograde peristalsis. Lepidopterans appear to have a valve between the thorax and abdomen and are able to isolate the hemolymph into whichever compartment it is being pumped. Thus, while the hemolymph is in the abdomen, the thoracic hemocoelic pressure is reduced, facilitating ventilation of this area. The reverse is true when the hemolymph is in the thorax. During this period, however, abdominal ventilation continues and actively facilitates gaseous exchange in this area. It would be interesting to know if during flight the heart reverses direction, putting

Figure 4.17 Oxygen uptake and carbon dioxide output in a diapausing cecropia moth, *Hyalophora cecropia*, pupa at 25°C. *Based on Schneiderman and Williams, 1955.*

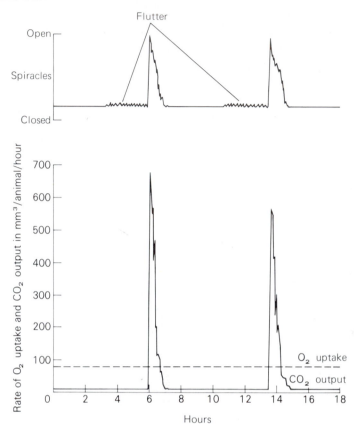

the hemolymph into the abdomen, thus facilitating thoracic gaseous exchange at a time when the flight muscles are most active and have the highest oxygen demand. This idea is partially supported by the new idea of "tidal flow," which proposes an antagonistic movement of air and hemolymph into the wings of a moth (Wasserthal 1982). According to this idea, as hemolymph leaves the hemocoelic spaces of the wings, air enters the wing tracheal system while the reverse is true when the hemocoelic spaces of the wing are filled with hemolymph.

*Passive Suction Ventilation and Elimination
of Carbon Dioxide*

The rate of diffusion of carbon dioxide through air is not too much different from that of oxygen, but in tissues carbon dioxide diffuses about 35 times more rapidly than oxygen. This being the case, carbon dioxide is much more likely to be eliminated from the body through the tracheal linings and integument than is oxygen to be absorbed along the same routes. Thus, although most of the carbon dioxide produced by respiration is eliminated via the tracheae and tracheoles, some of it may escape through the general body surface of soft-bodied insects and the intersegmental membranes of hard-bodied insects.

Many insects have been found to eliminate carbon dioxide through the spiracles in regular "bursts," while oxygen consumption remains constant (figure 4.17). Between these bursts the spiracles remain closed (but apparently not sealed) or "flutter" between slightly open and closed. The spiracles open completely during a carbon dioxide burst. As oxygen is removed from the tracheoles and tracheae by respiration, at least a portion of the carbon dioxide produced presumably goes into solution as bicarbonate in the hemolymph. This would cause the development of negative pressure in the tracheae and tracheoles. As a result, air would be sucked in (bulk flow) through closed (but not sealed) or fluttering spiracles, a process called *passive suction ventilation*. A carbon dioxide burst probably indicates the previous buildup of carbon dioxide (in the hemolymph and tracheae) to a threshold above which complete spiracular opening occurs. Spiracular opening would then allow equilibrium between the atmosphere, the tracheae, and the hemolymph to be reached.

Passive suction ventilation has been observed particularly in insects that are inactive owing to a low ambient temperature or to being in a dormant developmental stage. This phenomenon has also been induced experimentally in insects in which it does not normally occur by lowering the temperature and by injuring certain parts of the brain. The ability to release carbon dioxide periodically in this manner allows an insect to keep its spiracles partially or entirely closed most of the time and hence is thought to be an adaptation that favors the conservation of water by diminishing the rate of transpiration. In addition, the bulk flow of air into the tracheae would tend to retard the outward flow of water vapor.

Passive suction has also been observed in mosquito larvae and pupae and enables these aquatic insects to inspire air rapidly with only brief contact with the surface.

Ventilation in Aquatic Insects

Many insects spend all or part of their lives in an aquatic environment. These insects must either be able to utilize oxygen in solution or have some means of tapping a source of undissolved oxygen whether it be at an air–water interface (at the water's surface or in the form of submerged bubbles) or from aquatic vegetation. A wide variety of structural and physiological adaptations associated with ventilation in an aquatic environment have evolved. Mill (1974) considers aquatic insect respiration in detail.

Aquatic insects with closed ventilatory systems (which are found only in immature forms) depend entirely upon the diffusion of dissolved oxygen through some region of the integument. However, even in many of these insects the tracheal system is involved. These insects obtain oxygen in a variety of ways. Many possess *tracheal gills* (figure 4.18a), which are integumental evaginations covered by a very thin cuticle and are well supplied with tracheae and tracheoles. Such gills may be found anywhere on the body but are commonly found along the abdomen. In some insects they are located at the posterior end of the abdomen (figure 4.18b). Thin areas of the integument may be well supplied with tracheae and tracheoles and function in a similar fashion. Other aquatic insects with closed ventilatory systems possess *spiracular gills,* or *cuticular gills* (figure 4.18c; Hinton 1968). These are filamentous outgrowths, consisting mostly of very thin cuticle (about 1 micrometer thick) that open directly into the tracheae.

Many aquatic insects with closed ventilatory systems lack any specialized gill structures and depend upon diffusion of oxygen across the general body surface, a process called *cutaneous ventilation.* In fact, all forms with or without gills probably depend to a greater or lesser extent on cutaneous diffusion of oxygen.

Most aquatic insects have open tracheal systems and usually obtain oxygen at the water's surface. They are of two general types: those that must surface periodically and depend at least in part on atmospheric oxygen and those that may remain submerged for an indefinite period of time and are somewhat independent of atmospheric oxygen. These two groups are rather indefinite, with intermediates between them. In addition, cutaneous ventilation may under certain circumstances play a role in some of these insects.

Most of the members of the group of aquatic insects that must surface (e.g., most mosquito larvae; figure 4.15b) possess hydrofuge structures. These structures are generally associated with particular spiracles and are highly variable from insect to insect. However, all have essentially the same function, the breaking of the surface film of the water, thereby exposing associated spiracles to the atmosphere. Hydrofuge structures also serve to keep water out of the tracheae when the insect is submerged. Hydrofuge structures are usually made up of hairs and are resistant to "wetting" by water. Thus, when an insect approaches the surface, the cohesive properties of water cause it to be drawn away from the hydrofuge areas.

Many insects with open tracheal systems—aquatic Hemiptera and adult Coleoptera, for example—carry air stores in the form of bubbles or films into which spiracles open. These air stores are often held in place by a pile of erect hydrofuge hairs, but in some insects the body is so shaped that it forms a storage area without the use of hydrofuge hairs. Films or bubbles of air would obviously be a temporary source of oxygen if the insect were forced to remain submerged. How long an insect could survive beneath the surface by utilizing stored oxygen depends on a number of factors, but unless the insect has some way of replenishing the store from oxygen dissolved in the water, its survival time is not likely to be very long. The mechanisms involved in replenishing air stores vary. Water scorpions (Hemiptera, Nepidae), for example, use a caudal siphon, a long hollow tube extending from the rear of the body.

Many aquatic insects that carry stores of air are able to replenish the oxygen without surfacing. This is accomplished by the air store acting as a "physical gill." As the oxygen in reserve is used up, a point is reached where the partial pressure of oxygen is less in the air store than it is in the surrounding water. At this point oxygen diffuses from the water into the air store. Nitrogen in the air store does not readily diffuse into the water and hence tends to keep the air store from collapsing.

Air stores usually make an insect positively buoyant and may play a role in hydrostatic balance. By decreasing specific gravity air stores may also reduce the amount of energy expended in locomotion.

Insects that are able to remain submerged indefinitely usually possess a structure known as a *plastron* or obtain their oxygen from submerged vegetation. A plastron is a very thin layer of gas held firmly in place by tiny hydrofuge hairs or other very fine cuticular networks. The latter are typically associated with spiracular gills. Hair

Figure 4.18 Ventilatory structures in aquatic insects. (*a*) Lateral abdominal tracheal gills in a mayfly nymph. (*b*) Terminal abdominal tracheal gills in a damselfly nymph. (*c*) Cuticular gills on the thorax of a black fly pupa.
(*a–b*) *redrawn from Boreal Labs key card; (c) redrawn from Packard, 1898.*

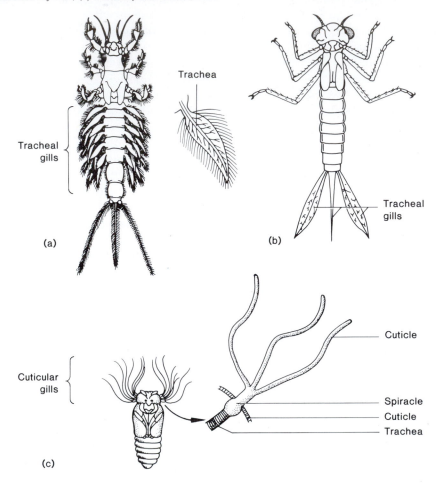

plastrons are found on adults of certain aquatic beetles (e.g., *Elmis*), nymphs and adults of the aquatic bug *Aphelocheirus aestivalis,* and adult females of the wingless lepidopteran *Acentropus niveus.* Plastrons composed of cuticular networks are found in the larval and/or pupal stages of certain beetles and flies. Unlike typical air stores, the gas layer held by a plastron cannot be displaced by water. The spiracles open into the plastron, and it functions in a manner similar to a physical gill except that it does not require repletion by a visit to the surface.

Insects that obtain oxygen from submerged vegetation may do so in a variety of ways. Many are able to capture bubbles on the surface of plants by means of hydrofuge structures. Others penetrate the tissues of submerged plants by biting into them or by inserting a specialized ventilatory structure into the intercellular air spaces. Examples include certain mosquito larvae (figure 4.19), other flies, and some larval beetles.

Aquatic insects may facilitate gaseous exchange by various movements such as rhythmically undulating the tracheal gills and/or undulating the body. Such movements bring oxygen-fresh water to sites of absorption.

Ventilation in Endoparasitic Insects

Endoparasitic insects are those that invade the tissues of their host, as opposed to external parasites. The great majority of the insects in this group are endoparasitic only in the immature stages. The environment of these insects presents problems similar to those of the aquatic environment in that the insects must either obtain oxygen in solution or from the atmosphere or both. In several insects, the tracheal system is nonfunctional and ventilation is cutaneous, gaseous exchange occurring directly between the tissues of the parasite and body fluids of the host. Some endoparasitic insects have tracheal gills similar to those found in insects inhabiting aquatic

Figure 4.19 Ventilatory apparatus adapted for penetration of aquatic plants in the mosquito *Mansonia perturbans*. The anal papillae are involved in osmoregulation.
Redrawn from Matheson, 1944.

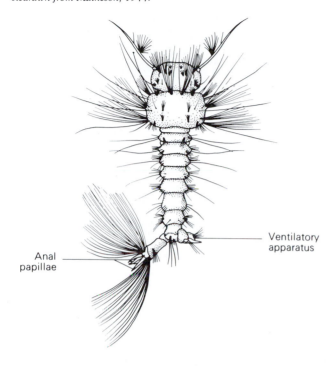

Anal papillae

Ventilatory apparatus

use as an anesthetic, CO_2 is used by many insects, especially hematophagous insects, as a cue for host location. Researchers have used CO_2 (i.e., usually in the form of dry ice) in traps to enhance both attractancy and trap catch. The use of CO_2 is gaining acceptance as a method to kill stored pest insects. Because levels of CO_2 higher than the atmospheric CO_2 concentration cause insect spiracles to open, CO_2 is often added to fumigants to increase their effectiveness. Finally, an interesting interaction has been shown where a virus-induced CO_2 sensitivity results in death when the insect experiences increased levels of CO_2. This phenomenon has only been shown in the Diptera (i.e., mainly *Drosophila*) and its mode of action is not fully understood. Nicolas and Sillans (1989) review the effects of CO_2 on insects.

Central Nervous Control of Ventilation

It is obvious that behavior involving the movement of body parts (e.g., ventilation and flying) must involve the muscular system for the expression of the movement(s) and the nervous system for activation of and coordination over the muscular movement(s). How the different muscles are activated and regulated to produce coordinated movements is an area of active research interest. Of special interest is how rhythmic outputs such as ventilation are generated and controlled. Figure 4.20 is a model showing how various arrhythmic and rhythmic coordinated motor outputs are believed to function. This model will be referred to again in chapter 8. Because ventilation is an event that involves an oscillating system (i.e., one that repeats itself at regular intervals), one wonders where the oscillating control center is located. In insects, as in other invertebrates, many events are controlled by specific ganglia. The regions of the ganglion having this control are termed *preprogrammed motor outputs* or *central motor pattern generators*. Regardless of their name, these regions of the ganglia contain the essential neural network and mechanisms for generating coordinated rhythmic outputs of the motoneurons that stimulate the muscles which ultimately results in expression of the desired behavior. These motor output programs are usually instinctive, stereotypic, and species specific. They consist of a neural network that, when activated, produces the self-contained program. The preprogrammed motor output or pattern generator can have its own built-in oscillator. In insect ventilation control it is believed that each ganglion, which registers some input to the muscles that produce either ventilatory movements or spiracular closure, has its own local control center; but one of these (maybe because it has the lowest threshold) serves as the pacemaker. A pacemaker is a neuron or group of neurons that undergoes intrinsic (usually without external influence from other neurons), rhythmic neuronal firing and

environments. Others depend, at least partly, on atmospheric air, obtaining it either by tubes or other structures that communicate with the tracheal system and that extend out of the host to the atmosphere. As with most aquatic forms, cutaneous ventilation plays a greater or lesser role in most endoparasites.

Ventilation during Molting

At each molt, the cuticle covering the insect, the cuticular lining of the foregut and hindgut, plus the tracheal and sometimes tracheole linings are shed. During this period, when the old tracheal lining is not functional and spiracular muscle control may also be impaired, very little is known about how the insect meets the high energy and oxygen requirements of this process. Mill (1985) presents some information but more research needs to be directed on this important process.

Carbon Dioxide and Insects

Carbon dioxide is the major anesthetic used to immobilize insects for study or surgery. Regardless of the technique used (i.e., CO_2, physical restraint, or low temperature), some side effect(s) will result. Changes in CO_2 levels have produced various effects on the behavior, physiology, and metabolism of insects. In addition to its

Figure 4.20 Schematic model of nervous control and coordination of preprogrammed motor outputs or central motor pattern generators.

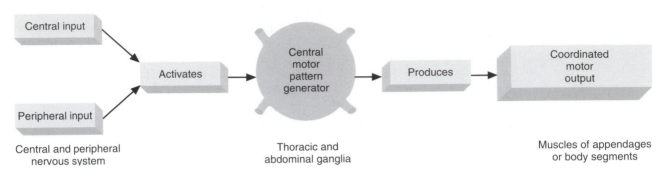

Central and peripheral
nervous system

Thoracic and
abdominal ganglia

Muscles of appendages
or body segments

may also drive other control centers. The ventilatory pacemaker in *Schistocerca gregaria* appears to be in the metathoracic ganglion, whereas in immature aeshnids (a family of dragonflies) it resides in the last abdominal ganglion. How an excess of carbon dioxide and/or low oxygen affects the overall neural control of ventilation is not known.

Oxygen Transport in the Hemolymph
Except over the very short distances between most tracheolar endings and the cells they oxygenate, the hemolymph does not generally function as an oxygen carrier. However, as mentioned earlier, certain chironomid (Diptera, Chironomidae) larvae (bloodworms), the endoparasitic bot fly larva (*Gasterophilus*), and the backswimmers *Anisops* and *Buenoa* (Hemiptera, Notonectidae) have the oxygen-carrying pigment hemoglobin in the hemolymph. Under conditions of high oxygen tension this hemoglobin is saturated with oxygen and thus does not serve as a carrier. However, under conditions of low oxygen tension, a common occurrence in the bloodworm's natural habitat (stagnant water), the hemoglobin is unsaturated and hence available to carry oxygen. Horse bot fly larvae ventilate periodically when they come into contact with an air bubble in the intestine of their host. Hemoglobin permits these insects to take in a larger supply of oxygen than they could otherwise. Through increasing the amount of oxygen that can be stored, hemoglobin enables the aquatic backswimmers to remain submerged for long periods of time. Mill (1985) discusses the differences between the hemoglobin's affinity for oxygen in chironomids versus the other insects with hemoglobin and presents a functional difference in life-styles for this difference.

The Excretory System

The function of the excretory system is to maintain *homeostasis*, i.e., a constant internal environment. Because the hemolymph bathes the tissues and organs of the insect body, it largely determines the nature of this internal environment. Thus, the excretory system is responsible for

the maintenance of the uniformity of the hemolymph. It accomplishes this by the elimination of metabolic wastes and excesses, particularly nitrogenous, and the regulation of salt and water (Maddrell 1971, Stobbart and Shaw 1974, Cochran 1985a, Taylor 1986, Bradley 1985). The Malpighian tubules are the major organs involved in filtration of the hemolymph, whereas the hindgut (especially the rectum) is involved in reabsorption of important ions and water.

Malpighian Tubules

Malpighian tubules (figure 4.21*a,b*) are usually long slender tubes closed at their distal ends and found in association with the posterior portion of the alimentary canal. They are named after their discoverer, Marcello Malpighi, a seventeenth-century Italian scientist. These tubules are commonly convoluted and vary in number, depending on species, from 2 to 250 or more, usually occurring in multiples of two. They are apparently lacking only in members of the order Collembola (springtails) and the family Aphididae (aphids; order Homoptera). Malpighian tubules lie at the junction between the midgut and hindgut. However, their embryonic origin is a matter of controversy, and they may open directly into the midgut or hindgut or more commonly into a dilated ampullar structure.

The Malpighian tubules are usually free in the body cavity and are bathed directly by the hemolymph. At least some are always in close proximity to the fat body and parts of the alimentary canal, other than when they open into it. In some groups of insects, for example, most Lepidoptera (butterflies and moths) and Coleoptera (beetles), the distal ends of the tubules are embedded in the tissues surrounding the rectum (figure 4.21*c,d*). This is the *cryptonephridial* ("hidden kidney") *arrangement*. In some insects, the distal ends of two tubules anastomose, forming a closed loop. In addition, there may be anatomical differences between the tubules within the same insect.

Figure 4.21 Major types of Malpighian tubule–hindgut systems. (*a*) Orthopteran type. (*b*) Hemipteran type. (*c*) Coleopteran type. (*d*) Lepidopteran type. Arrows indicate directions of movement of substances in and out of the tubule lumen. *Redrawn from Patton, 1963.*

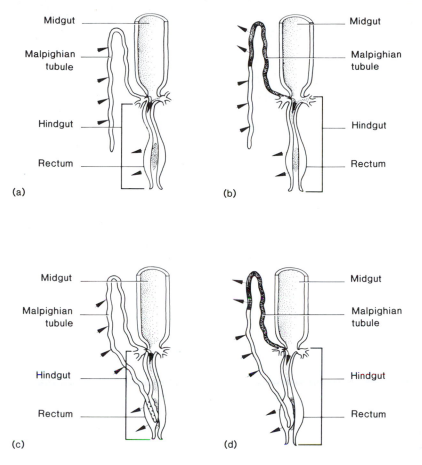

In many insects the Malpighian tubules are capable of a variety of movements by contraction of variously oriented muscles associated with the tubules. Apparently only members of the orders Thysanura (silverfish and relatives), Dermaptera (earwigs), and Thysanoptera (thrips) do not have muscles associated with the Malpighian tubules. The function of the tubule movement is currently a matter of conjecture. Suggestions have included such functions as propulsion of the contents of the lumen toward the opening into the alimentary canal, mixing of the luminal contents, and exposure of the tubules to more hemolymph. The tubules are usually well tracheolated and are one-cell thick. A basement membrane lies beneath the muscles and surrounds the tubule cells. The cytoplasm of these cells varies in appearance. It is usually colorless but may have a faint green or yellow appearance. It is generally filled with various refractile or pigmented inclusions and sometimes contains needlelike crystals but may be nearly clear. The tubule cells communicate with the lumen by means of a finely striated border. The electron microscope has revealed a very large number of mitochondria (consistent with the occurrence of active transport) in the cytoplasm of the tubule cells. The ultrastructure of the Malpighian tubules is covered in Bradley (1985) and in Martoja and Ballan-Dufrançais (1984). One ultrastructural fact worthy of mention here is that the microvilli of both the midgut digestive cells and the Malpighian tubules maintain their upright "stiffness" or support because of the presence of microfilaments composed of the single globular protein subunit called actin. However, when actin is put together with myosin, it serves as a contractile system within cells (e.g., muscle fibers, see figure 3.23).

Dietary Problems

Insect diets (whether liquid or solid, plant or animal) have greatly influenced the evolution of efficient digestive and excretory systems capable of handling the problems created by the type of food ingested. In table 4.4, one can see that insects feeding on vertebrate blood must actively conserve sodium by reabsorption of Na^+ ions from a food source low in that particular ion, whereas phytophagous

Table 4.4 The Ionic Composition of the Diet in Relation to That of the Hemolymph in Some Terrestrial Insects

Diet and insect	M.equiv./l. or /kg. wet weight			
	Na	K	Ca	Mg
human blood	87·0	51·1	3·0	2·5
Rhodnius	164·0	6·0	—	—
horse blood	84·8	31·4	1·7	3·3
Gasterophilus larva	175·0	11·5	5·7	32·0
lettuce leaves	13·0	86·2	—	—
Periplaneta	113·0	25·6	—	—
privet leaves	46·4	152·1	824·5	39·9
Carausius	8·7	27·5	16·2	142·0
carrot leaves	25·6	176·9	214·5	35·6
Papilio larva	13·6	45·3	33·4	59·8
potato leaves	trace	144·5	128·6	85·9
Leptinotarsa	3·5	65·1	47·5	188·3
Ribes leaves	trace	249·1	271·2	53·6
Pteronidea larva	1·6	43·4	17·5	60·5

Source: Data from various sources.

insects face a different problem (i.e., the food is high in both K^+ and Ca^{++} ions). Thus, they must excrete the excessive amounts of both ions to maintain homeostasis of the hemolymph.

In addition to the problems associated with differences in ionic concentrations between the food sources and the hemolymph, phytophagous insects face an additional problem (i.e., the noxious phytochemicals termed allelochemicals). How these are handled by the digestive system has been discussed earlier, but one must realize that, if absorbed into the hemolymph, the detoxified chemicals and/or the allelochemicals themselves must be excreted. One of the better studied cases is the grasshopper, *Zonocerus variegatus,* in which, following ingestion of plants containing the allelochemical cardiac glycosides, the transport mechanism of the Malpighian tubules is induced, thus facilitating rapid excretion of the toxin from the insect's hemolymph (Bradley 1985). Later, we will discuss how some insects have opted not to throw away but to retain and sequester these allelochemicals to their benefit rather than to excrete them.

Salt and Water Balance

Different environmental situations pose different salt and water problems for insects. Terrestrial forms are constantly faced with the tendency to lose water through transpiration and are generally dependent on ingested food for needed water and salt. Depending on the water content of their diet, the fecal material may be quite watery, as from plant-feeding insects that take in an excess of water, or a dry powdery pellet, as from those insects that feed on materials of very low water content. Freshwater insects must excrete the large amounts of water absorbed through the integument and by the gut along with ingested food and at the same time must conserve the inorganic ions. Saltwater forms, like terrestrial insects, must constantly overcome the tendency to lose water, in this case due to a difference in concentration between the insect's internal fluid and the surrounding medium.

In the majority of insects the regulatory problems outlined in the preceding paragraph are, at least in part, solved through the activities of the Malpighian tubules and the rectum in the hindgut. The Malpighian tubules are freely permeable to most small molecules. These molecules may be passed into the lumen of a tubule by simple diffusion or by active transport of potassium ions, which generates fluid flow. A more detailed presentation of the general scheme of how the excretory system functions is presented in figure 4.22. Materials in excess in the hemolymph are basically filtered through the Malpighian tubules and important materials that were lost in this process or from the feces are reabsorbed in the rectum. Secretion of water and ions is not a selective process. However, substances that are required are reabsorbed into the hemolymph in the proximal portions of the tubules and/or in the rectum. These reabsorption processes may also be active transport or simple diffusion. It has been suggested that it is more energy-efficient for organisms to develop, in the evolutionary sense, ways of reabsorbing needed substances than to develop ways of excreting every possible type of unneeded substance.

Figure 4.22   Schematic showing the movements of ions, water, and other molecules between the hemolymph and Malpighian tubules and the hindgut and the hemolymph in a generalized insect.
Taken from Gillott, 1980.

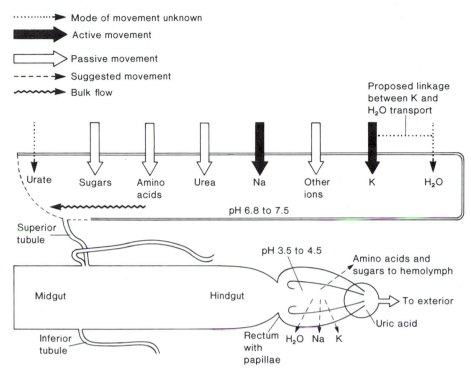

Absorption of materials in the rectum is considered to be a function of the anal papillae described earlier in this chapter (alimentary system). Ultrastructural and histochemical studies of these papillae have led to the development of a hypothetical explanation of the mechanism of absorption (Berridge 1970). The mechanism proposed involves the creation, by the active transport of ions, of gradients down which water, ions, and other molecules can passively diffuse. The proposed mechanism also provides a means for some molecules to move into the hemolymph against a concentration gradient.

Active transport is probably not always involved in rectal absorption. For example, in *Dysdercus* (Hemiptera, Pyrrhocoridae), absorption is entirely passive and occurs only when the rectal fluid is hypotonic relative to the hemolymph.

A number of Malpighian tubule–rectal cycling systems have been described. In the simplest situation, found in members of the order Orthoptera (grasshoppers and relatives), the tubules are composed of a single cell type and have only fluid in the lumen (figure 4.21*a*). This fluid passes down the lumen and mixes with the gut contents, and as it passes down the hindgut, particularly in the rectum, needed water and ions are reabsorbed into the hemolymph. A more complex cycling system is found in most Hemiptera (figure 4.21*b*). In this case movement of

materials into the lumen of a tubule occurs in the distal portion; reabsorption of needed water and ions into the hemolymph occurs in the proximal portion and in the rectum. A third system is that typically found in beetles (figure 4.21*c*), in which the distal portion of a Malpighian tubule is embedded in the tissues surrounding the rectum. This is the cryptonephridial arrangement mentioned earlier. Secretion into the lumen of a tubule apparently occurs in its exposed portion, whereas reabsorption of needed materials likely takes place through the portion embedded in the rectum. It has been suggested that cryptonephridial arrangement increase the efficiency of the cycling process.

A cryptonephridial arrangement is also typical of members of the order Lepidoptera (figure 4.21*d*) except that in these reabsorption of needed materials into the hemolymph occurs in the proximal portions of the tubules as well as in the rectum. Other situations have been described, but those mentioned here should give a general idea of the processes involved. It should be pointed out that factors such as spiracular control, integument permeability, food selection, and habitat selection are also involved with the regulation of salt and water in insects. Also, in certain aquatic insects there exist mechanisms for the uptake of ions other than via the gut. For example, chloride ions were taken into the hemolymph of mosquito larvae by way of four papillae, which surround the anus (figure 4.19). This is an active process, occurring against a rather severe concentration gradient. In addition, these papillae are responsible for sodium, potassium, and water uptake. Bradley (1987) provides detailed information about the osmoregulatory role of the anal papillae of mosquito larvae in various types of habitats. The methods for studying and measuring ion transport both in the Malpighian tubules and hindgut are given in Bradley and Miller (1984).

Control of Diuresis and Gut Motility

Diuresis, or the production of urine, in insects is controlled by diuretic or antidiuretic hormones. These substances have been isolated from and localized in the pars intercerebralis of the brain, the corpus cardiacum, and various ventral chain ganglia, including the subesophageal ganglia. Using immunological techniques (i.e., known vertebrate antibodies against vertebrate diuretic hormones) plus a bioassay, the diuretic peptide from *Locusta* was isolated and identified (Holman et al. 1990). An antidiuretic hormone, having its effect on the Malpighian tubules of the house cricket, *Acheta domesticus,* has been purified and awaits identification. At the same time, a chloride transport stimulating hormone has been isolated, which has been shown to regulate both ion and water balance in the rectum of the locust.

The topic of endocrine control of diuresis in insects is not only exciting and in a pioneer stage, but may offer a new, nonpesticidal control strategy against some insect pests. What would happen if we could force an insect to lose water? If the gene controlling the production of the antidiuretic hormone were genetically engineered into a carrier-infective vector (e.g., baculovirus), which would then carry the gene into the pest insect at an inopportune time, the insect's osmoregulative capabilities would be upset and should result in death when the gene is expressed. This is just one way in which basic research continually provides applied research with novel applications in insect biology (Menn and Bořkovec 1989).

What regulates hindgut motility and possibly defecation in insects? Proctolin, a neuropeptide that was isolated from the proctodeum or hindgut of *Periplaneta americana* was initially called the "gut factor" responsible for motility and was found to be absent from this region when the proctodael nerves were severed. This pentapeptide has now been well characterized, is widely distributed in the insect nervous system, and functions as an excitatory neurotransmitter (i.e., it produces a myotrophic effect on the visceral muscles of the hindgut). The other functions of proctolin within the insect have not been fully delimited (Huddart 1985). Evidence is accumulating that two excitatory neurotransmitters, L-glutamate and proctolin, control hindgut longitudinal muscles and that, in addition, proctolin also serves as a neuromodulator (Huddart 1985). More information about this exciting chemical is found elsewhere (Cook and Holman 1985, and O'Shea and Adams 1986).

Nitrogenous Excretion

Nitrogenous products of various types tend to accumulate in the hemolymph as a result of protein, amino acid, and nucleic acid metabolism. These materials are generally of no use to an insect and may be toxic. This being the case, they must either be excreted or stored in an inert state until they can be used for another function or be excreted.

Where an insect lives usually determines the type of excretory product it produces. In general, most terrestrial insects are uricotelic, whereas most aquatic insects are ammonotelic (i.e., excrete ammonia). This can be clearly seen by examining table 4.5, which lists the major nitrogenous excretory products of various insects. Notice that the aquatic immature dragonfly (*Aeshna cyanea*) produces ammonia as its major excretory product while a butterfly adult (*Pieris brassicae*) produces uric acid. Uric acid, however, is the major waste product formed and excreted, making up 80% or more of the nitrogenous end products found in the urine of most terrestrial and many aquatic insects. It does not require a large amount of water for its elimination because it is not very soluble in water. Spider feces also contain a lot of uric acid. Have you ever noticed the white fecal spots underneath a spider's web? Try washing them off the floor. Why are they so difficult to remove? Because of its insolubility, organisms that excrete uric acid are also interested in conserving water. On the other hand, ammonia is the major nitrogenous waste formed by many aquatic insects, as mentioned above, and by blow fly larvae (see table 4.5 for *Lucilia sericata* larva). Since ammonia is highly toxic it requires a large amount of water for its elimination, thus limiting its production as a major nitrogenous waste molecule to insects that live in an aquatic or semiaquatic environment. Other nitrogenous products that have been identified in insect urine are allantoin, allantoic acid, and urea. These arise from the breakdown of uric acid and are usually present in very small amounts. Amino acids are sometimes found but usually in very small quantities.

The Malpighian tubules are the major organs involved in the excretion of nitrogenous materials, but other tissues may be involved.

Even though cockroaches are uricotelic, they are extremely unusual, because unlike other insects, they normally do not excrete uric acid, but they store it. Cockroaches and a few other insects have specialized storage cells, usually located in the fat body, where they store uric acid in the form of spherules. These special storage cells are termed urocytes, or urate cells. In addition, males of several species have specialized storage glands, called uricose glands, that are part of the

Table 4.5 Nitrogenous Excretory Products of Various Insects[a]

	Uric acid	Allantoin	Allantoic acid	Urea	Ammonia	Amino acids
Odonata						
Aeshna cyanea (larva)	0.08	—	0.00	—	1.00	—
Dictyoptera/Phasmida						
Periplaneta americana	1.00	0.00	0.00	—	—	—
Blatta orientalis	0.64	0.64	1.00	—	—	—
Dixippus morosus	0.69	1.00	0.44	—	—	—
Hemiptera						
Dysdercus fasciatus	0.00	1.00	0.00	0.26	—	0.24
Rhodnius prolixus	1.00	—	—	0.33	—	trace
Coleoptera						
Melolontha vulgaris	1.00	0.00	0.00	—	—	—
Attagenus piceus	0.72	—	—	1.00	0.57	0.50
Diptera						
Lucilia sericata	1.00	0.30	—	—	0.30	—
Lucilia sericata (pupa)	1.00	0.00	—	—	0.15	—
Lucilia sericata (larva)	0.05	0.02	—	—	1.00	—
Lepidoptera						
Pieris brassicae	1.00	0.04	0.01	—	—	—
Pieris brassicae (pupa)	1.00	0.03	0.05	—	—	—
Pieris brassicae (larva)	0.28	0.16	1.00	—	—	—

[a]Quantity of nitrogen excreted in the different products is expressed as a proportion of the nitrogen in the predominant end product. Taken from Gillott (1980).

accessory reproductive gland system involved in spermatophore formation. When the male transfers the spermatophore (i.e., a specialized, usually proteinaceous structure housing the sperm) to the female, uric acid is deposited over the spermatophore. Once the sperm are released and the spermatophore is voided by the female, she usually eats it. Also, some females feed on a urate "slurry" found in the male's genital chamber.

If cockroaches don't excrete the uric acid, what is to be gained by storing it, and why would some species of males evolve specialized glands for its storage? Cockroaches evolved long before human beings entered the scene; in fact, most cockroaches do not live in human environments in which food or scraps of food rich in nitrogen are available. Rather, the lack of nitrogen-rich foods during the Pennsylvanian or Upper Carboniferous period, when cockroaches are believed to have entered the scene, possibly was important in cockroaches evolving the strategy of storing and not excreting uric acid. Researchers now believe that the "payoff" in storing uric acid is that it provides an alternate source of nitrogen. Studies, using radiolabeled precursors of uric acid, that showed that the male's uric acid passed to the female on the spermatophore reached either the female's store or in her embryos. Also, the female's own uric acid stores could be passed onto her embryos. Thus, it is suggested that not only can the male make a "nutritional investment" in the female but both sexes can make a "parental investment" in the offspring. One major problem, however, still exists. Most insects cannot use uric acid directly as a food source.

They lack uricase and must rely on bacteria or other microorganisms for breaking it down into a usable form. It is suggested, but not conclusively proven, that the bacteria stored in specialized cells called mycetocytes biochemically degrade uric acid, making it available to the cockroach. In fact, these cells are also located in the fat body in close association with the urate cells. Some female cockroaches possess specialized cells harboring symbiotic bacteria in their ovaries. Thus, when eggs are laid, these bacteria are transovarially transmitted by the female to the eggs. This would guarantee that the developing embryos could use the paternal uric acid as a nitrogen source. More information about this exciting evolutionary adaptation of cockroaches can be obtained from either Cochran (1985b) or Mullins and Cochran (1987).

Insect Urine

The nature and composition of insect urine is, not surprisingly, highly variable. Its physical appearance varies from the dry powdery material egested by terrestrial insects that inhabit dry environments to a clear fluid in those in which water conservation is not a problem (e.g., plant-feeders and freshwater forms). Its chemical composition is dependent upon the dietary substances that are in excess of an insect's needs or not usable by the insect, the nitrogenous wastes formed as a result of protein and amino acid metabolism, and the excesses of salts and water that may occur.

CHAPTER 5

Reproduction and Morphogenesis

In this chapter we consider the systems and processes involved in reproduction and then discuss the major events that begin with the newly fertilized egg (*zygote*) and terminate with the death of the insect. The behavioral aspects of reproduction are discussed in chapter 8.

Reproductive System and Gametogenesis

Reproduction in most insects is bisexual. The male reproductive system functions in the production, storage, and delivery of spermatozoa; in the production of seminal fluid, which nourishes and provides an appropriate environment for the sperm; and in providing the female with either chemical or physical signals to tell her that she has mated and that she should commence laying eggs or depositing larvae. In some cases the male may even transfer chemicals that aid the development of the embryo (e.g., the male cockroach transfers uric acid to the female, which may end up in the embryo and may serve as an additional source of nitrogen). The female system produces and stores eggs, provides the eggs with the necessary nutrients for embryonic development, receives and stores spermatozoa, is the site of fertilization, deposits eggs or larvae when appropriate, and may provide additional protection to the embryos in the form of a covering (e.g., ootheca in cockroaches and abdominal hairs in the gypsy moth).

General reviews of this topic include Davey (1965), de Wilde and de Loof (1973a,b), Engelmann (1970), Adiyodi and Adiyodi (1983), and Raabe (1986). Matsuda (1976) contains descriptions of both internal and external aspects of male and female reproductive systems in all orders of insects, and volumes 1 and 2 of Kerkut and Gilbert (1985) are devoted to reproduction, embryogenesis, and postembryonic development.

Male Reproductive System, Spermatogenesis, and Spermatozoa

The male reproductive system (figure 5.1) is located in the posterior portion of the abdomen and typically consists of paired gonads (*testes*) connected by various ducts, which ultimately open into the intromittent organ, the *penis,* or *aedeagus. Accessory glands* of various sorts are usually associated with the ducts. Although fundamentally similar, this system varies among the different species of insects. The following description is of a generalized male reproductive system with comments on variations.

Testes

Although the testes are usually bilateral, paired structures, they have undergone a medial fusion in some insects (e.g., Lepidoptera). Basically each testis is composed of a varying number of *testicular follicles* (sperm tubes), which are usually encased by a layer of connective tissue. Each follicle is in turn enclosed by a layer of epithelial cells, which are thought to serve a trophic (nutrient) function, absorbing nutrients from the hemolymph and making them available to the germ cells within.

Ducts

A tiny duct, the *vas efferens,* leads from each follicle to a common lateral duct, the *vas deferens,* and, finally, the vasa deferentia from each testis join to form the *ejaculatory duct,* which ends in the penis or aedeagus at the *gonopore.* The vasa deferentia are at least partly of mesodermal origin. In some insects (Protura, Ephemeroptera, and some Dermaptera) each vas deferens has its own opening to the exterior. Both the vas deferens and ejaculatory duct are invested with a layer of muscles, which

Figure 5.1 Generalized male reproductive system. (*a*) Principal male organs. (*b*) Detailed structure of a testis. (*c*) Section of a testis.

Redrawn from Snodgrass, 1935.

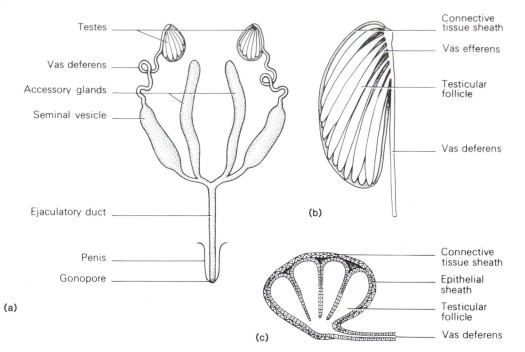

are involved in the propulsion of semen. Each vas deferens may be formed into a series of convolutions forming an *epididymis* or may have a dilated portion (i.e., *seminal vesicle*) in which the spermatozoa are stored in a quiescent state.

Accessory Glands

Various glandular structures are typically associated with the vasa deferentia or with the ejaculatory duct, although some insects (e.g., Apterygota and some Diptera) lack accessory glands. These accessory glands and their ducts are either of mesodermal origin (i.e., evaginations of the vasa deferentia) or of ectodermal origin (i.e., evaginations of the ejaculatory duct). They usually occur as a single pair, but in some insects there may be several in a cluster (e.g., the mushroom body in male cockroaches). In some insects, portions of the vasa deferentia or ejaculatory duct may also have glandular functions.

Among the known functions of male accessory glands is the secretion of seminal fluid, which may contain chemicals involved in activation of spermatozoa and also in the production of spermatophores. In addition, accessory gland secretions may influence an inseminated female in a variety of ways, including stimulation of oviposition, acceleration of oocyte maturation, stimulation of contractions of the genital ducts that aid in sperm movement, and inhibition of subsequent insemination by formation of vaginal plugs or by exerting an effect on the female's behavior. Hinton (1974) and Leopold (1976)

discuss functions of male accessory gland secretions at length. Davey (1985), Raabe (1986), and Kaulenas (1992) provide a general overview of the different functions of the accessory gland secretions, Happ (1984) discusses the ultrastructure, and Chen (1984) covers the biochemistry and hormonal and genetic aspects of synthesis. The biology of the accessory glands in other invertebrates is reviewed by Adiyodi and Adiyodi (1983). Evidence is accumulating that small peptides are the active factors in the male's seminal secretions and that these pass into the female's hemolymph where they are carried to specific target sites, which are ultimately responsible for mating inhibition or oviposition initiation (see table 3.2). These target sites are currently not known.

Seminal fluid production is tightly coordinated with mating behavior, which in many insects is influenced by adult nutrition and/or the environment (e.g., overwintering in diapause). These two variables (i.e., nutrition and environment) also influence the production and release of juvenile hormone. It ultimately is juvenile hormone that has been shown in most insects to regulate the development of and biosynthesis of secretions by the accessory glands. Thus, during periods of food shortage and/or severe environmental periods (i.e., winter) usually neither mating takes place nor the production of the accessory gland secretions. The influence of juvenile hormone on the accessory glands has mainly been shown by removing the corpora allata.

Figure 5.2 Section of a testicular follicle (diagrammatic).
Redrawn from Snodgrass, 1935.

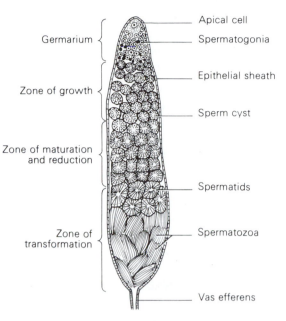

Germarium {
— Apical cell
— Spermatogonia

Zone of growth {
— Epithelial sheath
— Sperm cyst

Zone of maturation
and reduction {
— Spermatids

Zone of
transformation {
— Spermatozoa

— Vas efferens

Figure 5.3 Photomicrograph of a sagittal section of a testis in a male mosquito, *Aedes triseriatus*. 1, Germarium; 2, zone of growth; 3, zone of maturation and reduction; 4, zone of transformation; 5, spermatozoa; 6, epithelial sheath; 7, vas efferens. (Scale line = 50 μm.)

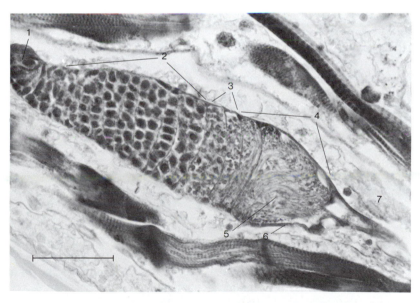

Spermatogenesis

The testicular follicles contain the *germ cells* and are hence the sites of the meiotic cell divisions that give rise to *spermatozoa*, the entire process being referred to as *spermatogenesis*. This process usually occurs during the last larval instar or pupal stage and in some species continues in the adult stage. Each follicle contains a large apical cell or complex of cells that apparently serve a trophic function, providing nutrients for the developing *spermatogonia*. Each follicle is divided apically to basally into zones that represent the different stages of spermatogenesis (figures 5.2 and 5.3). Apically the *germarium*, or *zone of spermatogonia*, is composed of the germ cells (spermatogonia) and somatic mesodermal cells. In the next region, the *zone of growth*, or *zone of spermatocytes*, the spermatogonia undergo several mitotic divisions, forming *primary spermatocytes* that become encysted in somatic cells. The primary spermatocytes

Figure 5.4 Generalized insect spermatozoon.

Redrawn from Breland, Eddleman, and Biesele, 1968.

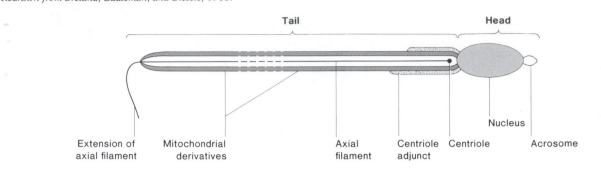

Tail | Head

Nucleus

Extension of axial filament · Mitochondrial derivatives · Axial filament · Centriole adjunct · Centriole · Acrosome

(diploid cells) undergo meiosis and produce haploid daughter cells in the next region, the *zone of maturation and reduction.* With the first and second meiotic divisions the primary spermatocytes become *secondary spermatocytes* and *spermatids,* respectively. In the basal *zone of transformation,* the secondary spermatids become transformed into flagellated spermatozoa. When the cysts in which they have been encased throughout spermatogenesis rupture, the spermatozoa enter the vas efferens and vas deferens and lodge in the seminal vesicles. Commonly, when the spermatozoa are released into the ducts, they remain in bundles (*spermatodesms*) held together by gelatinous material. The spermatozoa of most insects studied are filamentous with very narrow heads (figure 5.4). Why the head of insect sperm is so narrow as compared to other animal groups is not known. It may have something to do with the size of the micropyle, which itself must be small to prevent water loss by the egg or developing embryo.

The movement of spermatozoa within the male reproductive system is due not to their inherent motility but to contractions of the muscles associated with each vas deferens and the ejaculatory duct.

Further information on insect spermatozoa and spermatogenesis can be found in Baccetti (1972), King (1974), Phillips (1970), Raabe (1986), and Jamieson (1987).

Control of Spermatogenesis

In cell culture experiments a humoral factor has been implicated in the differentiation of spermatocytes in giant silkworm moths (Lepidoptera, Saturniidae). Ecdysone may play a role in influencing the permeability of the testis walls to this humoral factor. Juvenile hormone has been shown in some species to have an inhibitory effect on spermatogenesis.

Spermatophores

In some insects the spermatozoa are produced and transferred to the female in specialized packets held together by proteinaceous secretions of the accessory glands. These packets are called *spermatophores* and often assume

Figure 5.5 Spermatophore from a male silverfish, *Lepisma saccharina* (see also Fig. 8.7).

Redrawn from Sturm, 1956.

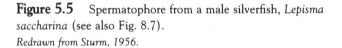
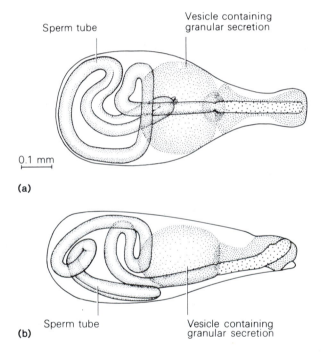

Sperm tube · Vesicle containing granular secretion

0.1 mm

(a)

(b) Sperm tube · Vesicle containing granular secretion

rather distinct forms (figure 5.5). Spermatophores are common in the lower orders (e.g., the apterygotes) and rare or absent in some of the higher orders, such as Hymenoptera. Once spermatophores are emptied of spermatozoa, they may be digested and absorbed within the genital ducts if placed there during copulation or be eaten by the female in those species in which the male places the spermatophore on the substrate.

An area of current interest is the importance of male secretions or spermatophores on female remating. This topic has both basic and practical implications and will be discussed later. In addition, considerable research is being directed at understanding the strategies used by males to ensure that their sperm fertilizes the female (Smith 1984).

Figure 5.6 Generalized female reproductive system. (*a*) Principal female organs. (*b*) Single ovariole. *Redrawn from Snodgrass, 1935.*

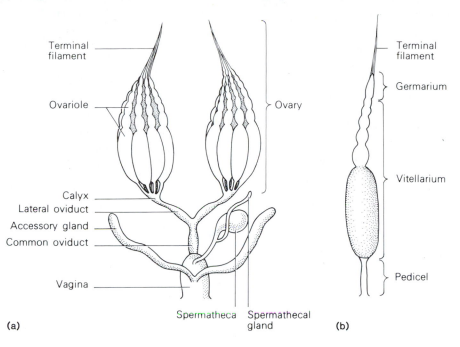

Terminal filament

Ovariole

Ovary

Calyx

Lateral oviduct

Accessory gland

Common oviduct

Vagina

Spermatheca

Spermathecal gland

Terminal filament

Germarium

Vitellarium

Pedicel

(a)

(b)

Female Reproductive System, Oogenesis, and Ova

Like the male reproductive system, the female system (figure 5.6) is located in the posterior part of the abdomen. It typically consists of paired gonads (*ovaries*) connected by a series of tubes to the *vagina,* which opens to the exterior and receives the penis during copulation. There are also a variety of *accessory glands* present. Although the female reproductive system is basically similar among different species, there is considerable variation. The system described below will be a generalized one with comments on some variations.

Ovaries

The ovaries are bilaterally located, mesodermal organs that produce eggs. They are composed of a number of functional units, *ovarioles,* which are invested in a layer of epithelial cells. This layer has muscles and tracheae associated with it. The large number of tracheae associated with the ovaries provides the developing follicles with the necessary oxygen to help meet the high energy demands of egg maturation. In addition, the muscle sheaths surrounding most ovarioles and ovaries contain neurosecretory cells, which presumably are involved in regulating muscle contractions.

The number of ovarioles per ovary varies greatly, from 1 in the tsetse flies, *Glossina* spp., and some aphids to over 2000 in the queens of certain termite species. In most insects a terminal thread from the cephalad portion of each ovariole joins those of its neighbors, forming a *terminal filament,* which attaches to the dorsal diaphragm.

Ducts

At the base of each ovariole is a small duct or *pedicel,* which joins those of the other ovarioles in a bulbous *calyx*; this in turn opens into the *lateral oviduct.* The lateral oviducts, which are, like the ovaries, of mesodermal origin, join to form the *common oviduct.* The common oviduct serves as a communicating tube between the lateral oviducts and the *bursa copulatrix* or *vagina,* which opens to the outside. The bursa copulatrix, when it occurs, is a saclike expansion of the vaginal region. The common oviduct and bursa or vagina are of ectodermal origin and, like the foregut and hindgut and tracheae, are lined with a modified form of cuticle. There is usually a single outpocketing from the bursa, vagina, or common oviduct in which spermatozoa are stored prior to fertilization. This outpocketing is called the *spermatheca,* and in some insects (e.g., certain flies) it is a paired structure.

Accessory Glands

There are generally one or two pairs of accessory glands, which usually open into the apical portion of the bursa copulatrix, vagina, or unpaired oviduct. These glands vary in structure and function. They are commonly involved in the secretion of adhesive materials (in which case they are called *colleterial glands*) and serve to cement eggs to the substratum or hold them together in masses. In cockroaches, secretions of the accessory glands form a capsule or *ootheca* around eggs that have accumulated in the bursa copulatrix. In many aquatic insects, mayflies, stoneflies, and so on, the gelatinous masses that surround the eggs are accessory gland secretions. The accessory glands of female tsetse flies are modified to form "milk glands," which produce the secretion utilized by the incubating larva for food while remaining and developing within the female's uterus. What is the exact chemical composition of this "milk," and what regulates its production? In comparison with male insects, relatively little is known about the biochemistry, regulation, and function of the accessory gland fluid secretions in female insects. A review of this topic in invertebrates other than insects is provided by Adiyodi and Adiyodi (1983).

Oogenesis

Under the heading oogenesis we include all those processes that ultimately lead to the development of a mature ovum, capable of being fertilized, and development of the nutritive capacity to support embryonic development. Oogenesis may be completed prior to or during the imaginal stage.

Each ovariole is divided into zones that contain germ cells or *oocytes* in various stages of development and maturation (figure 5.6*b*). There are two broad zones, the apical germarium and the basal *vitellarium* both of which are invested in an outer layer of cells, the *ovariole sheath.* The germarium contains the primary female germ cells, the *oogonia,* which divide mitotically and eventually become *primary oocytes.* The germarium also contains prefollicular tissue, which comes to form the *follicular epithelium* in the vitellarium. The vitellarium in an active ovariole is comprised of oocytes that are undergoing the uptake from the hemolymph and deposition into the oocyte of nutrients for the mature egg. One of these nutrients is yolk, which is composed of spheres of proteins, neutral lipids, and carbohydrates (e.g., glycogen). Vitellogenin or the female specific protein is generally produced in the fat body, released into the hemolymph and taken up against concentration gradients by the oocyte through the process of receptor mediated endocytosis (refer to figure 4.10). The process of vitellogenin biosynthesis and its uptake is referred to as *vitellogenesis* (yolk deposition). Some insects (e.g., mosquitoes, figure 5.7; and *Phormia regina,* figure 5.8) synchronously develop only the terminal oocytes. These species usually lay their eggs collectively in batches, whereas other species may have several oocytes within an ovariole developing at the same time and may lay eggs singly. In these species (e.g., moths), the terminal oocyte is generally the most well developed. The interactions among nutrition, endocrines, and oogenesis is one of the most active areas of reproductive research.

Each developing oocyte is surrounded by a follicular epithelium, and the oocyte and its associated epithelial layer comprise a *follicle.* The follicle cells are extremely important in that they are involved in the selective uptake of nutrients from the hemolymph and are also the cells that secrete and produce the eggshell. In asynchronous

Figure 5.7 Photomicrographs of vitellogenesis in the mosquito *Culex nigripalpus.* (*a*) Sagittal section of ovary in an unfed female. (*b*) Same as (*a*), higher magnification. (*c*) Oocyte, 48 hours following a blood meal. 1, ovary; 2, lateral oviduct; 3, Malpighian tubule; 4, follicle; 5, nurse cells; 6, follicular epithelium; 7, yolk. (Scale lines: A = 250 μm; B = 50μm.)

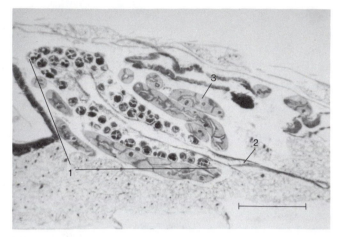

(a)

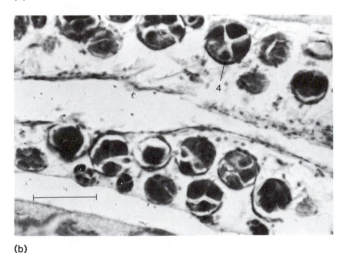

(b)

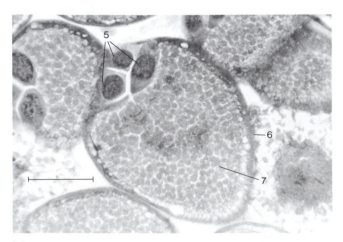

(c)

Figure 5.8 Electron micrographs of terminal follicles of *Phormia regina*. (*a*) Follicle cells (FC) closely appressed with no extracellular space between the cells of a sugar-fed female. (*b*) Spaces between adjacent follicle cells are evident in protein-fed females. Arrows show the direction vitellogenin (Vg) takes from the hemolymph into the developing oocyte. OC = oocyte; NC = nurse cell; OS = ovariole muscle sheath.
Courtesy of M. Mazzini, J. Stoffolano, and Chih-Ming Yin.

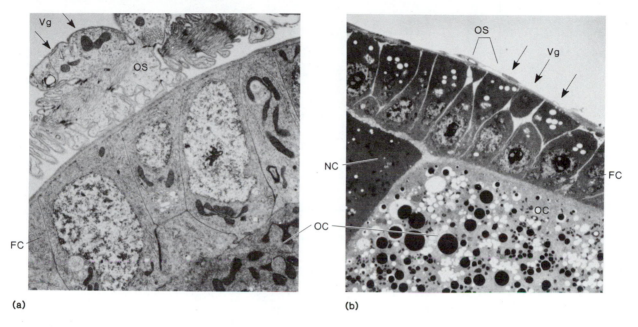

(a)

(b)

egg development, the oocytes in the vitellarium have progressively more yolk in an apical to basal sequence. The most mature basal oocyte is separated from the lumen of the pedicel by an epithelial plug, which ruptures when the oocyte is ready to leave the ovariole and move into the lateral oviduct. This process of exiting from the ovariole is called *ovulation*. Following ovulation, the follicular epithelial cells remain behind and eventually degenerate.

There are three major types of ovarioles, based on the method by which yolk deposition occurs (figure 5.9): *panoistic, telotrophic,* and *polytrophic.* Bonhag (1958), Mahowald (1972), and Telfer (1975) review ovarian structure and vitellogenesis.

The first and most primitive type of ovariole listed, panoistic, has no trophocytes. However, each developing oocyte is surrounded by the follicular epithelium, and follicular plugs exist between adjacent oocytes. Panoistic ovarioles are found in the apterygotes, Orthoptera, Isoptera, Odonata, Plecoptera (stoneflies), Siphonaptera, and some Coleoptera.

Unlike panoistic ovarioles, both telotrophic and polytrophic possess trophocytes. Telotrophic ovarioles differ from polytrophic ones in that the trophocytes are not directly associated with each developing oocyte but are all located in the apical region of the ovariole. The trophocytes are connected to the various oocytes by means of "nutritive cords." Otherwise, the two types are similar,

each oocyte being surrounded by a follicular epithelium and being separated from the others by follicular plugs. Telotrophic ovarioles are characteristic of Hemiptera and some Coleoptera.

Polytrophic ovarioles have nutritive cells, *nurse cells,* or *trophocytes,* associated with each developing oocyte. Trophocytes are within the follicular epithelium, which surrounds each oocyte. Characteristically, the oocyte and trophocytes originate from a single oogonium that undergoes a series of mitotic divisions. Incomplete cytokinesis results in the formation of cytoplasmic canals interconnecting the trophocytes and oocyte. For further information, see Telfer (1975). A follicular plug exists between each oocyte with its accompanying trophocytes. Polytrophic ovarioles have been described in Neuroptera, Lepidoptera, some Coleoptera, Diptera, and Hymenoptera.

Polytrophic and telotrophic ovarioles are sometimes collectively referred to as *meroistic ovarioles.*

In polytrophic and telotrophic ovarioles, the trophocytes furnish all or nearly all of the RNA contained in the mature egg. In the panoistic type, the oocyte nucleus produces all of the RNA in the mature egg. Much of the yolk protein originates in the fat body (as *vitellogenin*) and is transferred to the ovaries via the hemolymph (Hagedorn and Kunkel 1979). Yolk protein is called *vitellin.*

Figure 5.9   Ovariole types. (*a*) Panoistic. (*b*) Telotrophic. (*c*) Polytrophic.

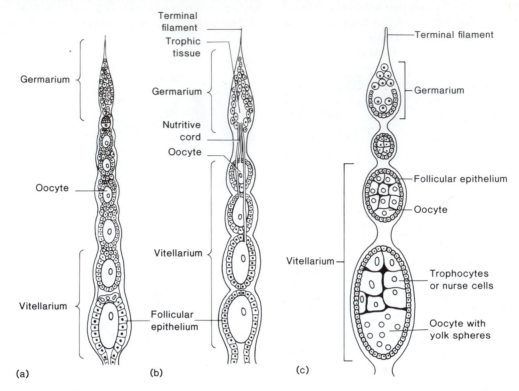

There are a few exceptional ovariole structures that do not fit into the previous classification. For example, in ovarioles from the beetle *Steraspis speciosa*, the germarium and vitellarium are very short, there are no nutritive cords or trophocytes, and there is a long region of glandular tissue proximal to the vitellarium (Engelmann 1970).

In the vast majority of insects, oogenesis occurs in the last larval instar, in the pupa, and in the adult stages. However, in some species immature stages are capable of producing mature oocytes that may commence and even complete embryogenesis. This phenomenon, *pedogenesis,* has been described in a number of insect species, among which are *Micromalthus debilis* (Coleoptera) and cecidomyiid flies in the genera *Miastor* and *Oligarces.*

See King (1974), volume 1 of King and Akai (1982) and volume 1 of Kerkut and Gilbert (1985) for more information on oogenesis and vitellogenesis. Invertebrates other than insects are covered by Adiyodi and Adiyodi (1983). A much broader coverage of these topics is given in Browder (1985). Browder's book treats oogenesis, which includes vitellogenesis, from a comparative viewpoint. Both invertebrates and vertebrates are covered. It must be emphasized that comparative approaches are extremely useful. A comparative study often provides new approaches or ideas to the problem at hand. In addition, comparative studies usually reveal new techniques or methods that may help solve a problem which, prior to this information, would remain unsolved.

Eggs

Mature insect eggs are typically elongate and oval in longitudinal section (figure 5.10*a*) although some assume other forms. In most instances the largest portion of an egg is filled with *yolk,* or *deutoplasm,* and the cytoplasm and nucleus occupy only a small portion. The yolk contains carbohydrates, protein, and lipid bodies, the protein bodies being the most abundant. Cytoplasm is located around the nucleus (*nuclear cytoplasm*) and around the periphery of the yolk (*periplasm,* or *cortex*). The egg may be encased by two layers: the *vitelline membrane* or *envelope*, and the *chorion,* or eggshell, which is secreted by the follicle cells. Several species lack a chorion (e.g., some viviparous species). When a chorion is present, it is nonchitinous, may be composed of two to several layers, varies considerably in thickness in different species, and may be smooth or sculptured in a variety of ways.

Because insect eggs are usually deposited outside the parent female, they are subject to a drying atmosphere, attacks by parasites and predators, and other dangers. The chorion serves the same functions as the cuticle. It serves as a protective coating (against physical damage, attack

Figure 5.10 (*a*) Sagittal section of a typical insect egg. (*b*) Scanning electron micrograph of an egg showing the five micropyles or holes at the apex. Note that sperm are located at five micropyles. The sculpturing on the eggshell is a result of the way in which species produce the chorion by the follicle cells.
After Harold R. Hagen, Embryology of the Viviparpus Insects. Copyright 1951 The Ronald Press Company, New York. Reprinted by permission of John Wiley and Sons, Inc., New York, NY. Courtesy of M. Mazzini.

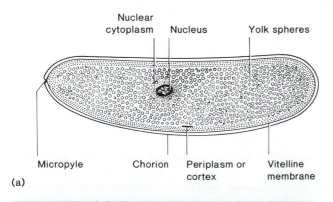

Nuclear cytoplasm   Nucleus   Yolk spheres

Micropyle   Chorion   Periplasm or cortex   Vitelline membrane

(a)

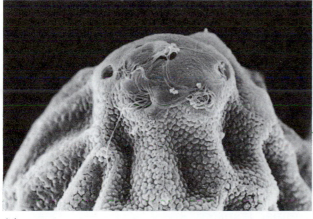

(b)

by parasites, etc.), a barrier against water loss and is important relative to ventilation of the egg. Coatings secreted by the accessory glands that cover eggs may aid in water conservation in addition to serving to hold eggs together or glue them to a surface. Some eggs laid in moist situations are capable of absorbing water from their surroundings. In insects in which the chorion is thin, ventilation may occur over the entire surface. In others the chorion may be lined with a porous, gas-filled layer that communicates with the outside of the egg by means of channels or *aeropyles*. Some eggs possess specialized structures that act as physical gills or plastrons, allowing them to obtain oxygen from water. Hinton (1969) reviews and synthesizes the major literature pertinent to respiratory systems of insect eggs. Miller (1974) also provides information on this topic. Spermatozoa gain entrance to an egg by means of one to several special channels or *micropyles* (figure 5.10*b*), which are perforations of the chorion and are located at various places on the eggs of different species.

Hinton (1979) provides a comprehensive three-volume treatise on insect eggs. Furneaux and Mackay (1976) consider the structure, formation, and composition of the chorion and vitelline membrane. King and Akai (1982) and volume 1 of Kerkut and Gilbert (1985) provide information on the biochemistry, ultrastructure, and molecular aspects of chorion formation.

Control of Oogenesis

The great diversity of insects is clearly expressed when one considers the different mechanisms (i.e., neural and/or hormonal) regulating oogenesis-vitellogenesis, in particular. The exact sequence of involvement and interaction of these events differs among species but generally involves four major control centers: head (i.e., neurosecretory cells), corpora allata (produces juvenile hormone), ovary (may produce ecdysone), and fat body (site of biosynthesis for vitellogenin). Recent research on midgut endocrine cells suggests that they may also be involved; however, definitive proof is still lacking. Some of the kinds of experiments and/or techniques that have enabled investigators to arrive at their conclusions are as follows:

1. Histological observations looking for correlations between neurosecretory cell(s) or corpora allata activity and ovarian development.
2. Ovarian activity being attenuated by the microsurgical removal of either the neurosecretory cells, corpora allata, or the ovaries and its restoration by tissue or gland implantation.
3. Allatectomy, either surgically or by using precocene (a plant derived chemical that specifically destroys the corpora allata) to attenuate ovarian activity followed by either hemolymph transfusion from a donor with active corpora allata or using the insect's own synthetic hormone or a juvenile hormone mimic.
4. Radiochemical analysis of the biosynthate produced by *in vitro* incubation of the corpora allata and/or ovary and their incorporation of the desired radiolabeled precursor into the biosynthate (JH or ecdysone).
5. Evaluation of the hemolymph or desired tissue extracts, using a radioimmunoassay for ecdysone.
6. ELISA (enzyme linked immuno-sorbent assay) to determine the concentration of vitellogenin in the hemolymph and/or the level of vitellin in the ovary (i.e., the eggs).
7. Using either a chromogen (color-producing) substance, a fluorescent material, or colloidal-size gold particles linked to an antibody that recognizes the antigen (i.e., either vitellogenin or vitellin).

It should be emphasized here that three major areas of biology are currently providing the driving force for new discoveries and breakthroughs in insect reproductive biology: immunology, computerization of optical instruments, and molecular biology. Breakthroughs using these techniques are also being used in other areas of entomology, for example, systematists are applying them in attempts to unravel problems at the species level of identification (see chapter 11).

The corpus allatum hormone may affect egg maturation by stimulating the incorporation of yolk into the oocyte and simultaneously regulating metabolism, particularly of proteins. In certain species (e.g., *Rhodnius* sp.) corpora allata from adults have been transplanted into larvae, and their secretions have had the same effect as the juvenile hormone. Conversely, larval corpora allata stimulate ovarian activity (gonadotropic effect) in adults.

Recently, certain substances (e.g., farnesol and farnesyl methyl ether) have been found that mimic the effect of the corpus allatum hormone, acting as "juvenile" hormone and having a gonadotropic action (i.e., stimulating the ovaries). In some species of insects allatectomy does not prevent egg maturation, but there is evidence that the corpus allatum hormone may be present in the hemolymph from earlier stages. In most species studied, secretions from the median neurosecretory cells in the pars intercerebralis in the brain are necessary for oogenesis. These secretions have been found to activate the corpora allata or stimulate protein synthesis (necessary for yolk formation) or produce a gonadotropic hormone. Recall that the corpora cardiaca serve as neurohaemal organs (see chapter 3), storing and releasing the products of some of the neurosecretory cells in the brain. This allows the accumulation of secretory products in the corpora cardiaca for some time prior to their release from this organ. Thus, the action of the median neurosecretory cells is a function of the release of their products from the corpora cardiaca. Such release of material is probably under neural control. The structure of the corpora allata and the role of juvenile hormone and ecdysone in reproduction are discussed in Tobe and Stay (1985) and in volumes 8 and 9 of Kerkut and Gilbert (1985).

Other factors, mediated hormonally or via the nervous system, that have been found to influence the activity of the corpora allata in different species include mating, light (photoperiod), chemical stimulation (pheromones from males), various nutritional factors, and the presence or absence of eggs in the brood chamber. These factors may have a stimulatory or inhibitory effect on the corpora allata.

In addition to secretions from the median neurosecretory cells and the corpora allata, ecdysone has been found to be involved in the control of oogenesis in female mosquitoes (Hagedorn and Kunkel 1979). In response to ingestion of a blood meal, secretions from the median neurosecretory cells (i.e., *egg development neurosecretory hormone*) induce the ovaries to release ecdysone,

which in turn triggers the synthesis of vitellogenin in the fat body (figure 5.11). Juvenile hormone secreted by the corpora allata "activates" the fat body and ovaries. Recent results, using some of the seven techniques listed, have shown that in the blow fly, *Phormia regina,* neither juvenile hormone nor ecdysone is produced or released into the hemolymph of sugar-fed flies in significant quantities (i.e., the critical titer for these hormones was not achieved). Also, it was shown that ecdysone produced by the ovary stimulates the fat body to produce vitellogenin, whereas the major role of juvenile hormone is in controlling the uptake of vitellogenin by the developing egg. In sugar-fed females, the follicle cells remain closely appressed and do not permit the vitellogenin molecules to enter the oocyte (figure 5.8*a*). In protein-fed females, however, juvenile hormone somehow is involved in creating the spaces between the follicle cells (figure 5.8*b*) and also in the development of the receptor mediated endocytotic apparatus involved in vitellogenin uptake. Thus, this fly is *anautogenous* (i.e., requiring an exogenous source of protein for oogenesis) rather than *autogenous* (i.e., being able to develop at least one batch of eggs without a protein meal).

Several factors probably interact in the control of oogenesis in any given insect species, and the present level of understanding does not permit broad generalizations. The reviews of Davey (1965), de Wilde and de Loof (1973b), Engelmann (1968), Riddiford and Truman (1978), Telfer (1965), and Raabe (1986) should be consulted for detailed information. Bownes (1986) covers the mechanisms influencing genes that code for vitellogenin production.

Seminal Transfer, Fertilization, and Sex Determination

Seminal Transfer

Internal fertilization, which prevents exposure of gametes to the drying atmosphere, is looked upon as one of the several prerequisites for the evolution of animal life in the terrestrial environment. Internal fertilization in insects is generally brought about by the act of copulation, during which time the *semen* (spermatozoa plus various glandular secretions) produced in the male reproductive system is transferred to an appropriate site in the female reproductive system, that is, by *insemination*. External genital structures are discussed in chapter 2 and copulatory behavior in chapter 8. Suffice it to say at this point that both genitalia and copulatory behavior vary greatly among the different groups of insects, but their common "goal," seminal transfer, is the same. Seminal transfer may involve the passage of either free semen or, in many insects, one or more spermatophores from the male to the female. The involvement of a spermatophore is considered to be the more primitive situation (Hinton 1964).

Figure 5.11 Control of oogenesis in the mosquito, *Aedes aegypti*. A "competent" fat body is able to produce vitellogenin when stimulated by α-ecdysone; a "resting" ovary is ready to secrete α-ecdysone and undergo maturation when stimulated by egg development neurosecretory hormone.

Modified from Riddiford and Truman, 1978; based on Hagedorn, 1974, and Flanagan and Hagedorn, 1977.

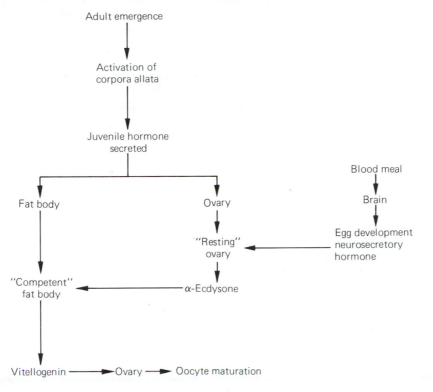

Free semen is usually deposited in the bursa copulatrix or vagina, but in some species it may be deposited in the common oviduct, in the lateral oviducts, or even directly into the spermatheca. Male dragonflies and damselflies deposit semen in a specialized organ on the venter of the second abdominal sternite. A portion of that organ is then placed in the female's vagina, and seminal transfer is accomplished. Many species in the superfamily Cimicoidea (e.g., bed bugs; family Cimicidae) have a rather bizarre method of seminal transfer. The males of these species inseminate the females by perforating the integument at a specialized site in their abdomen and ejaculating semen directly into the hemocoel. This method of semen introduction has been called *hemocoelic insemination,* or *traumatic insemination.* The seminal fluid and many of the spermatozoa are digested by the female, while some of the spermatozoa reach the ovaries. The adaptive value of this method of insemination is thought to be that it provides the female with additional nutritive substance (Hinton 1964).

Spermatophores are usually deposited by the male somewhere in the female reproductive system: the bursa copulatrix, vagina, or, rarely, the spermatheca. However, in the apterygotes (e.g., Thysanura), the male deposits a spermatophore on the substrate and the female picks it up and deposits it within herself. This method of indirect sperm transfer without copulation also occurs among the Diplura, Collembola, and in several noninsect arthropod groups (Schaller 1971). Gerber (1970) considers the methods of spermatophore formation in pterygotes.

Once a spermatophore is in the female reproductive system, various mechanisms may account for the release of the spermatozoa. In most insects the spermatozoa are either forced out by pressure applied by the female or by the mechanical perforation of the spermatophore. In the house cricket there is a significant difference between the osmotic pressure of a block of protein or pressure body within the spermatophore and the osmotic pressure of the evacuating fluid, which surrounds the spermatophore at the time of ejaculation. The evacuating fluid is absorbed by the protein body, which swells up and forces the exit of the spermatozoa (Hinton 1964). In some insects the spermatophore may be digested away, causing the release of the spermatozoa. After the spermatozoa have been released, the spermatophore is probably digested and absorbed by the female and thus may have some nutritional significance.

After release from the spermatophore or deposition in the form of free semen, the spermatozoa move to the spermatheca. This is apparently due to muscular contractions of the female ducts but may also be dependent upon sperm motility. In the honey bee the spermathecal duct has a "pumping" structure, which seemingly transports the sperm to the receptacle following copulation and subsequently releases them when it is time for an egg to be fertilized. In some insects the male's accessory gland fluid forces the sperm up the female's spermathecal ducts and into the spermathecae.

Multiple Matings and Sperm Precedence

As in many other organisms, insects generally have two sexes—males and females. Because of space constraints, it is impossible to cover the exciting controversies surrounding the evolution of sex (Stearns 1987; Michod and Levin 1988). Currently research emphasis is on the topics of sexual selection, reproductive competition, and the evolution of insect-mating systems (see chapter 8). More information can be obtained by reading Blum and Blum (1979), Thornhill and Alcock (1983), and Smith (1984). As one can imagine, insects—especially *Drosophila*—serve as the ideal models of choice to answer questions in these areas because of ease of rearing, short generation time, extreme diversity of mating systems, and, most important, because of our knowledge of their genetics.

Since the origin of insects, evolution has shaped mating strategies of the two major interest groups: males and females. Why do some female insects mate only once, whereas others may accept several matings from different males? Is there a benefit to the female to mate with several different males (multiple matings)? Some believe there is, and suggest that it takes the form of "investments" by the male to the female or "parental investment" by the male to developing embryos. Because it is in the best interest of the male to pass on his sperm, how do males ensure that their sperm fertilizes the eggs? We have already explained that in some species the females are rendered unreceptive to other mating attempts by the transfer of accessory gland fluid (i.e., small peptides) to the female. Other tactics to prevent multiple matings will be covered in chapter 8.

Research on multiply mated females, using radiation sterilized males, genetic markers, or parental identification using isozyme markers, has clearly shown that the sperm of the last male to mate (i.e., sperm precedence) generally succeeds in fertilization (Michiels and Dhondt 1988). Physiologically speaking, almost nothing is known about how this mechanism works. The older literature often uses the term *sperm competition,* which has now been replaced by the term *sperm precedence.* This change in terminology probably developed because of the inability of investigators to account for the mechanism of sperm competition. For more detailed information, consult Parker (1970), Ridley (1988), or Cheng and Siegel (1990).

How and Why Questions

Alcock (1975) notes that ". . . all biologically significant questions can be placed in two categories: how and why." How does the last male to mate ensure that his sperm fertilizes the eggs? The how questions address issues of proximate cause and usually have a physiological and/or genetic and developmental bases. Why then have males evolved mechanisms to ensure success of their sperm? Surely, the evolutionary bases or ultimate reasons for sperm precedence is to ensure the male that his genetic complement is passed on to the next generation. Students will find that researchers are generally involved in answering only one of these questions (i.e., how or why). Also, the way investigators approach their research is generally channeled the same way. Physiologists and biochemists are usually interested in the how questions, whereas behaviorists, ecologists, and evolutionary biologists are usually interested in the why questions. In fact, these two groups often study the same problem, approach it from different angles, and seldom talk to one another or make references to one another's research.

Male Contributions

In addition to passing on either a spermatophore, which is the proteinaceous package or housing for the sperm, or free sperm, males are reported to pass on other materials believed to provide nutrients for either the female or the developing embryos. Males of some species provide the females with nuptial gifts (e.g., in the balloon flies) that the female feeds on, thus improving her fecundity. The females of some species eat the spermatophores. In other cases, the female may feed on nonreproductive secretions (i.e., salivary gland or paranotal gland secretions) produced by the male. Recall the discussion concerning the fact that cockroaches do not excrete uric acid but store it in the accessory glands (see chapter 4). During mating this material is passed either onto the female or her eggs and is believed to provide either the female or the eggs with an extra source of nitrogen. The ecological correlates of paternal investment in cockroaches is discussed by Schal and Bell (1982).

Multiple mating is one way researchers believe females acquire additional nutritional resources. Otherwise, why remate? In addition to cockroaches, what other evidence shows that materials are transferred during copulation and that they are used as a nutrient source? Here we rely mainly on research on the Lepidoptera (and some on Orthoptera) and the major studies that used radiolabeled materials from the male. One major word of caution, however, is that there is no solid evidence showing that these materials are incorporated into the female's biochemical processes. Also, when calculated out, the amount of material transferred, in comparison to the amount of material in each egg, is almost insignificant and one questions the nutritional value (Kaulenas 1992). Parently investment is covered by Smith (1984) and Thornhill and Alcock (1983).

Fertilization

The processes involved in fertilization may be divided into three parts.

1. Release of spermatozoa from the spermatheca.
2. Entry of the egg by spermatozoa.
3. Formation and fusion of the male and female pronuclei.

As mentioned earlier, spermatozoa are stored in the spermatheca of the female until it is time for fertilization. The females of many insect species (e.g., *Apis,* honey bee; *Glossina,* tsetse fly; and *Rhodnius* and *Triatoma,* conenose bugs) mate only at a single time during their lives with one or more males, and the spermatozoa introduced at that time are stored and used to fertilize their eggs for the rest of the reproductive period. Spermatozoa stored in this fashion may survive for several months or years. Other insect species are inseminated periodically throughout their reproductive lives, and in these species the storage of spermatozoa in the spermatheca may only be for a short period of time.

The mechanisms for release of spermatozoa from the spermatheca are not clearly understood. Stimulation of sensory hairs by the passage of eggs in the oviducts, hemocoelic pressure forcing the exit of the sperm, and the inherent motility of spermatozoa that have been "activated" by some secretion have all been advanced as possible mechanisms in various species. Studies are needed to determine how sperm release is controlled by the female. These may show that the spermathecal ducts of many species are surrounded by muscles and that these muscles are regulated by neurosecretions from the neurosecretory cells embedded in these muscles. Davey (1985) notes that mechanisms which regulate the amount of sperm released for each egg have evolved, thus preventing sperm wastage. How the females accomplish this is not known. Many female Hymenoptera regulate the release of sperm to the eggs, thus controlling the sex ratio of their offspring. Because unfertilized eggs of several insect orders develop into males, whereas fertilized eggs develop into females, the females can control the sex ratio by regulating whether sperm is or is not released. How females evaluate the situation and make a decision to alter sex ratios is not clearly understood. Information about how this is accomplished may have important consequences in implementing control strategies using various hymenopterous (i.e., parasitic) biological control agents.

Following ovulation, the egg is oriented in the reproductive passage in such a way that the micropylar region is in rough proximity to the site of sperm release. The sperm migrate to the micropylar region of the egg, possibly responding chemotactically, and one or more enter the egg through the micropyle (figure 5.10). In the vast majority of insects more than one spermatozoon enters the egg, but usually only one fuses with the egg pronucleus. Excess sperm usually degenerate without disrupting the development of the zygote. Whether ovulation (i.e., the passage of the egg from the ovary into the oviducts) is regulated by the same factor(s) controlling ovipositioning (i.e., passage of egg to outside) is not known. In some insects these two events are closely linked and occur almost simultaneously. Are there insects where these two events are widely separated in time? The importance of a factor (i.e., probably a peptide), derived from the male accessory glands, and its effect on female readiness to mate has already been discussed. For the female, however, the involvement of the brain (i.e., especially neurosecretory products) in regulating these two events has been demonstrated for several insects (Davey 1985). Even though ovulation may be regulated by hemolymph-borne factors, oviposition usually is a complex behavior involving muscle coordination. Such an event surely involves the nervous system.

Shortly following the entry of sperm into the egg, the egg nucleus undergoes meiotic division, forming the female pronucleus (figure 5.12*a*). The spermatozoon that will fuse with this female pronucleus loses its tail, becoming the male pronucleus. The fusion of the two pronuclei forms the *zygote* and signals the commencement of morphogenesis.

Sex Determination and Parthenogenesis

Nearly all insects are bisexual; that is, male sex organs occur in one individual, and the female sex organs are in another individual. Several species in different groups are capable of reproducing *parthenogenetically* (i.e., they are able to produce individuals from unfertilized eggs). Insects in which both male and female sex organs occur in the same individual are said to be *hermaphroditic.* The cottony cushion scale (*Icerya purchasi*) and one or two relatives are the only insects in which true hermaphroditism has been established. Reproduction in these hermaphroditic insects usually occurs by self-fertilization.

Sex determination in bisexual insects is considered to depend upon a balance between genes for maleness and genes for femaleness. This balance is in most forms tipped in the direction of one sex or the other by a sex-chromosome mechanism in which one sex possesses two X (sex) chromosomes (i.e., the *homogametic sex,* XX) and the other a single X chromosome (X0) or a single X

Figure 5.12 Fertilization—germ band formation (chorion omitted; (*a*–*e*) sagittal sections). (*a*) Just prior to fertilization (at the stage of fusion of spermatozoon and female pronucleus). (*b*) Cleavage. (*c*) Blastoderm forming and germ cells differentiating. (*d*) Blastoderm formation complete. (*e*) Germ band and primary dorsal organ formation. (*f*) Cross section at same stage as (*e*). The location of the section is indicated in (*e*) by the line x—x.
Redrawn from Johannsen and Butt, 1941.

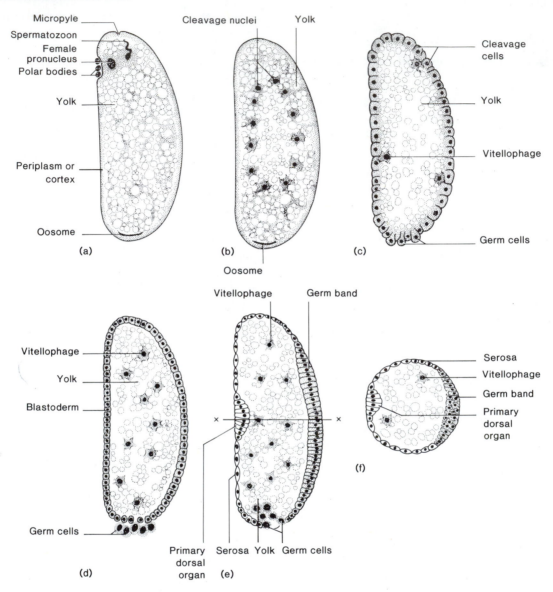

chromosome plus a smaller Y chromosome (XY). Individuals possessing the X0 or XY configurations comprise the *heterogametic sex*. In most insectan groups the males are heterogametic and the females homogametic. However, the reverse is true in the Lepidoptera and Trichoptera. In some insects (e.g., Hymenoptera, Thysanoptera, and certain Homopterous insects) males develop from unfertilized eggs, females from fertilized eggs. Recall that several species of Hymenoptera have been shown to regulate the sex ratios. How the females do this is not known. Also, whether this occurs in Thysanoptera and Homoptera remains to be demonstrated. White (1964), Bergerard (1972), and Raabe (1986) discuss sex determination in insects.

In some instances environmental factors have been shown to exert an influence on sex determination. For example, in the butterfly *Talaeporia* sp., more males than females are produced at high temperatures; the converse is true at low temperatures (Davey 1965).

Since extirpation of gonads or implantation of gonads into a previously castrated individual produces no effects, it is generally assumed that secondary sexual characters

Figure 5.13 An overall schematic view of the varied reproductive events and their regulation.
Taken from Raabe (1986).

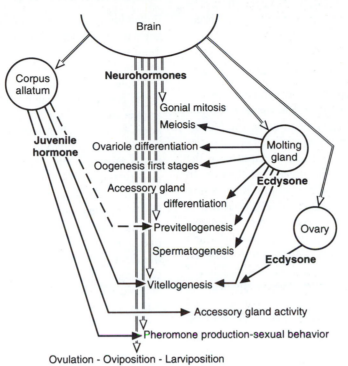

are not determined by humoral secretions associated with the gonads. However, in the beetle *Lampyris noctiluca*, there is evidence that special secretory tissue associated with the testis rudiment controls the development of male structures, and implants of this tissue masculinize females. The appearance of this secretory tissue is determined by the presence of active neuroendocrine cells in the brain (Naisse 1966, 1969).

Apparently every cell in the insect body is involved in sex determination, as is evident in the occasional occurrence of *gynandromorphs*. These individuals are literally sexual mosaics, some parts of the body possessing typically male traits and other parts typically female traits. This phenomenon is explained by differences in the sex chromosomes in the cells comprising the various tissues (i.e., some cells are "male" and others are "female"). These differences are known to occur by a number of mechanisms. One such mechanism is the loss of one X chromosome in the cleavage cells of a female (XX becoming X0, e.g., in *Drosophila,* vinegar or fruit flies). These cells thereafter give rise to male traits in whatever tissue they happen to form. Another mechanism occurs in the honey bee, in which the fusion nucleus and an extra sperm that has entered the egg both undergo cleavage. The cells that result from cleavage of the fusion nucleus (diploid) give rise to female traits, and those from the sperm nucleus (haploid) produce male traits.

Overall Integration of the Reproductive System

In order to coordinate the various control centers and the events they regulate, both the nervous and endocrine systems are involved. The major centers include the neurosecretory cells of the brain, the production of juvenile hormone by the corpora allata, and either ecdysone production by the molting or ecdysial gland in immature insects or ecdysone biosynthesis by the ovary in adult insects (figure 5.13). Both hormones may act either independently or together to produce their effect. Not discussed here are the potential effects of nutrition and environmental factors (i.e., especially photoperiod and temperature) on the overall process of oogenesis.

Embryogenesis

Embryogenesis includes those developmental events that occur between the formation of the zygote and the exit of the fully developed individual from the egg (*eclosion*). The changes that occur following hatching from the egg will be included in the discussion of postembryonic morphogenesis. *Morphogenesis* comprises all developmental events that occur between the formation of the zygote and the emergence of a sexually mature adult. Although there is much variation in detail from group to group, a general account of insect embryogenesis is possible.

Anderson (1973, 1979), Counce and Waddington (1972, 1973), Hagen (1951), Johannsen and Butt (1941), Sharov (1966), Snodgrass (1935), and Weygoldt (1979) are recommended for further reading in the area of insect embryogenesis.

Formation of the Blastoderm and Germ Cells

The first distinct layer of cells to form during the embryogenesis of any metazoan animal is the *blastoderm*, which is composed of a single layer of cells, the *blastomeres*. The means by which this layer of cells arises varies in different kinds of animals and is correlated with the quantity of yolk material initially present in the egg. In animals with little yolk material the zygote divides into two equal parts; each of these daughter cells divides into two equal parts; and so on. This is the process of *cleavage* and eventually leads to a ball of cells, the *morula*, which subsequently develops an internal cavity, the *blastocoel*, surrounded by the blastoderm. Because in this method of blastoderm development the entire zygote divides, the term *holoblastic* (holo = whole; blast = bud or sprout) *cleavage* is applied. However, in the vast majority of insects—Collembola (springtails) and certain parasitic Hymenoptera being the most outstanding exceptions—the eggs have large quantities of yolk. In most insects the fusion nucleus, with associated cytoplasm, behaves as though it were an individual cell and proliferates mitotically (figure 5.12b). The daughter (cleavage) nuclei eventually migrate to the periphery of the egg to the region known as the cortex (i.e., a thin region surrounding the egg and generally devoid of yolk) and there form the blastoderm (figure 5.12c,d). During the course of this process, the individual cleavage cells develop cell membranes. This form of cleavage is referred to as *meroblastic* (mero = part) *cleavage*. The eggs of some arthropods and a few insects (e.g., Collembola) undergo total cleavage initially and subsequently peripheral cleavage in the formation of the blastoderm. This is called *combination cleavage*.

During meroblastic cleavage, some of the cleavage cells remain in the yolk or return to it after reaching the periphery of the egg. These are the *vitellophages* (yolk cells; figure 5.12c,d) and are considered to be responsible for the initial digestion of the yolk, making it more readily assimilable by the other embryonic cells.

At the same time the blastoderm forms, some of the cleavage cells differentiate into *germ cells* (figure 5.12c–e), which will give rise to gametes or reproductive cells in the late larval, pupal, and adult stages. In many embryos the differentiation of germ cells is correlated with the passage of cleavage cells through a specialized region of the egg called the *oosome*.

Formation of the Germ Band and Extraembryonic Membranes

Following the completion of the blastoderm, the cells on one side of the egg become columnar along the longitudinal midline of the egg (figure 5.12e,f). In a lateral direction from this midline, the cells become successively less columnar, finally merging with the remaining cells of the blastoderm, which tend to become squamous. This thickened area of columnar cells of the blastoderm is the *germ band*, which subsequently elongates and develops into the embryo, while the remaining cells take part in the formation of the extraembryonic membranes. In most insects, folds from the area outside the germ band grow over the germ band (i.e., overgrowth; figure 5.14a), eventually meeting along the longitudinal midline. The inner and outer layers of one fold then merge with the respective layers of the other, forming an inner *amnion* around the developing embryo and an outer *serosa* around the yolk, amnion, and embryo. In some insects the extraembryonic membranes may form by invagination or involution of the embryo instead of overgrowth (figure 5.14b,c). The first method is found in the Apterygota; the second occurs among the orders Odonata, some Orthoptera, and Homoptera. As the extraembryonic membranes are forming, the germ cells become located at what will be the posterior end of the embryo.

As the germ band forms in some insects, particularly in Apterygota, but also in some Pterygota, a cluster of cells form the primary dorsal organ (figure 5.12e,f). This structure disappears at dorsal closure and may be glandular in function.

Differentiation of the Germ Layers

In animals with little yolk material and in which a blastocoel develops, an invagination eventually occurs that subsequently differentiates into well-defined mesodermal and endodermal layers. These three germ layers (*ectoderm, mesoderm,* and *endoderm*) then differentiate into the various tissues and organs of the fully developed organism. The formation of the mesoderm and endoderm is referred to as *gastrulation*. Unfortunately, this process in insects is not straightforward and clear. Unequivocal interpretation of insect morphogenesis in terms of this *germ-layer theory* has proved to be extremely difficult; insect embryologists hold widely divergent opinions. Fox and Fox (1964) point out that there is much to be said for not attempting to distinguish between mesoderm and endoderm in the insect embryo but for simply recognizing an outer ectoderm and inner layers. For convenience of discussion, we shall label those cells that give rise to the midgut as endodermal in origin, keeping the aforementioned qualifications in mind.

Figure 5.14 Formation of extraembryonic membranes (diagrammatic).

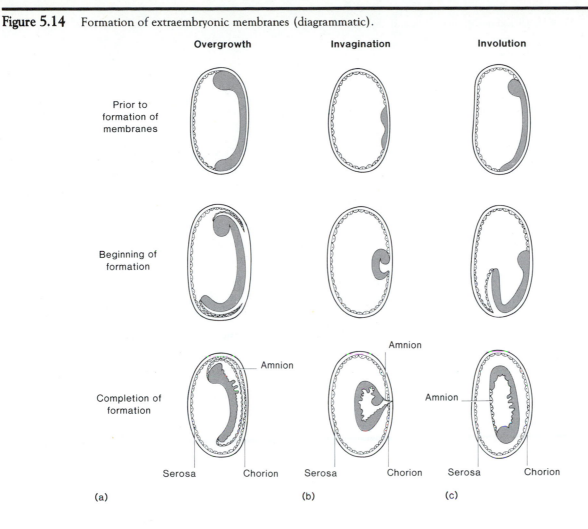

Gastrulation in most insects occurs as the amnion and serosa are forming. It begins, in most species, as a longitudinal, furrowlike invagination (figure 5.15a–c), which runs most of the length of the venter of the germ band. Eventually the invagination flattens out and the outer edges come together and fuse, forming a longitudinal band (inner layer, or *mesentoderm*) of cells surrounded by an outer layer, which at this point is appropriately referred to as *ectoderm*. Another type of inner layer formation (figure 5.15d,e) consists of a ventral longitudinal band of cells in the germ band sinking into the yolk and being overgrown by the remaining cells of the germ band. In a third and less common method of inner layer formation, a longitudinal band of cells proliferates from the germ band (figure 5.15f). Whatever the method of formation, the end result is essentially the same (an outer or ectodermal layer and an inner or mesentodermal layer). Eventually the inner layer differentiates into two lateral longitudinal bands (*mesoderm*) and a median strand with cell masses located at its anterior and posterior ends

(figure 5.16). For our purposes the median strand with its anterior and posterior cell masses will comprise the *endoderm*.

Segmentation, Appendage Formation, and Blastokinesis

Segmentation of the embryo begins very soon after the germ band has formed and originates in the mesoderm. Segmentation later becomes quite evident in nearly all the organs of mesodermal origin (dorsal vessel, muscles, etc.) as well as those of ectodermal origin (nervous system, tracheal system, etc.). The endoderm and the portions of the ectoderm that give rise to the foregut and the hindgut are unaffected by the segmentation process.

As segmentation proceeds, transverse furrows become externally evident in the ectoderm (figure 5.17). Soon after segmentation begins, bilateral evaginations of the ectoderm, which contain mesodermal tissue, begin to appear. These will form the various body appendages.

Figure 5.15   Germ-layer differentiation (yolk and chorion omitted). (*a–c*) Invagination. (*d*) and (*e*) Overgrowth. (*f*) Delamination.

Redrawn with modifications from Snodgrass, 1935.

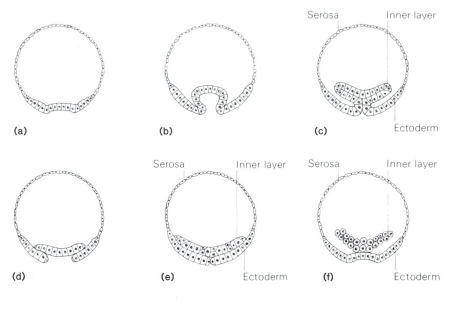

(a)   (b)   (c)

(d)   (e)   (f)

Figure 5.16   Sagittal section of an embryo (diagrammatic).

Redrawn from Snodgrass, 1935.

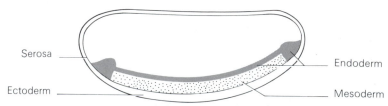

Initially the germ band (figure 5.17*a*) is composed only of the *protocephalon,* which is bilaterally expanded, and a narrow lobe extending posteriorly, from which the remaining body segments will develop. When the segmentation of the embryo is essentially complete (figure 5.17*b–d*) and all appendage rudiments have formed, the portions of the embryo that will form the three tagmata of the insect body can be discerned.

The protocephalon, the antennal and intercalary segments, and the following three segments with their paired appendages will form the definitive head. The antennal and intercalary segments usually cannot be clearly separated, but the next three segments, the appendages of which will form the mandibles, maxillae, and labium, respectively, are easily outlined. These last three segments comprise the *gnathal segments.* The region of the developing head anterior to the gnathal segments has been interpreted in several different ways regarding the true number of primitive segments involved. Rempel (1975) discusses the various interpretations and makes a strong case for the insect head having arisen from the amalgamation of an anterior body cap (*acron,* or *prostomium*) plus three segments (*labral, antennal,* and *intercalary*) plus three gnathal segments (see figure 11.13). This topic is considered further in chapter 11.

The three segments posterior to the gnathal segments will form the definitive thorax, and their appendages will form the thoracic legs. The remaining segments, which never number more than twelve, including the posterior telson, will compose the definitive abdomen. Pairs of limb buds appear on the abdominal segments, but the first seven pairs and the tenth pair are resorbed. Those on segments eight and nine form the external genitalia, and those on the eleventh segment form the cerci. As mentioned in chapter 2, the tendency in insect evolution has been toward a reduction in the number of abdominal

Figure 5.17   Segmentation and appendage formation. (*a*) Prior to segmentation. (*b*) and (*c*) Successive stages. (*d*) Segmentation complete.

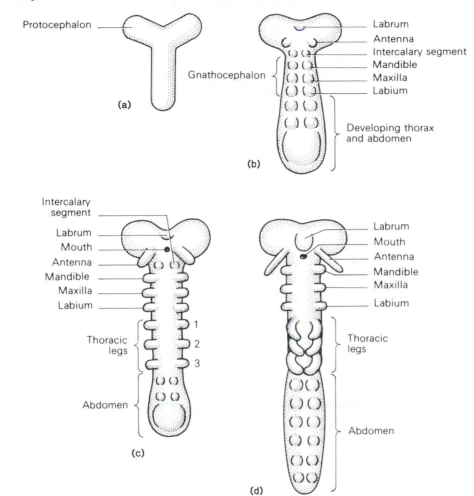

segments, the most primitive contemporary forms having only eleven, with the eleventh containing the anus, and the more advanced forms often having fewer than this number.

Blastokinesis is a term used to denote "all displacements, rotations, or revolutions of the embryo within the egg" (Johannsen and Butt 1941). These movements occur in a variety of predictable ways characteristic for each species. They may represent an adaptation to the large quantity of yolk characteristic of the vast majority of insect eggs and enable the embryo to make the most efficient use of the yolk material. These movements probably also result in protection from desiccation by causing the embryo to be surrounded by yolk.

Organogenesis

Subsequent to the formation of the three germ layers, each undergoes further differentiation, eventually forming the various tissues and organs of the fully developed embryo.

Fate of the Mesoderm

As segmentation occurs, the intersegmental portions of the mesoderm either become very thin or separate altogether. Eventually each segmental mesodermal layer develops two lateral lumens (figure 5.18*a,b*), the *coelomic sacs,* and finally the mesoderm differentiates segmentally into the following parts (figure 5.18*c*).

1. The *splanchnic,* or *visceral layer.*
2. The *somatic layer.*
3. The fat body.
4. *Cardioblasts.*
5. The *genital ridge.*

As development progresses, the splanchnic layer comes to form the visceral muscles, and the somatic layer, the skeletal muscles. At dorsal closure, to be discussed briefly later, the cardioblasts form the heart portion of the dorsal vessel. The aorta arises from the median walls of the antennal coelomic sacs (Johannsen and Butt 1941). The genital ridges are later suppressed in all but the eighth

Figure 5.18 Cross section of an insect embryo at successive stages of differentiation (chorion and most of yolk omitted); serosa omitted in (*b*) and (*c*).

Redrawn from Johannsen and Butt, 1941.

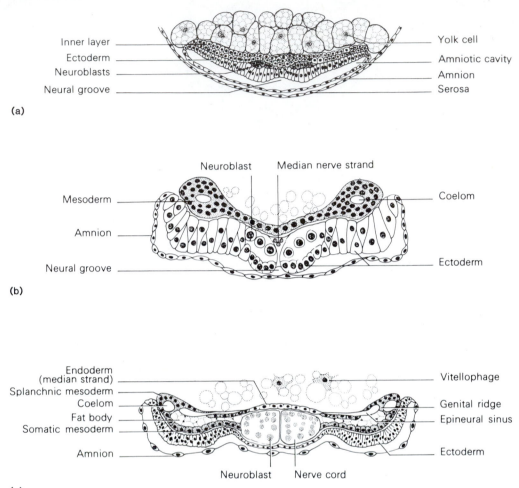

Fate of the Ectoderm and Endoderm

The alimentary canal begins its development as two invaginations of the ectoderm, one in the cephalic region (*stomodael*) and one in the last abdominal segment (*proctodael*; figure 5.19*a*). These two invaginations subsequently form the foregut (*stomodaeum*) and hindgut (*proctodaeum*), respectively (figure 5.19*b*). The midgut (*mesenteron*) arises from the proliferation of the cell masses at each end of the median strand of endoderm (figure 5.16). Some of the cells freed from the middle strand may also contribute to midgut formation in some species. The endodermal cells eventually envelop the remaining yolk material (figure 5.19*b*). The alimentary canal is completed when the membranes between the stomodaeum, mesenteron, and proctodaeum perforate. The Malpighian tubules develop as evaginations of the proctodaeum immediately posterior to the mesenteron.

The brain, subesophageal ganglion, segmental ganglia, and associated paired nerve cords are of ectodermal origin. The earliest evidence of their development is the presence of a longitudinal neural groove in the ectoderm and the differentiation of *neuroblasts* (figure 5.18). The neuroblasts subsequently give rise to the ganglia, nerve cord, and associated nerves. The ganglia and nerves of the stomatogastric nervous system develop from neuroblasts, which arise from the stomodaeum.

All the ventilatory structures of insects are of ectodermal origin. Spiracles, tracheae, and tracheoles arise

and ninth abdominal segments. In these segments, the germ cells, which, you will recall, differentiated while the blastoderm was forming, migrate into the genital ridges and together with these mesodermal cells form the *anlage* (precursors) of the gonads.

Figure 5.19 Sagittal sections of an insect embryo at successive stages in the development of the alimentary canal.
Redrawn from Johannsen and Butt, 1941.

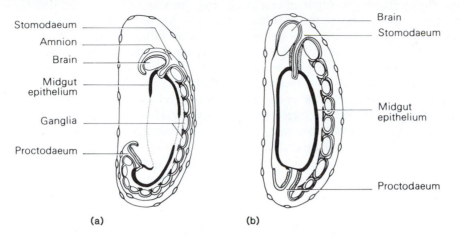

(a)   (b)

as invaginations of the ectoderm and eventually secrete a form of cuticle; the various "gill" structures develop as evaginations.

Oenocytes arise segmentally in the abdomen from the ectoderm.

Dorsal Closure and the Definitive Body Cavity

As the embryo develops, it spreads over the yolk material, and the extraembryonic area becomes smaller and smaller. The extraembryonic membranes, the amnion and serosa, usually disappear before the embryo has completed its development. Generally, they fuse at some point ventrally and a longitudinal cleft forms. The resultant folds, which consist of amnion fused to serosa, are drawn back over the embryo and the serosa condenses, forming the *secondary dorsal organ*. Eventually the dorsal organ sinks into the remaining yolk material, and the amnion provides a temporary dorsal closure. Subsequently, the ectoderm grows over the dorsal organ, which is detached and absorbed into the yolk. The ectoderm effects the final dorsal closure.

As should be evident from the preceding discussion, the definitive body cavity of insects is not a true coelom. The body cavity develops from the blastocoel, which is invaded by mesodermal tissue (Snodgrass 1935). Thus, as explained in chapter 3, the body cavity is correctly referred to as a hemocoel or blood cavity.

Polyembryony

Among several groups of parasitic Hymenoptera (e.g., Chalcidoidea, Proctotrupidae, Vespoidea, Braconidae, Ichneumonidae) and members of the Strepsiptera, a developing egg may divide mitotically and produce several embryos. This asexual multiplication of individuals is

called *polyembryony*. This phenomenon in the Hymenoptera is considered to be an adaptation associated with parasitism, whereas in the Strepsiptera, it is associated with ovoviviparity (i.e., "live birth"; offspring "born" as larvae instead of eggs). In the Strepsiptera the eggs develop in the mother's body cavity.

The eggs and embryogenesis in polyembryonic insects differ from those of other insects as follows:

1. The eggs are extremely small.
2. There is no yolk.
3. The chorion, if present, is very thin and permeable.
4. Cleavage is holoblastic.

These differences are not surprising when one compares the environments of polyembryonic and most other insect eggs.

Ivanova-Kasas (1972) reviews the literature on polyembryony in insects.

Control of Embryogenesis

The insect egg is a highly evolved, immobile package designed to withstand desiccation and other environmental insults. At the same time, it is extremely vulnerable to predation and parasitization. Thus, there is usually a premium placed on rapid embryonic development. In order to achieve this, however, the machinery for protein synthesis (i.e., ribosomes), the stored reserves (yolk), and the instructions (genetic information) necessary for directing and orchestrating this complex event must be in place. Some of the contributions to this package are maternal in origin (e.g., maternal genome); the remainder is produced within the egg by the zygotic genome (e.g., germ band and embryo).

Two major components of embryogenesis are *cellular differentiation* and *morphogenesis*. Cellular differentiation occurs when cells that are uncommitted become committed or specialized with respect to their ultimate fate. Thus, an undifferentiated cell keeps its option open and can become one of several different types of tissues when the time comes, whereas a differentiated cell has made the commitment to become only one specialized tissue type. Morphogenesis is the process in which primordial cells (i.e., undifferentiated) become increasingly differentiated in a stepwise fashion, finally becoming the specialized parts that comprise the total functional organism. Morphogenesis includes pattern formation, creation of body shape and appendage formation, plus the various morphogenetic movements that help shape the final embryo.

Experiments that have elucidated the control mechanisms of embryogenesis have been based on the removal, destruction, disruption, or separation of various parts of the egg or embryo at different times during development. Techniques used in these experiments include the use of microsurgery, ultraviolet light, X ray, cautery, and ligation. More recent techniques use various mutants; microsurgical lasers; and numerous molecular techniques such as visual molecular probes, cloning and labeling various transcripts, and nuclear transplantation techniques. Sander, Gutzeit, and Jackle (1985) and Weaver and Hedrich (1989) discuss modern techniques used to study insect embryogenesis.

Recent reviews pertaining to control of embryogenesis are those of Agrell and Lundquist (1973), Chen (1971), Counce (1973), Sander (1976), and Schwalm (1988). *Milestones in Developmental Physiology of Insects* (Bodenstein 1971) contains reprints of several classic papers in the control of embryogenesis.

Embryogenesis is a developmental process involving a symphony of gene action in which parts of the genome (whether maternal or zygotic) are expressed or repressed. The types of genes controlling embryogenesis fall into three major categories: *maternal effect genes, segmentation genes,* and *homeotic genes.* Experiments have shown that, prior to germ band formation, the pre-embryonic events are controlled by maternal mRNA that is either produced by gene amplification of the egg oocyte nucleus, as in the case of insects with panoistic ovarioles, or by the maternal contribution from nurse cells found in insects with meroistic ovarioles (Berry 1982). Protein synthesis, yolk utilization, and respiration are very low during pre-embryonic development. Two important questions concerning pre-embryonic development are: How is the egg compartmentalized? and What mechanism(s) accounts for the localization and/or movements of nuclei and cells during this period?

The egg package is basically divided into two major compartments: peripheral region or cortical cytoplasm and central region or yolky cytoplasm. Studies have recently shown that the cortical cytoplasm (often called periplasm) contains the major portion of the maternal mRNA. The female products that are contributed to the egg, such as rRNA, mitochondria, and ribosomes, are mainly located in the yolky cytoplasm.

Currently, it is believed that maternal effect genes produce *cytoplasmic determinants,* or chemicals that can act at a region or center of highest concentration, and that these chemicals also produce gradients that can influence cells in either adjacent areas or even cells at a distance. How these chemicals influence gene expression or repression is not known.

Cleavage and migration of the nuclei and their accompanying islands of cytoplasm begin at the *cleavage center,* which is located in the region where the future head will form. The earliest effect of the cytoplasmic determinants on cleavage cells occurs at the *activation center,* which is initially located near the posterior pole of the egg. The interaction of the cleavage cells and the activation center results in a change that initiates further growth and development. The effects of this change then proceed anteriorly. Blocking the effects of the cleavage center by removal or exposure to ultraviolet light results in a failure of germ-band formation. Further developmental direction comes from the *differentiation center,* located near the middle of the presumptive germ band, at which point the prothorax will eventually form. From this center all subsequent changes are induced, both in an anterior and posterior direction, and differentiation not only begins here but is always at an advanced stage relative to the rest of the developing insect. It has been shown by ligation experiments that the differentiation center does not release material out into the rest of the egg, but is the point of initiation of a contraction of the yolk from the chorion, an action that creates a space into which the cells forming the blastoderm can move. (It must be mentioned that the activation-center and differentiation-center descriptions are based on the study of comparatively few species of insects. Although there are probably similar centers in the embryos of most species, one should be cautious in applying these ideas to all species.)

Earlier in this section mention was made of the specialized region near the posterior pole of the egg. This is quite similar to the "centers" described above in that it determines which cleavage cells (those which migrate through it) will form the germ cells.

Insect eggs are quite variable as to when the cleavage cells become "determined" (i.e., when they become irreversibly destined to form a specific tissue). One extreme found, for example, in *Musca* and *Drosophila,* is the "mosaic" egg, in which determination is completed before the egg is deposited and the elimination of any part of it will result in the formation of an embryo missing whatever structure the excised or destroyed cells would have formed. In eggs at the other extreme, determination is not complete until well after deposition, the cells having retained the "potency" to form any tissue. Such eggs are referred to as "regulation," the capacity for a change in developmental fate (regulation) persisting through a much more advanced state than in the mosaic type.

Platycnemis (Odonata, Zygoptera) is an example of an insect with a regulation type of egg, and if a recently deposited egg is ligated in the middle, a "dwarf" embryo forms, demonstrating its well-developed capacity for regulation (Wigglesworth 1972). Many gradations between the extremes of mosaic and regulation have been described and seem to correlate with phylogenetic level, the mosaic condition occurring in such groups as Diptera and the capacity for regulation being found to an increasing extent among the "lower" orders, such as Odonata. A correlation also exists between the regulative capacity and the amount of cytoplasm; the greater the cytoplasmic volume, the less the regulative capacity (Agrell and Lundquist 1973).

How centers or gradients are formed is not clearly understood. New information has shown that the egg's cytoskeleton, especially actin filaments, may act as "anchors" for keeping the gradients in specific regions of the egg and that these areas somehow both attract and retain the maternal mRNA. Not only are the cytoskeletal elements (i.e., actins and microtubules) important for establishing centers and gradients, but they are also important for most movements within the egg.

Embryonic development begins with the formation of the germ band and the embryo itself. This period is characterized by an increase in cellular respiration, *de novo* synthesis of proteins, and increased utilization of the stored yolk reserves. A major shift occurs from maternal genetic control to zygotic or embryonic genetic control. The control of embryonic development appears to be under the influence of two categories of genes: segmentation genes and homeotic genes. What are segmentation genes and homeotic genes and how do they assist in the control of embryo formation?

Our understanding of both segmentation genes and homeotic genes comes from the important studies using *Drosophila* mutants. Using *Drosophila,* researchers demonstrated two major genetic effects. Some single gene mutations resulted in embryos missing complete segments; these developmental events are said to be controlled by segmentation genes. Other mutations showed major alterations in the fate or outcome of the segments and/or their appendages; thus these morphological events are said to be controlled by homeotic genes. Because a single gene mutation could produce these two events (i.e., segmentation and segmental or appendage formation), which are clearly under the control of many structural genes, these individual genes have been termed master genes, or regulatory genes. Thus, a segmentation regulatory gene can control the number and polarity of body segments, whereas a homeotic gene controls body part or appendage formation.

Homeotic mutants cause the transformation of morphological characters of specific body regions in the larva, nymph, or adult insect into characters appropriate for other regions. The best known homeotic gene is the antennapedia gene, in which the mutation of the single gene locus results in the transformation of cells that normally would be antennae into legs (figure 5.20*a*). Another interesting homeotic mutation occurs when the metathorax in *Drosophila,* which normally lacks a pair of membranous wings, is transformed into a segment having wings (figure 5.20*b*). The ability of various master or regulatory genes to produce such major changes via single gene mutations could provide a simple mechanism to account for the evolution of major morphological changes. An example of this could be the loss of one pair of wings in primitive Diptera, as fossil evidence now reveals they were once four-winged (Riek 1977).

Even more exciting than demonstrating the presence of homeotic genes is the evidence that most homeotic genes studied to date—whether they are from earthworms, centipedes, insects, frogs, or humans—have a highly conserved region of the DNA that encodes a sequence of 60 amino acids. This region is known as the *homeobox* and represents a highly conserved and uniform sequence of DNA. The reason this region has been so resistant to evolutionary change remains speculative.

In conclusion, insect embryogenesis and the mechanisms that regulate and control embryogenesis are areas of research that are rapidly evolving as new molecular and genetic techniques are being applied.

Symbionts in Eggs and Embryos

Many insects rely on intracellular symbionts for some aspect of their survival. Yet we know relatively little about how they are transferred from the female to the egg, where they reside in the pre-embryonic stage of the egg, and where they are located in the embryo. Their functions and effects during these periods remain unknown. One extremely important aspect of arthropods as vectors of plant and animal diseases, however, is the process of transovarial transmission of pathogens from the infected female to the next generation by the transfer of the pathogen to the egg. This process is also referred to as vertical rather than horizontal transmission.

Oviparity and Viviparity

Most insects are *oviparous;* that is, when they are released from the parent, they are surrounded by the egg-shell, or chorion. Depending on the species involved, the chorion ranges from very thin and delicate to very hard and thick, and the developmental stage of the insect within it varies from being in early cleavage to being ready to hatch as an active, free-living individual.

Figure 5.20 (a) The fruit fly, *Drosophila melanogaster*, showing the homeotic mutant, antennapedia. (b) The homeotic mutant, ultrabithorax.
(a) and (b) © David Scharf/Peter Arnold, Inc.

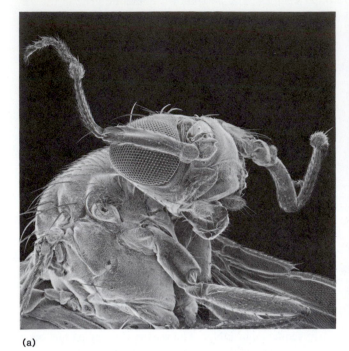

(a)

(b)

On the other hand, many insects are *viviparous* (i.e., they give "birth" to individuals having no chorionic covering). The terms *larviparous, nymphiparous,* and *pupiparous* are commonly used to refer to viviparous larvae, nymphs, and pupae, respectively. Although a group of Diptera are referred to as Pupipara, they are, in fact, larviparous; their name is really inappropriate. Members of the genus *Glossina* probably represent an extreme of viviparity in insects. One larva develops at a time in the highly specialized "uterus" and receives nutriment by means of specialized *uterine* (accessory) *glands*. When released from the parent insect, the tsetse larva is ready to pupate and does so within a few hours. While in the uterus, the larvae ventilate by protruding their posterior spiracles through the parent's genital opening.

The term *ovoviviparity* is sometimes used, and it refers to instances where an egg with a chorion develops, but the egg hatches within the parent before it is deposited (Hagen 1951). In some insects the embryo develops in the hemocoel of its parent (e.g., members of the Strepsiptera, as explained in the section on polyembryony).

Oviposition

Insects, among all other animals, are unique in having an ovipositor. Not all insects possess this structure, but those that do are able to exploit a multitude of niches that otherwise would be unavailable to them. The presence of an

ovipositor plus the uniqueness of the insect egg and its covering membranes have been major factors in the diversification of insects. This topic is reviewed by Zeh et al. (1989).

Egg laying in insects is often presented as a single act, when in fact it consists of two steps: ovulation and oviposition. The maternal parent insect typically deposits eggs in situations favorable for the survival of the offspring. Oviposition behavior is discussed in chapter 8.

In oviparous insects the eggs are propelled down the oviducts by peristaltic waves and may be deposited either in groups or singly, depending on the species. The ovipositing structure may be a well-developed appendicular ovipositor (see figure 2.44*a–d*), or the abdomen may be modified in such a way that it can telescope into a relatively long tube and hence function effectively as an ovipositor. The latter structure is commonly referred to as an *ovitubus* and is found among the Thysanoptera, Diptera (see figure 2.44*e*), and others. Ovipositors are reduced or lacking in the following orders: Odonata, Plecoptera, Mallophaga, Anoplura, Coleoptera, and the panorpoid orders.

Eggs are deposited in a variety of ways, some merely being dropped passively (e.g., walkingsticks, Phasmida) and others being "glued" singly or in masses to some substratum. Lacewings (Neuroptera, Chrysopidae) deposit their eggs at the tips of stiff stalks. Several orthopterous insects (e.g., locusts, mantids, and cockroaches) deposit their eggs in packets, or oothecae. The locusts secrete a frothy material that encases an egg mass, which is deposited in the ground. Mantids deposit their eggs on twigs in a foamy secretion that subsequently hardens to produce the ootheca. Cockroaches have oothecae with a cuticlelike surface; some species carry these around with them, whereas others glue them to a substratum. In most species the secretions involved with forming, covering, and gluing the egg masses to the substrate come from the accessory glands. Like the cuticles of the integument and chorion, oothecae of various sorts probably permit gaseous exchange without undue water loss. Insects such as parasitic Hymenoptera use their ovipositors to inject eggs deep into a host insect. Aquatic insects commonly surround their eggs with a gelatinous secretion. Insects that parasitize vertebrates often attach their eggs to hair or feathers.

The insect ovary is not innervated, whereas the oviduct is. Ovulation, or the passage of the eggs out of the ovarioles into the oviducts, is under the control of neurosecretory products from the brain. These secretions have a myotropic effect on the muscle sheaths of the ovarioles. Oviposition, or the passage of the eggs down the oviducts, and the placement of the eggs either into or onto a substrate is a complex behavioral event. It is believed to be under the control of the central nervous system and influenced by hormones. Recall that most insects will not lay eggs until they have mated. In some species, small molecules (probably neuropeptides) are transferred from the male's accessory reproductive glands to the female. How these molecules act on the female's nervous system is not known.

Eclosion

Eclosion is used here to denote the process of hatching or exiting from the egg (it is sometimes used to mean emergence of the adult insect). Although the details of this process vary from group to group, eclosion generally first involves the swallowing of amniotic fluid by the larva with the attendant diffusion of air into the egg. In addition to the amniotic fluid, some of this air may also be swallowed. The problems faced at eclosion are the rupture of the chorion and other embryonic layers and escape from the confines of the egg. This rupture may occur irregularly about the surface of the egg or along preformed lines of weakness. In some instances the embryonic membranes may be weakened by the action of digestive enzymes. The actual force involved in the perforation of the chorion and embryonic membranes may be applied via various structures, including spines or eversible bladders; or it may simply involve the forced expansion of one region of the body by contraction of another, a process aided by the previous swallowing of amniotic fluid and, in some instances, air. Some insects (e.g., Lepidoptera) chew their way out of the eggshell. In one instance [*Glossina spp.* (the tsetse flies)] in which the egg is retained and hatches within the "uterus" of the parent, the larva splits the chorion, but a parental structure removes it from the larva.

Postembryonic Morphogenesis

Here we include those events that occur between eclosion (or from the completion of embryogenesis in viviparous forms) and emergence of the adult insect.

Growth

Associated with the evolution of an external, relatively inexpansible exoskeleton has been the concurrent evolution of a mechanism that allows for increase in size, that is, the process of *molting*. Molting involves the periodic digestion of most of the old cuticle, secretion of new cuticle (usually with increased surface area), and shedding of undigested old cuticle. The last-mentioned step, shedding of undigested old cuticle (the *exuvia*), is commonly referred to as *ecdysis*. The term *molt* is also sometimes used synonymously with ecdysis.

As a typical insect progresses from the newly hatched immature form, it goes through a series of molts, generally increasing in size with each one. Each developmental stage of the insect itself is called an *instar*, and the interval of time passed in that instar is sometimes referred to as a *stadium*. For now, we will view an instar as the individual between successive ecdyses.

In many insects, especially those with a small number of instars (e.g., mosquitoes), it is possible to determine exactly the instar of a given individual larva by characteristic morphological traits. In others, there may be little or no change between instars other than growth, which may vary considerably with the availability of food and other environmental factors. The final instar, during which sexual maturity is reached and functional wings (if they are going to develop at all) appear, is the adult, or *imago*. A few insects (e.g., thysanurans) continue to molt after they reach the adult stage; however, it is unlikely that there is any growth associated with these molts. In members of the order Ephemeroptera (mayflies) there are two instars with wings, the first being the *subimago* and the second the adult. Other than mayflies, adult winged insects have never been known to molt under natural conditions.

The number of instars varies among insect species, the majority having between 2 and 20. In some insects the number of instars is constant; in others it may be variable, in response to environmental factors (e.g., availability of food and temperature). In some species the number of male instars may be different from the female number. The more specialized insects tend to have fewer instars.

Figure 5.21 Graph depicting Dyar's rule as it might apply to three hypothetical species of insects. A, B, and C are curves generated by plotting the log of a linear measurement against the number of a given instar. Given the information for the known instars in curve C, one can determine the total number of instars and the magnitude of the linear dimension pertinent to each.

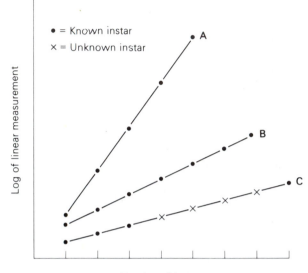

Growth in insects occurs as the result of an increase in the number of cells by mitotic division and/or an increase in cell size. The increase in weight between a newly hatched immature and a fully grown immature is usually quite pronounced. For example, a fully grown larva of *Cossus cossus* (Lepidoptera; Cossidae, the carpenter moths) weighs 72,000 times its first instar weight, and it takes three years to accomplish this growth (Richards and Davies 1977). In many other insects the magnitude of growth is somewhat smaller, varying from about 1,000 times on up. Increase in weight may approximate an S-shaped curve, although its continuity may be disrupted by the swallowing of water by aquatic insects or the ingestion of food.

Growth in length and surface area appears to be discontinuous, owing to the relative inflexibility of the cuticle. Based on observations of a degree of regularity in the extent of linear growth in various structures in several species, certain "growth laws" have been induced. The most important of these is *Dyar's rule,* which is based on the assumption that growth as reflected in various linear dimensions follows a geometric progression.

According to this rule, the ratio between a given dimension in one instar and the same dimension in the next is constant throughout all instars and constant for each

species. Dyar (1890) showed, for example, that in 28 species of lepidopteran larvae, the head capsule increases by a ratio of 1.4 at each molt. Thus if we were to plot the logarithm of the measurement of some linear dimension at each instar of a given species against the number of the instar, we would expect to obtain a straight line (figure 5.21). In cases in which this rule applies, it is useful in determining the number of instars of an insect, because if the dimensions of the final instar and the ratio of a given dimension between any two successive instars are known, the dimensions of all the other instars can be deduced by interpolation and their number found by counting the number of points on the generated growth curve. However, in instances in which the number of molts varies in response to certain environmental or other factors, Dyar's rule will not allow the determination of the number of instars.

An extension of Dyar's rule is *Przibram's rule,* based on the assumption that weight doubles during each instar, which says that all linear dimensions at each molt are increased by the ratio of 1.26 or the cube root of 2 (Wigglesworth 1972). Although this rule applied in the species studied when it was devised and in certain other species, it fails to apply in many other instances, for example, in instances in which weight more than doubles during each molt or in which growth is attained by increase in cell size instead of cell number, as, for example, in the larvae of certain true flies. In addition, Przibram's rule assumes *harmonic growth,* or *isogonic* growth (i.e., that all parts of the insect and the body as a whole increase by the same ratio during each molt). However, in the majority of insects, growth during each molt is *heterogonic,* or *allometric* (i.e., some parts of the insect body develop at different rates than other parts).

Metamorphosis

Most insects at eclosion are morphologically different from the adult. The degree of difference varies from relatively slight to extreme, with many intermediates. The developmental process by which a first-instar immature stage is transformed into the adult is called *metamorphosis,* which means literally "change in form." This process may take place gradually, with the immature being in general appearance comparatively similar to the adult, or it may be quite abrupt, the immature instars being drastically different from the adult with the transformation from the immature to the adult form occurring in a single stage.

The class Insecta can be divided into groups based on the type of metamorphosis, if it is present. Members of the Apterygota do not undergo any change in form, the immature instars differing from the adults only in size and the development of the gonads and external genitalia (figure 5.22a). The insects are sometimes grouped as Ametabola and said to undergo *ametabolous* development.

Members of the Pterygota can be divided into two groups relative to the degree of metamorphosis that occurs: *hemimetabola* and *holometabola*. Among the hemimetabolous (includes both hemimetabolous and paurometabolous forms as will be explained below) insects the immatures resemble the adults in many respects, including the presence of compound eyes, but they lack wings, gonads, and external genitalia. During the course of development (figure 5.22b,c) the wings become externally apparent as *wing pads*. Thus the orders that fall into this group are those that are classified as *exopterygotes* (in reference to wings developing externally). The hemimetabolous form of development is often called simple, direct, or incomplete metamorphosis. The immature instars in this group of insects are commonly known as *nymphs,* although they may also be correctly referred to as *larvae.*

In the past, insects with hemimetabolous development were subdivided further. Those insects (Odonata, Ephemeroptera, and Plecoptera) that pass the immature instars in an aquatic environment and the adult instar in a terrestrial/aerial environment, and the immatures of which at least superficially appear to be quite different from the adult stage (e.g., immature stoneflies, with highly specialized ventilatory gills, versus adults, which lack these structures and have well-developed wings) were classified as being *hemimetabolous,* or undergoing an incomplete metamorphosis (figure 5.22b).

Those insects passing both the immature and adult instars in essentially the same environment were classified as being *paurometabolous,* or undergoing gradual metamorphosis (figure 5.22c). The immature instars of hemimetabolous insects were called *naiads;* those with paurometabolous development were referred to as nymphs. Morphological studies have indicated that although Odonata and Ephemeroptera show some affinity with each other, they are not closely related to Plecoptera, and hence these three orders do not form a cohesive phylogenetic group. The paurometabolous and hemimetabolous groupings have therefore been abandoned and the two merged under the heading "hemimetabolous."

The remaining group of orders, those classified as *endopterygotes* (in reference to wings developing inside the body of the larva), all undergo what is referred to as *holometabolous* development or complete metamorphosis (figure 5.22d). In these insects the immature instars are quite dissimilar to the adults and generally are adapted to different environmental situations. Immature instars

in this group of insects are called *larvae.* Individuals in the larval stage typically lack compound eyes and usually have mandibulate mouthparts, whether or not they are mandibulate in the adult instar. They may or may not have thoracic or abdominal legs. Most of the changes in the transformation from the last instar larva to the adult are compressed into a single intervening instar, the *pupa.* The pupa is typically a resting stage protected in some way (by a silken cocoon, hidden in leaf litter, in a puparial case, etc.), but the pupae of some insects are quite active. For instance, the aquatic pupae of mosquitoes ("tumblers") are very active and dive in response to potentially threatening stimuli. In most endopterygotes the larval instars resemble one another except for a few minor morphological details, which are useful in distinguishing one instar from another. However, some holometabolous insects pass through one or more larval instars that are distinctly different from the others (figure 5.22e). This phenomenon is called *hypermetamorphosis* and has been described in certain species in the orders Neuroptera, Coleoptera, Diptera, and Hymenoptera, and in all species in the order Strepsiptera.

The grouping of insects based on metamorphosis as described is not without exceptions. For example, members of some orders that are included in Hemimetabola are secondarily wingless in the adult stage, and their development more closely resembles that of the apterygotes in that it is essentially ametabolous. More important, certain of the hemimetabolous insects in fact have a pupal stage. For example, thrips (Thysanoptera) and certain members of the Homoptera (whiteflies; and certain scale insects, e.g., *Pseudococcus* spp.) have a distinct resting "pupal" stage between the last "larval" and the adult instars (Hinton 1948). Even members of the "primitive" order Odonata undergo considerable morphological change during the transformation from the aquatic nymph to the terrestrial/aerial adult. These changes include loss of rectal gills, restructuring of the labium, modifications in the head and abdomen, complete reconstruction of the alimentary canal, and nearly complete replacement of the abdominal musculature.

It is difficult not to look upon the development of these insects as being "holometabolous," indicating that holometabolism has probably evolved on separate, independent occasions. However, Hinton (1963) recognizes a distinct difference between the holometabolous exopterygotes and the endopterygotes. He points out that since the wings develop internally in endopterygotes, there is insufficient room in the larval thorax for the development of both the wings and the wing muscles. Therefore, a molt is required to evaginate the wings (the larval–pupal molt) and provide space for the development of wing

Figure 5.22 Types of insect metamorphosis. (*a*) Ametabolous, silverfish. (*b*) Hemimetabolous, damselfly.
(*c*) Paurometabolous, leafhopper. (*d*) Holometabolous, house fly. (*e*) Hypermetamorphosis, beetle, *Epicauta cinerea* (Coleoptera; Meloidae). (A) adult; L_1, first instar larva; L_2, second instar larva; L_3, third instar larva; Ln, *n*th instar larva; N_1, first instar nymph; N_n, *n*th instar nymph; P, pupa.
(*c*) *redrawn from Essig, 1958; (d, e) redrawn from Packard, 1898.*

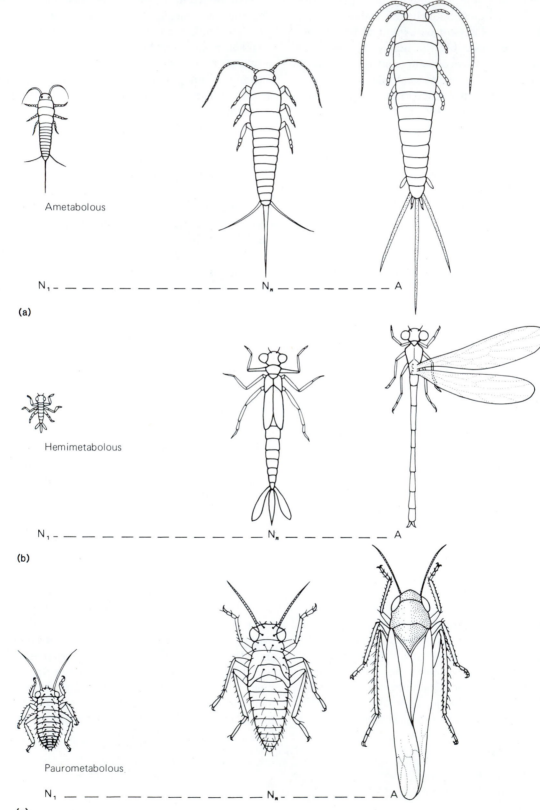

Figure 5.22 Continued

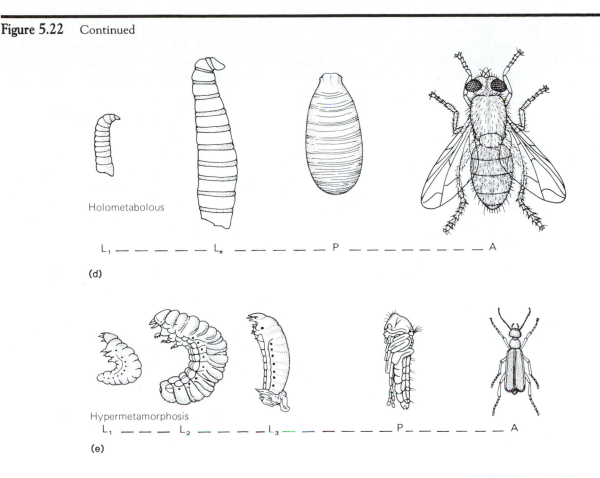

Holometabolous

L₁ — — — — — Lₙ — — — — — P — — — — — — — A

(d)

Hypermetamorphosis

L₁ — — — — L₂ — — — — — L₃ — — — — — — P — — — — — A

(e)

muscles. A second molt (pupal–adult) is then required to release the adult from the confines of the pupal cuticle. A pupal stage, in this sense, is unnecessary in exopterygotes since the wings develop externally. Hinton accounts for the "pupae" in certain exopterygote insects as follows.

In some exopterygotes such as the Aleyrodidae and male Thysanoptera and Coccoidea, the general structure of the feeding larval instars has departed very widely from that of the adult. The structural differences between the feeding larval instars and the adult are normally bridged in the last larval instars. As the differences between the two stages became greater and involved a greater degree of reorganization of the internal tissues, it would seem that the last or last two larval instars became more and more quiescent and eventually ceased to feed. It may be noted here that the structural reorganization required to bridge the gap between the feeding larval stages and the adult of some exopterygotes is greater than in the primitive endopterygotes, e.g., some Megaloptera. No difficulty necessarily arises if these quiescent or semiquiescent stages of exopterygotes are called pupae provided that it is recognized that their origin is quite independent from that of the endopterygote pupa and their initial functional significance is different.

Wigglesworth (1972) and others confine usage of the term "metamorphosis" to those changes that occur when an insect becomes an adult, regardless of whether the adult stage is reached by a single molt of the last immature instar or in two molts with a pupal stage in between. Wigglesworth (1954) points out that metamorphosis is commonly looked upon as a renewal of embryonic development. However, he regards the larvae and adults of holometabolous insects as essentially two organisms, which are latent within the genome of the embryo and which are expressed in sequence. He views this as a sort of temporal polymorphism. The evolution of a pupal stage in which comparatively drastic changes can occur has evidently enabled the divergent evolution of larvae and adults, which are usually adapted to radically different environmental modes of existence. As will be explained later, the same basic endocrine mechanism probably governs development in all insects.

The immature stages of insects take on a wide variety of forms. In most instances the nymphs of hemimetabolous species closely resemble the adults, but in the holometabolous species, the larvae are often drastically

Figure 5.23 Larval types. (*a*) Campodeiform, alderfly, *Sialis* sp. (Neuroptera; Sialidae). (*b*) Carabiform, ground beetle, *Harpalus* sp. (Coleoptera; Carabidae). (*c*) Elateriform, click beetle (Coleoptera; Elateridae). (*d*) Eruciform, clear-winged moth (Lepidoptera; Aegeriidae). (*e*) Platyform, aquatic beetle, *Eubrianax edwardsi* (Coleoptera; Dascillidae). (*f*) Scarabaeiform, branch and twig borer, *Polycaon confertus* (Coleoptera; Bostrichidae). (*g*) Vermiform, flesh fly, *Sarcophaga* sp. (Diptera; Sarcophagidae). (*a, b, d, g*) *redrawn from Packard, 1898; (c, e, f) redrawn from Essig, 1958.*

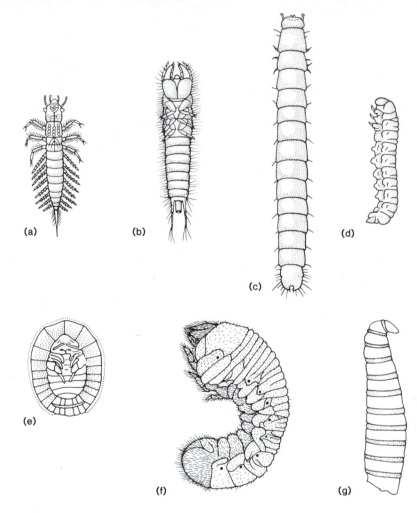

different from the adults. Although there is considerable variation in the appearance of larvae of the different holometabolous groups, there are sufficient similarities to allow recognition of distinct "larval types." Some types of larvae (figure 5.23) and terms commonly used to describe them are as follows:

1. *Campodeiform* larvae resemble diplurans in the genus *Campodea,* having flattened bodies, long legs, and usually long antennae and cerci (e.g., several beetles, Neuroptera, and Trichoptera).
2. *Carabiform* larvae resemble the larvae of carabid beetles, which are similar to the campodeiform type but have shorter legs and cerci (e.g., several beetles).
3. *Elateriform* larvae resemble click beetle larvae (Coleoptera; Elateridae) and have cylindrical bodies with a distinct head, short legs, and a smooth, hard cuticle (e.g., certain beetles).

4. *Eruciform* larvae are the typical Lepidopteran "caterpillars" or the caterpillarlike larvae of certain Hymenoptera and Mecoptera and have a cylindrical body with a well-developed head, short antennae, and short thoracic and abdominal legs (*prolegs*).
5. *Platyform* larvae have flattened bodies with or without short thoracic legs (e.g., certain Lepidoptera, Diptera, and Coleoptera).
6. *Scarabaeiform* larvae are the "grubs" and have a cylindrical body typically curled into a C shape, a well-developed head, and thoracic legs (e.g., several beetles).
7. *Vermiform* larvae have wormlike bodies with no legs and may or may not have a well-developed head (e.g., several Diptera, Coleoptera, Hymenoptera, Siphonaptera, and Lepidoptera).

Figure 5.24 Pupal types. (*a*) Exarate, ichneumon wasp. (*b*) Obtect, moth. (*c*) Coarctate, house fly. (*d*) Puparial case removed, exposing exarate house fly pupa.

(a) *redrawn from Boreal Labs key card; (b) redrawn from Packard, 1898; (c, d) redrawn from Wigglesworth, 1970.*

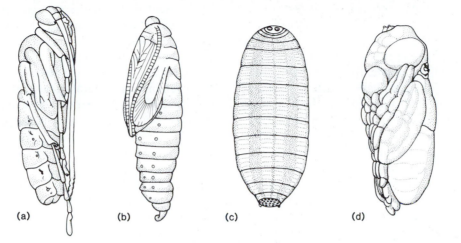

(a)   (b)   (c)   (d)

It must be emphasized that this classification of larval types is a pragmatic one reflecting only gross similarities and not phylogenetic affinities. Obviously one should expect to find numerous examples of larvae that display a mixture of characteristics of two or more of these types.

Berlese classified insect larvae based on the assumption that they are essentially free-living embryos that are released to the external environment in different stages of development in different species (Richards and Davies 1977). He recognized three such stages: *protopod, polypod,* and *oligopod.* Protopod larvae are very uncommon and represent a very early stage of development in which comparatively little segmentation has occurred. These larvae are found, for example, among certain parasitic Hymenoptera that larviposit in the hemocoel of other insects, placing the "embryo" in the only kind of environment possible for survival—a protected one in which it is surrounded by nutriment. Polypod larvae resemble caterpillars, and hence the eruciform type described would fall readily into this group. Oligopod larvae lack the abdominal prolegs and may or may not have well-developed cerci. The campodeiform, carabiform, elateriform, platyform, and scarabaeiform types listed fall nicely into this category. The vermiform (apodous) type of larva is considered to have developed secondarily from the polypod or oligopod type, depending on the kind of insect, and hence is not recognized as a separate group.

A subscriber to Berlese's view of holometabolous larvae must recognize a fundamental difference between these larvae and the nymphs of hemimetabolous insects, which in this context represent postoligopod stages of development. However, one can as easily look upon holometabolous larvae as specialized nymphs with a concurrent concentration of the changes to the adult form in a single stage, the pupa. From this point of view there are no fundamental differences between larvae and nymphs, and the terms could appropriately be used as synonyms. As will be explained subsequently, the results of comparatively recent investigations of the physiological mechanisms that control metamorphosis have supported this latter conception of the relationship between hemimetabolism and holometabolism.

As with larvae, pupae have been grouped according to similarities. Hinton (1964) classified pupae based on whether or not they have articulated mandibles that are used in escaping from a cocoon or pupal cell. Those pupae having such mandibles he described as being *decticous.* Examples of decticous pupae include members of the orders Neuroptera, Mecoptera, Trichoptera, and certain lepidopterous families. Pupae without functional mandibles used in escape from cocoon or pupal cell are termed *adecticous* and are found in all or some members of the orders Strepsiptera, Coleoptera, Hymenoptera, Diptera, and Siphonaptera. Another approach to grouping pupal types (figure 5.24) is based on whether the appendages are free or adherent to the body. *Exarate* pupae (figure 5.24*a*) have the appendages free and are usually not covered by a cocoon. In *obtect* pupae (figure 5.24*b*) the appendages adhere closely to the body and are commonly covered by a cocoon. All decticous pupae are exarate, whereas the adecticous types may be obtect, exarate, or *coarctate* (figure 5.24*c*). A coarctate pupa is encased in the hardened cuticle of the next-to-last (penultimate) larval instar, the *puparium.* However, the pupa itself is of the adecticous exarate form (figure 5.24*d*).

Many histological changes occur during the transition from immature to adult (Whitten 1968). In hemimetabolous insects these changes are comparatively gradual, being spread throughout the nymphal instars, although there are generally more changes during the last instar than during earlier ones. In holometabolous insects, these changes occur mostly in the pupal stage. The extent of change during the pupal period is variable; in some species a large amount of tissue change occurs. These changes are accomplished by means of tissue breakdown (histolysis) and tissue reorientation, growth, and differentiation (histogenesis). Histolyzed tissues may simply dissolve in the hemolymph or may be ingested by phagocytic hemocytes. Holometabolous insects have evolved a strategy of setting aside, early on in embryonic development, specific primordial or progenitor cells for later use during the pupal period. These undifferentiated cells will eventually develop into pouches or packages of epidermal cells termed *anlagen*. They are used for constructing adult organs or appendages (figure 5.25). From a developmental standpoint, it is an ideal evolutionary strategy for setting aside the building blocks early in the life cycle of the insect, which are carried by the insect through the embryonic and larval stages and are not used until the pupal stage.

Imaginal Discs

Many adult tissues are formed from masses of cells that have persisted in an undifferentiated state throughout the larval instars even though they increase in number by mitotic divisions. These masses of embryonic cells are generally referred to as *imaginal buds,* or *discs* (figure 5.25), and are involved in the formation of structures such as mouthparts, antennae, wings, and legs. Imaginal discs, especially from *Drosophila melanogaster,* have proved to be extremely valuable in studies of cellular differentiation (Gehring and Nöthiger 1973, Ursprung and Nöthiger 1972). In addition to imaginal discs, rings of embryonic cells are generally found at the posterior extremity of the foregut (*anterior imaginal ring*) and the anterior extremity of the hindgut (*posterior imaginal ring*). These undergo considerable growth and differentiation during the pupal stage and contribute significantly to the formation of the adult alimentary canal. For example, the ventral diverticulum in mosquitoes (see figure 4.2) develops as the result of mitotic division of cells in the anterior imaginal ring. In some instances, certain larval tissues persist into the adult stage, and these vary with the different groups of insects.

Imaginal discs are ideal for studying the process of determination and seeking answers to questions concerning when and how the progenitor cells or imaginal discs become programmed to express a specific phenotype at the time of differentiation. By using transplantation experiments of either whole or pieces of discs, investigators have learned at what stage the discs become committed and what some of the controlling factors are that govern their growth into specific organs or appendages. Generally, commitment occurs early in embryonic development, whereas the commitment to differentiate into an adult tissue is reached during the third stage of larval development. A high titer of ecdysone, which occurs at the end of larval development, is the signal triggering differentiation. Another aspect of imaginal disc biology is to examine what factors contribute to cellular recognition within the disc proper. By using enzymes to separate individual cells, disc cells can be dissociated and then left to reassociate. The mechanism(s) by which related cell types correctly reassociate and recognize one another is currently under investigation and is an important biological phenomenon. Finally, homeotic mutants and genetic markers are used to develop fate maps (figure 5.26) of individual discs. The *Drosophila* homeotic mutant, post-bithorax, results in the haltere imaginal disc (figure 5.20*b*) transforming into wing structures. Thus, homeotic and segmentation genes have a regulatory role controlling the ultimate fate of imaginal discs.

Why does the larva not recognize the imaginal discs as foreign and encapsulate them? Information in chapter 4 explained that they are not recognized and encapsulated because they are surrounded by a basement membrane, which contains surface molecules denoting "self." Experiments with larval *Drosophila* mutants, which encapsulate their own imaginal discs, reveal that the basement membrane is either abnormal or missing. The chemical identity of the basement membrane molecules remains an area of research in need of researchers to identify these molecules and determine exactly how they relate to recognition.

The use of genetic, molecular, and cellular probes to study imaginal disc determination is presented in Oberlander (1985) and Larsen-Rapport (1986), and the role of gene function in embryogenesis is the main topic of the 1990, volume 11, issue of the journal *Developmental Genetics.*

Also of interest in the development of insects are the changes and control of changes in integumental pattern (distribution of bristles, etc.) that occur during growth and metamorphosis. Lawrence (1970, 1973, 1992) and Waddington (1973) review the literature on the development of spatial patterns in the integument.

Pupa

Although some pupae (e.g., mosquitoes) are active and can evade potential predators or adverse environmental conditions, most are not and hence are quite vulnerable. A number of mechanisms have evolved that decrease this vulnerability. Probably the most common of these is the use of silk in constructing a cocoon of some sort (e.g.,

Figure 5.25 Schematic representation of the imaginal discs of a fly larva and the adult organ or appendage into which they differentiate.

Taken from Figure 1, Ursprung and Nöthiger (1972).

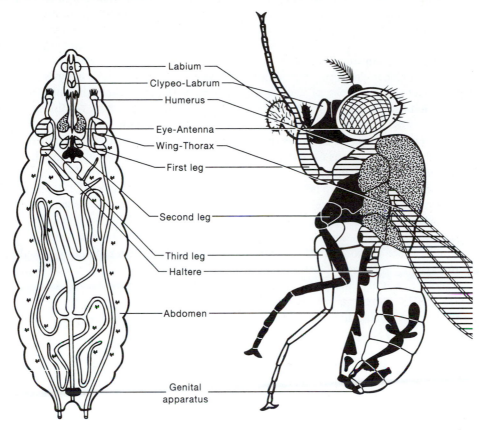

Figure 5.26 Fate map of foreleg or morphogenetic map of the male imaginal disc of a mature fly larva. Once the disc has reached this size and is in the proper stage of the larva, specific regions of the disc are committed to become specific parts of the adult organ or appendage.

Taken from Figure 1(a–e) from Ursprung and Nöthiger (1972).

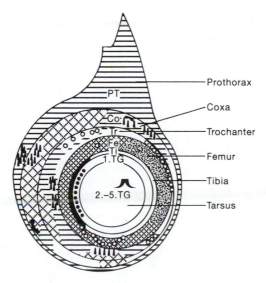

many Lepidoptera, Hymenoptera, Neuroptera, Trichoptera, and Siphonaptera). This cocoon may be composed solely of silk, or it may be basically silk but with bits of environmental debris incorporated into it. Recall the beetle, mentioned in chapter 4, that constructs a cocoon from collapsed peritrophic membrane. Many insects (e.g., Coleoptera and Lepidoptera) construct cells beneath the surface of the soil and pupate within them. Another example of a pupal protection mechanism is the puparium mentioned previously. Obviously the pupae of endoparasitic forms derive protection from being within the body of the host.

Concurrent with the evolution of pupal protection mechanisms was the development of means for escaping at the time of adult emergence. Decticous pupae chew their way out of the cocoon or cell and may be aided in this process by posteriorly directed spines. Escape methods for adecticous pupae include a variety of spines or other hard and sharp protuberances and an eversible bladder, the *ptilinum,* in the head that is used to force open the "cap" of a puparial case in the cyclorrhaphous Diptera. The labial glands of recently emerged adult saturniid and bombyliid moths secrete a solution containing the enzyme *cocoonase,* which digests the silken cocoon and allows easy exit (Kafatos et al. 1967, Kafatos 1972).

When an adult is nearly ready to emerge, the pupal cuticle splits in various places, particularly along the dorsal midline of the body. At emergence hemostatic pressure is exerted on the pupal cuticle by the contraction of body muscles (likely aided by the swallowing of air or water). The newly emerged adult is in a very vulnerable state until the cuticle hardens and the wings expand and harden. The term *teneral* is commonly used to refer to such a newly emerged, pale, soft-bodied individual regardless of stage.

In addition to morphological, histological, and cellular changes during growth and development, there are biochemical changes as well. Major reviews on this topic include Agrell and Lundquist (1973), Chen (1971), L'Helias (1970), and Wyatt (1968).

The Instar Definition Controversy

A controversy exists over the precise definition of an instar (Hinton 1946, 1958, 1971, 1973, 1976; Whitten 1976; Wigglesworth 1973). Traditionally, since Linneaus, an instar has been defined as the individual between successive ecdyses. However, Hinton (1958, 1971) argues that an instar should be defined relative to the cuticle to which the epidermis is actually attached. Thus he suggests that a new instar actually begins with each *apolysis* (the time of separation between the epidermis and old cuticle prior to secretion of new cuticle; Jenkin and Hinton 1966). He denotes the period of time between apolysis

and the beginning of new cuticle secretion as the *exuvial phase,* and the period of time between the first appearance of new cuticle beneath the old and ecdysis as the *cuticular phase.* Between apolysis and ecdysis, the insect is still within the confines of the old cuticle. Hinton refers to an insect in this state as being *pharate.*

By the traditional terminology the pupal instar, for example, would be an individual between larval–pupal ecdysis and pupal–adult ecdysis. By Hinton's terminology, on the other hand, the pupal instar actually begins with larval–pupal apolysis within the confines of the fully grown larva's cuticle and remains a pharate pupa until larval–pupal ecdysis, after which it is called a pupa, until pupal–adult apolysis. At pupal-adult apolysis, it becomes a pharate adult.

The controversy over the definition of an instar has not been resolved. It would seem that adherence to Hinton's definition would be especially important from the developmental, histological, physiological perspective, whereas the traditional definition would be of value from the point of view of behavior, ecology, and systematics in which the organism is studied as a whole.

Control of Growth and Metamorphosis

The search for the mechanisms controlling growth and metamorphosis has occupied the time of investigators for decades. As a result, a reasonably clear picture of patterns of tissue and hormone involvement has emerged. Major reviews include Doane (1973); Etkin and Gilbert (1968); Gilbert and King (1973); Highnam and Hill (1980); Kroeger (1968); Novák (1975); Riddiford and Truman (1978); Schneiderman and Gilbert (1964); Sláma, et al. (1974); Thomson (1976); Wigglesworth (1954, 1959, 1964b, 1970); and Willis (1974). Additional reviews are in Downer and Laufer (1984), volumes 7 and 8 of Kerkut and Gilbert (1985), Law (1987), and Gupta (1990).

Metamorphosis prepares the insect for major changes in both ecology and behavior. The morphological adaptations of the young or larvae of most animals usually permits them to focus on eating and growing while the adult form concentrates on dispersal and reproduction. This generalization, however, is reversed in many aquatic invertebrates (e.g., Annelids, Mollusks, and Echinoderms) where the larvae are the dispersal forms and the stationary adults carry on food acquisition and reproduction. It should be recognized that most events in the life of an insect are not unique to insects alone. A case in point is metamorphosis. Most animals exhibit some type of metamorphosis. The frog and the insect are the two most-widely studied examples. A comparative approach to this topic is presented by Gilbert and Frieden (1981).

Many different techniques have been employed in the elucidation of the mechanisms controlling growth and

Figure 5.27 Principal endocrine tissues in the giant silkworm moth, *Hyalophora cecropia*.
Modified from Gilbert, 1964 (Gilbert, personal communication, 1992).

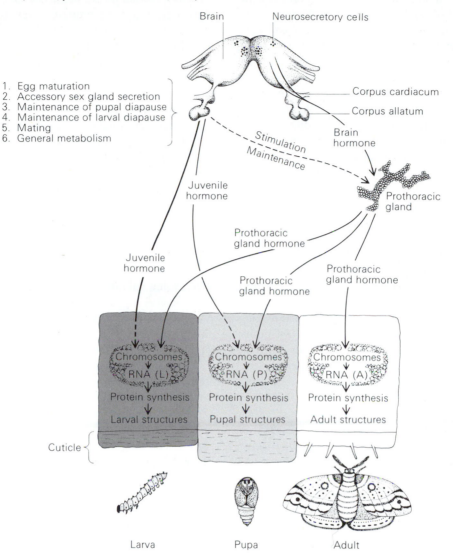

metamorphosis in insects. Early workers used ligation and decapitation extensively, and these techniques still find application today. Later techniques employed have included microsurgical removal of various tissues, implantation of various tissues, hemolymph transfusion, and the joining of the hemolymph circulation of one individual with that of one or more other individuals (*parabiosis*). Recently, sophisticated biochemical techniques have made possible the extraction and purification of the various hormones, and cytological techniques have brought a degree of understanding of the nature and modes of action of these hormones (Gilbert 1976, Menn and Beroza 1972, Morgan and Poole 1976, Rees 1977, Riddiford and Truman 1978). Gupta (1990) and Riddiford (1985) review the modes of action of hormones at the cellular level.

Among the structures and tissues involved in the control of growth and metamorphosis (figure 5.27) are

1. The *median neurosecretory cells* in the pars intercerebralis region of the brain.
2. The *corpus cardiacum.*
3. The *prothoracic glands* located in the prothorax in close association with tracheae.
4. The *corpora allata.*

The *neurosecretory cells* in the brain secrete the *brain hormone* (also called *prothoracicotropic hormone—PTTH* and *ecdysiotropin*), which accumulates in the *corpora allata* and is subsequently released into the hemolymph. Previously, brain hormone was thought to be stored in the corpus cardiacum, but at least in Lepidoptera recent studies employing sophisticated tracer techniques have cast the corpora allata in this role (Gilbert, personal communication). Brain hormone stimulates the secretory activity of the prothoracic gland. The stimuli that cause the secretion of brain hormone vary and in most cases are not known. In the blood-sucking bug *Rhodnius* (Hemiptera, Reduviidae), abdominal distension resulting from feeding can lead to brain hormone secretion. In the grasshopper *Locusta* (Orthoptera, Acrididae), stretch receptors in the wall of the pharynx are involved. In addition to stretch caused by the meal, recent evidence with *Rhodnius* suggests that a nutritional component of the diet may also be able to cause the secretion of the brain hormone. The prothoracic glands secrete *ecdysone* (also called *prothoracic gland hormone* and *molting hormone*), which initiates the growth and molting activities of cells. Ecdysone is also produced in other tissues, for example, the ovaries of mosquitoes, *Phormia,* and the locusts, but these sources are not considered to have a bearing on molting. The corpora allata secrete *juvenile hormone,* which promotes larval development and inhibits development of adult characteristics. It has been described as having a "status quo" effect.

During the larval instars both ecdysone and juvenile hormone are produced. However, during the last immature instar, juvenile hormone is not produced or decreases below some threshold; hence, the expression of adult characters is not inhibited, and metamorphosis to pupal and then the adult stage occurs. In hemimetabolous insects there is a gradual progression to the adult stage. On the other hand, in holometabolous forms the changes from immature to adult are compressed into the pupal stage, but the principle of discontinuation or reduction below some threshold of secretion of juvenile hormone during the last larval instar still applies. Thus, there appears to be no significant difference in the fundamental physiological mechanisms controlling growth and metamorphosis in either of these groups.

Ecdysone is actually a generic term for a number of similar steroid compounds, some of which have even been found in certain plants (*phytoecdysones;* Williams 1970). More on this topic will be discussed in chapter 9. Ecdysone apparently acts on genes through a receptor-mediated process, determining which genes are brought into action at a given time, and as a result influences the kinds of proteins (both enzymes and structural proteins) synthesized. Support for this idea of the action of ecdysone has been found in studies of *chromosomal puffs* in *giant chromosomes* (Ashburner 1970, 1972; Beerman 1972). These puffs, representing localized unravelling of the DNA molecule in the chromosomes, are taken to indicate heightened gene activity at the points of their occurrence. It has been shown, for example, that injection of pure ecdysone into *Chironomus* larvae produces puffing patterns in chromosomes identical to those patterns that occur during pupation. Additional support for the notion that ecdysone acts on genes comes from its involvement in sclerotization of the puparium in higher Diptera. Ecdysone is thought to activate a gene that directs the synthesis of a key enzyme in the sclerotization process (Karlson and Sekeris 1976). For detailed information on the extraction and analysis of ecdysone, see Gupta (1990) and for information on the mechanisms of sclerotization of dipterans, see Lipke et al. (1983).

The amino acid sequence and the gene for PTTH have been determined for *Bombyx mori.*

The chemical structure of juvenile hormone has been elucidated (Gilbert 1976). Actually, five different hormones have been found. They are terpenoids. The mode of action of the juvenile hormones is not known for certain, but they may activate or repress target genes (or gene sets) by binding through its association with a specific receptor to a regulatory sequence of the DNA (Gupta 1990). As with ecdysone, juvenile hormone and JH-mimics have also been found in plants. Menn and Beroza (1972) and volume 7 of Kerkut and Gilbert (1985) deal with several aspects of juvenile hormone.

At least two other hormones are involved in regulating processes associated with the growth and development of some insects, *eclosion hormone* and *bursicon.*

Eclosion hormone (Truman 1971–1973, Riddiford and Truman 1978), a neuropeptide, has been found in several Lepidoptera. It is synthesized in the brain. Release is controlled by a circadian clock. This hormone influences several aspects of pupal-adult ecdysis ("eclosion"), including the behavior associated with ecdysis, and the subsequent degeneration of the abdominal intersegmental muscles used in the act of ecdysis (see chapter 3). Eclosion hormone is apparently widely distributed throughout all insects and regulates ecdysis in all stages of development. The eclosion hormone gene of the tobacco hornworm, *Manduca sexta,* has been isolated and consists of 7.8 kilobases with three exons. The hormone itself is a 62-amino acid neuropeptide.

Bursicon (*tanning hormone;* Riddiford and Truman 1978, Cottrell 1964, Fraenkel and Hsaio 1965, and Downer and Laufer 1984) is a neurosecretory hormone synthesized and/or released from a variety of different sites according to species. It is commonly found in neurohaemal organs (similar to the corpora cardiaca) associated with the ventral chain ganglia. Bursicon stimulates tanning and sclerotization of the cuticle following ecdysis. It was initially discovered in blow flies (*Calliphora* spp., Diptera) and has subsequently been found in several different insects and at different life stages.

The results of a symposium (1989), which focused on various aspects of ecdysone research (e.g., ecdysone receptors, techniques for analysis, and gene regulation), are published in the *Journal of Invertebrate Reproduction and Development,* volume 18, pages 1–2, (1990).

Major reviews covering the four major hormones involved in insect metamorphosis are provided in volumes 7 and 8 of Kerkut and Gilbert (1985) and Gupta (1990).

Diapause

The ability of insects to survive unfavorable conditions (e.g., unsuitable weather or lack of food resources) in a state of arrested development is termed *diapause*. This physiological, biochemical, and behavioral adaptation is hormonally regulated and is usually primed by a series of recurring environmental cues (e.g., photoperiod). Diapause has permitted insects to extend their ranges and explore new niches. This aspect of development is considered further in chapter 8. The ecological aspects of diapause are covered by Danks (1987) and Denlinger (1986) and the biochemical and physiological aspects of embryonic diapause are presented in volume 1, with hormonal control treated in volume 8 of Kerkut and Gilbert (1985).

Polymorphism

The term *polymorphism* means literally "many forms." From this definition it follows that any number of phenomena could be cited as examples of polymorphism: differences between the sexes (*sexual dimorphism*), differences between individuals caused by the external environment, differences between developmental stages (i.e., the "temporal polymorphism" referred to earlier), and so on. However, the more restricted definition of Richards (1961) seems more useful, although there is still disagreement regarding the meaning of polymorphism. Richards' definition is as follows:

> Polymorphism exists when one or both sexes of a species occur in two or more forms which are sufficiently sharply distinct to be recognizable without a morphometric analysis; the occurrence is regular or recurrent; the rarer of the two forms makes up a reasonable proportion of the population (say, at least 5 per cent) or, as in some social species, the rarest type is at any rate essential to the survival of the species.

Actually, since form and function are inseparable, functional as well as morphological differences should be included in one's definition and concept of polymorphism.

Polymorphism is expressed in a variety of traits in different insects. Examples of these traits include coloration and patterns of color, as, for example, in many butterflies; the presence, absence, or attenuation of wings, as found in aphids and other insects; chromosomal differences; the relative size of different structures; and various integumental structures, including horns, spines, and other protuberances. Species in several orders exhibit one or more types of polymorphism, but polymorphism has not been found in the following orders (Richards 1961): Thysanura, Diplura, Protura, Ephemeroptera, Grylloblattodea (included with the Orthoptera in this text), Embioptera, Mallophaga, Thysanoptera, Mecoptera, and Trichoptera.

Following Richards' definition, there are basically two types of polymorphism, both of which are controlled by genes.

In one form of polymorphism (i.e., *balanced polymorphism*) the relative abundance of the different forms depends on variation in the selection pressures acting on members of the same species. Hence different populations may show significant differences in the relative abundance of the different forms. An example is found in the variations in relative numbers of different forms of the mimetic butterflies (*Papilio dardanus*) in different regions depending on the abundance of the species that they mimic (Sheppard 1961).

In the other type of polymorphism, expression of form is determined by various environmental factors, often influencing the endocrine system and ultimately the action of different genes, all individuals in a population having essentially the same genotype. A number of environmental influences have been shown to play a role in polymorphic expression in different species.

Nutrition, both in terms of quantity and dietary balance, plays a significant role in polymorphism in honey bees. Female larvae (from fertilized eggs) all have the potential to develop into either workers or queens, depending on the period of time they are fed royal jelly, a highly nutritious substance produced by the salivary glands of workers. If larvae are fed royal jelly for only two or three days, they develop into worker adults, but if they are fed royal jelly throughout the larval stadia, they develop into queens. By experimentally varying the length of time larvae are fed royal jelly, it has been possible to produce intermediates between workers and queens.

Variations in the quantity of food available can produce polymorphism in insects that grow heterogonically by inducing early or late pupation. For example, the mandibles in stag beetles (Lucanidae, Coleoptera) develop at a greater rate than other parts of the body. So a larger individual that had an abundance of food as a larva will have disproportionately larger mandibles in comparison with a smaller individual that had less food.

Invasion by certain parasites is known to influence morphologic expression. For instance, mermithid nematodes cause worker or soldier ants to show female traits. Parasitization of certain insects by strepsipterans may also cause such changes in form.

Several environmental factors are known to influence polymorphism in aphids (Homoptera). Aphids may exist as winged (*alate*) or wingless (*apterous*) parthenogenetic forms or as wingless or *brachypterous* (very short wings) sexual forms (see figure 12.32). These forms vary with the environmental situation, and the life cycle of a given species may be complex, including generations of different forms at different times. A variety of environmental factors have been found to influence polymorphism in different species, for example, photoperiod, temperature, changes in host plant (e.g., wilting and nutritive changes), and crowding. The changes in form induced by environmental changes are adaptive. For example, there may be several parthenogenetic generations during long photoperiods when host plants are abundant, and with the shortening photoperiods associated with autumn, sexual forms appear, mate, and produce eggs that will overwinter. Alate forms may be produced in response to crowding, thereby facilitating dispersal. As with termites, the endocrine system (especially juvenile hormone and ecdysone) is thought to be involved in the expression of polymorphism. The physiology of caste development in social insects is reviewed by deWilde and Beetsma (1982).

Crowding results in "phase changes" in some Orthoptera and Lepidoptera. For example, in the migratory locust, *Schistocerca gregaria,* there are changes in color pattern and behavior between solitary individuals and the same individuals when they enter the migratory (gregarious) phase in response to the constant agitation brought about by crowding.

Pheromones (chemicals produced by one individual that influence another individual of the same species; see chapter 8) may influence the expression of particular forms and functions, for example, among the lower termites. These insects typically have an elaborate caste system consisting of winged adults (kings and queens), wingless supplementary reproductive adults, larvae and nymphs that serve as workers and as a pool for transformation into other castes, and soldiers. This caste system is very flexible. Following hatching from the egg, there are several instars called larvae followed by several instars called nymphs. Older larvae have the capacity to develop into nymphs, soldiers, or supplementary reproductives. Nymphs can develop into winged reproductives,

supplementary reproductives, or soldiers and can even regress, becoming larvae again. The caste system is at least partly mediated by substances transferred from one individual to another by *proctodael feeding* (i.e., one individual ingests material from the anus of another individual). These substances regulate the numbers of a given caste in a colony (Lüscher 1961). For example, in many termite species supplementary reproductive castes do not appear as long as the original king and queen are present. The king and/or queen produce a substance that circulates throughout the colony and inhibits the formation of supplementary reproductives. Death of the royal pair would then result in the loss of this inhibitory substance, and hence the supplementary reproductives would appear, assuring the continued survival of the colony. Soldiers may also produce a substance that passes from individual to individual throughout the colony and inhibits development of more soldiers. The substances that inhibit development in larvae and nymphs apparently exert their effect on the endocrine system, in particular on the balance between juvenile hormone and ecdysone (Lüscher and Springhetti 1963). For example, the implantation of additional corpora allata can induce development of larva or nymph into a soldier. Juvenile hormone itself may be passed from individual to individual.

The Royal Entomological Society's symposium, *Insect Polymorphism* (Kennedy 1961), Lüscher (1976), volume 8 of Kerkut and Gilbert (1985), and volume 3 of Gupta (1990) should be consulted for further information.

Regeneration

Most, if not all, insects are probably capable of at least some regeneration, particularly with regard to wound healing. This process often involves hemolymph clotting and hemocytes, but the epidermal cells play the major role. Dead or injured epidermal cells in *Rhodnius prolixus* (Hemiptera, Reduviidae) apparently produce a substance that has an attractive effect on the surrounding epidermal cells, which migrate to a wound and lay down new cuticle (Wigglesworth 1972). Mitoses in the regions from which the epidermal cells migrate concurrently restore the original density of cells in those regions. Some insects (e.g., the walkingstick, *Carausius,* Phasmida) can be decapitated and the head replaced with the result that the epidermis and gut, but not neural tissue, grow together again. However, this is not the case in other insects that have been studied (e.g., *Cimex* and *Rhodnius,* Hemiptera) in which only the continuity of the integument (cells and cuticle) is reestablished.

If appendages of developing larvae are removed, they may be regenerated during later instars, indicating that at least some of the cells surrounding an appendage are sufficiently undifferentiated to retain the ability to reform that appendage. Some insects (e.g., walkingsticks, Phasmida) exhibit the ability to spontaneously amputate a leg, generally between the trochanter and femur. This phenomenon is called *autotomy* and has the obvious adaptive value of allowing escape from the grasp of a predator. Many immature insects (e.g., walkingsticks and mantids) can completely regenerate a lost appendage, and this regeneration usually requires a molt. Evidently the capacity for regeneration of the external form of an appendage is wholly within the epidermal cells, inasmuch as removal of associated ganglia has failed to block regeneration of form, at least in the species studied.

In some instances abnormal regeneration may occur. Abnormalities include the duplication or triplication of an appendage at its tip and *heteromorphous* regeneration, in which the regenerated appendage is like another appendage on the body but unlike the one it replaces. An example of the latter would be the growth of a leg where an antenna had originally been.

Current information on regeneration in insects can be found in volume 2 of Kerkut and Gilbert (1985), Treherne et al. (1988) cover the exciting area of neural repair and regeneration.

Aging

Brief consideration of those processes that ultimately bring about termination of the life of an insect is an appropriate way to complete a discussion of postembryonic morphogenesis. The reviews by Clark and Rockstein (1964), Rockstein and Miquel (1973), and Collatz and Sohal (1986) cover aging at length.

Aging includes all those changes in structure and function that occur from the beginning of life to its termination. These changes are predictable and reproducible. *Senescence* refers to all those changes in structure and function that decrease an individual's capacity for survival and ultimately lead to the death of the individual.

Theories on the causes of senescence are based in two areas, heredity and environment. Probably in the majority of situations, both heredity and environment act together to produce the observed senescence. Hereditary factors are as follows (Rockstein and Miquel 1973):

1. Aging as the result of programmed retardation or cessation of growth.
2. Programmed retardation and/or failure of some substance necessary for maintenance of the nonsenescing state.
3. Depletion of DNA, RNA, enzymes, coenzymes, and other substances essential for cell function.
4. Scheduled production or accumulation of an aging substance.
5. Scheduled accumulation of material(s) that may become harmful to an organism in time.

Environmental factors are as follows (Rockstein and Miquel 1973):

1. Cumulative radiation (ionizing, infrared, etc.) effect.
2. Cumulative pathological effects from parasitic invasions.
3. Cumulative effects of physical insults such as extremes in temperature or mechanical injury.

In instances in which information is available, species have a constant distribution of individual life spans under defined (genetically and environmentally) conditions. Among species, life spans are quite variable, and the overall range is probably fairly accurately reflected by 1 day for adults of certain mayflies to more than 25 years for certain termite species. Differences in mean life span have been found between different genetic strains, between populations with different diets, and between populations exposed to different temperature regimes. Some examples of mean adult life span in different insects are given in table 5.1.

Learning has been shown to be influenced by age in both ants and hymenopterous parasitoids (Papaj and Prokopy 1989). It appears that ants or wasps are more efficient at learning at particular ages.

Table 5.1 Mean Adult Life Span (in Days) of Insects[a]

Insect	Male	Female	Reference[b]
DIPTERA			
Drosophila melanogaster Wild (line 107)	38.1	40.1	Gonzalez (1923)
Vestigial mutant	15.0	21.0	
D. subobscura 9 inbred lines (average)	40.0	36.4	Maynard Smith (1959)
4 outbred populations (average)	56.8	60.0	
Musca domestica	17.5	29.0	Rockstein and Lieberman (1959)
M. vicina	20.8	23.3	Feldman-Muhsam and Muhsam (1945)
Calliphora erythrocephala	35.2	24.2	
Aedes aegypti		15	Kershaw et al. (1953)
LEPIDOPTERA			
Acrobasis caryae	6.5	7.3	Pearl and Miner (1936)
Bombyx mori (unmated)	11.9	11.9	Alpatov and Gordeenko (1932)
B. mori (mated)	15.2	14.2	
Fumea crassiorella (unmated)		5.5	Matthes (1951)
F. crassiorella (mated)		2.3	
Samia cecropia	10.4	10.1	MacArthur and Baillie (1932)
S. californica	8.7	8.8	
Tropea luna	5.9	6.0	
Philosamia cynthia	5.9	7.1	
Telea polyphemus	8.1	10.0	
Collasamia promethea	4.6	7.0	
Pyrausta nubilalis	13.0	17.4	
P. penitalis	6.8	7.7	
Carpocapsa pomonella	9.4	10.2	
HYMENOPTERA			
Apis mellifera Summer bees		35	Ribbands (1952)
Winter bees		350	Maurizio (1959)
Habrobracon juglandis Wild type	24	29	Georgiana (1949)
Small wings, white eyes, mutant	20	24	
H. serinopae	62	92	Clark and Rubin (1961)
ORTHOPTERA			
Blatta orientalis	40.2	43.5	Rau (1924)
Periplaneta americana	200	225	Griffiths and Tauber (1942)
Schistocerca gregaria	75	75	Bodenheimer (1938)

Table 5.1   Continued

Insect	Male	Female	Reference[b]
COLEOPTERA			
Tribolium confusum	178	195	Park (1945)
T. madens	199	242	
Procrustes	374	338	Labitte (1916)
Carabus	323	386	
Necrophorus	232	291	
Dytiscus	854	740	
Hydrophilus	164	374	
Melolontha vulgaris	19	27	
Cetonia aurata	57	88	
Lucanus cervus	19	32	
Dorcus	327	375	
Ateuchus	338	467	
Sisyphus	198	266	
Copris	497	623	
Geotrupes	700	642	
Oryctes	37	55	
Blaps mortisaga	848	914	
B. gigas	700	728	
B. magica	700	728	
B. edmondi	700	728	
Akis	854	951	
Pimelia	669	714	
Timarcha	135	182	

[a]From Clark and Rockstein (1964).
[b]Full citations can be found in Clark and Rockstein (1964).

CHAPTER 6

Sensory Mechanisms; Light and Sound Production

Sensory Mechanisms

In chapter 3 we discussed the system responsible for the collection, integration, and interpretation of information from the external and internal environment (the nervous system) and the systems that enable an insect to respond and hence adjust to environmental changes (the muscular and endocrine systems). We shall now consider the structures that carry on the actual collection of environmental information: the various sense organs or *sensilla* (sing. *sensillum*).

The basic function of any sense organ is to receive some form of energy (*stimulus*) from the environment and initiate a chain of events that ultimately results in a nerve impulse (Dethier 1963). The transformation of energy from one form to another is called *transduction*.

A large portion of the energy "sensed" by insects is in the form of various mechanical changes, either gross changes such as the bending of a hair or the stretching of a portion of the body, or molecular movements in the form of sound waves propagated through a solid, liquid, or gaseous medium. The sensation of these changes falls under the general heading of *mechanoreception*.

Another form of energy perceived by insects is "potential energy existing in the mutual attraction and repulsion of the particles making up atoms" (Dethier 1963). The perception of this form of energy is referred to as *chemoreception*. If the molecules are water, *hygroreception* is the appropriate term.

In other instances, the energy stimulating a given sense organ may be in the form of electromagnetic waves (or photons). The sensation of electromagnetic waves/photons is called *photoreception*. Or the stimulating energy may be the kinetic energy associated with the random motion of molecules, that is heat. Sensation of heat is called *thermoreception*.

Morphology of Sense Organs

The majority of sense organs are composed of two types of cells: *receptor cells* and *accessory cells*. Receptor cells are usually bipolar neurons that perform the actual detection of stimuli and generation of the nervous impulse. They are modified epithelial cells that during the development of an insect send out a process (the axon or neurite) that eventually communicates with the central nervous system. Accessory cells surround the receptor cells and are usually involved with or actually secrete the specialized cuticular structures that make up the most obvious parts of a sense organ. In addition to the type of sense organ just described, there are multipolar receptor neurons having no contact whatsoever with the cuticle. These multipolar neurons are associated with the various muscles, the alimentary canal, and entad surface of the integument.

Sense organs that communicate with the external world can be classified morphologically on the basis of the differences in associated cuticular structures. It is thought that most external sensilla were originally derived from setae and hence are homologous structures. However, photoreceptors do not fall into this grouping and will therefore be discussed separately.

Because the various sensilla are considered to be derived from setae, the hair sensilla (*sensilla trichoidea*) should be the first type discussed. Recall from chapter 2 that each hair is formed by two cells, the "hair-forming" trichogen cell and the surrounding "socket-forming" tormogen cell. The addition of one to several bipolar receptor cells to this picture produces the basic *trichoid sensillum* (figures 6.1*a* and 6.2). Other closely related sensilla include those with bristlelike processes (*sensilla chaetica*), scalelike processes (*sensilla squamiformia*), and peglike or conelike processes (*sensilla basiconica;*

Figure 6.1 Types of sensilla related to the trichoid type (diagrammatic). (*a*) Trichoid. (*b*) Coeloconic. (*c*) Ampullaceous. (*d*) Campaniform. (*e*) Placoid.

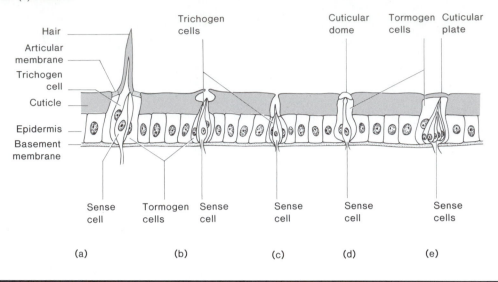

Figure 6.2 Scanning electron micrographs of sensilla on the antennal flagellum of a worker honey bee. 1, trichoid; 2, basiconic; 3, campaniform; 4, placoid; 5, coeloconic or ampullaceous. (Scale lines = 10 μm.)
Courtesy of Alfred Dietz and Walter J. Humphreys.

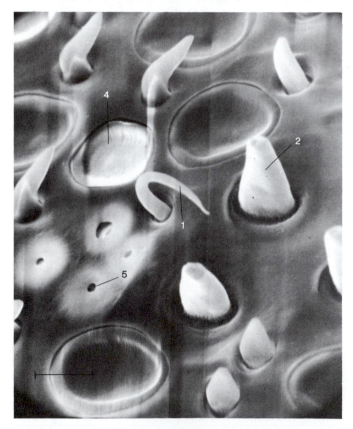

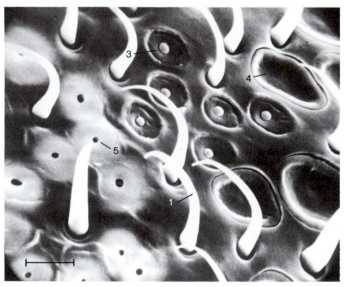

Figure 6.3 Simple chordotonal organ in a beetle larva. *Redrawn from Hess, 1917.*

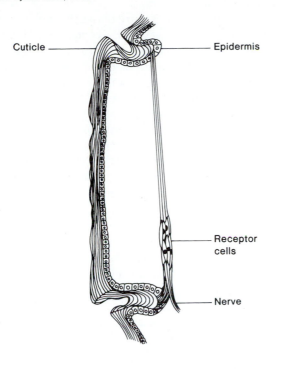

Cuticle

Epidermis

Receptor cells

Nerve

figure 6.2). A further modification consists of processes such as those just described, which are sunken in shallow pits (*sensilla coeloconica;* figures 6.1*b* and 6.2) or comparatively deep pits (*sensilla ampullacea;* figures 6.1*c* and 6.2).

Two other types, *campaniform sensilla (sensilla campaniformia;* figures 6.1*d* and 6.2) and *placoid sensilla (sensilla placoidea;* figures 6.1*e* and 6.2) lack hairs, pegs, cones, or bristles. Campaniform sensilla are shallow round or oval pits and in longitudinal section consist of a bell-shaped cuticular cap innervated by a single receptor cell. Placoid sensilla are platelike structures made up of a round or oval cuticular plate surrounded by a narrow membranous ring. In contrast to the campaniform type, a placoid sensillum is innervated by a number of receptor cells. A type of sensillum that is rather dramatically different from those already described is the *scolopophorous organ,* or *chordotonal organ* (figure 6.3). This kind of sensillum usually occurs in bundles of structures called *scolopidia* (or *scolophores*). Each scolopidium consists of a bipolar neuron enveloped by a scolopale cell, and an attachment cell. Scolopidia, and hence chordotonal organs, are usually stretched between two internal integumental surfaces. Burns (1974) describes the ultrastructure of a chordotonal organ in the femora of the locust.

A given morphological type of sensillum varies both in the appearance of the associated cuticular structures and the numbers of receptor neurons. In addition, a morphological type does not necessarily imply a particular function, because a given sensillum may have different functions in the same insect or may contain two or more receptors, which receive different forms of stimuli. For example, a single hair on the labellum of the blow fly, *Phormia,* may have four chemoreceptor cells and a mechanoreceptor cell associated with it (Dethier 1963; see figure 3.10). However, there are certain sensilla that always seem to be associated with the same general function. For example, the campaniform type has always been found to be a mechanoreceptor, which is stimulated by deformation of the cuticle. The same is true for the chordotonal sensilla, which are sound, vibration, and stretch receptors.

The current literature on insect sensilla may include the use of terms that are different from those presented here to identify the various types of sensilla. One newer classification scheme is based upon whether or not the sensillum has a cuticular pore and, if so, how many. This type of classification results in the terms *aporous* (i.e., without a pore and usually indicating a mechanosensillum), *uniporous* (i.e., a single generally terminal pore normally suggesting a contact, gustatory, or taste sensillum), and *multiporous* (i.e., having many pores and typically signifying an olfactory sensillum). More can be obtained on this newer classification scheme by consulting volume 6 of Kerkut and Gilbert (1985) and Zacharuk and Shields (1991).

Methods Used to Study Insect Sense Organs

Earlier work with insect sense organs generally consisted of a combination of behavioral and morphological approaches. Based on their gross appearance and appearance in stained sections under the light microscope, structures were identified as sensory or not sensory. This approach did not, of course, enable investigators to be certain of the function of a given sense organ, but it did allow (sometimes accurate) speculation. The use of behavioral criteria allowed the process to be carried somewhat further since innate or learned responses to various stimuli could be taken advantage of in different ways. For instance, an insect may be naturally attracted to, repelled by, or respond in some other way to a given stimulus. Once this fact is established, a structure or structures suspected of receptor activity can be removed or blocked in a variety of ways to determine if they are, in fact, active with regard to the stimulus being tested.

When a specific receptor or specific receptors are identified, further experiments can be made to determine their sensitivity by varying the concentration or intensity of the stimulus. An excellent example of this kind of work has been carried out on the tarsal receptors of the blow fly (Dethier 1955). The tarsal sense organs are "taste" or contact chemosensilla and are sensitive to various sugars, salts, and water. When a solution of a "tastable" sugar is applied to them, the fly extends its labellum in response. Using this labellar response, the sensitivities to various concentrations of several sugars have been worked out.

These earlier approaches are still used today but have been complemented by the much more detailed observation possible with the electron microscope (both transmission and scanning) and by the use of highly sophisticated electrophysiological techniques. The latter have enabled investigators to make critical studies of the functions of the sensory structures at the cellular level.

Sensory Fields

Insect sensilla of diverse functions may tend to be concentrated on specific body regions or appendages called sensory fields. In addition to obtaining food, the head serves as the insect's major source of sensory fields. The major sensory fields of most insects are the antennae, mouthparts, legs and wings, genitalia and cerci, and the ovipositor. For a particular species, the sensory fields usually are different for the immature stage than for the adult, especially among the holometabolous insects. Generally speaking, researchers tend to specialize and research one type of sensory field (e.g., the chemosensilla on the tarsi or the sensilla on the antennae).

Mechanoreception

Organs that are sensitive to the actions of stretching, bending, compression, torque, and so on applied to the integument or some internal organ are the *mechanosensilla*. These sense organs are responsible for the maintenance of posture, stability during locomotion, and body position with respect to gravity. In addition, many are designed so that they enable the insect to detect sound waves or vibrations in a solid substrate. In certain instances they provide information as to the state of certain internal organs (e.g., the alimentary canal).

Insects are known to possess the following mechanoreceptive senses: tactile (touch), proprioceptive, and sound or vibration. Mechanosensilla can also be grouped as Type I, those associated with the cuticle or cuticular invaginations, and Type II (*stretch receptors*), those associated with the viscera, but not the cuticle (McIver 1975, 1985).

McIver (1975, 1985) and Schwartzkopff (1974) provide reviews of the structure, physiology, and biological functions of mechanoreceptors.

The Tactile Sense

The external receptors involved in the sense of touch are typically hair sensilla. Movement of a hair affects the associated bipolar receptor cell(s). The distal process (dendrite) of a receptor cell is in very close contact with the base of a mechanoreceptive hair and contains an array of

microtubules, the *tubular body*. The mechanism by which deformation of the tubular body initiates a nervous impulse is not understood.

Many hair sensilla are strictly mechanoreceptive in function and have only a single receptor cell. However, others may possess a number of other kinds of receptor cells (e.g., chemical) in addition to the mechanoreceptive ones. An example of the latter was given earlier—certain hairs on the labellum of the blow fly, *Phormia*.

Mechanosensitive hairs appear to fall into two groups: velocity sensitive and pressure sensitive. In the first group neural impulses are generated only in the presence of constantly changing stimuli, for example, in an insect during flight. Not surprisingly, these hairs have been found on the anterior edges of the wings of the various insects. Hair sensilla that fall into the second category initiate a steady train of nervous impulses when statically deformed. This situation would occur, for example, when an insect is at rest on a solid substrate.

Hair sensilla are commonly found on the legs, mouthparts, and antennae, all of which frequently come into direct contact with the substrate or other surfaces. They are also found on the cerci and here may initiate an escape response resulting from air movement or something suddenly touching these appendages and consequently bending the associated hairs (e.g., in cockroaches and grasshoppers). Hair sensilla on the anal papillae of silkworm moths (*Bombyx mori*) provide information on the unevenness of the oviposition site and are thus important in regulating the position of eggs within egg clusters (Yamaoka et al. 1971). The hair sensilla described in this paragraph are usually velocity sensitive.

In addition to being found in locations in which they come into contact with other than the insect body itself, many hair sensilla are located between joints, between body segments, and in other areas in which there is direct contact between two body surfaces. Their function in these situations is proprioceptive, and they are generally of the pressure-sensitive type.

The Proprioceptive Sense

Proprioceptive structures are stimulated by changes in various parts of the insect body. The changes may be in length, tension, and so on. These receptors provide the insect with continuous information as to the position of the various body parts and the tensions of the various muscles. It follows, then, that these types of sensilla are of critical importance in the maintenance of "proper" orientation of the body parts with respect to one another or of the entire body with respect to gravity in both the stationary and the moving insect. You will find examples of this later in the section on locomotion. The role of proprioceptors in orientation is complemented by the actions of other receptors (e.g., tactile and photoreceptors). Mill (1976) provides a series of reviews on many aspects of

Figure 6.4 Transverse section of the head of the ant *Formica polyctena*, showing hair plates on the prothorax (diagrammatic).
Redrawn from Markl, 1962.

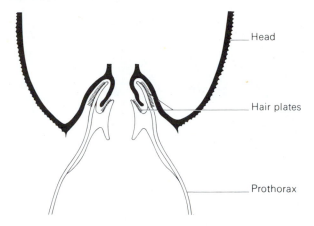

Figure 6.5 (*a*) Female apple maggot ovipositing into an apple. Notice the egg in the hole on the right. (*b*) Scanning electron micrograph of the ovipositor showing the numerous campaniform sensilla and some of the hair-shaped mechanosensilla.
Taken from Stoffolano and Yin, 1987; (b) courtesy of Stoffolano.

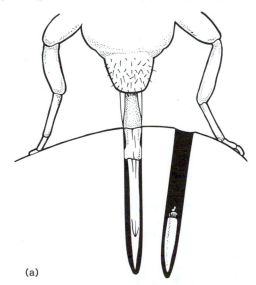

(a)

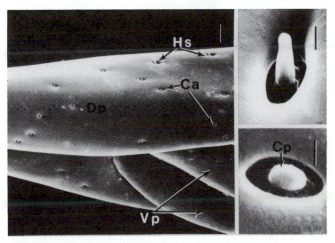

(b)

invertebrate proprioceptors, including those of insects. Weis-Fogh (1971) and Wendler (1971) consider proprioceptors in insects. Horn (1985) reviews the role of receptors in gravity perception.

A number of different types of sensilla function as proprioceptors. Among these are clusters of tactile hairs or hair plates, campaniform sensilla, stretch receptors, and chordotonal organs (including Johnston's organ).

Hair plates are very common in insects and appear as clusters of tiny trichoid sensilla. They may be looked upon as tactile receptors that respond to the insect "touching" itself since they are usually located in appressed or overlapping areas of the body. For example, in the ant, hair plates are located between certain segments of the antennae, in the neck region between the head and thorax (figure 6.4), at the bases of the coxae and the trochanters, in the petiolar region between the first and second abdominal segments, and in the ventral gaster region between the second and third abdominal segments. Because a given position of one body part relative to another would "stimulate" particular hairs of specific plates to specific degrees, a unique sensory pattern would impinge on the central nervous system for every possible body position and movement. These hair plates in the ant are then important in the maintenance of "proper" posture, whether the insect is stationary or moving. In the praying mantis, hair plates in the neck region function in the process of prey capture. As the mantis turns its head, visually following potential prey, the changing pattern of impulses from the hair plates is critical in determining the accuracy of the strike with its grasping forelegs. If the operation of the hair plates is experimentally interfered with by denervation or immobilization of the head, the accuracy of the strike is seriously impaired or destroyed altogether (Roeder 1967). Hair plates on the

vertex of the head of locusts (*Schistocerca* and *Locusta*, Orthoptera) sense airflow and are involved in the regulation of flight.

Campaniform sensilla serve as compression and stretch receptors and are located only in areas of the integument that are exposed to strains of various sorts. They are concentrated particularly in areas where compression and stretching occur as a result of muscular activity, for example, in the legs, wings, halteres, the bases of the mandibles, and ovipositors. Figure 6.5 shows a female apple maggot, *Rhagoletis pomonella,* ovipositing into an apple. Located all over the surface of the ovipositor are campaniform sensilla, which monitor the pressure necessary to penetrate the skin and pulp of the apple fruit

Figure 6.6 Photomicrograph of a sagittal section of Johnston's organ in the mosquito *Aedes triseriatus*. 1, Nerve; 2, scolopophores; 3, cell bodies of neurones; 4, pedicel. (Scale line = 50 μm.)

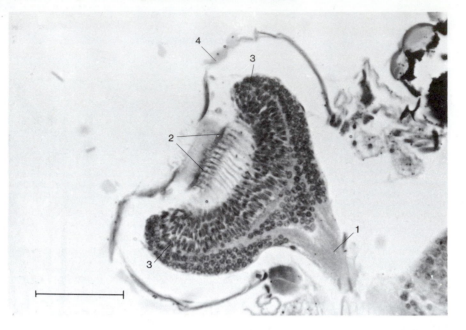

for placement of the egg. The role of campaniform sensilla in association with the legs, wings, and halteres will become clear when we discuss locomotion (chapter 7).

Multipolar neurons associated with muscles, the alimentary canal, and entad surface of the integument have been found to act as stretch receptors. They respond with a nervous impulse when the tissue in which they are embedded is subjected to a change in length. They have been found in dragonfly nymphs and members of the orders Orthoptera (grasshoppers and relatives), Hymenoptera (ants, bees, wasps, and relatives), and Lepidoptera (moths and butterflies) (Dethier 1963).

Chordotonal organs, as explained earlier, are usually stretched between two internal integumental surfaces. They have been found in the pedicel of the antennae in all insects studied, and are commonly found in the mouthparts, wing bases, halteres, legs, and abdominal segments. Chordotonal organs have also been found closely associated with tracheae and pulsatile structures and in the hemocoel. Early investigators ascribed an auditory function to them, and, in fact, many of them are auditory in function. However, a number are known to be proprioceptive, and as Dethier (1963) points out, very likely "all not associated with tympanic membranes or grouped to form subgenual and Johnston's organs . . . will eventually be proved to be proprioceptors." Several different proprioceptive functions have been suggested for chordotonal organs, for example, sensation of body orientation, passive body movements, and muscular movements. Their close associations with tracheae, pulsatile

organs, and the various hemocoelic cavities suggests that they may respond to changes in intertracheal air pressure and in blood pressure.

In virtually every insect studied, a specialized group of sense organs, which are quite similar in structure to chordotonal sensilla, are found in the pedicel (second segment from the base) of each antenna (figure 6.6). These structures are attached to the pedicellar wall and to the membrane between the pedicel and the third antennal segment and are in a radial arrangement. They make up the *Johnston's organ*. This structure varies in complexity depending on the insectan species and reaches its greatest development in two families of flies (Culicidae, mosquitoes; Chironomidae, midges). In these two dipteran families it completely fills the pedicel. Johnston's organ is known to function as a proprioceptor in some insects, although it has different functions in others (e.g., sound reception in mosquitoes and midges). An example of an insect in which it has a proprioceptive function is the honey bee. During flight, Johnston's organ responds to movements of the antennal flagellum and in this way provides the bee with a measure of the stream of air passing over it. The amplitude of the wingbeat is regulated on the basis of this measurement (Wigglesworth 1972).

Another example is the use of the antennal movements and hence Johnston's organ by certain aquatic Hemiptera (true bugs). Some of these (*Corixa* and *Naucoris*) swim dorsal side up, whereas others (*Notonecta* and *Plea*) swim dorsal side down. In either case, the proper body orientation during swimming is maintained

Figure 6.7 Section of a metathoracic tympanal organ of a noctuid moth. ✕ indicates a tracheal air sac. *Redrawn from Roeder, 1959.*

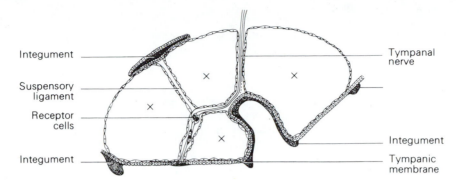

because the insect is able to sense when its dorsum is up or down. This is accomplished by the buoyant action of a small air bubble trapped between the ventral part of the head and each antenna. Any change in the position of the insect results in a change in the direction of the buoyant force of the bubble relative to the insect and hence results in a movement of the antennae, which in turn causes a change in the sensory patterns generated by each Johnston's organ.

Sound Perception

For the purposes of this discussion, we define sound as longitudinal waves of kinetic energy propagated through a continuous medium (gas, liquid, or solid), and divide perception of sound into two senses: the sense of vibration, perception of sound via the substratum, and the sense of hearing, perception of sound via air or water.

The perception of sound is important in a number of ways. Many of the stimuli that impinge on an insect from the external environment are in the form of sound. Some of these sounds are produced by other insects of the same or different species (see the section on sound production in this chapter), and other sounds come from a variety of environmental sources. Sound perception may be of value in sensing potential danger, a potential mate, prey, other members of the same species (e.g., when individual territories are maintained), and so on.

Sensilla Involved in Sound Perception
Only two basic types of sensilla have definitely been shown to be involved in sound reception—trichoid sensilla and specialized organs composed of chordotonal receptors. However, others, such as campaniform sensilla or stretch receptors, may also be involved. We shall consider first the organs composed of chordotonal sensilla and then the trichoid sensilla.

Those sound-sensitive structures composed of chordotonal sensilla are the *tympanic organs, subgenual*

organs, and *Johnston's organ,* which was described earlier. Most tympanic organs are composed of the same fundamental parts. However, the degree of development of these parts varies from group to group. The basic structures (figure 6.7) involved are a thin integumental area (the *tympanum*) and a group of chordotonal sensilla attached directly or indirectly to the entad surfaces of the tympanum. A tracheal air sac is usually closely associated with the tympanum and sensilla. In some insects (e.g., male cicadas) the tracheal air sac may serve to amplify certain frequencies. The number of chordotonal receptors varies from 2 in moths in the lepidopteran family Noctuidae to 1500 or more in cicadas (Dethier 1963).

Tympanic organs have been identified in a number of different locations in a variety of insects. In the order Orthoptera (grasshoppers and relatives), they are found on the tibiae of the forelegs in the families Tettigoniidae (long-horned grasshoppers; figure 6.8*a,b*) and Gryllidae (crickets) and on either side of the first abdominal segment in members of the family Acrididae (short-horned grasshoppers). Tympanic organs occur in the metathorax of noctuid moths and in the abdomen in geometrid and pyralid moths. In cicadas (Homoptera) these structures are located in the abdomen. Tympanic organs in nocturnal flying members of the Orthoptera, Neuroptera and Coleoptera are used to detect ultrasounds produced by echolocating, insectivorous bats. Once a bat is detected, the insect either remains silent, if stationary, or, if flying, goes into an evasive flight pattern. In addition to the role of hearing in defense, hearing is also often involved in reproductive communication. Spangler (1988) reviews the literature concerning the role of moth hearing and discusses both its significance to the insect and the possible origin of defensive flight behavior in moths. In addition, tympanic structures have been identified in a few members of the order Hemiptera (e.g., in water boatmen, family Corixidae). Examples of specific functions of hearing via the tympanal organs will be discussed in chapter 8.

Figure 6.8   Tympanic organ on the foretibia of a long-horned grasshopper. (*a*) Anterior view of the tibia. (*b*) Transverse section at the level indicated approximately by the line x—x in (*a*), epidermal cells omitted. (*b*) *redrawn with modifications from Schwabe, 1906.*

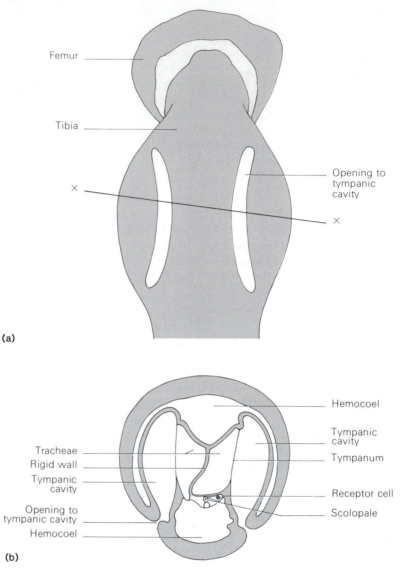

(a)

(b)

Subgenual organs (figure 6.9) are groups of chordotonal sensilla located in the basal portion of the tibial leg segment. They are not associated with any joints. They vary considerably in degree of development from group to group, being somewhat weakly developed in the true bugs, more developed in the members of the orders Lepidoptera and Hymenoptera, and most developed in the beetles and the true flies.

Vibration Perception

Among the insects in which vibration sensitivity has been measured, those showing the greatest sensitivity have been found to possess subgenual organs (Schwartzkopff 1974). These organs are thought to be specifically involved in the perception of vibration. Subgenual organs have not been found in the less sensitive species, in which the trichoid

and small chordotonal sensilla in the legs are the vibration receptors. In *Locusta* (Orthoptera, Acrididae) trichoid sensilla on the sternites are sensitive to substrate vibration (Dethier 1963).

A number of large insects (e.g., crickets, cicadas, and katydids) use loud airborne sounds that permit communication over a considerable distance. Because of their size limitations, smaller insects would have to use very high frequency sounds to communicate over comparable distances. Unfortunately, these high frequency sounds are not conducive for communication in the complicated habitats of most smaller insects, which so often involve complex plant surroundings. The communication option used by many of these insects to communicate is to use low frequency vibrations that are transmitted or propagated via the plants they inhabit. Claridge (1985) focuses on

Figure 6.9   Subgenual organ of an ant exposed by section of the tibia.

Redrawn with modifications from Schön, 1911.

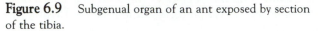

Nerve

Sense cells

Enveloping cells

Cap cells

Accessory cells

how leafhoppers and planthoppers produce and transmit sound (e.g., using plant substrates as the medium of transmission) and even mentions sound production in the aphid genus *Toxoptera*. The symbiotic association between some ant species and butterfly caterpillars, especially of the butterfly family Riodinidae, is apparently maintained by the caterpillars using their vibratory papillae to produce sounds that travel throughout the plant. These sounds are used by the ants to locate the caterpillars (DeVries 1990). In this mutualistic relationship, the ants receive food secretions from the caterpillars, which in return are protected from their predators by the ants. By examining butterfly species from diverse geographic regions, DeVries suggests that this ability to produce sounds has evolved independently at least three different times. Similar relationships between insects using sound vibrations traveling throughout plant materials, either to maintain symbiotic relationships or to locate prey, is probably more common than is currently known. A general review of acoustic and vibrational communication in insects is that of Kalmring and Elsner (1985).

Hearing

Schwartzkopff (1974) describes the use of aquatic surface waves by water striders (Hemiptera, Gerridae), backswimmers (Hemiptera, Notonectidae), and whirligig beetles (Coleoptera, Gyrinidae) as "preliminary stages of hearing." Water striders and backswimmers are predaceous and orient to their prey by the surface waves. These waves are sensed by receptors in the tarsi. The tarsal organ in the backswimmer *Notonecta glauca* is composed of scolopidia. Water striders maintain tarsal contact with the water's surface from above and backswimmers from below. Whirligig beetles use surface waves in navigation, avoiding moving and stationary objects,

as they dart to and fro along the water's surface. The surface waves are sensed by Johnston's organ in the antennae.

Tympanic organs, trichoid sensilla, and Johnston's organ have been shown to perceive sound in a number of insects. The tympanal organs are, as far as known, always used for hearing. However, trichoid sensilla and Johnston's organ have been demonstrated to have other functions. As already mentioned, tympanic organs in the nocturnal noctuid and related moths detect the ultrasonic emissions used by echolocating, predatory (insectivorous) bats (Roeder 1965, 1966, 1967). Detection of these sounds stimulates avoidance behavior of various sorts (see chapter 8). In the orders Orthoptera and Hemiptera tympana receive sounds produced by members of the same species, and these sounds are involved with sexual behavior. In the water boatmen, *Corixa* (Hemiptera, Corixidae), males produce waterborne sounds that are perceived and responded to by the females. Evidently males also hear each other produce these sounds, because they chirp as long as the tympanic organs are intact. These are the only insects known to be sensitive to waterborne sounds (Dethier 1963). Since tympanal organs occur bilaterally on some part of the insect body, they facilitate localization of a sound through differential stimulation.

Certain trichoid sensilla on exposed regions of the body in a number of insects have been shown to be sensitive to airborne sounds. Sensilla on the cerci of cockroaches (Blattaria) are especially sensitive, and stimulation by sound may elicit the characteristic "alarm" reaction mentioned earlier.

Johnston's organ in the antennae of some species of mosquitoes and midges has been shown to be a sound receiver. The males are able to detect the sounds produced by the rapidly beating wings of the females. Response to these sounds can be elicited by using a tuning fork that sounds at an appropriate frequency.

The green lacewing, *Chrysopa carnea,* has a swelling of the radius of each forewing that contains two scolopophorous structures. These sensilla respond to the ultrasonic chirps of echolocating bats (Miller 1970).

Hawk moths (Lepidoptera, Sphingidae, Choerocampinae) detect ultrasound by means of structures associated with the mouthparts (Roeder 1972). The second palpal segment is bulbous and is composed almost completely of an air sac. The medial region of this palpal segment rests against the distal lobe of the *pilifer,* a small appendage associated with the labrum. Ultrasonic vibrations are translated via the palps to the pilifer, which contains the sensory transducer.

Chemoreception

As defined earlier, chemoreception is the process by which the "potential energy existing in the mutual attraction and repulsion of the particles making up atoms" is detected (Dethier 1963). Thus chemoreceptive organs are responsive to direct contact with atoms and molecules. Chemical cues from the environment are useful to insects in several ways—for example, food (host plant or animal, prey, decaying organic material, and so on) procurement, mediation of caste functions in social forms, mate location, identification of noxious stimuli that are a potential threat to survival, selection of oviposition site, and habitat selection.

In general terms the chemoreceptive activities of insects may be divided into three "chemical senses": distance chemoreception, or *olfaction,* contact chemoreception, or *gustation,* and "general," or "common" chemical sensitivity.

Distance chemoreception is mediated by chemosensilla that are responsive to molecules or ions of a chemical in the gaseous state at comparatively low concentrations. These sensilla are very sensitive and may show a high degree of specificity with regard to the kind of chemical that elicits a response.

Contact chemosensilla are excited by direct contact with molecules or ions of a chemical in solution at a concentration usually somewhat higher than olfactory chemostimuli. Generally, these sensilla are less sensitive than the distance chemosensilla and are commonly associated with feeding activities.

The "general" chemical sense involves sensilla that are comparatively insensitive except to relatively high concentrations of a stimulating chemical. These sensilla are much less discriminating than either contact or distance chemosensilla and are usually associated with an avoidance or escape response. No general chemical sensilla have been positively located, and nonspecific effects on neurons may be involved (Hodgson 1974).

Although the classification of the "chemical senses" presented above is quite useful, it is not without difficulties. First, the distinction between distance and contact chemoreception must be qualified by pointing out that in both types the molecules or ions of the stimulating chemicals always come into direct contact with the receptor cell membranes. Second, in aquatic and subterranean insects the same receptors may respond to chemicals either in the gaseous state or in aqueous solution. Third, there are instances among terrestrial insects (e.g., blow fly taste sensilla) of contact sensilla reacting to certain volatile substances.

A wide variety of the morphological types of sensory structures may be involved with chemoreception. Among those for which there exists evidence of olfactory activity in certain insects are the sensilla trichoidea, sensilla basiconica, sensilla placoidea, and sensilla coeloconica. Sensilla trichoidea and sensilla basiconica have been identified as contact chemosensilla in a number of insects. Chemical stimuli reach the nerve endings via pores in the cuticle. Olfactory sensilla typically bear many of these pores (figure 6.10a), whereas contact chemosensilla usually have only one or two located distally (figure 6.10b). The number of receptor cells associated with a chemosen-

sillum varies from a very few—three to four neurons in *Phormia regina* (Diptera, Calliphoridae)—to fifty or more (Hodgson 1974).

Proved or suspected contact chemosensilla have been found in several parts of the insect body. They probably exist in the mouthparts of all insects, for example, the hypopharynx and epipharynx in caterpillars, the tips of the maxillary and labial palps and in the buccal cavity of the cockroach *Periplaneta* sp., and possibly the buccal cavity (cibarium) of mosquitoes. Chemosensilla probably also exist in the foregut; for example, sensilla in the pharynx of the house fly apparently are involved in the control of the passage of ingested food.

In honey bees and certain wasp species contact chemosensilla are located on the distal segments of the antennae, enabling the insects to differentiate between sweetened and unsweetened water (Wigglesworth 1972). The distal portion of the tibia and tarsi of the forelegs of many insects bear contact chemosensilla. In many butterflies, true flies, and bees the mouthparts are extended in response to stimulation of the fore tarsi with sugar water. Contact chemosensilla are located on the ovipositors of parasitic Hymenoptera, which deposit their eggs directly into host insects or in insects whose ovipositor penetrates either the vegetative parts or fruit of a plant, e.g., the apple maggot fly also has chemosensilla near the tip of the ovipositor (Stoffolano and Yin 1987). The wasp *Venturia canescens* (Hymenoptera, Ichneumonidae) can distinguish a parasitized host from an unparasitized host with its ovipositor (Ganesalingam 1974). Similarly, distance (olfactory) receptors have been identified in the antennae and mouthparts of a variety of insects and in the ovipositor of at least one. Bell and Cardé (1984) cover the ecological aspects of insect chemoreception.

Considerable work has been done with the contact chemoreceptive abilities of the honey bee. This insect is capable of differentiating the qualities sweet, bitter, acid, and salt (von Frisch 1950). The honey bee and human beings have been compared with regard to their sugar-tasting abilities. According to Wigglesworth (1972), "out of 34 sugars and related substances tested, 30 appear sweet to man, only 9 to the honey bee; all these nine being present in the natural food of the bee and capable of being metabolized by it." Honey bees are a little more sensitive to bitter substances such as quinine, whereas acetylsaccharose is exceedingly bitter to humans but not to the honey bee. This substance has been suggested for addition to cane sugar so that the sugar might be sold more cheaply to beekeepers and not used as human food (Wigglesworth 1972).

Von Frisch (1950) established two thresholds for honey bees relative to solutions of the sugar sucrose (cane sugar). The first he called the *threshold of acceptance,* which he defined as the minimum concentration of sugar that the bees could be induced to ingest. This threshold was established at about 40% when many of the bees' foraging plants were in bloom and approximately 5% in the

Figure 6.10   Chemosensilla. (*a*) Olfactory sensillum. (*b*) Contact chemosensillum. (Diagrammatic; based on several sources.)

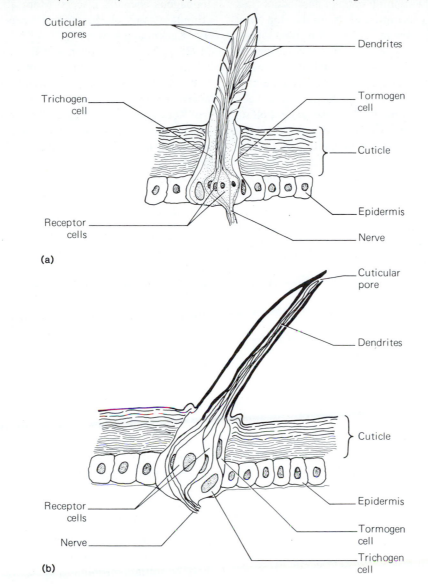

fall when flowers were scarce. The second threshold von Frisch measured was the *threshold of perception,* that is, the minimum concentration of sucrose that the bees could perceive. This threshold, if accurately determined, would not be expected to vary with the scarcity or abundance of flowers. By starving the bees for several hours and then determining the minimum concentration of sucrose they would accept, he established this threshold as somewhere between 1% and 2%. Contact chemoreception has also been extensively studied in a number of true flies (*Phormia* sp., *Calliphora* sp., *Musca* sp., *Drosophila* sp., and others). In those instances, both behavioral and electrophysiological techniques have been used to establish thresholds and to gain insight into the basic mechanisms involved. Dethier (1976) provides information on contact chemoreceptors and their role in the feeding behavior of *Phormia regina* (Diptera, Calliphoridae).

Olfactory sensilla may occur in tremendous numbers. For example, in the antennae of male polyphemus moths (*Antheraea polyphemus,* Lepidoptera, Saturniidae) there are more than 60,000 sensilla with about 150,000 sense cells. Most of these sensilla are receptors for the female sex pheromone (approximately 60%–70%; Schneider 1969).

Olfactory cues play a major role in the lives of many insects; they are important in all forms of behavior involving chemical communication (see chapter 8). The distance chemoreceptive abilities of some insects are fantastically acute. For example, the male silkworm moth, *Bombyx mori,* reacts to the sex pheromone (*bombykol*) produced by the female at a concentration as low as 100 molecules of attractant per cubic centimeter of air (Wilson 1970). A single molecule of female sex pheromone is sufficient to trigger an impulse in the male receptor cell (Schneider 1974). Considerable interest is being shown in identification of various host odors (i.e., for vertebrates and plants) in an effort to develop traps or better understand how the insect locates its food source. Visser (1986) focuses on host odors involved in phytophagous insects finding their food or oviposition sites.

As with contact chemoreception, the distance chemoreceptive abilities of the honey bee have also been widely investigated. One would, of course, expect olfactory abilities to be great in these insects since their life depends to a large extent on flowering plants, and it would be of obvious advantage for a bee to be able to visit repeatedly a particular kind of flowering plant that was currently in bloom. Evidently, the olfactory abilities of honey bees and humans are similar in terms of threshold concentration of various scents; however, honey bees seem to have a great ability to discriminate among many different scents (Wigglesworth 1972).

Sensory Coding: Phagostimulants and Phagodeterrents

When only one chemosensory hair on the tarsus of the fly is stimulated with a solution, how does the fly know whether to accept or reject it? If it accepts, it will extend its proboscis; if it rejects, it will fail to lower the proboscis. In other words, how does the fly interpret the nervous impulses sent from the 3 major incoming chemosensilla (i.e., from the salt-sensitive neuron, water-sensitive neuron, and sugar-sensitive neuron) of a tarsal hair to the central nervous system? Recall that the largest spike or action potential from an individual tarsal-hair sensillum was produced by the salt-best neuron in response to salts, whereas the sugar-best neuron produced a medium-sized spike with the water-best producing the smallest spike (figure 3.10). Thus, the peripheral message sent to the central nervous system of the fly comes from the tarsal chemosensilla and consists of a mixture of these three types of spikes. More simply stated, what the fly interprets as acceptance, or as a phagostimulant, is a message received within a few milliseconds, that contains more sugar-sensitive spikes than either the water or salt-sensitive neurons. On the contrary, rejection, or a phagodeterrent solution, would result in a message received within the same time interval that contains more of the salt-sensitive spikes. It is believed that 1 impulse from a deterrent cell (i.e., the salt-sensitive neuron) can counter 2.5 impulses from a stimulatory cell (i.e., the sugar-sensitive neuron). Information about phagostimulants is extremely important for control strategies that incorporate a feeding stimulant as well as a toxicant. A good example is the bait used for the Mediterranean fruit fly, which includes a sugar plus the insecticide malathion. Also, information concerning phagodeterrents is vital for plant breeders who hope to develop plants that are resistant to specific pests. In addition, current research (e.g., on the deterrent in the Neem seed) is being obtained to use the deterrent as a spray to protect crops from phytophagous chewing insects. More about coding, phagostimulants, and phagodeterrents can be found in volume 9 of Kerkut and Gilbert (1985), Dethier (1976), Dethier (1988), and Schoonhoven and Blom (1988).

For further information on chemoreception, see Hodgson (1958, 1965, and 1974), Lewis (1970), Schoonhoven (1977), Slifer (1970), Chapman (1982) and volume 6 of Kerkut and Gilbert (1985). Consult Kaissling (1986) for the latest information concerning transduction mechanisms in insect chemosensilla. Schneider (1969), Seabrook (1977), Payne et al. (1986) and Kaissling (1986) deal specifically with olfaction.

Thermoreception

Based on definite behavioral responses, it is well established that many insects are sensitive to changes in temperature. For example, honey bees trained to visit warm places are able to detect temperature differences as small as 2° C (Wigglesworth 1972). In some insects, heat sensitivity seems to be somewhat generalized over the entire body. In others, specific locations have been identified. Temperature receptors have been found on the antennae, maxillary palps, and tarsi of many insects. For example, in the blood-sucking bug *Rhodnius prolixus* (Hemiptera, Reduviidae) the antennae are extremely sensitive to small differences in air temperature. The sensilla presumed to be involved are the thick-walled sensilla present in very large numbers on the antennal segments. In *Rhodnius* and other blood-sucking insects (mosquitoes, lice, bed bugs, etc.), perception of warmth is important in host location. Unlike nonblood-sucking flies, which detect the food source using tarsal contact chemosensilla to initiate proboscis extension followed by sucking, blood-sucking insects feeding on warm-blooded vertebrates use the temperature differential between themselves and the host to initiate probing (i.e., using the mouthparts— mainly the mandibles and maxillae—to puncture the host's skin). Once they have pierced the skin, chemosensilla at the tip of the labrum monitor blood components. The mechanism of initiation of probing in blood-sucking insects which feed on cold-blooded vertebrates is unknown. A few insects may also have cold receptors. Altner and Loftus (1985) discuss thermoreceptors in insects.

All insects will move away from high temperatures, but this is probably a generalized sensitivity with no particular sensilla being involved. In addition, there is evidence that spontaneous activity within the central nervous system is influenced by temperature (Chapman 1982*b*). Some insects are evidently able to perceive the radiant heat of the sun or other light source; for example, stink bugs will turn their dorsal sides toward a light source when the ambient temperature is low. By thus exposing the largest surface to the light, they are able to receive the maximum possible radiant heat.

Behavior associated with changes in temperature is discussed further in chapter 8.

Hygroreception

As with thermoreception, the ability of insects to perceive moisture in the air is well-known through observation of specific behavioral responses. Springtails, like other small soil-dwelling insects, are very sensitive to moisture both in the air and the substratum. They are attracted to regions of high humidity. Other insects, such as earwigs and mealworm beetles, avoid very moist areas. Some insects, the honey bee and others, can perceive water from a distance. The sensilla that are sensitive to moisture have been identified in only a very few insects, and these have been found on the antennae and maxillary palps. In the human louse, *Pediculus humanus*, antennal "tuft organs" composed of several small hairs have been shown to be specific hygroreceptors (Wigglesworth 1941). Basiconic sensilla on the antennae and maxillary palps of the mosquito *Aedes aegypti* respond electrophysiologically to water vapor (Kellogg 1970). Altner and Loftus (1985) also discuss hygroreceptors.

Photoreception

Photoreception may be defined as the ability to perceive energy (light) in the visible or near visible (near ultraviolet) range of the electromagnetic spectrum. In order for an organism to perceive light, there must be a pigment capable of absorbing light of a given wavelength and a means of producing a nervous impulse as a result of this absorption.

Many different kinds of environmental information are available to an insect in the form of light stimuli. For instance, an insect may perceive, to a greater or lesser extent, form, pattern, movement, distance, color, relative brightness, the polarization plane of light, light versus dark, and the length of a light period.

Generally speaking, three types of photoreceptive structures have been found in insects. These are the *compound eyes* and *stemmata* (or *lateral ocelli*) and *dorsal ocelli*. Each of these kinds of photoreceptive structures will be considered in turn. Refer to chapter 2 for information regarding the location and numbers of compound eyes and ocelli. In addition, the larvae of certain higher dipterans (true flies) have specialized photoreceptive organs that do not readily fit into any of these three groups of photoreceptors. These consist of photosensitive cells (figure 6.11), which are located in small cavities in the anterior end of a larva (the larvae or "maggots" of higher Diptera do not have well-defined heads). The negative phototactic response (i.e., the tendency to move away from a source of light) is presumably accomplished by the larva orienting its direction of movement so that the light-sensitive cells receive minimal stimulation. This orientation would obviously be with the anterior end of the larva away from the light source, with most of the body interposed between the light source and the sensitive cells.

Figure 6.11 Photoreception in the house fly larva; cephalopharyngeal skeleton with pocket that contains photoreceptive cells, indicated by hatching.

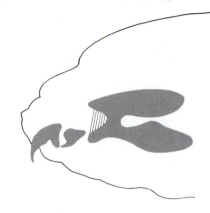

In addition to discrete organs associated with light perception, many insects apparently possess a light sensitivity over the general body surface. This is evidenced by the fact that certain insects will give definite responses to light even when the operation of the photoreceptive structures listed is disrupted in some way. For example, cockroaches continue to demonstrate a preference for dark situations even after being totally blinded. Similarly, decapitated mealworm larvae (*Tenebrio* sp.) continue to avoid light. The fact that extraretinal photoreception is present in insects has been established and is reviewed by Truman (1976).

Photoreceptors may be reduced or absent in cave-dwelling (*cavernicolous*), burrowing, and other species that live in dark situations.

Various aspects of insect vision are reviewed in the following: Bernhard (1966), Carlson and Chi (1979), Goldsmith and Bernard (1974), Horridge (1975, 1977), Mazokhin-Porshnyakov (1969), Ruck (1964), Wehner (1971), Carlson et al. (1984) and Blest (1988). Hoeglund et al. (1973) and Summers et al. (1982) discuss the biochemistry and physiology of insect photopigments. Volume 6 of Kerkut and Gilbert (1985) has 4 major chapters on insect vision.

Compound Eyes

The compound eyes (figure 6.12*a*) are the major photoreceptive organs of adult insects. When present, there are two, one located on either side of the head. Each is composed of a number of individual sensory units, or ommatidia (figure 6.12*b*). Externally, these ommatidia are marked by hexagonal cuticular facets. The facets, and hence the ommatidia, may vary in number from a very few to several thousand, for example, from 12 to 17,000 in some Lepidoptera and from 10 to more than 28,000 in some Odonata (Richards and Davies 1977).

Figure 6.12 Compound eye structure (diagrammatic). (*a*) Head with compound eye. (*b*) Four ommatidia removed and magnified. (*c*) Apposition ommatidium. (*d*) Superposition ommatidium.

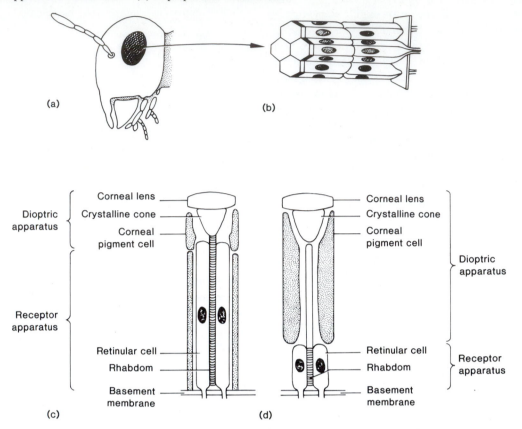

Structure of the Compound Eye

An individual ommatidium (figure 6.12*c,d*) is divisible into two parts: the *dioptric apparatus,* which acts as the "lens," and the *receptor apparatus,* in which the events leading to the initiation of a nervous impulse occur.

The dioptric apparatus is composed of the *cornea,* the *crystalline cone,* and *corneal pigment cells.* The cornea is a cuticular structure and is continuous with the cuticle of the integument. It has the shape of a plano-convex lens, the convex portion forming the outer surface. In many insects a hexagonal array of very small conical projections (approximately 0.2 nanometer from tip to base and center to center) may be found on the outer surface of the cornea. These "corneal nipples" are thought to act as an antireflection coating, which reduces the reflection for the air-cornea interface (Miller, Bernard, and Allen 1968). It is suggested that these "nipples" may also serve in insects active during light periods as camouflage by cutting out mirrorlike reflections from the cornea, which might attract predators. In insects active at night, they may help in some way to increase the sensitivity of the

eyes. In addition to corneal nipples, some insects, such as the house fly, *Musca domestica,* and the honey bee, *Apis mellifera,* have interommatidial hairs (i.e., hairs between the facets). In flying insects, these hairs are thought to function as aerodynamic organs. For example, removal of these hairs causes loss of flight control in honey bees. They may also provide sensory input that elicits cleaning behavior.

The crystalline cone lies immediately beneath the cornea and is composed of a translucent material. The darkly pigmented corneal pigment cells are usually located on the periphery of the crystalline cone, except in very primitive insects such as thysanurans (silverfish and relatives), in which they lie beneath the cornea. They are considered to be the cells that originally secreted the cornea.

The receptor apparatus (figure 6.12*c,d*) is composed of six or seven *retinular* (nerve) *cells* arranged in one (usually) or two layers. If arranged in two layers, one is proximal to the other. Like the crystalline cone, the group of retinular cells is usually surrounded by rather darkly pigmented cells. Each retinular cell usually gives rise to an axon that passes through a basement membrane and enters the brain and contributes to the formation of a centrally located retinal rod, or *rhabdom*. The contribution of each retinular cell to the rhabdom is called a *rhabdomere*. The rhabdomeres are considered to be the receptive surfaces of the retinular cells. Use of the electron microscope has revealed that the rhabdomeres are made up of tiny, closely packed fingerlike projections (microvilli) from the retinal cells. These microvilli project at right angles to the long axes of the retinal cells. It is generally thought that the microvilli contain the light-absorbing pigment(s), the *rhodopsins* and *metarhodopsins,* that are directly involved with photoreception.

There are a variety of different ways the dioptric apparatus and the receptor apparatus are associated with one another. However, based on the association of these components, ommatidia generally fall into one of two rather broad categories. If the retinal cells lie immediately beneath the crystalline cone (figure 6.12c), an ommatidium is of the "apposition" type, which is characteristic of diurnal insects (those active during the daylight hours). If there is a clear space between the retinal cells and the crystalline cone (figure 6.12d), an ommatidium is referred to as "superposition," and this type is characteristic of nocturnal or crepuscular insects (those active during dark or dusk periods, respectively). Currently, superposition eyes are referred to as "clear-zoned eyes," which refers to the clear space between the retinal cells and the crystalline cone.

In addition to the dioptric apparatus and receptor apparatus, groups of tracheal branches lie in the vicinity of the basement membrane. These form a surface from which light that has traversed the rhabdoms from distal to proximal is reflected back along the rhabdoms, giving these "receptors" a double exposure to the light and hence probably helping to increase the light sensitivity. These tracheal branches are sometimes referred to collectively as the *tapetum,* because their function seems analogous to this structure in the vertebrate eye. The eyes of many insects, particularly certain nocturnal Lepidoptera, when illuminated will appear to glow as a result of reflection from the tapetum.

Image Formation

The *mosaic theory* of insect vision initially proposed by Müller in 1826 and elaborated by Exner in 1891 is still generally accepted today. However, in light of more recent work, it has undergone considerable modification. According to the mosaic theory, each ommatidium "sees" only a small portion of the insect's surroundings. The combination of the images sensed by individual ommatidia supposedly together forms a composite or mosaic view of the external environment. This situation is somewhat analogous to looking at the surroundings through a handful of soda straws. Only a small part of the total view is seen through any one straw, but the combination of these "small parts" gives a mosaic image of the surroundings (Wigglesworth 1964).

There are two basic types of compound eyes, *apposition* (or *photopic;* Goldsmith and Bernard 1974) and *superposition* or *clear-zoned eye* (or *scotopic;* Goldsmith and Bernard 1974). Apposition eyes (figure 6.13a) are composed of ommatidia of the apposition type and are thus, as mentioned earlier, characteristic of diurnal insects. Similarly, superposition, or clear-zoned eyes (figure 6.13b,c) are made up of superposition ommatidia and are characteristic of nocturnal or crepuscular insects. In the apposition type of eye there is little or no movement of the pigment in the pigment cells surrounding the crystalline cone in response to changes from light to dark or vice versa. Thus the pigment remains rather uniformly distributed. However, in superposition eyes, there is considerable pigment movement in response to a light–dark change. In a lighted situation (figure 6.13b), the pigment in the pigment cells tends to migrate proximally, producing the light-adapted condition. On the other hand, in a dark situation (figure 6.13c), the pigment migrates distally (dark-adapted condition).

According to the mosaic theory, in the apposition eye and the light-adapted superposition eye, the rhabdom receives only light rays entering parallel (or nearly so) to the long axis of an individual ommatidium. Oblique rays are absorbed by the pigment, optically isolating adjacent ommatidia from one another. The image formed in this situation is an *apposition image,* because the light reaching a single rhabdom enters only via the dioptric apparatus of the same ommatidium. In the dark-adapted superposition eye, distal movement of the pigment has the effect of removing the optical isolation between adjacent ommatidia. In this instance, the light rays reaching the rhabdom of a given ommatidium enter via several ommatidia. The image thus formed is a *superposition image*. One would not expect a superposition image to be as sharply defined as an apposition one because of the increased amount of light reaching each rhabdom. Also, the formation of a sharply defined superposition image would require that all the light rays reaching a single rhabdom impinge at exactly the same point. It follows, then, that eyes in which pigment optically isolates individual ommatidia would probably have a greater resolving ability than those in which pigment does not form an optical boundary. Exner theorized that the dark-adapted condition in a superposition eye increases the light sensitivity since it enables more light to reach each rhabdom. The adaptation of the eye to light or dark situations is probably also partly due to a change in the retinal cells or in the central nervous system.

In recent decades details of the mosaic theory have been modified. Ommatidial visual fields have been found to overlap in all species studied. Few, if any, insects with

Figure 6.13 Image formation in compound eyes (diagrammatic). (*a*) Apposition type. (*b*) Superposition type, light adapted. (*c*) Superposition type, dark adapted. Dashed lines represent rays of light originating from a common point.

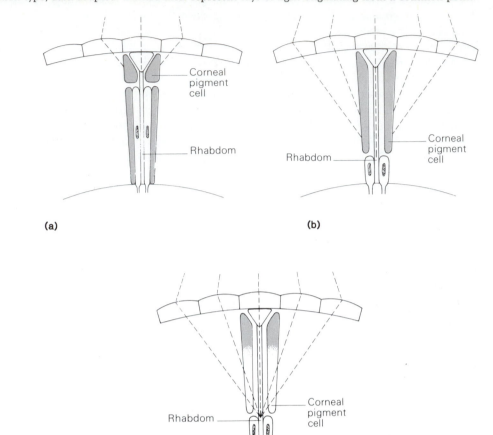

(a)

(b)

(c)

superposition eyes form superposition images. In addition, there is evidence that the resolving ability of eyes is, in fact, not decreased by the distal migration of pigment. However, Exner's idea that distal pigment migration (dark adaptation) in superposition eyes increases sensitivity still is thought to be true.

The crystalline tracts found in the eyes of many insects may act as light guides in that once light has entered them it does not escape but is carried directly to the rhabdoms. Besides acting as a light guide, a crystalline tract surrounded by pigment cells is thought to act as a longitudinal pupil (Miller, Bernard, and Allen 1968). When pigment envelops the tract (light adapted), it absorbs some of the light; however, when the pigment migrates distally and no longer envelops the tract (dark adapted), much more light is transmitted to the rhabdom. Other recent additions include the possibility that in at least certain instances the retinal cells are the individual functional units instead of the whole ommatidium. Also, images that are formed proximal to the retinal image have recently been discovered. They are essentially of the superposition type, and their significance is unclear.

Perception of Form and Pattern

There is strong behavioral evidence that insects are at least capable of simple perception of pattern and form. In particular, much work has been done with honey bees. For example, honey bees were trained to associate a particular pattern of colored paper with the presence of sugar water in a cardboard box. They were subsequently presented with several boxes marked with new pieces of colored paper cut in the pattern to which they had been trained and others with a rather similar pattern that did not contain sugar water. It was judged that the bees could discriminate between these two patterns since only a small number visited the boxes marked with the second pattern (von Frisch 1950). Other work, again with training to a particular pattern, has demonstrated that bees can discriminate between the shapes shown in the upper row of figure 6.14 and those in the lower row but cannot distinguish among the shapes within each row. This suggested that the bees perceive form on the basis of the "brokenness" of pattern (von Frisch 1950). It is supposed that broken or interrupted patterns produce a flickering visual impression as the bee passes during flight. It is not sur-

Figure 6.14   Figures used to study form perception in honey bees.

Redrawn from Hertz, 1929.

prising, therefore, that bees tend to visit flowers that are being shaken by the wind more readily than those that are not.

It is well known that bees, ants, and wasps are able to locate their nests on the basis of various landmarks. In addition, flying bees have been shown to be able to distinguish right from left, before from behind, and above from below (Wigglesworth 1972). However, the form and pattern perception abilities of insects are considered to be less than those of humans.

The honey bee, because of its ability to associate forms and patterns with a carbohydrate reward, has been the insect of choice for studying visually mediated learning. In order to evaluate other insect systems, greater effort must be made to understand the insects' visual environment. This same concern is aptly expressed by Järvilehto (1985), "The basic problem is to find ways of defining the clusters of stimulus parameters in the animal's multidimensional space that appear to be relevant to the animal." Researchers currently using this philosophy of close scrutiny of the animal's visual space have created an area of study termed "visual ecology" (discussed in this chapter), which in many cases has led to practical implementation of the information to integrated pest management (IPM) programs.

Intensity and Contrast Perception

Optomotor responses have been used to study intensity perception. An optomotor response is behavior stimulated by a moving visual pattern. When an insect flies, the apparent movement of the surroundings tells it the direction and rate of movement. Certain orientation movements, maintenance of flight, and changes in velocity of flight and landing may be in response to a change in the rate or direction of the moving visual pattern. In the laboratory, a moving visual pattern can be produced by surrounding a stationary insect with a cylinder on which are painted vertical stripes. Rotation of the cylinder gives the insect the sensation that it is moving.

In the study of perception, optomotor responses to patterns of rotating stripes have been used to determine the ability of insects to discriminate between different levels of intensity of adjacent stripes. Measurements indicate considerable variation among different insects. Discrimination of different intensities apparently depends on whether those being compared are relatively high or low. For example, in the house fly, at low intensities the more intense light must be 100 times brighter than the less intense, whereas at high intensities the magnitude difference may decrease to 2.5 times (Wigglesworth 1972).

The brightness of the background may influence intensity perception. For example, the hummingbird hawk moth will favor a dark disc on a white background over an equally dark disc on a gray background. In other words, the moth "prefers" the situation in which there is greater contrast. It is interesting that under natural circumstances these moths enter dark crevices to spend the night.

Movement Perception

As was the case for form and pattern perception, there is behavioral evidence for the perception of movement. In fact, certain responses are elicited only by movement. For example, dragonfly nymphs will not attack prey with their labial jaws unless it is moving. The optomotor response elicited by the moving pattern of stripes is based on the perception of movement by an insect.

Distance Perception

It is not difficult to think of several instances in which it would be essential for an insect to possess the ability to judge the distance of an object from itself. An excellent example would be in prey capture. To catch prey, particularly on the wing, distance perception must be especially acute. One needs only to watch a dragonfly capture its prey in flight to be convinced. Binocular vision must be involved since the ability to judge distance accurately is lost when one compound eye is blocked in some fashion. However, unlike those of humans, the eyes of insects are fixed, and depth perception depends not on convergence of the eyes on a fixation plane, but upon the equal, simultaneous stimulation of corresponding retinal points. The distance of an object will then be determined by the location of the object relative to the points of intersection of projections of axes of the corresponding ommatidia (Dethier 1953). A schematic representation of how depth perception is accomplished by a dragonfly nymph is shown in figure 6.15. For this means of depth perception, the insect must, of course, face the object perceived.

Motion Detection

As will be discussed in chapter 7 regarding insect flight, the compound eyes are important for several aspects of flight. Landing and takeoff responses in many insects are initiated by the motion of an object, whether animate or inanimate, approaching the insect. Once in flight, many

Figure 6.15 Depth perception in a dragonfly nymph. The lines represent the visual axes of selected ommatidia. The distance and position of objects are determined by the points of intersection of the visual axes. The extended labium is shown in gray. Potential prey at point A is within reach of the extended labium, but not at points B, C, D, and E. *Redrawn from Baldus, 1926.*

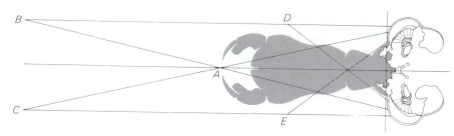

insects use the apparent movement of the visual environment (i.e., trees or other ground reference points) to make both height and speed compensations. Such compensatory adjustments to visual input are referred to as *optomotor responses*. Researchers have used this particular phenomenon to study insect flight in wind tunnels. In most cases, the base of the wind tunnel has a moveable floor with stripes painted on it. The stripes are usually perpendicular to the insect's direction of flight and provide the flying insect with a visual reference point as to its flight speed. By moving the floor slowly, an insect may compensate by flying faster while rapid movement causes the insect to slow down. Can the researcher manipulate the floor speed so that the insect remains in stationary flight? Also, how do night flying insects evaluate their flight speed? More information about the importance of vision to insect flight can be found in Horridge (1975) and Goldsworthy and Wheeler (1989).

Color Vision

The range of the electromagnetic spectrum perceived by insects is from about 253 nanometers (near ultraviolet) to approximately 700 nanometers (infrared; Dethier 1963). Although there is a large amount of variation in the sensitivities of different insects to different wavelengths, insects in general are particularly sensitive to the ultraviolet and blue-green regions of the spectrum. Some insects, such as the nocturnal stick insect, *Dixippus* sp., are apparently color-blind. In insects that do possess color vision, part of an eye may be color sensitive and another part color-blind [e.g., in the water boatman, *Notonecta* (Hemiptera)]. Evidence exists that there is more than one type of color receptor in the cockroach eye (Dethier 1963). The dorsal area of the eye is more sensitive to ultraviolet than to blue-green light; the reverse is true in the ventral area. The spectral sensitivity of an insect may vary with its physiological state. For example, cabbage butterflies (*Pieris brassicae*) seem to prefer blue or yellow flowers; gravid females ready to oviposit seem to prefer green and blue-green. Likewise, nongravid apple maggot adults (*Rhagoletis pomonella*) prefer a yellow trap, which mimics foliage color, whereas gravid females prefer spherical red traps, which mimic the apple fruit in color. Whether the spectral sensitivity changes observed in an insect at different physiological states, and measured only at the behavioral level, are due to changes at the peripheral receptor level and/or are modulated at the central nervous system can only be determined by monitoring the changes at the receptor using electrophysiological and neuropharmacological techniques. Researchers interested in behavior and ecology must become aware of the limitations imposed on research results obtained only by measuring behavioral parameters. Without investigating the underlying mechanism(s) of behavior, incorrect conclusions can be made concerning either the ecological and/or evolutionary adaptation(s) of the specific behavior(s) being investigated. More will be said about peripheral versus central modulation of behavior in chapter 8.

Both electrophysiological and behavioral techniques have been used to establish the color-perceiving abilities of insects. Different wavelengths have been tested to see if they elicit a particular electroretinogram (ERG) pattern or a specific behavioral response. Both the various optomotor responses and training experiments have been utilized. For example, von Frisch (1950) used training techniques to demonstrate the color-perceiving capacities of the honey bee. He first placed a dish of honey on a blue card where bees could get to it. After several hours the bees were "trained" to associate the color blue with the presence of honey. The bees were then presented with a fresh (unscented) blue and a red card where only the blue one had been previously. The bees visited the blue card and ignored the red one. Although this experiment shows that bees can distinguish between blue and red, it does not prove that they actually perceive color, since red and blue differ in relative brightness. To determine whether the bees were distinguishing brightness or color, several cards of different gray shades between white and black and a blue card were placed where only the blue card had previously been. In addition, to discount the possible role played by scent the cards were covered with

Figure 6.16   Comparison of the spectra perceived by humans and honey bees. *Redrawn from von Frisch, 1950.*

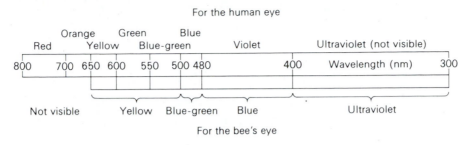

a glass plate. In this situation the bees still visited only the blue card; therefore, they are able to perceive color.

Similar results were obtained when the bees were trained to orange, yellow, green-violet, and purple. However, the color red was confused with the black and dark gray cards; therefore, the bees are red-blind. Further experiments in which bees were presented with several different-colored cards showed that they were unable to distinguish certain colors from others. As a result of these and other experiments, it has been demonstrated that bees can accurately distinguish only 4 colors as opposed to the 60 or so that can be distinguished by the human. The portion of the spectrum in which we can recognize several distinct colors between orange and green appears yellow to bees, and so on (figure 6.16). These characteristics of color vision in the honey bee have very interesting ramifications when we consider the relationship between bees and flowers.

We have mentioned that the spectral range in which insect eyes are generally most sensitive is in the ultraviolet and blue-green regions. We also established that bees are red-blind. However, some insects are highly sensitive to wavelengths in the red region. For example, certain butterflies are capable of recognizing red flowers or red models of flowers, and the firefly, *Photinus* sp., is able to perceive flashes of light up to 690 nanometers, which is well into the red region of the spectrum.

Inasmuch as insects can perceive light into the ultraviolet zone of the spectrum, it is of significance that many flowers display patterns based on differential reflectance and absorbance of ultraviolet light. In fact, many bee pollinated flowers have a special area of the flower (i.e., leading to the source of nectar, termed *nectar guides*) that absorb ultraviolet or UV light while the rest of the petal reflects UV. The nectar guides are used by the insect as visual cues to direct them to the source of nectar. To us, however, these flowers appear uniform in coloration. The same is true for some insect body coloration; that is, the different patterns displayed by insects (e.g., butterfly wings) may be due to differential reflectance and absorbance of UV light. Matthews and Matthews (1978) and Barth (1985) discuss UV and its importance as a species-specific signal for mating and for pollination.

Polarized Light Perception

That insects could recognize the direction of polarization of light was first discovered in honey bees and ants. These insects were found to be able to detect the polarization pattern of the sky, which varies with the position of the sun and enables them to determine direction. This directional ability is, of course, important in finding the hive or nest after a foraging or hunting trip. According to von Frisch, it is also of importance in the orientation of the bees' communication dances that are used to inform other members of the hive as to the location of food. The ability to detect the polarization plane of light has been found in all insects examined in this regard.

The discovery that crickets return to a "home" base, possess the ability to use polarized light (i.e., diurnal species), and have a time compensated celestial orientation (i.e., nocturnal species) ability (Huber et al. 1989) indicates that insects, other than social insects, that return to a "home" base probably will also be shown to have the ability to use either polarized light, if they are diurnal, or celestial cues, if they are nocturnal, for orienting and returning to these refuges.

In the desert ant (*Cataglyphis bicolor*) and the honey bee, the receptors involved in the perception of polarized light are sensitive to ultraviolet light. This is significant because the ultraviolet wavelengths are much less affected by atmospheric conditions (e.g., clouds) than the longer wavelengths visible to humans. Thus sensitivity to the polarization pattern of ultraviolet light gives a particularly stable navigational point of reference. For a fascinating account of polarized light navigation, see Wehner (1976), and for astronavigation in insects, see Wehner (1984).

Stemmata

Stemmata (or *lateral ocelli*) are structurally variable. Some types are similar in structure to an individual ommatidium of a compound eye. For example, in larval butterflies and moths (figure 6.17*a*), each eye consists of a cornea, a crystalline body, and a number of retinal cells forming a rhabdom.

Figure 6.17 (*a*) Section of a stemma (diagrammatic). (*b*) Section of a dorsal ocellus (diagrammatic).

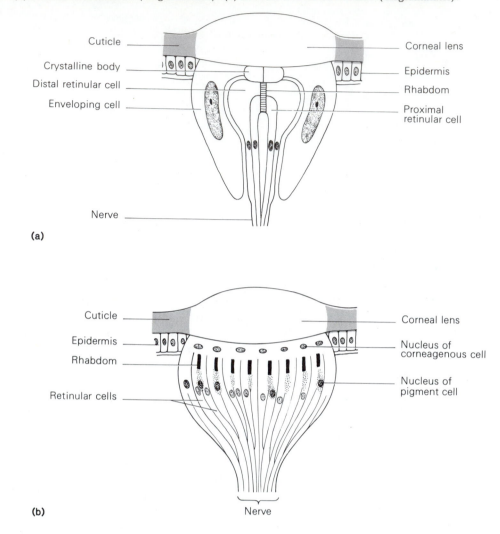

Cuticle — Corneal lens

Crystalline body — Epidermis

Distal retinular cell — Rhabdom

Enveloping cell — Proximal retinular cell

Nerve

(a)

Cuticle — Corneal lens

Epidermis — Nucleus of corneagenous cell

Rhabdom

Retinular cells — Nucleus of pigment cell

Nerve

(b)

Stemmata function in the manner of eyes. Typically they are the only eyes found in holometabolous larvae. In various insects they have been shown to be involved with color, form, and distance perception. Like compound eyes, they receive nerves from the optic lobes of the brain.

Although a detailed pattern of the external surroundings is not likely to be perceived at any one time by an insect possessing only a few stemmata, the movement of the head back and forth, "scanning," may allow much greater detail perception than would otherwise be possible. For this type of activity to be effective, the insect must be able to convert spatial patterns into temporal patterns. In other words, the external scene is viewed as a sequence of events in time.

Dorsal Ocelli

Dorsal ocelli vary somewhat in structure. They are typically made up of a *corneal lens;* a layer of *corneagen cells,* which secretes the lens; and the *retina,* which is composed of up to 1000 photosensitive cells depending on the species (figure 6.17*b*). In some species, pigment cells are associated with the dorsal ocelli. Dorsal ocelli have been shown to be light sensitive, but are apparently not important in image perception. They may be "stimulatory organs" that increase the sensitivity of the compound eyes to light, for when they are blocked, reactions of the insect to light are diminished somewhat. Alternatively, they may be involved along with the compound eyes in orientation behavior relative to light. However, according to Carlson and Chi (1979), "The ocellus remains an enigma as to its specific role in insect behavior."

There is a correlation between the presence of dorsal ocelli and flight, because they are not found in apterous insects. On the other hand, there are strong fliers that lack dorsal ocelli, for example, horse flies (Diptera, Tabanidae) and relatives and sphinx moths (Lepidoptera, Sphingidae). However, recent histological and electrophysiological studies provide evidence for the presence of an internal ocellus in sphinx moths (Eaton 1971).

Goodman (1970) reviews the literature on the structure and function of dorsal ocelli.

Magnetic Field Reception

The earth is invested in a magnetic field and depending on where you are on its surface, the magnetic field lines (i.e., also called dip angles) intersect the earth at different angles. The closer you get to the poles, the steeper these dip angles become. Thus, if an animal were able to perceive these lines and also calculate the angle of intersection, it would know where it was located. There is little doubt, especially for birds, that some organisms are able to use the earth's magnetic field for orientation. Finding the bacterium, *Aquaspirillum magnetotacticum*, which is literally a "swimming magnet," is evidence that magnetic detection may be widespread among living things. Long (1991) presents a stimulating article on animal navigation and includes information on the monarch butterfly, ants, and bees. The major question at the moment, however, is how insects perceive the magnetic field. Gould (1980) presents experimental evidence for magnetic field detection by honey bees. Gould (1982) believes that insects and other animals may even have sensory systems capable of detecting stimuli that humans will find difficult to imagine. Exactly how bees perceive the magnetic field still remains a mystery.

Visual Ecology

This currently active area of research has been succinctly defined by Allan et al. (1987) as ". . . the description and analysis of an animal's natural optical environment in terms of the animal's visual system and the relationship between environment, visual perception, and behavior." In an effort to develop nonchemical methods of pest control based on the insect's behavior and also to develop techniques for monitoring pest populations (e.g., with traps), researchers are examining significant visual features of the pest's environment that can be mimicked and incorporated into the development of a visual panel, model, or trap. One question that should always be addressed in studies concerning visual ecology is: "What are the visual components of the environment that are most significant to the organism for food, shelter, egg laying, or mate finding?" Once understood, a trap or panel can be made to intercept pests seeking one of these specific requirements. By using this approach, Prokopy and Owens (1983) have used a red-sphere with sticky material on it to mimic a ripe apple to intercept gravid females ready to oviposit whereas a yellow trap attracts nongravid females. These fruit and foliage mimics or visual models are also used in basic behavioral studies as well as in monitoring populations of the apple maggot, *Rhagoletis pomonella*, for integrated pest management programs of apples. At the same time, researchers studying biting flies have used cut-out or three dimensional models of cows, horses, and deer to attract biting flies. The reviews by Prokopy and Owens (1983) and Allan et al. (1987) provide an excellent entry point into the topics of visual ecology of herbivorous insects and biting flies, respectively.

The practical and commercial exploitation of insect visual ecology as a means of developing traps for either monitoring or reducing pest populations continues to be an active area of development. Are the UV-electrocuting "bug-zappers" effective in reducing mosquito populations, thus annoyance? Maybe they are just a waste of electricity? There is little doubt that many night-flying insects are attracted to UV light and are destroyed by these traps; but, to date, no convincing evidence exists to support the advertiser's claim that they reduce mosquito populations (Nasci et al. 1983).

Light Production

The production of light by living organisms, *bioluminescence*, has fascinated scientific investigators and curious lay observers for centuries. However, it has been only during the last hundred years that a fundamental understanding has begun to develop. The phenomenon of bioluminescence has been described in several groups of plants, microbes, and animals. For example, it has been found in the clam *Pholas dactylus*, the crustacean *Cypridina hilgendorfii*, several marine annelids, a number of insects, marine dinoflagellates, and several fungi and bacteria. Luminous bacteria commonly exist symbiotically with animals. For example, they are responsible for the luminescent body regions of several marine fish.

Insects with specific light-producing mechanisms are found in a few species among the Collembola (springtails), Homoptera (cicadas, leafhoppers, and relatives), Coleoptera (beetles), and Diptera (true flies). Bioluminescence in other insects has been found thus far to be due to the presence of bioluminescent bacteria.

When certain rare species of springtails (e.g., *Achorutes muscorum*) are stimulated, luminescence occurs over the whole body.

There is only one well-established case of bioluminescence among the Homoptera (McElroy et al. 1974). This is in the species *Fulgora lanternaria*. Luminescence of the head in this species has apparently been observed only when males and females are together and evidently has something to do with mating behavior.

Bioluminescence has been described in more beetles than any other group of insects. Several families are known to be involved, among them Lampyridae (fireflies), Elateridae (click beetles), Drilidae, and Phengodidae. Members of the family Lampyridae have been the most extensively studied. The light-producing organ is in the abdomen and may occur in both sexes, or only in females, and in the larval stage. Luminescent larvae and some females in this family are often called "glowworms." Not all species of lampyrids are luminescent, but in those that are, the light flashes are usually involved in attracting members of the opposite sex. The flashing patterns are species specific. A well-known American species of fireflies is *Photinus pyralis,* the mating behavior of which has been extensively studied. In the family Elateridae, those in the genus *Pyrophorus* are especially well-known. Some of these beetles have luminous green spots on the prothorax and orange ones on the abdomen. The orange spots are visible only during flight.

Pyrophorus is particularly significant historically because these were the first insects in which the biochemical nature of the light-producing reaction was studied. In the late nineteenth century, DuBois ground up their light-producing organs in cold water. The homogenate glowed for a short period of time, and then the glow gradually faded away. Addition of an extract obtained by boiling light-producing organs in water briefly restored the glow. DuBois later carried out similar experiments with a luminescent clam and named this reaction the *luciferin-luciferase reaction.* More recently, the active principle in the hot water extract has been found to be the heat-stable substance adenosine triphosphate (ATP).

A rather bizarre example of the luminous beetles is in the family Drilidae among larvae in the genus *Phrixothrix.* These larvae are commonly known as "railroad worms" because of the eleven pairs of green luminous spots on the thorax and abdomen and the pair of red luminous spots on the head. The movement of these larvae when stimulated to luminesce gives the impression of a string of railroad cars at night.

Truly luminescent dipterans are found among larvae of the families Platyuridae and Bolitophilidae. The members of both these families are called *fungus gnats.* The larvae of the New Zealand glowworm, *Bolitophila luminosus,* occur in well-shaded humid areas, for example, the environmental situation found in certain caves. The light-producing organ has been found to be modified portions of the Malpighian tubules, which are located near the end of the abdomen. Adult female flies deposit eggs in a mucous glue on the ceiling and larvae spin silken sheaths from which they hang. The network of sticky threads traps small flying insects, which are eaten by the larvae. The prey are presumably attracted by the bioluminescent light. Luminescent larvae of the species *Platyura fultoni* have been found in moist environments in the Appalachian Mountains.

The morphology of the light-producing, photogenic organs is extremely complex and variable and will not be considered here. Similarly, the biochemical aspects are also very elaborate and far from being completely understood. For further information on bioluminescence, see the reviews by McElroy et al. (1974), Lloyd (1983), McElroy and DeLuca (1985), and the article by McElroy and Seliger (1962). The role of bioluminescence in behavior is discussed in chapter 8.

Sound Production

The ability to produce sound is widely distributed among insect groups and has evolved independently many times. These sounds are commonly correlated with well-developed organs of hearing and often play an important role in various types of behavior. Sound as a mode of insect communication is discussed in chapter 8.

A useful classification of sound-producing mechanisms has been presented by Haskell (1974).

1. Sounds produced as a by-product of some usual activity of the insect.
2. Sounds produced by impact of part of the body against the substrate.
3. Sounds produced by special mechanisms.

In the first category no specifically adapted structures are involved in sound production. Sounds of this type include by-products of flight produced by wingbeats, vibration of the thoracic sclerites, the wings striking one another, and similar mechanisms. This category also includes sounds produced as a result of movement during copulatory behavior, cleaning, feeding, and so on (Haskell 1961). In many instances the sounds have no specific function. On the other hand, some do have behavioral significance. A good example is found among the mosquitoes, where the sound produced by the wingbeat of the flying female elicits a mating response in males. Another possible example is the "piping" of queen honey bees, which may result from the vibration of the thoracic sclerites. This sound is produced when a colony possesses a number of virgin queens and has been suggested to be a sound of challenge (Butler 1963).

A number of insects are known to produce sound by tapping the substrate with some part of the body. The "death watch" beetles (*Anobium* and *Xestobium,* Anobiidae) tap the walls of their galleries in wood (unfortunately sometimes furniture) with their heads, producing a characteristic sound. Other insects in this category include the subterranean termite (*Reticulitermes flavipes*), book lice (Psocoptera), and several members of the Orthoptera.

Among the special sound-producing mechanisms are frictional mechanisms, vibrating membrane mechanisms, and mechanisms involving air movement.

Frictional mechanisms are found among several groups of insects, and although these mechanisms are structurally diverse, they consist of similar parts. The production of sound by means of a frictional mechanism is commonly called *stridulation.* Frictional mechanisms (figure 6.18a,b) are located in areas where two surfaces (two wings, a leg and a wing, etc.) may be rubbed together. One of these surfaces, the "file," bears a row of regularly spaced ridges, and the other bears the "scraper,"

Figure 6.18 Examples of sound-producing mechanisms. (a) File and scraper on underside and posterior edge, respectively, male katydid, *Neduba carinata*. The file of one wing and scraper of the other are rapidly rubbed together, producing a chirping sound. (b) Hind femur of a short-horned grasshopper, *Stenobothrus* sp., with a portion of the file enlarged. In sound production the file is rapidly rubbed against a thickening (the scraper) of the basal portion of the forewing. (c) Section of a tymbal of a cicada. The dashed lines and arrows indicate the pattern of movement of the tymbal during sound production.
(a) redrawn from Essig, 1942; (b) redrawn from Comstock, 1940; (c) redrawn with modifications from Haskell, 1961.

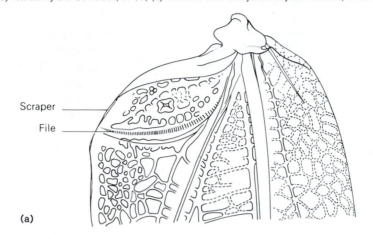

(a)

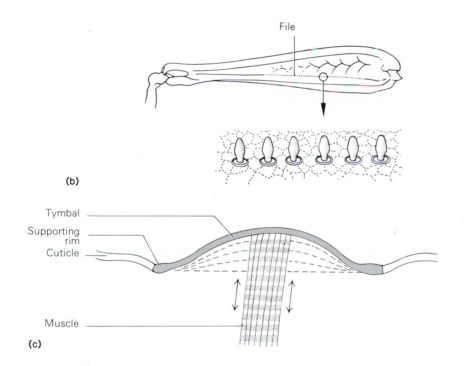

(b)

(c)

a ridge or knoblike projection. When the file and scraper are rapidly rubbed together, a sound is produced, the quality of which is based on the rate of rubbing, the spacing of the ridges on the file, and the resonance characteristics of the surrounding cuticle. Frictional mechanisms have been described in several life stages of several orders and involve many different combinations of body parts.

Sound production in the mole cricket, *Gryllotalpa nivae* (Orthoptera, Gryllotalpidae), is of particular in-

terest. This insect constructs a burrow that acts as a resonating chamber (Bennet-Clark 1970).

Vibrating-membrane or tymbal mechanisms have been found only in members of the orders Homoptera, Hemiptera and Lepidoptera. Although tymbal mechanisms have been described in several homopterans, the best-known examples are found among the cicadas (figure 6.18c). In these insects the tymbal organs are paired structures on the dorsolateral surface of the base of the abdomen. The tymbal organs of Lepidoptera have been found in certain arctiid moths (tiger and footman moths) and others. In the moths these organs are paired and are located on either side of the metathorax. In addition, tymbal mechanisms have been identified in the hemipteran family Pentatomidae.

Mechanisms involving the movement of air are uncommon and poorly understood. One well-established example is the death's head hawk moth, *Acherontia atropos,* a European species, which produces a sound by the forcible inhalation and exhalation of air through the proboscis by means of the pharyngeal muscles. Passage of air over the epipharynx apparently produces the sound. Air released forcibly through the spiracles may result in sound production in certain Diptera, Hymenoptera, and Blattaria. The cockroach *Gromphadorhina portentosa* of Madagascar is capable of producing an audible hissing sound via the spiracles when disturbed.

Sound and light are not used by insects solely for intraspecific communication. Instead, parasites and predators often use these stimuli to locate a host or prey. Firefly predators attract their prey by mimicking their flash signal (Lloyd 1975), whereas parasitic flies locate their hosts by eavesdropping on their "love songs" (Cade 1975, and Soper et al. 1976). People are also attempting to use these stimuli to their benefit in I.P.M. programs or for collecting particular species.

Wherever there is a market, there are usually entrepreneurs to seize the opportunity to exploit the market. One such area is the use of sound—whether ultrasound or sound audible to the human ear—to rid cats, dogs, and even houses of fleas. Similar devices are being sold to repel mosquitoes. Dog and cat collars and even devices to put onto the ceiling of flea-infested rooms are marketed. To date, however, there is no scientific evidence that these devices work (Hinkle et al. 1990). In fact, there is no experimental data demonstrating that fleas "hear" these sounds (Rust and Parker 1988). The biophysical aspects of sound communication in insects is dealt with by Michelsen and Nocke (1973).

CHAPTER 7

Locomotion

The ability to change position within the environment is of essential importance to the survival of all nonsessile organisms. Escape from predators, feeding, dispersal, mate finding, and adjustment to temperature and humidity changes all depend to some degree upon the ability of an organism to move about. Insects were originally terrestrial organisms that subsequently invaded both the aerial and aquatic environments. They are the only group of invertebrates with members capable of flight. Recent treatments of various aspects of insect locomotion may be found in Kerkut and Gilbert (1985).

Terrestrial Locomotion

Walking and Running

Walking and running are accomplished by the six thoracic legs. Actually these limbs, unless modified for some function other than walking, serve two basic purposes; they suspend and support the insect body off the ground, and they exert the necessary forces to propel the insect. The body literally "hangs" close to the ground, resulting in a low center of gravity and thus a high degree of stability. Also, the tip of the abdomen may be in contact with the ground, providing additional stability. The use of legs as a mode of terrestrial locomotion is found in the adults of nearly all flying and nonflying insects and many nymphal and larval forms as well. In many insects walking legs are the only source of locomotion or the only source used to any great extent. For example, apterygote insects such as silverfish are solely walkers and runners; cockroaches, although they possess functional wings, seldom fly. In several beetles the elytra (forewings) are fused; hence these insects are grounded for life. Larval and nymphal forms are entirely dependent upon legs for locomotion.

Although several insects are rather sluggish in their walking and, in fact, may use their legs more as organs for clinging to a surface than for actual locomotion, many are rapid, agile runners. For example, cockroaches are well-known for their running ability. The fastest speed measured for the cockroach *Periplaneta americana* is 2.9 mi/hr or 130 cm/sec (McConnell and Richards 1955). In the absolute sense this is a much slower speed than many vertebrates are able to attain. However, in relation to body size, it is remarkably fast. If speed were increased proportionately, an American cockroach the size of a lion could run about 70 miles per hour (Delcomyn 1985). Insects also have the ability to accelerate rapidly over short distances and to change direction suddenly. The speed of walking and running is proportional, within certain limits, to temperature.

Functional Morphology of the Insect Leg

In the discussion of the insect skeleton the leg has already been described as being composed of a series of segments that articulate with one another. The nature of these intersegmental articulations is of paramount importance in the capabilities for movement of the entire leg. Two basic types of articulations are found: *monocondylic* and *dicondylic*. A condyle is a prominence of exoskeleton upon which an adjoining segment articulates. A monocondylic joint (figure 7.1*a*) is one that contains a single condyle. This type of joint allows considerable movement and has been said to be the nearest single-articulation equivalent to the ball-and-socket joint found in vertebrate animals. A dicondylic joint (figure 7.1*b,c*) consists of two condyles and restricts movements to a single plane.

Figure 7.1 Types of joints (diagrammatic). (*a*) Monocondylic. (*b*) Dicondylic (end view). (*c*) Dicondylic (lateral view).
Redrawn from Snodgrass, 1935.

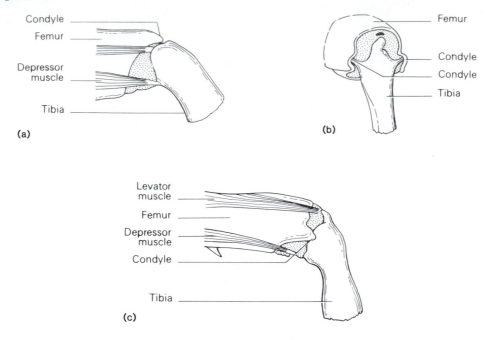

The insect leg is composed of four main regions (Hughes and Mill 1974).

1. The coxa, which articulates with the thorax by a dicondylic joint or sometimes a single pleural articulation.
2. The trochanter and femur, which are fused, the trochanter articulating with the coxa by means of a dicondylic joint.
3. The tibia, which forms a dicondylic joint with the femur and a monocondylic joint with the tarsus.
4. The tarsus, which is composed of tarsomeres joined by monocondylic joints.

The dicondylic joints between the coxa and thorax and between the femur and tibia are oriented such that they enable movements in different planes and therefore allow the entire leg to move in all directions about the articulation with the body (i.e., the functional equivalent of a ball-and-socket joint).

It was mentioned in chapter 3 that leg movements are accomplished both by intrinsic muscles, which originate within the leg itself, and extrinsic muscles, which are attached between the leg and the thoracic wall. The movement of the leg segments relative to one another is possible because the muscles cross one or more joints between their origins and insertions.

Patterns of Leg Movement during Walking and Running

The sequence of movements or gait of the legs during walking and running has been a topic of considerable interest. Attempts to detect details of gait by direct observation are extremely difficult, if not impossible, because of the general rapidity of leg movements. Even if one were able to discern the gait of a slow-moving insect, the results would not necessarily apply to the same insect moving more rapidly or to other species. The use of cinematography has afforded an excellent solution to this problem. Insects may be filmed at whatever speed or under whatever conditions the experimenter desires; then the film may be projected at reduced speed, and enlarged prints of different frames may be produced.

The order or sequence of leg movement during walking seems to be fairly constant at a given speed for a given insect. However, variations occur at different speeds and between different species. The classical description of forward walking of insects is that of alternating tripods of support (figure 7.2*a*). The first and third legs on one side and the middle leg on the other advance while the other three legs remain stationary and provide a tripod of support. The cycle then repeats itself with the previously stationary three legs becoming the moving ones. An insect progressing in this fashion follows a zigzag course. Many deviations from this classical description are known; thus it is useful only in a very general way. Probably the most common pattern is a modification of the alternating triangles of support where the three legs of each triangle do not move simultaneously but in sequence (figure 7.2*b*). In addition, accidental or experimental amputation of one or more legs immediately results in a new coordinated pattern of walking suitable for the remaining legs. Also, not every insect uses all six legs in walking. For example, the praying mantis, when walking slowly, uses only the mesothoracic and metathoracic legs. The review by Delcomyn (1985) considers the topic of gait in detail.

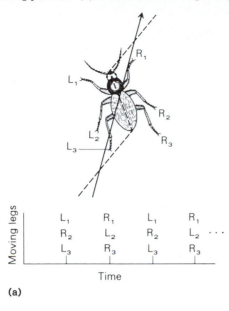

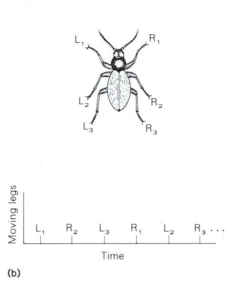

(a)   (b)

Muscle Contraction and Coordination of Legs during Walking and Running

The muscle action that produces leg movement is complex in that individual muscles associated with a given joint differ in size, orientation, and the kinds of excitatory axons (slow, intermediate, fast) they receive (Delcomyn 1985). "Slow" axons are involved in slow walking, whereas increasing numbers of "fast" axons are recruited as the frequency of motor axon firing increases with increasing speed. Inhibitory axons may also be involved, facilitating rapid relaxation of a muscle after contraction.

The coordination of insect walking and running may be resolved into two basic components: the movements of each leg and the collective movements of all the legs. It is well established that each thoracic segment contains afferent and efferent pathways necessary for the individual movement of the two legs associated with it. However, since cutting the longitudinal connectives disrupts walking, the coordination of all the thoracic legs is considered to be intersegmental. This coordination lies within the thoracic ganglia, because experiments such as decapitation (removal of brain and subesophageal ganglion) do not disrupt the ability of an insect to walk. It is thought currently that the locomotor rhythms for each leg are centrally generated by networks of neurons called central pattern generators and that these central patterns are modified, according to circumstance, via sensory feedback. However, the sensory feedback from a given leg appears to influence the nearest motor axon or central pattern generator, whereas the central pattern generators in the individual legs are linked by pathways within the central nervous system (see chapter 3).

Comparatively little work has been done on control of actual walking behavior (Delcomyn 1985). Although, as previously mentioned, decapitation does not affect the coordination of walking, it does influence locomotor activity. A decapitated insect (i.e., one with both the brain and subesophageal ganglion removed) may live for a long time but not walk spontaneously; however, an insect with just the brain removed may display an increase in walking behavior. How these two cephalic ganglia control walking behavior is not known.

Aids to Walking and Running

For the propulsive force applied by the legs to move an insect, there must be a certain amount of friction between the leg and the substrate. Although tarsal claws generally suffice on rough or dirty surfaces, which afford ample points for grasping, there are situations in which tarsal claws fail—for example, the very smooth surface of an inclined piece of glass or a window pane. Yet many insects, such as the house fly, are able to walk with ease on such surfaces. This ability is due to the presence of a variety of adhesive structures: the pulvilli and tarsal pads described in chapter 2 or pads at the base of the tibiae. These adhesive structures are usually covered with dense mats of tiny hairs with expanded tips. The expanded tips are covered with a secretion from glands located at the bases of the hairs and are the parts of the adhesive pads that come into direct contact with the substrate. Apparently, molecular forces among the expanded tips, the glandular fluid, and the substrate account for the adhesion of the pad. Because it is the tiny hairs that are ultimately responsible for the "clinging" ability, they are

Figure 7.3 Climbing organ of *Rhodnius prolixus* (Hemiptera; Reduviidae). (*a*) and (*b*) Proximal portion of tibia and tarsus showing position of climbing organ. (*c*) Tent hairs that compose the climbing organ. *Redrawn from Gillett and Wigglesworth, 1932.*

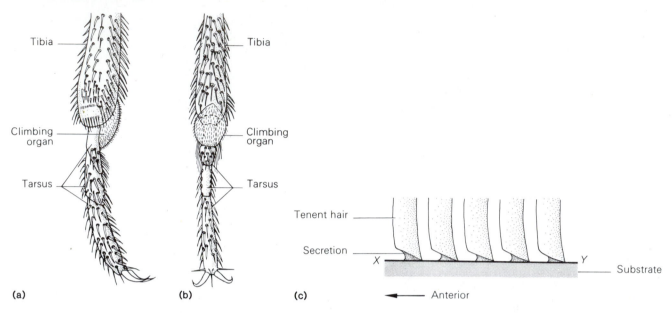

commonly referred to as *tent hairs*. Gillett and Wigglesworth (1932) advance the following explanation of the functioning of tent hairs in *Rhodnius prolixus* (Hemiptera; Reduviidae).

The tent hairs that compose the climbing organ (figure 7.3) on the distal end of the tibia are wedge-shaped at their tips (figure 7.3*c*), and only the hindmost part comes into contact with the substrate. When the insect moves in an anterior direction (toward *X*, figure 7.3*c*), or an external force (e.g., gravity) pulls the insect in an anterior direction, the secretion between the hairs and the substrate acts as a lubricant, reducing the friction. However, if the insect is moving or being forced to move in the opposite direction (posterior or toward *Y*, figure 7.3*c*), the surfaces of the hair tips come into very close contact with the substrate, and "seizure" occurs. The surfaces are held together by the adhesive forces of the molecules of the glandular secretion. Thus *R. prolixus* can walk up a pane of carefully cleaned glass at an incline of 80° but slips at an angle of 22° when walking down the same pane. The holding power of adhesive pads is well illustrated by the fact that one bug in the same family as *R. prolixus* can support a tension of greater than 50 grams (Wigglesworth 1972).

Jumping

Under this heading we include all the means, other than wings, by which insects are able to propel themselves through the air. Many insects are capable of jumping, but it has become a pronounced specialization only in certain groups. The suggestion has been made that perhaps wings first appeared on a jumping insect that used the jumping ability for a propulsive force and early wings merely as gliding surfaces. Indeed, flight typically begins with a jump (Möhl 1989). Mechanosensilla on the tarsi of most flying insects provide inhibitory feedback to the flight control system and without the "jump" to remove this inhibition, flight does not occur.

Most jumping insects use modified hindlegs in jumping (see figure 12.15), although other specialized mechanisms occur in some species. Those that use their legs in jumping typically have enlarged, muscular femora on the hindlegs. Examples are the members of several orthopteran families (e.g., grasshoppers and katydids), flea beetles (Coleoptera, Chrysomelidae), and fleas (Siphonaptera). Relative to their size, these insects are capable of rather astounding leaps. For example, the trajectory of a jumping flea may reach a height of 8 inches and cover a distance of 13 inches; if proportionally carried out by a human being, this would result in a jump of 800 feet high (Wigglesworth 1965)! However, as explained in chapter 3, strength of contraction is a function of cross-sectional area of muscle (linear dimension squared), and increase in mass is a function of volume (linear dimension cubed). Therefore a flea the size of a human being would have relatively much more mass per unit cross-sectional area of muscle than a normal-sized flea and would thus probably still be able to jump only 8 inches high. Grasshoppers have been reported to leap up to 30 centimeters high over a distance of 70 centimeters.

Springtails (Collembola), click beetles (Elateridae), and cheese maggots (Diptera, Piophilidae) have unique methods of jumping.

Springtails (see figure 2.42b) are minute wingless insects that possess a fork-shaped structure, the furcula, on the posterior part of the abdomen and a structure called the tenaculum on the anteroventral portion of the abdomen. The furcula is bent anteriorly, and its tip is engaged by the tenaculum. Prior to a jump, tension is built up in the muscles of the furcula. When released by the tenaculum, the furcula "snaps" posteriorly against the substrate, propelling the insect into the air.

Click beetles utilize a similar principle, an elongated spine on the posterior part of the prosternum being inserted into a receptacle on the mesosternum. The jump produced by the sudden release of this prosternal spine is elicited only when the beetle is on its back and hence serves as a righting mechanism.

Cheese maggots use their entire body in jumping. The last abdominal segment is held by the mouth hooks, muscular tension is built up within the body, and the last abdominal segment is suddenly released against the substrate to propel the insect.

It has been assumed for a long time that direct contraction of leg muscles accounts for the force released when an insect jumps. However, the role of cuticular elasticity has come to be appreciated (Bennet-Clark 1976). Cuticular elasticity is associated with regions of the cuticle that contain high concentrations of rubbery protein, resilin. The Oriental rat flea, *Xenopsylla cheopis*, has been studied extensively in this regard (Rothschild et al. 1973; Bennet-Clark 1976). In this flea, resilin is concentrated in the pleural arch of the metathorax. It has been known for a long time that the large metathoracic legs are the most important in jumping. The pleural arch is considered to be homologous with the wing hinge ligament of winged insects. The metathoracic leg is so constructed that it can be "cocked." When this leg is in the cocked position, the femur is raised up and engaged by a catch and the trochanter is in contact with the substrate. A catch mechanism engages plates of the mesothorax to the metathorax. Also, when the metathoracic leg is in the cocked position, the resilin in the pleural arch is compressed and elastic energy is stored in it. When the catches are released, the elastic energy in the resilin is released and translated to the trochanter via a tendon. The trochanter thrusts against the substrate, accelerating the flea from the substrate. Then the thrust of the tibia against the substrate provides additional acceleration. Jumping in locusts (Orthoptera, Acrididae) and click beetles is also thought to involve energy stored in resilin.

Crawling

Crawling (or creeping) is generally associated with the locomotion of larval insects, particularly those that propel themselves by means other than the six thoracic legs alone.

However, many larval and most nymphal insects walk like adults. Some of these hexapodous immature forms (for example, beetle larvae like mealworms, *Tenebrio molitor*) are aided in their locomotion by eversible *pygopodia*, which arise from the terminal abdominal segment. A pygopodium is everted by means of hemolymph pressure and aids the thoracic legs in the progression of the insect.

The immature forms of butterflies and moths (caterpillars) and certain wasps (Hymenoptera, Symphyta) bear, in addition to the three pairs of thoracic legs, accessory legs, or prolegs, on certain of the abdominal segments. The movement of the prolegs depends on the integrated activity of the body musculature of the abdomen. Since these insects are typically soft-bodied, hemolymph pressure is important in maintaining a "hydrostatic skeleton." The pressure of the hemolymph is maintained by the turgor muscles of the insect body. Puncturing one of these insects results in an immediate shriveling, owing to the contraction of the turgor muscles. This indicates that these muscles respond to a reduction in hemolymph pressure by contraction and gives some insight as to how they function in an intact larva. In addition to the turgor muscles, there are locomotor muscles that fail to contract when a larva is punctured and are involved with the actual movement of the prolegs. These muscles are arranged in longitudinal, dorsoventral, and transverse patterns.

Walking patterns in larvae with prolegs have been best described in caterpillars. As in walking solely with thoracic legs, the rhythms involving prolegs are probably generated in the central nervous system. Single peristaltic waves of contraction usually pass anteriorly from the posterior end of the insect. As each wave passes, the two legs of each segment always move simultaneously. Each wave involves three main phases. Hughes and Mill (1974) describe these three phases as follows (figure 7.4).

In the first, the dorsal longitudinal and dorsal intrasegmental muscles contract, together with the large transverse muscles. This results in the segment becoming shortened dorsally and consequently its posterior end becomes inclined forward so that the segment behind is lifted from the ground. In the next phase the segment contracts dorsoventrally and its feet are released simultaneously from the substratum. After the legs have been moved forward, the ventral longitudinal muscles contract so as to bring the segment down toward the ground and the feet become fixed. The wave of peristalsis passing over the body is therefore not limited to a single segment but involves a simultaneous contraction of muscles in at least three segments. Contraction starts at the last segment with the release of the terminal appendages. They are lifted and placed on the ground at varying distances forward.

Figure 7.4 Caterpillar walking. Arrow indicates direction of peristaltic waves and direction of progression. *Redrawn from Barth, 1937.*

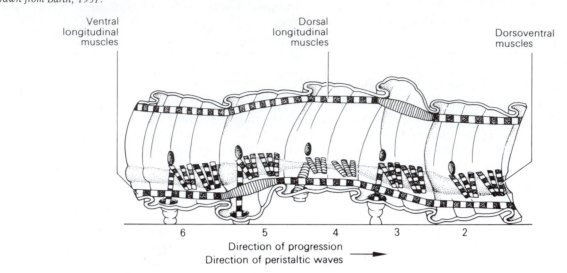

Direction of progression
Direction of peristaltic waves →

Figure 7.5 Locomotion of "inchworm," geometrid moth caterpillar. d = distance advanced in one cycle. *Redrawn from Snodgrass, 1961.*

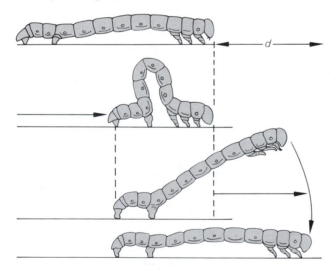

Figure 7.6 Locomotion in a crane fly larva. *Redrawn with modifications from Kevan, 1962, after Gilyarov.*

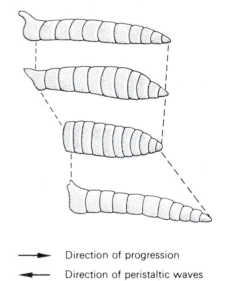

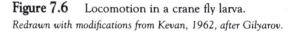
→ Direction of progression
← Direction of peristaltic waves

This description is simplified, since other sets of muscles (e.g., the transverse muscles) are no doubt involved. In addition, contractions of the turgor muscles must be in coordination with the contractions of the locomotor muscles. An interesting variation of this type of crawling is shown by the larvae of geometrid moths. These larvae are often referred to as inchworms, or loopers. In these larvae much of the trunk is out of contact with the substrate as they move along (figure 7.5).

Terrestrial larvae having no legs whatsoever (for example, fly maggots) depend upon peristaltic waves of contraction for their progression. Many possess spines or other structures that increase the friction between their bodies and the substrate. In most apodous larvae, waves of peristaltic contraction, like those in larvae with pro-legs, pass from posterior to anterior, that is, in the direction of progression. However, in certain ones adapted to burrowing through the soil (for example, burrowing crane fly larvae and bibionid larvae (Diptera)), the waves of contraction proceed from anterior to posterior, or in the direction opposite that of progression. In this method, the insect's posterior serves as a stable point while the anterior portion of its body is pushed through the soil (figure 7.6).

Aquatic Locomotion

We shall first examine the locomotion of aquatic insects that propel themselves on the surface of the water and then those that "swim" beneath the water's surface.

Nachtigall (1985) includes literature citations on aquatic locomotion and provides somewhat detailed explanations of the mechanics and functional anatomy of aquatic locomotion in insects.

Surface Locomotion

To appreciate adaptations associated with insect locomotion on the water's surface, we must first consider some of the characteristics of that surface and its interaction with other surfaces. At temperatures and pressures typical of the environments in which insects are found, water molecules are strongly attracted to one another, a phenomenon known as *cohesion*. In the absence of outside forces upon the molecules, a droplet of water tends to contract so that it has the smallest possible surface area. This theoretically would result in a sphere. The molecules of the gases that compose air have little attraction for water molecules and little attraction for one another. Hence any surface of water in contact with air tends to contract. This force of contraction is called *surface tension*, and its presence makes the water's surface behave somewhat like a thin, elastic membrane. Some surfaces or surface coatings attract water molecules, and when this attractive force is equal to or greater than that between water molecules themselves, the water spreads out on the surface, which is said to be wettable or hydrophilic (water loving). Other surfaces have the opposite effect, exerting little or no attractive force on water molecules. Such surfaces are water repellent or hydrophobic (water fearing).

The magnitude of the forces associated with surface tension seems rather small when we think of large animals, which have a comparatively small ratio of surface to volume and hence comparatively little surface area per unit body mass. However, surface-tension forces are quite significant in small organisms, such as insects, that have relatively large ratios of surface to volume and thus a comparatively large amount of surface area per unit body mass. Because in general the insect cuticle is hydrophobic, many small insects are able to be supported against the force of gravity by the forces of surface tension. However, if the cuticle is made hydrophilic by the application of a "wetting agent," such as a detergent, the insect will immediately sink or become trapped by the water. Many surface-dwelling forms secrete a waxy material that makes their tarsi hydrophobic and allows these insects to walk on the water's surface. The surface-dwelling bugs (several families of Hemiptera) are able either to walk on the water as other insects would walk on the ground (e.g., the water measurer, *Hydrometra*, Hydrometridae) or to use the middle legs as synchronous oars, rowing from place to place (e.g., water striders, Gerridae and others). The collophores and tarsal claws of certain springtails (e.g., *Podura aquatica*) are hydrophilic, but the rest of the integument is hydrophobic. This enables these insects to be anchored to the water and yet easily move from place to place by jumping with their furculae. Beetles in the genus *Stenus* (Staphylinidae) secrete a substance from anal glands that lowers the surface tension of the water behind the insect. As a result they are drawn forward by the effects of the contractive forces of the water in contact with their bodies. These beetles are able to attain speeds of 45–70 cm/sec. Water striders in the genus *Velia* (Veliidae) are similarly propelled by a surface-tension-reducing substance discharged through their posteriorly directed mouthparts. This occurs as an escape response (Blum 1978).

Subsurface Locomotion

Many insects live beneath the surface of the water. These insects can be conveniently divided into four groups.

1. Those utilizing appendages in swimming.
2. Those moving by various undulations of the body.
3. Those that are able to produce jets of water through the anus.
4. Those that walk over the substrate like terrestrial insects.

Many of these insects have air stores of various sorts, which increase buoyancy and aid in the maintenance of body position in the water (*hydrostatic balance*). In addition, in many, particularly the rapid-swimming, hard-bodied forms, the body is flattened and tapered, a hydrodynamic adaptation that minimizes friction in the form of drag during movement through the water.

Insects that utilize appendages in swimming include representatives from the orders Coleoptera (Dytiscidae, predaceous diving beetles; Gyrinidae, whirligig beetles; Haliplidae, crawling water beetles; Hydrophilidae, water scavenger beetles; and others), Hemiptera (Notonectidae, backswimmers; Corixidae, water boatmen; Belostomatidae, giant water bugs; and others), Trichoptera (caddisflies), Neuroptera (dobsonflies, lacewings, and relatives), and even Lepidoptera. Other than insects that merely crawl about on the bottom of a body of water or on submerged vegetation, the vast majority of the insects in this category have legs that are flattened like oars or that are covered with "swimming hairs," or commonly both. For example, in whirligig beetles in the genus *Gyrinus* the rowing legs are so flattened that the broad side has five times the area of a comparable round leg.

The swimming legs of adult dytiscid beetles bear large numbers of hairs. These hairs (figure 7.7) are movable; when the insect swims, they open out in response to water pressure, increasing the surface area applied against the water during the thrust stroke, and they fold down on the legs during the recovery stroke. Rowing or swimming legs operate in various combinations. Whirligig beetles swim with the oarlike mesothoracic and metathoracic legs. Their forelegs are long and slender and are used in prey capture. These beetles are the ones commonly seen darting to and fro, the hydrophobic dorsum of their bodies breaking the water's surface. Beetles in

Figure 7.7 Swimming movements of a predaceous diving beetle *Acilius sulcatus*. Numbers indicate the temporal sequence of leg positions during the swimming stroke. Top, thrust stroke; bottom, recovery stroke. Note the positions of the hairs at various points in the stroke cycle.
Redrawn from Nachtigall, 1960.

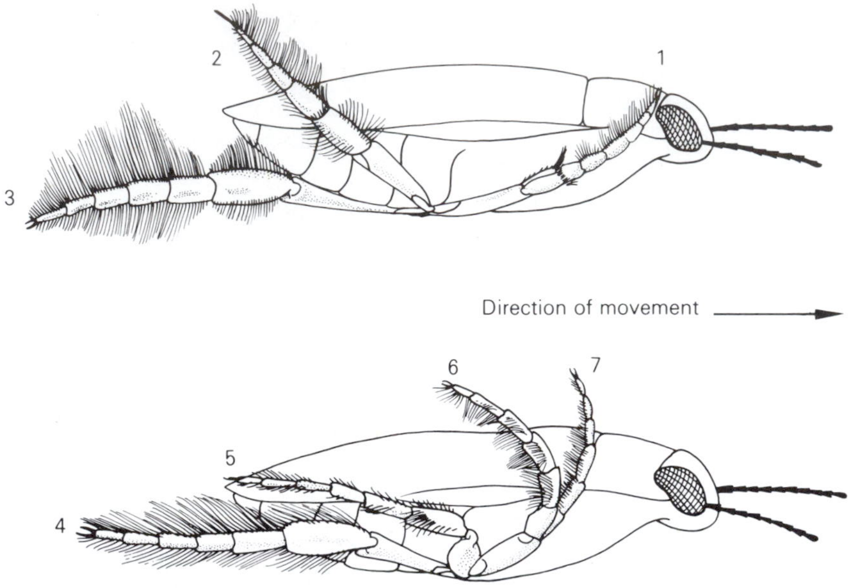

the family Dytiscidae may use all three pairs of legs in locomotion. However, in the large members of this family the forelegs are used in prey capture and the midlegs and hindlegs for propulsion. Hydrophilid and haliplid beetles swim with all three pairs of legs, but in contrast to the beetles just mentioned, the thrust and recovery strokes of legs on the same thoracic segment are in opposite phase.

Among the aquatic Hemiptera members of the families Nepidae (water scorpions), Naucoridae (creeping water bugs), and Belostomatidae (giant water bugs) swim with the last two pairs of legs. The strokes of the two legs of each segment are out of phase by 180°. The forelegs of these insects are generally used in grasping prey, and swimming legs are modified in ways similar to those of the aquatic beetles. Backswimmers have a rather different manner of swimming. They swim beneath the water with the ventral side up and the head directed downward. Swimming is accomplished with mesothoracic and metathoracic legs, which move in the same phase. In water boatmen the third pair of legs provides propulsion, the second pair act as grasping and steering organs, and the first pair scoops up food-containing materials.

Steering, braking, and curving while swimming are achieved by various differential movements of the legs. Whirligig beetles in the genus *Orectochilus* are purported to use their external genitalia as rudders (Nachtigall 1985).

In addition to legs, certain insects, such as the adult females of *Hydrocampa nymphaeata* (Lepidoptera, Pyralidae) and a few hymenopterous insects (*Caraphractus cinctus*, Mymaridae; and *Limnodite*, Proctotrupidae), are propelled through the water by their wings.

The legless aquatic larvae and pupae of several Diptera are propelled through the water by twitches and undulations of their bodies. Figure 7.8 shows some of the variations in this type of locomotion. Certain of these insects are buoyed by air stored in "air bladders" (e.g., *Chaoborus* larvae) or gas in external spaces resulting from the shape of the body [e.g., the ventral air space formed by developing legs, wings, and mouthparts in mosquito pupae Romoser and Nasci (1979)].

The nymphs of most dragonflies and the nymphs of the mayfly genus *Cloeon* draw water into, and expel it from, the rectum as part of the ventilatory process. If alarmed, these insects can expel water from the rectum with considerable force. The expulsion of water propels the insect several centimeters, and speeds of 50 cm/sec have been recorded. When the insects propel themselves in this fashion, the thoracic legs are held close to the body, which has the effect of reducing the friction between the water and the insect. In addition, the forcibly expelled water may disturb bottom material (especially silt) and throw a protective "smoke screen."

Bottom-dwelling insects, dragonfly and damselfly larvae, larval caddisflies, and so on, walk over the substrate with thoracic legs much as the terrestrial forms do.

Aerial Locomotion

As mentioned earlier, insects, alone among the invertebrates, possess the ability to fly. This ability is perhaps one of the most important reasons for their tremendous success relative to the rest of the animal kingdom. Flight

Figure 7.8 Swimming movements of legless aquatic larval and pupal Diptera.
Redrawn from Nachtigall, 1963.

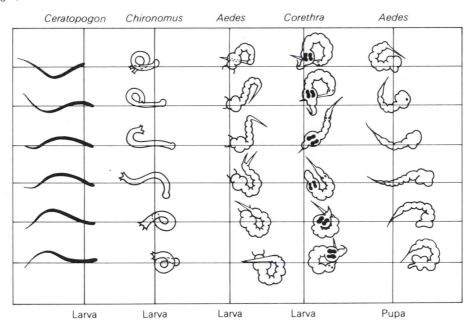

Figure 7.9 Generalized wing-bearing segment. (*a*) Lateral view. (*b*) Cross-sectional view.
Redrawn from Snodgrass, 1935.

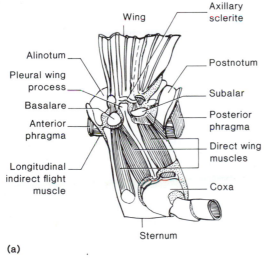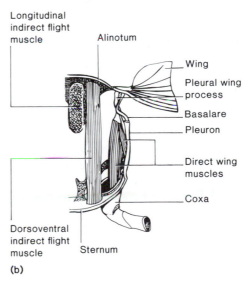

has enabled insects to take advantage of environmental situations that are virtually untouchable by their non-flying rivals. Goldsworthy and Wheeler (1989) are the coeditors of a multiauthored comprehensive treatment of insect flight that includes chapters on flight behavior and pest control, in addition to morphologically/physiologically oriented chapters.

Functional Morphology of the Flight Mechanism

General discussions of skeletal structure, thoracic musculature, and insect wings may be found in chapters 2 and 3. As explained in chapter 3, a generalized wing-bearing segment contains both indirect and direct wing muscles (figure 7.9). There are two sets of indirect wing muscles, neither of which attach directly to the wing

Figure 7.10 Generalized wing base.
Redrawn from Snodgrass, 1935.

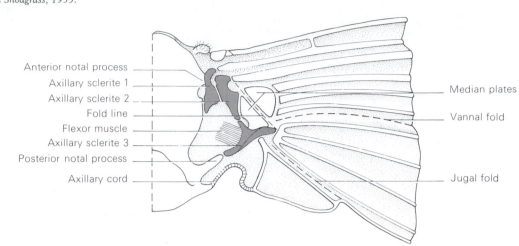

Anterior notal process
Axillary sclerite 1
Axillary sclerite 2
Fold line
Flexor muscle
Axillary sclerite 3
Posterior notal process
Axillary cord

Median plates
Vannal fold
Jugal fold

bases. The longitudinal indirect muscles run between dorsal apodemes (phragmata), and contraction causes arching of the tergum with the consequent depression of the wings. The dorsoventral (tergosternal) indirect muscles run between the notum and sternum. The structural relationship between the notum and pleura (see figure 3.21) is such that contraction causes elevation of the wings.

Direct wing muscles attach to various regions on the pleuron and coxa and to the axillary sclerites (figure 7.9) and basalar and subalar sclerites (figure 7.9). Contractions of these muscles are involved with wing movement during flight and with *wing flexion*. The muscles associated with the basalar and subalar extend and depress the wings. Contraction of muscles attached to the third axillary sclerite (figure 7.10) results in wing flexion (the opposite of extension); that is, the wing is folded posteriorly over the abdomen. Folds in the wing (figure 7.10) facilitate this process. With various modifications of the generalized scheme presented here, winged members of all pterygote orders, except the Odonata (dragonflies and damselflies) and Ephemeroptera (mayflies), are able to flex the wings over the abdomen. The origin of wings and evolutionary significance of wing flexion are discussed in chapter 11.

The wings themselves show some remarkable adaptations. The longitudinal wing veins are tubular thickenings of cuticle, which provide strength for the wing and tend to be heavier along the leading edge and anterior region of the wing (i.e., the part of the wing that must "cut" through the air). Longitudinal veins in the trailing part of the wing tend to be smaller and less rigid. Cross veins oriented at various angles relative to the longitudinal veins further strengthen the wings. In some insects (e.g., locusts) pleating of the wings provides additional stiffness without increasing the weight. Channels along the sides of wing veins, between the dorsal and ventral

epidermal layers that form the wings, are continuous with the hemocoel. Hemolymph circulates through these channels. At their bases, wing veins form rather complex articulations with thoracic sclerites. Wooton (1990) provides an excellent introduction to the "engineering and design" of insect wings.

There has been a tendency during insectan evolution toward the reduction of the wings to a single unit. The pressures in this direction may have been due to the inefficiency of a hind pair of wings operating in the turbulence produced by the movements of the forewings. The wings of certain groups of insects, such as the Orthoptera, Megaloptera (dobsonflies, fishflies and alderflies), Isoptera (termites), and others, operate separately, and these insects are rather poor fliers as a result. One evolutionary solution to this problem is found in dragonflies and damselflies. Although their pairs of wings operate separately, their flight is quite efficient because the sequence of wingbeat is reversed, the hindwings moving before the forewings during each cycle.

Single unit functioning of wings has been achieved in two ways. One has been by the development of a variety of mechanisms that enable the forewings and hindwings to be coupled to one another (see figure 2.40*a–c*). Correlated with the presence of wing-coupling mechanisms is a reduction of the size of the hindwings. Coupling mechanisms are found among the Hymenoptera, Trichoptera, Hemiptera, many Lepidoptera, and others. In contrast to wing coupling, many insect groups have undergone an evolutionary loss or extreme modification of one pair of wings so that they no longer function as propulsive organs of flight. For example, in the true flies, reduced hindwings form the halteres (balancing structures). The forewings of beetles have for the most part lost their function as organs of flight, although in some species they do move at low amplitude frequencies and may contribute modestly to flight (Nachtigall 1989).

Flight and Its Control

As mentioned previously, flight typically begins with a jump initiated by contraction of "jump" muscles (e.g., the tergotrochanteral muscle of flies). Following the initial jump, flight ensues.

Both "direct" and "indirect" muscles are associated with wing movement. Probably the muscles attached directly to the wings were the original sources of the major propulsive force of wing motion. Direct muscles are still dominant in cockroaches (Blattaria) and very important in members of the Orthoptera as well as Odonata and many Coleoptera. In other insects the main propulsive forces are furnished by the indirect muscles of each wing-bearing thoracic segment. In these insects the indirect muscles are quite large relative to the direct ones, although the direct muscles remain important in wing motion. There is a direct correlation between the weight of flight muscle as a percentage of body weight and the strength of flying. For example, according to Wigglesworth (1972),

> In the relatively weak flying orthopteran *Oedipoda* the flight muscles comprise only 8% of the total body weight, but in strong fliers they make up far greater proportions: *Musca* [fly] 11%, *Apis* [honey bee] 13%, *Macroglossa* [sphinx moth] 14%, *Aeshna* [dragonfly] 24%.

Wing-movement patterns in flying insects have been analyzed by several techniques. Earlier methods included attaching light-reflecting devices to the wing tips (which made the trajectory of each wing tip visible) and observation of wing positions in freshly killed insects. More recently, high-speed cinematography, coupled with microcomputers, has proved very useful. Another important technique has been the use of the strobe light, a device that can produce regularly intermittent light flashes over a wide range of rates. At appropriate rates of flashing the entire wing is illuminated regularly at various points in its cycle, which makes the wing appear to be moving very slowly. This makes possible direct observation of the changes in wing tilt and so on as they occur. When the flash frequency is a whole-number multiple of the wingbeat frequency, the wing is caught in the same position during each cycle and appears to stand still. This is one of the methods for determining wingbeat frequency. Recording and analysis of the pitch of the sound produced during wingbeat is another important method for determining frequency of wingbeat. Highly sophisticated mechanical and electronic recordings of neuromuscular activities during flight have also been used extensively.

In modern insects, the wings are responsible for both lift and propulsion and usually are of little importance in gliding. Naturally, for bilateral appendages to produce these effects, the movements must be complex. These movements consist of "elevation and depression, promotion and remotion (fore and aft movement), pronation and supination (twisting) and changes of shape by folding and buckling" (Pringle 1957).

The velocity of wing motion varies in different parts of the cycle. In many insects movements like wing twisting are entirely under the control of direct wing muscles. However, in some of the better fliers, such as members of the orders Diptera and Hymenoptera, the basal wing articulations are designed mechanically such that the appropriate twisting of the wing occurs at the proper time during a wingbeat cycle. This is not to say that direct muscles do not exert some degree of influence, even in these insects. The propulsive action of the wings is quite efficient. According to Wigglesworth (1972), "the flying insect produces a polarized flow of air from front to rear during approximately 85% of the cycle." In typical forward flight, a wing traces out a figure eight relative to the body at its base (figure 7.11). Relative to a point past which an insect is flying, the wing traces out a pattern resembling a sinusoidal curve. Many insects can hover by changing the inclination of the figure eight (i.e., the *stroke plane*) relative to the body (figure 7.11). Many of the especially good fliers (Diptera, Hymenoptera, and Lepidoptera) are able to fly backward (figure 7.11), sideways, or rotate around the head or tip of their abdomens. Rotation and flight sideways are apparently produced by unequal activity of the wings on either side. Besides being able to vary the patterns of wingbeat, insects may vary the power of wingstroke by varying the number of actively contracting muscle fibers.

To get a visual feel for the movements of insect wings during flight, see the superb stop-action photos in Dalton (1975).

The number of wingbeat cycles per second varies from one group of insects to another (table 7.1). Not only does wingbeat frequency vary in different insectan groups, but it also varies with age, sex, season, temperature, humidity, load, air resistance, air density, air composition, wing inertia (moment of inertia), muscle tension, and fatigue. However, variation in frequency of wingbeat is apparently not generally used by insects to control flight. Assuming a single series of muscle contractions for each wingbeat cycle, many frequency measurements seemed in the past to be fantastically high. Electrophysiological recordings from flight muscle and associated nerves showed that in certain insects the number of nervous impulses impinging on a wing muscle were much fewer than the number of muscle contractions observed. Thus the high wingbeat frequency observed could not be explained on the basis of a one-to-one relationship between nervous impulse and muscle contraction. In other cases there was

Figure 7.11 Wingbeat patterns traced by the tip of a wing during flight. Dashed lines indicate longitudinal axis of body. (*a*) Forward flight. Numbered lines indicate position and inclination of the wing at different stages of the wingbeat cycle: (*b*) Hovering. (*c*) Backward flight.

(a) redrawn with modifications from Mangan, 1934; (b–c) redrawn with modifications from Stellwaag, 1916.

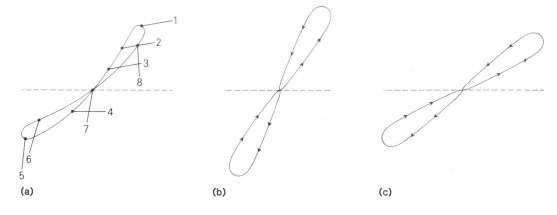

(a) (b) (c)

Table 7.1 Representative Wingbeat Frequencies[a]

Insect	Wingbeats per second[b]
Odonata	
Libellula	20
Aeshna	22, 28
Coleoptera	
Melolontha	46
Coccinella	75–91
Rhagonycha	69–87
Lepidoptera	
Pieris	9, 12
Colias	8
Saturnia	8
Macroglossa	72, 85
Acidalia	32
Papilio	5–9
Diptera	
Tipulids	48, 44–73
Aedes, male	587
Culex	278–307
Tabanus	96
Musca	190, 180–97, 330
Muscina	115–220
Forcipomyia	988–1047
Hymenoptera	
Apis	190, 108–23, 250
Bombus	130, 240
Vespa	110

[a]Data from Wigglesworth (1965) based on various sources.
[b]Two or more sets of numbers indicate data obtained by different investigators.

a one-to-one relationship between nerve impulses and muscle contraction. These differences remained a puzzle until the elastic nature of the insect thorax and the action of resonating flight muscles were discovered.

In addition to the actions of direct and indirect muscles, it has been found that in many insects the elasticity of the thoracic skeleton plays a significant role in the movement of the wings. As the wings move in one direction, some of the energy of muscle contraction may be stored as elastic energy in the thoracic skeleton and antagonistic muscles and used in the movement of the wings in the opposite direction. This can be seen in the locust *Schistocerca*. When the major flight muscles are relaxed, the wings are completely depressed. This is the stable position for them in relation to the elasticity of the thoracic skeleton. Any movement away from this stable position is opposed by the elastic forces.

In several insects, such as certain Diptera and Coleoptera, the wings have two stable positions: completely elevated and completely depressed. Movement of a wing away from one of these stable positions is opposed by the elasticity of the thorax to a point at which the direction of the elastic force reverses. At this point the wings are suddenly driven by the elastic energy into the opposite stable position. This has been called the click mechanism. In chapter 3 it was pointed out that insects possess both resonating and nonresonating muscles and that, under appropriate mechanical conditions, the resonating type of muscle is capable of undergoing several successive contractions when stimulated with a single nervous impulse. The insect thorax with sufficient elasticity to produce a click mechanism is an example of such an appropriate mechanical situation. The resonating muscles involved are the mutually antagonistic dorsal longitudinal and dorsoventral flight muscles. As one set of these muscles is stimulated by a nerve impulse to contract, a point is reached where the "click" occurs (i.e., where the elastic forces reverse). When this happens, the tension on the contracting muscles is suddenly released and the antagonistic pair is suddenly stretched. This sudden stretch stimulates the antagonistic pair to contract and the cycle then repeats itself. Thus several oscillations of the wing mechanism are produced before another nervous impulse initiates the series all over again. These oscillatory contractions have been identified in many of the higher Diptera and Coleoptera, and others.

Information regarding the mechanics of insect flight can be found in Nachtigall (1989). Aidley (1989) deals with the structure and function of insect muscle, and Kammer (1985) covers both muscular and neurobiological aspects of insect flight.

The speeds attainable by different insects during flight cover a broad range (table 7.2) with weak fliers like mayflies on one end and strong fliers like deer bot flies on the other.

The sensory cues pertinent to flight control undoubtedly come from a number of different sources, both from sensilla in the wings themselves and from sensilla located in other parts of the body. However, no sensory endings have been found in insect flight muscles. Wing sensilla fall into three main categories.

1. Bristles and hairs.
2. Campaniform sensilla.
3. Chordotonal sensilla.

The bristles and hairs may be widely distributed over the wing's surface and respond to tactile stimulation. Thus they may respond to the movement of air over the wings during flight. Campaniform sensilla are arranged in definite patterns in the wings. Apparently different ones respond to different torques acting on the wings during flight, depending on their specific orientation. Since chordotonal sensilla are sensitive to changes in length, those in the wings probably respond to displacements of the cuticle at the wing base resulting from various torques. Like the campaniform sensilla, they are arranged in definite patterns. In certain instances they may actually respond to changes in air pressure or even have an auditory function.

Sensilla that collect information useful in flight control have been described not only in the wings, but also in the head, neck, halteres, legs, and other locations. Sensilla in the head are commonly in the form of hairs on the frontal region and Johnston's organ in each antenna. Both kinds of sensilla are sensitive to airstreams across the head. The compound eyes must not be overlooked as

T able 7.2   Average Speed of Sustained Flight in Various Insects[a]

Insect	Speed, km/h
Mayflies, small field grasshoppers	1.8
Bumble bees, rose chafers	3.0
Malaria mosquitoes	3.2
Stag beetle, damselfly, *Ammophila* (a fossorial wasp)	5.4
House fly	6.4
Cockchafer, cabbage white butterfly, garden wasp	9.0
Blow fly	11.0
Desert locust	16.0
Hummingbird hawk moth	18.0
Honeybee, horse fly	22.4
Aeschna (a big dragonfly), hornet	25.2
Anax (one of the biggest of European dragonflies)	30.0
Deer bot fly	40.0

[a]Data from Nachtigall (1974) based on various sources.

sense organs important to flight control. Dragonflies have hair plates on the neck that respond to changes in head position.

Probably any stimulus that in some way indicates danger would cause an insect to take flight as a means of escape. However, a number of specific stimuli that induce flight have been identified. The most widespread and important of these is loss of contact with the substrate. This may be easily demonstrated by suspending a cockroach or other insect from a string by means of a drop of paraffin. As long as the legs of the insect are in contact with a surface, there will be no flight movements, but if the insect is suddenly lifted into the air and surface contact lost, the wings will immediately commence to beat. This is the tarsal reflex because mechanosensilla on the tarsal segments are involved. Other flight-initiating stimuli that have been identified include change in the relative positions of the prothorax and mesothorax of the cockroach *Periplaneta* and acoustic stimuli in locusts.

Certain stimuli—for example, a dimming of light or the visual sensation of a moving object approaching the flying insect—may initiate a landing response.

External environmental factors may influence whether or not an insect takes flight. For example, moths in the family Spingidae (sphinx moths) are unable to fly until the wing muscles reach a temperature of 30° C. If the ambient temperature is 30° C or above, they have no trouble taking flight, but if it is below, these moths must rapidly vibrate the wings until the necessary temperature is reached.

During flight, constant corrections must be made for any forces that tend to disrupt the stability of the insect. Instabilities can be resolved into tendencies of the insect to move in one or a combination of three planes: movement about the longitudinal-horizontal axis, or *roll*; movement about a transverse-horizontal axis, or *pitch*; and movement about a vertical axis, or *yaw*. An unstabilizing force or combination of such forces undoubtedly affects different sensilla to different degrees and in different ways and thus results in some change in the pattern of stimuli that impinge on the insect's nervous system during stable flight. In response, the insect modifies its flight movements in such a manner that the impinging stimuli again are in the pattern that indicates stability. These stabilizing responses can probably be explained by simple reflexes, but flight muscle and wingbeat rhythms are probably in many cases determined by central pattern generators in the central nervous system (Möhl 1989).

The halteres of true flies are a specific example of stabilizing organs (see figure 2.39e). These "modified hindwings" move in synchrony with the forewings. Groups of campaniform and chordotonal sensilla are arranged in specific patterns at the base of each haltere. Because the halteres are constantly vibrating during flight, they act like a gyroscope, which generates forces when there is any tendency to change direction. You will probably recall from basic physics the simple demonstration of gyroscopic action in which a rapidly rotating bicycle wheel held by two hands exerted resistive forces when you attempted to turn it away from its original plane of rotation. The forces generated by the vibrating halteres when subjected to turning forces are "sensed" by the campaniform and chordotonal sensilla in a manner similar to the way the resistive forces were "sensed" by the sensilla in your hands and arms. The halteres apparently monitor changes in all three of the planes described previously.

Other sensilla that respond to unstabilizing forces during flight are the hairs on the neck of the dragonfly. During stable flight a characteristic distribution of pressure from the head is exerted on these sensilla and causes a "normal" pattern of signals. However, if a force tends to turn the insect in some direction, the pressure on these hairs becomes changed as a result of the inertia of the head, causing the insect to tend to resist the directional change. This change in pattern of stimulation of the hairs results in compensatory changes in the wing movements and hence correction for the unstabilizing forces. In addition to the neck hairs, the flight stability of the dragonfly is maintained by a dorsal light response, by which the insect moves to keep the source of illumination dorsal to itself, and an optomotor reaction to the general visual pattern. Johnston's organ in each antenna and the sensitive hair plates on the head may also detect changes in airflow and thus collect information pertinent to guidance as well as flight maintenance.

Möhl (1989) presents an excellent discussion and review of the sensory aspects of flight control.

CHAPTER 8

Behavior

In the preceding chapters organ systems have been discussed more or less individually. In this chapter, insects are considered as whole organisms, and the results of the very complex, integrated actions of the various systems in response to changes in the external and internal environments are examined.

Before beginning, however, it is significant for students of behavior to know that three behaviorists, Karl von Frisch, Konrad Lorenz, and Nikolaas Tinbergen, shared the 1973 Nobel Prize for Physiology and Medicine. When presented with the award, it was said that they were "eminent founders of the new science called the comparative study of behavior, or *ethology*." They not only shared the award but also shared the experience of studying insects at one time in their lives. Von Frisch is known for his classic studies that deciphered the language of honey bees (von Frisch 1967). Although Lorenz made his mark by his important research on duckling and gosling imprinting behavior, he reported on observations concerning the predatory behavior of dragonfly larvae and predaceous diving beetles (i.e., family Dytiscidae) in his well-known book *King Solomon's Ring* (1952). Tinbergen (1951, 1972) demonstrated that the bee wolf wasp, *Philanthus triangulum* (family Sphecidae), learned to use visual landmarks to locate its nest entrance. By using bee models, he also showed that "bee odor" was an essential stimulus to elicit attack. Without the odor they ignored the models.

Insect behavior has been extensively investigated under natural and laboratory conditions, both approaches being valid, appropriate, and useful. Since the founding of ethology, which represents the more objective study of animal behavior as compared to the more subjective area of animal psychology, new techniques have pushed the study of animal behavior into more comprehensive studies of the biochemistry (i.e., neuropharmacology), the genetics, and the cellular circuitry underlying behavior (i.e., using electrophysiological and other methods to study the "wiring of the system"). This new approach to the analysis of behavior is termed *neurobiology*. It is an extremely popular discipline and one in which major breakthroughs are being made. More will be said about how this new approach has answered questions that previously went unanswered or, as in many cases, resulted in the correct interpretation of results that previously were misinterpreted. Regardless of the level at which one studies behavior, the necessary first step is to describe an insect's behavior as accurately and completely as possible (*ethogram*). This should include consideration of the time of appearance of specific behavioral variations associated with physiological state (nutrition, etc.). During this process of developing an ethogram, specific questions will emerge concerning the functional, ecological, and evolutionary significance of the behavior(s) to the individual or the species.

Following description, there are three primary objectives.

1. Determination of the control of behavior from the nervous, endocrine, and genetic points of view.
2. Elucidation of the function of the various behaviors in the insect's life, that is, the adaptiveness of behavior.
3. Determination of the probable phylogenetic origin of behavior, the rationale being that behavioral traits, like morphological and physiological traits, are genetically based and represent adaptations that have arisen through natural selection.

Kandel (1970) refers to the effect of natural selection on behavioral traits when he states that "the application of biological techniques to behavior has its roots in the writings of Charles Darwin, who argued that since man had evolved from lower animals, human behaviors must have parallels in the behaviors of lower forms. Darwin's radical insights stimulated studies of animal behavior, opening the way to experimentation that was not feasible in man." As you will soon see, *Drosophila* has become the "workhorse" for laboratories studying the genetic and developmental aspects of behavior.

One of the pitfalls encountered during the study of insect behavior has been the temptation to ascribe human purposiveness or goal seeking to many behavior patterns. This is commonly referred to as *anthropomorphism* and should be avoided. Anthropomorphic usage tends to obscure the fact that adaptive behavior is the result of natural selection and anticipation of a goal is unnecessary.

The approach of this chapter is to consider the basic kinds of behavior characteristic of insects, the control of behavior (nervous, endocrine, and genetic), and the biological functions of behavior.

The texts by Atkins (1980), Matthews and Matthews (1978) and the review by Markl (1974) deal specifically with insects. In addition, volume 9 of Kerkut and Gilbert (1985) is devoted exclusively to insect behavior. Carthy (1965) considers the behavior of arthropods, including insects. Richard (1973) traces the historical development of studies of insect behavior.

Kinds of Behavior

In very broad terms, insects exhibit two kinds of behavior, innate and learned. *Innate behavior* to a large extent consists of a more or less fixed response or series of responses to a given stimulus or pattern of stimuli. The "more or less" in the preceding sentence should be emphasized, since innate behavior is usually somewhat flexible and may be modified by experience (learning). Innate behavior is generally considered to be based upon the inherited properties of the nervous system. On the other hand, *learned behavior* is not inherited but is acquired through interaction with the environment (including conspecifics) during the life of the individual. Obviously, although specific patterns of learned behavior are not inherited, the potential for learning is. In many instances it is difficult to determine whether an observed behavior pattern is inherited or learned. There is no evidence of the ability to reason among insects, the observed behavior patterns being explainable on the basis of innate patterns and learning. Hansell (1985) presents a brief historical account concerning the usage of the term instinct.

Innate Behavior

Some innate behavior patterns seem to be comparatively simple, for example, the *reflexes*. Reflexes may involve only a part of the body, as in proboscis extension, or the whole body, as in the righting reflex, which occurs when an insect is placed on its dorsum. Reflexes can be grouped into two classes, *phasic* and *tonic*. Phasic reflexes are comparatively rapid and short-lived and are involved in rapid movements such as proboscis extension. Tonic reflexes are slow and long-lived and are involved with the maintenance of posture, body turgor (see the discussion of crawling in chapter 7), muscle tone, and equilibrium. Reflexes vary in complexity, the simplest being mediated by a single afferent impulse from a receptor to an interneuron and then along an efferent neuron to an effector (see figure 3.1), the *reflex arc*. Probably all reflexes are more complex than this, involving many more neural connections. Individual segmental ganglia may show considerable reflex autonomy. For example, in the silkworm moth (*Bombyx mori*) the oviposition reflex, which results from the ovipositor making contact with a surface, resides entirely in the caudal (last abdominal) ganglion.

More complex innate behavior includes the various orientation patterns. Orientation may be defined as "the capacity and activity of controlling location and attitude in space and time with the help of external and internal references (i.e., stimuli)" (Jander 1963). Birukow (1966), Fraenkel and Gunn (1961), Jander (1975), Markl (1974), and Matthews and Matthews (1978) are all useful in their consideration of this topic. The terminology used in describing the kinds of orientation is confusing. Fraenkel and Gunn (1961) offer a classification scheme for the various kinds of orientation. They first recognize two broad kinds of orientation, primary and secondary. *Primary orientation* is the assumption and maintenance of the basic body position in space—that is, the normal stance—either in a stationary or in a moving insect. For example, reflexes that keep a flying insect on an even course are responsible for the primary orientation. *Secondary orientation* is superimposed upon primary orientation and has to do with the positioning of the insect in response to various stimuli. These stimuli may be external (e.g., light, humidity, temperature) or internal (e.g., the presence of a particular hormone). Secondary orientation may be of value to an insect by leading it toward potential prey or a potential host, or away from potential danger or an unsuitable environment, or toward a favorable one.

In addition to the primary and secondary divisions, Fraenkel and Gunn (1961) also offer a classification based on the mechanisms of orientation, dividing them into *kineses, taxes,* and *transverse orientations*.

Kineses are undirected locomotor reactions to stimuli, and hence there is no particular orientation of the long axis of the body relative to a stimulus source. For example, in tsetse flies, *Glossina* spp., there is an increase in activity in an arid atmosphere relative to activity in a humid one. This behavior would result in the insect "finding" and remaining in a humid atmosphere. The body louse, *Pediculus,* makes fewer and fewer directional changes as temperature, humidity, and odor increase as the louse moves closer to a potential host.

Figure 8.1   Relationship between magnitude of turning tendency and angular deviation.
Redrawn with modifications from Jander, 1963.

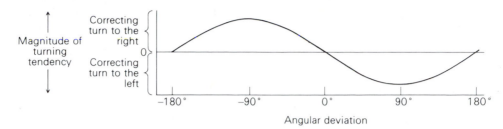

Taxes are different from kineses in that they are directed responses relative to a stimulus source. The long axis of the body takes on a definite orientation relative to the stimulus. The insect may move toward (positive) or away from (negative) the stimulus source. Theoretically, when an insect is maintaining a course relative to a stimulus, angular deviations from this course result in the initiation of a "turning tendency," that is, the tendency for the insect to correct for angular deviation by turning back to its original course. The magnitude of this turning tendency increases with increasing angular deviation. Hence, if the magnitude of this turning tendency is plotted as the ordinate and angular deviation as the abscissa, a sine curve is generated (figure 8.1).

Taxes may be classified according to the stimuli involved (e.g., *phototaxis,* light; *geotaxis,* gravity; *anemotaxis,* air currents; *rheotaxis,* water currents; and *thigmotaxis,* contact).

A good example of a taxis is *klinotactic orientation.* An insect orienting klinotactically swings all or part of its body back and forth across a stimulus field and moves toward or away from the region of maximum stimulation, depending upon whether it is attracted or repelled by the particular kind of stimulus. For example, when fly maggots have completed feeding and will soon pupate, they move the head region back and forth until they are heading directly away from a source of light. You will recall from chapter 6 that this is the position in which the light-sensitive cells are the least affected by the light stimulus. These fly maggots are then photonegative.

Transverse orientations are those in which the body is oriented at a fixed angle relative to the direction of the stimulus. Locomotion may or may not be involved. Fraenkel and Gunn (1961) discuss two types of transverse orientations found in insects: the *light-compass reaction* and the *dorsal light reaction.* The light-compass reaction, currently known as *time-compensated sun compass orientation,* consists of an orientation such that locomotion occurs at a fixed angle relative to the sun. It has now been demonstrated that all vertebrates and invertebrates that rely on some sort of navigational sense to find "home" or nest sites are able to use the sun's position as a reliable indicator of direction. In fact, these organisms also have evolved sophisticated ways of compensating for the sun's movement if they are gone for a long time from their "central place" (for example, back to the nest for ants and the hive for bees). In a similar fashion, crickets return to a "home base" and use time-compensated celestial orientation or polarized light to do this (Huber et al. 1989). Because these insects return to this central place, they have been termed *central place foragers.* Both bees and ants use the sun's position or skylight, as we will soon discuss, as the main directional cue to locate their nesting site. Once in the area of their nest or hive, however, they often rely on landmarks to locate the exact position of their home. If both of these systems fail, they engage in a systematic searching behavior. The research by Wehner (1976) on the desert ant, *Cataglyphis bicolor,* in an area usually devoid of reliable and often shifting landmarks showed that they used either the sun's position on clear days or, on cloudy days, they used the pattern of polarized ultraviolet light produced by the sun. Von Frisch, using honey bees, was the first to demonstrate that insects could use the pattern of polarized light, because on cloudy days bees navigate perfectly well. Thus, insects have evolved the ability to use the pattern of polarized light as their main backup navigational point of reference when the sun is concealed by clouds. Polarized light perception is discussed in chapter 6 and appears to be a common feature of the compound eye of diurnal insects.

A topic currently being debated and researched by neurobiologists is the way animals make cognitive maps of their environment. This same debate has now entered the insect arena. Wehner and Menzel (1990) present an update on this topic and conclude that no evidence exists that insects use landmarks to make cognitive maps; but instead they use vector information based on their position to the sun as their major reference point. There is little doubt, however, that certain species of bees use landmarks to establish specific foraging routes that they use daily to collect food (Wehner and Menzel 1990).

In the dorsal (or ventral) light reaction, both the long and transverse axes of the body are kept perpendicular to a directed source of light (Fraenkel and Gunn 1961). Although symmetrical receptors (the eyes) are involved, this type of reaction is different from phototaxis because orientation is transverse to the light rays and not parallel to them. Examples of both dorsal and ventral light responses are found in certain aquatic members of the order Hemiptera. When the air bubbles trapped between the head and the antennae are removed, backswimmers, *Notonecta* and *Plea,* depend entirely upon a ventral light reaction for the maintenance of their normal swimming position, whereas the converse is true for the water boatman, *Corixa,* and the creeping water bug, *Naucoris* (Rabe 1953). Some caterpillars are shaded in such a way that, provided they are oriented properly with respect to light, they are effectively camouflaged. Dorsal and ventral light responses come into play in these insects, ensuring "proper" orientation. The dorsal (ventral) light response is a good example of a means for maintaining primary orientation.

Jander (1975) discusses the ecological aspects of orientation behavior and provides a classification of these behaviors developed on ecological bases.

The most complex innate behavior patterns (i.e., the *fixed action patterns*) are more commonly known as *stereotypic, preprogrammed motor programs*. This is an ideal point in the text to alert students that what often appears to be scientific progress is really controversy over the precision of terms. The term fixed action pattern or FAP was developed by ethologists who, at the time, did not have at their disposal the new information the science of neurobiology has provided on cellular circuits, hormones, and behavior. Scientists often fail to move forward in their research programs because they become embroiled in controversies over terminology and often refuse to let go of the concepts or terms that new research has altered or given a new meaning.

One book that discusses a controversy in the general biological arena is that of James Watson (1968), who provides his account surrounding the discovery of the structure of DNA in his book entitled *The Double Helix.* Wenner and Wells (1990) present the history surrounding the controversy over the exact nature of the "language of the bee." Both books provide a look at the personal interactions between scientists, how they handle controversy, and how "truth" is reached. Graham Hoyle (1976) presents a historical account of the usage of the term FAP and nicely points out that any term used to describe behavior—whether it be "reaction-specific energy," "releaser," "drive," "motivation," or "appetitive"—has a physiological basis. The noted insect behaviorist, J. S. Kennedy, wrote an article in a book (Beament and Treherne 1967) dedicated to the father of insect physiology, Sir V. B. Wigglesworth, entitled "Behaviour as Physiology." FAPs or preprogrammed motor outputs are patterns that are usually characterized by being unlearned, species specific, and adaptive. They differ from reflexes and orientation mechanisms in that they frequently require an internal readiness before they

can occur. This internal readiness (i.e., of the nervous system) is set or determined by hormones, which are often termed *primers.* Another difference is that the eliciting stimulus (*releaser*) is commonly not required to act during the entire course of the pattern. The releasing stimulus may be internal in the form of hormone or neuropeptide secretion or an excitatory signal from a higher neural center. The stimulus may also be external in the form of a physical factor-light, temperature, mechanical contact, and so on. Or the behavior, appearance, odor, and so on, of another animal may be involved (more on this in the section on communication).

A variety of factors may affect the development of a state of internal readiness or responsiveness necessary for a given behavior pattern to occur. These may be either internal or external in origin (Markl 1974, Markl and Lindauer 1965). If one were to read the 2nd edition of this text, one would notice that the term *motivation* was equated with the term internal state of readiness. As we previously stated, precision in terminology and the avoidance of anthropomorphism is highly recommended. Thus, the term *motivation* has been removed from this text. The historical debate over the use of the term *motivation* for animals, other than human beings, has been long and, in many cases, vigorously debated. For those interested, Kennedy (1987) reviews this debate and presents strong evidence for the removal of this term from the literature when defining specific qualities of an insect's behavior. Internal factors influencing an insect's state of readiness or responsiveness are neural, hormonal, and neuromodulatory. For example, the relative concentrations of the juvenile hormone and the molting hormone (ecdysone) determine the spinning behavior of the caterpillars of certain moths (*Galleria* or wax moths and others). A hormone secreted by the ovaries is necessary for the female grasshoppers (Acrididae) in the subfamily Truxalinae to sing (Markl and Lindauer 1965, citing the work of Haskell 1960).

Internal receptors are commonly involved when a behavior pattern appears spontaneously without any discernible external stimulus. An excellent example of this has been found in the blow fly, *Phormia,* in which neural stretch receptors associated with the crop carry stimuli that inhibit food ingestion as long as the crop is full. No inhibitory stimuli arise from these receptors when the crop is empty and feeding can occur. In other instances of spontaneous behavior, regular cycles of behavioral activity of internal origin (endogenous rhythms) determine when a particular pattern may be released. For example, crickets are active only during a certain period each day and this daily rhythm continues for several days when the insects are kept in constant darkness in the laboratory. Periodicity in behavior is considered later in this chapter. Growing support for the role of biogenic amines, such as dopamine and octopamine, in modulating the "central excitatory state," arousal level, or level of responsiveness of insects, is being obtained. Evans (1985) discusses several of these cases for octopamine, especially the work in which octopamine addition increased the arousal level of honey bees to incoming olfactory stimuli. This neuropharmacological approach to insect behavior is an area that will have major impact on our understanding of this area of insect science.

External factors—light, temperature, substances produced by other insects of the same species, activities of other insects of the same species, and others—have definite effects on readiness in addition to serving as releasers. Light and temperature commonly determine whether an insect is active or inactive. Recall that when we discussed the dorsal ocelli as stimulatory organs, we gave examples in which insects failed to demonstrate certain light responses when the ocelli were blocked. Every insect has a temperature below which and one above which it becomes inactive. Within the range of these temperatures, specific temperatures may be necessary for particular activities. For example, the readiness of certain butterflies (e.g., *Pieris*) to copulate is determined by temperature. Adult reproductive *diapause* (i.e., a state of no egg development) is an example in which external factors (i.e., photophase length and temperature) have a profound impact on behavior. Adults in diapause usually do not mate, are photonegative and hide, and do not feed.

Many instances are known in which a substance secreted by one insect influences the behavior of another of the same species. The term *pheromone* is commonly used to denote such a substance. A good example of a pheromone affecting behavior is found in the honey bee. The queens produce a substance known as the queen substance in the mandibular glands, which is passed from worker to worker orally and which may inhibit ovarian development as well as the behavior involved with queen-cell construction. Queen cells, in which new queens develop, are constructed when a hive has lost its queen and hence the effects of the substance of her mandibular glands. Pheromones will be discussed in the section on communication.

When an insect is in a state of readiness or responsiveness, it may begin to behave in a manner that increases the probability of exposure to a releasing stimulus. This has appropriately been called *appetitive behavior*. Probably the most common appetitive behavior is increased locomotion. Appetitive behavior does not reduce the insect's responsiveness and is likely to continue until the insect comes into contact with the appropriate releasing stimulus. At this time the specific behavior pattern involved with the ensuing act is manifested. This occurrence has been called the *consummatory act* and it reduces or lowers the initial level of responsiveness. Male cicadas that have been sexually aroused move about randomly (appetitive behavior), thus increasing the probability of encountering a female. This random movement

will continue until contact is made with a female. At this time copulatory behavior (consummatory act) is released and sexual arousal reduced. There are probably a number of behavioral steps between the initial behavior and the consummatory act. It is generally felt that behavior is organized into a hierarchical sequence of patterns ultimately based on a hierarchical organization of nervous centers. From this point of view a given behavioral pattern (e.g., sexual behavior and food getting) then consists of a definite sequence of patterns, each with a specific releaser, the final pattern being the consummatory act that somehow reduces the internal state of readiness or responsiveness. A sequence of behavior patterns may be aborted at any time if a given releaser is missing along the way. Markl and Lindauer (1965) cite an excellent example from the work of Baerends, who studied the sequence of patterns involved in brood care in the digger wasp, *Ammophila adriaansei*. In a definite sequence, three groups of activities follow one another: digging of a nest, hunting for a caterpillar, and oviposition; bringing more caterpillars and temporary closing of the nest; fetching still more caterpillars that are required and the final closing of the nest. The process is complicated by the fact that several nests are simultaneously cared for. At a second nest, work can only proceed in the spaces between the three phases of supplying the first. Therefore, a characteristic pattern of activities appears in the supplying of the nests. The single parts of the process are also organized hierarchically. In the process of bringing in the first caterpillar, this sequence is always as follows: the depositing of the caterpillar in the front of the entrance of the nest; digging up of the entrance; turning around at the open nest; and pulling in of the caterpillar. The sight of the nest situation releases the dropping of the caterpillar and scraping, the half-open nest site releases digging, the open nest releases the turning around, and the sight of the caterpillar releases the pulling in.

Learned Behavior

Although there are a number of possible definitions of learning, we shall use that of Thorpe (1963): learning is "that process which manifests itself by adaptive changes in individual behavior as a result of experience." Thus learning involves the accumulation and storage (memory) of environmental information and the subsequent effects of this stored information on an animal's behavior. Extreme care has been necessary in calling a given pattern of behavior learned, because an innate pattern that appears at some point during the development of an animal may give the distinct impression of having been learned.

Space limitations unfortunately prevent us from presenting a complete picture of this most important aspect of insect behavior. This topic is too broad, especially with the recent emphasis on foraging strategies, which usually focus on nectar gathering and involve learning. The major breakthroughs in molecular biology have made *Drosophila* an even more powerful model for studies examining questions related to the molecular, cellular, and genetic basis of behavior.

There is no generally agreed upon classification of the kinds of learning. However, Markl (1974) outlines the following designations, which have been described among insects and which are useful for our purposes: *habituation, associative learning* (including *classical conditioning, instrumental conditioning,* and *shock-avoidance conditioning*), and *latent learning.* Carew and Sahley (1986) separate learning into two categories: non-associative and associative. Nonassociative learning includes habituation and sensitization. These authors also present an overview of invertebrate learning and discuss the principles and paradigms of learning in *Apis mellifera,* the locust and cockroach, and *Drosophila.* Students of behavior should realize that the kinds of stimuli an animal can learn depends both on the animal group being studied and on the context within which the stimuli are presented. One would not expect *Drosophila* to be instrumentally conditioned to push a lever when a bell rings. Because of their evolutionary history, phylogenetic constraints exist for each animal. Thus, not all stimuli are meaningful, especially when presented in a completely unnatural environment.

Habituation has occurred when a stimulus that initially evoked an escape or avoidance response no longer elicits such a response. This type of learning is common in insects and has been demonstrated relative to substrate vibrations, noise, visual stimulation, chemical repellents, etc. The ability of insects to learn in this fashion avoids unnecessary, energy-consuming behavioral responses.

Sensitization results when there is an increase in the animal's response to a variety of stimuli as a result of presenting another stimulus that can either be intense or noxious to the insect.

Classical conditioning was originally discovered by the Russian scientist Pavlov. In this type of learning, two stimuli are involved. One of the stimuli (the unconditioned stimulus, UCS) elicits a response (the unconditioned response, UCR). The other, the conditioned stimulus (CS), prior to conditioning, does not elicit the UCR. However, when these two stimuli are repeatedly presented to an animal in very close succession (CS briefly preceding the UCS), the animal gradually begins to respond to the CS in the absence of the UCS.

Classical conditioning has purportedly been shown to occur in honey bees. In these insects, the proboscis extension reflex can supposedly be conditioned to respond to the essence, coumarin, in the absence of sugar. In this case, proboscis extension is the UCR, the sugar is the UCS, and the odor of coumarin the CS. However, Alloway (1972) explains that the experiments carried out to demonstrate the conditioning of the proboscis extension reflex were insufficiently controlled. Today, there is

little doubt that insects exhibit both nonassociative and associative learning (McGuire 1984, Carew and Sahley 1986).

In instrumental (trial and error) learning, a response or series of responses is induced or inhibited by the presentation of a stimulus or pattern of stimuli [reinforcer(s)]. Maze learning is a good example of instrumental learning. A maze is placed between an insect and its nest, food, or some other strong positive reinforcer; or a negative reinforcer, such as an electric shock, is applied for an incorrect response. If an insect is capable of learning the maze, the number of errors (wrong turns) will decrease with the number of trials. For example, ants are capable of learning a rather complex maze placed between themselves and their nest (positive reinforcer). Another example of instrumental learning is the "training" of honey bees to associate a particular color with a sugar (reward) source (see chapter 6). Honey bees can also learn to associate odors with a food reward.

There are many instances in which an insect has been conditioned in an inhibitory way. For example, a mantis can learn not to strike an object it associates with an electrical shock.

Horridge (1962a, b) showed that a single leg of a decapitated cockroach suspended above a container of electrified saline solution can be trained to stay above a level where it would receive a shock (shock-avoidance learning). The experiments were designed so that a test animal received shocks according to the level of the test leg, while a control animal received the same number and intensity of shocks, but in a random fashion, without regard to leg position. Later investigators (Eisenstein and Cohen 1966) demonstrated that even a single isolated ganglion in the ventral nerve cord could mediate the shock-avoidance learning of a single leg. Eisenstein and Reep (1985) provide a review on this fascinating leg-learning preparation, including some of its pharmacological bases. It appears that the prothoracic ganglion is essential in the leg-learning preparation and that isolated preparations showed that an intrinsic set of motoneurons (pacemaker) had an intrinsic frequency of firing and that what is modified (i.e., modulated) during learning is the frequency, which is ultimately controlled by the K^+ conductance values.

Latent learning occurs without apparent reward or punishment. Social insects learn characteristics of the immediate vicinity of their nest such that they are able to find their way back from foraging flights. The predatory wasp, *Ammophila,* previously mentioned, can find its nest despite forced detours and displacements (Thorpe 1950).

Olfactory conditioning may be a form of habituation or latent learning. An example of this phenomenon is a situation in which exposure to an ordinarily repellent odor during a certain period of development results in the loss of repellency of this odor to adults. This is the case in

Drosophila melanogaster, the adults of which are normally repelled by the odor of peppermint oil. However, if the larvae have been reared in the presence of peppermint, they are strongly attracted to it as adults.

Although the ability to learn is widespread in the animal kingdom, the presence of either short-term or long-term memory has not been demonstrated for all groups. There is no doubt, however, that insects possess short-term memory. Long-term memory is relative and usually depends on the life of the organism being studied. Some researchers believe that insects lack long-term memory and that both their short lives and size of brain have placed constraints on this aspect of memory. *Drosophila* has recently been studied to dissect the neurogenetic components of both learning and short-term memory (Dudai 1988). Lynch et al. (1984) provide an overview of the neurobiology associated with both learning and memory.

Alloway (1972, 1973) and Markl (1974) review learning in insects. Wells (1973), Erber (1975), and Menzel and Erber (1978) deal specifically with learning in the honey bee. Tully (1984) reviews the behavior and biochemical bases for behavior in *Drosophila* learning, and Papaj and Prokopy (1989) review the ecological and evolutionary aspects of learning in phytophagous insects. This area of behavior (i.e., learning and memory) is one in which students will see an ever increasing amount of research.

Periodicity in Behavior

Insects, along with all other organisms, have evolved in an environment characterized by regularly occurring, cyclic changes. Many of the activities of insects (locomotion, feeding, mating, oviposition, eclosion, etc.) also occur at regular intervals. Further, the timing of these recurring activities is adaptive; that is, insects, under natural circumstances, behave in ways that maximize chances for survival. Thus herbivorous (plant-eating) insects hatch from eggs in synchrony with the availability of host plants; insects become dormant and cold-hardy or migrate at a time that favors survival of an upcoming cold or dry season, and so on. The term *periodicity* is applied to such recurrences of behavior. The patterns of recurrence of particular activities may be every 24 hours, monthly, annually, in coordination with the lunar cycle, and so on. Periodicity in behavior involves both endogenous and exogenous components. Endogenous components arise from within the insect and are ultimately related to its heredity. These include, in particular, rhythms, but the stage in a given physiological-behavioral cycle (e.g. gonotrophic cycle), developmental stage, and other endogenous factors may also play a role. Exogenous components arise from the external environment.

Rhythms are processes that are controlled by an "innate time-measuring sense" or "biological clock." They characteristically continue even when external conditions (temperature, light, etc.) are kept constant. Most rhythms that have been identified have a period of approximately 24 hours and are referred to as *circadian* ("about a day") *rhythms*. Several examples of circadian rhythms have been identified in insects. For example, cockroaches are nocturnally active; *Drosophila* emerge from the pupal stage at dawn; katydids and other orthopterans sing at a particular time of day or night.

A number of noncircadian rhythms, so called because they occur only once in the life of an insect (e.g., pupariation and adult eclosion), are known for insects. Another example, the lunar emergence of certain chironomid flies and mayflies and the annual emergence of a species of dermestid beetles (Corbet 1966). A well-known example of noncircadian rhythmicity is found in the species of periodical cicadas (*Magicicada* spp.), the various broods of which emerge en masse at 13-year or 17-year intervals. These rhythms are sometimes called *gated rhythms* and are important for population survival, as opposed to individual survival. Thus the members of a population are coordinated and reach and pass through a behavior "gate" at nearly the same time. If an individual misses the gate on a given day, it awaits until the next gate for activation of the behavioral pattern. This way, population events that are related to survival, such as group mating-swarming, are synchronized. The appearance of the gate is usually under the control of a circadian rhythm. The best-known example is the adult eclosion behavior of two silk moths, *Hyalophora cecropia* and *Antheraea pernyi*. This behavior associated with the adult leaving the pupal case is an example of gated-regulated behavior whose expression is under the control of a biological clock located in the brain. Using these two species, Truman (1973) and his colleagues showed that one moth (*H. cecropia*) ecloses in the morning, whereas the other (*A. pernyi*) ecloses just prior to darkness. To locate the clock, Truman did several manipulations of the nervous system (including brain removals) and concluded that the clock was located in the brain (figure 8.2). In unoperated specimens, the two adult species eclosed at their normal time, but by debraining them, the eclosion rhythm was random and scattered. By reimplanting an active brain into the debrained specimen of the same species, Truman showed that the rhythm could be reinstated and reaffirmed that the clock was located in the brain. The next experiment was the transfer of the brains into debrained specimens of the other species. Surprisingly, the specimens now eclosed at the time set by the clock located in the implanted brain. Most interesting, however, was that the preprogrammed motor output associated with each species remained virtually

Figure 8.2 Operations used to locate the site of the clock responsible for the gated rhythms in two species of silk moths. (a) Experiment performed to locate the site of the gating rhythm within each species. (b) Experimental data showing that the clock located in the brain is species specific. Black bar represents darkness. *Taken from Truman, 1973.*

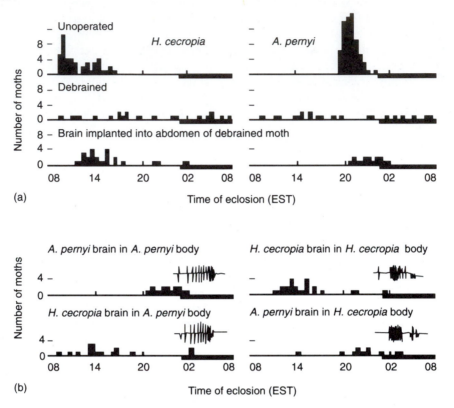

unchanged. Thus, the clock located in the brain and responsible for gating the eclosion behavior is species specific, whereas the normal program controlling the behavior appears to be species independent. More will be said about this preprogram when we discuss hormonal control of behavior.

Some investigators have maintained that rhythms are under the control of undetected exogenous factors rather than endogenous biological clocks. However, evidence favors an endogenous source, particularly the fact that the phase of a rhythm does not change even if the animal is transported to a place on the earth far from the locality of entrainment ("clock setting"). A gene controlling the period of locomotor activity or circadian rhythm in *D. melanogaster* has been identified and was termed the "*per* locus" because it affects the locomotor period. Several allelic forms of this *per* gene have been identified: *per+* is the normal, wild-type; *per^s* has a shortened period; *per^L* has a lengthened period, whereas *per^o* has an abolished rhythm. The development, function, or both of special cells in the brain are believed to represent the circadian pacemaker cells, and it is these cells that are affected by the mutation. The use of genetics and molecular biological techniques is moving us closer to understanding the mechanisms regulating circadian rhythms. This "*per* locus" not only regulates the circadian rhythm in *Drosophila* but has parallel effects on the interpulse interval modulation rhythm, which is so important in the species-specific courtship song. This effect will be discussed in greater detail in the section that deals with the genetic control of behavior. Interested students are referred to Hall and Rosbash (1988) for more information concerning the genetic basis for rhythms. A central theme concerning the location of the clock controlling circadian rhythms in all animals is its location in the brain. As previously discussed, the work with silk moths by Truman showed that the clock not only resides in the brain but its expression is also regulated by the genetics of the donor brain—a theme also found for the *per* gene for *Drosophila*. This same theme is found in mammals in which similar brain transplant experiments confirmed the finding of Truman (cf., Rusak and Bina 1990). Again, the insect model is shown not to differ drastically from the vertebrate model. Research on the role of neurotransmitters in mammalian circadian rhythms (Rusak and Bina 1990) appears to be somewhat more advanced than that of the insect models.

Although rhythms are not controlled by exogenous factors, some exogenous factors are known to influence rhythms. Particularly significant are the external time cues (*Zeitgebers*), which set the phase of the rhythmic patterns of the individuals in a population. In a population under artificially constant environmental conditions (e.g., constant illumination) the rhythmic activities of the individuals may be out of phase with one another, but with the introduction of an appropriate time cue, they are set in phase and the activities of the individuals become synchronous. For example, mosquitoes such as *Aedes aegypti* held under constant illumination oviposit arrhythmically, but when exposed to a dark period, even a very brief one, oviposit synchronously at 24-hour intervals thereafter. Under laboratory conditions, the phase of a rhythmic activity can be shifted about at the investigator's will. Among the exogenous factors that may act as time cues are length of the light or dark phase of a photoperiod ("a cycle consisting of a period of illumination followed by a period of relative darkness"—Beck 1968), light intensity, temperature, time at which food is available, and others.

Exogenous factors may influence rhythms in ways other than as time cues. For example, in the dragonfly *Anax imperator,* as described by Corbet (1966):

> Emergence (probably rhythmic) usually occurs abruptly soon after sunset; if, however, the air temperature falls below 10° C., after larvae have left the water but before ecdysis has begun, some of them will return to the water and emerge the next morning after temperatures have risen again. For these individuals the diel periodicity of emergence has been changed by a short-term response to unfavorable exogenous factors, but the phase-setting of their rhythms has remained unaltered (as evidenced by their attempting to emerge at the normal time).

Many external environmental factors display a 24-hour periodicity, which may result in corresponding periodicity in insects without the involvement of biological clocks. Among these are light intensity, moisture, temperature, wind velocity, and various possible biotic factors. An example of how one of these factors may cause periodic behavior is the swarming of many species of mosquitoes, which is determined by light intensity. These insects continually possess the "drive" to swarm, but do so only under appropriate conditions of light intensity and can be induced to swarm for hours if the light intensity is held artificially at the appropriate level. Under natural conditions, the appropriate light intensity may be periodic (e.g., that which occurs during crepuscular periods). In this situation the mosquitoes would swarm "periodically."

For more information on biological clocks and insect periodicities, see Brady (1974), Corbet (1966), Danilevskii (1965), Hall (1990), Page (1985), and Saunders (1974, 1976a, b). Bünning (1967) discusses biological clocks in general.

The Control of Behavior

An insect is continually bombarded with multitudinous stimuli in space and time. Many stimuli provide information pertinent to the insect's life; many do not. In order to survive, an insect must respond to the "right" stimuli and do the "right" thing at the "right" time: escape, locate a food source, feed, locate a mate, copulate, oviposit, migrate, become dormant, and so on. That is, an insect's behavior must be controlled.

How the environment is perceived and what behavior is exhibited is a function of the nervous and endocrine systems. The functional characteristics of the nervous and endocrine systems are in turn determined by hereditary mechanisms. In this section, the involvement of the nervous and endocrine systems in behavior and the role of heredity are discussed.

Nervous Control

Insects are especially good subjects for studying nervous function and behavior because their behavior tends to be stereotyped, but with some learning; and their nervous systems contain a relatively small number of neurons. For detailed information on the nervous control of insect behavior see Barton Browne (1974), Howse (1975), Hoyle (1970), Huber (1974), Miller (1974), and Roeder (1967, 1970, 1974).

Aspects of nervous control of several motor patterns (especially flight, walking, and sound production) have been studied, but very little is known about the neural control of complex sequences of behavior. Techniques that have been used include observation of behavior before and after decapitation, severance of nerves, ablation or destruction of all or parts of ganglia; electrophysiological recording from various nerves and ganglia *in vivo* and in isolated preparations; and use of various stains to elucidate structural connections. A rather ingenious approach has been addressed to the establishment of the location of the sites within the nervous system directly associated with various abnormal (mutant) behavior patterns. This approach involves the use of genetic mosaics and the establishment of "embryonic fate maps" (Hotta and Benzer 1972; Benzer 1973).

The operation of the nervous system is dependent upon the properties of individual neurons and the ways these neurons are associated with one another both spatially and temporally. Behavior is the result of interactions between sensory input, reflexes, and centrally generated neural patterns. More on reflexes and central patterns will follow. See chapter 3 for a discussion of the properties of neurons and the organization of the nervous system and chapter 6 for information on sensory mechanisms.

Behavioral studies provide strong evidence for the filtering of sensory input (*stimulus filtering*); that is, from the many, diverse stimuli impinging on an insect, only certain ones or only certain aspects of a given stimulus are involved in eliciting a particular behavior.

Reflexes depend on sensory input for their occurrence. At one time, insect behavior was thought to be explainable solely on the basis of reflexes in which one reflexive response provided the stimulus for the next reflex in line. For example, the repetitive motor patterns associated with flight and sound production were viewed as being controlled by cyclic reflexes, which would continue until inhibited by some other reflex. Although reflexes are undoubtedly important in behavior, more recent research has revealed evidence for the generation of patterns within the central nervous system independent of sensory input.

It is known from behavioral and neurophysiological studies that patterns of behavior may occur spontaneously without any external sensory triggering. Thus the nervous system must be capable of spontaneously generating organized patterns of behavior from within (*endogenously*). A decapitated male mantid will continuously turn and perform copulatory movements. The central nervous system in a locust with all sensory input from the head and wings removed still produces motor output nearly identical to the motor output recorded during normal flight (Wilson 1972). These activities are thought to arise from the spontaneous, preprogrammed activity of *neural pattern generators* in the central nervous system.

If copulatory behavior arises endogenously in the mantis, why doesn't the mantis behave this way all the time? How does the locust control flight if environmental conditions (wind, obstacles, etc.) vary? Why don't insects display all of their behavior patterns simultaneously? Obviously endogenous activity must be regulated. Endogenous activity is thought to be regulated through excitatory and inhibitory stimuli from "higher centers" and by modulating input from peripheral sensilla, and probably directly or indirectly by hormones. In particular, the brain and subesophageal ganglion have both been found to exert excitatory and inhibitory influences on endogenously controlled motor patterns. These patterns seem to reside largely in the particular segments involved (e.g., the thoracic ganglia in flight). Thus the mantid brain (which was removed with the head in the decapitated individual mentioned above) sends out impulses that inhibit the endogenous activity of the ventral chain ganglia relative to copulation. In the locust, there are wind-sensitive sensilla on the head, which help sustain flight, and also tarsal receptors, which stimulate flight when contact with the substrate is lost. In addition, there are stretch receptors, which aid in the regulation of flight, on the wing hinges and other parts of the wings and pterothorax. Thus, although locust flight pattern is centrally generated, the basic pattern is modifiable by peripheral, sensory input.

Something like neural pattern generators is probably involved in sequences of stereotyped behaviors more complex than simple motor patterns. For example, courtship in the grasshopper *Gomphocerus rufus* occurs in a very definite sequence of simple behavior patterns and this sequence is not interrupted by surgical action designed to block potentially controlling sensory feedback. Similarly, in *Drosophila,* an excited male will display all the intricacies of courtship in the presence of an anesthetized, unresponsive female. Thus complex behaviors appear to be programmed into the central nervous system as "tapes" of behavior sequences and a given insect must have a complete bank of "tapes" to be played as appropriate. According to Hoyle (1970):

> For purposes of functional description the nervous system can be considered to have three major functional divisions. In one of these the enormous mass of incoming sensory input is reduced to a small series of sets of information which can serve to drive motor output. The second 'selects' from the sets the one which shall at a given time be operative, and inhibits others. The third division contains neural generators whose purpose is to produce sequences of motor nerve impulses.

If the brain is involved in excitation and inhibition of the rest of the central nervous system, how is the brain, in turn, excited? Although little is known, there is evidence for excitatory input to the brain from the ocelli (mentioned earlier in this chapter and in chapter 6), compound eyes, and antennae. Spontaneous, endogenous activity and endocrine factors probably also play a role in the levels of excitation of the brain.

Tully and Hirsch (1982) reported a single gene effect regulating the central excitatory state of a fly, whereas research on biogenic amines points to octopamine or enhanced responsiveness in certain insect studies. Regardless of the mechanisms of control, research into the neural, genetic, and chemical basis for the regulation of responsiveness in insects will be a fruitful area of investigation.

How learning provides a way to modify or "rewrite" behavioral "tapes" through experience is not exactly understood, but some basic principles seem to be emerging (Carew and Sahley 1986):

1. Learning can generally be traced to individual neurons.
2. Learning involves changes in either the intrinsic cellular properties or existing synaptic connections of the neurons associated with learning.

3. Learning does not involve the appearance of new or novel synapses.
4. Modulation of K^+ conductance is involved in memory storage.
5. Second-messengers, such as cAMP and Ca^{2+}-calmodulin, are also involved in learning.

How the endocrines, nervous system, and genetics of an organism help "rewrite" behavioral tapes is a future priority of insect neurobiology.

Endocrine Control

Hormones play major roles in insect behavior. Through acting directly on the nervous system they can cause major overall changes in patterns of behavior. Whereas strictly neural mechanisms usually mediate the short term, sometimes split-second, behavioral responses to rapidly changing external and internal environmental demands (potential attack by a predator, an obstacle encountered during flight, etc.), endocrine mechanisms typically exert their influence over the longer term, coordinating behavior with less immediate environmental demands (changing seasons, changes in the availability of food, etc.). Thus strictly neural mechanisms are involved with escape responses, maneuvering during flight, and the like; endocrine mechanisms are involved with development of migratory behavior or dormancy, changes in feeding behavior, development of adult behavior patterns, development of sexual receptivity, and so on. In the terminology of the section on the nervous control of behavior, hormones are commonly involved in determining which reflexes can be elicited and which behavioral "tapes" are played in a given circumstance.

Interfacing of the nervous and endocrine systems in which the latter acts in a "switching" fashion on the former may explain, at least in part, how the very large array of behavior patterns displayed by insects can be packed into their comparatively small, simple nervous systems.

The determination of hormonal influences on behavior has usually been based on one or more of the following criteria:

1. Correlation of hormone secretion and a particular behavior pattern.
2. Application of hormone or implantation of active glands inducing a behavior pattern sooner than it would ordinarily occur.
3. Correlation between removal of an active gland and disappearance of a specific behavior.
4. A combination of the criterion in number 3 and subsequent restoration of the specific behavior by hormone application or implantation of active glands.

Obviously, the fourth criterion provides the strongest evidence for hormonal control of a given behavior.

Riddiford and Truman (1974) recognize two categories of hormonal effects on behavior: hormones can act as *modifiers* or as *releasers* of behavior.

When a hormone acts as a modifier, it alters the responsiveness of an organism. Where a given stimulus previously elicited one kind of behavior, a different behavior occurs. The hormonal effects on behavior as described earlier are as modifiers. An excellent example of a hormone acting as a modifier is the influence of ecdysone on the biting behavior of female *Anopheles freeborni* (Culicidae, mosquitoes, Diptera) (Beach 1979). This mosquito does not take blood meals once ovarian development has been initiated. Ecdysone, which is secreted by the ovaries, inhibits biting. This was concluded on the basis that biting inhibition does not occur after removal of the ovaries and is restored by either replacing ovaries or injecting ecdysone. Bowen (1991) reviews the literature concerning the sensory physiology of host seeking in mosquitoes and points out that a blood-borne factor somehow affects the peripheral sensitivity of the antennal sensilla to host derived odors. She also notes, as did Stoffolano (1974), that several behaviors (i.e., feeding, mating) of insects during diapause are modified by the existing endocrine picture, which is usually different in diapausing versus nondiapausing adults.

When a hormone acts as a releaser of behavior, a specific behavior pattern occurs within a few minutes of secretion. This situation is analogous to the action of nervous input eliciting a behavioral response, but is much slower in occurring (minutes vs. milliseconds). An example is the phallic nerve stimulating hormone in the cockroach *Periplaneta americana* (Milburn, Weiant, and Roeder 1960; Milburn and Roeder 1962). This hormone can be extracted from the corpora cardiaca and when injected into male cockroaches it causes the abdominal movements characteristic of copulation.

The "eclosion hormone" (eclosion meaning pupal–adult ecdysis in this context) secreted by the brain (neurosecretion) and corpora cardiaca in the moth *Antheraea pernyi* acts both as a releaser in triggering pre-eclosion behavior and as a modifier by "turning on" adult behavior (Truman 1973). In addition to showing that the clock for gating the eclosion behavior resided in the brain, Truman also showed that the preprogrammed motor output resided outside the brain. The eclosion hormone has three major effects: first, it signals the onset of the eclosion behavior; second, it sets the stage for increased hemolymph flow to help expand the wings once the moth has escaped the pupal case and cocoon; and third, it silences the motor neurons going to the abdominal intersegmental muscles. The motor "tape" or program (which lasts about 1.5 hours) for the eclosion behavior is located in abdominal segments 3–6. Once activated, each individual ganglion sends a message to the intersegmental muscles involved in both the rotary and peristaltic movements used by the adult to exit the pupal case. For each abdominal segment to act in concert with the others, co-ordination exists between the segments so that they act together to produce the desired behavior. Once the adult moth has exited and expanded its wings, the silenced motor neurons somehow activate a preprogrammed cell death of these muscles, which are no longer needed by the adult moth.

Truman and his colleagues have since moved to a better insect model, the tobacco hornworm (*Manduca sexta*), and have purified, characterized, and isolated the gene (it is 7.8 kilobases and consists of three exons) for the eclosion hormone. The eclosion hormone itself is a 62-amino acid neuropeptide. As in other vertebrate and invertebrate nervous systems, this neuropeptide mediates its effect through the action of second messengers or cyclic nucleotides such as cAMP and cGMP. This research of Truman, using an insect model, has added greatly to our understanding of how the nervous systems of both insects and vertebrates are "turned on." The research of Truman also represents an excellent approach: one involving integrative studies of the nervous system, circulatory system, muscular system, and endocrine system. By reading all of Truman's and his colleagues' papers one can also appreciate how their research has progressed from classical gland or organ removals and implants to the use of molecular techniques to identify the eclosion gene.

Many examples and discussions of endocrine mechanisms and behavior can be found in Barth and Lester (1973), Truman and Riddiford (1974), and Riddiford and Truman (1974). Additional examples may be found in other sections of this chapter.

Genetic Control of Behavior

There are many examples of the demonstration of genetic control of insect behavior. Not surprisingly, a great deal of work has been done with *Drosophila melanogaster* and other *Drosophila* species as well as a number of other kinds of insects. An extensive treatment of the genetics of behavior is far beyond the scope of this text, and the reader is referred to Dudai (1988), Ehrman and Parsons (1976), Ewing and Manning (1967), Fuller and Thompson (1960), Hirsch (1967), Huettel (1986), McClearn and DeFries (1973), Quinn and Greenspan (1984), Rothenbuhler (1967), and Tully (1984) for more information on this subject.

It is hoped that one day we will understand the chain of events occurring between the action of genes and the expression of behavior. Although we do not know all the answers yet, considerable progress has been made, especially with the advent of molecular techniques combined with genetics and biochemistry. Investigators are approaching genes and behavior in a number of different ways. A few examples follow.

In rare instances, it has been possible through breeding experiments to account for particular behavior patterns on the basis of the action of single genes. An outstanding example of this is the work of Rothenbuhler (1964). Rothenbuhler studied two inbred lines of honey bees, one that had been selected for resistance to American foulbrood (a disease of honey bee larvae) and another that was susceptible. He noted a distinct behavioral difference between the two lines; the resistant line removed foulbrood-killed larvae relatively soon after death (hygienic behavior), and the susceptible line did not (nonhygienic behavior). Genetic analysis revealed that the action of two sets of dominant/recessive genes on different chromosomes would account for the results of breeding experiments between the two lines. One set of alleles involved uncapping of the cells in which diseased larvae were developing: recessive allele, uncapping and dominant allele, no uncapping. Thus homozygous recessives displayed uncapping behavior. Homozygous dominants and heterozygotes did not. The other set of alleles involved removal of diseased larvae: recessive allele, larvae removed and dominant allele, diseased larvae not removed. Homozygous dominants and heterozygotes did not remove diseased larvae. Figure 8.3 shows the two genes assorting independently and the resulting behavioral phenotypes. The "remove only" group was identified by artificially removing the caps from cells containing diseased larvae and subsequently observing the removal of diseased larvae.

Most behavior patterns are probably under polygenic control, and *directional selection experiments* have been useful in establishing whether a given behavior has a genetic basis (Ehrman and Parsons 1976). In selection experiments, "individuals are selected at high or low extremes of a distribution in the hope of forming separate high or low lines in subsequent generations" (Ehrman and Parsons 1976). If a trait does not have a genetic basis, then attempts to select for extremes would fail. A good example of a selection experiment is the work of Hirsch and Erlenmeyer-Kimling (1961). They showed that with selection, *Drosophila melanogaster* that had previously been assumed to be negatively geotactic could be separated into both positively and negatively geotactic lines. Following establishment of "extreme" strains, it has been possible in some cases to evaluate the degree of involvement of different chromosomes. For example, in the case of geotaxis in *Drosophila,* genes on each of the three major chromosomes were found to exert an influence (Hirsch and Erlenmeyer-Kimling 1962). In *Drosophila,* several other behavioral traits have been shown to be under polygenic control, including locomotor activity, chemotaxis, duration of copulation, optomotor response, phototaxis, and preening.

Figure 8.3 Genetic hypothesis to explain hygienic and nonhygienic behavior in the honey bee. u = gene for uncapping (recessive) and r = gene for removing diseased larvae (recessive). + = dominant alleles; that is, no uncapping or no removal of diseased larvae.
From Rothenbuhler, 1964.

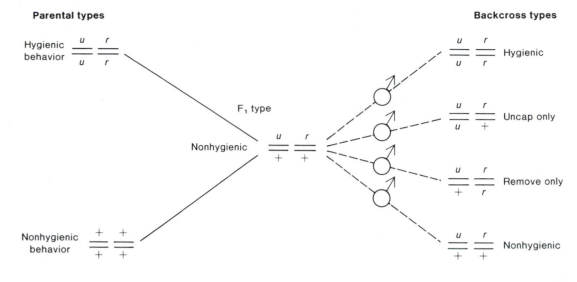

There are several examples of gene mutations in *Drosophila* that change a behavior pattern (Benzer 1973). An early example was a sex-linked mutant with yellow body color instead of the normal gray (Bastock 1956). Males of these yellow mutants were less successful in mating than the gray wild type. The mutation rate for most organisms is normally low and varies from 1×10^{-8} for a bacteriophage to $4 - 12 \times 10^{-5}$ for achondroplasia (i.e., a kind of dwarfism in humans). Regardless, these rates are low and researchers using mutants as genetic models must increase these rates in the laboratory by using various mutagens (e.g., ethylmethane sulfonate or EMS). Only by increasing the natural rate of mutation have geneticists been able to increase their supply of mutants.

Another example of a behavioral mutant is the *per* mutant in *Drosophila*. As you may remember, the *per* mutant not only affects the circadian rhythm of the fly but has parallel effects on the interpulse modulation rhythm of the courtship song (i.e., performed by the male vibrating its wings). The values for the *Drosophila* songs for two species and the *per* mutants are listed in table 8.1. Using molecular techniques, Wheeler et al. (1991) tested the hypothesis that *per* contains the instructions for the species-specific song. Since *D. melanogaster per*[o] is arrhythmic, this mutant could be used as the test recipient animal to see if they could take a cloned copy of the donor or, in this case, *D. simulans per*[+] (i.e., with song 30–40s) and introduce it into the *D. melanogaster per*[o] (i.e., with song arrhythmic), thus restoring the song rhythm. They cloned the DNA segment from the *per*[+] locus of *D. simulans* and, using a P-element construct, were able to

Table 8.1 Average Period Length for the Interpulse Song Interval for *Drosophila melanogaster per* mutants and *D. simulans*[a]

Species	Secs (Range)
D. melanogaster	
per[+] or normal, wild-type	50–65
per[s] or shortened	35–45
per[L] or lengthened	75–95
per[o] or abolished	Arrhythmic
D. simulans	
per[+] or normal, wild-type	30–40

[a]Data from Wheeler et al. (1991).

introduce the cloned gene into the *D. melanogaster per*[o]. Using this molecular approach, they were able to rescue or restore the *per*[o], thus producing a genetically engineered fly that initially was arrhythmic but now had a pulse rhythm identical to the recipient-donor gene, (i.e., in this case of another species). It is suggested by these authors that for *D. melanogaster* and *D. simulans* that maybe only 4 or fewer amino acid substitutions in a specific region of the *per* locus could account for the species-specific courtship song. The importance of this *per* gene in speciation of *Drosophila* species is currently being studied. This example of the *per* gene and its control of a specific behavior shows the direction future studies aimed at understanding the behavioral genetics of an organism are taking.

Some researchers are approaching behavioral genetics from a neurophysiological point of view. For example, Bentley and Hoy (1972, 1975), Hoy and Paul (1972), and Hoy (1974) have found that the sounds produced by adult male crickets are not learned, but genetically controlled. Further, hybrids produced by crossing *Teleogryllus* and *Gryllus* species exhibit song patterns that are intermediate between those of the parents. *Teleogryllus* songs are controlled by several different genes on different chromosomes. The books by Huber et al. (1989) on the cricket, Menzel and Mercer (1987) on the honey bee and Huber et al. (1990) on the cockroach are excellent sources for information on the behavior and neurobiology of these model systems.

Communication

Communication in one form or another is part of every organism's life at all levels of complexity. Thus, for example, the nucleus communicates with the cytoplasm in directing cellular activities; cells communicate with one another as cooperators in a multicellular organism both during development and in the fully developed organism; tissues, organs, and organ systems communicate with one another via nerves and hormones; and finally, organisms communicate with one another. At this point we are specifically interested in communication between organisms. At this level, we may define *communication* as the influence of signals from one organism on the behavior and/or physiology of another organism, with the outcome being beneficial to the sender, receiver, or both. In its broadest sense, then, communication involves members of the same (*intraspecific*) or different (*interspecific*) species. Thus a worker honey bee dancing on the hive to give information about the location and direction of a nectar source is communicating, and so is a bug when it releases a repugnant chemical in response to a threatening predator. One could say that virtually any aspect of an organism, even something as basic as color or form of the body, can be involved in influencing the behavior and/or physiology of another organism and is hence a means of communication. Communication often involves the use of specialized structures or methods by the sender and/or the receiver. Communication systems have evolved independently many times in insects.

Although systems of communication can be grouped in various ways, such as the type of information conveyed (Otte 1974), we will discuss this topic from the standpoint of modes of communication used by insects, specifically *tactile, acoustical, visual,* and *chemical.* Sensory mechanisms (discussed in chapter 6) are channels of reception. Mechanisms of sending (signal output) include sound and light production (see chapter 6) and chemicals produced by exocrine glands (see chapter 3). Discussions directly pertinent to insect movement, also a form of signal output, may be found in chapters 2, 3, and 7. Books and reviews that consider communication in general include Frings and Frings (1977), Lewis (1984), Montagner (1977), Otte (1974), Sebeok (1968, 1977), Smith (1977), and Wilson (1975). Reviews that deal with communication in particular groups include Hölldobler (1977), Ishay (1977), Lindauer (1974), Stuart (1977), and Wilson (1965) on social insects; and Attygalle and Morgan (1985), Ewing (1977, 1990), McNeil (1991), Otte (1977), Schneider (1975), and Silberglied (1977) on various other insect groups.

As previously mentioned, communication may be between members of the same or different species. Intraspecific communication is involved in species recognition, sexual recognition, alarm, and social coordination. Interspecific communication is also involved in species recognition (e.g., signals from a plant attracting a pollinating insect) as well as aggression, defense, etc. Many examples are included in the following section on the biological functions of behavior and also in chapter 9.

Tactile communication has been the least studied. This form of communication is limited to situations in which direct contact is possible, as during courtship and copulation. Wilson (1975) describes a form of communication to explain nest building in termites. Instead of direct touch, information is passed from one individual to another by a product of work. Thus a termite responds to a portion of already constructed nest by adding more material. This response is released by the presence of a physical structure. Wilson includes this form of communication as part of a broader category, *sematectonic communication,* which he defines as "the evocation of any form of behavior or physiological change by the evidence of work performed by other animals, including the special case of the guidance of additional work."

The ability to perceive sounds is found in several groups of insects (see chapter 6). However, members of at least five orders are known both to "hear" and to produce distinctive sounds: Orthoptera, Hemiptera, Homoptera, Lepidoptera, Coleoptera, and Diptera. Interspecific acoustical communication tends to be mediated by unpatterned sounds, such as "warning" sounds, whereas intraspecific communication typically involves highly patterned sounds (Matthews and Matthews 1978). As intraspecific signals, sounds are involved in aggregation, sexual behavior, aggression, alarm, and social interactions. Acoustical signals are very effective in that they can operate over relatively long distances, can be modulated in many different ways, can operate in the dark, can provide directional information, and so on. A new discovery of infrasound communication has been made in a species of cricket, *Phaeophilacris spectrum,* whose males lack sound producing devices on the wings but use the wings to produce a series of wing flicks leading to the movement of air at 8–12-Hz (Huber et al. 1989).

Research on elephants has also shown that they communicate using infrasound. In addition, some orthopterans use plants as "transmission channels" to produce vibrational songs (Michelsen et al. 1982). Acoustical signals in insects are monotonal, but vary in intensity and pattern of pulses. Useful references on acoustical communication include Alexander (1967, 1975), Claridge (1985), Haskell (1974), Huber and Thorson (1985), Ishay (1977), Kalmring and Elsner (1985), Otte (1977), and Schwartzkopff (1974).

Visual communication is common in insects. Visual signals have a great advantage in that they can provide a large amount of information. Recall the various aspects considered under photoreception in chapter 6: perception of form and pattern, intensity and contrast, movement, distance, color, and polarized light. Consider the possibilities for variation in all of these parameters. Some insects are bioluminescent and use light production and reception as a mode of communication. However, most insects use light indirectly, that is, reflected or transmitted by something else. Visual signals are involved in such things as reproductive behavior, aggressive displays (e.g., in territorial behavior of dragonflies), and in defensive startle displays. For further information on visual communication, see Hailman (1977) and Lloyd (1971, 1977, 1983).

Chemical communication is by far the most common form of communication among insects (in fact among animals in general). Chemical signals act over large distances, but they are not easily modulated and thus are used mainly as on-off signals, for example, to indicate sexual readiness in a female. They may also give directional information as in the case of a chemical trail. A given chemical signal may be a single compound or a mixture of compounds and may have been synthesized entirely by the insect or may have been obtained from the environment, especially from food.

Chemical signals can be classified on the basis of whether they mediate intraspecific or interspecific communication. Chemicals that function intraspecifically are called pheromones. For example, when ants are threatened or injured, they may release chemicals that induce dispersal from an area among foragers or aggressive behavior in the vicinity of the nest. These chemicals are acting as *alarm pheromones*. Chemicals involved in interspecific communication are commonly grouped as *allomones* and *kairomones*. An allomone is emitted by a member of one species and induces a response that is adaptively favorable to the emitter. For example, the repugnatorial substance released by a bug in response to a predator would be acting as an allomone. A kairomone is adaptively favorable to the receiving organism. For example, the mosquito *Aedes aegypti* is attracted by lactic acid (as well as other compounds) in human perspiration. In this situation, the mosquito (receiver) benefits, and

hence the lactic acid is acting as a kairomone (Galun 1977b). In many ants, bees, wasps, and termites the same chemical can function both as an allomone and a pheromone. For example, in ants formic acid can act as a defensive allomone and alarm pheromone that alerts nestmates to an intruder (Blum 1978). Likewise a volatile or mixture of volatiles may act as both a pheromone or a kairomone. In Scolytidae (Coleoptera), for example, the aggregating pheromone may also act to attract predators and parasites of the beetles. As stated in chapter 3, cuticular hydrocarbons produced by exocrine glands of insects have been shown to be extremely important in intraspecific communication, as well as taxonomy (see chapter 11). For a review see Howard and Blomquist (1982).The following references deal with chemical communication: *General*—Beroza (1970), Ebling and Highnam (1970), Pasteels (1977), Roelofs (1975), Shorey and McKelvey (1977), and Wilson (1965, 1970, 1975); *Pheromones*—Attygelle and Morgan (1985), Birch (1974), Blum and Brand (1972), Jacobson (1972, 1974), Karlson and Butenandt (1959), Law and Regnier (1971), McNeil (1991), Roelofs and Cardé (1977), Seabrook (1978), Shorey (1973, 1976, 1977a, b), Weaver (1978a), and Wilson (1963); *Allomones and Kairomones*—Blum (1978), Brown, Eisner, and Whittaker (1970), Rosenthal and Janzen (1979), Weaver (1978b), and Whittaker and Feeney (1971). Volume 9 of Kerkut and Gilbert (1985) is devoted to behavior and has chapters on neurobiology of pheromone perception, sex pheromones, alarm pheromones, allomones and kairomones, and aggregation pheromones, and Bell and Cardé (1984) contain several chapters on various aspects of chemical communication in insects.

The Biological Functions of Behavior

This section considers how insects are behaviorally adapted for the maintenance of the individual through sexual maturity and for those activities that culminate in the production of offspring. Gregarious, subsocial, and social behavior is discussed at the end of this section.

Feeding Behavior

Here we consider the sources of food, how insects are able to locate these sources, and the control of feeding.

What an insect eats is intimately associated with where it is found. It is usually the feeding of an insect that makes it a pest. Caterpillars eat large quantities of vegetation, blood-feeding insects transmit parasites and other microorganisms while attempting to get a blood meal, termite feeding causes untolled structural damage, and stored-grain insects spoil and deplete food stores. Finally, many insects depend on a meal prior to being able to mate and produce eggs.

Insects may be broadly grouped on the basis of the biological source of nutriment as follows: *detritivores* (*saprophagous*); *herbivores* (*phytophagous*, "plant-eating"); *mycetophagous*, ("fungus-eating"); and *carnivores* (*zoophagous*, "animal-eating"). Insects that fall into more than one of these categories are called *omnivores*. A qualification from Dethier (1976) before proceeding:

> No two animals have precisely the same diet. Each species possesses some of the qualities of a gourmet. The common categorization as herbivore, carnivore, and omnivore depicts only the outline of the story. The details of diet reveal the enormous complexity of the feeding processes and diverse and precise specifications to which the physiological and biochemical machinery must be designed.

Slansky and Rodriguez (1987) provide extensive coverage of the nutritional ecology of the above categories based on source of nutriment (e.g., herbivores, etc.).

Detritivores

Detritivorous insects feed on decaying organic materials: leaf litter, carrion, dung, and so on. Such materials were probably the basic food source for the most primitive insects, which lived on the forest floor. Many contemporary insects use these materials throughout all or a part of their lives. These insects are important in the progressive degradation of decaying organic material. Cockroaches, for example, subsist on a wide variety of dead plant and animal matter. They are commonly described as scavengers. Other examples of detritivores are dung beetles, carrion beetles, and many of the soil-inhabiting apterygotes (e.g., Collembola). Actually many insects that upon cursory examination appear to be detritivores are found upon closer scrutiny to be dependent upon the microorganisms associated with decaying organic material. The fruit fly, *Drosophila melanogaster,* is an example of such an insect. The larvae cannot be maintained on a sterile culture medium. Thus, although one associates them with the decaying organic matter, they are actually "microphagous" (Brues 1946).

Herbivores

Living plants are the source of nutriment for the greatest number of insects, and the habits of many insects have evolved closely with plants (coevolution is discussed in chapter 9). Members of the orders Orthoptera, Hemiptera, Homoptera, Lepidoptera, Thysanoptera, Isoptera, and some other smaller orders are predominately or partly herbivorous. In addition, large numbers of insects in the orders Coleoptera, Hymenoptera, and Diptera are also herbivores. Although flowering plants are by far the most common plant hosts, there are many insects that feed on fungi, algae, ferns, and so on.

Some herbivorous insects may show very little feeding preference, whereas others show some degree of preference, and still others are very specific as to food choice. Insects such as the migratory locust (*Schistocerca gregaria*) are *polyphagous* and will eat almost any green plant in their path. On the other hand, the milkweed bug and many others are *monophagous*, feeding only on plants of a single species or genus. Most phytophagous insects lie somewhere between these two extremes (*oligophagous*). Variation also exists with regard to what part of a plant is used as food. No major part of plant anatomy has been ignored by insects. There are those that feed in or on roots, stems, woody parts, buds, flowers, and fruits. However, a given insect is usually rather specific as to which part of a plant it eats. Thus there are leaf rollers, leaf miners, rootworms, root borers, stem borers, and so on. In addition to providing nutriment for the phytophagous insects, plants usually serve as a place to live, particularly in the case of larval and nymphal forms.

Many fungus-eating insects maintain fungus gardens within their nests. These are mutualistic relationships in that both the fungi and the insects derive benefit (see chapter 9). Outstanding among these are the fungus-growing ants in the tribe Attini (Formicidae; Weber 1972). Weber (1966) provides a brief description of their activities in the summary of a paper dealing with these insects.

> Fungus-growing ants (Attini) are in reality unique fungus-culturing insects. There are several hundred species in some dozen genera, of which *Acromyrmex* and *Atta* are the conspicuous leaf cutters. The center of their activities is the fungus garden, which is also the site of the queen and brood. The garden, in most species, is made from fresh green leaves or other vegetable material. The ants forage for this, forming distinct trails to the vegetation that is being harvested. The cut leaves or other substrates are brought into the nest and prepared for the fungus. Fresh leaves and flowers are cut into pieces a millimeter or two in diameter; the ants form them into a pulpy mass by pinching them with the mandibles and adding saliva. Anal droplets are deposited on the pieces, which are then formed into place in the garden. Planting of the fungus is accomplished by an ant's picking up tufts of the adjacent mycelium and dotting the surface of the new substrate with it. The combination of salivary and anal secretions, together with the constant care given by the ants, facilitates the growth of the ant fungus only, despite constant possibility for contamination. When the ants are removed, alien fungi and other organisms flourish.

Certain species of scale insects (Homoptera, Coccidae), a wood wasp (*Sirex*), ambrosia beetles (families

Scolytidae, Paltypodidae, and Lymexylomidae), and termites (*Macrotermes* and *Odontotermes*) also carry on mutualistic relationships with fungi (Batra and Batra 1967).

Carnivores

Carnivorous insects are those that use living animals as a source of food. They may be divided somewhat arbitrarily into *predators* and *parasites,* although there is no sharp line of demarcation between the two. Predatory insects capture and eat on the spot living animals, usually other insects. Predators are usually larger than the captured individual, the *prey,* and many prey individuals are eaten during the life of the predator. Most parasites differ from predators in several respects. They live on or within the bodies of their food sources, or *hosts.* Parasites often do not immediately kill the hosts, and if the host is killed, only one host is usually required during the life of the parasite. A parasite is smaller than its host, but an immature insect parasitizing another insect may grow and attain the same size as the host.

Predatory behavior is widespread among the insects, having arisen independently on many different occasions in widely divergent groups. The orders Odonata, Hemiptera, Neuroptera, Coleoptera, Diptera, and Hymenoptera include large numbers of predatory forms, and many are found scattered among other orders, for example, praying mantids (Mantodea).

Predatory insects usually possess speed and agility and/or great stealth. For example, adult robber flies and dragonflies are noted for their impressive demonstrations of agility, capturing prey in midair. The praying mantid and many dragonfly nymphs provide examples of stealth in prey capture. These insects remain motionless and strike at passing prey with incredible speed and accuracy. A dragonfly nymph can only extend its prehensile labium directly in front of itself and must do so completely or not at all because the force for extension is hydrostatic. Thus potential prey must be a certain distance directly in front of the nymph in order to be caught. The compound eyes serve as the ranging sensors in these insects (see figure 6.15). The praying mantid has a more flexible prey-capture mechanism, being able to judge the distance and position of prey relative to the longitudinal axis of its body. Thus prey can be caught in many different positions relative to the mantid. This mechanism is based on an interaction between the compound eyes, sensory spines in the neck region that monitor head position, and the muscles of the raptorial forelegs.

Certain insects, such as antlion larvae (Neuroptera, Myrmeleontidae) and the larvae of some caddisflies (Trichoptera), literally trap their prey. The antlion forms a shallow cone-shaped pit in a sandy area and buries itself just beneath the surface at the bottom of the pit. When an ant or other small insect passes near enough to the pit to cause grains of sand to roll down the sides, the antlion responds by creating a miniature landslide. This brings the prey within reach of the antlion's prehensile jaws, through which it injects enzymes and then sucks out the digested insides of the prey. The aquatic larvae of certain species of caddisflies construct silken nets in which they are able to catch small organisms.

The eggs of some carnivorous insects are placed by the parent insect in proximity to prey species or the offspring may depend upon sluggish, easily captured prey. Many examples of parent insects locating prey for their offspring are found among the Hymenoptera (e.g., the case of the female *Bembix* wasp described in the section on reproductive behavior). In the higher Diptera, the larvae have poorly developed mouthparts and move about very slowly. Thus predaceous species like many syrphid flies (Syrphidae) depend on weaker, more sluggish prey, such as aphids, scale insects, mites, other dipteran larvae, and spider and insect eggs.

Most predaceous insects eat other insects, particularly herbivorous species. This is not surprising inasmuch as herbivorous insects (primary consumers) are the most abundant. Small crustaceans (especially aquatic), mites, and occasionally spiders may also fall prey to insects. Certain members of several families of beetles (Carabidae, Silphidae, Drilidae, and others) and some species of giant water bugs (Belostomatidae) feed upon snails. Protozoans compose a significant proportion of the diets of many aquatic insects. Members of the more specialized groups (e.g., solitary wasps) tend to be restrictive in prey choice, whereas members of the more generalized, primitive orders exhibit a wider range of prey choice.

Parasitic behavior, like predatory behavior, has arisen independently on several occasions among insects. Parasitic insects are probably best viewed on the basis of hosts, in particular as parasites of vertebrates and as parasites of other insects and related arthropods. Askew (1971) provides a comprehensive treatment of parasitic insects.

Most insects that are parasitic on vertebrates are found in the orders Anoplura (sucking lice), Mallophaga (chewing lice), Hemiptera (true bugs), Diptera (true flies; figure 8.4), and Siphonaptera (fleas). *Ectoparasitic* insects derive all or part of their sustenance from the host and live entirely on its external surfaces. *Endoparasitic* insects invade the tissues of the host. The majority of insect ectoparasites of vertebrates, save most Mallophaga, are blood feeders (hematophagous). Among the members of this group, some remain on the host throughout their life cycle, some live on the host only during a particular stage of the life cycle, and others visit the host intermittently during a particular stage of the life cycle and are otherwise free-living. Parasitic insects that remain on the host (e.g., the lice) tend to be host specific. They have close evolutionary ties with the host

Figure 8.4 Blood feeding by *Phlebotomus longipes* (Diptera; Psychodidae) on a human host. This species is a vector of *Leishmania aethiopica*, a protozoan parasite of humans in Ethiopia. (Length of fly approximately 3.0 mm.)
Courtesy of Woodbridge A. Foster.

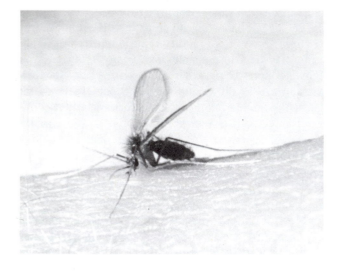

species and are most likely to be entirely dependent on it. Fleas spend only one stage of their life cycle on the host. Flea larvae are free-living and feed on organic debris that accumulates in the host nest or burrow. Adults tend to remain on the host, taking intermittent blood meals. However, adult fleas are active and commonly move from host to host. They are less host specific than the lice.

Hematophagous insects like adult female mosquitoes, "no-seeums" (Diptera, Ceratopogonidae), and some other adult Diptera, a few dipteran larvae, and nymphal and adult bed bugs and conenose bugs (Hemiptera: Cimicidae and Reduviidae, respectively) visit host vertebrates only to take a blood meal and exhibit varying degrees of host specificity. Most feed on a single or very few vertebrate species and usually cause little or no harm. This is not meant to imply that they never harm the host, as they obviously can when they occur in large numbers or introduce disease-causing organisms. Insects that parasitize vertebrates are discussed by specialists in Smith (1973). The feeding behavior of mosquitoes is described by Jones (1978). Nelson et al. (1975) discuss several aspects of vertebrate-host–ectoparasite relationships. Almost all insects that are endoparasites of vertebrates are dipteran larvae (maggots), in several families (Zumpt 1965), which may invade, for example, the alimentary canal, nasal cavities, and open sores. These larvae attack a wide variety of vertebrates, including human beings. An animal infested with fly larvae is said to be suffering from *myiasis*.

Some insects act as ectoparasites of other insects, taking intermittent blood meals from host insects in the same way those mentioned above do from vertebrate hosts. In fact, many members of the same orders and families are involved. For example, "no-seeums" have been observed feeding on caterpillars, along the wing veins of lacewings, dragonflies, and so on. Insects that feed in this fashion probably individually do little damage to their host.

Most insects that parasitize other insects do so as larvae and eventually destroy their host. These insects begin life as "typical" parasites in that they are much smaller than the host and live in intimate association with it. However, they eventually grow, sometimes nearly as large as the host itself, and the host is eventually killed. Thus they lie somewhere between parasites and predators. For this reason, the term *parasitoid* has been suggested to distinguish these insects from the more typical parasites and predators.

Most parasitoids live and feed as larvae within or on the host and become free-living adults. Exceptions include most female Strepsiptera (twisted-wing parasites), which spend their entire life within the host insect. The majority of parasitoids are in the order Hymenoptera in the superfamilies Ichneumonoidea, Chalcidoidea, and Proctotrupoidea, although a few occur scattered among other hymenopteran groups. A number of dipteran families also contain parasitoid forms (e.g., Tachinidae and Sarcophagidae). These flies attack a wide variety of insects, including grasshoppers, caterpillars, true bugs, beetle grubs and adults, and hymenopterous larvae. Rhipiphorid beetles attack such insects as hymenopteran larvae and cockroaches.

Some parasitoids exhibit a high degree of host specificity, attacking a single or very few species of insects. In most instances only a single parasitoid can exist in a host. However, there are instances where several and, rarely, several hundred or a thousand can develop in a single host. The parasitoids presumably feed on nonvital tissues until they are ready to pupate, and then they devour vital tissues. Some parasitoids attack other parasitoids. Thus one may find a caterpillar harboring a larva of a tachinid fly, which in turn harbors a chalcid wasp larva. This phenomenon is referred to as *hyperparasitoidism* (*hyperparasitism*). During hyperparasitism one of the attackers kills the other parasite. Superparasitism is when the parasites or parasitoids both share the host and do not kill one another. Both hyperparasitism and superparasitism are reviewed by Sullivan (1987) and van Alphen and Visser (1990), respectively. Both of these aspects of parasitism (i.e., hyperparasitism and superparasitism) are currently receiving much research attention because of the implications to biological control programs. Some of the topics being addressed are: How do parasites or parasitoids use marking pheromones to avoid either superparasitism or hyperparasitism? How do parasites "count" the number of eggs they lay into a host, and can they regulate the sex of their offspring? and finally, How do parasites and parasitoids use learning either to find a suitable patch (i.e., the area in which a resource, whether food, mate, etc., is located) or host? Many *entomophagous* ("insect-eating") insects, especially the hymenopteran parasites, have been utilized in attempts to control various pest-insect species (see chapter 15).

The feeding behavior of certain insects falls under the heading *social parasitism.* Brues (1946) describes the behavior of several species of wasps in the genus *Vespula,* which parasitize other members of the same genus, as follows.

All the parasitic species have lost the worker, or infertile female, caste and consist only of males and fertile females, the latter corresponding to the "queens" in the host species. The parasitic females enter the nests built and maintained by their hosts where they lay their own eggs in the paper cells already provided. Their young are then fed and reared in cells, intermingled with those containing the brood of the host, all of the feeding being done by workers of the host species.

Several other social Hymenoptera display this kind of parasitism.

Certain beetles in the family Meloidae invade the nests of solitary bees (Askew 1971). The beetle larvae destroy the bee's eggs and consume the food stored in the cells of the nest. Such nest invasions are also found among the Hymenoptera, for example, the cuckoo wasps (Chrysididae). This phenomenon has been called *cleptoparasitism* (Matthews and Matthews 1978).

Many insects may be found living and feeding as more or less permanent residents in the shelters of other insects (both social and nonsocial). These forms are sometime called *inquilines.* For example, certain galls "constructed" by one species are shared by larvae of another species without any apparent damage to the original owner. For social integrity and organization to be maintained it is extremely important for nest mates to recognize conspecifics. How do they do this? Do they "card" them? This aspect of social organization is discussed for wasps by Gamboa et al. (1986), for ants by Hölldobler and Wilson (1990) and for social insects in general by Wilson (1971). Because most of these social insects live in enclosed areas, they have evolved pheromones of low volatility, thus minimizing sensory habituation. Because social insect societies represent "closed clubs," one wonders how any social parasite or inquiline can gain entrance without the proper "ID card." One interesting example of how this is done is with the termitophilous beetle, *Trichopsenius frosti,* because its "ID card" (i.e., its cuticular hydrocarbon pattern-contact pheromone) is identical to "club members." Other examples of this "ID card falsification" are presented by Howard and Blomquist (1982).

Location of Food Sources

Dethier (1966) describes feeding behavior as "a complex and interacting sequence of responses to a variety of stimuli culminating in ingestion to repletion." He outlines the basic sequence of events involved in feeding: "locomotion bringing the insect to its food, cessation of locomotion on arrival, biting or its equivalent (probing, sucking, etc.), continued feeding, termination of feeding." In this and the following section we shall consider some of the factors involved in food location and initiation and termination of feeding.

In many herbivorous insects, the female parent oviposits on or in the vicinity of a food source for her offspring. The same is true for most parasitic insects. In these cases location of oviposition site, host location, and food location are not distinctly separate activities.

Much research has been carried out to determine the stimuli involved in food location by different insects. Visual and olfactory stimuli are probably the main ones used in food location by insects in general.

As an example of the use of visual cues, recall the use of color, form, and movement in the location of flowers by honey bees (see chapter 6). Many butterflies also respond to these stimuli. As mentioned in the preceding section, dragonfly larvae and adults and praying mantids rely heavily on visual stimuli in prey capture. Visual stimuli are important in host location in mosquitoes, tsetse flies, and other hematophagous flies and likewise in insects that are parasites of other insects.

Olfactory stimuli (volatile chemicals) are involved in food location in dung-feeding insects such as dung beetles (*Geotrupes* and others) and certain flies (e.g., many in the family Calliphoridae). These insects are attracted by the volatile substances skatol and ammonia, which are found in their food. These substances are referred to as *token stimuli* or *sign stimuli,* because they serve as indicators of substances with nutritive value but have none themselves. Token stimuli are very important in the attraction of herbivorous insects to their host plants. For example, the butterfly *Papilio polyxenes* is attracted to several of volatile oils that impart an odor to their characteristic food plants, members of the Umbelliferae. However, these oils have no food value for the butterflies. Also, they have no known role in plant metabolism. These oils and many others are called *secondary plant substances* (more on these in chapter 9). Secondary plant substances may also act as repellents to some insects. Thus some act as kairomones and some as allomones. Olfactory stimuli play a role in host location by mosquitoes. Carbon dioxide, steroids, amino acids, and other volatile substances characteristic of vertebrates are involved.

In addition to visual and olfactory cues, tactile, thermal, and hygrostimuli are involved in host location by the ectoparasites of warm-blooded vertebrates. For example, human body lice, which live between the skin and clothing of their host, show a marked preference for rough-textured materials. They show an increased frequency of turning on smooth surfaces, and this frequency is inversely proportional to the stimulus strength. Lice also demonstrate a preference for the 26.4° C–29.9° C range in a temperature gradient and 76% relative humidity in a humidity gradient.

Backswimmers and whirligig beetles respond to vibrations produced by the swimming or struggling of potential prey. The backswimmer locates prey by orienting itself in such a way that it receives equal stimulation of the sensory hairs on its oarlike legs. The larvae of ladybird beetles catch their prey, aphids, by crawling along the veins of a leaf and periodically stopping and "scanning" the vicinity by turning their body to and fro about the temporarily secured abdomen. When they are successful in capturing an aphid, they show an increase in the frequency of turns, a response that keeps them in the immediate locality. This behavior increases the probability of encountering additional prey and provides a good example of *klinokinesis*.

A considerable amount of attention is being given to the practical and theoretical aspects of how animals search and forage for resources (e.g., food) (Bell 1990, 1991; Bowen 1991; Traniello 1989; and Mitchell 1981). The stimuli used by the insect in locating food varies from either the animal (Allan et al. 1987) or plant (Prokopy and Owens 1983) being sought. One important aspect of studying food location in visually oriented insects is to duplicate, as closely as possible, any visual model being used to study visual orientation to food. The use of red spheres coated with a sticky material and hung in apple trees to attract gravid females of the apple maggot and the use of deer, cow, or oxen models to study biting fly orientation are just two examples of the use of artificial models to monitor or remove individuals from the population and to study food-location behavior The use of olfactory stimuli (e.g., CO_2 and 1-octen-3-ol, both breath mimics from ruminants) combined with a visual model usually enhance the traps' effectiveness.

Once potential food is located, specific stimuli are involved in the induction of feeding. These stimuli are often referred to as *phagostimulants* (a category of kairomones). They may be purely token stimuli, having no nutritional value, or they may be substances that have definite nutritive value. Many of the volatile oils (secondary plant substances) characteristic of certain species of plants are phagostimulants. For example, the cabbage aphid, *Brevicoryne brassicae,* is induced to feed even on an abnormal host when the substance sinigrin, a material extracted from Brassicaceae (mustard family) and other plants, is present. Sucrose, a common sugar in the food of many insects, is an effective phagostimulant in the majority of insects. Amino acids, glucose, certain proteins, ascorbic acid, and several other chemicals have also been shown to be feeding stimulants.

For more information on food location in herbivorous insects, see Ahmad (1983), Bell and Cardé (1984), Dethier (1970), Kennedy (1977), Kogan (1977), Schoonhoven (1972), Staedler (1977), and Visser (1986). Galun (1977a, b), Friend and Smith (1977), Hocking (1971), and Bowne (1991) deal specifically with host location in hematophagous insects. Askew (1971) contains much information on host location in parasitic insects in general, and Traniello (1989) covers ant foraging. Vinson (1976) deals specifically with host selection in parasitoids.

The larval and/or adult stages of some parasitic, parasitoid, and predatory insect species are transported to the host or prey species' nest by adults of the host or prey species (Clausen 1976). The insect doing the transporting is unharmed. Such biological "hitchhiking" is called *phoresy.* For example, a female of the Australian wasp, *Synoditella fulgidus* (Hymenoptera, Scelionidae), clings to the abdomen of an ovipositing host grasshopper. When host oviposition is complete, the wasp descends into the hole made by the host and deposits her eggs on the grasshopper's egg mass. These wasps may be carried by the grasshoppers when they swarm.

An example involving phoresy by larvae is found among the meloid beetles previously mentioned. Depending on the species of beetle, a female may oviposit in the soil near a host nest, in flowers, or at the entrance to the host nest. The first instar beetle larva, *triungulin,* is very active. Those species hatching in the soil climb vegetation and tend to aggregate in flowers. These, along with species that hatch on flowers, respond to the vibration produced by a visiting bee and some crawl onto the bee's body. If the bee is of the "correct" species, the beetle larva, which is transported by the bee to the bee's nest, will enter a larval cell. Triungulins that hatch near the entrance to a host nest climb onto passing bees and gain entry to the nest. The larvae do not harm the adult bees, but simply gain transportation and access to a food source.

Control of Feeding

The control of feeding has been studied extensively, but there is still a great deal to be learned (Barton Browne 1975; Bernays and Simpson 1982). Insects deprived of food typically behave differently from those that have recently obtained food; that is, they behave in a fashion that increases the probability of locating food. Such behavior usually includes increased locomotor activity and increased responsiveness (i.e., decreases in thresholds of response) to various stimuli. For example, the tendency for the desert locust, *Schistocerca gregaria,* to move upwind in the direction of the odor of grass increases with the duration of food deprivation. Similarly, the response of tsetse flies, which feed on vertebrate blood, to a slow-moving visual stimulus increases with duration of food deprivation. Responsiveness in food-deprived insects also increases to stimuli that do not emanate from food. For

Figure 8.5 Regulation of uptake of sugars in the blow fly *Phormia regina*. "Slug" refers to a mass of food that moves as a unit. *Slightly modified from Gelperin, 1971.*

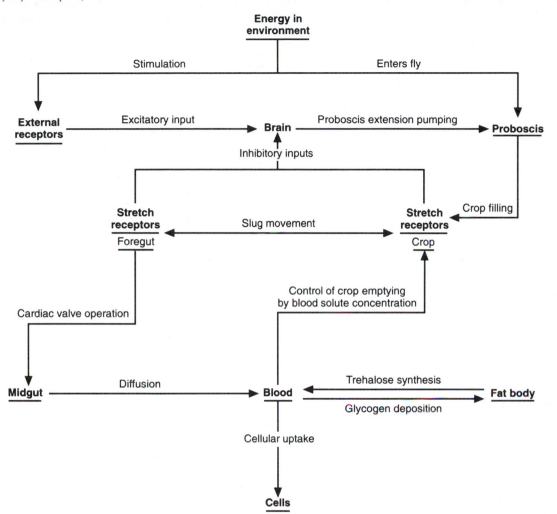

example, the photonegative (tending to move away from light), nocturnally feeding larvae of the cutworm, *Tryphaena pronuba,* tend to become less photonegative with starvation. Such modification of behavior may result in these larvae feeding above ground in the daylight.

Contact chemoreceptive sensilla are important once contact is made with food. These sensilla are responsible for the perception of stimuli that induce subsequent contact between the mouthparts and food. For example, stimulation of the chemoreceptors on the tarsi of the blow fly, *Phormia regina,* and many other flies and also on the tarsi of butterflies results in extension of the proboscis and contact with food. Sensilla on the mouthparts are then involved with the initiation and maintenance of feeding. The thresholds of responsiveness of sensilla on the tarsi and mouthparts decrease with food deprivation.

In many insects, the size of a meal is a function of the state of food deprivation. As food is ingested and assimilated, excitability (evidenced by increased locomotion and responsiveness) decreases, and inhibitory in-

fluences increase. These inhibitory influences may include adaptation of external sensilla (i.e., receptors become progressively less responsive to continuous stimulation), internal monitoring of nutritional state, and the action of stretch receptors associated with the alimentary canal or body wall of the abdomen. External receptors responding to feeding-deterrent molecules may also exert an inhibitory influence from the very first, even to the point of inhibiting feeding.

The control of feeding has been especially well studied in the blow fly, *Phormia regina* (Dethier 1976). Gelperin (1971) devised a very useful scheme for modeling the regulation of uptake of sugars (potential energy) from the environment. This scheme (figure 8.5) summarizes nicely much of the information presented here.

Several factors other than food deprivation also influence feeding behavior including molting, temperature, dormancy (to be discussed later in this chapter), the presence of other insects, food previously ingested, and reproductive state.

Escape and Defense Behavior

Throughout the evolutionary history of insects there has been a significant selective advantage in favor of the development of protective mechanisms against predators and parasites (see chapter 9). We shall examine several examples of such mechanisms from two standpoints: escape and defense.

Escape

Many insects will exhibit an escape reaction when threatened. Recall the alarm response of cockroaches mentioned earlier. Even those insects that have protective means other than escape will usually attempt to flee if the stimulus is great enough. A coupling of defensive behavior with escape is shown by some. For example, the beetle *Chlaenius* (Carabidae) discharges a secretion that is repellent to ants and remains invulnerable to them for 8–13 minutes following a discharge. This beetle can run an estimated 100 meters during this period of time!

Particularly interesting examples of insect escape reactions are those that are elicited in certain moths (especially in the family Noctuidae) in response to ultrasonic impulses from bats (Roeder 1965, 1966, 1967; Spangler 1988). The tympanic organs of hearing have already been described (see chapter 6). Insectivorous bats locate nocturnally airborne insects by emitting ultrasonic impulses and locating the sources of echoes. The moths detect these impulses beyond the range of the bats' echo-locating sensitivity and evasive behavior is initiated. Patterns of evasive behavior were studied by the use of an electronic sound source that produced sound in the range of the bat "chirps." The sound was emitted from a loudspeaker mounted on an upright pole. The paths of moths arriving in the vicinity of the sound source were recorded photographically by means of a stroboscopic (regularly blinking) flash that illuminated the moths at regular, very closely spaced intervals of time. Each time an approaching moth was illuminated, the reflected light, indicating the position of the moth at that moment, was recorded on the same photographic plate. With this technique a number of different evasive tactics were detected. These included turning and flying directly away from the sound source, oblique turns upward, "power dives," "passive drops," complex "looping dives," and zigzag movements. Some individuals landed on the ground and remained motionless for a period of time. When moths were tossed into the air in the vicinity of bats, similar responses were recorded with the same technique.

Defense

This broad topic is reviewed by Evans and Schmidt (1991). Insects display great variation in their form, coloration, and patterns of coloration. Instances in which one or all of these characteristics provide protection of the insect are legion. The close resemblance of wings and other body parts to leaves, twigs, thorns, and other plant parts; insects with "eyespots" or bright colors on the wings, which are suddenly displayed in response to a threat; *mimicry* of species that vertebrate predators learn to avoid, through experience with poisonous stings, irritating sprays, disagreeable tastes, and so on; and insects that blend in so well with their normal surroundings that they become nearly invisible to predators (*crypsis*) are but a few examples.

An insect must behave in an appropriate way if it is to gain any protective benefit from its form and coloration. Thus an insect that closely resembles a leaf will tend to orient itself on a stem in an appropriate manner, one with "eyespots" or other brightly colored markings will display these characteristics when threatened by a potential predator, and one that resembles other species may also adopt similar behavior. Insects that blend with their surroundings tend to remain motionless when threatened. Protective form and coloration will be discussed further in chapter 9. For more information on warning coloration and mimicry read Huheey (1984).

Skeletal structures, although not always used for defense, may function in this capacity. The raptorial forelegs of a praying mantis, although functioning primarily in prey capture, are sometimes used to strike out at an attacker. Many, if not all, of the sound-producing orthopterans produce sounds of warning and aggression. Any skeletal structure of sufficient strength and/or hardness can serve in a passive defense capacity. Cleptoparasitic cuckoo wasps (Hymenoptera, Chrysididae) roll up into a ball when threatened. In this position, the exposed areas of exoskeleton apparently provide protection from their hymenopteran host (Matthews and Matthews 1978).

A large number of insects produce nonskeletal barriers of various types. Nests, cocoons, cases, webs, and so on no doubt function in defense. The cast skins and fecal material of the larvae of the beetles in the genus *Cassida* (Chrysomelidae) remain on a "fork" of posterior abdominal appendages that are highly maneuverable. This waste material dries and forms a sturdy "fecal shield," which is used by the larva as a protective device that is said to be highly effective in blocking the bites of ants (Eisner, van Tassell, and Carrel 1967). The predaceous larvae of the green lacewing, *Chrysopa slossonae,* feed on the wooly alder aphid, *Prociphilus tesselatus.* The lacewing plucks the waxy "wool" from the bodies of the aphids and applies this material to its own body, coming to look like its prey. This "disguise" protects it from the ants that "herd" the aphids. Artificial removal of the waxy "wool" results in attack by ants (Eisner et al. 1978).

The primary defense mechanism of many insects is chemical. A large variety of noxious substances are secreted by the repugnatorial glands associated with the integument (see chapter 3). These substances represent "the means by which predators and other potential enemies are 'told' to desist or withdraw'' (Eisner and Meinwald 1966). The glands producing repugnatorial substances are variable in number, location, and morphology and have no doubt arisen independently many times during the course of insect evolution. Repugnatorial material is discharged from storage reservoirs associated with the glands by a variety of mechanisms. Basically these mechanisms fall into three categories (Eisner and Meinwald 1966).

1. The secretion oozes onto the integumental surface.
2. The gland is evaginated and the secretion allowed to volatilize.
3. The secretion is forcibly discharged.

Caterpillars of the species *Papilio machaon* (and probably all swallowtail butterfly larvae; Papilionidae, Lepidoptera) possess glands behind the head, osmeteria, which they evert when threatened, that produce a secretion effective against ants (Eisner and Meinwald 1965).

The forcible discharge of material is effected in a variety of ways. Contraction of muscles associated with the glands, hydrostatic pressure of the hemolymph, and air pressure are among them. Examples of insects that are capable of forcible discharge of odoriferous material are particular species of cockroaches, stink bugs, earwigs, stick insects, caterpillars, and carabid and tenebrionid beetles. Some of these are able to spray several feet, and many have a very accurate aim. A particularly impressive discharge mechanism is found in the bombardier beetles (*Brachinus* spp.). These beetles eject a hot spray, containing quinones, from glands in the posterior part of the abdomen with a distinctly audible "explosion." This is the result of a chemical reaction in which oxygen is liberated. The beetle's raised posterior at the time of the explosion adds to the impressiveness of this display. One may wonder why the defensive secretions liberated by insects do not harm the insects themselves. Apparently the gland cells allow two or more harmless precursors to be mixed in tubes or chambers isolated from the rest of the cell. This principle operates at the multicellular level in the repugnatorial glands of the bombardier beetle (figure 8.6). Each gland, which is two-chambered, contains phenolic precursors of quinones and hydrogen peroxide in an inner chamber and the enzyme catalase in the outer chamber. When these substances come into contact with one another, oxygen is liberated as a result of the mixing of catalase and hydrogen peroxide, and the phenols are oxidized to quinones, producing the reaction described above (Eisner and Meinwald 1966).

Blum (1978), Duffey (1977), Roth and Eisner (1962), and Eisner and Meinwald (1966) contain tables listing many specific chemicals and the arthropods in which they have been found. Besides functioning as defense allomones, some of these chemicals are thought to have antimicrobial effects. Some also act as alarm pheromones (Blum 1969). For example, soldiers of certain termites discharge a spray that not only works against

Figure 8.6 Defensive gland of a bombardier beetle, *Brachinus* sp.; E = enzyme catalase.
Redrawn from Schildknecht and Holoubek, 1961.

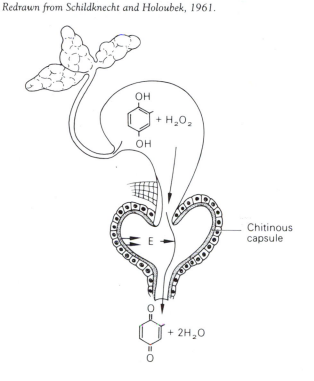

nest invaders but also alerts other soldiers. Alarm pheromones are also found in several groups of Hymenoptera (ants, bees, wasps, etc.) and are reviewed by Blum in volume 9 of Kerkut and Gilbert (1985). Eisner (1970), Blum (1978), Pasteels et al. (1983) and Prestwich (1984) are good sources for further information on the chemical defenses of arthropods.

Some insects are behavioral mimics of those that produce defensive secretions. For example, *Eleodes longicollis,* a species of tenebrionid beetle that possesses defensive glands, stands with its posterior well raised and discharges material from the glands. A second species, *Megasida obliterata,* mimics the stance of the first but lacks the defensive glands.

The products of venom glands associated with the hymenopteran stingers (modified ovipositors) and with the stinging hairs of many lepidopteran larvae (and some pupae and adults) and the toxic saliva of some Hemiptera may also be placed in the chemical defense category.

Not all defensive chemicals are directly associated with integumental glands. Some are found in the hemolymph and impart a bad taste or render an insect indigestible. For example, the Monarch butterfly, *Danaus plexippus,* possesses such substances, called *cardenolides,* in its body tissues (Reichstein et al. 1968). Cardenolides are derived from the characteristic food plants (i.e., milkweeds) of this species. Some lygaeid bugs (e.g., the milkweed bug, *Oncopeltus fasciatus*) also obtain cardenolides from their food sources (again milkweeds).

Rothschild (1972) and two chapters in Barbosa and Letourneau (1988) provide many more examples of insects that have been found to use secondary plant substances (see "Adaptations Associated with Interspecific Interactions" in chapter 9) as defensive chemicals.

Many species of beetles in the families Coccinellidae, Chrysomelidae, Meloidae, Lampyridae, and Lycidae discharge blood (*reflex bleeding*) in response to a threat. The blood may serve as a physical deterrent by clotting and entangling an attacker or, like the blood of many "blister" beetles (Meloidae), it may contain noxious substances. Many grasshoppers and other herbivorous insects may regurgitate or defecate when disturbed.

Many insects that produce repugnatorial substances, taste bad, or sting are brightly colored. This *aposematic coloration,* or *warning coloration,* "advertises" their "undesirability" as food (more on this in chapter 9).

Behavior and the Fluctuating Environment

Most habitats are characterized by both short-term and long-term (seasonal) fluctuations in their suitability for living organisms. Although few habitats are constant, many are "predictable"; that is, they vary, but with a degree of consistency. Others vary with little consistency. This variation in suitability is related directly and indirectly to abiotic factors, especially temperature and humidity. Thus for a given insect, a given habitat at one time may provide adequately for its life requirements (food, shelter, favorable temperature, favorable moisture, etc.) and at another time be entirely unsuitable. In this section, the role of behavioral adaptations associated with variation in environmental suitability is discussed.

Dispersal and Migration

Dispersal has been defined in a number of ways. For our purposes it can be looked upon as the result of several mechanisms, probably the most important of which is migration. Other probable mechanisms of dispersal include those associated with the search for food or a mate, phoresy, and responses to gradients of environmental factors (e.g., temperature, moisture, and CO_2). Insects may be carried passively by the wind or by water currents. As Johnson (1966) points out, "It is extremely difficult to measure how much dispersion occurs accidentally, incidentally, or adaptively."

Migration may be viewed as an active mass movement adapted to displace populations. Kennedy (1975) used the term "adaptive traveling." Migration is characterized behaviorally "as an accentuation of locomotor function with a depression of vegetative function" (Johnson 1966). In other words, migrating insects are persistent in their movement and usually do not respond to "vegetative" stimuli (stimuli associated with food, potential mates, etc.). Kennedy (1975) adds to increased locomotor function and decreased vegetative function the

following characteristics of migratory behavior: the straightening out of "the flier's track over the ground so that it travels, in the sense of traversing new ground instead of frequently changing its direction in a series of 'trivial,' station-keeping flights." In addition, the female sex is always present in a migration and may or may not be accompanied by males. Migrant females are generally sexually immature, but there are exceptions. Typically, many individuals migrate simultaneously, and migrant species tend to produce large numbers of offspring.

Migration is accomplished mainly by flight, and the direction of displacement for many is influenced by the wind. Even among migrants in which wind plays a major role, there is evidence that behavior plays a role, that is, behavior associated with straightening out the track over the ground (Kennedy 1975). Migration may occur by means other than flight. For example, army ants (*Eciton hamatum*) migrate on the ground (*pedestrian migration*) (Kennedy 1975, Schneirla 1971).

Insect migration has been studied by a variety of techniques including nets; airborne suction traps; aerial balloons; aircraft; photography; and marking with dyes, paint, radioisotopes, or genetic markers. Since 1966, when radar began to be used to track the migration of insects, that technique has contributed a great deal to our understanding of insect migration (Schaefer 1976). Incredibly, in addition to being able to obtain data on sizes of migrating populations and their flight performance, "radar entomologists" have also been able to determine species and sex (up to a distance of 1.5 km) by echo "signature" analysis.

Three types of migration may be recognized on the basis of adult life span (Johnson 1966).

1. Short-lived adults that emigrate and die within a season.
2. Short-lived adults that emigrate and return.
3. Long-lived adults that hibernate or estivate.

Members of the first group usually leave the breeding site, oviposit elsewhere, and die. Examples include locusts, termites, aphids, thrips, and many butterflies. The distance covered is variable; in locusts and butterflies it may be several hundred miles, in termites only a few feet or yards. The flights of many of these are windborne. Relatively short-lived adults, which emigrate and return, depart from the breeding site to feeding sites, where the eggs mature. The females then fly back to the vicinity of the original breeding site and oviposit. This emigration and return may be repeated in a given season by the same individual. For example, such behavior is characteristic of many dragonfly species. Insects in the third category fly to hibernation or estivation sites and return to the original breeding site the following season. The monarch butterfly, *Danaus plexippus* (Urquhart 1976, Brower 1977), many noctuid moths, and several beetles fall into this category.

The result of migration is the transference of a group of insects from an old to a new site. This is of particular importance when the "old" site was only temporarily suitable for survival of the insects. Temporary sites include pools of water that tend to dry out, seasonal food plants, and so on. Schneider (1962) points out, "In the course of evolution, a low level of dispersive movement has been associated with the colonization of permanent habitats and a high level has been closely correlated with the adoption of a temporary one." He further explains that this idea also applies to species in which the adults and larvae feed in quite different habitats (e.g., mosquitoes) and to those that hibernate or estivate in habitats different from the breeding sites. Thus migration has come to be viewed as an adaptive strategy of insects that use temporary habitats. It was formerly thought that only a few species of insects migrated. Now it appears that some migratory behavior is probably characteristic of most species (Johnson 1976). The details of migratory behavior vary greatly among species.

A couple of examples will illustrate how migratory behavior is adaptive. A particularly well-known migrant is the desert locust, *Schistocerca gregaria* (Johnson 1976). These insects breed in an area from North Africa to East Africa to Pakistan in the spring. As these areas dry out in the summer, young locusts begin to migrate in swarms (some estimated to contain from 10^5 to 10^{10} individuals). Their migration paths lead to a zone where the generally westerly winds of Africa collide with the moisture-laden monsoon winds from the Indian Ocean (i.e., the intertropical convergence zone). In this zone there is much rainfall, and lush vegetation results. Later in the season, when the convergence zone moves northward, new swarms migrate northward. Clearly, the migratory behavior of the desert locust is adjusted to ecological changes in a way that favors survival.

The monarch butterfly, *Danaus plexippus,* is a well-known migrant in Canada and the United States (Urquhart 1976; Brower 1977). Monarchs breed far into the north in British Columbia, Alberta, Quebec and Ontario during the summer, feeding and ovipositing on several species of milkweed plants. In the fall, they migrate southward. In the west, they fly to southern California, and in the midwest and east, they fly into central Mexico and Florida. The evolution of migratory behavior in these insects has enabled them to expand their feeding and breeding range in the summer and avoid the cold of winter.

Two hypotheses have been advanced to explain the cause of migration (Johnson 1966). The first one suggests that insects respond to the onset of adverse conditions by flying away. The second explains migration on the basis of endocrine changes correlated with particular environmental effects, such as crowding, food deficiency, and short days. For example, crowding in aphids results in the production of winged (alate) forms instead of the non-migrant wingless (apterous) forms, and this seems to be associated in some way with activity of the corpus allatum. Available evidence pertinent to the cause of migration favors the second hypothesis. It has been suggested that migratory behavior might depend on a particular balance between ecdysone and juvenile hormone in the blood.

The movement or dispersal of insect pests is a topic of great concern to individuals developing various integrated pest management programs. Not only is the pest a problem, but pathogens and parasites are often transported from field to field or even state to state. This topic is reviewed by Stinner et al. (1983).

For further information on various aspects of insect migration consult Dingle (1972, 1978a), Drake and Farrow (1988), Goldsworthy and Wheeler (1989), Johnson (1963, 1969, 1974, 1976), Kennedy (1975), Kring (1972), Rainey (1976), and Schneider (1962).

Dormancy

Insect development (including embryogenesis and larval and ovarian development) and general activity are subject to two extreme kinds of suppression (*dormancy*) relative to changes in the abiotic environment. At one extreme, an insect responds to adverse environmental conditions by a slowdown in metabolism and development. When conditions are no longer adverse, metabolism and development resume immediately. This type of dormancy is commonly referred to as *quiescence*. At the other extreme, an insect enters into a state of metabolic and developmental arrest in response to certain environmental conditions. These conditions may or may not be adverse, but serve as indicators of the imminent onset of adverse conditions. Development does not necessarily resume immediately with the return of favorable conditions. This type of dormancy is called *diapause*. Cold-hardiness, increased resistance to desiccation, and so on (see chapter 9) may be associated with quiescence and diapause.

Diapause is looked upon as a physiological timing mechanism that provides for

1. The induction of a comparatively resistant state during periods of adverse environmental conditions, particularly low temperatures.
2. The resumption of development when adverse conditions have disappeared and food is available (e.g., synchrony with host plants).
3. Synchrony of adult emergence in species with a short adult life.

Diapause may occur in any of the life stages in different insect species. In the adult stage it is characterized by the failure of the gonads to enlarge.

In many insects, diapause is *obligatory* (i.e., insects enter this state during every generation in spite of variation in environmental conditions). Other species display facultative diapause and can go on for several generations before entering diapause. Species or strains that exhibit obligatory diapause have only one generation per year (i.e., they are *univoltine*). On the other hand, those with facultative diapause may complete two or more generations per year (i.e., they are *bivoltine, trivoltine, quadrivoltine,* etc.). As in the milkweed bug, *Oncopeltus fasciatus,* diapause may occur in association with migration (Dingle 1978b).

The stage at which an insect enters diapause is a species characteristic, and even closely related species may enter diapause at different stages. Usually diapause occurs only during one life stage, but there are exceptions where diapause is entered twice. In a given species, there may be strain differences with regard to diapause. One strain may show an obligatory diapause, another facultative diapause, and still another not enter diapause at all. For example, univoltine, bivoltine, and quadrivoltine races of the silkworm, *Bombyx mori* (Lepidoptera, Bombycidae) are known in addition to races that undergo no diapause whatsoever. It is not uncommon for geographical races to show differences in the number of generations per year, depending on locality. An example of such racial differences has been found in the European cornborer, *Ostrinia nubilalis* (Lepidoptera, Pyralidae). Several years after the introduction of this species into the eastern United States it was discovered that the race in the Great Lakes states was univoltine with an obligatory diapause, whereas the New England race was bivoltine.

As mentioned above, the environmental conditions responsible for the induction of diapause are not necessarily adverse, but may signal the imminent onset of adverse conditions. The most important environmental factors that induce diapause appear to be those associated with the daily alternation of light and dark and the relative durations of light and dark periods. Compared to other environmental factors, these factors are the most accurate and invariable cues relative to seasonal changes in temperate zones. The mosquito *Aedes triseriatus* completes several generations during the summer, but in the fall the eggs (i.e., embryos) of the last generation enter diapause in response to shortening day length despite the fact that temperatures may actually be higher than in the spring when diapause is terminated and development resumes. In a few insect species the length of the dark period is the determining factor in the initiation of diapause.

Temperature is another important factor in the induction of diapause in certain species. Generally, low temperatures favor the initiation of diapause, although this is not always the case (e.g., high temperatures favor diapause in *Bombyx mori*). Temperature usually acts in association with photoperiod in producing diapause. Other environmental factors that may be involved include unfavorable nutritional conditions, low moisture content of food, desiccation, crowding, maternal diet, maternal age

at oviposition, and maternal exposure to low temperatures during oogenesis. Diapause in parasitic insects may be induced in synchrony with diapause in the host. In some parasitic species development is closely attuned to the hormonal changes in the host, and in some host species the presence of a parasite disrupts a diapause that would otherwise have occurred.

Various factors may be involved in the termination of diapause. For example, several species (e.g., *Bombyx mori*) must be exposed to low temperatures for certain intervals of time before diapause can be terminated. When diapause occurs during a hot dry season, exposure to high temperatures for a period of time may be necessary before diapause can be terminated. Wounds and various kinds of shock, such as pricking with a needle or electrical stimulation, may cause the termination of diapause in individuals of certain species.

Diapause is probably controlled by the endocrine system. Williams (1946–1953) showed that the activity of the prothoracic glands is necessary for the termination of postembryonic diapause in the silk moth *Platysamia cecropia*. In some other moths (*Bombyx mori,* Fukuda and Takeuchi 1967; *Phalaenoides glycinae,* Andrewartha et al. 1974) a secretion of neurosecretory cells in the subesophageal ganglion of the female parent induces diapause in her embryos.

Danks (1987), Denlinger (1986), Müller (1970), Mansingh (1971), and Thiele (1973) consider various classifications of types of dormancy, dormancy in tropical insects, and the ecological aspects of dormancy. Danks (1987), Dingle (1978a), volumes 1 and 8 of Kerkut and Gilbert (1985), Masaki (1980), Stoffolano (1974), Tauber and Tauber (1976), and Lees (1955) review various aspects of diapause.

Behavior, Temperature, and Humidity

It has been known for a long time that insects placed in a temperature gradient will demonstrate a "preferred" temperature by locating themselves at a particular point along the gradient and, further, that this preferred temperature is roughly correlated with habitat preference. Otherwise insects have generally been viewed as *poikilothermic* ("cold-blooded"). However, evidence has been accumulating for the existence of *thermoregulation*—maintenance of body temperature different from ambient temperature—in insects (May 1979). Behavior, along with physiological mechanisms, plays an important role in thermoregulation. Mechanisms of thermoregulation involve heat exchange with the environment (via conduction, convection, radiation, and evaporation) or variation in the production of metabolic heat. *Ectotherms* rely on heat from the external environment, in particular directly from the sun. *Endotherms* generate most of the required heat metabolically. Some insects may be both ectotherms and endotherms.

Common mechanisms of ectothermic regulation include control of solar input posturally, selection of "appropriate" microhabitats (see chapter 9), and shifting of activity rhythms (May 1979). Many insects, for example, several grasshoppers and locusts, butterflies, and cicadas, bask in the sun, assuming postures that facilitate heating. Conversely, ground-dwelling insects may "stilt," raising their bodies away from hot ground. Insects commonly move about and come to rest in a microhabitat with appropriate temperatures. For example, many insects tend to remain preferentially in shaded areas. Many desert insects burrow in response to high temperatures. In several diurnal species, which display midday activity peaks under cool conditions, the activity peaks may be shifted to early morning or evening in hot weather. Color may not be especially important in thermoregulation in most insects, but a few are known to undergo color changes in association with changes in temperature.

A good example of endothermy is found in the sphinx moth *Manduca sexta* (Lepidoptera, Sphingidae), which has been found to regulate body temperature just prior to and during flight (Heinrich and Bartholomew 1972; Heinrich 1974). These insects, which bear a striking resemblance to hummingbirds, cannot fly unless their bodies are at a temperature between 35° C and 38° C, but they have been observed flying when the ambient temperature was as low as 10° C. When ambient temperature is less than 35° C, these moths heat the thorax by assuming a characteristic stance and vigorously vibrating the wings through a small arc. This "shivering" is accomplished by the simultaneous stimulation of both the tergosternal and longitudinal flight muscles. The duration of this warm-up period varies indirectly with ambient temperature. These moths are also able to prevent "overheating" (i.e., reaching temperatures above 46° C). This is accomplished by increased heartbeat rate, which increases hemolymph circulation from the thorax to the abdomen. Loss of excess heat occurs across the abdominal wall, which is not insulated against heat loss as is the thorax.

Raising body temperature by fluttering the wings without flight has been observed in several insects in addition to *Manduca sexta*. There is also evidence that some "singing" insects warm up before singing.

A few insects are known to cool their bodies by enhancing evaporation. For example, honey bees may release fluid from their mouthparts and spread it on the venter (May 1979).

Thermoregulation in many insects is probably under central nervous control, but little is known about this topic. Insects may cooperate in thermoregulation. Many insects are gregarious and tend to form clusters under conditions of low temperature; overwintering honey bees form such clusters within their hives. The combined effects of the metabolic heat of the component individuals may actually accelerate the rate of development, as in the case of certain caterpillars (Bursell 1974a). The nests of social insects provide good examples of temperature control. For instance, the nests of some species of termites consist of cells having a structure that permits them to act as an air blanket, the effect being analogous to that of storm windows. During a hot day, the outer surface of a nest may be too hot to be touched by a human hand, whereas deep within the nest the temperature may be somewhere around 30° C. On a cold winter day, the internal temperature of the same nest may be 9° C or 10° C warmer than the air surrounding it (Skaife 1961). Lüscher (1961) describes the evolution of such "air conditioning" in termite nests. The nests of some wasps (e.g., hornets) are constructed with an outer "paper" envelope covering the layers that contain the rearing cells, resulting in an air-blanket effect. Other social wasps (e.g., *Polistes*), do not construct such an envelope. If a nest gets too hot, a number of its inhabitants will fan their wings vigorously at the nest entrance and may even bring in drops of water, which has the effect of increasing the cooling from fanning (Richards 1953b). The behavioral, evolutionary, and ecological aspects of thermal responses by aquatic insects are covered by Ward and Stanford (1982). Beck (1983) reviews insect thermoperiodism.

It is well established that insects will respond to a moisture gradient by moving into an "optimal" zone (Bursell 1974b). This behavior obviously promotes survival. Other factors, of course, may influence exactly how an insect behaves in a moisture gradient. If the flour beetle, *Tribolium,* has been starved, it will show a preference for moist air, but if it has been given access to food and water, it prefers dry air. Most insects offset the influence of dryness and other factors promoting water loss by drinking water or taking in food.

Reproductive Behavior

Reproductive behavior involves first the location of a mate, followed by courtship and mating, oviposition, and sometimes brood care. Territoriality may arise out of competition for mates or food. These aspects of reproductive behavior are considered in this section. Examples have been drawn primarily from the reviews of Carthy (1965), Markl (1974), and Matthews and Matthews (1978). The literature cited in the section on communication contains much information on reproductive behavior. Reproductive behavior is typically complex and highly variable. The evolutionary significance of this complexity and variability is generally assumed to be reproductive isolation.

Mate Location
A wide variety of mechanisms function in bringing together the sexes. Initially, over relatively long distances, these mechanisms usually involve the visual, olfactory, and auditory modes of communication, singly or in combination.

The visual stimuli associated with mate location and the responses they elicit vary in complexity. The simplest pattern, observed among water striders (Gerridae, Hemiptera), involves a male insect approaching any moving object of appropriate size that happens to enter its visual field. Other insects require more specific stimuli to induce approach. For example, male damselflies in the family Lestidae (Odonata) approach any insect that has transparent wings and flies in a manner similar to that of damselflies. Damselflies in the genus *Calopteryx* (Calopterygidae) require even more specific stimuli, the males of different species being able to recognize members of their own species by the amount of light allowed to pass through the wings.

More complex visual stimuli involved in mate location include the use of luminescent organs in signaling for a mate. This is best exemplified by the fireflies in the beetle family Lampyridae (see Lloyd, 1966, 1971, 1977, 1983). In some species both the males and females produce light and have a very complicated signaling system; in other species only the males produce the light signals. In either case the light-signaling systems are species-specific. In fact, several species have been discovered through observation of differences in signaling systems. The mating behavior of *Photinus pyralis,* a well-known American species, is described in the following quote from McElroy and Seliger (1962).

At dusk the male and female emerge separately from the grass. The male flies about two feet above the ground and emits a single short flash at regular intervals. The female climbs some slight eminence, such as a blade of grass, and waits. Ordinarily she does not fly at all, and she never flashes spontaneously. If a male flashes within three or four yards of her, she will usually wait a decorous interval, then flash a short response. At this the male turns in her direction and glows again. The female responds once more with a flash, and the exchange of signals is repeated—usually not more than five or 10 times—until the male reaches the female, waiting in the grass, and the two mate.

In some tropical species of fireflies, several males congregate in a single tree and flash synchronously (Buck and Buck 1968, 1976). McElroy et al. (1974) describe this phenomenon as follows.

In Burma and Siam and other eastern countries, all the fireflies on one tree may flash simultaneously, while on another tree some distance away this same synchronous flashing would be apparent, but out of step with those of the first tree. Observers have been particularly impressed by the display, which is one of the interesting sights of the Far East.

In some cases visual markers in the environment may be involved in mate location. For example, among the true flies males may swarm or sit near a marker and wait until a female approaches the marker (Fletcher 1977).

The use of olfactory cues in mate location is widespread among the insects. Sex pheromones are produced by specialized glands in males, females, or both sexes of a given species. Sex pheromones are very potent substances; "Detected by the insect in fantastically minute amounts, these attractants are undoubtedly among the most potent physiologically active substances known today" (Jacobson 1965). The sex pheromone *bombykol,* produced by the female silk moth, *Bombyx mori,* can elicit a response from a male in a concentration of 1000 molecules per cubic centimeter. Electrophysiological studies of antennal receptors have revealed that a single molecule of bombykol can initiate a single nerve impulse (Schneider 1974). Most sex pheromones are apparently species-specific; however, there are several known examples of nonspecificity.

The production by female insects of pheromones that attract males has been observed and described for several species in a number of orders. Some examples follow. In the Gypsy moth, *Lymantria dispar* (Lymantriidae), virgin females produce a pheromone that is capable of luring males over a distance of 1000 meters. The containers in which virgin females have been held apparently absorb some of the odorous substance, because they remain attractive to males for 2 to 3 days following the removal of females. Virgin females of various species of silk moths (Saturniidae) "call" males by protruding posterior segments of the abdomen and exposing pheromone-secreting glands to the atmosphere. This "calling" posture only occurs at certain times of day or in response to certain stimuli and is controlled by the release of a "calling" hormone from neurosecretory cells in the corpora cardiaca (Riddiford and Truman 1974). Female cockroaches produce a sex pheromone that stimulates male alertness, antennal movement, searching locomotion, and vigorous wing flutter (Jacobson 1965). Filter paper that has been in contact with virgin females has the same effect.

As mentioned earlier, males in many species produce sex pheromones. The pheromones produced by the males of some of these species serve both as attractants and excitants or excitants alone. Two examples of male production of sex pheromones follow. Males of the greater wax moth, *Galleria mellonella,* secrete from glands on their wings an odorous substance that is very attractive to females. The odor is dispersed by the male vibrating its wings and dancing around. Bumble bee (*Bombus terrestris*) males produce an attractant in their mandibular glands that lures females. This substance has been extracted from bumble bee mandibular glands with pentane and identified as farnesol, a substance present in the flower oils of many plants. These flower oils may be the bee's source of farnesol (Jacobson 1965).

Three species of beetles in the genus *Dendroctonus* (Scolytidae) provide an interesting variation in the function of sex pheromones. In these species a substance that is attractive to both sexes is produced by sexually mature, unmated females feeding on fresh Douglas fir phloem. Since both sexes are attracted by this odorous substance, it serves to bring them together, which eventually leads to courtship and copulation. *Trypodendron lineatum,* an ambrosia beetle, produces a substance that has a similar effect. Such substances have been called *assembling scents* or aggregation pheromones.

As previously mentioned in this chapter, neurobiologists now have many new techniques and instruments to probe the underlying neural mechanisms of behavior. This new approach to studying behavior has helped researchers understand how specific stimuli are received, integrated, and how they ultimately influence subsequent behavior. Just consider the specialized topic of sexual communication in moths. Fabre in his 1879–1908 treatise *Souvenirs Entomologiques* reported that male moths were attracted from a considerable distance to a screen-caged female moth, but when covered with a glass top so that no "odor" was released into the air of the room, the males were not attracted to her any longer. It was not until 1959 that Butenandt and his coworkers (Butenandt et al. 1959) reported the structure for the sex pheromone of *Bombyx mori,* which was called bombykol. With the advent of electronic technology, researchers were able to record the response of the antennal sensilla (i.e., referred to as an EAG, or electroantennogram) to purified sex pheromone (see figure 3.8). The question was asked, "Why don't female moths respond to their own pheromone or that of other females?" The only answer given at that time was placed in an evolutionary context without any understanding of how pheromones were received or how the input was integrated in the central nervous system. We will use this moth-pheromone system as just one example to illustrate how neurobiology has provided answers to questions that previously remained either unanswered or incorrect answers were given. The first step came when researchers, using an isolated antenna and purified pheromone, were unable to obtain any EAG response from the female antenna while the male's antenna responded. What did this mean? Either the peripheral sensilla and/or the central integration centers were lacking in females. It was later shown that female moths of these particular species lack the antennal sensilla that specifically respond to the pheromone and that females also lack a region of the central nervous system known as the macroglomerular complex. By using intracellular dyes, researchers traced the sensilla trichodea nerves of the males into the macroglomerular complex. It was shown later, using very fine microelectrodes, amplifiers, and an oscilloscope, that when the sensilla trichodea responded to stimulation by the sex pheromone, the electrical impulses were sent to this region of the brain for

integration. Thus, the response to the original question of why female moths do not respond to their own pheromones can be answered on the basis of morphology—the female lacks both the peripheral and central equipment to receive and integrate information relative to sex pheromones. It should be emphasized that some species of female moths do respond to sex pheromones produced by the males. In these species, one would expect both the peripheral and central systems to process the incoming stimuli.

Shorey (1977a, b) and volume 9 of Kerkut and Gilbert (1985) consider sex pheromones as well as pheromones in general. Howard and Blomquist (1982) cover the contact sex pheromones or cuticular hydrocarbons. Fletcher (1977), Borden (1977), and Bartell (1977) include discussions of the roles of pheromones in the sexual behavior of Diptera, Coleoptera, and Lepidoptera, respectively.

Acoustic signals serve to bring the sexes together for a large number of insects—many Orthoptera, Homoptera, and Diptera; a few Lepidoptera, Hemiptera, and Coleoptera. Calling songs may be quite elaborate and are usually species-specific. In general, a response to one of these songs depends in part on the state of internal readiness for courtship and copulation. In the majority of cases, specialized structures are involved in the production of sounds (see chapter 6), but several insects produce sounds as a direct result of wing movements in flight. Examples of sounds produced in this fashion that serve to bring the sexes together have been found in members of the Diptera, Hymenoptera, and Orthoptera. For example, the flight sound of a mature female mosquito is very attractive to a male and elicits copulatory behavior. The males react similarly when the frequency of the flight sound is produced by a tuning fork. In some insect groups the acoustical counterpart of the "assembling scent" has been found. For example, cicadas in the genus *Magicicada* sing an aggregating song in chorus that is responsible for assembling both males and females (Alexander and Moore 1962). Fly parasites of both crickets and cicadas use sound to locate their host. Some species of crickets have males, called sentinels, that remain quiet around singing males thus avoiding parasitism but benefiting from the other male's song to lure the female.

In many cases, aggregation of males and females may occur in association with biological activities other than sex, for example, at emergence, oviposition, or feeding sites. Such aggregation can subsequently lead to sexual activities.

Once an insect has been induced to approach a possible mate, other stimuli are usually required to release further pursuit. These are generally olfactory, but they may also be a characteristic behavior on the part of the pursued. An insect that recognizes a member of its own species may or may not be able to distinguish the sex of the individual it pursues. Males of *Pyrrhocoris* and *Gerris*

(Hemiptera) and *Leptinotarsa* (Coleoptera, Chrysomelidae) recognize members of their own species by specific odors but will attempt to copulate with either sex. On the other hand, some fruit flies are able to recognize the sex of an individual by its odor.

Courtship and Mating

Following mate location, variable and often elaborate behavior is released, which leads ultimately to copulation. Courtship displays function in species and sex identification, the meeting of solitary individuals, and in the stimulation and maneuvering involved in copulation. During courtship, escape and attack responses are usually inhibited, at least to the extent that copulation may occur.

Many of the cues that serve in mate location act further as releasers of courtship and copulation. For example, the flight sound of the female mosquito not only attracts males but also releases copulatory behavior. In many insects the sex pheromone produced by one of the sexes also serves as an excitant (aphrodisiac), stimulating courtship or attempts at copulation. For instance, the sex pheromone produced by female American cockroaches (*Periplaneta americana*) not only attracts males but induces a wing-raising display and attempts to copulate. In the absence of a female, males may attempt to copulate with one another if the female odor is present. This female sex pheromone is also capable of eliciting courtship behavior in males of other *Periplaneta* species and males of *Blatta orientalis* (oriental cockroach). In some insects, the sexually excited male or the female may produce a substance that acts as an excitant for the opposite sex. Such is the case in *Lethocerus indicus,* a giant water bug, and a family of beetles (Jacobson 1965).

> During sexual excitement the male is easily recognized by its odor, for its abdominal glands secrete a liquid with an odor reminiscent of cinnamon. . . . This substance, produced in two white tubules 4 cm long and 2–3 mm thick, occurs to the extent of approximately 0.02 ml per male and is used in southeast Asia as a spice for greasy foods. The female does not secrete the substance, which is believed to act as an aphrodisiac to make her more receptive to the male.

> . . .

> Males of Malachiidae, a family of tiny tropical beetles, entice females first with a tarty nectar and then expose them to an aphrodisiac. The males possess tufts of fine hair growing out of their shells (in some species on the wing covers, in others on the head). These hairs are saturated with a glandular secretion that the females cannot resist. During the mating season, the male searches for a female; when he finds one he offers his tuft of hair, which the female then accepts and nibbles upon. In so doing, her antennae come in contact with microscopic pores in his shell, through which the aphrodisiac substance is excreted, thus putting her in a state of wild excitement.

Insects show a seemingly endless variety of sexual patterns. Some are simple, the male and female simply coming together and copulating with little or no courtship maneuvers. On the other hand, many have elaborate and sometimes bizarre patterns of courtship. Two examples will give you an idea of how elaborate some of these courtships are. The silverfish, *Lepisma saccharina,* guides the female to an externally deposited spermatophore by spinning a series of threads that restrict her movements to those which bring her closer and closer to making contact with the spermatophore (figure 8.7). Jacobson (1965) cites the work of Bornemissza (1964) in the following description of courtship and copulation of two species of scorpion flies (*Harpobittacus,* Mecoptera).

> Males of both species hunt for the soft-bodied insects on which they feed; females have never been observed hunting, capturing, or killing prey in the field. When the male holds its prey and begins to feed, two reddish-brown vesicles are everted on the abdomen between tergites 6–7 and 7–8 and begin to expand and contract, discharging a musty scent perceptible to humans. This scent attracts females to the vicinity of the male, moving upwind in his direction. As soon as the female is within reach, the male retracts his vesicles and brings the prey to his mouthparts. The female attempts to get hold of the prey but is prevented by the male, whose abdomen seeks out the tip of the female's abdomen and copulation takes place. Once in copula, the male voluntarily passes the prey with his hind legs over to the female.

Darwin proposed that it is the female of most species that ultimately makes the decision to accept or to reject a male suitor. This has been termed "female choice." How the female makes decisions about which male to copulate with is generally not understood. For some insects, however, there is evidence indicating how the female might make this decision. For example, many predatory male insects present the female with nuptial gifts in the form of prey-food. While feeding on these, the male usually copulates with her. Clearly, a female could easily visually distinguish a larger prey-gift from a smaller; and, the larger prey-gift would make a greater contribution to the reproductive success of the female. The selective forces acting on insect mating systems are aptly covered in Blum and Blum (1979), Smith (1984), and Thornhill and Alcock (1983).

Figure 8.7 Courtship in the silverfish, *Lepisma saccharina*. (*a*) Approach. (*b*) Male affixes threads to wall and floor and deposits spermatophore (only the main thread and one secondary thread are shown, although many secondary and irregular threads are produced). (*c*) Female is guided to the spermatophore by the threads. *Redrawn from Sturm, 1956.*

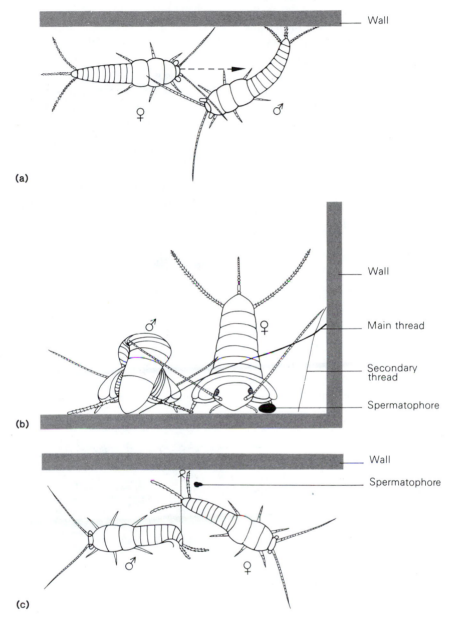

Rivalry and Territoriality

Rivalry between males may arise over the courtship of the same female and may result in direct physical aggression or displays. Reactions to members of the same species are different from those involved with escape or defense, which are stimulated by threats of danger. When a male grasshopper approaches another male that is serenading a female with a courtship song, they face one another and sing a characteristic "rivalry" song (Haskell 1974). Eventually one of the males leaves, and the other continues the courtship. Male silverfish, *Lepisma* (Lepismatidae, Thysanura), will fight over a female.

Territorial behavior is not common among insects, but there are some very definite examples. Males of certain odonates in the genera *Calopteryx* (damselflies) and *Pachydiplax* (dragonflies) defend territories against other males. A male entering another's territory is met with an attack. Females are recognized by sight since there is distinct sexual dimorphism in these particular species. Instead of aggression, these female are met with attempts to copulate. In some other genera of dragonflies (e.g., *Aeshna* and *Libellula*), when the sex of an approaching member of the same species is not recognized, males will attempt to copulate with males. The tendency for males to avoid such encounters results in their spreading out into individual territories. The suggestion has been made that such sexual encounters between males preceded the evolution of territorial behavior (Manning 1966). Male crickets are territorial; when one male enters another's territory, he is greeted by the "rival" song of the other. This is followed either by the exit of the intruder or holder of the territory or by fighting. A rank order may become established among males whose territories are in close proximity to one another. A female entering a male's territory is also greeted with the "rival" song, and will either leave or remain quietly. Price (1984) provides a list of examples of territoriality in several insect orders and Baker (1983) updates the topic.

Competition in mating often occurs even after insemination (Matthews and Matthews 1978). It is evolutionarily adaptive for males to evolve mechanisms that ensure that it is their sperm that fertilizes the eggs and not that of a competitor. The various strategies used by males to ensure success of their own sperm and the evolution of mating systems within the Insecta are discussed by Blum and Blum (1979) and Thornhill and Alcock (1983). Because sperm are commonly stored in a spermatheca, it would seem likely that more than one male could successfully inseminate a given female—multiple mating. A number of different mechanisms that help prevent additional insemination have evolved. For example, male dung flies "protect" or "guard" females from the copulation attempts of other males (figure 8.8). Other examples include the tandem flight of dragonflies (Odonata); prolonged copulation characteristics of the lovebug, *Plecia nearctica* (Diptera, Bibionidae), and mating plugs formed in the genital tract of the female from secretions of the male's accessory reproductive glands. Waage (1979) reported that some male damselfly species have the penis or aedeagus modified for removing sperm of rival males.

Many male insects during copulation transfer an unidentified substance in their accessory gland to the female that enters her hemolymph, acts on the nervous system, and renders her unreceptive to other mating attempts. In many species this fluid also contains a substance, which may or may not be the same molecule as just mentioned, that "informs" the female that she has mated and initiates oviposition. These chemicals are believed to be neuropeptides, but their exact nature and how they act on the female's central nervous system remains unknown.

If a female mates more than once (referred to as *multiple mating*), which sperm fertilizes the eggs? Studies have demonstrated that it is usually the sperm of the last male to copulate. This phenomenon is referred to as sperm precedence, or sperm competition. The former term is preferred because there is no evidence that competition is the mechanism that explains sperm prece-

Figure 8.8 Copulation and oviposition in the yellow dung fly, *Scatophaga stercoraria* (Diptera: Anthomyiidae). (*a*) Copulation. Female on substrate; copulating male above; guarding male thwarting takeover by another male (partly out of picture). (*b*) Oviposition with the male protecting the female from other males. This behavior prevents further mating activity from interfering with oviposition. *Courtesy of Woodbridge A. Foster.*

(a)

(b)

dence. A considerable amount of research attention is being given to the topic of single versus multiple mating because of its implications in understanding not only population dynamics but also its importance to control strategies. Unlike the screwworm, *Cochliomyia hominivorax*, which mates only once, insects that multiply mate are not good candidates for using the sterile male release technique to reduce pest populations (see chapter 15).

More information on reproductive strategies of insects is found in Blum and Blum (1979), Thornhill and Alcock (1983), and Eberhard (1985).

Males of some species—certain *Drosophila* (Diptera, Drosophilidae), dragonflies, and others—congregate and display at a particular place. These aggregations of males are presumably more effective in attracting females than isolated males.

Oviposition

In addition to wings, the ovipositor makes insects unique. As far as is known, no other animal group possesses an ovipositor. Thus, its uniqueness to this successful group may have important evolutionary implications. The major characteristics used in the past to explain the biological success of insects are "small size, rapid development, high fecundity, efficient terrestrial sperm transfer, the cuticle, tracheal system, wings, wing folding, and complete metamorphosis" (Zeh et al. 1989). The majority of these traits, however, are adult. Little emphasis has been given to what are termed "insect-egg" traits. These include the presence of the amnion, other aspects of the complex architecture of the egg shell (see chapter 5), and the ovipositor. Zeh et al. (1989) present the case for the "insect-egg" hypothesis as another way to explain insects' expansion into new niches and their rapid and extensive phyletic divergence. The survival, growth, and development of immature insects depends to a great extent on oviposition in an appropriate environment. This is especially important to insects with specific diets (e.g., a particular plant, or in the case of many parasitic insects, a particular host). For example, female mosquitoes that are ready to oviposit must do so in a place where the eggs are in water or will eventually be. Another example is the female apple maggot (see figure 6.5) that uses her long ovipositor to penetrate the skin of the apple fruit and to deposit an individual egg. The highly evolved ovipositor of fruit flies contains both mechanosensilla and chemosensilla that provide sensory input with respect to fruit hardness and quality. After laying, the female drags the ovipositor over the fruit surface while simultaneously depositing an oviposition deterrent pheromone. This pheromone prevents other females from ovipositing into the limited larval resource, thus avoiding larval competition for food. Apparently, females have to learn the "odor" of this pheromone. Although some insects merely drop their eggs wherever they may happen to be (e.g., some mayflies), more often they are specific as to their choice of oviposition site, locating it by means of a variety of stimuli, depending upon the kind of insect. For example, the beetle *Hylotrupes* (Cerambycidae) is attracted to the terpene odor of the wood in which it deposits its eggs. The parasitic wasp *Nasonia* locates the puparial cases of host blow flies (Diptera, Calliphoridae) by the odor of the decaying flesh in which blow fly larvae and pupae are commonly found. Insects ready to oviposit may respond to stimuli that previously elicited no response whatsoever. For example, *Pieris* (Pieridae) butterflies show a definite preference for objects with a green color when they are ready to oviposit, but when they are searching for food, they demonstrate no such preference. When an insect is highly selective in the choice of a site for oviposition, this does not mean that it has foreknowledge of the needs of its offspring. Oviposition in response to specific stimuli that are associated with an environmental situation in which the young can survive has, during the course of evolution, no doubt been selected for. Thompson and Pellmyr (1991) and Bentley and Day (1989) present information on the sensory inputs received by female Lepidoptera and mosquitoes, respectively, concerning oviposition site selection and evaluation. An amazing engineering feat of surveying and measuring is accomplished by the minute parasitoid wasps of the genus *Trichogramma*. How they do this was masterfully uncovered by Jonathan Schmidt. Measurement and evaluation of the host egg is achieved by the female walking from one side of the egg, across the apex, and down the other side. While taking this egg walk, females continuously drum the egg surface with their antennae. The egg-walk measurement and antennal drumming provide her with information about both the size of the food resource and quality (Schmidt and Smith 1989). Using these inputs, she regulates both the number of eggs laid and the sex ratio. More will be said about regulation of sex ratios by females in the discussion of social behavior. In general, more information on how insects evaluate resource quality and quantity is needed. Once this is known, manipulation of important factors for such things as developing plants that are resistant to specific pests can begin.

Brood Care

Once a parent insect has fulfilled its responsibility for placing the egg (or larva, in some instances) in an appropriate environmental situation, it may simply leave. However, some continue an association with the eggs and immature stages. This association is most highly developed in the social insects (ants, bees, wasps, and termites) in which the brood form a core around which all activity is centered, and these brood are "reared" from egg to adult by workers. Social insects differ from social vertebrates in that a vertebrate colony is composed of a number of mating pairs and offspring, whereas an insect colony is usually the product of a single female or single male–female pair.

Many nonsocial insects also do more for their offspring than merely deposit the egg. Earwig females (Dermaptera) lay their eggs in burrows in the ground and guard them until they hatch. Female beetles in the genus *Omaspides* protect their brood from the ravages of invading ants. Most sphecid and eumenid wasps, and solitary bees go to the extent of preparing elaborate nests and stocking them with food for their young. For example, a female solitary wasp in the genus *Bembix* digs a nest in the soil with her legs and mandibles. She then captures an adult fly (or sometimes another insect) and brings it to the nest. Evans (1957) describes the prey capture.

> The capture and stinging of the prey occur with great rapidity. The female wasp proceeds slowly through the air or hovers over a source of flies, pouncing upon the flies either in flight or at rest; then she descends to the earth or to some other solid object, where she quickly bends her abdomen downward and forward, inserting the sting on the ventral side of the thorax or in the neck region of the fly. . . .

Following capture, the wasp returns to the nest and deposits an egg on the fly. Thereafter, a number of flies are brought to the nest as the larva grows. Eventually, the larva spins a cocoon, pupates, and emerges from the ground as an adult. Depending on the species and the time of year, it may remain over the winter in the cocoon.

Insects in Groups

Insects display many gradations between solitary behavior and complex, organized social behavior. However, comparatively few insect species are truly social (*eusocial*), a few thousand perhaps, and these are found in only two orders, Isoptera (termites) and Hymenoptera (ants, bees, wasps, and relatives).

Matthews and Matthews (1978) provide a useful classification of intraspecific insect associations other than sexual interludes. They divide interactions into aggregations, simple groups, primitive societies, and advanced societies (*eusocial* insects).

Group behavior is thought to provide a number of different benefits; including protection as a result of such things as collective displays and more efficient detection of potential predators, increased efficiency in detection and utilization of food, and moderation of adverse physical environmental factors. Insects that produce defensive secretions and/or display warning coloration no doubt derive increased protection by pooling their defensive capabilities. Wooly apple aphids, *Eriosoma lanigerum* (Homoptera, Eriosomatidae), secrete waxy material, which gives their bodies a whitish, bushy appearance. Disturbance of aggregations of these insects causes them to rhythmically move their bodies at the same time. This collective display gives the impression of a much larger organism. Honey bees are very efficient food users by virtue of their ability to communicate the locations of food sources to one another. During the winter, honey bees form into tight clusters, which provide protection from freezing temperatures.

Many insects form into loose, temporary *aggregations* under certain circumstances. These aggregations may result from common attraction to a particular habitat. Attracting stimuli include a particular temperature, light condition, food, and particular chemicals. Aggregations may also result from mutual attraction independent of external conditions. Examples of aggregations are hibernating groups of ladybird beetles (Coccinellidae) commonly found in human habitations during the winter; cockroaches, which are often found in feeding aggregations; and nighttime clusters of usually conspecific bees or wasps.

Simple groups are characterized by coordinated movements. Examples are migrating groups of butterflies and locusts.

Groupings in the *primitive society* category range from simple parental care of offspring beyond oviposition (*brood care*) to interactions that border on true social behavior. Reciprocal communication is typical, and cooperation in nest construction and defense is common. Several examples of brood care and nesting behavior are given in the preceding section. Additional examples of communal nesting include web spinners (Embioptera), which construct and inhabit networks of silken tunnels, and tent caterpillars (Lepidoptera, Lasiocampidae), which cooperate in the construction of silken shelters in which they pass the night. During the daytime, tent caterpillars forage along the branches of the host tree. They communicate by depositing chemical trails in association with feeding success. Other individuals follow these trails preferentially.

Three traits are generally agreed upon as characterizing eusocial insect behavior (Wilson 1975).

1. Conspecific individuals cooperate in brood care.
2. There is a division of labor based on reproduction; that is, one or more fecund individuals reproduce while essentially sterile individuals serve as a labor force.
3. At least two generations serve in the labor force; that is, offspring assist parents during part of their life.

Behavior that incudes only one or two of these characteristics is viewed as *presocial*. Many of the examples of primitive societies given earlier can be viewed as presocial.

Prior to discussing eusocial behavior, it is worth mentioning that B. Hölldobler and E. O. Wilson shared the 1991 Pulitzer Prize in nonfiction literature for their 1990 book on ants. In addition to this, both individuals have made significant contributions to our understanding of the ecology, behavior, and evolution of social insects.

A requisite for the development of social behavior has been the evolution of systems of communication between the members of a society. In most cases communication in social insects involves chemical signals, although other sensory cues may also be important. Communication makes possible complex levels of behavior involving groups of individuals, such as construction of complex nests. However, these complex patterns of behavior result from the integration of relatively simple individual behavior patterns, which are evident in the behavioral repertoires of solitary and presocial insects. Wilson (1975) outlines nine categories of responses to communicatory signals found in social insects.

1. Alarm; 2. Simple attraction (multiple attraction = "assembly"); 3. Recruitment, as to a new food source or nest sites; 4. Grooming, including assistance at molting; 5. Trophallaxis (the exchange of oral and anal liquid); 6. Exchange of solid food particles; 7. Group effect: either increasing a given activity (facilitation) or inhibiting it; 8. Recognition, of both nestmates and members of particular castes; 9. Caste determination, either by inhibition or by stimulation.

Communication in social insects is discussed in Attygalle and Morgan (1985), Hölldobler and Wilson (1990), Lewis (1984), von Frisch (1967, 1971), and Wilson (1971, 1975).

The members of insect societies are divided into forms specialized for different functions within the colony. Differences among the various specialized forms, or *castes,* vary from solely behavioral (e.g., many bees and wasps) to both behavioral and morphological. Morphological differences, in turn, range from slight to extreme. Further, the particular role played by a given individual may vary with age, *temporal polyethism.* For example, an individual might be involved with brood care and nest maintenance during the first part of its life and with foraging during the later part (e.g., honey bee workers).

In the broadest sense, a social insect colony may be divided into reproductives and nonreproductives. Reproductives may include one or more fecund females (*queen* or *queens*) and males. One or more *worker* castes are the numerically most abundant nonreproductives. In addition, forms specialized for colony defense, *soldiers* may be present.

In the termites, there are typically the colony-founding primary reproductives, the king and queen, workers (immatures in the lower termites and fully differentiated forms in the higher termites), and soldiers. The soldiers show a wide array of variation in the structure of the head, especially the mandibles, depending on species. In addition, termites usually have the potential for producing *supplementary reproductives* should the king and queen be removed.

Whereas the termite worker castes are composed of both males and females, those of the social Hymenoptera are all females. Unlike the termites, male hymenopterans contribute little to the colony save insemination of the queen. In the ants, reproductives (in particular, queens),

workers, and soldiers are morphologically well-defined. In a few species workers are divided into morphological subcastes, but in most they are divided into temporal subcastes (i.e., temporal polyethism). Castes are only behaviorally differentiated in most bees and wasps, but in a few species, for example, the honey bee, *Apis mellifera,* there are clear morphological differences between queens and workers.

Details on the biology of termites and social Hymenoptera may be found in chapters 12 and 13, respectively. Caste differentiation is discussed in chapter 5. Table 8.2 compares some of the aspects of the social biology of termites and social Hymenoptera. Further information on social insects may be found in Andrews (1971), Evans and Eberhard (1970), Free (1977), Gamboa et al. (1986), Goetsch (1957), Hermann (1979), Hölldobler and Wilson (1990), Krishna and Weesner (1969, 1970), Menzel and Mercer (1987), Michener (1974), Oster and Wilson (1978), Plowright and Laverty (1984), Prestwich (1984), Richards (1953a), Schneirla (1971), Skaife (1961), Spradbery (1973), Sudd (1967), Traniello (1989), von Frisch (1950, 1967), Weber (1972), Wheeler (1910), Wilson (1971), and Winston (1987).

Origins of Social Behavior

Eusociality is considered to have evolved once in the termites and at least eleven times within the Hymenoptera. Although there are many differences between the various groups, there are also many similarities as a result of convergent evolution (table 8.2).

The termites (especially the "lower" termites) have many traits in common with the cryptocercid cockroaches. Outstanding among these similarities is the fact that the lower termites and cryptocercid cockroaches are the only wood-eating insects that depend on cellulase-producing symbiotic intestinal protozoans. These symbionts are passed from one generation to the next by means of anal *trophallaxis.* This behavior involves interaction between members of overlapping generations and hence rudimentary social behavior. Termites may be viewed as "social cockroaches," bound together originally by the contact required to pass along the symbiotic protozoans and subsequently forming into complex societies. Without the intestinal symbionts, the termites would be unable to digest cellulose and would starve to death.

Two possible evolutionary pathways to eusocial behavior among the Hymenoptera have been envisioned (Wilson 1975). The *parasocial* sequence (which may have been followed by most social bees) began with *communal* behavior involving cooperation in nest construction, but with separate brood rearing. Subsequently, cooperation in brood care was added (i.e., *quasisocial* behavior). This was followed by the development of a nonreproductive worker caste (*semisocial* behavior). With the development of cooperation of two or more overlapping generations the path to eusocial behavior was complete.

Similarities	Differences	
	Termites	**Eusocial Hymenoptera**
1. The castes are similar in number and kind, especially between termites and ants.	1. Caste determination in the lower termites is based primarily on pheromones; in some of the higher termites it involves sex, but the other factors remain unidentified.	1. Caste determination is based primarily on nutrition, although pheromones play a role in some cases.
2. Trophallaxis occurs and is an important mechanism in social regulation.	2. The worker castes consist of both females and males.	2. The worker castes consist of females only.
3. Chemical trails are used in recruitment as in the ants, and the behavior of trail laying and following is closely similar.	3. Larvae and nymphs contribute to colony labor, at least in later instars.	3. The immature stages (larvae and pupae) are helpless and almost never contribute to colony labor.
4. Inhibitory caste pheromones exist, similar in action to those found in honeybees and ants.	4. There are no dominance hierarchies among individuals in the same colonies.	4. Dominance hierarchies are commonplace, but not universal.
5. Grooming between individuals occurs frequently and functions at least partially in the transmission of pheromones.	5. Social parasitism between species is almost wholly absent.	5. Social parasitism between species is common and widespread.
6. Nest odor and territoriality are of general occurrence.	6. Exchange of liquid anal food occurs universally in the lower termites, and trophic eggs are unknown.	6. Anal trophallaxis is rare, but trophic eggs are exchanged in many species of bees and ants.
7. Nest structure is of comparable complexity and, in a few members of the Termitidae (e.g., *Apicotermes, Macrotermes*), of considerably greater complexity. Regulation of temperature and humidity within the nest operates at about the same level of precision.	7. The primary reproductive male (the "king") stays with the queen after the nuptial flight, helps her construct the first nest, and fertilizes her intermittently as the colony develops; fertilization does not occur during the nuptial flight.	7. The male fertilizes the queen during the nuptial flight and dies soon afterward without helping the queen in nest construction.
8. Cannibalism is widespread in both groups (but not universal, at least not in the Hymenoptera).		

[a]From Wilson (1971).

The *subsocial* sequence of evolution of eusocialism has been envisioned for ants, social wasps, and a few species of social bees (and also applies to termites as described above). In this pathway, the female first remained with brood for a time following oviposition, but departed before eclosion (*primitively subsocial*). Subsequently, the female remained with a newly hatched brood, and an overlap in generations occurred (*intermediate subsocial I*). In the next stage, the first generation offspring aid in the rearing of the next generation of brood (*intermediate subsocial II*). Differentiation into reproductive and non-reproductive castes then led to complete sociality.

A major factor in the evolution of social behavior among Hymenoptera is considered to be *haplodiploidy* (Hamilton 1964, Trivers and Hare 1976, Wilson 1975). The term *haplodiploidy* denotes that female Hymenoptera arise from fertilized eggs and are diploid, whereas males arise parthenogenetically and are haploid (see chapter 5). As previously mentioned when discussing how an egg parasitoid "measures" and "evaluates" the egg resource (i.e., prey) prior to ovipositing, females of *Trichogramma* and *Nasonia* species are able to regulate the sex of the offspring by controlling the release of sperm.

Thus fertilized eggs become females, whereas males develop from unfertilized eggs. How the female controls which eggs are fertilized has not been determined. By consulting Blum and Blum (1979), Smith (1984), Suzuki et al. (1984), and Thornhill and Alcock (1983), more information can be obtained on the benefits of sex ratio manipulation and its relationship to kin selection. As a result of haplodiploidy, female Hymenoptera are more closely related to their sisters than to their daughters. That is, on the average, sisters share 75% of their genes with each other but only 50% with their daughters. This process has been termed *kin selection* and has been advanced as the method to account for the tendency for Hymenopteran societies to be characterized by *altruistic behavior* between sisters. An example of altruistic behavior is a honey bee "sacrificing" its life to protect the hive. When a honey bee stings, the barbed sting apparatus remains embedded in the victim and is ripped away from the bee's body; the bee subsequently dies. The notion of altruistic behavior is linked to *inclusive fitness,* a central concept in the rapidly developing area, *sociobiology* ("the systematic study of the biological basis of all social behavior"—Wilson 1975). *Fitness,* in Darwin's original view, is a measure of the reproductive success of an individual, that is, an individual's relative contribution of genes to the next generation. Inclusive fitness is more encompassing than Darwinian fitness in that the contribution of an individual to the reproductive success of close relatives (*kin*) is considered in addition to an individual's own reproductive success. In this view, then, it is possible for an individual to be fit in the evolutionary sense without reproducing directly. Thus, the honey bee worker is not really making such a "sacrifice" after all, because the act of apparent altruism actually enhances the reproductive success of her younger sisters, with whom she shares most of her genes and from among whom future queens will arise. An important issue in sociobiology is the ability of nestmates (i.e., workers in eusocial societies) to recognize nestmate and non-nestmate kin and nonkin (Gamboa et al. 1986).

The study of insect societies is playing a major role in the development of sociobiology. For more information on this topic see Andersson (1984), Barash (1977), Dawkins (1976), Hamilton (1964), Hölldobler and Wilson (1990), Jeanne (1980), Plowright and Laverty (1984), Ruse (1979), Tallamy and Wood (1986), Trivers and Hare (1976), and Wilson (1975).

PART TWO # Insects and Their Environment

The study of organisms in relation to their environment is the modern science of ecology, which has developed in recent years from the earlier, essentially descriptive endeavor, natural history. Ecological studies of insects are important in that they provide insights into insect evolution, enable us to develop ways to manage both beneficial and pest species, and form the basis for assessment of the potential negative and positive effects of manipulation of an ecosystem.

The following two chapters examine the multitudinous relationships between insects and the components of their environment. In chapter 9, we consider some of the basics of insect population biology, the physical/chemical, abiotic, components of the environment, and begin an examination of the biotic components of the environment by discussing the various relationships between insects and other animals, and between insects and microbes. Chapter 10 is devoted entirely to the manifold relationships between insects and plants.

CHAPTER 9

Insect Populations and the Physical-Chemical and Biotic Environment

Before we discuss insect populations, a brief review of a basic functional unit of the environment, the ecosystem is needed. An *ecosystem* is composed of a community of living (biotic), interacting organisms (plants, animals, and microorganisms) and the nonliving (abiotic) components of their environment. Ecosystems vary in size and complexity, ranging anywhere from a woodland pond to the entire *biosphere* (that portion of our planet in which living forms exist). The biotic part of an ecosystem is made up of producers, consumers, and decomposers. *Producers* are primarily green plants, which synthesize food (carbohydrates and other compounds) from carbon dioxide and water by using energy derived from solar radiation (i.e., *photosynthesis*). *Consumers* are those organisms, the vast majority of which are animals, that obtain energy and molecular building blocks by ingesting organic matter from plants (herbivores and detritivores) or from other consumers (carnivores and detritivores). *Decomposers,* mainly bacteria and fungi, break complex organic molecules down to simple inorganic nutrients usable by producers, facilitating the continuous cycling of inorganic nutrients throughout an ecosystem.

A number of general ecology texts are available, among them Begon et al. (1990), Ricklefs (1990), and Smith (1990). Price (1984) and Huffaker and Rabb (1984) deal specifically with insect ecology. Andrewartha and Birch (1973) trace the historical development of insect ecology.

The place where an organism lives is its *habitat*. This "place" may be an entire field or forest, the shoreline of a stream, or the undersides of leaves of a particular kind of plant. A given species is typically found living in the same kind of surroundings throughout its distribution, although there may be variation within the limits of tolerance of the species. The degree of suitability of a given habitat will vary with the extent to which it satisfies the needs within these limits of tolerance.

Each species of the community of organisms living in a given habitat requires certain resources (e.g., food, shelter, breeding sites, favorable temperature, favorable humidity) in sufficient amounts, of sufficient quality, and at appropriate times in order to survive and reproduce. That is, every species occupies a particular physical location at a given time and does a given thing (eats, copulates, rests, etc.) at that place and time. A compilation of all needed resources, both in qualitative and quantitative terms, describes the *ecological niche* of a species. A community is composed of individuals of different species segregated into different niches.

Insects occupy a wide variety of niches in almost every kind of terrestrial and freshwater habitat. About the only habitats not broadly colonized by insects are those within the ocean; although many insects from different orders are found in intertidal marine habitats, and some are found on the open ocean (e.g., several species of water striders, *Halobates,* Hemiptera, Gerridae; Cheng 1976). However, there are no known marine insects that spend their entire life cycle submerged. Janzen (1977) provides an interesting discussion of the possible reasons why insects in particular have been able to occupy so many different niches.

The Life-System Concept

Clark et al. (1967) introduce a particularly useful concept for our purpose, the concept of the life system (figure 9.1). A life system is "that part of an ecosystem which determines the existence, abundance, and evolution of a particular population." In other words, the life system of an insect is that part of the environment (*effective environment*) that directly influences the fate of a given population plus the population itself. Not all parts of an ecosystem necessarily directly influence a given animal population, although the situation becomes hazy when one

Figure 9.1 Components of a life system.
Redrawn from Clark et al., 1967.

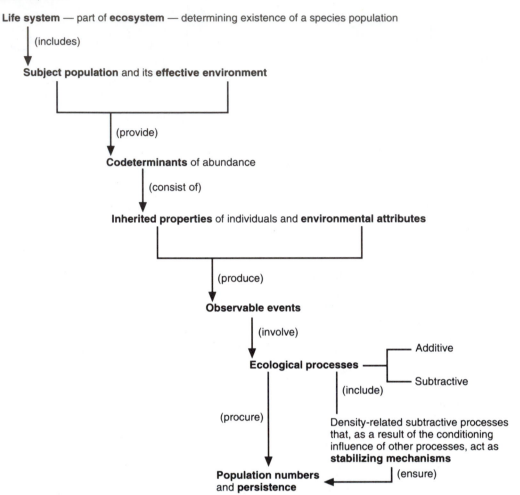

considers indirect influences. In view of the life-system concept, an ecosystem becomes a series of interlocking life systems.

Various components of the environment (both abiotic and biotic) have been found to be significant in their influence on insect or other animal or plant populations. Any combination, or all of these components, may at some time act to varying degrees on a given population. When one of them approaches or exceeds the species-specific limit of tolerance, whether expressed in terms of survival, development, fecundity, and so on, it becomes a limiting factor. That is, it becomes the environmental component that is directly responsible for limiting the extent of survival, degree or rate of development, fecundity, and so on. In the field it is often very difficult to pinpoint a given environmental component as a limiting factor since, obviously, several components are likely to be interacting to produce the "effective environment" of any population. For the same reason, it is difficult to relate studies of environmental effects under the carefully controlled conditions of a laboratory to what actually occurs in the field.

Another problem further complicates the picture. Environmental components (e.g., temperature, moisture, light, and particular resources) are not uniform throughout an ecosystem. For example, the temperature of one part of a plant may be quite different from that of another part. Thus, at least in many instances, to obtain a truly accurate picture, one must measure the temperature in specific parts of an ecosystem (e.g., bottom side of a leaf). Realization of this problem has led to the development of the concept of "microenvironment," which implies the recognition of distinct horizontal and vertical (spatial) and temporal differences in environmental components within an ecosystem.

This chapter and the following chapter deal with aspects of the major functional components of a life system—population and effective environment. In this chapter, we consider some of the basics of insect population biology, the physical/chemical, or *abiotic,* components of the "effective environment," and we begin an examination of the biotic components of the "effective environment" by discussing the various relationships between insects and other animals and between insects and microbes. Chapter 10 is devoted entirely to relationships between insects and plants.

Populations

It is the goal of population studies to understand the composition and behavior of populations, the environmental determinants of this behavior, and ultimately to develop predictions about these populations.

Members of a given insect species are typically separated into more or less discrete groups, or *populations*. A population may also be a group of individuals of the same species delineated, for purposes of study, by a biologist.

Like individuals, populations have characteristics that can be measured and described: for example, *genetic composition* (the individuals of most natural populations vary phenotypically and genotypically), *sex ratio* (the proportion of females to males), *age composition* (most populations are composed of adults and immatures of varying ages), arrangement in space, or *dispersion* (most natural populations tend to be composed of clumps of individuals due to irregularities in the distribution of food and shelter and the fact that individuals often attract one another), *population size* (expressed as total number of individuals), *population density* (the number of individuals per unit area or volume), *biomass* (the total weight of a population), and *dynamics* (changes in numbers and/or density over time).

It has been long recognized that organisms have the capacity under ideal conditions to increase by a geometric progression (exponentially). Since this capacity for exponential increase (*biotic potential*) is never fully realized under natural circumstances, it is clear that there are factors in the environment (predators, limited resources, etc.) that act to prevent such increases. The collective action of these factors has been called *environmental resistance*. In the broadest terms, one may view the dynamics of a given population as representing the outcome of the interaction between biotic potential and environmental resistance.

Three basic approaches are used to study populations.

1. Analysis of population performance under controlled, but artificial, conditions in the laboratory.
2. Evaluation of populations in the field.
3. Development of theoretical, mathematical models that describe population dynamics.

Each of the three approaches to studying populations has some value and definite limitations. Our overall understanding of populations results from an amalgamation of ideas generated by all three approaches.

Population studies in the laboratory are of necessity too simplified and only remotely representative of what must actually occur under the complexities characteristic of field conditions. On the other hand, the complexities of natural conditions make it difficult to assess the action of particular environmental factors. Further, such studies must rely on *sampling*—studying a comparatively small number of individuals from a population with the hope that the characteristics exhibited by a sample accurately represent those of the whole population. Whether this hope is realized depends upon the adequacy of sampling. Sampling methods vary with insect species, habitat, and the kind of information to be derived from the sample. Many sophisticated sampling techniques have been developed, and elaborate statistical analyses have been made easier by the use of computers. It is necessary that population studies be made over several generations. Theoretical studies are especially valuable in developing descriptive and predictive models of population performance; however, as with laboratory studies they tend to represent oversimplifications and must rely on assumptions that may or may not apply in natural situations. Fortunately, modern computers allow evaluation of increasingly complex models.

For information on population biology, see Clarke et al. (1967), Dempster (1975), Kuno (1991), McDonald et al. (1989), Pielou (1974), Price (1984), Seber (1982), Solomon (1969), Southwood (1966, 1975), Varley and Gradwell (1970), Varley et al. (1973), Wilson and Bossert (1971), and Wilson (1975).

Relationship between Environmental Components and Populations

Population Growth

This section establishes a simple, but useful model, which helps in thinking about population growth. Assume we are evaluating the repopulation of an area from which all members of a given species have been previously removed. Because populations under ideal conditions have the capacity for exponential increase, the first step is to consider an equation that describes exponential increase. Such an increase in a population is described by the equation $dN/dt = (b - d)N$, where $dN/dt =$ instantaneous rate of increase; $N =$ population; $t =$ time; b and $d =$ average birth and death rates per individual per unit. The average birth (natality) and death (mortality) rates, b and d, can be represented as $(b - d) = r$, where r is the *intrinsic rate of increase* of a population. The population equation then becomes $dN/dt = rN$. If b is greater than d, then the population is growing; if b is equal to d, the population is stable ($r = 0$); if b is less than d, the population is decreasing in size. The graph depicted in

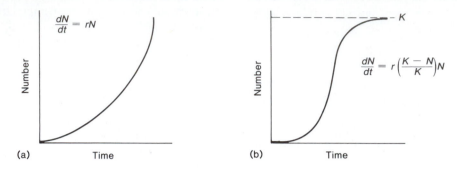

figure 9.2*a* shows the shape of a curve of the changes in the instantaneous rate of increase dN/dt that would be generated if r remained at a constant positive value. The population size increases at an increasing rate.

Under natural conditions, r does not remain constant, being subject to strong environmental influence. For example, r may well decrease with increasing population density. Competition for food and shelter, rates of predation, and invasion by parasites and parasitoids are examples of environmental factors that may increase as population density increases. Looking again at the equation for exponential growth, but with r decreasing at a constant rate as population size increases, one finds that a curve like that shown in figure 9.2*b* would be generated.

This curve is commonly called the *logistic model* of population growth and the changes in the instantaneous rate of increase are described by the equation $dN/dt = r(K - N/K)N$. It is evident that as N approaches K, the term $(K - N/K)$ approaches zero and hence the instantaneous rate of increase approaches zero, resulting in a leveling off of the population. The population size at which the leveling off occurs is called the *carrying capacity* (K) of the environment.

Although the logistic model may come close to describing population growth in some cases, it is severely limited because the assumptions on which it is based are rarely true in reality. Among these assumptions are equality of reproductive potential among all individuals of the population, an even age distribution with a constant proportion of individuals breeding all the time, reproduction uninfluenced by climate or other factors, and no changes in carrying capacity (Price 1984).

In addition to birth and death rates, emigration and immigration also influence population growth. Birth rate, death rate, and emigration-immigration rates are themselves influenced by a variety of factors. For example, weather, in particular temperature, influences all these rates. In addition, it can be argued that any factor that exerts an influence on one of these rates influences the others as well. For example, an environmental change, say in temperature, might cause higher mortality among

insects in one stage than in another or in insects of a certain age. If this age-specific mortality happened to occur in ovipositing females, one could say that both the birth and death rates were being affected.

Among the major factors that influence birth rate are average fecundity of the females, average fertility of the females, and the sex ratio. Average fecundity represents the average number of offspring that would be produced by each female under ideal environmental conditions; average fertility represents the average number of offspring actually produced, the difference between fecundity and fertility being related to such environmental factors as quality and quantity of available food, weather components, and population density. The sex ratio is the fraction of the total population that is female. In most insects the sex ratio is 0.5, but in species where parthenogenesis occurs there may be drastic deviations from this 50–50 balance of sexes. As with fecundity and fertility, the sex ratio may be influenced by several environmental factors.

The death rate is affected by such factors as adverse weather conditions, temperature extremes in particular; predators, parasites, and pathogens; accidents; low vitality; food shortage; and lack of adequate shelter (Clark et al. 1967).

The carrying capacity (K) of the environment also varies with many factors (e.g., availability of food, extent of predation).

More Regarding the Life-System Concept

Several different models have been proposed to explain the numerical behavior of insect populations. As previously mentioned, Clark et al. (1967) and more recently Hughes et al. (1984) provide a useful integrating concept (i.e., the "life system"). Our objective here is to elaborate somewhat on this concept, which is presented diagrammatically in figure 9.1.

Two kinds of "ecological events" occur as a result of the interaction of the "codeterminants of abundance" (i.e., the "subject population" and its "effective environment"): *primary* and *secondary* (figure 9.1). Primary events are those directly involved in the demographic

Table 9.1 Life Table for Second Generation of the Diamondback Moth on Early Cabbage, Merivale, Ont., 1961[a]

| Age interval | Numbers per 100 plants | Mortality | | |
		Causative factors	Per 100 plants	Percent
Eggs	1580	Infertility	25	1.6
Larvae				
Period 1	1555	Rainfall	1199	77.1
Period 2	356	Rainfall	36	10.1
		Parasitism by *M. plutellae*	52	14.6
Period 3	268	Parasitism by *H. insularis*	69	25.7
Pupae	199	Parasitism by *D. plutellae*	92	46.2
Moths	107	Sex (49.5% ♀♀)	1	1.0
Females × 2	106	Photoperiod	78	73.6
"Normal" females × 2	28	Adult mortality	20	71.4
Generation totals			1572	99.5

Trend index,[b] 0.55. Stable number rate,[c] 99.1%

[a]From Harcourt and Leroux (1967).

[b]Trend index $= \dfrac{\text{eggs laid in new generation}}{\text{eggs laid in old generation}}$

[c]Stable number rate = the mortality rate necessary for the population to remain the same size from generation to generation. Since the percent mortality for the generation represented in this table was 99.5%, it is apparent that the population was declining.

equation previously explained (i.e., births, deaths, emigrations, and immigrations). Secondary events are those that exert an influence on the magnitude, extent, frequency, or duration of primary events (quality and quantity of available food, weather components, etc.). Primary and secondary ecological events may be additive or subtractive. Processes that act in a positive way on a population are additive, for example, immigration and weather conditions that reduce populations of natural enemies. The planting of a monoculture, such as corn, is an additive process relative to species, such as the European corn borer, that thrive on this crop, as well as to insects that are parasitoids or predators of the corn-eating species. Subtractive processes (e.g., emigration and adverse weather conditions) have a negative effect on a population, causing the death of individuals and/or a decrease in the number of progeny produced. The planting of corn would constitute a subtractive process for insects that were originally present, if they could not survive in a corn agroecosystem.

Additive and subtractive ecological events may be either density-independent or density-dependent (i.e., the effects of the action of a given ecological process on a population may or may not be related to the density of that population). Subtractive density-related processes are of special interest because they may act, in some instances, as regulators of the level of abundance; that is, as a given population increases, a point is reached where further growth is inhibited by some factor.

Huffaker et al. (1984) provide a good overview of the actions of various factors in the regulation of natural populations.

Numerical Changes in Populations

As populations interact with their effective environments, there are fluctuations in population size. Some populations fluctuate irregularly, apparently in response to changes in environmental components such as food supply and weather. Other populations tend to fluctuate regularly about a mean level of abundance. As previously mentioned, this may be a reflection of the action of some subtractive density-related process that is serving to regulate the level of abundance. Insects with a reproductive period restricted to a particular time or times of year tend to display regular, seasonal population peaks associated with reproduction.

It has become more and more apparent that, in a given life system, certain "key" ecological processes with age-specific effects determine major population trends, whereas other processes exert somewhat lesser effects. Elaborate "multifactor" studies have been carried out for a number of insect species. The objective of these studies has been to derive "life tables" that contain information as to population densities at different times during the life cycle and age-specific information regarding the "key" ecological processes that account for mortality. Such a life table is presented in table 9.1. The development of a truly representative life table requires the sampling of a population over many generations.

For additional information on the preparation and use of life tables, see Harcourt (1969) and Hughes, Jones, and Gutierrez (1984).

Another approach that has been used in which preliminary studies have indicated a few key influences involved in a life system is the *key-factor method*. In this type of study, population density is measured at only one point in each generation. This approach requires much less sampling and enables investigators to concentrate on critical periods of time in the life cycle when the "key" process or processes are active without spending much time on other periods.

Insects and the Abiotic Environment

In this section, we consider the influences, both direct and indirect, that abiotic factors (i.e., the physical and chemical features of the environment) exert on insect populations.

Abiotic factors include temperature, moisture, light, and several other physical/chemical parameters. Although these factors will be treated individually in sequence, it should be borne in mind that at any given moment in an insect's life all factors are acting in concert.

The combined action and influence of factors such as atmospheric temperature, pressure, moisture, and movement at any given moment in time result in what we call *weather*. It varies continually throughout days, weeks, months, and years and exerts an influence on insect abundance, longevity, rate of development, and so on, from one year or season to the next. *Climate,* on the other hand, is the average course or condition of the weather in a locality over a period of several years. Weather changes rapidly, often violently, whereas climate tends to remain pretty much the same or change very slowly over a period of many years. The main elements of weather are temperature, moisture, and light, although several other physical environmental factors are known or thought to exert a degree of influence on insects. Various materials and methods used in measuring and interpreting physical aspects of the environment are described in Platt and Griffiths (1964). Insect behavior in relation to changes in the environment is discussed in chapter 8.

A thought-provoking discussion of the effects of weather on insects may be found in Wellington and Trimble (1984).

Temperature

Insects are basically poikilothermic—that is, their body temperature tends to be the same as ambient temperature. However, this does not necessarily mean that an insect's body temperature is always the same as that of the environment.

For every insect species there is a fairly well-defined range of temperature within which it is able to survive. Exposure to temperatures above the high or below the low extremes of this range results in death. The range of tolerable temperatures varies from species to species, within a species, and with the physiological state of an individual. Thus there are times or stages during the life cycle when an individual may be able to survive exposure to high or low temperatures that at other times would kill it. For example, many insects are able to survive much lower temperatures in the fall and winter than in the spring or summer. Tropical species are generally less tolerant of cold than those in temperate zones. Terrestrial insects usually have a somewhat wider range of temperature tolerance than do aquatic insects, and, not surprisingly, the range of temperature variation in terrestrial habitats is usually substantially greater than that in aquatic habitats.

The range of survival relative to temperature for most insects probably lies somewhere between 0° C and 50° C, although it is likely that no one species can thrive throughout this entire range. The optimal temperature range for most species is 22° C to 38° C (Taylor 1981). There are, however, exceptional species that are able to survive at temperatures well beyond these ranges. For instance, some dipteran larvae apparently thrive at temperatures of 55° C or higher, while certain species of beetles go through their entire life cycle in ice grottoes at temperatures slightly below and slightly above 0° C (Andrewartha and Birch 1954). The optimal temperature range for one species of rock crawlers, *Grylloblatta* sp., is 3° C–12° C (Taylor 1981). These insects live at high elevations (e.g., the edges of glaciers). The firebrat, *Thermobia domestica,* can live indefinitely at temperatures of 42° C and higher (Andrewartha and Birch 1954). It typically inhabits such places as ovens, hotwater pipes, and similar "hot" environments.

If insects of a given species are exposed to a temperature gradient, they will move until they reach the "preferred temperature," at which point they will tend to congregate. Near the upper and lower tolerable limits of temperature, insects become dormant. In the range between these limits, they are active.

The actual cause or causes of death at the limits of the temperature range are not clear. At the lower limit the submicroscopic structures of cells may be disrupted by the formation of ice crystals or the metabolic balance may be thrown off. At the upper limit protein denaturation, metabolic imbalance, disruption of ordered molecules (as, for example, in the wax layer of the epicuticle), and desiccation are likely to be involved.

The phenomenon of insect cold-hardiness has been reviewed by Salt (1961), Downes (1965), Asahina (1966), and most recently and comprehensively by the chapters in Lee and Denlinger (1991). Many insects that become dormant in temperate regions are able to survive low temperatures for considerable periods of time. Most of these insects are dormant in a stage that is more cold-hardy than the preceding one. Some are capable of long exposure to low temperatures and display a certain amount of cold acclimation, but succumb to freezing of the body fluids. For example, *Aedes aegypti* larvae reared at 30° C are killed by exposure to −0.5° C for 17 hours, but survive such an exposure if preconditioned for 24 hours at 18° C or 20° C (Bursell 1974a).

Other insects are able to avoid freezing because they can withstand supercooling (i.e., being cooled below the point of freezing, but without freezing actually taking place). Individual species vary as to the temperature to which they can supercool. Cryoprotective compounds such as glycerol, sorbitol, and erythritol have been found in the tissues of overwintering insects (Baust and Morrissey 1977). Salt (1959) showed that glycerol depresses the hemolymph freezing point of *Bracon cephi* larvae as much as 17.5° C and that this compound plays a key role in cold-hardiness. Glycerol begins to appear in the tissues in correlation with the advent of dormancy (see "Behavior and the Fluctuating Environment" in chapter 8) and usually disappears rapidly when the insects come out of this state (Salt 1961). A number of insects (e.g., many caterpillars) are able to survive the formation of ice within their bodies. Members of this group of "freezing-tolerant" insects have been found to possess higher concentrations of cryoprotective compounds in their hemolymph. Apparently these compounds act to prevent tissue destruction by influencing how and where ice crystals form as well as depressing the hemolymph freezing point.

The lethal effects of high temperatures are difficult to study because of the influence of moisture changes. The interaction of moisture and temperature is considered later. It has, however, been established that lethal temperatures may vary depending upon the temperature to which an insect has been exposed previously. Such high temperature acclimation has been demonstrated in *Calliphora and Phormia* (Diptera, Calliphoridae), and others (Bursell 1974a). Acclimation to both high and low temperatures may or may not occur in the same species. Like cold acclimation, high-temperature acclimation is likely to be of value to insects under natural circumstances because seasonal and daily high temperatures are usually preceded by a gradual transition from somewhat lower temperatures. Acclimation at both ends of an insect's tolerable range of temperatures may be looked upon as promoting the insect's survival against the effects of extreme daily and seasonal temperature fluctuations.

Temperature also affects the duration of life, for example by influencing the rate of utilization of food reserves when food is present in limited quantities (Bursell 1974a). This is particularly evident in blood-sucking insects such as the tsetse flies, *Glossina* spp. (Diptera, Muscidae), which depend on stored food reserves between meals. Increases in temperature shorten the survival period between meals. Obviously, if a fly uses up its reserves from one meal before it is able to obtain another, it will perish.

As with lethal limits of temperature, insects also have definite tolerable ranges of temperature relative to reproduction and development beyond which neither will occur. For example, the temperature range in which the eggs of the beetle *Ptinus* (Coleoptera, Ptinidae) will develop is between 5° C and 28° C (Bursell 1974a). *Pediculus* (Anoplura, Pediculidae) fails to lay eggs below 25° C (Wigglesworth 1972). For a given species the range in which development will occur is probably somewhat broader than that in which reproduction will be successful. Within the tolerable ranges, egg development, oviposition rate, and the rate of larval and pupal development usually increase with increasing temperature. For example, the duration of pupal life of the mealworm beetle, *Tenebrio mollitor* (Coleoptera, Tenebrionidae), is decreased by 180 hours (from 320 to 140) as the temperature is increased by 12° C (from 21° C to 33° C) (Wigglesworth 1972).

The specific ranges of tolerance and influence of temperature on rates of reproduction and development vary among members of the same species. In addition, the span of ranges varies from species to species. Within each range of tolerance is an optimum zone in which the rates of reproduction and development are maximal. For example, the oviposition rate of *Toxoptera graminum* (Homoptera, Aphididae) increases with increasing temperature to a maximum of approximately 25° C and then falls off. Most laboratory studies of the effects of temperature on reproduction and development have been carried out at constant temperature. The results of such studies do not necessarily reflect what would occur under uncontrolled field conditions. Periodically fluctuating temperature, which is characteristic of field conditions, tends to induce higher rates of development than would occur at constant temperature. For example, "grasshopper eggs kept at a variable temperature showed an average acceleration of 38.6% and nymphs an acceleration of 12%, over development at comparable constant temperature" (Odum and Odum 1959).

Because temperature exerts such a strong influence on the rate of development, it influences the timing of the various life stages of an insect. Knowledge of the timing and duration of these life stages is of great importance in applied entomology. If the temperature history in a given region is known along with the quantitative effects of temperature on the development of a given pest insect, predictions as to when to expect the destructive stage to be reached, or perhaps the stage most vulnerable to a particular control measure to be reached, can be made and appropriate, timely control measures applied. Under usual circumstances, the temperatures in a given region do not remain constant, but fluctuate widely on a daily basis, from day to day and from season to season.

Figure 9.3 Tolerable temperature ranges for various aspects of a hypothetical insect's life. The left-hand thermometer of each pair registers the maximum tolerable temperature; the right-hand one, the minimum tolerable temperature. (*a*) Survival. (*b*) Mobility. (*c*) Development. (*d*) Reproduction.

Redrawn from Rolston and McCoy, 1966.

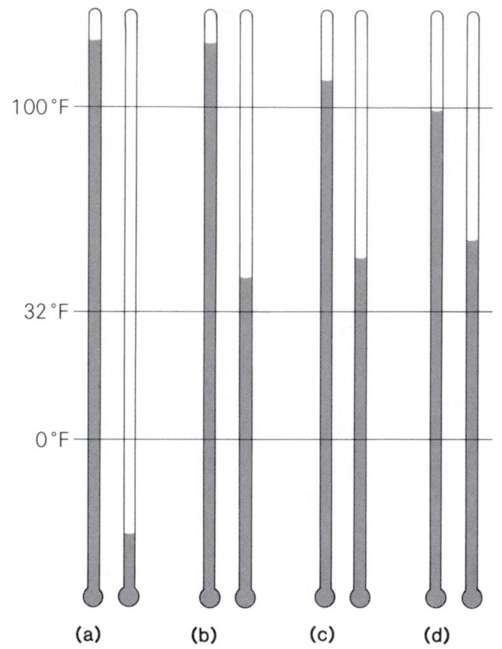

A useful method for the prediction of the timing of developmental events (eclosion, larval molts, pupation, adult emergence, etc.) is the use of *degree-days*. This topic is explained in chapter 16.

Figure 9.3 illustrates the idea of tolerable temperature ranges relative to survival, mobility, development, and reproduction.

The distribution, horizontal and vertical, of an insect species is often greatly affected by temperature. In temperate zones the northern extreme of a given insect's distribution is commonly determined by low-temperature extremes. When northern limits are determined in this manner, there is usually a zone somewhat below the extreme limits in which the overwintering stage is killed but which is repopulated during the warm season. Proceeding southward from such an area, a greater and greater percentage of the overwintering individuals survive. For example, the corn earworm, *Heliothis zea* (Lepidoptera, Noctuidae), in eastern North America must become completely reestablished during the warm season each year in Canada and to a progressively lesser extent in the United States proceeding southward (Andrewartha and Birch 1954). The ability to become cold-hardy also plays an important role in determining the northernmost limits of distribution. The number of generations per year often varies between the northern and southern regions of an insect's range. For example, the corn earworm completes a generation approximately every 36 days as long as weather permits (Rolston and McCoy 1966). Thus in its northern range it may only complete two or three generations per year, whereas in the southern portion it may complete several more, breeding the year round. One would expect a gradation in the actual number of generations between the northern and southern extremes of the range.

Temperature also affects, to a greater or lesser extent, the rate of dispersal of an insect species.

For a review of the influence of temperature on aquatic insects, see Ward and Stanford (1982).

To this point we have been discussing the various ways temperature influences insects. We now want to consider the actual temperature of an insect and the factors that may influence it. We have established the idea that the temperature, or any of several other parameters, may vary substantially even within a small area (e.g., different parts of a plant). Likewise, the temperature of an insect in one part of a habitat may not be anywhere near the same as one in another and may not be the same as the ambient temperature. On the other hand, the temperature of an insect under controlled laboratory conditions usually reflects the ambient temperature.

The temperature of an insect depends on the sources of heat gain and loss operating under a given set of circumstances. The major sources of heat gain are solar radiation and metabolic heat. Solar radiation may cause the temperature of an exposed insect to be significantly different from the ambient temperature (figure 9.4). The effect of solar radiation on the temperature of an insect is influenced by such factors as size, larger insects being more affected than smaller ones; color, darker colors absorbing more radiation than lighter ones; shape, the more surface directly exposed to radiation, the greater the absorption; and orientation with respect to the sun, some insects orienting in such a way that a large or small amount of body surface is exposed.

Metabolic heat results from the breakdown of complex organic molecules. Part of this energy is stored in the high-energy bonds of ATP; the remainder is released as heat. In the absence of solar radiation, this is the sole source of heat and can be quite significant in heat balance, particularly during flight or in clusters of gregarious forms. Sources of heat loss from an insect include evaporation, convection, conduction, and long-wave radiation. Evaporation of water from an insect has a cooling effect since heat is required to propel a molecule of water from the body surface. Evaporation is the major cause of heat loss in the absence of solar radiation. You will recall from chapter 2 that the rate of evaporation of water from an insect is dependent partly on the size of the insect.

Figure 9.4 Effect of radiation on the body temperature of selected insects. (*a*) Thermometer painted black. (*b*) Air temperature. (*c*) *Bombus* (bumble bee). (*d*) *Xylocopa* (carpenter bee). (*e*) *Apis* (honey bee). (*f*) *Anisoplia* (beetle). (*g*) *Asilus* (robber fly). (*h*) Damselfly. (*i*) Butterfly. (*j*) Butterfly. (*k*) Damselfly. Light gray, sunlight; dark gray, shade.

Based on data from Masek Fialla, 1941.

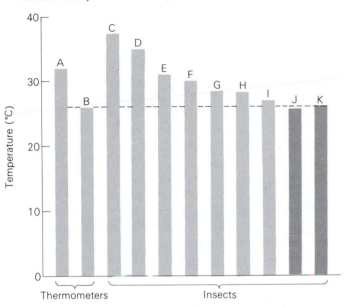

Because smaller insects have a larger ratio of surface area to volume, they have a greater tendency to lose water through evaporation than larger ones. The other causes of heat loss—convection, conduction, and long-wave radiation—are significant in the absence of solar radiation at which time the temperature differential between an insect and its surroundings is usually greater than in the presence of solar radiation. Air movement may accentuate the heat loss by contributing to the maintenance of a steep gradient between an insect and its surroundings. Dense coverings of hairs and scales, on the other hand, may serve as an insulating layer and retard heat loss.

As discussed in chapter 8 (see the section "Behavior, Temperature, and Humidity"), many insects probably exert a degree of control of heat loss and gain (i.e., thermoregulate). Heinrich (1981) provides an excellent discussion of insect thermoregulation from ecological and evolutionary points of view.

Moisture

The water content of insects varies from less than 50% to more than 90% of total body weight (Wigglesworth 1972). Variation occurs both between different species and between different life stages of the same species. Soft-bodied insects such as caterpillars tend to have comparatively large amounts of water in their tissues, whereas many insects with hard bodies (i.e., relatively thick cuticles) tend to have somewhat lesser amounts. Active stages commonly have a higher water content than dormant stages. In most instances it is critical that water content be maintained within certain limits, which are influenced by several other environmental factors (e.g., temperature, pressure, air movement, available surface water). If the limits of tolerance under a given set of circumstances are exceeded, an insect either perishes or many of its activities are seriously impaired.

Environmental moisture factors of major significance are precipitation, humidity, condensation, and available surface water. Rainfall is the most common and widespread form of fluid precipitation. Snow is the most common form of solid precipitation; hail and sleet are less common forms. The annual and seasonal amounts of precipitation are primarily determined by the movements of large masses of air and by topographical characteristics. Thus there are very "wet" and very "dry" regions and innumerable gradations between. Humidity refers to the amount of water vapor in the air and depends on temperature and atmospheric pressure.

Condensation (dew, fog, and white frost) occurs when the atmosphere becomes saturated with water vapor. Saturation is the result of the relative humidity approaching 100% or the temperature dropping below the dew point (the point at which the relative humidity becomes 100%).

Available surface water is related to all the other moisture factors and to the nature of the substrate (soil, leaf surface, bark, and so on). It is well known that soils vary in their water-holding capacity and the rapidity with which water runs off or soaks in.

All these moisture factors influence the water balance of terrestrial insects. What, then, are the environmental factors that influence the water balance of aquatic insects? The "wetness" or "dryness" of an aquatic environment is a function of the osmotic pressure. Thus insects living in fresh water must cope with a comparatively "wet" environment; those in brackish and salt water (e.g., many of the salt-marsh forms, such as the mosquito *Aedes sollicitans*) are living in a very "dry" environment (see "The Excretory System" in chapter 4).

Compounding the water-balance problem in insects is the fact that under other than carefully controlled laboratory conditions the moisture factors continually change. Obviously, an insect living in a given locale must be able to survive the extremes of these changes. As with temperature and the other weather parameters, one must think in terms of microenvironments. For example, the amount of rainfall measured at the edge of a forest hardly reflects the actual amount of water reaching an insect living in a tree hole or on the underside of a leaf. All the moisture factors vary both temporally and spatially. For example, relative humidity varies with location, time of

Figure 9.5 Changes in relative humidity (*a*) and water vapor pressure (*b*) at different levels above the ground and at different times of day.

Redrawn from Chauvin, 1967.

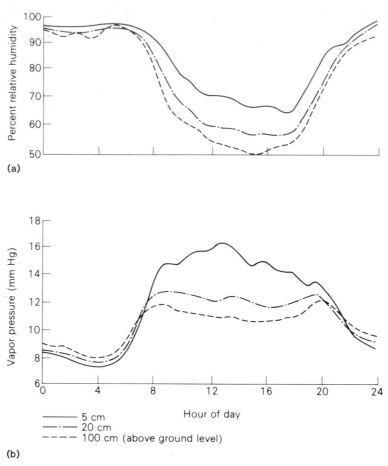

day or year, topography, vegetation, and so on, and commonly tends to be comparatively high during the night and lower during the day. It may also be different at different heights above the ground (figure 9.5).

As with temperature, there is an optimal moisture range in which a given species thrives. Mortality may occur under conditions of excessively low environmental moisture content for the active stages of most insects, and under conditions of excessively high moisture content for many. Death under very dry conditions is generally due to excessive water loss, but it can also result from indirect causes. For example, reduction of the moisture content of food can interfere with an insect's feeding, as would be the case in an insect with piercing-sucking mouthparts feeding on wilted leaves. Under conditions of excessive moisture the causes of death are more variable and may be direct or indirect. For example, drowning may be the result of excessive moisture, as is the case with overwintering pupae of the moth *Heliothis zea* during wet years in the southeastern United States (Andrewartha and Birch 1954).

Survival is indirectly affected by very wet conditions that favor the spread of viral, fungal, and bacterial diseases and by any negative effects the excessive moisture might have on the food of a given insect species. For example (N. A. S. 1969),

The fall webworm, *Hyphantria cunea,* the armyworm, *Pseudaletia unipuncta,* and the gypsy moth, *Lymantria dispar,* succumb most readily to viruses when the weather is warm and the humidity is high. Whether high humidity affects the host or the development of the pathogen is not clear. However, many viruses do not develop rapidly unless temperatures of 21 to 29.4° C and relative humidities of 50 to 60% are experienced.

If excessive moisture does not kill an insect, it may seriously affect the length of its life. For example, newly emerged adult migratory locusts, *Locusta migratoria,* live longer the lower the humidity (Andrewartha and Birch 1954). Because environmental moisture content may determine survival, it is commonly a major factor along with

Figure 9.6 Influence of humidity on various aspects of a hypothetical insect's life. (*a*) Insect not harmed by high humidity. (*b*) Insect adversely affected by high humidity. Zone 1, lethal dryness; Zone 2, favorable moistness; Zone 3, lethal wetness. *Redrawn from Andrewartha and Birch, 1954.*

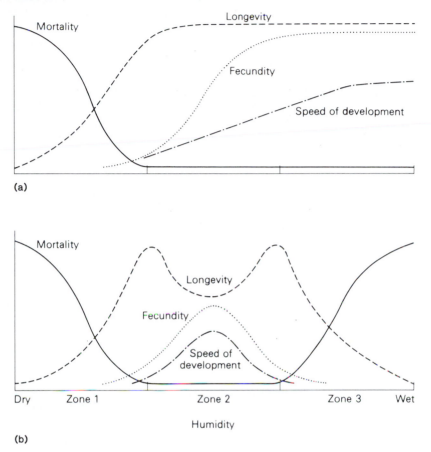

temperature in determining the geographic distribution of an insect species. For example, according to Rolston and McCoy (1966),

> The High Plains grasshopper reproduces and develops in an area of about 50,000 square miles, and from this center the adults may spread over an area perhaps twice as large. The region of endemic infestation lies in the short-grass belt of the High Plains where the average winter temperature generally falls between 28° and 38° F and the average annual precipitation ranges from about 15 to 18 inches. That the High Plains grasshopper flourishes in only a part of the short-grass belt indicates that climatic conditions, rather than host availability, limit the range of the insect.

Extremes of environmental moisture content directly influence many of the activities of insects, including feeding, reproduction, and development. Spruce budworm larvae stop feeding when the air becomes saturated with water (Graham and Knight 1965), and the tsetse fly, *Glossina tachinoides,* does not feed on its vertebrate hosts when the relative humidity is above 88% (Andrewartha

and Birch 1954). Newly emerged adult migratory locusts do not produce eggs below about 40% relative humidity (Andrewartha and Birch 1954). Generally low humidities adversely affect the rates of oviposition, which increase as humidity increases. The rate of development may be decreased by extremes of moisture content or development may be halted altogether. Under very moist conditions silkworm larvae fail to pupate (Andrewartha and Birch 1954). According to Bursell (1974b) the incubation time for eggs of the spider beetle *Ptinus,* under a constant temperature of 20° C, is 15 days at 30% relative humidity and 10 days at 90%. He further points out that generally higher humidities are more favorable for embryonic stages than low humidities. Two sets of hypothetical curves summarize the points made in this and the preceding paragraphs (figure 9.6).

Heavy rainfall (and hail) must surely be destructive to insects by direct mechanical damage (e.g., aphids; Hughes et al. 1984) or by drowning.

The effects of environmental moisture content are often strongly modified by other weather factors. Temperature and moisture, in particular, interact to a large

Figure 9.7 Interrelationship of temperature and humidity as they affect the rate of development of a hypothetical insect. (*a*) Region of most rapid development. (*b*) Region of favorable development. (*c*) Region of retarded development. (*d*) Region of no development.
Redrawn from Graham and Knight, 1965.

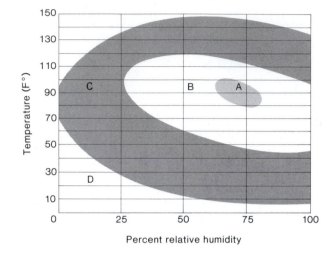

extent in their effects. Thus temperature exerts a relatively great effect on insects at the extremes of moisture conditions and vice versa. For example, the boll weevil is more tolerant of higher temperatures at comparatively low humidities than at high humidities (Graham and Knight 1965). Another good example of the interactions of temperature and moisture is their combined effect on the rate of development of insects (figure 9.7). As mentioned earlier, air movement, atmospheric pressure, availability of surface water, and other factors may also influence the effects of moisture conditions on insects.

Insects are adapted in a number of ways and to varying degrees to cope with changes in environmental moisture conditions. Dormant stages are often well suited for exposure to drought. The eggs of many species of mosquitoes (in the genera *Aedes, Psorophora,* and others) can withstand prolonged drying in air (Clements 1963). Many insects become dormant in response to different environmental conditions and in this state are able to withstand long periods of drought. For example, the potato beetle, *Leptinotarsa decemlineata,* becomes quiescent in response to dryness and can survive in this resting, desiccated state for months. Such a response ensures that when eggs are laid, the larvae will be exposed to environmental conditions more favorable for survival (Andrewartha and Birch 1954). Hinton (1951, 1960) describes a chironomid (Diptera, Chironomidae) larva, *Polypedilum vanderplanki,* that is able to survive almost complete dehydration for several years. These insects show no visible signs of metabolic activity and are described as being in a state of *cryptobiosis.*

Most insects offset the influence of dryness and other factors promoting water loss by drinking water or taking it in with food. Time is important in the survival of active insects under adverse moisture conditions. Many can survive extensive desiccation for extended periods. Some insects that live under extremely dry conditions, such as mealworm beetle larvae (*Tenebrio molitor,* Tenebrionidae, Coleoptera), are able to use metabolic water (i.e., water that becomes available as a result of the metabolic breakdown of food materials). Oxidation of 0.5 ug of fat requires 1 mm^3 of oxygen and produces 0.56 ug of water (Gordon 1984). Insects such as some Thysanura (silverfish and relatives) are able to absorb moisture directly from the atmosphere (Wigglesworth 1972). See chapter 2 for a discussion of the role of cuticle in water balance, chapter 4 for examples of several adaptations that help insects conserve water, and chapter 8 for behavior associated with variations in humidity. Cloudsley-Thompson (1975) and Tauber et al. (1984) discuss adaptations associated with arid environments.

Light

Light is of direct importance to insects more as an environmental point of reference than as a survival factor because its parameters (photoperiod, illuminance, wavelength, etc.) are more or less constant and it is seldom, if ever, directly lethal under natural conditions. In the nonequatorial region of the earth there is a regular change in photoperiod due to the tilt of the earth's axis of rotation 23½° from vertical to the imaginary plane that passes through the sun and the earth's orbit. Regular change in photoperiod serves as an annual clock for insects and is used by many to maintain synchrony with the seasons and their host plants. Photoperiod is one of the major stimuli that induces diapause (see "Behavior and the Fluctuating Environment" in chapter 8). The daily cycle of dark and light with crepuscular (dawn and dusk) periods in between also serves as a clock by which the feeding, mating, etc., are regulated (see "Periodicity in Behavior" in chapter 8). As with temperature and moisture, the reactions of insects to photoperiod and other light parameters vary both among different species and among different life stages of the same species.

Different wavelengths of reflected light are commonly used by plant-feeding insects in host location. The position of the sun and degree of polarization of light in different parts of the sky are important to many insects in orientation and navigation. More information on the response of insects to various light parameters may be found in chapters 6 and 8. The effect of light on both aquatic and terrestrial plants may indirectly influence the activities of insects. For example, the amount of light reaching submerged aquatic vegetation will affect oxygen-generating photosynthesis and in turn affect the oxygen concentration in the water, which then influences aquatic insects.

Cloud cover may diminish light intensity and thereby reduce the activity of some insects. For example, flight and oviposition by the cabbage butterfly are reduced by overcast skies, and this may diminish population growth (Dempster 1967, Gossard and Jones 1977).

Other Factors

Other environmental factors that under some circumstances influence insects include currents in air and water, gases dissolved in water, air composition, electricity, ionizing radiation, and soil composition.

Air and water currents are determined to a large extent by physiographic conditions. Air movement is modified by trees and other vegetation and by anything else that may block or redirect it. Currents in water are influenced by such factors as water volume, slope of stream bed, and temperature differences between the surface and various depths. Wind is a very effective agent in the distribution of insects—aphids, leafhoppers, and others being blown for hundreds or even thousands of miles (see "Dispersal and Migration" in chapter 8).

Air movement may be directly responsible for the death of insects in two ways. First, severe wind and heavy rain together may cause mortality. Second, movement of air above a surface where evaporation is occurring (e.g., insect cuticle) increases the gradient of water vapor concentration and hence tends to increase the rate of evaporation. Other factors being constant, the rate of evaporation is proportional to air movement. On the other hand, air movement may be beneficial if humidity is high.

High winds probably inhibit the activity of most flying insects. In response to strong air movements, many insects tend to hunker down and hold on. Strong downdrafts over marshes and lakes can result in flying insects dropping to the water surface, making them vulnerable to the adhesive forces of water and potential predation by water-dwelling predators.

Air movements can influence insect distribution and dispersal both locally and on a grand scale. For example, locally, airborne insects, especially those affected by relatively weak air currents, tend to congregate where there are updrafts. Prevailing winds can exert an influence on the pattern of movement of migrating insects. Some, for example the Monarch butterfly, have been shown to make efficient use of rising air currents, thermals, by soaring and thus conserving energy (Gibo and Pallett 1979). For recent discussions of the effects of atmospheric movements on insects, see Wellington and Trimble (1984) and Drake and Farrow (1988).

Water currents often determine which species of insects will live in a given area. For example, the various genera of mayflies (Ephemeroptera) may be classified into still-water or rapid-water forms (Needham, Traver, and Hsu 1935). The legs and bodies of these insects are appropriately adapted (e.g., legs capable of clinging and streamlined bodies associated with fast-moving water). Black fly larvae fasten themselves to stones or other stationary material in the water. Several caddisflies (Trichoptera) attach their cases to submerged objects. Many aquatic insects (e.g., mosquito larvae) are unable to survive in moving water. Another important aspect of currents in the aquatic environment involves the circulation of dissolved gases, salts, and nutrients.

The insect fauna in a given location are often determined to a large extent by the amount of dissolved oxygen. For instance, caddisfly and mayfly larvae may be found under conditions of relatively high oxygen concentration, midge and black fly larvae (Diptera; Chironomidae and Simuliidae, respectively) at somewhat lower concentrations, and certain mosquito (Diptera, Culicidae) and other fly larvae at very low concentrations (Usinger 1956). Dissolved oxygen is determined by water movement, wind, water splashing, photosynthesis in aquatic plants, and so on. Other dissolved gases that may be of importance are carbon dioxide (a highly soluble waste product of cellular respiration) and nitrogen.

In the absence of pollutants the gaseous composition of air is remarkably constant and is probably not a significant limiting factor for terrestrial insects and other animals. However, the gas content in cavities in flowers may be quite different from the surrounding air, and a large number of insects live in such cavities (Chauvin 1967). Further, the composition of air in the middle of a thick canopy of vegetation is not necessarily the same as in open air, nor is it necessarily the same throughout the day because temperature and the photosynthetic activities of plants vary during the course of a day. For example, measurement of the carbon dioxide concentration above fields of wheat and clover at different times of day revealed that carbon dioxide decreased toward the middle of a day and reached a maximum between midnight and 6 A.M. (Chauvin 1967).

Air pollutants, such as fluoride, sulfur, ozone, lead, and dust may have significant direct and indirect effects on insect populations (Alstad and Edmunds 1982; Riemer and Whittaker 1989). Direct effects involve the toxic action of pollutants on insects and the associated decline in insect populations. Indirect effects may cause an increase or a decrease in a given insect population. For example, if parasitoids and predators are more susceptible to a given pollutant than their host/prey insects, the latter may experience an increase in population size. The classic example of natural selection, industrial melanism in moths, was brought about by the action of soot causing the darkening of light-colored lichens and ultimately a decline in the light-colored moths that rested cryptically upon them. The darker moths experienced a population increase because they were more successful at avoiding predatory birds. See boxed reading 10.1 for additional information on the effects of air pollution on insects.

Electrical factors have seldom been taken into account, although they may have a direct effect on insects (Folsom and Wardle 1934). Under natural conditions, the ionization of air and atmospheric potential may vary and may affect certain activities of insects. For example, the ionization of air has been found to modify the flight activity of a blow fly, *Calliphora* sp. (Chauvin 1967). An increase in the number of ions causes a temporary increase in activity; for example, flight in *Drosophila* is abruptly reduced for a short period of time by sudden exposures to a potential gradient of 10–62 volts/cm (Chauvin 1967). Under natural conditions the atmospheric potential falls rapidly between the ground and a few meters above the ground (Chauvin 1967).

Electromagnetic radiation bombards the earth continuously. Radiation in the visible and near-visible region plays many important roles in life processes; for example, it affects photosynthesis and is involved in vision. Radiation of shorter, more energetic wavelengths also influences living organisms. This highly energetic radiation arises from cosmic, solar, and terrestrial sources. Most radiation from extraterrestrial sources is absorbed or reflected by the atmosphere and deflected by the magnetic fields, but a significant amount gets through nonetheless. Cosmic radiation probably varies considerably from time to time as a result of changes in the earth's magnetic fields, the occurrence of nearby supernovae (exploding stars), and so on. Such changes in the intensity of cosmic radiation may have played (and may continue to play) a major role in biological evolution by inducing drastic increases in mutation rates at various times during the earth's history. Most terrestrial sources of highly energetic radiation occur in trace amounts throughout the earth's crust. However, since the beginning of this century, human beings have learned how to release immense quantities of radiation by "splitting the atom." The inherent dangers of this activity are well-known.

Entomological interest in high-energy radiation stems from

1. The usefulness of radiation as a basic research tool, for example, radioisotopes are used as tracers in the study of biochemistry, and radiation is used to induce physiological, morphological, and genetic changes with the hope of further understanding biological processes.
2. The usefulness of insects as model biological systems in evaluating the harmful effects of radiation.
3. The applied value of high-energy radiation in inducing sterilization or deleterious genetic changes in insect pests (see "Genetic Control" in chapter 15). For information regarding the influence of high-energy radiation on insects, see Grosch (1974).

Soil composition may have an effect on insects that burrow and construct nests in it (e.g., ants and termites).

For example, the clay content in certain soils may have an adverse effect on termite activities (Lee and Wood 1971). On the other hand, soil insects and other soil arthropods exert a profound effect on the formation and nature of various soils (Chauvin 1967, Schaller 1968, Seastedt 1984).

Insects and the Biotic Environment

In this section and the following chapter, we examine the diverse relationships between insects and the biotic components of their environment. Relative to a given insect, these biotic components include individuals of the same as well as different species (i.e., both intraspecific and interspecific interactions).

Intraspecific Interactions

Interactions between members of the same species may be directly beneficial with some sort of cooperation involved. For example, large aggregations of individuals may by their overall appearance deter predation (Vulinec 1990). Intraspecific cooperation reaches its zenith among the various groups of social insects (see chapter 8, "Insects in Groups").

The ways in which members of the same species interact are related in part to population density. There seem to be advantages associated both with relatively low and relatively high densities. Examples of circumstances in which high population density may be advantageous include mate finding and survival of potential predation. In a situation in which the probability of finding a mate is decreased owing to low population density, the result would be fewer fertilizations and hence a lower average fecundity.

Whether a given population density is advantageous or disadvantageous is relative and depends upon other environmental influences. This is particularly apparent with regard to such available resources as food and shelter. The maximum population of a given species an area can support depends on the abundance of these resources in that area. When resources are limited, *intraspecific competition* may become increasingly pronounced as the maximum population is approached and may result in the death or emigration of individuals. Thus, intraspecific interactions may play a major role in the regulation of population size. However, other environmental influences may serve to keep a population well below a level in which competition for limited resources would occur.

Interspecific Interactions

Interspecific interactions may be conveniently divided into competition, symbiosis, predator-prey interactions, herbivore-plant interactions, and indirect interactions. Herbivore-plant interactions as well as all other insect-plant interactions are considered in chapter 10.

Interspecific competition comes about when the needs of two or more different species for a given resource (food, shelter, etc.) coincide—when their niches overlap. In situations in which two species with essentially identical life needs are brought together, one species may be expected to have a competitive advantage and eventually to eliminate the other. This is the phenomenon of *competitive displacement* (or *competitive exclusion*). The segregation of the different species in an ecosystem into different niches is explained on the basis of the operation of competitive displacement. A good example of competitive exclusion is the outcome when two species of flour beetles (*Tribolium castaneum* and *T. confusum*) are placed in the same container of flour (Park 1962). If the container is maintained under conditions of high humidity and temperature, *T. castaneum* invariably wins out. However, maintaining the beetles under conditions of low humidity and temperature results in a "victory" for *T. confusum*. In the absence of the other, either species can be maintained indefinitely under wet and warm or dry and cool conditions.

Detailed treatments of interspecific competition and the phenomenon of competitive displacement, including theoretical considerations and extensive literature citations, may be found in Lawton and Hassell (1984) and Price (1984) and also in the general ecology texts cited at the beginning of this chapter.

Symbiosis is used here to mean a close association between two different species. This association may be mutually advantageous (*mutualism*), disadvantageous for one of the species (*parasitism*), or advantageous for one without harm to the other (*commensalism*). General treatments of the phenomenon of symbiosis can be found in Ahmadjian and Paracer (1986), Goff (1985), Boucher (1982, 1985), Margulis (1981), Starr (1975), Schmidt and Roberts (1985), and Smith and Douglas (1987).

There are many examples of mutualistic associations between insects; one is seen when ants actively care for and protect aphids, which in turn excrete honeydew that is ingested by the ants (Way 1963). The relationships between fungus-growing ants and fungi (see "Feeding Behavior" in chapter 8 and the next section of this chapter) are good examples of mutualism between insects and other kinds of organisms. Additional examples include the relationships between microbes and insects relative to digestion and nutrition (see chapter 4). Dasch, Weiss, and Chang (1984) provide an excellent overview of ". . . procaryotes that are regarded as true insect endosymbionts, because their association with their hosts is so intimate that it is thought to be beneficial to both insect and symbiont." A review that deals specifically with the intracellular symbiotes (i.e., endocytobiosis) of the Hemiptera and Homoptera is provided by Houk and Griffiths (1980). Schwemmler and Gassner (1989) includes several detailed treatments of the endocytobiotic relationship found among insect groups.

Among the organisms that parasitize insects are insects themselves (figure 9.8; see also "Feeding Behavior" in chapter 8), microorganisms, mites, and nematodes. In some instances microbes and nematodes parasitize both insects and vertebrates. In these cases the insects may act as *vectors,* carrying parasites to their vertebrate or plant hosts (Harwood and James 1979; Maramorosch and Harris 1979; Kettle 1987; Mayo and Harrap 1984; Reeves 1990; Fulton, Gergerich, and Scott 1987; Pfadt 1985).

Well over 1,000 microbes, most of them pathogens (disease-causers), have been described in insects, and new ones are being added regularly. These microbes may gain entrance into insects orally or via wounds in the integument, or they may be capable of actively penetrating the integument (particularly fungi). In some instances pathogens that have gained entrance may be passed *transovarially* (via the egg) to offspring, as are certain arboviruses in mosquitoes.

Interest in pathogenic microbes associated with insects has largely been focused on their potential use in biological control (see chapter 15). During the past 30 years or so, insect microbiology has grown rapidly as an entomological specialty. For information on this topic, see Steinhaus (1949, 1963), Cantwell (1974), Fuxa and Tanada (1987), and references cited under "Biological Control" in chapter 15.

Many species of bacteria have been identified as insect pathogens. Their pathogenicity for insects is usually due to the production of host-specific toxins. Among the bacteria that infect insects, the best known is *Bacillus thuringiensis,* a spore-former that has been used with some success in biological control. It is pathogenic for nearly 200 species of pest insects, particularly Lepidoptera, and has been developed as one of the major agents used in the biological control of insect pests. The variety *B. thuringiensis israelensis* is widely used against mosquito larvae. Another member of the same genus, *Bacillus larvae,* causes American foulbrood, a serious disease of honey bee larvae. In addition to members of the genus *Bacillus*, certain species of *Clostridium, Pseudomonas, Serratia, Streptococcus,* and others include insect pathogens. For more information on bacterial diseases of insects, see Faust (1974) and Krieg (1987). Many bacterial diseases of plants are mechanically transmitted by insects. Certain bacterial pathogens of vertebrates are also transmitted by insects; for example, those that cause plague (*Yersinia pestis*) in humans.

Microbes somewhat similar to bacteria, but lacking cell walls, spiroplasmas, and mycoplasma-like organisms, have been found in a number of insect groups. Spiroplasmas were initially found in association with the yellows disease and witches broom diseases of plants (Krieg 1987; Tiivel 1989). At least 100 plant diseases associated with mycoplasma-like microbes or spiroplasmas are vectored by leafhoppers (Homoptera). Spiroplasmas have been found in other insects, including fulgorids and

Figure 9.8 Scanning electron micrographs. (*a*) Bee louse, *Braula coeca* (Diptera, Braulidae), an example of an insect that parasitizes another insect. (*b*) Dorsal view of the tarsus, showing the comblike claws used to cling to the characteristic branched hairs, (*c*) of the honey bee. The larva of this insect lives beneath the honey cappings; the adult lives on the body of an adult bee and takes food directly from the mouth of its host. An adult bee louse is about 1.5 mm long.
Courtesy of A. Dietz, W. J. Humphreys, and J. W. Lindner.

(a)

(b)

(c)

plantlice (Homoptera), fruit flies (*Drosophila,* Diptera), and honey bees (*Apis,* Hymenoptera). The spiroplasmas in *Drosophila* spp. are passed maternally and kill embryonic male flies, thus affecting the sex ratio.

Some rickettsiae are insect pathogens. In size they lie somewhere between viruses and most other kinds of bacteria, but, unlike viruses, they are capable of independent metabolism *in vitro*. Those rickettsiae that are pathogenic for insects are very slow in killing their hosts. They have been grown in mammalian-tissue culture and have killed white mice upon injection. Because of the slow kill of insects and the potential danger for mammals, rickettsiae are unlikely ever to be used in insect control (N.A.S. 1969). *Rickettsia prowazekii* (which causes epidemic typhus) and *R. prowazekii mooseri* (which causes murine and endemic typhus), which are pathogenic for humans, are vectored by body lice and fleas, respectively.

Rickettsiae in the tribe Wolbachieae infect a number of mosquito species, attacking the gonads in particular. *Wolbachia pipientis* infects the mosquito *Culex pipiens* and is responsible for sexual incompatibility between geographic populations (Yen and Barr 1973). Similarly, *Wolbachia* and *Wolbachia*-like rickettsiae have been found in other mosquitoes and in certain Lepidoptera (e.g., *Ephestia cautella*) and may cause the sexual incompatibility observed in *Culex pipiens* (Krieg 1987).

Literature in which rickettsiae associated with insects are considered includes Dasch et al. (1984), Krieg (1987), and Schwemmler and Gassner (1989).

Viruses (figure 9.9) are obligate cellular parasites composed of DNA or RNA, protein, and often surrounded by a lipoprotein envelope. They are classified on the basis of the part of the host cell in which they develop, the presence or absence and morphology of an inclusion body, and the kind of nucleic acid (DNA or RNA) present. Some insect viruses produce granular inclusion bodies; many produce crystalline polyhedral bodies in the nucleus or cytoplasm of an infected cell. Virus particles may be alone in an inclusion body or may be in packets. Biochemical and biophysical characteristics are increasingly being determined and incorporated into classification schemes (Evans and Entwistle 1987).

Several viruses that are pathogenic for insects have been identified and described. Most have been found among Lepidoptera, although several have been found among the Hymenoptera, Diptera, Coleoptera, and Neuroptera. It is interesting that a virus may act in conjunction with a bacterium to produce a disease syndrome not produced by the virus or bacterium separately. Such is the case in two silkworm diseases, gattine and flacherie, caused by *Streptococcus bombycis* and *Bacillus bombycis,* respectively (Rolston and McCoy 1966). The review by Evans and Entwistle (1987) and the multiauthored volume edited by Schwemmler and Gassner (1989) should be consulted for broad treatments of insects and viruses.

Figure 9.9 Electron micrograph showing the virus of Rift Valley fever in the tissue of the mosquito *Culex pipiens*. The virus is approximately 90 Å in diameter and is composed of RNA surrounded by a protein coat. The arrow indicates a single virus.
Courtesy of K. Lerdthusnee.

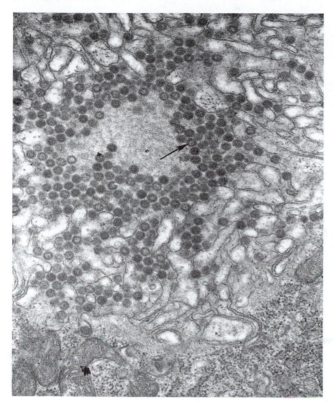

Figure 9.10 Microsporidian *Thelohania minuta* (indicated by arrows) infecting the fat body of the mosquito *Culex erraticus*.
Courtesy of H. C. Chapman and J. J. Petersen.

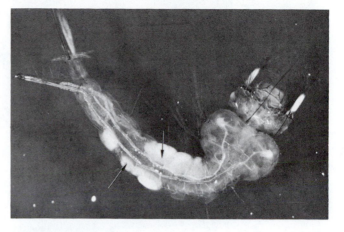

Insects, especially the aphids and leafhoppers (Homoptera), are important vectors of viruses pathogenic to plants (Harrison and Murant 1984, Adam 1984, Pfadt 1985). Insects also serve as vectors of several viruses that attack humans and other vertebrates, notably the *arboviruses* (arthropod-borne viruses), including those that cause yellow fever, dengue, and encephalitis. Kettle (1987), Horsfall and Tamm (1965), Theiler and Downs (1973), David (1975), Gibbs (1973), Reeves (1990), Smith (1967, 1977), and Vaughn (1974) provide extensive information on insects and viruses. The multivolume series by Monath (1988) provides a recent and comprehensive treatment of arboviruses.

There are more species of fungi that are known to be insect pathogens than of any other group of microbes. These *entomogenous* fungi have been described from several groups of fungi. Species in the genera *Beauveria, Metarhizium,* and *Paeciolmyces* have been used in attempts at biological control (Keller and Zimmermann 1989). However, the success of fungus infections in insects (i.e., whether they become established and kill an insect) depends to a great extent on weather conditions, and hence these microbes would likely be undependable

as agents of biological control. Carruthers and Soper (1987) provide a recent review of fungal diseases of insects from an ecological perspective.

Although insects are apparently not vectors of fungi that cause human or other vertebrate diseases, they do serve as mechanical vectors of fungi pathogenic to plants (Webber and Gibbs 1989). An example is the fungus that causes Dutch elm disease, which is vectored mainly by the European elm bark beetle, *Scolytus multistriatus.*

Many insects use fungi as food, and in some cases a highly developed mutualistic relationship has evolved. For example, termites in the subfamily Macrotermitinae (Family Termitidae, the "higher" termites) cultivate fungus "gardens" within their nests (Wood and Thomas 1987). Likewise, leaf-cutting ants in the Tribe Attini (Subfamily Myrmicinae) tend fungus gardens as food (Cherrett et al. 1987). The Royal Entomological Society of London in collaboration with the British Mycological Society published a symposium (Wilding, Collins, Hammond, and Webber 1987) that provides reviews of several aspects of insect-fungus interactions.

A large number of protozoa, especially *Microsporidians* (figure 9.10), are known to kill insects (Brooks 1974), but, like rickettsiae and fungi, they act very slowly. However, many of those slow-acting microbes probably weaken their hosts and make them more susceptible to the effects of other environmental components. The microsporidian *Nosema bombycis* (Sporozoa) causes the "pebrine" disease of silkworms. The elucidation of this relationship by Louis Pasteur in the nineteenth century stands as a classic in microbiological research. Other well-known microsporidian insect pathogens include *Nosema apis* and *N. pyraustai,* which attack the honey bee and European corn borer, respectively. A mosquito, *Culex erraticus,* parasitized by the microsporidian *Thelohania minuta* is shown in figure 9.10. Protozoans are common partners in mu-

Figure 9.11 Scanning electron micrograph of a mite (Arachnida; Acari) on the head of a termite (highly magnified). The mite is indicated by the arrow and is shown at a still higher magnification in the inset.
Courtesy of Walter J. Humphreys.

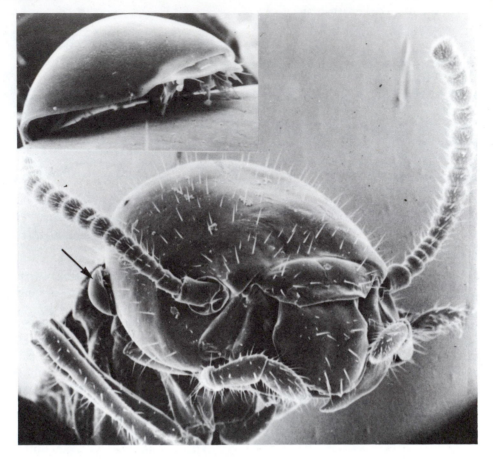

tualistic relationships with insects. We have already discussed those found in the gut of termites, which provide cellulase and enable their hosts to digest cellulose (Cleveland et al. 1934). Several protozoa pathogenic to humans and other vertebrates are vectored by insects. Examples include the malarias (*Plasmodium* spp.) by *Anopheles* spp. mosquitoes (different forms of malaria occur in birds, monkeys, and humans), American trypanosomiasis (Chagas' disease) caused by *Trypanosoma cruzi* and vectored by blood-sucking bugs (Hemiptera, Reduviidae), and the two forms of African sleeping sickness caused by *T. rhodesiense* and *T. gambiae* and vectored by tsetse flies.

Mites (Arachnida, Acari) are very small arthropods, many species of which parasitize insects (figure 9.11). For example, the larvae of members of the family Erythraeidae are parasites of insects and other arthropods and have been found attached to wing veins and other locations on the insect body (Baker et al. 1956). *Acarapis woodi* (Scutacaridae) causes the acarine disease of bees. This species either lives in the tracheae or in the external

parts of the bee's body and weakens the host to the point where it can no longer fly and eventually succumbs (Baker et al. 1956, Gochnauer 1963).

About 100 species of nematodes (phylum Aschelminthes, roundworms; figure 9.12) have been described as being associated with approximately 1,500 insect species (N.A.S. 1969). In most instances, these "entomophilic" nematodes damage and eventually kill their host. Nematodes show some promise in the biological control of insects. A number of species that attack humans— for example, filarial worms (*Wuchereria bancrofti* and others)—and other vertebrates—for example, dog heartworm—spend part of their life cycles in insects (in these cases, mosquitoes), which serve as their vectors. See Gangler and Kaya (1990), Nickle (1984), Poinar (1972, 1975), and Welch (1965) for reviews in the area of "entomophilic nematology."

Many relationships between two or more insect species or between insects and others organisms are commensal. For example, probably many of the microbes associated with insects neither harm nor particularly help

Figure 9.12 Nematode parasite, *Romanomermis culicivorax*, in the thorax of the mosquito *Culex quinquefasciatus*. (*a*) Several infected larvae. (*b*) Single infected larva. (*c*) Postparasitic stage of the nematode escaping. *Courtesy of H. C. Chapman and J. J. Petersen.*

(a)

(b)

(c)

their host but gain food and a place to live. Phoretic relationships (see also chapter 8) are good examples of commensalism. In a phoretic relationship an individual of one species attaches to an individual of another and gains a mode of transportation. Insects may be transporters or riders or both. Some chewing lice (Mallophaga) attach themselves by their mouthparts to louse flies (Diptera, Hippoboscidae), some of which are also parasites of birds, and in this way move from one host to another. The torsalo fly, *Dermatobia hominis* (figure 9.13), deposits eggs on other flies, such as mosquitoes, black flies, and house flies. When these insects come into contact with a potential vertebrate host (including humans), larvae emerge from eggs and penetrate the skin (Harwood and James 1979). Clausen (1976) discusses phoresy among entomophagous ("insect-eating") insects.

Predators that feed on insects include insects themselves; other Arthropoda, such as mites, spiders, scorpions, and pseudoscorpions; and vertebrates, including birds, reptiles, amphibians, fish, and mammals. Insects as predators are discussed in chapter 8, and comments regarding vertebrate predators are included in chapter 15 ("Biological Control"). Sweetman (1936a, b) discusses predatory invertebrates in some detail. In addition to animals, a few plant species—such as the Venus-flytrap, pitcher plant, and the aquatic *Utricularia vulgaris*—trap, digest, and assimilate insects (Frost 1959). Heslop-Harrison (1978) discusses carnivorous plants.

Insects may be indirectly influenced by other organisms. No doubt the greatest effects are those produced by humans purposely against pest populations and inadvertently against the vast majority of insects that are "innocent" (see chapter 15). The effects of modern agriculture, urban sprawl, and superhighways on various habitats are obvious, and the far-reaching and unfortunate effects of highly residual pesticides have become well-known. The effects of air and water pollution on the environment are beginning to get the kind of attention they

Figure 9.13 Mosquito, *Psorophora* sp., carrying the eggs (indicated by arrow) of the torsalo fly, *Dermatobia hominis*. *Redrawn from USDA photo.*

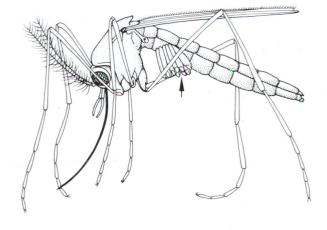

deserve. Insects themselves are sometimes polluters: "Predatory anthocorid bugs enter open galls and feed on a few of the aphids therein. The aphids which are not attacked die from the fumigant effect of chemicals produced by the stink glands of the bugs" (Clark et al. 1967, citing the work of Dunn 1960).

Adaptations Associated with Interspecific Interactions

The "goal" of every organism is to obtain enough food (matter and energy) to maintain itself long enough to reproduce and to pass its genes to the next generation. As an organism proceeds from conception to reproductive maturity, it must obtain food and at the same time avoid becoming food for some other organism. Organisms that are unable to do this successfully, for whatever reason, are maladapted and are, in the long term, selected against. Although a number of environmental factors are no doubt involved in the selection process, examination of specific adaptations often leads to the identification of specific selecting agents. Organisms that are closely associated ecologically often act as agents of selection reciprocally; that is, they act as selecting agents on one another. There is good evidence that many organisms have done so for long periods of geologic time. These organisms are said to have *coevolved* or to be *coadapted*.

Two (or more) different species may have coevolved to the extent that they benefit one another in the quest for reproductive success; their relationship thus is mutualistic. In this case the partners have become (or are becoming) more and more successful in their interaction, leading to sustained or improved reproductive success for both. Alternatively, the participants may have engaged and continue to engage in a kind of genetic warfare or "arms race" in which one member "evolves" a mechanism of offense providing increased access to the other as food, only to have the other "respond," in the evolutionary sense, with a defense, and so on through time. Such is the case in interactions between parasites (and parasitoids) and hosts, predators and prey, and herbivores and plants (chapter 10).

In this section, examples of coevolutionary adaptations are examined. For a comprehensive introduction to coevolution, see Gilbert and Raven (1975) and pertinent sections in more recent evolutionary biology texts such as Futuyma (1986).

A good example of the evolutionary "fine tuning" that can occur between mutualists is seen in the association between attine ants and their fungus gardens (see "Feeding Behavior" in chapter 8). As ants prepare a fungus garden, they add their own fecal material, which benefits the fungus (Martin 1970). The fecal material contains proteolytic enzymes and various nitrogenous materials that play vital roles (i.e., protein digestion and enhancement of growth) in the metabolism of the fungus. In turn, the fungus contributes cellulase, which benefits the ants by catalyzing the digestion of cellulose. Such close metabolic interdependency is characteristic of many mutualistic associations.

Parasites and hosts apparently pass through successive cycles of "parasite attack—new host defense," and the parasite may become increasingly benign and the host more tolerant, perhaps eventually leading to a commensalistic or even mutualistic outcome. For example, parasites that invade the insect hemocoel are commonly encapsulated (see "Circulatory System" in chapter 4). The presence of such a capsule around a parasite may severely limit the parasite's activities by inhibiting mobility or interfering with feeding and/or gaseous exchange. On the other hand, some parasites secrete materials that block capsule formation (Jones 1977). Whitcomb et al. (1974), Duffey (1977), Dunn (1986), Bowman and Hultmark (1987) consider insect defenses against microorganisms and parasitoids.

A remarkable example of an extremely close adaptive adjustment between parasite and host is the relationship between the rabbit fleas *Spilopsyllus cuniculi* and *Cediopsylla simplex* and their rabbit hosts (Rothschild 1965a, b; Rothschild and Ford 1966, 1972). The reproductive behavior and physiology of these fleas is strongly influenced by the rabbit's reproductive hormones. Egg maturation only occurs on pregnant doe rabbits and has been shown to be stimulated by the host's corticosteroids. Estrogens and other hormones also contribute to egg development and influence oviposition. Larvae develop in the rabbit nests in which the young rabbits are born. Linkage of the sex cycles of the fleas and rabbit hosts ensures that the reproductive activity of the fleas occurs such that newborn flea larvae end up in the host's nest.

That predators of insects and their insect prey have exerted strong selection pressures on one another is clearly evident in the means predators have evolved to capture prey and the avoidance mechanisms of potential prey. Several examples of coevolutionary adaptation involving insects as predators and insects as prey have already been given in earlier chapters (see, especially, "Escape and Defense Behavior" in chapter 8). Here we consider examples of insect adaptations in response mainly to selection pressures by insectivorous vertebrates, in particular protective coloration, form, and pattern; and coevolutionary adaptations in bats and moths.

The multiauthored book in insect defenses edited by Evans and Schmidt (1990) provides an excellent overview of the evolutionary interactions between insects and their predators.

Examples of protective form, coloration, and pattern include crypsis, aposematic coloration, mimicry, and spots. As explained in chapter 8, such protective devices would be of diminished value (or even detrimental) without appropriate behavior. Thus, although we will emphasize form, coloration, and pattern in this discussion, it must be remembered that "appropriate" behavior has also come about as a result of selection pressure applied by predators. General treatments of protective form, coloration, and pattern include Cott (1957), Portmann (1950), Wickler (1968), Edmunds (1990), and Guilford (1990).

Crypsis refers to various combinations of form, color, and pattern that facilitate "hiding" from potential predators. Resemblance to various inedible, even repulsive, objects like thorns or bird droppings; coloration and shading that cause an insect to blend into the background; and transparent wings that allow the background to show through all belong in the crypsis category.

A classic example of crypsis (and one that has contributed significantly to our understanding of natural selection) is *industrial melanism* (Kettlewell 1959, 1961, 1973; Bishop and Cook 1975). Industrial melanism refers to a phenomenon first observed in the peppered moth,

Figure 9.14 The monarch butterfly, *Danaus plexippus* (a) serves as the model for its mimic, the viceroy butterfly, *Limenitis archippus* (b).

(a)   (b)

Biston betularia, in England. These moths are nocturnally active and during the day they rest on lichen-covered tree trunks. Prior to 1845, collection records of peppered moths described them as light-colored, an appearance that provided them with camouflage when they rested on the tree trunks. In 1845 and thereafter, dark (*melanic*) forms began to appear around the industrial center of Manchester and eventually came to be more common than the light-colored forms. It was hypothesized that the change in predominant phenotype (from light to dark) was associated with the overall darkening of tree trunks due to industrial pollution. Such darkening was thought to lessen the selective advantage of light-colored moths while favoring the survival of dark forms. Birds were assumed to be the agents of selection. Kettlewell demonstrated experimentally that birds did, in fact, prey more on uncamouflaged forms; light-colored forms had a definite selective advantage in unpolluted areas. He also showed that moths tended to rest on a substrate that matched their body coloration, indicating that they could discriminate between light and dark backgrounds. With the advent of pollution controls, light moths are beginning to appear again in industrial regions.

Aposematic coloration, or warning coloration, is sometimes called "advertising" in the sense that an insect tastes very bad (i.e., is unpalatable), stings, or does something similarly disagreeable to a predator and communicates this ability visually by colors and patterns that contrast blatantly with the background. Vertebrate predators soon learn to avoid these insects on the basis of a few unpleasant encounters. An aposematically colored insect may also be advertising falsely in that although it looks like (mimics) a harmful insect, it is really palatable and harmless. Aposematically colored insects are usually boldly patterned in shades of orange, red, or yellow contrasting with black and stand out against the greens and browns that are characteristic of the environment. Unpalatability is commonly due to the presence of chemicals in the hemolymph that have been derived from host plants. For example, the monarch butterfly (*Danaus plexippus;*

figure 9.14*a* and certain true bugs and beetles, which feed on various species of milkweed plants (Asclepiadaceae), accumulate chemicals (specifically *cardiac glycosides*) from these plants. These chemicals not only render the insects unpalatable to vertebrate predators (especially birds) but also act as emetics (i.e., they make a predator vomit) (Brower 1969, Brower and Brower 1964, Brower and Glazier 1975, Reichstein et al. 1968, Rothschild 1972, and Duffey 1980). Although all monarch butterflies are aposematically colored, an individual may or may not be unpalatable depending on the level of cardiac glycoside in the particular milkweed plant it fed upon as a larva (Brower et al. 1968). In an interesting coevolutionary turnabout, certain insectivorous birds have circumvented the monarch's chemical defense by learning to reject or ingest individuals on the basis of taste (Calvert et al. 1979). In this instance aposematic coloration would tend to be disadvantageous for the insects because it would render them more visible to predators.

Mimicry, as used here, refers to the close resemblance (in form, color, pattern, or otherwise) between one insect, the *mimic,* and another insect or a plant part, the *model* (insects may mimic thorns, leaves, etc.). Various forms of mimicry have been identified in insects, among them Batesian (Bates 1862) and Müllerian (Müller 1879). A *Batesian mimic* is palatable (and otherwise harmless) but is protected somewhat by virtue of its similarity to an unpalatable (or otherwise harmful) model. An oft-cited example is the viceroy butterfly (*Limenitis archippus*), which has until recently been viewed as a Batesian mimic of the monarch butterfly (figure 9.14*b*). However, recent "taste-testing" experiments with birds have shown that viceroys are also unpalatable (Ritland and Brower 1991).

Because viceroy caterpillars feed on nontoxic willows, they are apparently able to synthesize their own toxic chemicals. Batesian mimics are commonly less numerous than their models. This is interpreted as advantageous on the basis that large numbers of aposematically colored but palatable individuals would reduce the avoidance

Figure 9.15 Io moth (female), *Automeris io,* with forewings thrown forward, displaying eyespots.

learning in predators. *Müllerian mimics* differ from Batesian in that both mimics and models are unpalatable (or otherwise harmful). Several species that bear close resemblance to one another may comprise a *Müllerian mimicry complex,* all members gaining advantage by virtue of the fact that the number of different colors and patterns predators are required to learn to avoid are reduced; that is, predators need learn only one pattern to result in protection for all the insects in the complex.

For additional information on insect mimicry, see Rettenmeyer (1970) and Guilford (1990).

Many insects (e.g., certain moths, butterflies, and caterpillars) bear spots at various locations on their body. These spots often closely resemble eyes of vertebrates and perhaps, in some instances, eyes of other insects. Many moths, which have eye spots on the hindwings, are otherwise cryptically colored when resting on their usual substrate. If the first line of defense, crypsis, fails and a predator strikes, the forewings are thrown forward, suddenly displaying the eyespots (figure 9.15). Such behavior has been shown experimentally to frighten and repel insectivorous birds (Blest 1957). The eyespots resemble eyes of birds. Sargent (1990) discusses startle as an antipredator mechanism.

Small spots are commonly found along the edges of the wings of many butterflies (Wickler 1968). These have been shown to serve as pecking targets and by virtue of their location away from vital parts of the body may provide protection by diverting the peck of a predator and allowing time for escape.

A superb example of evolutionary interaction between predator and prey is the association between insectivorous bats and members of certain families of nightflying moths, in particular Noctuidae, Arctiidae, and Geometridae (see "Escape and Defense Behavior" in chapter 8). Bats locate prey by emitting pulses of sound beyond the range of human sensitivity (ultrasonic). As these sounds bounce off various objects, living and nonliving, the bats hear the echo and use this information to guide their flight and to locate flying prey insects. Details of the discovery and elucidation of this echo-locating ability are recounted in Griffin (1959). Tympanic organs that perceive the echo-locating cries of bats have evolved in several moth families (Roeder 1974, Fullard 1990; see "Sound Perception" in chapter 6). As described in chapter 8, the moths display various escape maneuvers in response to bat cries. The coevolutionary "battle" has proceeded a step further in certain Arctiid moths, which possess microtymbal organs on the metathorax. When a bat approaches one of these moths, trains of sound pulses are generated by the microtymbal organs. These sounds have been shown to, in effect, "jam" the bat's echolocating system and hence interfere with its ability to catch the moth (Dunning and Roeder 1965).

Because the earlier work with moths, insects from several other groups (e.g., green lacewings, praying mantids, crickets, locusts, and tiger beetles) have been found to display evasive maneuvers in response to artificial pulses of ultrasound (May 1991).

CHAPTER 10

Insects and Their Environment: Plants

Lowman / Morrow

I've watched you now a full half-hour,
Self-poised upon that yellow flower;
And, little butterfly! indeed
I know not if you sleep or feed.
　　William Wordsworth, "To a Butterfly"

Humans have always been fascinated by insects, and the majority of their observations occur on the substrate of plants. Plants form the basis of all food chains on earth because they are the only organisms that capture energy from sunlight. In turn, herbivores (of which many are insects) are plant consumers, thereby transferring energy to higher trophic levels (figure 10.1). Insects and plants have evolved together for millions of years, including both antagonistic and mutualistic interactions. Most early observations were simply accounts of plague insects (e.g., the Bible), whereas it is only during the last 75 years that insects on plants have been more carefully studied. Why has the biology of insects on plants received such extensive recent attention? There are a number of reasons for this:

1. Insect population fluctuations have a direct impact on human health and food supplies.
2. Insects are abundant, small and easy to raise, observe, and manipulate in scientific experiments.
3. Unlike many parts of ecosystems, insect-plant relationships have at least one (often two) sessile components, making the study of their relationships very convenient.
4. Insects and plants have evolved in response to each other, and scientists are curious about how these relationships developed.
5. Insects and plants together are perhaps three quarters of the known species on Earth, and the specific insects that feed on plants may represent one quarter of life on Earth (figure 10.2).

During the last 75 years of research on insects and plants, science has advanced enormously. Even as late as 1982 it was hypothesized that nearly 1,000,000 species of insects inhabited the earth (Strong et al. 1984; Richards and Davies 1977). But recent research in tropical rain forest tree canopies has led to a drastic revision of this estimate. Only 10 years later, in 1992, we now think that there may be as many as 30,000,000 insects, a 30-fold increase! (Erwin 1982, Wilson 1988, but see Gaston 1991). Most of that increase appears to be a consequence of insect communities in forests (both in trees and in soils), which may be much more complex and abundant than previously thought.

In this section, we shall look at the ecological relationships of insects and plants, comparing the abundance of insects on structurally simple plants and working up to the more complex distributions of insects in forests. In ecology, scientists often test *hypotheses,* which is an assumption based on careful deduction. Experiments to test an hypothesis comprise observations, experiments in the field or in the lab, and the interpretation of results that leads to either an acceptance or rejection of the original hypothesis. We will look at several examples of experimental ecology in our discussion of plants and insects.

Insects are associated with plants for a number of reasons. The most important relationship involves food supply: many insects on plants are herbivorous or phytophagous (*phyto* = plant, *phagous* = feeding). Others, however, use plants for territory, protection, mating, traveling, communal roosting, and other activities.

The colonization of habitats by arthropods and their active (but sometimes passive) abilities to colonize a site have been the subject of extensive research in the last decade. It is assumed that most arthropods colonize a habitat primarily because conditions are suitable for survival, and that species will co-exist if other populations

Figure 10.1 The energy pyramid of a land-based community. Note that a large proportion of the energy is lost between trophic levels. Trophic levels are pictured and labeled on the left.

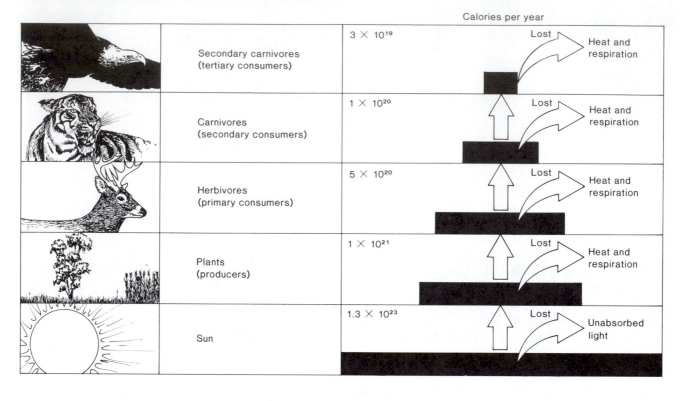

Figure 10.2 The number and proportions of species in major taxa, excluding fungi, algae and microbes. The proportion of phytophages is based on estimates for British insects, assuming that proportions in the world as a whole are broadly similar. *After Strong et al., 1984.*

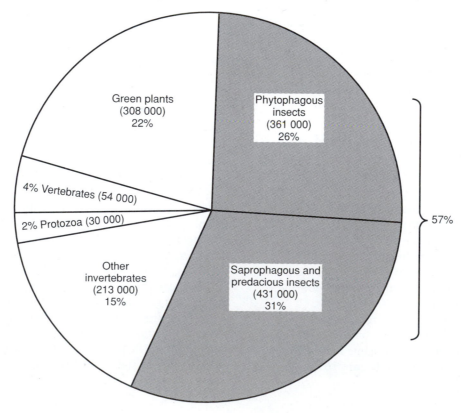

do not inhibit them or outcompete them. This perhaps originates from the original views by a plant ecologist named Gleason (1926) who advocated that communities are comprised of individual populations, with a continuum of species along an environmental or habitat gradient. Gleason's definition of a community is in contrast to another biologist, Clements (1916), who advocated communities as "organisms" comprised of highly interdependent populations. The individualistic view of Gleason and the organismic view of Clements both have come in and out of fashion over time, but draw attention to the importance of populations and their interactions in any community.

Populations of insects in communities fluctuate more extensively than do populations of longer-lived organisms. These oscillations impart both positive and negative aspects to field studies on insects: insects are easier to study because several generations and population cycles can occur during one season, but population dynamics may be more difficult to interpret because numbers may fluctuate wildly over a relatively short period of time. Functional groups of insects in a community exist at two levels: populations of a species, and guilds of similar, but different species that perform the same role (e.g., sap-sucking insects).

It is interesting to note that one plant can comprise a community with its resident populations of insects, whereas one ecosystem can also comprise a community, with many plants housing many populations of insects. For our purposes, an *insect community* is defined as populations of insects that interact on or in the vicinity of a common host plant.

The reasons for differences in numbers of both populations and guilds of insects on plants are many. We discuss three in this chapter: structural complexity and age of host plant, nutrition and toxicity of plant tissues, and evolution of mutualistic relationships.

Structural Complexity and Age of Host Plants

Insects on Structurally Simple Plants

It is well established that plant-dominated ecosystems have more insects than landscapes that do not. For example, volcanic peaks and alpine aeolian systems (i.e., regions that depend upon wind to import both energy and nutrients), have very sparse arthropod faunas (Edwards 1987), and as areas develop soils and then vegetation (termed *succession,* a classic sequence whereby a region undergoes development of soils and vegetation, often bare soil to field to shrubs to scrub to forest), a larger variety of plants is established, followed by a greater fauna of insects. As succession of the landscape leads to greater complexity, it is hypothesized that increased complexity

of plants results in greater insect fauna. The numbers of insects increase as plant size increases: monocots to weeds to perennial herbs to shrubs to trees (Lawton 1983).

Monocultures (single-dominant groups of a species versus polyculture) of cereal grains or other crop plants are both structurally and taxonomically simple, but their resident populations of insects have extremely complex behaviors (reviewed in Liss et al. 1986). On the one hand, populations of insects in monocultures tend to be taxonomically less diverse than insect communities in more diverse habitats; but on the other hand, their populations may fluctuate enormously with subtle changes in the quality and availability of food. Price (1976) examined seasonal changes in arthropods on soybean plants, with predictable increases in numbers as the season progressed, followed by decreases in numbers as the plants underwent senescence. The number of species fluctuated from 0 to 30 during a three-month period. Bach (1980) examined the population dynamics of insects on cucumbers in monocultures and in polycultures. She found that the populations of striped cucumber beetles (*Acalymma vittatum*) were higher in monocultures than in polycultures (figure 10.3). This is an example which supports the *resource concentration hypothesis* (Root 1973), which predicts that insect herbivores will reach larger per plant populations in areas with high densities of host plants than in regions where host plants are patchy or rare. The *enemies hypothesis* is an alternative idea that predators and parasitoids should control numbers of insects more effectively in mixed stands than in pure stands.

Both of these hypotheses are pertinent in recent experimental ecology. The relationships between insects and crop plants are studied extensively in a discipline called integrated pest management (IPM; see chapter 16). This topic is of obvious importance to the survival of human beings because we depend upon crop plants for our nutrition and try to minimize the losses of cereal crops to insect epidemics. Because crop plants are usually herbaceous and grown in monocultures, their structural simplicity poses a challenge to agriculturalists to protect the crops from excessive build-up of insect populations.

Insects on Structurally-Complex Plants

Lawton (1983) further hypothesized that plants offering a greater variety of resources (e.g., more leaves or branches) will support more individuals and more species of insects. Very little experimental evidence exists to support this hypothesis, however, because it is difficult to test. In one experiment, Bach (1981) found higher numbers of specialist chrysomelid beetles on vertically-grown (versus horizontal) cucumbers. This suggests that insects prefer the structurally vertical plant with its more complex arrangement of stems. With larger populations, species are more likely to persist on a plant despite predation,

Figure 10.3   Mean number of striped cucumber beetles (*Acalymma vittata*) on cucumber plants (*Cucumis sativus*) in garden plots in monocultures, or in polycultures planted with corn (*Zea mays*) and broccoli (*Brassica oleracea*). Density of cucumber plants (289 or 144/100 m²), by itself, has little influence on beetle density. *After Bach (1980)*.

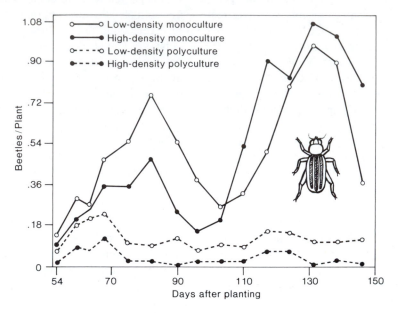

inclement weather, and so forth. Following Lawton's (1983) original progression, shrubs, saplings and ground vegetation are slightly more structurally complex than low-growing herbaceous monocultures.

Some studies indicate that numbers of individuals (not just species) also vary with increasing structural complexity of the host plant. Saplings of eucalypts and closely related angophoras in Australia were examined for abundance of arthropods over time, with the result that numbers fluctuated seasonally from 0 to 20,000 (figure 10.4). Some of these insects (e.g., Diptera and other flying insects) appeared to be situated on a plant as a result of chance, and were travelers; others such as the ants and aphids were residents and also very prolific. This combination of transient and permanent residents on a plant accounted for the high variability in numbers of insects on the eucalypt saplings. Other factors such as foliage quality and *phenology* (seasonal patterns of growth-related activities) of the leaves affected the numbers on a seasonal basis. These numbers appear to be extremely high, but eucalypts seem adapted to relatively large insect populations as compared to their northern temperate forest counterparts (see section on forests). Mature eucalypts have been observed to have many tens of thousands of Christmas beetles (*Anoplognathes* sp. Scarabaeidae) feeding in their canopies (Heatwole and Lowman 1986).

Studies comparing the abundance of insects on seedlings, saplings, and adult trees are rare because of the obvious time and energy required. It is possible to infer relationships of insect abundance from studies on her-

bivory, however, which is an indirect measure of the numbers of insects on plants. Coley (1983) examined the herbivory of rain forest saplings in Panama. Persistent, late successional species had greater defenses of leaf material and subsequently lower numbers of herbivores per tree than pioneer or early successional species of saplings. Similar results were obtained for herbivory of mature rain forest canopy trees in Australia, whereby the gap-colonizing giant stinging tree (*Dendrocnide excelsa*) had higher herbivory than the late successional red cedar (*Toona australis*) and sassafras (*Doryphora sassafras*) (42%, 5%, and 13% annual leaf area losses, respectively) (Lowman 1985).

According to the structural complexity hypothesis, trees and forest communities house the most abundant and diverse insect populations. In his classic work, Southwood (1961) (see also Southwood et al. 1982a, b) made extensive measurements of the arthropod fauna of various deciduous trees. He used an insecticide spray which harvested all insects within the canopy and essentially provided a "snapshot" of each tree at one point in time. This technique, called *fogging*, is useful for obtaining a comprehensive count of all insects within a complex structure such as a tree canopy. Comparisons among trees and continents showed a fairly consistent proportion of different trophic levels of insects: one-quarter phytophagous by species but over one-half by weight, one-quarter parasitoids by species but less than one-sixteenth by weight, one-quarter predators by species and the same proportion by weight, one-sixth scavengers by species and

Figure 10.4   Seasonal changes in numbers of insects on 10 individual eucalypti saplings growing in close proximity in an old field near Armidale, NSW, Australia. Arthropods responsible for peaks include ants (*A*); psyllids (*P*); Christmas beetles (*C*); leafhoppers (*L*). Sapling species are *Angophora floribuda (a, b)*; *Eucalyptus caliginosa (c, d)*; *E. blakelyi (e, f)*; *E. melliodora (g, b)*; and *E. viminalis (i, j)*.

Data from Heatwole, Lowman, Donovan, and McCoy.

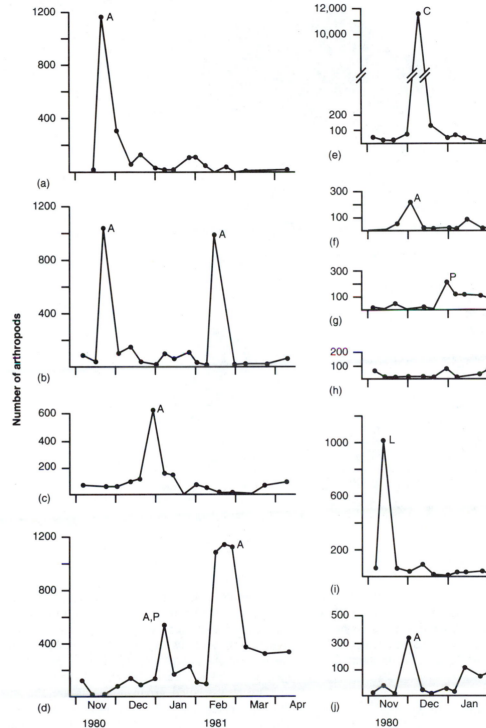
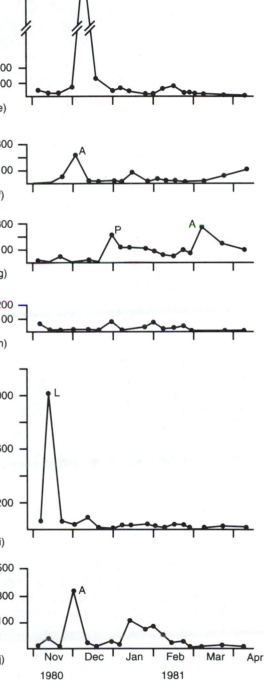

Figure 10.5 Composition of the arthropod fauna of trees in terms of major guilds, expressed as mean number of species, number of individuals, and biomass, based on comparable samples from six species of tree in Britain and six in South Africa. C = chewers, S = sap-suckers, E = epiphyte fauna, Sc = scavenging (dead wood, and so on) fauna, PR = predators, P = parasitoids, A = ants, T = tourists.

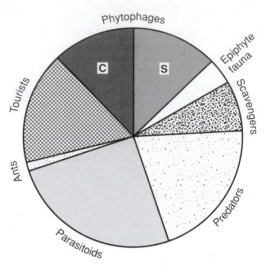

Species

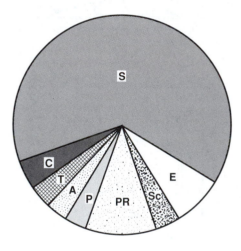

Individuals

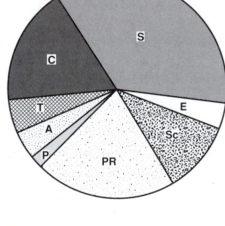

Biomass

by weight, and the remainder split amongst tourists (visitors to the tree), epiphyte fauna, and ants (figure 10.5). In total, they found between 180 and 425 species of insects for *Salix alba* and *Quercus* canopies, respectively, using a composite of both fogging and observation methods.

Insects on Plants of Different Ages

It is assumed that insects have coevolved with plants for a long time, so the older, more complex and more highly-evolved ecosystems may have richer insect faunas and more complex relationships between plants and insects. Once again, these hypotheses are difficult to test because they involve evolutionary time and cannot be manipulated experimentally. Biologists can look at different members of one group, however, and compare the differences of the plant-insect relationships they observe.

One well-studied group of plants are the thistles. Europe, especially the Mediterranean region, has a large and ancient thistle flora used by more than 50 genera of insects. Most of these species are *host-specific* (feed just on thistles) and an especially large guild feeds inside the flower heads. One tribe of thistles migrated to North

America and speciated extensively. But the diversity of insects on these thistles is low, apparently because few thistle-feeding taxa accompanied their ancestors as they migrated from Europe over the Bering Strait land connection in the late Miocene (Zwolfer 1988).

A perhaps older and more complex web of relationships exists between *Heliconius* butterflies and plants in neotropical rain forests. These butterflies have a variety of interactions with their sessile hosts; adults are brightly colored, a warning to would-be predators that they retain the toxins their larvae obtained from their passion vine hosts. Warning is enhanced by gregarious feeding of larvae, communal roosting of adults and participation of adults in mimetic complexes with other distasteful butterfly species (there are two kinds of mimics: Batesian mimics are palatable species that look like distasteful, or dangerous, model species; Müllerian mimics are distasteful and look like other distasteful species; see chapter 9). Butterflies eat pollen of widely scattered *Anguria* vines; young butterflies learn *trap lines* (regular routes traveled by pollinators to visit the same plants repeatedly) by following older adults when they leave the roost. Some passion vines make fake insect eggs and attract ants to escape larval attack (Turner 1981). Figure 10.6 illustrates the complex interactions and the range of relationships that have evolved between *Heliconius* and their hosts.

Other studies have shown that older trees have more phytophagous insects than younger ones (Banerjee 1981 on tea trees) and that older forests have a richer fauna than younger ones (Southwood et al. 1982a, b). Southwood (1961) compared the numbers of insects between dominant native trees and introduced trees in Britain, Sweden, Russia and Cyprus. They found that the older tree community averaged higher numbers of resident insects, as compared to introduced, less well-established tree species (presumably where the trees and insects have had less time to reach an equilibrium).

Plant Nutrition and Herbivory

Plants are the largest and most readily available source of food in terrestrial communities. Phytophagous insects have the capacity under ideal conditions to increase by a geometric progression (exponentially). But this capacity is rarely realized in nature. In an influential paper, Hairston et al. (1960) observed that plants seem to be damaged extensively only when natural enemies (parasitoids and predators) of herbivores are absent, due, for instance, to the greater sensitivity of enemies to pesticides or introductions of herbivores without their enemies. From this they deduced that herbivore populations are controlled by their natural enemies rather than by availability of food, in contrast to plants and carnivores, which they argued are controlled by competition for resources.

In some cases predation, parasitism, and disease control the abundance of herbivores, but experiments to test these regulatory effects are difficult to conduct (Price 1987, Price et al. 1990). Why then is the world green? Hairston et al. (1960) assumed that anything green is food for herbivores just as different antelope species are equally good for large predators. As we shall see, this assumption has proved false.

Herbivorous insect species are very abundant, comprising perhaps half of all individuals found in nature. However, out of 32 orders of insects, only 9 feed on living plants (figure 10.7). This suggests that plants are a formidable evolutionary barrier that most groups have not been able to overcome. But the high diversity of insect species within these orders suggests that, once these barriers are surmounted, insect groups radiate extensively (Mitter et al. 1988).

Ways of Eating Plants

Plants are complex and variable structures. The nutritious parts of plants, cytoplasm and fluids, are surrounded by wooden walls that insects cannot directly digest (see chapter 4), so they must mechanically disrupt or otherwise circumvent them. Insects have evolved specific morphological, physiological, behavioral and other adaptations to cope with and use plants. The large size of plants in comparison to insect herbivores has been associated with the evolution of insect specialists that feed on a particular plant part. Insects can be classified into several feeding *guilds,* or groups, that have similar styles of exploiting plants for food.

Chewers

Chewing is the most common way in which insects process plant material, including leaves, stems, flowers, pollen, seeds, and roots. In the orders Orthoptera, Coleoptera, Hymenoptera and Phasmida both juvenile and adult stages have chewing mouthparts. In the Diptera and most Lepidoptera, only the larvae chew their food. Because most of the material eaten by chewing insects includes tough cell walls, the mandibular surfaces are exposed to considerable wear and are replaced at each molt.

Miners and Borers

Larvae of many insects and some adult beetles feed within plant tissues. Leaf miners are chewing insects that eat one or more of the tissue layers between the intact upper and lower epidermis of leaves. As the insect tunnels its way through a leaf, it leaves a mine whose pattern is often characteristic for the species. Except in the Diptera, miners tend to be flattened to accommodate the dimensions of the leaf. Like other insects that feed inside growing tissues, miners often feed on one or a small

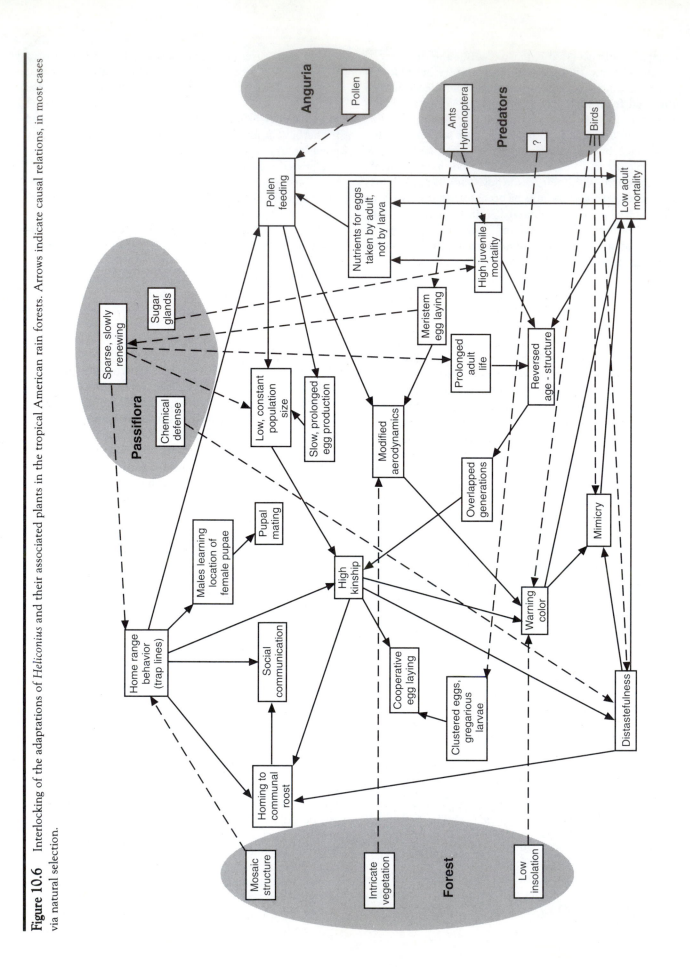

Figure 10.6 Interlocking of the adaptations of *Heliconius* and their associated plants in the tropical American rain forests. Arrows indicate causal relations, in most cases via natural selection.

Figure 10.7 The size of the insect orders containing significant numbers of herbivores.

The size of the order taken from Borror et al., 1976 and Richards and Davies, 1977. Proportion of herbivores in each order taken from Price, 1977, except for Thysanoptera, which was estimated from Lewis, 1973; as appeared in Weis and Berenbaum, Plant-Animal Interactions, *1989.*

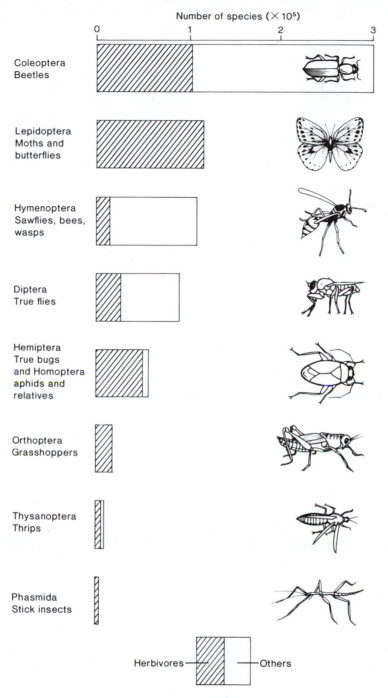

number of related plant species. Leaf-mining insects are found in the Lepidoptera, Diptera, Hymenoptera, and Coleoptera. The greatest diversity of forms and number of leaf-mining species are in the Lepidoptera, followed by the fly family Argromyzidae (Hespenheide 1991).

Boring insects live in the woody tissues of plants. The abundance and long-term availability of food provided by tree trunks permits growth to a large size. The larvae of *Syleutes boisduvale* (Cossidae), for instance, attain a length of 18 cm (Waterhouse 1970). In addition, the low nutritive value of wood tends to result in long life cycles. As an aid in using wood for food, some borers inoculate the wood with fungi that attack the plant and upon which the larvae feed. Dutch elm disease is caused by a xylem-blocking fungus introduced by the elm bark beetle, *Scolytus scolytus*. Other borers attack the living cambium

Figure 10.8 Insects galls. (*a*) *Andricus pattersonae* (Hymenoptera; Cynipidae) on leaf of blue oak. (*b*) Pine-cone willow gall caused by *Rhabdophaga strobiloides* (Diptera; Cecidomyiidae). (*c*) *Pemphigus betae* (Homoptera; Aphididae) on *Populus angustifolia*. (*d*) Longitudinal section through the goldenrod ball gall caused by *Eurosta solidaginins* (Diptera; Tephritidae). (*a*) *After Essig; (b) Redrawn from Boreal Labs key card; (c) from Whitham 1978; (d) from Abrahamson, 1989.*

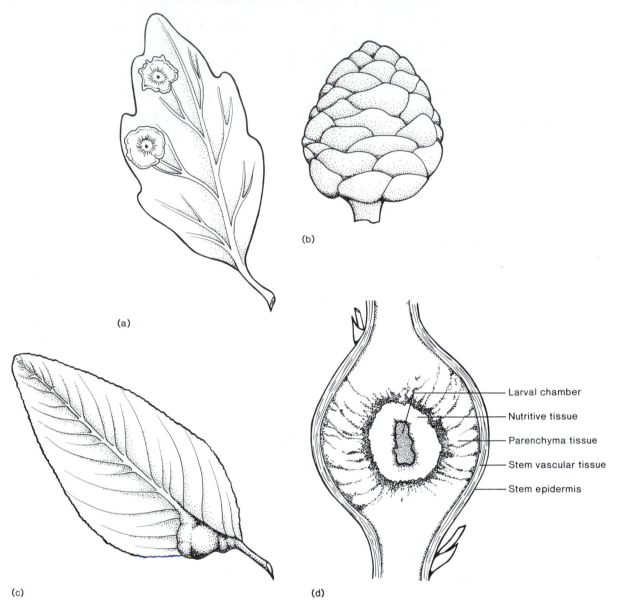

(a)

(b)

(c)

(d)

Larval chamber
Nutritive tissue
Parenchyma tissue
Stem vascular tissue
Stem epidermis

and vascular tissues, which can devastate the host. Bark beetles in the family Scolytidae are examples (Mitton and Sturgeon 1982).

Gall-Forming Insects

Many phytophagous insects induce the production of abnormal growth reactions, *galls,* in the tissues of their host plants, inside which they live and feed. Galls are found in buds, leaves, stems, flowers, or roots and their shape and location are often characteristic of the plant and insect species concerned (figure 10.8 *a–d*). In galls initiated by

species with chewing larvae the insect resides in an inner chamber lined by nutritive tissue upon which the insect feeds. The gall is entirely a product of the plant, developing in response to a chemical stimulus from the secretions of the insect. The diversity of gall types and the biology of host and gall can be explored in books by Meyer (1987) and Ananthkrishnan (1984).

Gallmakers (also called gallers) reach their peak of diversity on dicots, the broad-leaved flowering plants. In North America and Europe, 85% of gall-forming species are found on the oak (especially *Quercus*), rose, and sunflower families. In South America, galls are especially common on legumes, and in Australia half of all known gallmakers are found on plants in the Myrtaceae, especially on *Eucalyptus* trees (Meyer 1987). Gallers are found primarily in the Diptera (Cecidomyiidae) and Hymenoptera (Cynipidae), but there are some representatives in the Hymenoptera (Chalcidoidea, Agaonidae), Coleoptera, Thysanoptera and Lepidoptera and in the fluid-feeding Homoptera (Aphididae, Apiomorphinae) (Waterhouse 1970).

Sucking Insects

Some insects avoid the need to fragment plant tissues by using highly modified mouthparts (see chapter 2) to pierce the plant epidermis and suck plant fluids. In this way, true bugs (Hemiptera), aphids and their relatives (Homoptera), and Thysanoptera avoid consuming indigestible structural components of the plant as well as toxic materials produced by the plant.

Sucking insects survive on several types of plant fluids. Thrips pierce and macerate the contents of individual cells with their stylets and then suck the liquified material through their proboscis. Other species (e.g., the meadow spittlebug, 13- and 17-year periodical cicadas) tap xylem vessels, from which they get a very dilute nutrient solution. But most hemipteran species insert their beaks directly into phloem cells to tap the more nutritious fluid there. Phloem fluid is a sugar solution with low concentrations of necessary nutrients. The specialized filter chamber digestive system that allows phloem feeders to make use of this fluid is described in chapter 4.

The feeding of Hemiptera and Homoptera involves negligible mechanical injury to plant tissues. However, saliva injected by some species may be toxic and sucking mouthparts are often involved in the transmission of viral diseases. So some insects are of more importance as plant-disease vectors than for the direct damage they inflict by feeding.

Seed-Eaters

Seed-eaters (and seedling-eaters) are the only true plant predators among insects because they kill plants by consuming them. Herbivores that consume stems, leaves, buds, or fluids are likened to parasites because the host is not killed (Price 1980); rather a growth module of the plant (either branch or leaf) may be removed, and the plant can grow a replacement part. Seeds are a rich source of protein and are used by both sucking and chewing insects. Seed beetles (Bruchidae) are one of the largest groups infesting seeds. Janzen's (1971a) studies of this group in the neotropics led to the development of some important ideas in ecology (e.g., predator satiation, seed shadows).

Plants As Food

Plants, permanently rooted to a spot, cannot flee from their consumer. Therefore, it is easy to assume that plants are an available and abundant food. But their immobility is the only "advantageous" aspect of herbivory! The structural parts of plants vary extensively and require different ways of feeding. Plants are only marginally nutritious, and most plants have a variety of both toxic and physical defenses against consumers. Further, the effects of air pollution on plant quality may exert indirect effects on phytophagous insects (boxed reading 10.1). The wide variety of characteristics plants use to defend against herbivory have been the subject of extensive and exciting research over the past decade, both in natural ecosystems and in crop plants. Some of these characteristics, including structural materials, nitrogen, water, chemical and physical defenses, are described below.

Structural Materials

The rigidity of plants is imparted by the structural carbohydrates cellulose and hemicellulose. As plant tissue ages, these carbohydrates accumulate and tissues become tougher and more difficult for insects to chew or pierce. In studies of Australian rain forest leaves, herbivory of young leaves was 15–40 times greater than of old leaves, and leaf toughness increased similarly (Loman and Fox 1983). There is a large amount of energy bound up in cellulose and hemicellulose, but insects cannot digest them since they do not produce cellulase, which is necessary to break down cellulose. The bulk of what almost all chewing insects eat is consequently indigestible. Important exceptions to this are insects such as termites, leaf cutting ants, and bark beetles, which have internal or external mutualistic associations with microbes or fungi that can produce cellulase (see chapter 4). Plants also produce waxes, cuticle, and other indigestible structural materials and concentrations of these, in addition to cellulose and lignin, increase as tissues age. They are especially high in tissues of plants growing in high light, dry or nutrient poor habitats where the vegetation is referred to as sclerophyllous (hard leaf) (Morrow 1983).

Nitrogen

Insects are more than 50% protein and 7% to 14% nitrogen by dry mass because they use protein as a major component of their structural compounds (e.g., arthropodin, sclerotin; see chapter 2). In contrast, plants have nitrogen levels that are much lower, from 0.5% to 8% nitrogen (figure 10.9), so insects must concentrate large amounts of protein from a very dilute source. As the nitrogen content of their food increases, insects become more efficient at extracting it (figure 10.10a). Many studies demonstrate increased growth (figure 10.10b),

Figure 10.9 Variation in nitrogen content of animals, fungi, and different plant tissues. Bars span the range of typical values. *Data from Strong et al., 1984.*

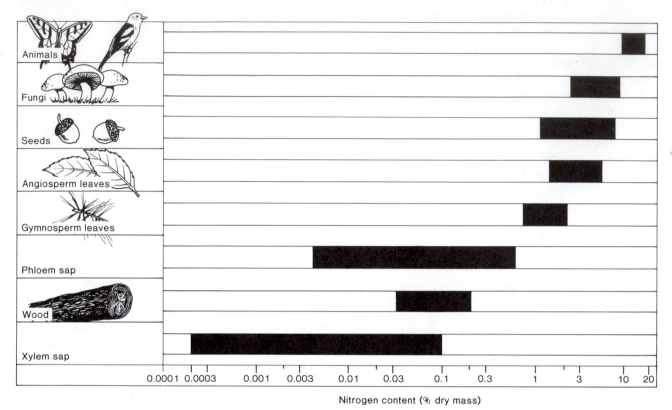

Nitrogen content (% dry mass)

Figure 10.10 (*a*) Efficiency of conversion (biomass gained/biomass consumed) for various invertebrate phytophages as a function of the nitrogen content of their food (Mattson 1980). Low-nitrogen diets are used inefficiently because the rate of feeding and of throughput in the gut are increased to compensate for low food quality; therefore, digestion is less efficient. (*b*) Effect of leaf nitrogen content on development rate of *Paropsis atomaria* larvae (Coleoptera; Chrysomelidae). The number of days from hatching until most larvae in a cohort become prepupae decreases as nitrogen content increases (Morrow and Fox 1980). Percentage nitrogen $\times$ 6.25 gives an estimate of protein content (Waldbauer 1968).

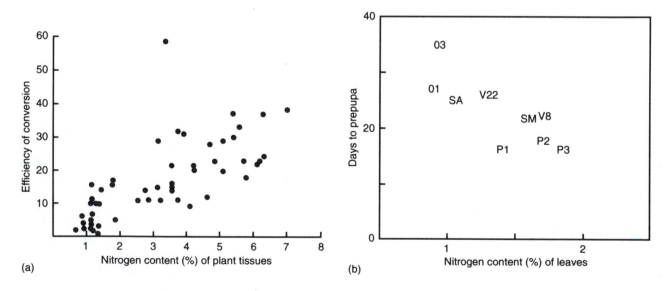

Pollution Induced Changes in Plant Quality

There is growing evidence that air pollution changes insect performance and levels of attack on plants (Riemer and Whittaker 1989). At the broadest level we might suspect that the nitrogen compounds in air pollution and increased CO_2 would affect levels of structural carbohydrates and concentration and composition of leaf proteins. Air pollution has resulted in dramatic inputs of nitrogen to the environment over the last 30 years. Emissions of NO^3 come largely from fossil fuel combustion and NH^4 from fertilizer applications (which have quadrupled since 1945) and intensive livestock operations (Wedin and Tilman 1992). Insect responses to fertilized and to pollution stressed plants are variable and we are not now able to predict how these factors alone, much less together, will influence insect populations.

The consequences of increasing atmospheric CO_2 levels on interactions between plants and their herbivores have been examined in a small number of studies. Some plants grown at high CO_2 levels had lower nitrogen levels, presumably because of dilution by cellulose and other carbon-based compounds, and insects increased their consumption of high CO_2 plants by 20% to 80%. Despite this, they grew more slowly than larvae on plants grown at ambient CO_2 levels (Bazzaz 1990). In contrast, sagebrush grown at high CO_2 was more readily digested by grasshoppers (Johnson and Lincoln 1991). Clearly, the direct and indirect effects of air pollution are not going to be easy to predict.

survival, fecundity, and longevity with increases in nitrogen. It would appear to be advantageous for insects to maximize their consumption of this limiting resource.

One way for insects to increase nitrogen intake is to eat more, but there is a limit to how much they can compensate for low quality food in this manner. *Eucalyptus* leaves must have at least 1% nitrogen in order for larvae of the chrysomelid beetle *Paropsis atomaria* to gain more nitrogen than they excrete in their frass (Fox and Macauley 1977). For grasshoppers and larvae of *Prodentia eridania,* the break-even level of plant nitrogen is 3% (Mattson 1980). But nitrogen content is more complex than nitrogen concentration alone implies. For example, when nitrogen levels are elevated by fertilization, insects do not benefit as predicted from studies of plants that have naturally high nitrogen levels or levels elevated as a result of stress (Morrow 1983).

The relative proportions of amino acids that make up plant and insect proteins are different. In artificial diets, Karowe and Martin (1989) found that the respiration rate of *Spodoptera eridania* larvae depended upon the amino acid composition, but not upon the quantity of protein. Southwood (1973) concluded that predatory insects convert animal proteins to biomass more efficiently than herbivores convert plant protein. One major difference is in the ratio of phenylalanine to other amino acids. Plants produce relatively small amounts, whereas insects require large amounts to make chitin. Grasshoppers, for example, have significantly increased growth rates when phenylalanine is added to their diet (Bernays 1982).

The nitrogen content of plants varies considerably between plant species, tissues (figure 10.9), and with season. Actively growing tissues require high levels of nitrogen to support protein synthesis, but once the tissue is

fully expanded, synthesis shifts to structural carbohydrates and nitrogen levels become diluted. The deciduous oak, *Quercus robur,* has leaf nitrogen levels of about 5% at bud burst in late April, 2.5% nitrogen when fully expanded and 1.5% just before leaves are shed in October (Scriber and Slansky 1981). The equivalent values for *Eucalyptus pauciflora,* an evergreen tree in nutrient poor, high altitude sites are 2%, 1.5%, and 1.2% nitrogen (Morrow 1983).

Nitrogen levels in phloem fluids also are variable. When the plant is actively growing and later when nitrogen is being retrieved from leaves before they are dropped, the concentration of nitrogen transported in the phloem may reach 0.6%. In between these periods, levels may drop to .03%, and the proportions of amino acids change (Mattson 1980). Phloem-tapping insects such as the green spruce aphid, *Elatobium abietinum*, respond to this nitrogen change by producing winged alates that disperse from the plant in late spring to find alternative hosts (McNeill and Southwood 1978).

Water

The water content of foliage varies from 45% to 95% of fresh weight. All else being equal, leaves with high water content (e.g., expanding leaves or leaves of annual plants) are digested most efficiently and support the highest rates of growth (Scriber and Slansky 1981). Leaf water content is correlated with nitrogen content, and both decrease with increasing levels of structural materials.

Adaptations have evolved in higher plants that enable these plants to deal with the harsh realities of extracting water and nutrients from dry land. The specialized ways of plant feeding that have evolved in phytophagous insects provide evidence that plants have passed difficult nutritional problems for insects to overcome.

Insects that have evolved the ability to use particular host plants exert selection pressures that favor the evolution of defenses in those hosts. We will consider these "co-evolutionary" interactions later. Here we simply outline the main characteristics of plant defenses.

Plant Defenses

Plant defenses against phytophagous organisms of all sorts fall into two basic categories: physical defenses and chemical defenses that are toxic or deterrent, or actively interfere with digestion.

Physical Defenses

Plants commonly have tough surfaces that make them hard to pierce, bite, and chew. Toughness is endowed by thick cell walls with high levels of lignin and by deposits of surface waxes. Silica crystals are common in epidermal cells of grasses and horsetails, *Equisetum,* and abrade the mandibles of herbivores.

Many plants are covered by small hairs called *trichomes,* or *pubescence.* These hairs come in many different forms: long, short, thin, sharp, flattened, slippery, and with or without glands (Levin 1973). Leaves and stems of *Passiflora adenopoda* are covered with small hooked hairs that protect it from larvae of *Heliconius* which are specialized feeders on passion vines. Where hooked hairs pierce the larva's skin, hemolymph drains out and the larvae dries up and dies (figure 10.11*a,b*). Trichomes are an important defense in some crop plants. For example, soybeans without trichomes supported an average of 30 leafhoppers per plant, and those with trichomes had just 2.7 (Singh et al. 1971).

In contrast, the leaves of the giant stinging tree (Urticaceae, *Dendrochide excelsa*) in tropical Australian rain forests are densely covered with stinging hairs, as the name implies. Observations by Lowman (1985) have shown that a host-specific beetle, *Hoplostinis viridipenis,* is undeterred by these defenses. The leaves lost an average of 32% leaf surface area during a lifetime. Stinging hairs may have evolved as a defense against mammalian herbivores in Asia, but are less effective against insects (figure 10.11*c*). While defenses can be ineffective against some herbivores, they may deter other potential herbivores, and so the plant has won at least those battles.

Chemical Defenses

Plants are the original pharmacies. Among the compounds they produce are vanilla, salicylic acid (aspirin), caffeine, nicotine, morphine, pyrethrum and tannins. Frankel (1959) argued convincingly that the diverse array of compounds produced by plants are defenses against herbivores. An enormous literature now exists on relationships between plant toxins and herbivores that support his view. This literature can be tapped through review

volumes such as Rosenthal and Janzen (1979), Rosenthal and Berenbaum (1992), and Harborne (1988) and journals such as *Chemical Ecology.*

Defensive chemicals, also called *allelochemics,* or *secondary compounds,* are produced from biosynthetic pathways related to the primary metabolic pathways (figure 10.12) common to all or large numbers of organisms. Plants produce an incredibly diverse array of these compounds and all of them can be traced back to a very limited number of precursors, acetyl coenzyme A, mevalonic acid, and shikimic acid (Waterman and Mole 1989). Allelochemics can poison the physiological systems of herbivores, many of which are the same as those of plants. Plants avoid poisoning by isolating allelochemicals from physiologically active sites, for instance in glands, resin ducts, or vacuoles. Or they may be stored as inactive precursors that become damaging only when metabolized by the herbivore or when the plant is damaged. For instance, free hydrogen cyanide, which poisons mitochondrial respiration, is released from damaged clover leaves when cells are ruptured and cyanogenic glycosides and glycosidases come together (Jones 1973). Free cyanide can deter insects and is even more toxic to mammalian herbivores (Bernays 1982).

We can roughly categorize allelochemicals into two groups, those that exacerbate the already low nutritional value of plants and those that are toxic to basic metabolic processes. The first group are variously referred to as *digestive reducers, quantitative, immobile,* or *carbon-based defenses.* Passively or actively, they make food less nutritious (table 10.1). Eating becomes a more time consuming process and increases the length of time an insect is exposed to predators, parasites and environmental hazards. These chemicals generally are present in high concentrations (e.g., up to 23% of the dry mass of Sitka spruce bark is lignin and 21% of eucalyptus leaves may be essential oil) and their effectiveness is dose dependent (Wainhouse et al. 1990). Quantitative compounds accumulate as tissues age and the plant does not scavenge their largely carbon components before shedding the structure (hence, *immobile* compounds). Compared to the next group of allelochemics, digestion reducers are not very diverse.

The second group of allelochemics are referred to as *toxic, qualitative,* or *mobile compounds.* Qualitative compounds are present and effective in small concentrations (generally less than 5% leaf dry weight) and target essential functions of the herbivorous animals such as respiration, DNA repair, transmission of nerve impulses, and hormone production (table 10.1). Often they are synthesized and degraded over fairly short periods, hours to weeks, and their component parts end up in other compounds. In the plant they may have nondefensive as well as defensive functions. For instance, nicotine, a potent insecticide, also is used by tobacco plants to transport nitrogen to the shoot from uptake sites in the roots, a

Figure 10.11 Larvae of *Heliconius melpomene* eat many species of *Passiflora* but not *P. adenopoda* for the reason illustrated in these scanning electron micrographs. (*a*) Third instar larva caught on leaf petiole (85×). Proleg marked x is enlarged in (*b*). (*b*) Proleg has hooked trichomes embedded at a, b, and c causing the larval skin to split and hemolymph to drain from the wound (85×). (*c*) Host-specific chrysomelid (*Hoplostinis viridipenis*) feeds on the giant stinging tree in Australian rain forests, obviously undeterred by the hairs. Hairs have a sharp point that on contact detach, injecting poisons whose effects may be felt by humans for up to a month.

(a) and (b): Gilbert, L. F., and SCIENCE: 172, 585–6 (1971), © AAAS. (c): © M. Lowman.

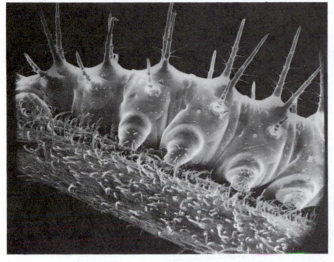

(a)

(b)

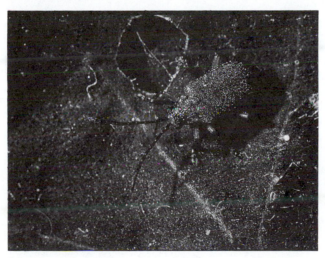

(c)

Figure 10.12 The ent-kaurene to gibberellin (a hormone) primary pathway probably present in all higher plants. Diterpenoids indicated to the right of the dotted line are secondary metabolites characteristic of goldenrods, *Solidago*, which inhibit feeding by many herbivores.

Modified from Graebe, 1987.

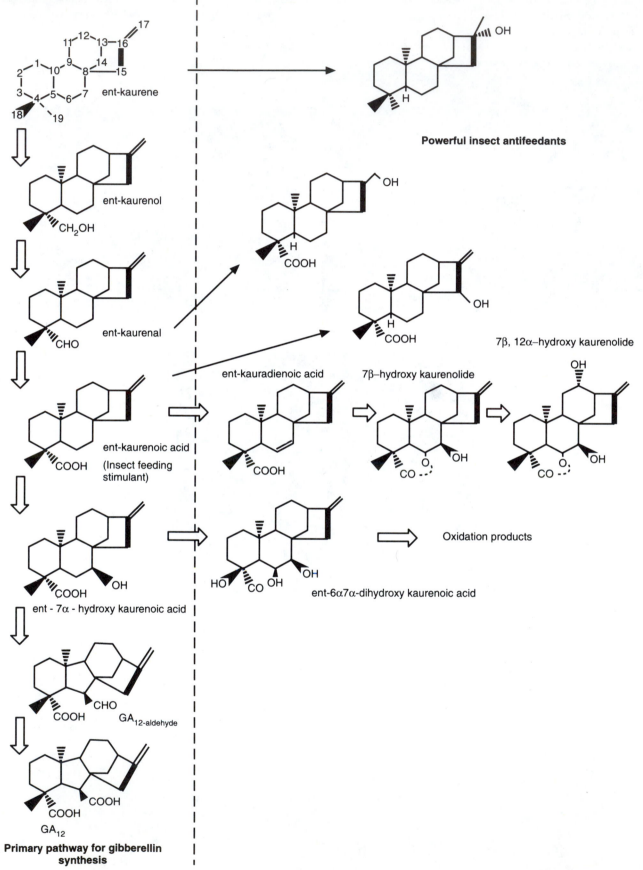

Table 10.1   Selected Plant Products that Reduce Herbivory[a,b]

Chemical group (number identified)	Description	Defensive role
Quantitative Digestibility Reducers		
Cellulose (1 basic type)	Sugar polymer	Requires gut flora for digestion
Hemicellulose (1 basic type)	Sugar polymer	Requires gut flora for digestion
Lignins (indefinite)	Phenolic polymers	Bind with proteins and carbohydrates
Tannins (indefinite)	Phenolic polymers	Bind with proteins
Silica (1 basic type)	Inorganic crystals	Indigestible
Qualitative Toxins		
Alkaloids (20,000)	Heterocyclic N-containing	Many; some stop DNA and RNA production
Toxic amino acids (260)	Analogues to protein amino acids	Compete with protein amino acids
Cyanogens (23+)	Glycosides that release HCN	Stops mitochondrial respiration
Glucosinolates (80)	N-containing K salts	Many; endocrine disorders
Proteinase inhibitors (indefinite)	Proteins or polypeptides in subunits	Bind with active site of enzymes
Terpenoids (100,000+)	Polymers of C_5 units	Many; some stop respiration

[a]See Rosenthal and Janzen (1979) for a discussion of other secondary compounds.
[b]C, carbon; H, hydrogen; K, potassium; N, nitrogen; HCN, cyanide.
Some compounds have characteristics of both groups. Silica, for instance, can be viewed as both a physical and a quantitative defense and the defensive properties of cellulose may be coincidental to its primary support function (Howe and Westley 1988).
From *Ecological Relationships of Plants and Animals* by Henry F. Howe and Lynn C. Westley. Copyright © 1988 by Oxford University Press, Inc. Reprinted by permission.

function performed in most plants by other nitrogen-rich but nontoxic compounds such as glutamine and asparagine and amino acids closely related to them (Pate 1973).

Plant defenses may be present in a plant all of the time or they may be produced in response to tissue damage. *Constitutive defenses* are a plant's permanent protection. They include most of the physical defenses like silica crystals and trichomes as well as many digestion reducing and toxic compounds (table 10.1). They help plants survive first encounters with herbivores. *Inducible defenses* are produced after the plant begins to experience damage (Tallamy and Raupp 1991; Haukioja 1991; Weins et al. 1991). The best documented inducible allelochemicals are proteinase inhibitors. These polypeptides and proteins block the catalytic activity of digestive enzymes in the gut by binding to the active site of the enzyme molecule (Ryan 1979, 1983). Proteinase inhibitors are extremely common in the plant kingdom, including the legume family. They are denatured by heat, one of the reasons we cook beans and some other vegetables. See chapter 4 for a brief discussion of how genetic engineers have used the proteinase inhibitor genes to develop plants that are resistant to certain pest insects.

Catching Food: Temporal and Spatial Availability

Plants are dynamic. All of the properties so far discussed change with time and in response to biotic and abiotic environmental challenges. The parts of a single plant can vary enormously in their potential to support the growth and reproduction of an insect. These differences have causes ranging from the divergent physical properties of various structures such as leaves and stems, the unequal production, transport and storage of nutrients, and fluctuations in the concentration of toxins and other compounds that interfere with feeding. Plant variation imposes severe constraints on many insects (Whitham 1983; Schultz 1983).

The boundaries of those constraints have been studied for the aphid, *Pemphigus betae,* by Whitham and his colleagues. In the spring, females emerge from eggs that overwintered in bark fissures of their cottonwood host tree, *Populus angustifolia* (figure 10.8c). Galls started within three days of bud break have the greatest chance of being successful. By day seven, leaves are fully expanded and lignified and cannot be galled. Bud break is nearly simultaneous throughout a tree so there is little room for error in female emergence time. This is not an easy problem to deal with because date of bud break varies between trees within a population (3–7 days) (Whitham 1978) and for individuals from year to year (e.g., figure 10.13a).

The gall aphid's temporal problem is exacerbated by spatial problems. On leaves less than 5 cm² in area, most galls fail. Above this size, the number of offspring that emerge from the gall increases with leaf size, which is related to the amount of nitrogen the tree sends to a leaf; larger leaves get more than smaller ones. Most leaves are too small and the large leaves are scattered throughout the crown. Females crawl from branch to branch searching for unoccupied large leaves. The females actually fight one another for possession of the best leaves (Whitham 1978). Not unexpectedly, the number of galls on a tree varies enormously from year to year (Moran and Whitham 1988).

Figure 10.13   (a) Twenty-six year record of the date of bud burst for tagged branch of an aspen (*Populus*) tree. (b) Relative abundance and duration of leaf production on five saplings of *Eucalyptus blakelyi* at a single location in one year. The asynchrony in leaf initiation, which increases as the (southern hemisphere) growing season progresses, is caused by individual responses of trees to different levels and timing of insect defoliation.

(a) *Data from Hodson, 1991. (b) Data of L. R. Fox.*

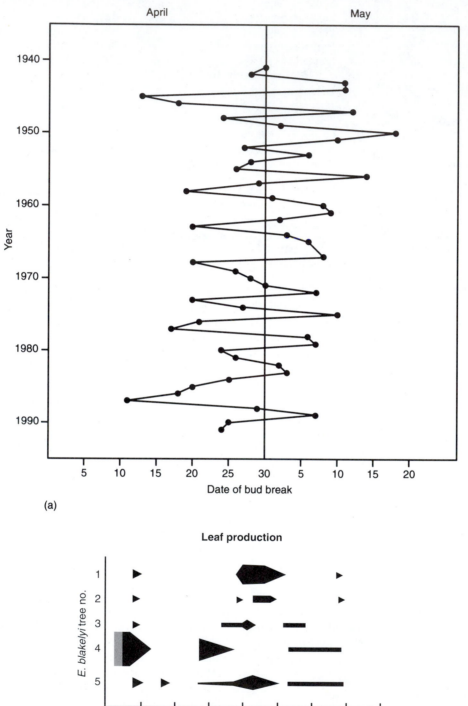

(a)

(b)

Trees in the large genus *Eucalyptus* have a very different leafing phenology. They initiate new leaves whenever it is sufficiently warm and wet and they are especially prone to do this in response to damage, for example, fire or insect attack. This behavior can result in continuous growth or in bursts of growth (figure 10.13*b*). Unlike the narrow window of time available for *P. betae* to exploit cottonwood, eucalypt insects can have multiple large windows. This is probably one of the factors responsible for the high levels of damage eucalypts sustain most years (Fox and Morrow 1983, 1986; Morrow and Fox 1989). Furthermore, insect attack opens these windows wider because trees respond to attack by producing new flushes of growth, which the insects then eat. This is a major contributing factor in the widespread eucalypt dieback in rural regions of Australia (Heatwole and Lowman 1986), widespread tree decline that has ranked amongst the most severe in the world. Outbreaks of psyllids, beetles, and other insects repeatedly defoliate the canopies of eucalypts until mortality occurs. Defoliation can reach 275% per year, because the insects defoliate the canopy each time it refoliates. The stress of repeated defoliation is all the more significant because the eucalypts already face other stresses, for example, widespread clearing of adjacent forest, soil compaction from livestock, recurrent droughts, and depletion of many insectivorous bird populations.

Recently, Whitham (1989) has shown that hybrid cottonwood trees support several orders of magnitude more gall aphids than the parent species. A contributing factor may be the increased time over which bud burst occurs in the hybrid zone, 21 days versus 3–7 days in the parent zone (T. G. Whitham, unpublished data).

Mutualism and Coevolution

Every organism must obtain enough food to maintain itself long enough to reproduce if it is to pass its genes to the next generation. As an organism proceeds from conception to reproductive maturity, it must constantly obtain food and at the same time avoid becoming food for some other organism. Organisms that are unable to do this successfully, for whatever reason, are maladapted and leave few descendants. Although a number of environmental factors are undoubtedly involved in the selection process, examination of specific adaptations often leads to the identification of specific selecting agents. Such is the case when we examine adaptation associated with many interspecific interactions. Organisms that are closely associated ecologically often act as agents of selection reciprocally; that is, they act as selecting agents on one another. There is good evidence that many organisms have done so for long periods of geologic time. These organisms are said to have coevolved.

Examination of several examples of coevolution reveals two types of interactions. Two different species or guilds of species that have coevolved to the extent that they benefit one another are termed *mutualistic*. In this

Figure 10.14 Bull's horn acacia branches (*Acacia cornigera*) are occupied by an aggressive stinging ant (*Pseudomyrmex ferruginea*). Ants live in the thorns and feed on starchy Beltian bodies and sugar-rich secretions from extrafloral nectaries. The ants clear vegetation around host trees and attack insect and mammalian herbivores that might eat the foliage.
From Ecological Relationships of Plants and Animals by Henry F. Howe and Lynn C. Westley. Copyright © 1988 by Oxford University Press, Inc. Reprinted by permission.

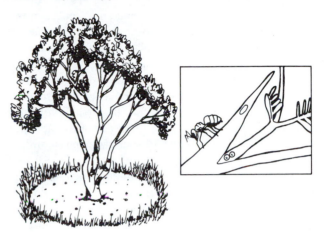

case, the presence of both enhances the reproductive success of both. Alternatively, two species may engage in a kind of genetic warfare whereby one adapts a mechanism of offense providing increased access to the other as food, only to have the other "respond," in the evolutionary sense, with a defense. Examples include predators and prey, parasites and hosts, and herbivores and plants. For a comprehensive introduction to coevolution, see Thompson (1982, 1989).

We give two examples of mutualisms here: ants and plants and plants and their pollinators.

Ants and Plants

Some species of ants in the neotropics live in acacia trees. An aggressive stinging ant (*Pseudomyrmex ferruginea*) nests in the hollow thorns of the bull's-horn acacia (*Acacia cornigera*) and feeds on nutrient-rich Beltian bodies and sugar secretions from extrafloral nectaries present on all the leaves. In return the ants attack herbivores and clear vegetation surrounding their acacia tree (figure 10.14). Acacias without ants have lower growth rates and are often killed by herbivores, fire, or crowding from other plants (Janzen 1966). In this case, the mutualism is *obligate*, neither the ant nor the acacia survives without the other.

More common than obligate mutualism is *facultative mutualism* (i.e., the interaction occurs under some conditions but not others). One of the most common kind of facultative mutualism is between the many species of plants with extrafloral nectaries and the many ant species that seek nectar. While searching for nectaries, ants may prey upon herbivores they encounter. Both mutualists benefit from the interaction but neither is dependent upon it. Tilman's (1978) study of ants, cherries, and tent caterpillars is an excellent example of facultative mutualism.

Plants have a mutualistic relationship with many ants for the purposes of seed dispersal. Ants benefit by gaining a food source, and plants benefit by having their distributions extended (because ants invariably drop some seeds on their way back to the colony). There are two major variations in this phenomenon: some ants feed upon the seeds (and so plant dispersal depends upon the frequency of seeds dropped); and some ants feed upon an *elaiosome* (fat body) that is attached to a seed, thereby ignoring the seed which is left to germinate (Handel and Beattie 1990). This latter relationship, termed *myrmecochory* (from the Greek for "ant," *myrmex,* and for "dispersal," *kore*) insures complete dispersal of all the seeds of a plant that are harvested by ants. The ants benefit from consumption of the elaiosome, and the plants benefit from effective seed dispersal. Elaiosomes have been found in taxonomically diverse plants, including northern temperate wild flowers (e.g., *Trillium* and *Viola*), tropical epiphytes and vines, and Australian arid shrubs (e.g., *Acacia*). The morphologically and taxonomically diverse origins of elaiosomes indicates convergent evolution: originally they protected seeds from predation but gradually changed into food lures for ants. Even the biochemical attractants contained within the elaiosomes have been identified, with some proving to be the same from plants found on different continents!

Other ants have special relationships with plants whereby they cultivate fungal gardens (providing nutrients for the ants) in exchange for shelter and a supply of plant material that the ants harvest from outside the nest. Huxley (1978) studied ant-plants in New Guinea that have evolved bulbous chambers to house their ant partners and Thompson (1981) explores how these associations may have arisen. Huxley and Cutler (1991) provide a review of anti-plant interactivities.

Plants and Pollinators

Flowers and their pollinators have developed specialized relationships over evolutionary time. The first winged insects were present 300 million years ago in the Carboniferous, long before the development of flowers (Smart and Hughes 1972). The first flowers appeared in the fossil record approximately 225 million years ago and were small (those of the gymnosperm order Bennettitales), so presumably they had insect pollinators. They are termed *entomophilous* (pollinated by insects that feed on pollen, nectar, or on the flower itself). At about the same time the first holometabolous insects appeared. The adults of these insects were adapted to feed on different foods from the larvae. This alteration of food supply between different life stages greatly enhanced the opportunities of insects to specialize upon leaves as larvae and upon pollen as adults. The radiation of Lepidoptera, an insect order whose coevolution in pollination ecology is well studied, began during the early Cenozoic, and today we appreciate a wide variety of specific relationships between pollinators and plants.

Sexual reproduction in flowering plants is accomplished by the transfer of pollen from the anther of a male flower to the stigma of a female flower. When a pollen grain contacts a stigma, the male germ cell in the pollen grain eventually unites with the female germ cell or egg, and a fertile seed develops. When the seed is exposed to the proper conditions in or on an appropriate substrate, it will give rise to a new plant. The transfer of pollen from a male to a female flower is accomplished primarily by the wind or by the activities of insects that associate with plants. Examples of wind-pollinated plants include cereal plants such as wheat and corn, ragweed (of hay fever fame), and many species of trees. The flowers of these plants, which are generally small with weakly developed petals, do not produce nectar, and produce dry pollen grains that are easily picked up by the wind.

The evolution of pollination involved the processes of *divergence* (whereby organisms become different) and *convergence* (whereby organisms become more similar over time). Fossil evidence indicates that animal mutualists changed more significantly over evolutionary time as compared to flowering plants, which have remained remarkably stable over the last few million years. However, some plant families exhibit classic specialization of reproductive parts that attract specific pollinators. A good example is the phlox family, which evolved from one bee-pollinated flower to a variety of flower heads that have different types of pollinators (figure 10.15).

Many factors affect the complex relationships between flowers and their insect pollinators. These include the nutritional rewards of pollen and nectar, as well as the physical features such as shape, color, and scent of flowers; the phenology (i.e., seasonal availability) of flowers in relation to population dynamics of insects; the growth structure of flowers (clusters versus solitary); the plant community surrounding a flower (i.e., competition from other plants present); and, of course, competition from other pollinators. With all these factors to account for, it is obvious how complex plant-pollinator relationships are for biologists to unravel!

In pollination, both partners enjoy a mutual benefit from the relationship. The plant is propagated, and the pollinator enjoys a calorific reward. An interesting aspect of the energetics of the plant-pollinator mutualism was investigated by Baker and Baker (1973). It has long been known that nectar contains various sugars and is used as an energy source for pollinators, but insects also require

Figure 10.15   The phlox family (*Polemoniaceae*) illustrates divergence of flower form due to selection from different pollinators. Self-pollination (S) evolved independently in several of these adaptive branches. *From Barth, 1985; after Grant and Grant, 1965.*

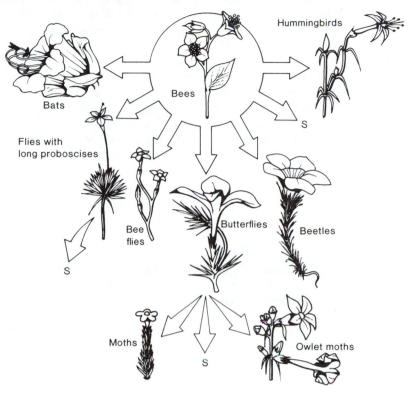

the molecules with which to synthesize proteins (amino acids) which presumably had to be found elsewhere (e.g., in Lepidoptera, from larvae feeding upon leaves). The Bakers surveyed 266 species of flowering plants in California and found that nectar contained amino acids in many species, with relative amounts of amino acids highest in specialized lepidopteran-pollinated flowers.

Plants and pollinators not only have complex relationships in terms of energy exchange, but their structures ensure long-term consequences such as regulation of the pollination process. The promotion of *outcrossing* (whereby a plant gets pollen from another individual, not from itself) was studied by Heinrich (1979), who found several interesting patterns: flowers may evolve to restrict visits to pollinators, excluding "robbers" that take the rewards without transferring pollen; plants evolve different levels of nutritional reward and stand densities to regulate the number of flowers that insects are encouraged to visit; plants can control the amount of the calorific reward in relation to the size and behavior of the pollinator (e.g., bumble bees require more energy than a small fly).

Distinct correlations can be made between the anatomical and physiological characteristics of the flowers of a given species and the anatomy, physiology, and behavior of their insect pollinators. Among the characteristics of flowers attractive to insects are:

1. The production of particular scents.
2. Color, size and shape of petals.
3. Patterns of stripes or spots on petals.
4. Separation or nonseparation of petals.
5. Shape of flower.

Scent production is often important in attracting a pollinator and in determining behavior. Flowers pollinated by flies, which breed in dung and carrion, may actually mimic the odor of dung or rotten meat, so "fly flowers" commonly have a disagreeable odor to humans. Conversely, "butterfly and moth flowers" usually have a sweet smell. Honeybees are attracted to blue, purple, and yellow flowers, but cannot "see" red flowers. Many insects are attracted to patterns of ultraviolet light. The shape of nectaries are often suited to restrict undesirable species and to encourage the pollinator species.

Over the course of evolution, many pollinators have adapted to using particular flowers. A good example of this is the orchid family, the majority of whose members have specific pollinators and structurally complex flowers to pollinate. Some orchids produce scents that mimic the effects of sex pheromones, stimulating the insects both visually and tactilely. Another interesting family is the figs with 600 species, each of which has a specialized wasp pollinator. Pollinators that visit only one taxon of plant are termed *monolectic* (versus *polylectic* pollinators that visit many different taxon). What is the evolutionary advantage of visiting only one taxon? In the case of fig wasps, they not only pollinate the fig flowers but also lay eggs in the floral receptacle called a *synconium*. This structure provides a safe home for the wasp larva, which consumes the ovule in which it was placed. Upon emergence, it can easily find a mate and fly to another ripe fruit, thereby completing its life cycle within one tree canopy. In this sense, monolectic behavior offers a secure existence to the host-specific fig wasp.

Even the foraging dynamics of pollinators can be a complex coevolved process. For example, euglossine bees forage for food throughout kilometers of tropical forest and frequently return to the same plants (Janzen 1971b). This trap-lining behavior is advantageous because outcrossing occurs despite the very low plant densities characteristic of many tropical tree species. Only one of several flowers must produce nectar each day, and floral morphology can become specialized for pollinators that can learn their way around a forest and that are usually also efficient (and do not waste pollen). Because smaller amounts of energy are required to ensure pollination, tropical trees can reproduce much earlier in their life span and also save energy for other more competitive aspects of survival (e.g., getting light and nutrients). Floral visibility is also less important and can be effective even under the shaded, complex tropical canopy.

The most common insect pollinators are members of the orders Coleoptera, Lepidoptera, Diptera, and Hymenoptera. Their most common adaptations include elongated mouthparts that allow them to get nectar from flowers with deep nectaries, plumose (featherlike) hairs on the body to which pollen clings, and various specialized pollen-collecting and/or transporting structures such as the corbiculae (pollen baskets; see figure 2.37a) on the hind tibiae of honeybees. The relationships of plants and insect pollinators are an exciting aspect of ecological research, because coevolution is a dynamic process and there are so many groups of insects and plants yet to study. The topic of insect-plant interactions is covered by Prince et al. (1991) and Abrahamson (1989).

Insects in Forests as an Example of Community Aspects of Insects and Plants

Insects form a community within a plant, and similarly insects and plants together form part of a larger community or ecosystem. Although insect epidemics can occur anywhere (even in deserts), forests generally contain the most diverse and abundant fauna of invertebrates. This is because forests contain a wide diversity of plants (trees, shrubs, herbs, mosses, lichens, ferns, and algae, each with its own insect fauna); are slow-growing and older than many surrounding early successional ecosystems; and offer a large selection of food sources of different ages, palatabilities, and nutritive qualities. The relatively long time scale involved in forest succession allows for colonization and more complex levels of interactions of organisms to become established.

Insects in forests are difficult to study, but represent a great challenge due to the anticipated diversity that will be found there. Whereas biologists can count and measure the number of mollusks on the intertidal rock platform (a relatively two-dimensional habitat), it is nearly impossible to view the invertebrates of a forest due to its structural density, complexity, and height. These logistic problems have led many scientists to study insects in other habitats; hence, a paucity of literature exists on forest insects, except in outbreaks in which numbers are easier to obtain.

In recent years, biologists have become particularly interested in insects of rain forests for two reasons. These habitats are endangered by human activity, and the need to study them before they are reduced to fragments or to clearings is paramount. Second, rain forests contain the greatest wealth of insect species anywhere in the world (Wilson 1988). Despite the fact that they occupy less than 7% of terrestrial habitats, they have a variety of plants and a relatively homogeneous wet climate that has enabled the evolution of a wealth of species. Wilson (1987) found 43 species of ants in one leguminous tree in the tropical rain forests of Peru, equivalent to the ant fauna of the entire British Isles! And Erwin (1982) found 1,080 species of Coleoptera by fogging four rain tree forest canopies in Brazil. What was more astounding than the diversity, however, was the fact that 83% were endemic to only one site. Only 3.2% of the species were shared among all four sites, indicating that almost every tree canopy in the tropics may have a specialized insect fauna!

In subtropical understory vegetation in Australia, Lowman (1982) sampled insect densities that ranged from 65 to 650 per 100 m^3 throughout different months of the year. So, even in the relatively benign climates of the tropics and subtropics, insect seasonality is very pronounced. This suggests that biologists studying tropical insects need to consider many factors in their sampling design and replication: numbers and diversity of tree species, age of forest, season of year, day versus night, phenology of foliage, and weather, to name a few.

Herbivory levels are often more commonly measured in vegetation than are the insects that cause the damage. This is partly because herbivores are often hard to see, feeding only intermittently or at night. Herbivory serves as an important indicator of insect abundance, particularly in such ecosystems such as rain forests where the herbivores are especially difficult to observe and count.

Figure 10.16 Herbivory in canopies of cool temperate rain forest, dominated by Antarctic beech (*Nothofagus moorei*). Each number represents mean amount of leaf surface area removed annually by insects (%). *Data from M. Lowman.*

D.a. = *Dicksonia antarctica* C.v. = *Cuttsia viburnea*
N.m. = *Nothofagus moorei* Q.s. = *Quintinia sieberi*
D.s. = *Doryphora sassafras* C.q. = *Coprosma quadrifolia*
C.s. = *Callicoma serratifolia*

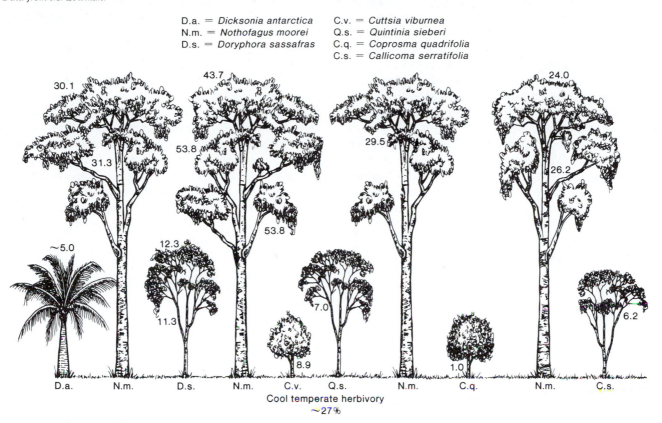

Cool temperate herbivory
~27%

For example, high levels of herbivory in the cool temperate rain forests of Australia led Lowman to look for the insect(s) responsible for this high defoliation. Like many aspects of field biology, searching for an insect in a complex mosaic of green foliage is equivalent to finding a needle in a haystack. Fortuitous observations led to the discovery of a new genus of chrysomelid beetle whose larvae are prolific but only during a 10-day period in spring (October) every year. This herbivore, *Novocastria nothofagi,* eats only the Antarctic beech trees that dominate the cool temperate rain forests in New South Wales. The beetle larvae emerge synchronously with the leaf emergence and defoliate over one half of the new leaves, contributing significantly to the overall 30.5% annual leaf surface removal for beech canopies (figure 10.16).

This new beetle is just one of a vast fauna of yet-to-be-discovered insects in this rain forest type alone! No one can predict the actual numbers of insects and plants becoming extinct as rain forests are continually cleared, but the urgency to document, study, and preserve the ecology and biodiversity of what remains has become an important issue of international priority today.

How many insects are found in a tree? This question appears simple and straightforward, but is actually very complicated and not well-defined. There are many logistic aspects (previously mentioned) challenging biolo-

gists who aspire to find all the insects in a tree. In addition, factors that lead to variability in the numbers of insects inhabiting a tree, as previously mentioned, the methods of counting insects in trees are not well established, and ecologists face challenges to merely locate all insects in a tree, much less count them! Some insect orders characteristically feed on leaf surfaces (Coleoptera, Hemiptera, Phasmida, and Lepidoptera larvae) and are best sampled with sweep nets or with beating trays. Other orders are often involved with temporary feeding on trees (e.g., pollinators) and occupy the air spaces between trees (Diptera, Lepidoptera, some Hymenoptera). Some insects are nocturnal and may be collected with light traps. Insects associated with the soils and roots of trees are difficult to quantify. Pitfall traps offer opportunities to collect ground insects, but once again they are not comprehensive in their ability to quantify the abundance of organisms. Insects in the wood and root systems of trees usually go uncounted. There are no comprehensive methods to measure insect abundance, although the use of sweep nets is perhaps the most widely-used technique. It is obvious that numbers of insects in trees are vastly underestimated.

Fogging is perhaps the most comprehensive method of documenting the numbers of insects on trees. It involves the dispersal of an insecticide throughout the upper tree canopy, usually with a spraying device. The insects are collected on sheets at ground level after the insecticide has been effective. The sampling technique requires windless conditions with careful attention paid to season, time of day, and forest structure and composition for replication. This method fails to collect insects in the bark and roots, and may underestimate numbers of large insects that may be more resistant to insecticide.

The paradox of insect outbreaks and forest trees is a vast and complex topic, summarized in an entire volume (Barbosa and Schultz 1987). Trees are sessile, long-lived, and extremely reliable food sources for insects, particularly in forests where the amount of foliage ranges from 1–9 t ha^{-1} yr^{-1} (Van Cleve et al. 1983). More than 120 lepidopteran species feed upon *Quercus robur,* yet very few reach outbreak levels (Feeny 1970). Insects are very fecund and are renowned for their ability to adapt to large persistent food supplies (agricultural fields, trees), so it is surprising that outbreaks are not more commonplace. Of the vast numbers of herbivores on western American trees, only 31 species are considered pests (Furniss and Carolin 1977). Only 16 of these are defoliators (with 15 as feeders in bark or wood), and three of the 16 are introduced (e.g., gypsy moth). The number of native pests on this apparently abundant food supply is extremely low!

Whether or not insects are considered pests is an *anthropocentric* (human-centered) view, and oftentimes the fluctuations in insect numbers are a consequence of human activities. The eucalypt dieback in Australia and the gypsy moth outbreaks in the eastern United States are good examples. As long as insects continue to fluctuate in numbers over time and space and as long as they continue to compete with humans as herbivores, we will continue to study the complex interactions of insects on plants.

PART THREE # Unity and Diversity

$\mathbf{One}$ need only look around to appreciate the
tremendous diversity of living things, and insects are
certainly no exception. Yet a close look reveals that
despite the diversity, many organisms (insects) are very
similar to one another. In fact, organisms can be
grouped on the basis of similarities. This has, of course,
been recognized for centuries. Aristotle (300 B.C.) was
among the early scientists who devised a grouping
(classification) of organisms. Modern classification
begins with the work of Carl Linné (Linneaus) and
others in Europe during the eighteenth century. In the
nineteenth century, Charles Darwin and Alfred Russel
Wallace developed a paradigm, evolution by the process
of natural selection, which, to date, stands as the best
explanation of both the diversity and underlying unity
of organisms.

   In the following three chapters we examine unity
and diversity as exemplified by insects. Chapter 11,
"Insect Classification and Evolution," first deals with
the methods and principles involved in describing,
naming, and classifying insects and then considers the
evolution of insects beginning with the origin of the
phylum Arthropoda. Chapters 12 and 13 provide an
overview of the class Insecta, including a framework of
classification and information on the biology and
importance of each order.

CHAPTER 11

Insect Classification and Evolution

McPheron/Romoser

Systematics

Having gained some idea of insect morphology, physiology, and behavior from the preceding chapters, you are in a position to consider the threads of unity that run through the diversity of insects. Finding these threads of unity—morphological, physiological, or otherwise—and using them to produce meaningful and useful groupings of insects is the objective of *insect systematics*. Some authors define systematics simply as the science of classification, making it synonymous with another widely used term, *taxonomy*. Most, however, expand the definition of systematics to include the study of the evolutionary history of organisms.

Regarding the place of systematics among the other specialties in the biological sciences, the following quotations are appropriate.

> Systematics is at the same time the most elementary and the most inclusive part of zoology, most elementary because animals cannot be discussed or treated in a scientific way until some systematization has been achieved, and most inclusive because systematics in its various guises and branches eventually gathers together, utilizes, summarizes, and implements everything that is known about animals, whether morphological, physiological, psychological, or ecological. . . (Simpson, 1961)

> Systematics, the study of biological diversity, is sometimes portrayed as the mere classification of organisms, but in fact its range and challenge are among the greatest in biology. . . . If a biologist is well trained in classification of the organisms encountered, the known facts of natural history are

an open book, and new phenomena come more quickly into focus. (Wilson 1985)

The major tasks of systematics are

1. Identification.
2. Description.
3. Concern with the proper application of the rules of nomenclature.
4. Study of speciation.

These tasks are typically carried out in sequential order for a given group of organisms. For a few groups of insects, systematics has reached the speciation level, but most groups are still in the identification, description, and classification stages of development. Danks (1988) and Kim (1989) clearly articulate the central role of systematics relative to nearly all other entomological pursuits. Tuxen and Lindroth (1973) trace the historical development of insect systematics.

Obviously, collection and preparation of specimens for study must precede identification and the other steps. Discussion of collection and preparation techniques would be inappropriate here, but see Bland and Jaques (1978), Borror, Triplehorn, and Johnson (1989), Borror and White (1970), Cummins et al. (1965), Knudsen (1966), Peterson (1959), Ross (1962), and Stehr (1987). In addition to collection of one's own specimens for study, extensive collections, housed in museums throughout the world, are usually available for serious study.

It is often desirable to culture certain species. Such cultures can provide a wealth of useful biological data (e.g., biochemical, behavioral, physiological, developmental, genetic), that would be unavailable from preserved specimens. Information on culturing insects can be found in Needham et al. (1937), Peterson (1959), Siverly (1962), and King and Leppla (1984).

Identification

Purpose of Identification

The purpose of identification is to determine what kind of organism a given specimen is. The meaning of "kind" depends largely upon one's objectives. A student in a general zoology course may be happy to identify a given organism as a fly, whereas a professional entomologist may need to know the species of fly. Professional reasons for wanting to know the species of an organism (insect) fall into two categories: the nonsystematist's reasons and the systematist's reasons. The nonsystematist, for example, needs to identify a specimen in order to find pertinent literature and to have a name under which to publish his or her findings. Thus the species name of an organism serves as the "file category" under which any and all information pertinent to this species is stored.

The systematist, in addition to using the species name as a key to the literature and a category for data storage, needs to identify an organism to species or at least to the point where it is realized that the organism has not been previously described. This has been called *taxonomic discrimination* (Mayr, Linsley, and Usinger, 1953). Identification of a specimen must necessarily precede fitting this specimen into a scheme of classification.

Methods of Identification

Ultimately, all methods used to identify organisms are based on comparison. One of the most common methods of identification is the use of a key. A key is a printed information-retrieval system into which one puts information regarding a specimen-in-hand and from which one gets an identification of that specimen to whatever level the key is designed to reach. Most keys are *dichotomous* (double branching; figure 11.1). At the beginning of a key, one is presented with two alternatives (a *couplet*), each of which leads to another pair of alternatives. At each point the user of a key picks the alternative that best describes some aspect of the specimen being identified. Finally one reaches a terminal pair, one member of which presents the identity of the organism. The simplest keys are designed for the layman and may offer only superficial identification or deal only with insects found in a specific habitat, such as a garden. More complex keys are designed for the nonsystematist professional biologist who needs to identify an organism to a more specific level than the layperson. These keys are more technically oriented and require some training in morphology before they can be used effectively. Finally, there are keys designed for the specialist, which are very technical and require detailed knowledge of the morphology of a particular group of organisms. These keys usually carry identification to the species level.

Another method of identification is the use of pictures in the form of color plates, black-and-white photos, or line drawings. These pictures may be of entire organisms or parts of organisms and are often used in conjunction with keys. "Pictorial keys" offer alternatives in the form of pictures with or without accompanying verbal alternatives. Pictures are particularly useful when an organism is highly patterned or characteristically colored.

One of the best means for identifying a specimen is to compare it with specimens that have previously been identified. This approach can be used to obtain an identification at any level. *Type specimens* are those on which an original species description is based. These specimens may be extremely useful in making species identifications. Original written descriptions of species based on type specimens are the permanent records of the attributes of a given species. Such records are particularly useful when the type specimens are available, and they constitute the only original record if type specimens are destroyed, lost, or unavailable.

Combinations of the above methods are usually used in identification. Although the layperson or nonsystematist professional biologist can—in many, if not most, instances—identify a specimen to family, or in some cases below the family level, the best and safest way to obtain a species identification is to consult a specialist. From experience and familiarity with pertinent literature, the specialist is in the best position to use the highly specialized keys, original descriptions, and type specimens. In view of the vast number of insect species already described, a given individual is likely to specialize in the systematics of only a very few families or even a single family.

Problems Encountered in Identification

Unfortunately, identification is not without problems. Keys may not include the organism being identified, fail to include the stage or sex of the unknown organism, or fail to take into account any of a large number of possible variations in morphological or other characters used. Pictures can be misleading, especially if an unknown organism very closely resembles the one depicted. Written descriptions can be ambiguous, particularly when such traits as color and texture are described, and such descriptions commonly require knowledge of specialized terminology to be of use. Original species descriptions are sometimes in early volumes of obscure journals that may be difficult to obtain and may be in a foreign language. A collection of accurately identified specimens is not always available. Type specimens may be very difficult or impossible to obtain and may have been lost or destroyed. The major difficulty with consulting a specialist on a given group of organisms is that such a person may be dead or perhaps have never existed. Even if a specialist does exist, there is still the problem of contacting this person.

An additional problem, first recognized by students of insect natural history but greatly expanded by the application of population and molecular genetics principles

Figure 11.1   Sample dichotomous key that differentiates the insects from centipedes and millipedes.

1(a) two body regions (head and trunk).. 2
1(b) three body regions (head, thorax, and abdomen)..an insect, CLASS INSECTA

2(a) one pair of legs per body segment; poison jaws present...................a centipede, CLASS CHILOPODA
2(b) two pairs of legs per most body segments; poison jaws lacking..........a millipede, CLASS DIPLOPODA

to systematics, is that insect groups contain numerous *sibling species*. A sibling species group is an assemblage of morphologically indistinguishable forms that actually represent two or more biologically distinct species. Unfortunately, morphology is sometimes the only avenue for identification, and sibling species remain undetected. Pashley (1986) provides a discussion of the biological consequences of failure to recognize sibling species.

Description

The systematist, having completed an identification, will follow one of two courses. If a species has already been described, it will probably be put aside for possible future study with other specimens of the same species. If not, it will be described as a new species and placed in whatever grouping the systematist sees as appropriate. This description will serve to identify this species for other investigators. The production of such descriptions is the job of descriptive systematics.

Subjects of Description
Ideally, specimens to be described are obtained by a carefully planned and executed procedure such that they constitute a statistical *sample* of a population (i.e., they display the same distribution of traits as do the members of the entire population). The other, and more common, alternative is that a group of specimens is obtained in a more or less erratic fashion; perhaps several individuals were taken in the same locality, or perhaps different individuals were taken from different localities at different times and under different circumstances. Such a group is termed a *series*. Although many species descriptions are based on series and samples, several are based on a single specimen or on a series of a few individuals.

Characters Described
A *character* is any possible trait an individual might possess. This could be anything from the shape of a particular sclerite (morphological character), to a particular kind of amino acid in the hemolymph or cuticular hydrocarbon profile (biochemical characters), to a particular mechanism for excretion (physiological character), to a specific way of responding to a change in photoperiod (behavioral character), to the nucleotide sequence of a particular piece of DNA (genetic character). However,

all characteristics of a given organism are not used in a description of that organism. It follows, then, that any species description plus all information gained since the discovery of that species is only a partial picture and that a species description in this sense is never really completed. In reality the vast majority of species descriptions are based on selected morphological features, and although the modern trend is definitely in the direction of including other sorts of data (particularly physiological, biochemical, genetic, and behavioral), morphological characters will likely continue to be central to most descriptions, at least of insects.

There is a wide variety of characters to choose from in a description (Mayr, Linsley, and Usinger, 1953).

I. Morphological characters
   A. General external morphology
   B. Special structures (e.g., genitalia)
   C. Internal morphology
   D. Embryology
   E. Karyology (and other cytological differences)
II. Physiological characters
   A. Metabolic factors
   B. Serological, protein, and other biochemical differences
   C. Body secretions
   D. Genic sterility factors
III. Ecological characters
   A. Habitats and hosts
   B. Food
   C. Seasonal variations
   D. Parasites
   E. Host reactions
IV. Ethological characters
   A. Courtship and other ethological isolating mechanisms
   B. Other behavior patterns
V. Geographical characters
   A. General biogeographical distribution patterns
   B. Sympatric–allopatric relationship of populations
VI. Molecular genetic characters
   A. Isozymes
   B. Nucleic acid sequences
   C. Gene expression and regulation

Phenomenal advances in biology since 1973, when it first became possible to clone DNA, necessitate a brief digression into the role of molecular genetic characters in insect systematics. One major approach to the study of molecules has been the application of isozyme techniques (reviewed in Murphy et al. 1990). This is a method for separating proteins on the basis of their electrical charge. Protein charge is a function of amino acid composition, which in turn reflects the DNA sequence of the gene encoding that protein. Thus, isozyme analysis permits, within certain limits, an estimate of the genetic relationships among organisms. This approach has been widely applied to the study of insect relationships, especially at the level of the genus and species (Berlocher 1984, Menken and Ulenberg 1987).

As techniques were developed to uncover the variation among organisms at the DNA level, these methods were adapted to estimate phylogenetic patterns among organisms. Indirect or direct methods for measuring nucleotide sequence diversity include restriction fragment length polymorphism (RFLP) (Dowling et al. 1990) and DNA or RNA sequencing (Hillis et al. 1990), respectively. RFLP studies rely on bacterial enzymes that cut DNA at specific 4 or 6 nucleotide sites, generating pieces of DNA that may be visualized as bands on a gel matrix after exposure to an electric current. The presence or absence of DNA bands in different organisms is used to establish the average similarity of the total DNA sequence in the organisms. Sequencing, on the other hand, actually gives the scientist a list, base by base, of the DNA structure. Thus, changes among organisms may be directly compared. These DNA-based techniques are becoming more common in insect systematics (DeSalle et al. 1986, Hendriks et al. 1986, Vossbrinck and Friedman 1989, Wheeler 1989).

While not every systematist may use these approaches, it is vital that a researcher understand the basis of the techniques in order to evaluate the importance of such studies. The best chance to resolve issues in systematics will be to investigate every character available (Patterson 1987, Hillis and Moritz 1990). Continuing advances in molecular biology provide an ever-changing opportunity for systematists. It is now possible to recover DNA sequence information from museum specimens, even of extinct organisms (Thomas et al. 1989) and fossils (Golenberg et al. 1990). Our museum collections may now provide an even richer repository of systematic information than previously realized.

No matter what type of character is used, it is extremely important to know the extent to which that character varies within a species.

For characters to be of value in identification and classification they must be reasonably constant or vary predictably. Failure to recognize variation can lead to confusion. There are any number of potential kinds of variation that can occur within a population of a given species (Linsley and Usinger 1961):

I. Extrinsic or noninherited variation
   A. Variation due to age
   B. Seasonal variation
   C. Castes in social insects
   D. Variation due to habitat
   E. Variation due to crowding
   F. Climatically induced variation
   G. Heterogonic variation
   H. Traumatic variation
II. Intrinsic or inherited variation
   A. Primary sex differences
   B. Secondary sex differences
   C. Alternating generations
   D. Gynandromorphs
   E. Intersexes
   F. Mutations resulting in continuous variation
   G. Mutations resulting in discontinuous variation
   H. Genetic polymorphism

Classification

The many hundred thousands of species descriptions would be practically impossible to deal with if they were not organized in some fashion. Such organization is necessary because of the differences between organisms and, at the same time, is made possible by the similarities among them. By grouping organisms based on degrees of similarity, one can arrive at a system of classification.

Classifications serve several purposes (Warburton 1967):

1. Classifications provide intellectual satisfaction for the taxonomists who make them.
2. They make possible the identification of species and higher groupings.
3. They ". . . provide a convenient, practical means by which zoologists may know what they are talking about and others may find out" [Warburton quoting G. G. Simpson].
4. They provide a system of information retrieval.
5. They may reflect phylogenetic relationships [to be discussed later].
6. They may serve as summarizing and predicting devices.

The first purpose listed requires no elaboration and is probably the primary motivation for the efforts of most scientists. The ability to place a given organism at some point in a classification facilitates further identification by ruling out myriads of other organisms. For instance, if one can determine the class, order, or even family of an organism by determining where it fits in an already established classification, one is several steps closer to a complete identification.

A classification's value as a basis for communication is obvious. As mentioned earlier, an organism's name is the "file category" under which all information pertaining to that organism is stored. Within a classification one has not only the name of an organism but also the names of the successively more inclusive groups of which it is a part and hence the "file categories" of the literature pertinent to these groups.

A classification, by the nature of the way it is constructed—a descending or ascending hierarchy of successively less or more inclusive groups—is a summarizing device, since all categories are arrived at by the grouping of organisms on the basis of similarities. Thus a great deal of information is summarized merely by recognizing that a given organism is an insect. Similarly, an organism's position in a classification enables one to predict certain additional things about that organism.

Characters Used in Classifying Organisms

When developing a classification, only *homologous* characters—those that could theoretically be traced back to a common ancestor—are used as comparative bases. For example, the wings of all insects, although they vary considerably in both structure and function, are fundamentally similar and are assumed to have been derived from an ancestor common to all winged insects. Hence insect wings are considered to be homologous among the various insectan groups. In practice, homology is not always easy to establish. For example, one might note the rather close similarity between the raptorial forelegs of a praying mantid (see figure 12.21) and those of a mantispid (see figure 13.8*a*) and conclude that these two insects share a common ancestor with raptorial forelegs. However, such is not the case. Based on comparison of many other characteristics, mantids and mantispids are placed in widely divergent orders (Mantodea and Neuroptera, respectively). Thus the character "raptorial" is not homologous in these two insects and must have arisen independently in each. Similarities between nonhomologous characters are considered to be due to *convergent evolution,* that is, evolutionary responses (adaptation) to similar selection pressures. In the preceding example, then, we would view the raptorial legs as representing closely similar, but independent, adaptations associated with prey capture.

Since common ancestry is nearly impossible to prove directly, a number of pragmatic criteria for judging homology exist. For example, two characters are considered to be homologous on the basis of minutely detailed resemblance, close similarity in embryological origin, and the occurrence of many other homologous features in the same organism. Obviously, the existence of a complete

fossil lineage back to a common ancestor would clinch a decision of homology. Such complete lineages unfortunately do not exist.

Kinds of Classification

There are basically three schools of thought regarding classification, *phenetics, cladistics,* and *evolutionary systematics* (Mayr 1982). These approaches differ philosophically over which evolutionary features of organisms can or should be reflected in their classification. Entomologists were instrumental in the early development of these various approaches. Michener (a bee taxonomist) and Sokal (1957) pioneered phenetic classification, whereas writings by the fly expert Willi Hennig initiated the cladistic method of classification (Hennig 1965, 1966).

The phenetic school (Sokal and Sneath 1963) argues that phylogeny cannot be deduced with any certainty, so classifications must rely on observing the similarity (or dissimilarity) of as many different features as possible; characters are given equal weight, and a phenetic distance between pairs of organisms results from this comparison. These distance measures are used to generate a *phenogram* (or *dendrogram*), generally accomplished with computer assistance (this dependence upon computers is arguably one of the most important contributions of the phenetic school). A phenetic classification results from analysis of the patterns of connection within the phenogram.

In contrast to phenetics, cladistics is based on the notion that it is possible to reconstruct phylogeny and that the classification of organisms must reflect the underlying pattern of evolutionary divergence. The cladist attaches special significance to unique characters that are shared among different organisms because of common ancestry. The "family trees" obtained from such an analysis are called *cladograms.* One drawback of pure cladistics is the difficult requirement of assigning a taxonomic category (see next section) to each branch in the cladogram. This makes a very unwieldy classification scheme.

Much space in scientific journals has been devoted to arguments over the merits of phenetics versus cladistics. Indeed, new scientific societies have been created to defend one or the other approach. Both schools have contributed to modern classification, and debate, while often philosophical and sometimes caustic, has been healthy (see Felsenstein 1982 or Swofford and Olsen 1990 for reviews of phylogeny reconstruction).

What have taxonomists been doing while the debate raged? Most have combined the most useful points of both phenetics and cladistics and think of themselves as evolutionary systematists. Such scientists examine as many characters as possible. They use patterns of shared characters to define the phylogeny of the organisms under study whenever possible. In addition, the evolutionary systematist recognizes that, when a new species arises, it accumulates many unique features through evolutionary processes, and these features help the scientist assign this organism to its proper category.

Insect classification reflects the influence of all three schools. Kamp (1973) provides one example of the use of numerical phenetics in a group of insects. Many analyses of molecular characters (e.g., isozymes and nucleic acids) have employed phenetic approaches (Berlocher 1984). Boudreaux (1979) interprets arthropod phylogeny (insects included) using a cladistic approach. Examples abound of the influence of evolutionary systematics in insect classification (Kim and Ludwig 1978, Kristensen 1989).

Components of Biological Classification

Four components form the basis for most systems of biological classifications (definitions from Simpson 1961).

1. Hierarchy—"a systematic framework for zoological classification with a sequence of classes (or sets) at different levels in which each class except the lowest includes one or more subordinate classes."
2. Taxon (pl. taxa)—"a group of real organisms recognized as a formal unit at any level of a hierarchic classification."
3. Category—"a class, the members of which are all the taxa placed at a given level in a hierarchic classification."
4. Rank—a category's "absolute position relative to other categories."

Thus Insecta (taxon) is a class (category) that in a widely accepted biological hierarchy (figure 11.2) is ranked between superclass and subclass.

Taxonomic Categories

The *species* is the basic unit of taxonomic classification, referring directly to real populations of real animals. The *subspecies* is a subdivision of a species and also refers to real animals. On the other hand, *higher categories* (all those categories in the biological hierarchy above the species level) are essentially subjective, their nature depending upon the professional opinions of the taxonomists who "recognize" them. They refer to real animals indirectly in that they ultimately include one to several species.

Despite its objective reality, a number of different concepts revolve around the term "species." Early pre-Darwinian biologists considered each species to have arisen as a result of special creation and to be composed of nonvarying individuals. With this *type concept* of species, a single specimen was all that was necessary to describe a series, and morphological characters, sometimes supplemented with geographical data, were generally the

Figure 11.2 Hierarchy of generally accepted taxonomic categories.

Kingdom
Phylum
Subphylum
Superclass
Class
Subclass
Cohort
Superorder
Order
Suborder
Superfamily (*-oidea*)
Family (*-idae*)
Subfamily (*-inae*)
Tribe (*-ini*)
Genus
Subgenus
Species
Subspecies

only ones used. Note that the type species and the use of "types" in species description mentioned earlier are very different. Types in the latter sense serve a pragmatic purpose, but modern concern is with series or samples of individuals of a species, and great stress is placed on identifying the kind and degree of variation of any character used.

With modern understanding of organic evolution and heredity, a new concept of species, the *biological species,* has developed. According to this concept, "Species are groups of actually (or potentially) interbreeding natural populations which are reproductively isolated from other such groups" (Mayr, Linsley, and Usinger 1953). However, according to Blackwelder (1967):

> Morphogeographical species are those of the ordinary taxonomist, from Linneaus to modern times. Although data other than "morphology" and geography have been increasingly used in recent years, these still remain the basic species of taxonomy.

A subspecies may be defined as ". . . an aggregation of local populations of a species, inhabiting a geographical subdivision of the range of the species, and differing taxonomically from other populations of the species" (Mayr 1963). In other words, subspecies are geographical races within the same species that are sufficiently different in some regard for them to be classified in this manner. Based on the biological species definition, subspecies should be able to interbreed if given the opportunity to do so.

As mentioned earlier, the higher categories are subjective, depending on the opinions of taxonomists. This subjective nature is apparent when one realizes that a taxon that was once an order is now a family (or vice versa) or that a given taxon may be at the same time recognized as a family by some systematists and as an order by others, for example, order Blattaria (cockroaches) versus family Blattidae (cockroaches when included in the order Orthoptera). Further, the taxa of one group of animals usually do not correspond to the taxa of another group relative to the diversity in a given category. For example, the orders of birds are much less distinct than the orders of insects.

A higher category may appropriately be defined in relation to the category immediately below it in the hierarchy. For example, a genus may be defined as a group of species that have a sufficient number of features in common to warrant inclusion within a single group. In practical terms the decision regarding "a sufficient number of features" is up to the systematist making the decision. Generally, the higher the category, the greater the stability with regard to changes. In other words, genera are changed by systematists much more frequently than families or orders.

Future of Classification

Classifications should be looked upon as organizational systems that are continually undergoing a sort of "evolutionary" development in light of new information about organisms and their relationships to one another. "The well-classified parts of the animal kingdom form no more than a hundredth part of the known fauna and probably only a thousandth part of what actually exists" (Blackwelder 1967). Since the largest portion of the animal kingdom is held by insects, at least a major part of the job will have to be done by insect systematists.

Nomenclature

Nomenclature is involved with the naming of organisms and groups of organisms and the rules and procedures to be followed in such naming. The rules of nomenclature have been developed and gradually changed and modified over the years. The current rules are in the 1985 revision of the International Code of Zoological Nomenclature.

Scientific Names

Every species and higher category has a Latinized scientific name, and the binomial system of nomenclature developed by Linneaus is still used today. At the species level, the scientific name consists of two parts, the genus and the *specific epithet,* which comprise the *binomen.* An example would be *Aedes occidentalis,* in which *Aedes* is the genus and *occidentalis* the specific epithet. These two words together comprise the scientific name of a species of mosquito. The epithet is not the species name and should not be used as such. By convention, all species names are printed in italics (or underlined when written or typed without italics), the generic name is capitalized, and the epithet is all lowercase. Actually a complete species name can be more complex than a simple binomen, although the binomen is still valid to use. For example, the complete scientific name of one of the subspecies of

the mosquito species just mentioned is *Aedes* (*Finlaya*) *occidentalis occidentalis* (Skuse 1889) (*Culex*). The first and third terms are the binomen. The second term, in parentheses, is the subgenus and the fourth indicates the subspecies. Skuse is the taxonomist who first described this species, and his description was published in 1889. The parentheses indicate that Skuse placed this species in a different genus than *Aedes.* The last name, *Culex,* in parentheses, is the genus in which Skuse placed this species. Sometimes a generic name is written followed by sp. or spp., for example, *Aedes* sp., or *Aedes* spp., if the specific epithet is unknown. These abbreviations indicate a particular species or several species within a given genus, respectively.

The names of families are uninomials that end in -idae, and name changes are strictly regulated. All family names are based on a type genus; for example, the family name of tiger beetles, Cicindelidae, is based on the generic name *Cicindela.*

The names of categories above family are all uninomials and are not extensively considered by the international code. They are used in the form of plural Latin nouns (e.g., Pterygota, Insecta), and no type categories are involved. The categories used in a given classification depend entirely upon the opinion of the taxonomist preparing the classification. Like family, several other categories have standardized endings (figure 11.2).

Common Names

Insects that are particularly common in a given locality may be given vernacular names. For example, dragonflies have been variously referred to as "mosquito hawks," "snake feeders," and "devil's darners." Such names have little value other than in the context of local color. However, certain common names are useful when applied to agricultural or medical pests since they facilitate communication between the professional and layperson.

The Entomological Society of America has formalized rules by which certain arthropod species may be given unambiguous common names (E.S.A. 1989). The E.S.A. and the Commonwealth Agricultural Bureaux International each publish separate checklists of preferred common names for insects. Most orders and families and many species have common names that over the years have become well established. Two particularly outstanding common names are "bugs" and "flies." Although these words are found in several ordinal and familial names (butterflies, mayflies, doodlebugs, etc.), they apply specifically to the Hemiptera (true bugs) and the Diptera (true flies). When used in reference to members of these orders, "fly" and "bug" are written as separate words (e.g., house fly and conenose bug). However, if they are used for members of any other order, they are written with a modifier of some sort as a prefix (e.g., dragonfly or ladybug). When there is no common name, or

Figure 11.3 Representatives of four classes of segmented worms, phylum Annelida. (*a*) Chaetogaster (Oligochaeta). (*b*) Leech (Hirudinea). (*c*) Earthworm (Oligochaeta). (*d*) *Dinophilus* (Archiannelida). (*e*) Lugworm (Polychaeta). *Redrawn from Stiles, Hegner, and Boolootian, 1969.*

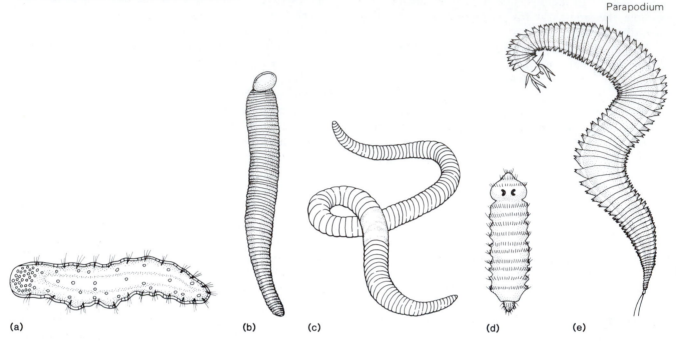

Parapodium

(a)   (b)   (c)   (d)   (e)

when it is preferable not to use a common name, the scientific name may be used as a noun or modifier (e.g., mosquitoes may be called culicids, derived from the family name Culicidae).

Insect Evolution

Evolutionary change may be viewed as occurring in three successively longer time frames: *microevolution,* the changes that occur in populations usually in response to changes in selection pressures, such as industrial melanism (see chapter 9) and the development of insecticide resistance in a population (see chapter 15); *speciation,* the changes that typically occur over much longer time spans than microevolution and that result in reproductively isolated, new species; and finally *macroevolution,* the major phylogenetic patterns that develop over wide spans of geological time.

The objective of this section is to outline the major macroevolutionary events that are thought to have occurred among insects and their relatives. The discussion includes consideration of the origin of the arthropods, origin of the insects, and phylogeny within the class Insecta.

At least a rudimentary knowledge of evolutionary theory is assumed in this discussion. Readers unfamiliar with natural selection, speciation, the concept of geological time, and so on, should consult a good general biology or zoology text or any of several excellent texts available which deal specifically with organic evolution such as Futuyma (1986) and Mayr (1963).

Arthropods and Relatives

Comparison of present-day (extant) forms has led to the recognition of structural and developmental affinities among the Annelida, Arthropoda, and Onychophora.

Annelida

Annelida, the segmented worms, apparently evolved from the ancient protostomes, which also gave rise to the mollusks (Meglitsch 1967). Annelids (figure 11.3) are soft-bodied, elongate, cylindrical, bilaterally symmetrical animals. Their bodies are composed of repeating segments or metameres capped anteriorly by the prostomium or acron and posteriorly by the periproct or telson. The mouth is on the venter between the prostomium and first metamere. The anus opens from the periproct. Some annelids have several bilateral pairs of segmental appendages called *parapodia,* which arise as evaginations of the body wall. Segmentation of the body is evident in nearly all tissues except the alimentary canal, which is essentially a longitudinal tube running from the mouth to the anus. The circulatory system is a closed system of tubes, usually one dorsal longitudinal and one ventral longitudinal vessel, connected by lateral trunks that provide vascularization for the various body tissues. The blood commonly carries a respiratory pigment. The body cavity is a coelom divided into segmental compartments. Excretion is accomplished by paired segmental *nephridia,* tubes that communicate between the coelom and the exterior. Ventilation is cuticular or by means of gills evaginated from the body wall. The nervous system is

Figure 11.4   Scorpion, illustrating the major chelicerate features.

After U.S. Public Health Service.

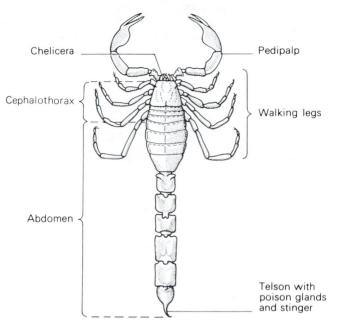

Chelicera —————

Cephalothorax

Abdomen

Pedipalp

Walking legs

Telson with
poison glands
and stinger

composed of an anterior prostomial ganglion that lies dorsal to the alimentary canal. It is connected by a pair of bilateral connectives to a ventral chain of paired segmented ganglia linked by paired longitudinal connectives.

Arthropoda

Arthropods, the joint-footed animals, are a highly successful diverse group of animals unrivaled by other animals in diversity of structure and function. They are found in nearly every imaginable ecological niche. A feature that has played a central role in their success is their relatively impermeable exoskeleton. Such an impermeable covering was a prerequisite for invasion of the terrestrial environment, which poses a continuous challenge to the water-holding capabilities of all animals and of arthropods in particular because of their generally small body size (see "Permeability Characteristics" in chapter 2). Additional adaptations are discussed later in this chapter.

Arthropods (figures 11.4 through 11.7) have segmented bodies composed of chitinous exoskeletal plates of varying degrees of hardness, which are separated by less-hardened "membranous" areas. Paired segmental appendages are usually present on at least some body segments. Arthropods share several fundamental traits with annelids. Particularly outstanding are their segmented bodies and the ventral chain of segmental ganglia. On the other hand, the arthropods have an open

circulatory system, and the body cavity is a hemocoel; coelomic sacs appear, during embryogenesis, but do not form the body cavity. Also most arthropods ventilate by means of a tracheal system and have well-developed appendicular mouthparts. For more information on the arthropods, see Boudreaux (1979), Clarke (1973), and Manton (1977).

Extant arthropods include the chelicerates, crustaceans, myriapods, and insects.

The chelicerates (figures 11.4 and 11.5) include the spiders, mites and ticks, scorpions, solpugids, and their relatives (class Arachnida). Horseshoe crabs (Xiphosura, e.g., *Limulus*), sea spiders (Pycnogonida), water bears (Tardigrada), and tongueworms (Linguatulida or Pentastomida) are variously classified, some authors including them with the chelicerates. Chelicerates are characterized by the presence of six pairs of appendages; the anterior pair are called *chelicerae* and are jawlike in more primitive forms. They have been modified in a variety of ways in more advanced forms. The second pair of appendages are called *pedipalps* and also vary in structure. The remaining four pairs of legs generally have a locomotor function. Antennae are lacking, and there are usually two body regions: a fused head and thorax (*cephalothorax*) and an abdomen. The appendages are borne by the cephalothorax.

Figure 11.5 Representatives of the subphylum Chelicerata. (*a*) Spider, *Loxosceles* sp. (Arachnida; Araneida). (*b*) Mite (Arachnida; Acari). (*c*) Tick (Arachnida; Acari). (*d*) Wind scorpion (Arachnida; Solpugida). (*e*) Harvestman or daddy longlegs (Arachnida; Phalangida). (*f*) Horseshoe crab, *Limulus* sp. (Xiphosura).
(a–e) *After U.S. Public Health Service.*

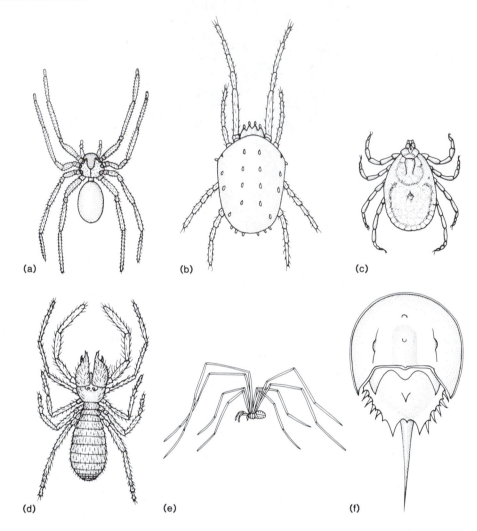

(a) (b) (c)

(d) (e) (f)

The crustaceans (crayfish, lobsters, crabs, shrimp, barnacles, copepods, sowbugs, and so on; figure 11.6) have two pairs of antennae and a pair of mandibles. The crustaceans commonly have two body regions: a cephalothorax, which is covered by a *carapace,* and an abdomen. These arthropods are mostly aquatic animals.

Myriapods are cylindrical, have a well-defined head bearing primitive mandibulate mouthparts, and have a trunk with nine or more bilateral pairs of legs. There are four myriapod classes: Diplopoda (millipedes), Chilopoda (centipedes), Pauropoda (pauropods), and Symphyla (symphylans). All are terrestrial animals.

As adults, the insects possess a head, thorax (usually with wings and legs), and an abdomen. This is the only group of invertebrates that contains members capable of active flight. Insects are primarily terrestrial with many secondarily aquatic forms.

A major group of extinct arthropods, the trilobites, have played an important role in the study of arthropod evolution. These animals (figure 11.7) were extinct or very rare in the Permian period (see figure 11.14) and therefore are known only from the fossil record. They were marine animals, and their bodies were made up of three tagmata: head, thorax, and pygidium. The name Trilobita refers to these three tagmata. The head and first few thoracic segments were covered by a carapace, most of the body segments had a bilateral pair of jointed legs, and the head carried a single pair of antennae.

Figure 11.6 Representative mandibulates. (*a*) Crayfish (Crustacea). (*b*) Sowbug (Crustacea). (*c*) Centipede, *Scolopendra heros* (Chilopoda). (*d*) Eastern house centipede, *Scutigera coleoptrata* (Chilopoda). (*e*) Millipede, *Narceus americanus* (Diplopoda).

After U.S. Public Health Service.

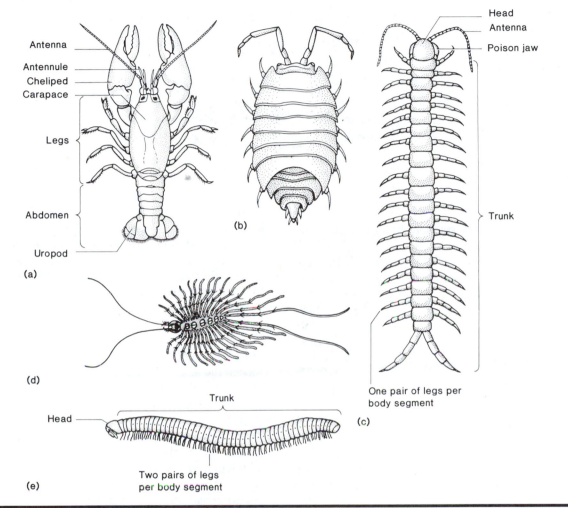
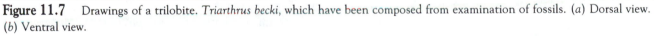

Figure 11.7 Drawings of a trilobite. *Triarthrus becki,* which have been composed from examination of fossils. (*a*) Dorsal view. (*b*) Ventral view.

Redrawn from Ross, 1965 (from Schuchert after Beecher).

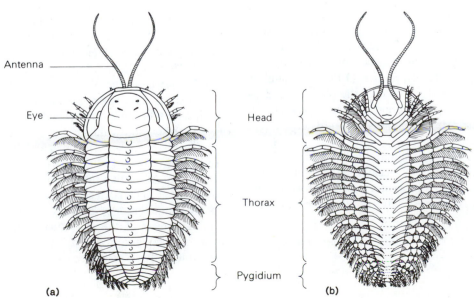

Figure 11.8 Onychophoran, *Peripatus*. (*a*) Dorsal view. (*b*) Ventral view.
Redrawn from Boreal Labs key card.

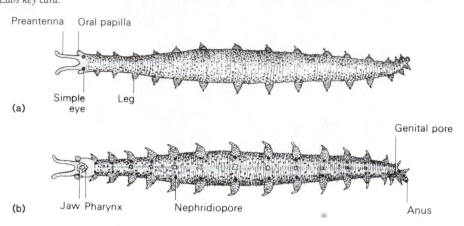

Onychophora

Onychophorans are a very small group of terrestrial animals, but are significant in that they show morphological affinities with both annelids and arthropods. These animals (figure 11.8) have elongate, wormlike bodies with several bilateral pairs of lobelike legs. They are not segmented as adults, but are clearly segmented during their embryogenesis. They have paired nephridia (excretory organs) along the body that open near the leg bases. This fact, together with other morphological characters, suggests an evolutionary relationship with the annelids. On the other hand, onychophorans share several characters with arthropods, including a chitoproteinous cuticle (Hackman and Goldberg 1975), an open circulatory system with a hemocoel, tracheal ventilation, and appendages associated with the mouth. Onychophorans have been variously classified as annelids, arthropods, or as a separate phylum. Recent analysis based on ribosomal RNA sequences provides evidence that the onychophorans are modified arthropods (Ballard et al. 1992).

Origin of Arthropoda

The basic similarities among annelids, arthropods, and onychophorans strongly suggest common ancestry at some point in metazoan evolution. Some authors have, in fact, classified all three groups as members of a superphylum, Annulata (Snodgrass 1952). Sharov (1966), in a similar vein, supports Cuvier's* concept of the phylum Articulata with annelids, onychophorans, and arthropods being recognized as separate subphyla. The details of the evolutionary relationships among these three groups are purely hypothetical. Unfortunately, knowledge of the fossil record is of little help, because each group is distinctly defined in the Cambrian strata, from which the oldest "good" fossils come. This, of course, means that

the supposed common ancestor arose in Precambrian times. There is, however, general agreement that the common ancestor was an annelid or annelidlike creature.

Some have suggested that the primitive arthropod evolved directly from a primitive polychaete annelid (Meglitsch 1967). Snodgrass (1952), later supported by Sharov (1966), conceived of a segmented wormlike animal that produced two branches, one ultimately giving rise to the polychaete annelids and the other to a form with paired, undifferentiated lobelike legs (*lobopod*). He envisioned the lobopods as subsequently branching into arthropodan and onychophoran lines.

Once established, members of the phylum Arthropoda are viewed by some as having branched out into four groups (subphyla): the trilobites (Trilobita), the chelicerates (Chelicerata), the crustaceans (Crustacea), and the myriapods and insects (Uniramia) (figure 11.9).

There is disagreement as to whether the arthropods actually constitute a single phylum (are *monophyletic*), sharing a common arthropodan ancestor, or are actually two or three groups (are *polyphyletic*) of distantly related animals in which major arthropodan traits, especially jointed appendages but also many other structures, evolved convergently (Scudder 1973, Boudreaux 1979, Willmer 1990). Recent cases for the latter point of view have been made by Tiegs and Manton (1958), Manton (1964, 1970, 1972, 1979), Cisne (1974), and Anderson (1973, 1979). On the basis of mandibular structure and movement and locomotor limbs, Manton recognizes three groups that became "arthropods" convergently (figure 11.10): the trilobites and chelicerates; the Crustacea; and the Onychophora, myriapods, and hexapods (insects). Anderson (1973, 1979) provides embryological corroboration. According to this point of view, the trilobites, chelicerates, and crustaceans have primitively biramous, double-branching, *appendages* and *gnathobasic mandibles*. The form of the gnathobasic mandibles and biramous appendages in the trilobites and chelicerates is fundamentally different from those of the crustaceans.

Georges Cuvier (1769–1832) was among the founders of paleontology, the study of the fossil record.

Figure 11.9 Hypothetical phylogeny of arthropods, annelids, and onychophorans.

Redrawn from data of Snodgrass, 1952, and Sharov, 1966.

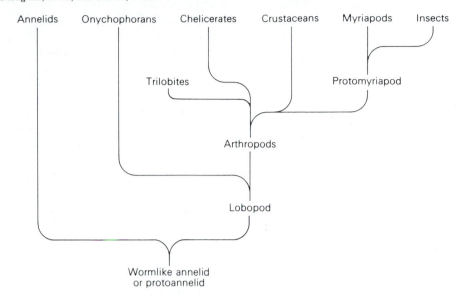

Figure 11.10 Hypothetical phylogeny of arthropods based on Tiegs and Manton (1958).

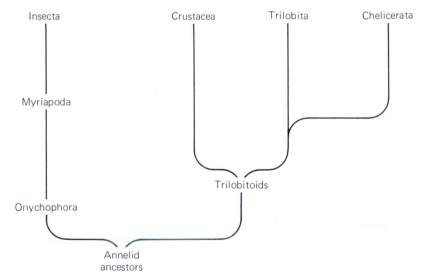

On these and other bases, the trilobites and chelicerates are viewed as evolving independently. The Uniramia all have similar uniramous, unbranched, limbs and bite with the tips of the mandibles; thus they are considered to constitute a monophyletic group. Cisne (1974) studied parts of the internal anatomy of pyritized trilobites using X rays and provides evidence in support of a polyphyletic origin of arthropods. However, he views the trilobites as a link between the chelicerates and crustaceans and hence recognizes two arthropod lines (figure 11.11).

Although there is substantial evidence for a polyphyletic origin of arthropods, acceptance of this point of view requires acceptance of the independent, convergent origin of several characters: a hemocoel; paired, jointed locomotor appendages; tracheae; Malpighian tubules; compound eyes; a chitinous cuticle, and so on (reviewed by Willmer 1990). Some experts find this very difficult to accept. In a multiauthored volume entitled *Arthropod Phylogeny* (Gupta 1979), a monophyletic origin of arthropods is supported by studies of the embryology of the head, olfactory mechanisms and sensilla, eye structure, visceral anatomy, intersegmental tendon systems, and sperm ultrastructure, although Anderson (1979) supports a polyphyletic origin based on embryological features. Boudreaux (1979) and Weygoldt (1986) also favor a monophyletic explanation of arthropod origin; the

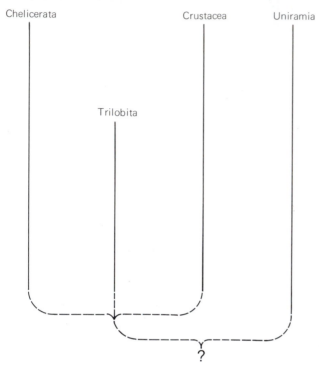

Figure 11.11 Hypothetical phylogeny of arthropods based on Cisne (1974). Uniramia here includes onychophorans, myriapods, and insects.

former provides a very detailed discussion of arthropod phylogeny. Until recently, the fossil record has not provided sufficient information to resolve the issue of a monophyletic versus polyphyletic origin of the arthropods (Bergström 1979). Recently, however, Kukalová-Peck (1992) has presented compelling fossil evidence that insect and other uniramian appendages were primitively branched and that the unbranched (uniramous) condition is a recent development. On this basis the taxon "Uniramia" is no longer valid and the monophyletic interpretation is well-supported and gains ascendence over the polyphyletic view (Shear 1992). Additional support for the monophyletic interpretation comes from an analysis based on ribosomal RNA sequences (Ballard et al. 1992).

Origin of Insects

The class Insecta is generally considered to have evolved from a myriapod or protomyriapod of some sort prior to the Devonian period. Based on differences in mandibles and mandibular movement, Manton (1964) concluded that insects are not direct descendants of myriapods and that these two groups are best looked upon as sharing a common ancestor. The molecular data of Ballard

et al. (1992) do not support the close relationship between insects and myriapods hypothesized by earlier investigators.

A series of diagrams originated by Snodgrass (1935) and modified by Ross (1965) is helpful in a very simplified way in visualizing the hypothetical origin of the insects. The first (figure 11.12a) represents the segmented, legless, wormlike annelid or annelidlike stage. The undifferentiated body is composed of a series of somites or metameres capped anteriorly by the prostomium (acron) and posteriorly by the terminal body segment, the periproct (telson). The mouth is located between the prostomium and the first body segment; the anus opens in the periproct. Figure 11.12b represents the evolutionary stage in which were developed paired, bilateral, lobelike appendages on the somites as well as a pair of simple eyes on the prostomium and antennae on the second somite. The appendages on the first somite have been lost altogether. This level of organization is onychophoranlike. The next stage (figure 11.12c) represents a protomyriapod/protoinsect in which arthropodization has occurred; that is, the bilateral appendages of each body segment have themselves become segmented. The appendages of somites 4, 5, and 6 have been reduced and have moved into close association with the primitive head, becoming involved with the manipulation of food. The appendages of the hindmost somite have become sense organs and no longer function in locomotion. As mentioned earlier, there are varying opinions as to the origin of the myriapod and insect groups. Figure 11.12d can be looked upon as representing a myriapodlike level of organization, bilateral appendages being retained on most body segments and the appendages of the 2nd, 3rd, and 4th body segments becoming the typical, primitive mandibulate mouthparts. The insectan level of organization is represented by figure 11.12e. The body has been differentiated into the three tagmata (head, thorax, and abdomen) characteristic of insects. The appendages of segments 7, 8, and 9 have been retained as locomotor structures, but locomotor appendages have disappeared from the remaining body segments. The appendages of abdominal segments 8 and 9 have been modified as external genitalia, and the cerci, appendages of segment 11, have been retained. The telson has been lost, and the anal opening is now within the 11th abdominal segment.

The evolution of the insect head from the prostomium plus four somites, which is implied in figure 11.12, has been disputed. Rempel (1975) reviews the evidence regarding various theories of origin of the insect head and supports the one depicted in figure 11.13 where the head originates from the fusion of the prostomium and six somites.

Figure 11.12   Hypothetical stages in the evolution of the insect form.
Redrawn from Ross, 1965.

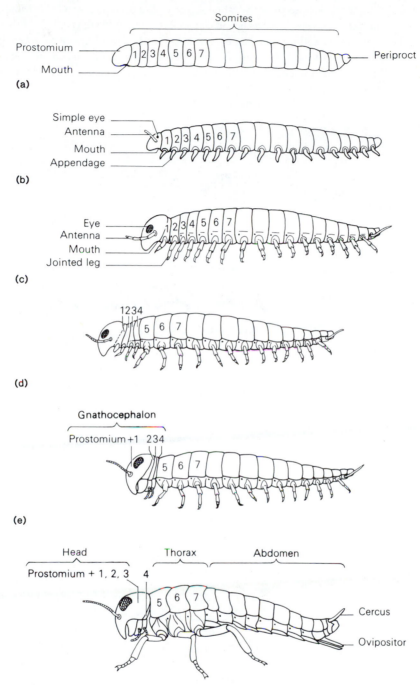

Figure 11.13   Origin of the insect head from fusion of the prostomium and six somites. (*a*) Annelidlike ancestor with prostomium containing a ganglionic mass (archicerebrum). (*b*) Cuticle, segmental appendages, and apodemes have evolved. (*c*) Hypothetical arthropod; ganglion of first somite (now the labral segment) has become preoral. (*d*) More advanced arthropod; ganglion of second somite (now the antennal segment) has become preoral. (*e*) Insect head; ganglion of the third somite (now the intercalary segment) has become preoral, and the ganglia of somites 4, 5, and 6 have formed the subesophageal ganglion. According to this theory the protocerebrum of the brain arose from the fusion of ganglionic masses associated with the prostomium (archicerebrum) and first somite; the deuto- and tritocerebrum, from ganglionic masses in the second and third somites, respectively.

Redrawn from Rempel, 1975.

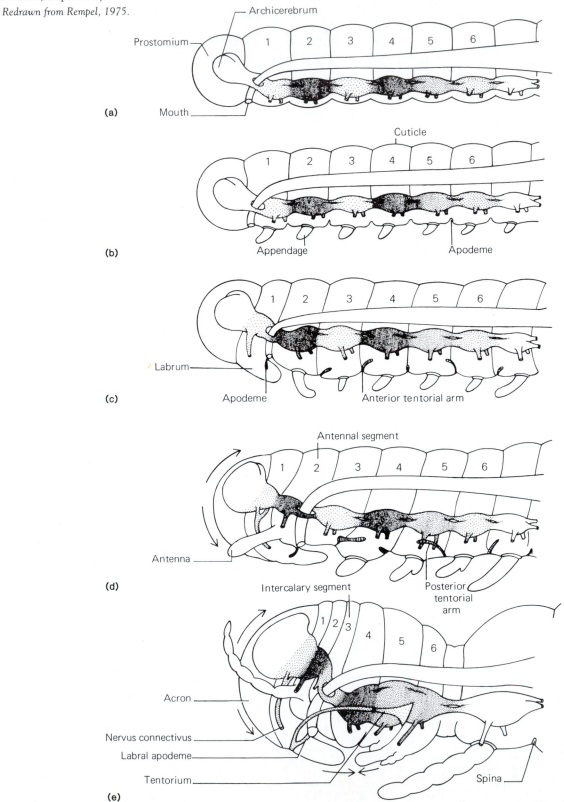

Figure 11.14 Major stages in insect evolution superimposed on a geological time table. Thick solid lines run from earliest known fossil representative of an extant group to the present. Dashed lines represent hypothetical branches. Thin solid lines bounded by circles represent the following extinct orders: a, Monura; b, Palaeodictyoptera; c, Megasecoptera; d, Diaphanopterodea; e, Protodonata; f, Protorthoptera; g, Caloneurodea; h, Miomoptera; i, Protelytroptera; j, Glosselytrodea. *Based on information from Carpenter, 1977; Sharov, 1966; Smart and Hughes, 1972; Kukalová-Peck, 1987. Geological eras and periods based on Villee and Dethier, 1976.*

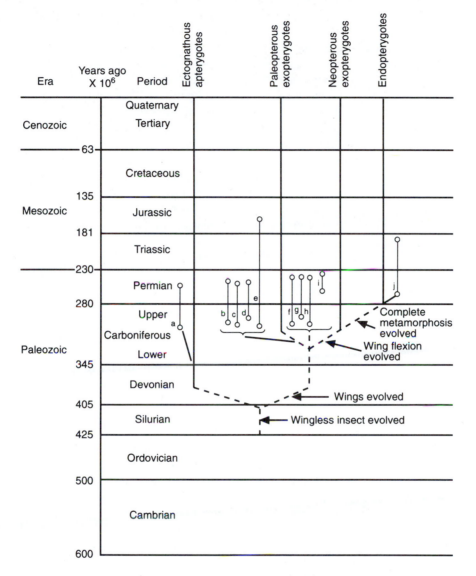

Insect Phylogeny

Carpenter (1953, 1977) recognizes four major stages in the evolution of insects. The first was the appearance of primitive wingless insects, which probably resembled contemporary Archeognatha and Thysanura. These primitive *apterygotes* are thought to have arisen during the Silurian period or before (figure 11.14). Fossil Archeognatha (Shear et al. 1984, Labandeira et al. 1988) and the entognathous order Collembola (Whalley and Jarzembowski 1981) are known from Devonian-age rocks.

This evidence indicates that the divergence of ectognathous and entognathous insects (see chapter 12) took place prior to the Devonian. The earliest myriapod fossils are found in the Silurian (Mikulic et al. 1985, Robison 1990). Thus, no matter whether insects evolved from some group of myriapods or a protomyriapod ancestor gave rise to both modern myriapods and insects, it seems likely that the six-legged condition we associate with insects existed in the Silurian or, perhaps, even earlier (Kukalová-Peck 1987).

The second major step was the development of wings. When *pterygote* (winged) insects first evolved is a puzzle. Insects with fully-formed, obviously functional wings first appear suddenly in the Carboniferous period (figure 11.14). Scientists are confident that these complex structures and the physiological and behavioral requirements to use them evolved through a long series of intermediate forms. However, what that evolutionary process must have been like remains unclear.

Insects were the first animals to invade the air, doing so some 50 million years before the reptiles and birds. Since predatory amphibians, reptiles, and arthropods were abundant at this time, wings no doubt provided a strong selective advantage by facilitating escape. The early winged insects had very simple wing articulations that allowed flight, but at rest the wings were held out from the body, not being flexed out of the way over the abdomen. These *paleopterous* ("primitive-winged") insects became the dominant group of insects during the Carboniferous and Permian periods but are represented today by only two orders, Odonata and Ephemeroptera.

A number of hypotheses have been advanced to explain the origin of wings in insects. It is generally agreed that wings and their control and propulsive mechanisms have been superimposed on segments originally adapted to walking. Wings have been added without the loss of legs. This is different from the vertebrates, in which walking appendages and the associated nerves and muscles have become the organs of flight.

A widely discussed hypothesis of the origin of insect wings views these structures as having arisen as bilateral expansions (*paranotal processes*) of the thoracic nota (Hinton 1963, 1977, Flower 1964, Hamilton 1971). Hinton (1963) and Flower (1964) envision paranotal expansions serving first as surfaces that enabled insects to control body attitude when falling from vegetation, ensuring that they would land in an upright position and could immediately run to shelter if being pursued by a predator. Subsequently, with further lateral expansion of the paranotal processes, gliding from high vegetation or even a running jump became possible. However, it is difficult to imagine a subsequent gradual development of an articular and neuromuscular arrangement by which the tilt fore and aft of the paranotal processes could be varied, thereby exerting a degree of glide control. The next step would have been the development of basal articulations and neuromusculature that allowed the wings to exert both controlling and propulsive forces. It has been suggested that these lateral expansions were originally involved in courtship displays by male insects (Alexander

and Brown 1963) or may have been important for thermoregulation (Douglas 1981, Kingsolver and Koehl 1985). These explanations seek to define a scenario in which selection for nonflight uses gave rise to the complex machinery that was then adapted to flight.

Wigglesworth (1976) renewed support for a theory, advanced in the late 1800s, that views insect wings as having originated from tracheal gills in a primitive aquatic form. Gills in aquatic insects are commonly articulated, contain branching tracheae, and are movable. For example, mayfly larvae are able to wave their gills to and fro, a movement that brings oxygenated water closer to the body (see "Ventilatory System" in chapter 4). Gills may also aid in swimming.

Kukalová-Peck (1983, 1987) has proposed that wings evolved from lobes originally between segments of the primitive arthropod leg (lobes are common in arthropods; as mentioned earlier, crustaceans, some chelicerates, and the extinct trilobites have biramous legs, one branch being an elaboration of a lobe between two leg segments). The leg segments surrounding this lobe are no longer visible as such in modern insects. Rather, they have evolved into the body wall and form the upper and lower articulations for the wings. This explanation accounts for not only the complex wing articulations but also provides an origin for innervation and musculature for the novel structures. Her analysis of early apterygote fossils leads Kukalová-Peck to believe that the term "paranotal" is erroneous, and these lateral expansions never were, in fact, associated with the notal surface of the insect body.

Kukalová-Peck (1987) discusses the ecological conditions under which these structures may have first been selected and elaborated. She favors the interpretation that the lobes originally had a respiratory function (moving water over the body surface). As insects evolved a more integrated control system, these structures may have been used more and more for locomotion, particularly escape from predators and dispersal. Based on her evaluation of fossil insects, Kukalová-Peck proposes that the evolution of the ancestral pterygote form was occurring at the same time as the colonization of the terrestrial habitat, rather than much later in the Devonian, when tall land plants were already present. This view is a stimulating twist on a salient problem in insect evolution and will likely generate considerable debate (Trueman 1990).

The third major evolutionary advance appears to have occurred prior to the Lower Carboniferous (figure 11.14). This was the development of a wing-flexion mechanism, which enabled insects to flex the wings posteriorly over the abdomen, the *neopterous* ("new winged") condition.

This added the advantage of being able to run and hide from predators and to move into niches inappropriate for forms with continuously outstretched wings. The neopterous insects subsequently radiated rapidly and became the dominant group of insects, as they are today. They comprise 90% of the contemporary orders and 97% of the total number of species (Carpenter 1953). Among extant orders, only apterygotes and the Odonata and Ephemeroptera are not neopterous. The extinct order Diaphanopterodea, although closely related to the paleopterous Megasecoptera and Paleodictyoptera, was able to fold its wings. Kuklalová-Peck (1974) has shown that the detailed mechanisms of wing flexure were different than is found in the true neopterous insects. This apparent convergent evolution of wing folding should reinforce the need for careful evaluation of homology, even in the case of very complex characters.

The fourth major evolutionary step was the development of complete metamorphosis (see chapter 5), also prior to the Upper Carboniferous (figure 11.14). The evolution of complete metamorphosis enabled insects to benefit from the favorable aspects of different habitats. For example, an insect with aquatic larval stages and terrestrial-aerial adults would benefit from the abundant and readily available nutrients of the aquatic habitat and later from the potential for more efficient mate location and dispersal offered by the terrestrial-aerial habitat.

Hinton (1977) discusses several insectan features, in addition to those recognized by Carpenter (1977), the origin of which he considers as major events in the evolution of insects. Among these are the evolution of a tracheal system, a relatively impermeable cuticle (mentioned earlier), and the fat body. These major adaptations are characteristic of most terrestrial arthropods and likely preceded the evolution of insects as such. They are therefore perhaps best viewed as preadaptations, which set the stage for the major evolutionary events just outlined.

The evolution of the tracheal system enabled insects (and other terrestrial arthropods) to localize respiratory and moisture exchanges with the environment as opposed to these exchanges occurring over the general body surface. This was followed by the development of a relatively impermeable cuticle (see "Permeability" in chapter 2 and "Arthropoda" earlier in this chapter), a feature that enabled insects (and other arthropods) to radiate widely into terrestrial habitats. Evolution of spiracular closure mechanisms also deserves mention in this context (see "Ventilatory System" in chapter 4). These developments are related to the evolution of flight because they counteract the drying tendency of the ambient air. Hinton views the fat body (see chapter 4), particularly with regard to the storage of fat, as a major terrestrial adaptation that allows insects and other arthropods to withstand comparatively long periods without water. He points out that the total amount of fat in arthropods is greater in terrestrial forms than in aquatic forms.

Hinton (1977) also views the size range of insects as a particular advantage, pointing out: "Insects are large enough so that they are not imprisoned by surface tension forces; many make great use of these during locomotion. At the same time they are small enough so that they can fall to any distance without injury." Janzen (1977) sees a great ecological advantage to the typical small size of insects: "I think that there are so many species of insects because the world contains a very large amount of harvestable productivity that is arranged in a sufficiently heterogeneous manner that it can be partitioned among a large number of populations of small organisms."

In the view of Zeh, Zeh, and Smith (1989) the ovipositor and egg characteristics have played major roles in the diversification of insects, roles that heretofore have been unrecognized: ". . . egg-stage characters . . . reduced constraints on suitable sites for egg deposition, and enabled insect lineages to diversify into previously inaccessible niches. In addition, the self-sufficient insect egg, resistant to osmotic rupture, desiccation, and drowning, may explain the low incidence of postzygotic parental investment among insects relative to other terrestrial arthropods. . . . We suggest that insects would have been unable to exploit the potential of holometaboly and flight without the capacity to ensure egg survival in the wide range of oviposition substrates provided by terrestrial environments."

During the course of their evolution, insects have clearly influenced and been influenced by other organisms (see chapters 9 and 10). See chapter 8 for comments on the evolution of behavior.

Based on Carpenter's four major evolutionary stages in insects, the class Insecta may be divided into groups as follows:

Apterygotes—primitively wingless insects

Pterygotes—winged insects

  Paleopterous exopterygotes—wing-flexion mechanism lacking; simple metamorphosis

  Neopterous exopterygotes—wing-flexion mechanism present; simple metamorphosis

  Neopterous endopterygotes—wing-flexion mechanism present; complete metamorphosis

Figure 11.14 displays a widely held concept of the phylogenetic relationships between these various groups. This cladogram is superimposed on a geological time scale to indicate approximately when each group originated. The grouping of insects as listed above has served as the basis of organization for chapters 12 and 13. Additional information on these categories and on the possible phylogenetic associations of the various insect orders is included in chapters 12 and 13. Boudreaux (1979), Hamilton (1972), Hennig (1969), Jeannel (1949), Kristensen (1975, 1981, 1989), Mackerras (1970), Rohdendorf (1969), Ross (1955), Scudder (1973), and Wille (1960) all consider insect phylogeny in some depth. Handlirsch (1908) and Wilson and Doner (1937) should be consulted for early interpretations of insect phylogeny. Ross (1973) traces the historical development of concepts of insect phylogeny.

Although it has not provided definitive answers, the geological record has been of value in piecing together many aspects of insect phylogeny and in evaluating the evolutionary associations between insects and other organisms. Since fossilization occurs only under certain very uncommon circumstances, the fossil record for most organisms is fragmentary. Insect fossils consist mainly of wings, but various other body parts are also commonly represented. In addition to representation of most extant orders (save Zoraptera, Grylloblattodea, Mallophaga, and Anoplura) in the fossil record, 52 extinct orders of insects have been described by paleoentomologists (see Carpenter 1977, Smart and Hughes 1972). Many of these orders are based on very limited data, such as isolated wings or wing fragments. Carpenter (1977) reduces the number of recognized extinct orders significantly by applying the following criteria: "My acceptance of an extinct order requires the knowledge of both fore- and hindwings (in the case of Pterygota) and the nature of the head, including mouthparts." On these bases 10 extinct orders can be recognized.

Apterygotes
    order Monura
Pterygotes
  Paleopterous exopterygotes
    order Palaeodictyoptera
    order Megasecoptera
    order Diaphanopterodea
    order Protodonata
  Neopterous exopterygotes
    order Protorthoptera
    order Caloneurodea
    order Miomoptera
    order Protelytroptera
  Neopterous endopterygotes
    order Glosselytrodea

The spans of geological history known to have been occupied by members of these extinct orders and their possible relationships to main insectan lines are depicted in figure 11.14. Information on each extinct order listed above may be found in chapters 12 and 13. For further information on insect fossils see Callahan (1972), Carpenter (1977), Riek (1970), Smart and Hughes (1972), Wootton (1981), Kukalová-Peck (1987), and Schlüter (1990). Rohdendorf (1973) outlines the history of paleoentomology.

CHAPTER 12

Survey of Class Insecta: I. Apterygota and Exopterygota

D u r d e n / R o m o s e r

The objective of this chapter and the following one is to provide a survey of the orders of insects. A comprehensive survey is beyond the scope of this text. This chapter will introduce some basic concepts and provide a survey of the more primitive insect groups (the Apterygota and Exopterygota) and chapter 13 will treat the more advanced orders (the Endopterygota).

The original divisions of the class Insecta are a matter of contention, and it is not easy to decide which divisions to use. Many different ordinal groupings are recognized by different experts, and which one is "best" will no doubt remain a problem. Because there are different opinions regarding the category levels of various insect groups, especially between class and order, categorical names, such as subclass and division, have purposely been omitted. Thus the taxa between class and order described here, for example, Apterygota and Pterygota, should simply be viewed as groups within the Insecta that have certain characters in common. In most cases Boudreaux (1979) has been used as a basis for recognizing groups above order. However, the taxon names used by Boudreaux are not used in every case. See "Insect Phylogeny" in chapter 11 for literature citations pertinent to various attempts at phylogenetic classification of insects. General references on the class Insecta include the following: Arnett (1985); Askew (1973); Bland and Jaques (1978); Borror, Triplehorn, and Johnson (1989); Borror and White (1970); Cheng (1976); Chu (1949); C.S.I.R.O. (1991); Davies (1988); Edmondson (1959); Merritt and Cummins (1984); Pennak (1978); Richards and Davies (1977); Stehr (1987, 1991).

To facilitate comparison of the sizes of the different orders, the number of species described in each order is presented in graphic form (figure 12.1). Figure 12.2 depicts the size standards used throughout this chapter and the following one.

Apterygota (Ametabola)

The members of the apterygote orders of insects are all primitively or primarily wingless (i.e., none of their ancestors possessed wings). Except for size differences and the presence of genitalia and gonads, there are no major changes in morphology between immature and adult forms of these insects; they are consequently stated to have an "ametabolic" or "ametamorphic" development. Apterygote insects frequently possess small, paired styli (fingerlike projections) on some of the pregenital abdominal segments; such appendages are absent in adult pterygote insects. Based on the ordinal divisions recognized in this text, this group of insects includes the orders Protura, Collembola, Diplura, Archeognatha, and Thysanura.

There is disagreement as to whether these orders represent a monophyletic or polyphyletic group. Boudreaux (1979), Hennig (1969), Kristensen (1975), Tuxen (1970), and others view them as monophyletic. However, Manton (1970), Sharov (1966), and others support a polyphyletic interpretation, elevating the Protura, Collembola, and Diplura each to the rank of class. In any event it is generally agreed that members of Protura, Collembola, and Diplura do not show as close affinities

Figure 12.1 Relative sizes of the orders of Insecta (for described species).

Data from Bland and Jaques, 1978; Borror, Triplehorn, and Johnson, 1989; C.S.I.R.O.; 1991; Gaston, 1991.

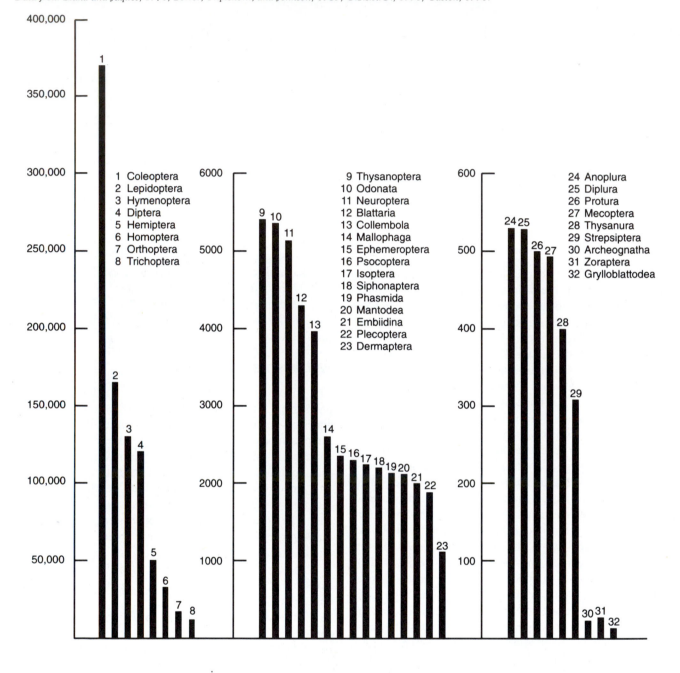

1 Coleoptera
2 Lepidoptera
3 Hymenoptera
4 Diptera
5 Hemiptera
6 Homoptera
7 Orthoptera
8 Trichoptera

9 Thysanoptera
10 Odonata
11 Neuroptera
12 Blattaria
13 Collembola
14 Mallophaga
15 Ephemeroptera
16 Psocoptera
17 Isoptera
18 Siphonaptera
19 Phasmida
20 Mantodea
21 Embiidina
22 Plecoptera
23 Dermaptera

24 Anoplura
25 Diplura
26 Protura
27 Mecoptera
28 Thysanura
29 Strepsiptera
30 Archeognatha
31 Zoraptera
32 Grylloblattodea

Figure 12.2 Size standards used in this chapter and the following one. Mi = minute; S = small; Me = medium; L = large; VL = very large. (Size abbreviation in each of the following figure captions pertains to all insects depicted, unless otherwise specified.)

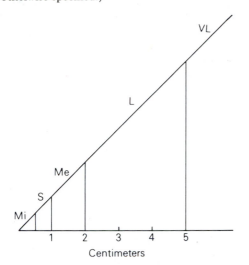

with the rest of the class Insecta as do the Archeognatha and Thysanura. The Protura, Collembola, and Diplura all have the mouthparts somewhat pulled into the head (*entognathous*). Those who recognize these groups as separate classes view the entognathous condition, absence or weak development of compound eyes, thin and pale cuticle, general small size, and several other traits as convergent adaptations to life within the soil. On the other hand, the presence of *ectognathous* (not pulled into the head) mouthparts, both simple and compound eyes, well-developed ovipositors, and a number of other characteristics clearly supports the hypothesis of a close relationship among the Archeognatha, Thysanura, and pterygote (winged) insects. The Thysanura, in particular, are considered to be descendents of primitive insects from which the pterygotes evolved. The Archeognatha have a single point of articulation between the mandible and head capsule, whereas the Thysanura and pterygotes have two such points of articulation.

Figure 12.3 is a cladogram depicting one possible interpretation of apterygote phylogeny (i.e., as a monophyletic group).

General information on the Apterygota can be found in Eisenbeis and Wichard (1987); Kevan (1962); Künnelt (1961); Schaller (1968); and Wallwork (1970).

Entognathous Apterygotes

Order Protura

Prot, first; *ura,* tail; Myrientomata; proturans (figure 12.4*a*).

Body characteristics minute, elongate, whitish.

Mouthparts entognathous, sucking.

Eyes and ocelli lacking.

Antennae lacking; "pseudoculi" may represent vestigial antennae.

Wings lacking.

Legs first pair carried in an elevated position suggestive of antennae; tarsi 1-segmented with a simple claw; an empodium may be present.

Abdomen short bilateral styli on first three segments; 12 segments in adults; 8–11 segments in immatures; cerci lacking.

Comments tracheae usually lacking; Malpighian tubules are small papillae.

Proturans are inhabitants of soil and leaf litter and require moist conditions. They apparently feed on decaying organic matter. Protura are worldwide in distribution and are common, but easily overlooked.

Proturan development is ametabolous. The adults are essentially identical to the immatures except adults have fully developed gonads and a larger number of abdominal segments. During the course of development abdominal segments are added anterior to the telson, and reach a total of twelve in the adult. The progressive addition of abdominal segments during development, termed *anamorphosis* (1 segment per molt in Protura) does not occur in any other groups of insects although it is known to occur in some other arthropods. Largely because of this, some zoologists do not consider proturans to be true insects.

Three families of proturans are currently recognized. Representatives of the family Eosentomidae are the only proturans that possess tracheae; also, all of the abdominal styli terminate in a vesicle in members of this family. In the Protentomidae, at least two of the three pairs of styli end in a vesicle, whereas in the family Acerentomidae, only the first pair of styli possess vesicles. Further information on the Order Protura is presented by Ewing (1940) and Tuxen (1964).

Figure 12.3 Hypothetical phylogeny of the apterygotes and their relationship with the pterygotes. *Based on Boudreaux, 1979.*

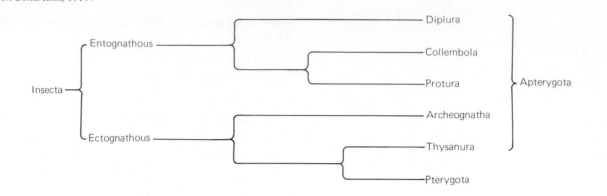

Figure 12.4 (a) Proturan, *Acerentulus barberi barberi* Ewing. (b) Diplura, *Japyx diversiungus.* (Mi.) *(a) redrawn from Ewing, 1940; (b) redrawn from Essig, 1942.*

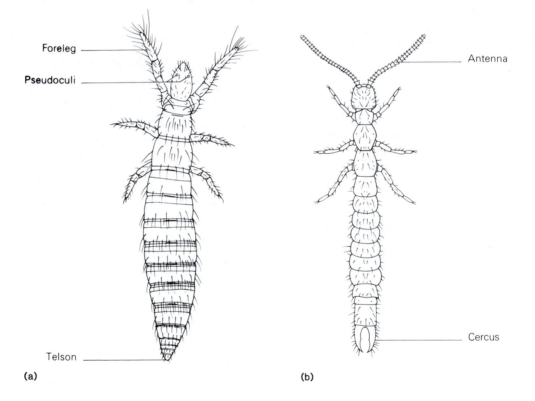

(a)   (b)

Order Collembola
Coll, glue; *embol,* a wedge; springtails (figure 12.5).

 Body characteristics minute, somewhat tubular (sub-order Arthropleona) or globose (suborder Symphypleona).

 Mouthparts entognathous, chewing.

 Eyes and ocelli compound eyes and dorsal ocelli absent; eye patches consist of 1 to several lateral ocelli comparable to those present in holometabolous larvae.

 Antennae short to long; usually 4-segmented, occasionally 5- or 6-segmented.

 Wings lacking.

 Legs claw and empodium at distal end of tibia; tarsi fused to tibia (tibiotarsus) or 1-segmented.

 Abdomen 6 segments; lobelike organ on venter of first segment, the *ventral tube,* or *collophore,* which functions in water uptake and as an "adhesive organ," and possibly also has an osmoregulatory function; forked

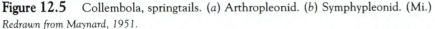

Figure 12.5   Collembola, springtails. (*a*) Arthropleonid. (*b*) Symphypleonid. (Mi.)
Redrawn from Maynard, 1951.

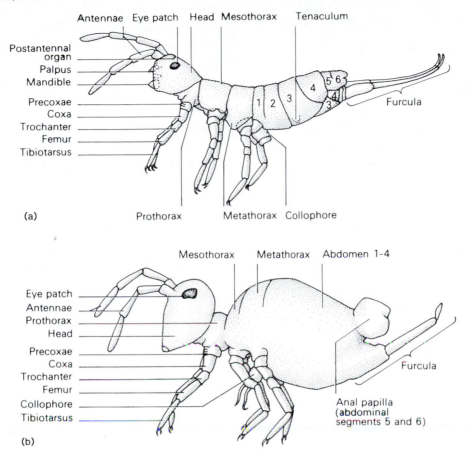

structure, the *furcula,* on venter of fourth segment of most species; *tenaculum* on venter of third segment in these species. The elongate Arthropleona have 6 distinct abdominal segments while in the oval to globular Symphypleona, the basal abdominal segments are partially fused.

Comments postantennal organs in many species; furcula directed anteroventrally and secured by tenaculum; "springing" in jumping species accomplished by sudden release of furcula from tenaculum; tracheae present or absent; Malpighian tubules lacking. It is in the possession of only 6 abdominal segments that the members of the order Collembola differ from all other insects. The thoracic segments are not readily distinguishable from those of the abdomen except that they carry legs. Springtails pass through 6–8 instars although some may attain reproductive maturity prior to a final molt. Springtails are so named because of their habit of jumping by means of the furcula, although jumping ability is reduced or absent in a few species.

This is by far the largest order of apterygote insects. Collembola are widely distributed throughout the Arctic, temperate, and tropical regions and are commonly found at very large population densities of several million in a single hectare. They are found in diverse, but always moist, habitats ranging from intertidal zones (e.g., *Anurida maritima*), the surface of snow (e.g., the "snow flea," *Achorutes nivicolus*) and glaciers to high mountains above the snow line. They are common in soil, leaf litter, and other decaying organic matter and may be found on the surface of water (e.g., *Podura aquatica*). Some species are inquilines inside ant, termite or vertebrate nests, and others are found in caves. They feed on decaying organic matter, which is usually abundant in the habitats mentioned. Species above the snow line apparently feed on sparse algae and pollen deposited by wind currents.

Although they are of no medical importance, certain species of Collembola have been identified as pests of mushrooms, greenhouse and truck garden crops, forage and cereal crops, sugarcane, and in households (Maynard 1951). Such species occasionally "swarm" in population explosions. The most notable pest is *Sminthurus viridis,* the "Lucerne flea," which is now established in numerous countries and may damage crops such as alfalfa, clover, and peas. For additional information on the Order Collembola, reference should be made to the following works: Butcher et al. (1971); Christiansen (1964, 1978); Christiansen and Bellinger (1978); Maynard (1951); Mills (1934); Salmon (1964, 1965); Scott (1961).

Order Diplura

Dipl, two; *ura,* tail; Entotrophi, Aptera, Entognatha; two-pronged bristle tails (figure 12.4*b*).

Body characteristics minute to medium—most are minute or small but one Australian species of *Heterojapyx* may reach a length of 5 cm; slender and elongate; whitish.

Mouthparts entognathous; chewing.

Eyes and ocelli lacking.

Antennae long; filiform.

Wings lacking.

Legs 1-segmented tarsi with 2 claws (movable spurs).

Abdomen 10 visible segments; cerci usually either forcepslike or long caudal filaments; styli present on segments 1–7 or 2–7.

Diplura are found in damp situations in caves, under tree bark, in the soil, and in similar habitats. They are found in temperate and tropical climates. Development is ametabolous. The Diplura share some traits with centipedes and they are considered by some entomologists to be the closest living relatives of the prototypical insects from which the class Insecta arose.

Four families are usually recognized within the Diplura. The largest and most frequently encountered family is the Campodeidae, the members of which possess long multi-segmented cerci. The Japygidae are armed with short 1-segmented forcepslike cerci. Members of both the Procampodeidae and the Anajapygidae have short but many-segmented cerci (these two families are distinguished by the position of the abdominal styli which occur on segments 2–7 in the Procampodeidae and on segments 1–7 in the Anajapygidae). The following references provide further information on the order Diplura: Remington (1954), Smith (1969, 1970).

Ectognathous Apterygotes

Order Archeognatha

Archeo, ancient; *gnath,* jaw; Microcoryphia; jumping bristletails (figure 12.6*a*).

Body characteristics small, elongate; many with scales on part of body—these scales often are arranged in distinctive patterns.

Mouthparts ectognathous, chewing; single point of articulation between mandibles and head capsule; elongate, 7-segmented maxillary palps.

Eyes and ocelli large contiguous compound eyes; 3 ocelli.

Antennae elongate, filiform.

Wings lacking.

Legs coxae of mesothorax and metathorax often with styli; 2- or 3-segmented tarsi with 2 claws.

Abdomen small bilateral styli on venter of segments 2–9; 3 ventral sclerites (*coxopodites*) and a central sternal plate on segments 2–7; elongate multisegmented cerci; elongate median caudal filament; 1 or 2 pairs of eversible vesicles on many abdominal sterna (usually on segments 1–7) that function to absorb water; females with elongate, jointed ovipositor.

Comments The name "Archeognatha" refers to the presence of a single articulation between the mandible and head capsule in contrast with two such points in the Thysanura and pterygotes. Members of the Archeognatha also are more cylindrical and have a more arched thorax than members of the Thysanura.

Jumping bristletails are commonly found in forest litter, under decaying wood, rocks, and so on. Most are nocturnally active and feed on lichens, algae, and decaying vegetation. Jumping, achieved by flexing the abdomen downward, is part of their escape behavior; a single leap may extend up to 30 cm.

These insects undergo ametabolous development and, prior to each molt, they cement themselves to the substrate apparently using fecal material to accomplish this. The Archeognatha are included by some with the Thysanura. For further discussion of this order, reference should be made to the following articles: Remington (1954, 1956); Sharov (1966); Smith (1969, 1970).

Order Thysanura

Thysan, fringe; *ura,* tail; Ectognatha, Ectotrophi, Zygentoma; bristletails, silverfish, firebrats (figure 12.6*b, c*).

Body characteristics small, elongate (occasionally oval); often dorsoventrally flattened; head and thorax broader than the tapering abdomen; usually covered with scales.

Mouthparts ectognathous, chewing; two articulations between each mandible and head capsule.

Eyes and ocelli compound eyes usually present but these are small and widely separated; 0 or 3 ocelli.

Antennae elongate, filiform, multisegmented.

Wings lacking.

Legs 2- to 5-segmented tarsi; 2 or 3 claws.

Abdomen 11 segments but terminal segment often reduced; segments 2–7 bear sternites or a pair of coxopodites; small bilateral styli on venter of several segments; females have elongate joined ovipositor; cerci elongate with many segments; elongate median caudal filament present.

Figure 12.6 (*a*) Archeognatha, jumping bristletail, *Machilis* sp. (*b*) Thysanura, common silverfish, *Lepisma saccharina.* (*c*) Thysanura, firebrat, *Thermobia domestica.* (S to Me.)

(a) redrawn from Essig, 1942, after Lubbock; (b, c) after U.S. Public Health Service.

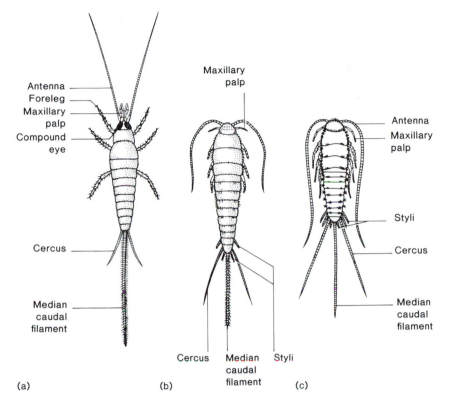

(a) (b) (c)

Comments styli interpreted as vestiges of locomotor appendages inherited from myriapodlike ancestors (same for Archeognatha); Malpighian tubules present.

Most species of Thysanura live in soil, leaf litter, rotting wood, nests of ants, termites, or mammals, and similar damp habitats. They typically feed on a variety of starchy substances.

Bristletails undergo ametabolous development. These insects differ from the pterygotes in continuing to molt periodically after they have reached sexual maturity. Longevity may be as long as five years. There are six or more instars; externally, early instars differ from later instars principally in their smaller size and the absence of scales.

Some thysanuran species, for example, the common silverfish, *Lepisma saccharina* (figure 12.6*b*), and the firebrat, *Thermobia domestica* (figure 12.6*c*) (both are representatives of the largest family, the Lepismatidae) are domesticated and share human habitations.

Silverfish are found in cool, damp locations; firebrats frequent warmer locations and are generally found around steam pipes, furnaces, and similar heat-producing objects. Both species can become pests because they feed upon such things as book bindings, starched clothing, and cloth of various sorts. They have no medical significance.

For additional information on the Thysanura, see Remington (1954); Slabaugh (1940); Smith (1969, 1970); Wygodzinsky (1972).

Fossil Apterygotes

Distinct fossil apterygotes have been found in beds ranging from the Upper Carboniferous through the Lower Permian in Europe and North America. These fossils are various species of *Dasyleptus* in the extinct order Monura (see figure 11.14). These insects resembled modern Archeognatha in a number of ways, but lack cerci. For more information on the Monura, see Sharov (1966) and Boudreaux (1979).

Pterygota (Hemimetabola and Holometabola)

It is generally agreed that the pterygotes comprise a monophyletic group. Although some insects classified as pterygotes are wingless (e.g., lice, bed bugs, and fleas), these insects are considered on the basis of other morphological (and developmental) grounds to have arisen from winged ancestors.

As explained in chapter 11, the four major evolutionary steps described by Carpenter (1977) are used as

Figure 12.7 Hypothetical phylogeny of pterygote insects.
Based, with modifications, on Boudreaux, 1979.

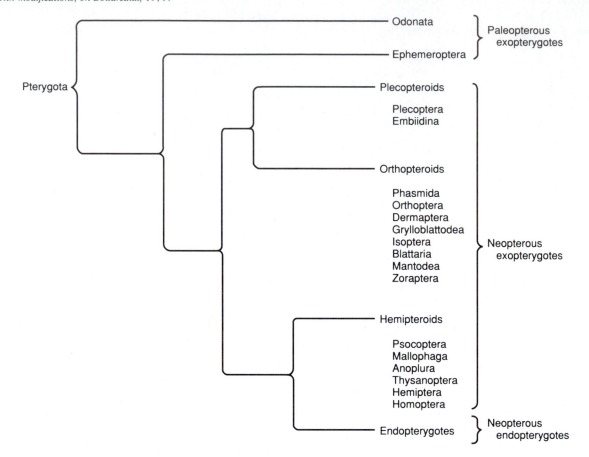

a basis for classifying insects in this text. Thus the subgroups of the Pterygota are the paleopterous exopterygotes, neopterous exopterygotes, and the neopterous endopterygotes. Figure 12.7 is a cladogram depicting a recent interpretation of the possible phylogenetic relationships among these groups. Note that although the pterygotes as a group and the neopterous groups within the Pterygota are viewed as monophyletic, the paleopterous exopterygotes are not. Some view the Ephemeroptera, on bases other than wing flexion, as being more closely related to the neopterous insects than to the Odonata (Matsuda 1970, Boudreaux 1979). On the other hand, Kristensen (1975) suggested closer ties between the Odonata and neopterous exopterygotes than between Ephemeroptera and the latter group.

The Exopterygota (Hemimetabola) undergo *hemimetabolous* or *hemimetamorphic* development. Therefore, except for the gradual external development of wings in winged Exopterygota, the immature instars outwardly resemble the adults. However, the type of habitat occupied by the immatures may differ greatly from that of the adults of the same species (as in the orders Ephemeroptera, Odonata, and Plecoptera).

Paleopterous Exopterygota

As mentioned earlier, the wings in this group are primitive in that they cannot be flexed and laid down over the abdomen and hence must, at rest, be held laterally or together above the thorax and abdomen. All other pterygotes with wings have a wing-flexion mechanism or are descendants of insects that possessed such a mechanism. Although the fossil record indicates the past existence of other paleopterous orders, Odonata (dragonflies and damselflies) and Ephemeroptera (mayflies) are the sole survivors.

Paleopterous insects undergo simple metamorphosis (i.e., they are hemimetabolous).

Order Odonata
Odon, a tooth; dragonflies and damselflies (figures 12.8 and 12.9).

Body characteristics medium to very large; elongate.

Mouthparts chewing; larva with prehensile labium.

Eyes and ocelli large compound eyes; 3 dorsal ocelli.

Antennae short; bristlelike.

Figure 12.8   Odonata, Anisoptera, dragonflies. (*a*) Adult, *Anax junius.* (*b*) Nymph. (VL.)
Redrawn from Garman, 1927.

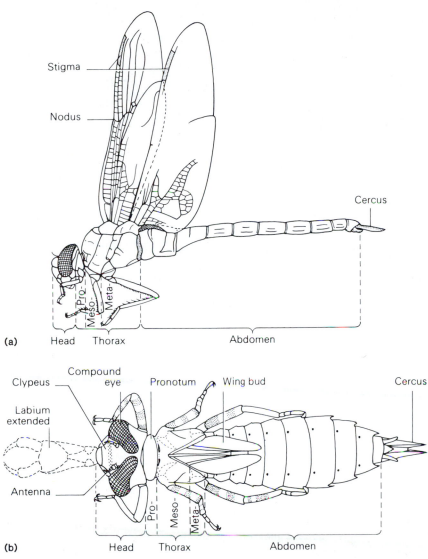

Stigma

Nodus

Cercus

Pro-
Meso-
Meta-

(a)   Head   Thorax   Abdomen

Clypeus   Compound
eye   Pronotum   Wing bud   Cercus

Labium
extended

Antenna

Pro
Meso-
Meta-

(b)   Head   Thorax   Abdomen

Wings 2 pairs; membranous; netlike venation; characteristic pigmented cell, the *stigma,* immediately posterior to the costal vein near the apex of each wing; well-developed cross vein, the *nodus,* near the middle of the leading edge of each wing.

Legs adults: basketlike arrangement, adaptation for prey capture; 3-segmented tarsi.

Abdomen elongate; adult males with gonopores on ninth segment and complex penis on venter of second segment; 1-segmented cerci in males serve as claspers during copulation.

Comments larvae: closed tracheal system, gill ventilators; adults: open tracheal system.

Odonata can be divided into three suborders, Anisoptera, the dragonflies (figure 12.8), Zygoptera, the damselflies (figure 12.9), and Anisozygoptera. Members of the first two suborders are easily distinguished from each other both in the larval and adult stages (see table 12.1). The Anisozygoptera share characters with both the Anisoptera and Zygoptera. Although there are many extinct forms, this group is represented today by only two species, one in Japan and one in India. Some other fossil dragonflies from the Carboniferous (= Mississippian and Pennsylvanian) period were relatively immense (with a wingspan of up to 64 cm or 2.5 feet) and represent some of the largest insects known to have existed.

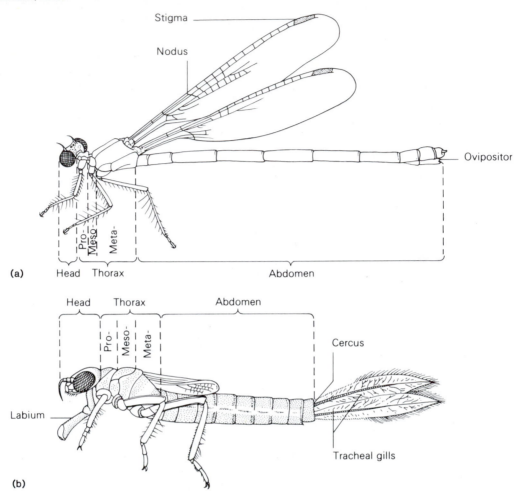

Stigma

Nodus

Ovipositor

Pro- Meso- Meta-

(a)   Head   Thorax   Abdomen

Head   Thorax   Abdomen

Pro- Meso- Meta-

Cercus

Labium

Tracheal gills

(b)

Immature dragonflies and damselflies are aquatic, whereas the adults are terrestrial-aerial but are generally found in the vicinity of water. Nymphs usually reach the adult stage in a single season; however, some species require two or three years to develop into adults. Eggs may be dropped into the water or attached to aquatic vegetation or other partially submerged objects. Rarely, eggs are deposited in gelatinous strings or masses. In a few species, females go beneath the surface of the water to oviposit. See "Seminal Transfer" in chapter 5 for the remarkable method of insemination characteristic of the Odonata.

Dragonfly and damselfly larvae prey on a wide variety of aquatic organisms. Some of the larger species may capture tadpoles and small fish. Adults prey on other flying insects (e.g., mosquitoes and midges, moths, bees, and other dragonflies). They are able to catch, hold, and devour prey in flight.

Although these insects have been recorded as occasional pests in apiaries and sometimes attack trout fry, they are generally considered to be very beneficial. They are harmless to man.

Some of the principal families of the Odonata are discussed in the following paragraphs. Separation of families within this order is largely based on the morphological attributes of the wings.

More information on the Odonata may be found in the following references: Corbet (1963); Gloyd and Wright (1955); Merritt and Cummins (1984); Needham and Westfall (1955); Pennak (1978); Smith and Pritchard (1956); Walker (1953, 1958); Walker and Corbet (1975); Westfall (1978).

Suborder Zygoptera (damselflies)   *Family Calopterygidae.* Often referred to as broad-winged damselflies these odonates are relatively large with the wing-bases narrowing gradually (the wing-bases are stalked in other zygopterans). They usually occur along clear fast moving streams.

Family Coenagrionidae. Members of this large family are commonly called narrow-winged damselflies. At rest, adults typically hold the wings together over the body with the body itself held in a horizontal plane. Most

Table 12.1 Outstanding Differences between Suborders of Odonata

Anisoptera	Zygoptera
1. Forewings and hindwings unequal in size; hindwings basally broader than forewings.	1. Forewings and hindwings approximately the same size and shape.
2. Wings laterally spread at rest.	2. Wings typically held together dorsally over the thorax and abdomen.
3. Strong, agile fliers.	3. Comparatively weak fliers.
4. Compound eyes in close proximity or meeting dorsally.	4. Compound eyes widely separated.
5. Males have 3 terminal abdominal appendages.	5. Males have 4 terminal abdominal appendages.
Nymphs	
1. Ventilatory gills in rectum, not externally visible.	1. Ventilatory gills are 3 terminal abdominal appendages, externally visible.
2. Stout, robust body.	2. Relatively slender, fragile body.
3. Able to propel themselves short distances by forcibly ejecting water from the rectum.	3. Lack "jet-propulsion" mechanism.

Family Cordulegastridae. These dragonflies are commonly called biddies and typically are brownish-black with yellow markings. They usually frequent small woodland streams where the adults patrol back and forth just above the water surface.

Family Corduliidae. In life, the eyes of most adult corduliids are bright green, which gives rise to their common (vernacular) name of green-eyed skimmers. Adults are black or metallic and usually are devoid of markings on the thorax or abdomen. Many species alternate periods of hovering with episodes of direct flying. The genus *Somatochlora* includes the bog skimmers, which are metallically-colored and associated with bogs or woodland streams.

Family Gomphidae. Many of the representatives of this family have clearly swollen terminal abdominal segments, which gives rise to their vernacular name of clubtails. Most adults are darkly-colored with yellow or green markings and have a tendency to alight on the surfaces of stones or rocks.

Family Libellulidae. This is a large and widely distributed family, the representatives of which are often referred to as common skimmers (in North America) or darters (in Europe). They are typically abundant around swamps and ponds and vary in size from about 20 mm to 75 mm. These dragonflies are usually robust and rapid fliers, although flight behavior is often erratic with frequent changes in direction. The wings are marked with (usually black) spots or bands in many species and the abdomen is typically flattened dorsoventrally so that it appears stout. The large genus *Libellula* is widely distributed and common near ponds, swamps, and lakes.

Family Macromiidae. Commonly called belted skimmers, or river skimmers, these dragonflies typically occur along pond and lake shores although some are associated with large streams and rivers. They are swift fliers and brownish in color often with characteristic thoracic and abdominal markings.

Order Ephemeroptera

Ephemero, for a day; *ptera,* wings; Ephemerida, Plectoptera; mayflies (figure 12.10).

Body characteristics small to medium; fragile; soft-bodied.

Mouthparts nymphs: chewing; adults: vestigial.

Eyes and ocelli compound eyes present; 3 dorsal ocelli.

Antennae short; setaceous.

Wings adults and subimagines; 2 pairs; membranous; forewings larger than hindwings; few species with only mesothoracic wings; held vertically at rest; typically many cross veins and pleated wing appearance.

Legs 3- to 5-segmented tarsi.

species are weak or inefficient fliers. A wide range of habitats is exploited including swamps and ponds, but swift streams are usually avoided. Males are frequently more brightly colored than females of the same species. Coenagrionids are well represented in both North America and Europe.

Family Lestidae. These insects are commonly called spread-winged damselflies because of their habit of holding the wings partly outspread and the body in an almost vertical plane when alighting on vegetation. Lestids are largely associated with swamps. Females often attach eggs to aquatic plants well beneath the water surface. *Lestes* is the largest and most widespread North American genus; this genus also occurs in Europe.

Suborder Anisoptera (dragonflies) *Family Aeshnidae.* Commonly referred to as darners (in North America) or hawkers (in Europe), the family Aeshnidae includes the largest and most robust dragonflies. Adults of most species measure about 75 mm in length and are green or blue. Aeshnids are common in or close to a variety of aquatic habitats including ponds, marshes, and canals.

Figure 12.10 Ephemeroptera, mayflies. (*a*) Adult. (*b*) Nymph. (S to Me.)
Redrawn from Burks, 1953.

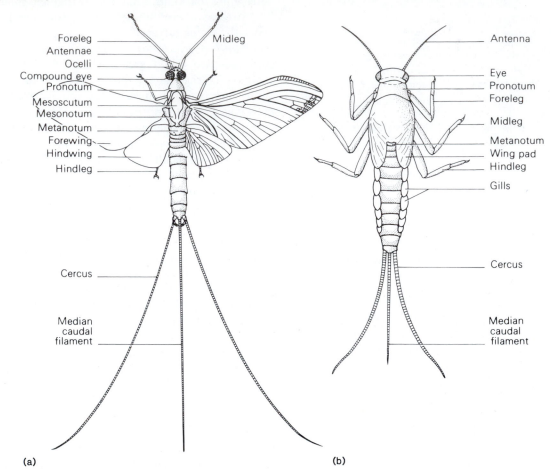

Foreleg
Antennae
Ocelli
Compound eye
Pronotum
Mesoscutum
Mesonotum
Metanotum
Forewing
Hindwing
Hindleg
Midleg
Cercus
Median caudal filament

(a)

Antenna
Eye
Pronotum
Foreleg
Midleg
Metanotum
Wing pad
Hindleg
Gills
Cercus
Median caudal filament

(b)

Abdomen nymphs: bilateral ventilatory gills on the first 4 to 7 segments; adults: pair of long, filamentous cerci; some with additional long median caudal filament.

Comments nymphs: closed tracheal system, gill ventilators; adults: open tracheal system.

The members of this order are aquatic in the nymphal stages and terrestrial-aerial as adults.

Mayflies are unique in the class Insecta in passing through a winged, subadult (subimaginal) stage that is essentially identical to the adult stage except that it lacks functional genitalia. The subimaginal and adult stages live for a brief period of time (a few hours to a few days), and their apparent sole function is to survive long enough to copulate and deposit their eggs, which are often simply dropped into the water and subsequently sink to the bottom. Some species attach their eggs to partially submerged rocks or aquatic vegetation. The larval period is generally fairly long, a year or more, although some species may develop in a few months.

Mayflies commonly emerge from lakes, streams, and rivers in fantastically large numbers, and in areas near towns or cities the shed subimaginal cuticles and dead adults may literally accumulate in piles on streets and sidewalks. This was once a common occurrence in towns along the shores of Lake Erie until pollution took its toll. In fact, mayflies are generally very good bioindicators (index species) of water pollution.

Larval mayflies (figure 12.10*b*) are solely responsible for the intake of nutrients because the adults do not feed. Their food consists of aquatic vegetation and small aquatic organisms.

Mayfly larvae are important food for aquatic predators, particularly fish, and provide a periodic feast for insectivorous birds and dragonflies when they emerge in vast numbers. They have been likened to terrestrial herbivores in that they are close to the bottom of several food chains, converting plant material into animal material. They also provide excellent models for many of the common "tied flies" cherished by avid fly fisherman.

In areas where large populations of adult mayflies invade towns and cities, there have been reports of severe allergic responses (asthma, conjunctivitis, and so on) to airborne detritus composed of the broken casts of subimagines and dried bodies of adults.

As in the order Odonata, family classification within the Ephemeroptera is based primarily on wing venation. Terminology used here for wing veins and wing regions is as shown in figure 2.23. A brief discussion of the characteristics of the main ephemeropteran families follows. Most of the families considered here are represented in both North America and Europe. The following references provide additional information on mayflies: Alba-Tercedor and Sanchez-Ortega (1991); Burks (1953); Day (1956); Edmunds (1959a, b, 1978); Edmunds et al. (1976); Hubbard (1990); McCafferty (1975, 1991); Merritt and Cummins (1984); Needham et al. (1935); Pennak (1978).

Family Baetidae. This is a large family of mayflies with a wide geographical distribution. Adults are small (wingspan, less than 25 mm) and hindwings may be present (as in the genera, *Baetis* and *Centroptilum*) or absent (as in *Cloeon* and *Procloeon*); if present, hindwing venation is reduced. Adults possess two caudal filaments. Nymphs occupy a variety of aquatic habitats and generally are good swimmers; they are cylindrically shaped.

Family Baetiscidae. These are small- to medium-sized mayflies (wingspan, 16–25 mm). Adults possess two caudal filaments that typically are shorter than the body. The forewings are devoid of cubital intercalary veins and vein 1A extends to the outer wing margin. Nymphs occur principally in cool, fast-flowing streams.

Family Caenidae. The mayflies belonging to this family are minute or small (wingspan, 4–12 mm) and lack hindwings. The forewings are fringed with setae posteriorly and venation is reduced with a few cross-veins arranged in a zigzag fashion. Adults have three caudal filaments. Nymphs are bottom-dwellers, partly carnivorous, and possess an enlarged second gill that acts as a protective shield covering the remaining gills.

Family Ecdyonuridae. These are small- to medium-sized mayflies (wingspan, 10–35 mm) that are typically associated with fast-flowing streams and rivers. Nymphs are dorsoventrally flattened and cling to stones on the stream or river bed to avoid being swept along with the current. Adults have two caudal filaments.

Family Ephemerellidae. These mayflies are small- to medium-sized (wingspan, 12–38 mm) with brown or yellow bodies and clear or smoky-brown wings. The family is common and widespread, occupying numerous aquatic habitats. Adults possess three caudal filaments. Marginal veinlets are present on the outer margin of the forewings and costal cross-venation is reduced.

Family Ephemeridae. These small- to large-sized insects (wingspan, 10–50 mm) are commonly referred to as burrowing mayflies because nymphs tunnel into silt or sand at the bottom of swift-flowing, slow-moving, or still water where mud accumulates. Presumably because of their burrowing lifestyle, the nymphal gills are bent back over the abdomen. Adults may have two (e.g., *Hexagenia*) or three (e.g., *Ephemera*) caudal filaments. Swarms of adult ephemerids may sometimes emerge from lakes and rivers.

Family Heptageniidae. This family of mayflies is widely distributed in North America. Like those of the Ecdyonuridae and Siphlonuridae, the nymphs are dorsoventrally flattened presumably to enhance attachment to stones in the fast-flowing streams that they occupy. Adults possess two caudal filaments and two pairs of cubital intercalary veins that are more or less parallel.

Family Leptophlebiidae. The leptophlebiids are small- to medium-sized mayflies (wingspan, 8–28 mm) with brown or yellow bodies and clear or smoky-brown wings. Adults have three caudal filaments and a fairly complete wing venation. Males have divided eyes with the upper eye sections possessing significantly larger facets than the lower sections. Nymphs typically occur in still water or in slow-flowing streams.

Family Siphlonuridae. This is another family of mayflies with streamlined (dorsoventrally flattened) nymphs in accordance with their preference for fast-flowing streams and rivers. Most adults are medium-sized (wingspan, 16–40 mm). The family is widely distributed in North America but less so in Europe.

Family Tricorythidae. This is principally a tropical family but a few species extend into parts of the United States. Adults are minute or small (wingspan, 6–18 mm) and possess three caudal filaments. The forewings are elongate-oval and lack marginal veinlets. Hindwings are absent except in the males of one genus (*Leptohyphes*). Nymphs frequent rivers and streams.

Fossil Paleopterous Exopterygote Orders

Carpenter (1977) recognized four extinct paleopterous exopterygote orders: Palaeodictyoptera, Megasecoptera, Diaphanoptera, and Protodonata (see figure 11.14). The first three listed are apparently closely related, sharing such traits as piercing-sucking haustellate mouthparts, long cerci, and an external ovipositor. They were probably terrestrial (fossil larvae lack gills) and presumably herbivorous. The Diaphanoptera were able to flex the wings over the abdomen, but the flexion mechanism differed from that found in modern neopterous insects. Protodonata were similar to modern Odonata in appearance and predaceous feeding behavior. Among the Protodonata were the largest known insects, one genus, *Meganeura,* having a wing span of 75 cm.

Neopterous Exopterygota

This taxon includes three groups of orders (figure 12.7): the plecopteroids, the orthopteroids, and the hemipteroids.

Plecopteroid and Orthopteroid Orders

The plecopteroid orders as described here include the Plecoptera and Embiidina, and the orthopteroid orders include Phasmida, Orthoptera, Dermaptera, Grylloblattodea, Isoptera, Blattaria, Mantodea, and Zoraptera (see figure 12.7). There are differences of opinion about the groupings of these orders, especially with regard to the plecopteroid insects. Both Plecoptera and Embiidina have been included as orthopteroids, and the Plecoptera are sometimes listed as being separate from all other neopterous exopterygote orders. Among the orthopteroid orders, there is strong evidence and much agreement that the Isoptera, Blattaria, and Mantodea are closely related. Thus the Blattaria and Mantodea are sometimes grouped as the order Dictyoptera. Lumping all three orders into a single order has also been suggested.

The plecopteroids and orthopteroids are similar relative to a number of features including: (1) primitive, "generalized" mandibulate mouthparts; (2) complex wing venation with hindwings typically larger than forewings owing to enlarged anal areas (hindwings fold along vannal and jugal folds when wings are flexed); (3) male intromittent organ formed from everted ejaculatory duct; (4) cerci present; (5) typically many Malpighian tubules; and (6) ventral chain ganglia distributed segmentally.

Both plecopteroid and orthopteroid orders undergo hemimetabolous development.

Among the bases for grouping Plecoptera and Embioptera as plecopteroids are: (1) external appendages associated with male genitalia reduced or absent; (2) abdominal styli lacking; (3) accessory clasping structures in males from various sources including extension of tenth abdominal tergum and modified paraprocts; (4) forewings not leathery or sclerotized; and (5) trochantin attached to epimeron with no suture between.

Traits common to the orthopteroid orders include: (1) external genital appendages present in males; (2) abdominal styli on ninth abdominal segment of males in many species; (3) vannus of hindwings forming pleats when folded; (4) forewings leathery or somewhat sclerotized; and (5) trochantin separated from epimeron by suture or membrane.

Order Plecoptera

Pleco, pleat; *ptera,* wing; Perlaria, stoneflies (figure 12.11).

Body characteristics small to medium; soft-bodied; elongate, flattened; body nearly parallel-sided.

Mouthparts chewing in nymphs; commonly vestigial in adults.

Eyes and ocelli well-developed compound eyes; usually 3 dorsal ocelli, rarely 2.

Antennae long and tapering.

Wings 2 pairs; membranous; males of some species apterous or brachypterous (short wings); at rest, hindwings folded in "pleats" beneath forewings.

Legs 3-segmented tarsi.

Abdomen ovipositor lacking; many-segmented cerci.

Comments nymphs: closed tracheal system, gill ventilators, gills usually on venter of thorax, but sometimes on other parts of body; adults: open tracheal system.

Stonefly nymphs (figure 12.11*c*) are aquatic and are usually found around and beneath stones in fast-moving, well-aerated water. The adults (figure 12.11*a*) are terrestrial-aerial and are found in the vicinity of nymphal habitats. The nymphs are campodeiform and resemble the adults in general appearance except for the absence of wings and the presence of tracheal gills.

Stoneflies are mostly herbivorous as nymphs, feeding on various forms of small aquatic plant life; a few are carnivorous and feed on other aquatic insects. The adults with well-developed mouthparts feed on algae and lichens; those with vestigial mouthparts do not feed at all. Although most species complete the nymphal stage in a year or so, some species require two, three, or four years to develop into adults.

Comstock (1940) mentioned adults of *Taeniopteryx pacifica* as fruit tree pests in the Wenatchee Valley, Washington. However, the value of stoneflies as fish food far outweighs this isolated instance of pestiferous behavior. Plecoptera are often good bioindicators of aquatic pollution.

Not surprisingly, the family classification of stoneflies is based primarily on wing venation. The families discussed below are representative of the order. All but one of these families occur in both North America and Europe (the exception, the Pteronarcidae, is not found in Europe).

For additional information on the Plecoptera, see Alba-Tercedor and Sanchez-Ortega (1991); Claassen (1931); Frison (1935); Harper (1978); Hitchcock (1974); Hynes (1976, 1977); Jewett (1956); Merritt and Cummins (1984); Needham and Claassen (1925); Pennak (1978); Ricker (1959).

Family Chloroperlidae. Commonly called green stoneflies, these insects are usually greenish or yellowish. Adults are characterized by having either no anal lobe or a small anal area without branched anal veins on the hindwings. There are no vestiges of nymphal gills on the ventral surface of the adult thorax. Body length varies from 6 mm to 24 mm. Nymphs usually are found in small streams, and adults appear during the spring.

Family Isoperlidae. The hindwings of these stoneflies have forked anal veins. Adults are devoid of remnants of nymphal gills on the thorax and are commonly observed running on vegetation adjacent to streams. The genus *Isoperla* occurs in both North America and Europe.

Figure 12.11 Plecoptera, stoneflies. (*a*) Adult male, *Pteronarcys californica*. (*b*) Terminalia of adult female. (*c*) Generalized nymph. (L.)

(a,b) redrawn from Essig, 1942; (c) redrawn from Claassen, 1931.

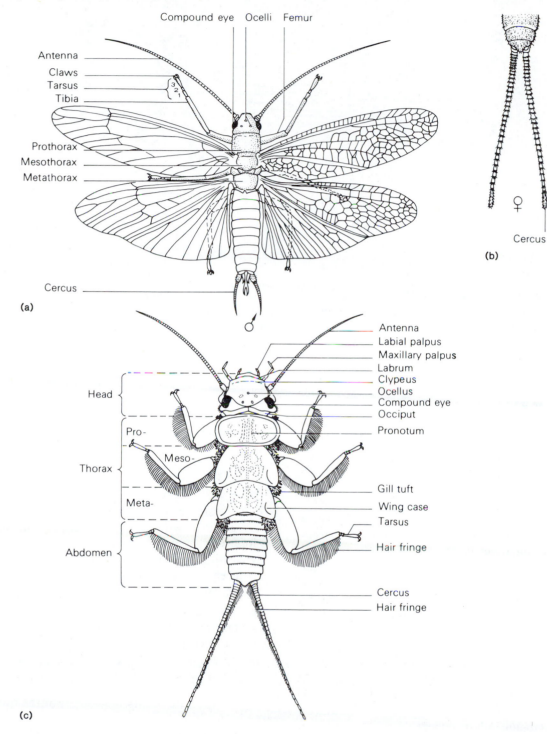

Figure 12.12 Embiidina, webspinners. (*a*) Male, *Oligotoma saundersii.* (*b*) Female. (Mi.)
Redrawn from Essig, 1942.

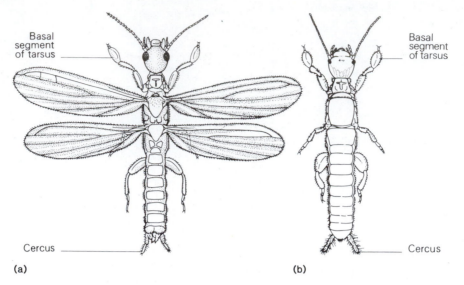

(a)   (b)

A common vernacular name for this group is green-winged stoneflies. Some authors include the isoperlids in the family Perlodidae.

Family Nemouridae. This family comprises four subfamilies, the Capriinae (small winter stoneflies), Leuctrinae (rolled-winged stoneflies), Nemourinae (spring stoneflies), and Taeniopteryginae (winter stoneflies), each of which may be elevated to familial rank by some authors. All of these stoneflies are brown or black in coloration and are small with a body length of 13 mm or less. Many species emerge as adults during the winter or early spring. Some male capriines are brachypterous (short-winged). Nymphs of most species are phytophagous.

Family Perlidae. Often referred to as common stoneflies, these insects collectively comprise the largest family within the order Plecoptera. Adults are medium to large in size (body length, 20–40 mm) and possess remnants of nymphal gills on the ventral surface of the thorax (these are immediately posterior to the leg bases in most species). Adults are usually on the wing during the summer and do not feed. Most nymphs are predaceous.

Family Perlodidae. These stoneflies are similar to the isoperlids but generally are larger (body length, 10–25 mm). Remnants of nymphal gills are either absent in adults or represented by a single gill on each side at the base of the labium. Adults appear during the spring or early summer close to the medium-sized to large streams inhabited by the nymphs.

Family Pteronarcidae. This group of giant stoneflies includes the largest members of the order Plecoptera with females of some species attaining a body length of 65 mm. Adults are brown or gray and the anal area of the fore-

wing has two or more rows of cross-veins. Adults appear in spring and early summer and do not feed. Nymphs are phytophagous and inhabit medium-sized or large rivers.

Order Embiidina

Embio, lively; Embioptera; webspinners, embiids (figure 12.12).

Body characteristics minute to small; brown to yellow; elongate.

Mouthparts chewing.

Eyes and ocelli compound eyes present; no dorsal ocelli.

Antennae filiform; 16–32 segments; shorter than body.

Wings males in some species with 2 pairs of membranous wings with hindwings slightly smaller than forewings; "smoky" in appearance and usually covered with fine setae; except for radial vein, venation reduced with each vein usually in the center of a brown band; females and some males wingless.

Legs short; basal tarsal segment of foreleg enlarged and contains silk-producing glands; 3-segmented tarsi; hind femur thickened.

Abdomen pair of 1–2 segmented cerci present; in males, these are asymmetrical and unequal in size.

Embiidina are mostly inhabitants of the tropics but are also found in temperate zones. About 2,000 species have been described worldwide and at least 9 species occur in the southern United States. Webspinners are so named because they live in a network of silken tunnels beneath stones, bark, and so on. Tunnels are constructed by both sexes and by immatures from the secretions of glands in the tarsi of the forelegs. These insects are gregarious and a tunnel network may contain many individuals. They are able to run quickly forward and backward within these tunnels. In addition to serving as a habitation, the tunnels probably protect the inhabitants from predators and may play a role in the maintenance of a humid atmosphere. Wings are soft and rollable, which allows the males to run forward or backward inside their tunnels without damaging the wings or becoming stuck. Each wing possesses a large hemocoel that can fill with hemolymph and render the wings sufficiently turgid for a clumsy mode of flight. During cooler weather, webspinners often retreat beneath the soil. They may play dead when disturbed. Webspinners feed mainly on decaying plant matter and lichens. In some species the male mouthparts differ in shape from those of the female suggesting that the sexes of these particular species use contrasting food resources. Winged male webspinners are poor fliers but are occasionally attracted to outside lights. At rest, the wings are held flat over the body.

These insects reproduce within their silken tunnels. Eggs are relatively large and oval and deposited in scattered groups; they may be covered with chewed food particles, which suffice as the first meal for the emerging nymphs. Webspinners are like the Dermaptera in that the females actively attend the eggs. Some species are parthenogenetic. Ross (1940, 1944, 1970, 1984) provides additional information on the order Embiidina.

Order Phasmida

Phasm, phantom; Phasmatodea, Phasmodea, Phasmoidea; stick and leaf insects; walkingsticks (figure 12.13).

Body characteristics large; commonly elongate, cylindrical with short prothorax and usually elongate meso- and metathorax; some flattened and leaflike (leaf mimics).

Mouthparts mandibulate.

Eyes and ocelli compound eyes present; some winged species have 2 ocelli.

Antennae filiform or moniliform; variable length.

Wings many wingless (all in the U.S. are wingless, save one species in southern Florida that has very short wings); forewings (*tegmina*) often reduced; hindwings sclerotized anteriorly with membranous anal area; in leaflike forms venation mimics leaf venation.

Legs generalized cursorial; tarsi usually 5-segmented.

Abdomen ovipositor small and somewhat concealed; short, unsegmented cerci.

Walkingsticks are rather sluggish, solitary, herbivorous insects. Although several species occur in temperate zones, they are predominantly tropical with the greatest diversity in the Oriental region. Some species grow quite large, with certain tropical forms ranging up to a foot

Figure 12.13   Phasmida. A stick insect, *Diapheromera femorata.* (VL.)
Redrawn from Hebard, 1934.

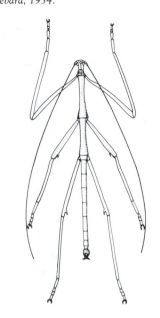

(30 cm) in length. *Megaphasma dentricus* is the largest insect (by length) in the United States, reaching 7 inches (18 cm).

Walkingsticks are somewhat protected from predators by their resemblance to stems or leaves of plants. Some, for example, *Carausius morosus,* are capable of physiological color changes effected by pigment granule migration in epidermal cells. Many phasmids produce repugnatorial substances from glands in the prothorax. *Anisomorpha buprestoides,* an aposematically colored species (see "Adaptations Associated with Interspecific Interactions" in chapter 9), is able to control the direction of its spray and will spray in response to an approaching potential predator such as a bird (Eisner 1965). If a nymph happens to be caught by a leg, *autoamputation* (autotomy) may result, allowing the insect to escape. Autoamputated legs subsequently regenerate.

Walkingsticks undergo hemimetabolous development. Their eggs, which may resemble host plant seeds, are generally dropped singly to the ground. Both obligate and facultative parthenogenesis is found among members of this order.

Leaf insects are more robust than walkingsticks and usually are dorsoventrally flattened and resemble leaves. Numerous representatives of the leaflike genus *Phyllium* inhabit tropical Asia and Australasia.

Some authors treat the phasmids as a suborder (Phasmatodea), or a family (Phasmatidae) within the order Orthoptera. However, phasmids do not have enlarged hind femora and do not jump. The order Phasmida is divided into two families, the Phasmatidae and

Figure 12.14 Orthoptera, a short-horned grasshopper (Acrididae), showing major external structures. (L.)
Redrawn from Essig, 1942.

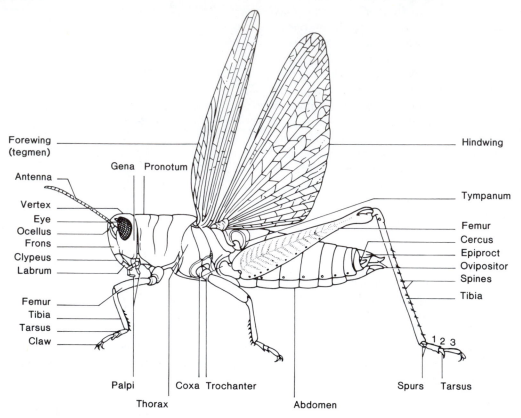

the Tememidae (=Phyllidae). The Phasmatidae (walk-ingsticks/stick insects) are elongate and never leaflike; the tibiae lack a triangular apical area. The Tememidae (leaf insects, timemas) typically are leaflike and stoutly set; tibiae have a small ventro-apical triangular area. For additional information on this order, the following references should be consulted: Blatchley (1920), Helfer (1963), Ragge (1973).

Order Orthoptera

Ortho, straight; *ptera,* wing: Saltatoria; grasshoppers, locusts, crickets, katydids, and related forms (figures 12.14 and 12.15).

Body characteristics minute to very large; body form variable.

Mouthparts chewing; all gradations between hypognathous and opisthognathous.

Eyes and ocelli well-developed compound eyes; 0, 2, or 3 dorsal ocelli.

Antennae various; filiform in many (e.g., katydids and crickets); somewhat shorter in many grasshoppers and others.

Wings most with 2 pairs; forewings parchmentlike tegmina; at rest, hindwings folded pleatlike beneath forewings.

Legs hindlegs adapted for jumping (the hind femora are enlarged, accommodating the muscles that produce the jumping force); tarsi 3- or 4-segmented.

Abdomen females in several families with well-developed ovipositors; cerci variable: short, long, clasperlike, segmented, unsegmented.

The Orthoptera includes grasshoppers, locusts, crickets, katydids, and relatives. This order can be divided into two suborders, Caelifera (grasshoppers, pygmy grasshoppers, etc.; figures 12.14 and 12.15 *f*) and Ensifera katydids, crickets, mole crickets, etc.; (figure 12.15 *a–e*). The Caelifera have comparatively short, filiform antennae (fewer than 30 segments), short ovipositors, tarsi with 3 or fewer segments, and tympana (if present) situated laterally on the first abdominal segment. The Ensifera typically have long hairlike antennae (with more than 30 segments), 3- or 4-segmented tarsi, and comparatively long cylindrical or sword-shaped ovipositors; tympanic organs (if present) are located at the bases of the tibiae of the forelegs.

Members of this order are typically terrestrial and are found in trees, bushes, and other vegetation; on the surface of, or burrowing into, the ground; and occasionally in caves. A few species are aquatic or semiaquatic.

Figure 12.15 Representative Orthoptera. (*a*) Field cricket, *Gryllus assimilis* (Gryllidae). (*b*) Bush katydid, *Scudderia furcata* (Tettigoniidae). (*c*) Jerusalem cricket, *Stenopelmatus longispina* (Gryllacrididae). (*d*) Tree cricket, *Oecanthus niveus* (Gryllidae). (*e*) Mole cricket, *Gryllotalpa hexadactyla* (Gryllotalpidae). (*f*) Pygmy grasshopper, *Telmatettix hesperus* (Tetrigidae). (*a*, *d*, and *f*), Me; (*b*, *c*, and *e*), (L.). *Redrawn from Essig, 1942.*

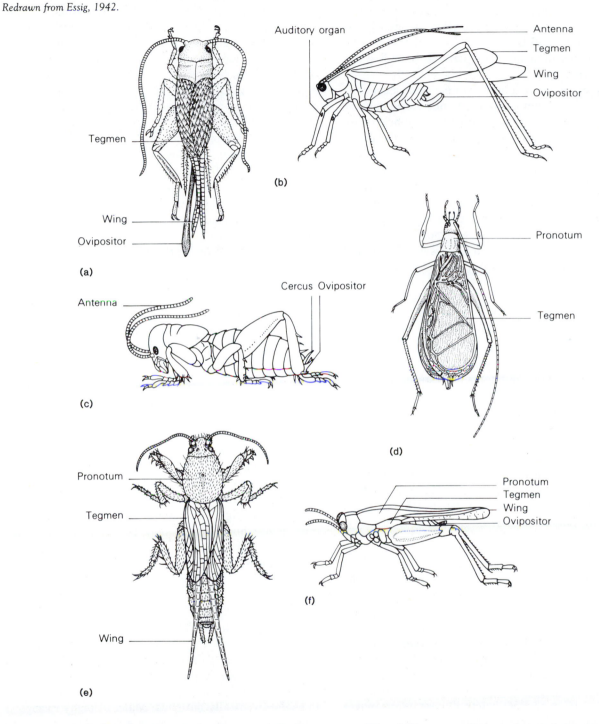

Orthoptera includes general scavengers, many voracious herbivores, and some carnivorous species.

There are generally five or more nymphal instars in an orthopteran life cycle. Eggs are deposited singly or in masses, in or on vegetation (e.g., most katydids), or in the soil (e.g., most grasshoppers).

The ability to produce and perceive sound is found among many orthopterans. Although sound production is effected in a variety of ways, it generally involves the rubbing of one part of the body against another. The females of some species are capable of producing sound, but it is typically the males that are the accomplished singers. The songs are most commonly involved with bringing members of the opposite sexes together. The snowy tree cricket, *Oecanthus fultoni,* is purported to be a "living thermometer" because adding 40 to the number of chirps in 15 seconds approximates the temperature in degrees Fahrenheit. This, of course, assumes that one is able to recognize the song of the snowy tree cricket. A phonograph record (Alexander and Borror 1956) is available that includes the songs of many orthopterans, including the snowy tree cricket.

Orthoptera includes a large number of economically important species. Members of the family Acrididae, the short-horned grasshoppers, are serious pests of field crops, particularly small grains. Some of the long-horned grasshoppers (family Tettigoniidae) are also serious pests of field crops. For example, the Mormon cricket, *Anabrus simplex,* attacks several crops in western North America. In Salt Lake City, Utah, a statue commemorating the sea gulls that saved the early Mormon settlers' crops from destruction has been erected in Mormon Square. The coulee cricket, *Peranabus scabricollis,* is a common pest of field crops in the northwestern United States. In many parts of Africa and the Middle East, various species of locusts (particularly *Locusta migratoria* and *Schistocerca gregaria*) have, throughout human history, caused crop damage, often on a devastating scale. Migration of locust swarms can be aided by storms and wind patterns so that, under certain conditions, these insects may occasionally appear in large numbers in Europe and other parts of Asia.

Familial classification within the Orthoptera is based largely upon the morphology of the legs and antennae although wing and general body structures are also considered in some groups. Some of the more important orthopteran families are discussed in the following paragraphs. The following references provide additional information on this order: Blatchley (1920); Chapman and Joern (1990); Dirsh (1975); Helfer (1963); Otte (1981, 1984); Uvarov (1966, 1977).

Suborder Caelifera (short-horned grasshoppers) *Family Acrididae* (figure 12.14). This family includes the short-horned grasshoppers and locusts. The tarsi are 3-segmented, and the ovipositor and antennae are comparatively short. The wings are usually well developed in adults and are brightly colored in some species. The pronotum does not extend posteriorly over the abdomen. Representatives of this large and widely distributed family are often abundant in meadows during the summer and early fall. The phytophagous habits of these insects makes some of them pests of various crops.

Family Eumastacidae. The monkey grasshoppers are similar to the acridids but lack wings. Antennae are shorter than the fore-femora and tympanal organs are usually absent. Most species are small and brownish and inhabit arid regions (particularly in the southwestern United States and Central America). They are found in bushes where they are remarkably agile as their vernacular name indicates.

Family Tanoceridae. Commonly referred to as desert short-horned grasshoppers, these insects are similar to the eumastacids in that they lack wings, inhabit bushes where they are agile, and are found in similar arid environments. The antennae in the Tanoceridae are longer than the fore-femora however.

Family Tetrigidae (figure 12.15*f*). These insects, commonly called pygmy grasshoppers, grouse locusts, or ground hoppers, are easily characterized by a posterior elongation of the pronotum, which extends as far as the tip of the abdomen. Hind-tarsi are 3-segmented, whereas fore-tarsi and mid-tarsi are 2-segmented. The tegmina are reduced to scales. Some species also have reduced hindwings and may be flightless. Adults often overwinter and are usually encountered during the spring and early summer.

Family Tridactylidae. The pygmy mole crickets are minute or small (body length, 4–10 mm). They excavate tunnels typically along stream and lake shores but they also are active jumpers. Tympanal organs are absent and the cerci are paired so that four of these structures appear to be present at the tip of the abdomen. The fore-tibiae are enlarged for digging, the antennae have 11 segments, the fore-tarsi and mid-tarsi are 2-segmented, and the hind-tarsus is either 1-segmented or absent.

Suborder Ensifera (long-horned grasshoppers) *Family Gryllacrididae* (figure 12.15*c*). These are the wingless long-horned grasshoppers. Two well-known subfamilies are the Rhaphidophorinae (cave crickets or camel crickets) and the Stenopelmatinae (Jerusalem crickets, sand crickets, or stone crickets). Gryllacridids generally lack sound-producing and tympanal organs and the wings are either vestigial or absent. Cave crickets or camel crickets are humpbacked in appearance and their antennae are contiguous (or nearly so) at the bases; they inhabit caves, cellars, and similar habitats. Jerusalem crickets (figure 12.15*c*) are large, robust, nocturnally active insects.

Family Gryllidae (figure 12.15*a,d*). This family constitutes most of the crickets. Members of this family have 3-segmented tarsi, long cerci, and long, cylindrical ovipositors. They generally differ from other orthopterans in having the forewings positioned horizontally (rather than vertically) over the body and in the tendency for the body to be dorsoventrally flattened. Males of many species of crickets are renowned songsters and each species typically has a characteristic song. Gryllids are omnivorous, but vegetable matter predominates in the diet of most species.

Figure 12.16 Dermaptera. (*a*) Suborder Forficulina; European earwig, *Forficula auricularia*, male (Me.) (*b*) Same, right hindwing unfolded. (*c*) Suborder Hemimerina; *Hemimerus talpoides*. (*d*) Suborder Arixeniina; *Arixenia jacobsoni*. (*a*) *and* (*b*) *redrawn from Essig, 1942;* (*c*) *and* (*d*) *after Askew, 1973.*

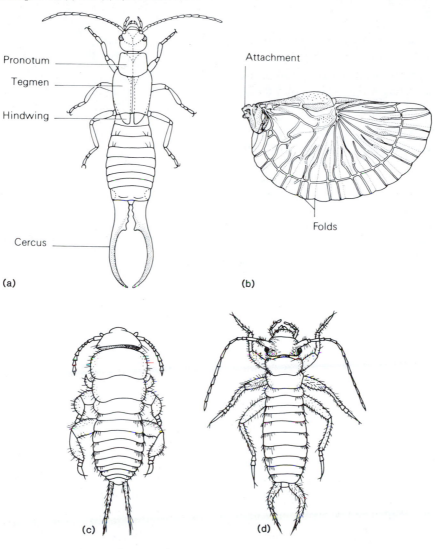

Family Gryllotalpidae (figure 12.15*e*). This family comprises the mole crickets, which characteristically have brown-black pubescent bodies, short antennae, and forelegs modified into broad spadelike organs. The ovipositor is not visible externally and the tegmina are short. These insects tunnel into moist ground, often close to streams or ponds. Mole crickets are widespread in some regions and may occasionally be turf grass pests (e.g., when large numbers burrow under turf on golf courses, etc.). Males are renowned songsters.

Family Tettigoniidae (figure 12.15*b*). This is a large family that includes the long-horned grasshoppers (or bush crickets) and katydids. These insects, in addition to having long hairlike antennae, possess 4-segmented tarsi, a bilaterally flattened bladelike ovipositor, and wings with less than 8 longitudinal veins. Except for a small portion that is held horizontally and dorsally over the body, the wings, at rest, slope vertically. Tettigoniids are usually greenish and cryptically colored; males of most species are songsters. Although the majority of tettigoniids are phytophagous, a few species are predaceous.

Order Dermaptera
Derma, skin; *ptera,* wing; Euplexoptera; earwigs (figure 12.16).

Body characteristics small to medium; narrow; elongate.

Mouthparts chewing.

Eyes and ocelli compound eyes present in most, but vestigial or absent in some; dorsal ocelli lacking.

Antennae long; slender.

Wings forewings short, leathery tegmina; hindwings semicircular in shape and membranous with radially arranged veins; hindwings folded fanlike beneath forewings; wingless species common.

Legs 3-segmented tarsi.

Abdomen forcepslike, unsegmented cerci in most species; abdomen not covered by wings.

Comments cerci apparently used in prey capture, defense, folding and unfolding wings, and possibly during copulation.

Most earwigs are nocturnal and prefer damp situations (e.g., soil, under bark, under stones, among vegetation, and near bodies of water). They are mostly omnivorous and feed upon small living or dead insects, decaying plant material, and tender parts of living plants. Some earwigs are associated with mammals (e.g., *Hemimerus* spp. which inhabit the fur of African rats).

Earwigs display parental care of offspring, which is rather uncommon among insects and may represent an early stage in the evolution of social behavior (see chapter 8). Eggs are deposited in the soil and the females literally "roost" on them until they hatch and then care for the newly hatched young.

When disturbed, an earwig may raise the abdomen with its imposing cerci, somewhat resembling the behavior of a scorpion. An additional defense is the ability of some species to squirt (sometimes over several centimeters) a repugnatorial substance produced by glands on the dorsum of the second or third abdominal segments.

Some species (e.g., *Forficula auricularia*), when abundant, may damage flowers and tender foliage. The common name "earwig" may have come from the notion that these insects enter the ears of sleeping persons, but this has little basis in fact, and they are considered to be harmless. Alternatively, their common name may be a corruption of "earwing," in reference to the hindwing resembling a human ear (Richards and Davies 1977).

The order Dermaptera is divided into three suborders: Arixeniina, Hemimerina, and Forficulina. Members of the first two suborders are associated with mammals and are represented by very few species. The Arixeniina (figure 12.16*d*) consists of a single family, Arixeniidae, the members of which are wingless with vestigial eyes and arched, hirsute cerci. Two arixeniid genera are known (*Arixenia* and *Xeniaria*) all representatives of which inhabit bat fur or bat guano in Southeast Asia. The Hemimerina (figure 12.16*c*) also consists of just one family, Hemimeridae, and has one genus, *Hemimerus*. Hemimerinids are morphologically similar to arixeniinids but possess long, straight, unsegmented cerci. Species of *Hemimerus* live in the fur of central African rats (*Cricetomys* and *Beamys*), subsisting on skin and body secretions of their hosts. The suborder Forficulina (figure 12.16*a,b*) includes the vast majority of earwigs and six

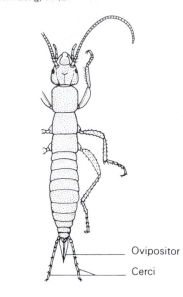

Figure 12.17 Grylloblattodea. A rock crawler, *Grylloblatta campodeiformis*. (L.)
Redrawn from Essig, 1942.

Ovipositor
Cerci

families (Carcinophoridae, Chelisochidae, Forficulidae, Labiidae, Labiduridae, and Pygidicranidae) are usually recognized. Forficulinids have well developed eyes, forcipate cerci, and the first tarsal segment is longer than the second segment; wings may be present or absent.

Further information on the Dermaptera can be found in the following references: Blatchley (1920); Helfer (1963); Popham (1965); Steinman (1989).

Order Grylloblattodea
Gryll, cricket; *blatta,* cockroach; Notoptera; rock crawlers or ice bugs (figure 12.17).

Body characteristics somewhat elongate and slender; medium to large; light brown to gray color.

Mouthparts mandibulate.

Eyes and ocelli compound eyes reduced or absent; ocelli lacking.

Antennae filiform; 28–50 segments.

Wings lacking; all rock crawlers are secondarily apterous.

Legs generalized cursorial; 5-segmented tarsi.

Abdomen well-developed sword-shaped ovipositor; long 8-segmented cerci.

This is the smallest order of insects with only 17 known species, 11 represented in North America (Bland 1978). Rock crawlers are active at low temperatures (range ca. −2.3° C to 5° C) and are found on snow, at edges of snow fields, in moss, under rocks, and so on. Some species are cavernicolous (found in caves). They appear to be omnivorous, feeding on mosses and dead or sluggish

Figure 12.18 Isoptera, termites. (*a*) Winged form, *Zootermopsis angusticollis*. (*b*) Eggs. (*c*) Third instar nymph. (*d*) Last instar nymph. (*e*) Soldier. (*f*) Head of *Nasutitermes* sp. (*a*) and (*e*), Me. (*a–e*) *redrawn from Essig, 1942.*

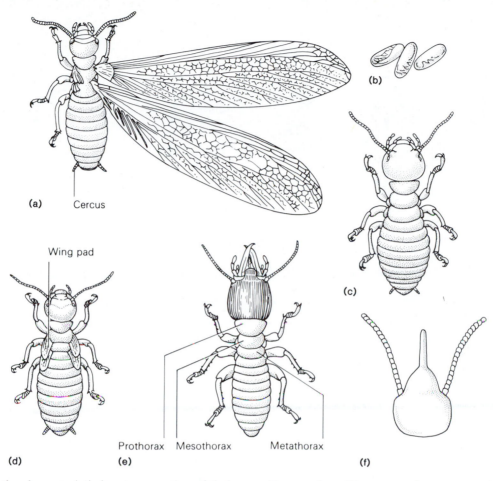

insects. Given the characteristic low temperature of their habitat, it is not surprising that they have low rates of metabolism and are slow in their development. They may take as long as 7 years to complete one life cycle.

The Grylloblattodea is included as a suborder of the Orthoptera by some authors, and its members may represent the sole surviving remnants of a primitive stock of insects from which the orthopteroid orders evolved. The three known genera of Grylloblattodea (*Grylloblatta* in western North America, *Galloisiana* in Japan, and *Grylloblattina* in far eastern Siberia) are usually included in one family, the Grylloblattidae. Gurney (1948) and Helfer (1963) should be consulted for more information about these insects.

Order Isoptera
Iso, equal; *ptera,* wings; termites "white ants" (figures 12.18 and 12.19).

Body characteristics minute to large.

Mouthparts chewing.

Eyes and ocelli compound eyes present in all winged forms, present or lacking in apterous forms; 0 or 2 dorsal ocelli.

Antennae short, moniliform or filiform.

Wings when present, 2 similar pairs; membranous; shed by breakage along a basal fracture line.

Legs 4- to 6-segmented tarsi.

Abdomen genitalia lacking or weakly developed; pair of short cerci.

Comments characteristic depression on dorsum of head called *fontanelle* in many species.

The termites are a very primitive order of orthopteroid insects, being most closely related to the cockroaches.

They are mainly tropical and subtropical, but also occur in temperate regions. Some termites live in dry wood (e.g., that frequently used in the construction of buildings and furniture), whereas some others inhabit damp wood (rotting logs, etc.). Others are subterranean and require

Figure 12.19 Subterranean termite, *Reticulitermes*. (*a*) Eggs. (*b*) Nymph. (*c*) Female winged form. (*d*) Male winged form. (*e*) Worker. (*f*) King. (*g*) Secondary reproductive (abdomen distended with enlarged ovaries). (*h*) Queen (abdomen distended with enlarged ovaries). (*i*) Soldier. (*j*) Earthen tubes from soil across surface of concrete foundation to wooden structures. (Mi to S.)

Redrawn from Boreal Labs key card.

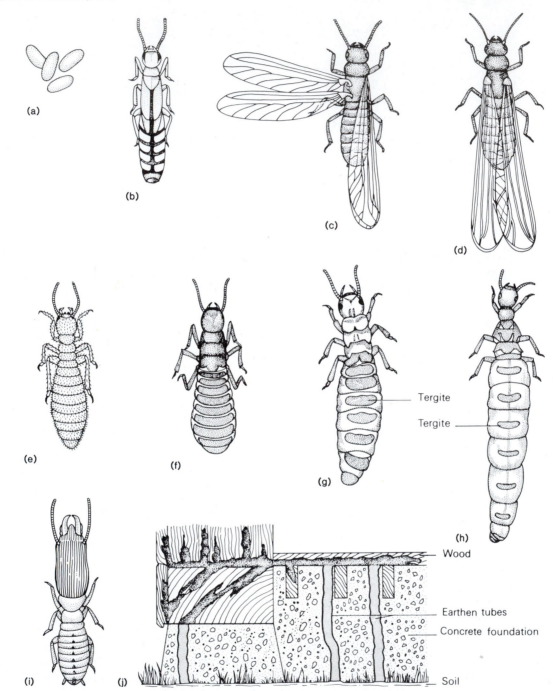

very humid conditions. Subterranean forms probably play an ecological role similar to earthworms in that they aerate and add nutriment to the soil. Most groups of termites can conveniently be separated according to their ecological requirements, as either "dry wood" (or "powder post"), "damp wood," or "subterranean" forms; these divisions can be especially helpful when one is considering pestiferous species. In addition to wood, termites feed on a wide variety of cellulose-containing materials, fungi, and dried animal remains. The digestion of cellulose is carried out by flagellate protozoans or bacteria, which appear to be mutualistic inhabitants of the gut.

Termites are commonly confused with ants. This is probably the reason that termites are sometimes referred to as "white ants." However, there are several distinct differences between these two groups of insects. Some of the more obvious include the following. In ants the hindwings are smaller and have fewer veins than the forewings and at rest are often held vertically over the abdomen, whereas in termites the wings are similar in size and venation and are flexed horizontally over the abdomen at rest. In wingless forms, ants have relatively dark, sclerotized bodies, whereas termites are pale and soft-bodied. The abdomen of ants is basally constricted in both winged and apterous forms, but this is not the case in termites.

All termites are social insects. A colony's population is initiated and maintained by a queen that may live for as many as 50 years in some species. Two to several castes may be present, depending on the species, and all castes are composed of both males and females. Conveniently, castes can be divided into reproductives and nonreproductives. There may be two kinds of reproductives present in a colony of a given species: primary reproductives and secondary reproductives. The primary reproductives are the queen (figure 12.19c,h) and king (figure 12.19d,f) and are thought to comprise the original caste in isopteran phylogeny. They typically have dark, sclerotized (at least compared to other castes) bodies with completely developed wings and compound eyes. Secondary reproductives (figure 12.19g) occur in a variety of forms (e.g., ones with shorter wings, less pigmentation, and smaller compound eyes than the primary reproductives or ones with no wings and pale bodies resembling workers).

The nonreproductive castes are the workers and soldiers. The workers (figure 12.19e) are typically wingless, unpigmented, soft-bodied forms, without compound eyes or well developed mandibles, that function as the labor force of the colony, maintaining and adding to the structure of the nest, tending fungus gardens, and feeding members of other castes and immatures. Soldiers (figures 12.18e and 12.19i), on the other hand, may or may not have compound eyes and generally have enlarged, sclerotized heads with well developed mandibles. They function as defenders of the colony either with their mandibles or in some species by plugging up holes with their heads. In the soldier caste of *Nasutitermes* spp. (called *nasuti* or *nasutes;* figure 12.18f), the duct from a gland in the head opens at the end of a long narrow snout through which a sticky secretion from the gland can be forced. This secretion is squirted as a means of defense and, in some species, also functions as an alarm pheromone. In the simplest social structure, characteristic of certain primitive termite species, there are reproductives and soldiers, and the immatures of both these castes function as workers (figure 12.18c,d).

The control of caste differentiation has long been a topic of interest and investigation and is still a long way from being understood completely. There is strong evidence that complex interactions of pheromones are involved (see "Polymorphism" in chapter 5).

New colonies are formed when at certain times of the year, winged primary reproductives called *alates* appear and swarm from the nest. Males and females pair off during this swarming, but mating usually occurs on the ground. Subsequently, their wings break off along preformed basal fracture lines, and the pair locates a site for nest construction. Another method of nest foundation is *sociotomy* (Lindauer 1965), the process by which colonies divide by the separation of immatures and secondary reproductives from the parent colony or by division of a migrating colony into two daughter colonies. Colony size may vary from a few individuals to over 100,000.

Termite nests vary from simple cavities in soil or wood to vast subterranean complexes or elaborate edifices that project well above the ground. A nest of a species in northern Australia, *Nasutitermes triodidae,* has been reported (Richards and Davies 1977) to reach a height of 20 feet (6 m) and a diameter of 12 feet (3.7 m) at its base! Very elaborate ventilation systems, designs that provide for maintenance of constant temperature, canopies that deflect rainwater, and other structural adaptations of nests have been described in various termite species. The means by which the behavior of individual members of a colony are coordinated to produce such complex structures has long been a source of amazement (see chapter 8).

Termites are very significant structural pests, damaging wooden structures (e.g., furniture, building timbers, and wooden floors). Some of the most destructive termites in the United States are subterranean and in the genus *Reticulitermes*. They can construct earthen tubes on concrete and are thus able to invade a structure even though it is not in direct contact with the soil. The presence of earthen tubes (figure 12.19j) is one of the characteristics used to "diagnose" a termite infestation. Not all destructive termite species are subterranean; for example, colonies of *Kalotermes* spp. (dry-wood termites) and *Zootermopsis* spp. (damp-wood termites) exist entirely in wood. These termites are found along the Pacific Coast of North America and in the southern United States.

Fossil termites are known from at least 250 million years ago and one primitive Australian genus, *Mastotermes,* is morphologically similar to cockroaches (Blattaria). The order Isoptera is usually divided into six families and four of these (Hodotermitidae, Kalotermitidae, Rhinotermitidae, and Termitidae) occur in North America. Termites are absent from the British Isles but several species occur in continental Europe. The families of termites that are generally recognized are briefly treated in the following paragraphs. The following references should be consulted for additional information on termites: Behnke (1977); Krishna and Weesner (1969, 1970); Lee and Wood (1971); Weesner (1965); Wilson (1971).

Family Hodotermitidae (figure 12.18). This family includes rotten-wood, damp-wood, and harvester termites. Adults lack a fontanelle and are somewhat flattened dorsoventrally. Many species attack dead wood and a ground contact is not required but moist wood is necessary for prolonged colony survival. Although many species inhabit damp, rotting logs, they may also damage wooden buildings, lumber, etc., especially in moist or foggy areas. The genus *Zootermopsis* (figure 12.18) occurs along the Pacific coast of North America.

Family Kalotermitidae. These dry-wood, damp-wood, and powderpost termites have a flat pronotum that is usually broader than the head, 4-segmented tarsi, no anal lobe on the hindwing, and no fontanelle or ocelli. Over 250 species are known worldwide. There is no true worker caste, and immatures of other castes fill this role. Members of this family are often of great economic importance, particularly in the United States where 16 species are found.

Family Mastotermitidae. This primitive family is represented by just one extant species, *Mastotermes darwiniensis,* from northern Australia. Tarsi are 5-segmented, the hindwings are large and have an anal lobe, and ocelli and a fontanelle are lacking. There is no true worker caste.

Family Rhinotermitidae (figure 12.19). These damp-wood, and subterranean termites are usually minute (adult length, 6–8 mm) and have a flat pronotum, 4-segmented tarsi, lack an anal lobe on the hindwings and lack ocelli. Members of this family are often economically important (e.g., the eastern subterranean termite, *Reticulitermes flavipes,* in the United States). A contact to the soil is always maintained and earthen tubes are constructed to connect wooden structures to the soil. Wingless forms are typically pale, whereas winged forms are black.

Family Termitidae. This large family has an almost global distribution and includes about 75% of all extant termite species. There is no anal lobe on the hindwings, wing venation is reduced, tarsi are 4-segmented, and both ocelli and a fontanelle are present. The pronotum (of workers and soldiers) is narrow and has a raised anterior lobe. Representatives of this family build a wide range of colony structures, use diverse food resources, and have various caste systems. Nasute-caste termites are included in this family. A ground contact is maintained, and some species are economically important.

Family Serritermitidae. This family includes a single species, *Serritermes serrifer,* from Brazil, which has bizarre falcate mandibles in both worker and adult castes. Various authors have placed this species in the Kalotermitidae, Rhinotermitidae, or Termitidae.

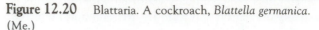

Figure 12.20 Blattaria. A cockroach, *Blattella germanica.* (Me.)

After U.S. Public Health Service.

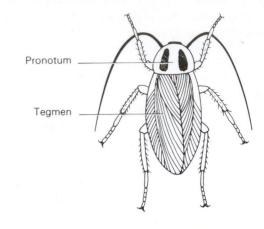

Order Blattaria

Blatta, cockroach; Blattodea; cockroaches (figure 12.20).

Body characteristics somewhat flattened dorsoventrally; small to very large; generally dark colored, but some tropical species brightly colored.

Mouthparts mandibulate.

Eyes and ocelli compound eyes usually well developed; 2 ocelli in some; fenestrae (pale regions with nerves to brain) in place of ocelli in most.

Antennae long; filiform.

Wings forewings tegmina; hindwings membranous, broader than forewings and with enlarged anal area.

Legs generalized, cursorial; large coxae.

Abdomen pair of styli on 9th sternum of males; ovipositor reduced and concealed by 7th abdominal sternum.

Thorax highly compressible.

Comments large shieldlike pronotum nearly covers dorsum of head.

The cockroaches are a predominately tropical group, but several species are found in temperate zones. They are basically omnivorous insects, mostly feeding on decaying animal and plant matter. Certain species, for example, *Cryptocercus* spp., feed on dead wood and are able to digest cellulose by virtue of the presence of cellulase produced by intestinal symbionts (protozoans in the case of *Cryptocercus,* bacteria in some others). Cockroaches are generally found in litter, among low vegetation, or on or in the ground. Some species inhabit caves, and a few are associated with ant colonies (*myrmecophilous*). Cockroaches are very agile runners and usually depend on this ability, instead of flying, to escape potential predators. The cockroach *Gromphadorhina portentosa* produces a hissing sound through the spiracles when threatened. Cockroaches tend to be nocturnally active.

Cockroaches undergo hemimetabolous development. There are many oviparous and ovoviviparous species, and one species is viviparous. In the oviparous forms, the eggs are lined up in an egg purse or ootheca, which is sometimes cemented to the substratum (e.g., the American cockroach, *Periplaneta americana*).

The cockroaches are common household pests and feed on a wide variety of household goods, but the major indictment against them is that they are dirty, distasteful, and odoriferous creatures and are attracted to such materials as garbage, feces, and foodstuffs consumed by humans. The most common household-invading species (domiciliary) in the United States are (1) *Periplaneta americana,* the American cockroach; (2) *Blattella germanica,* the German cockroach (figure 12.20); and (3) *Blatta orientalis,* the Oriental cockroach. Although there is little conclusive evidence that points to cockroaches as disseminators of pathogenic organisms, the circumstantial evidence is strong, and it has been suggested that they may, in fact, rival house flies in their capacity for disease transmission. Their nocturnal, secretive habits have perhaps been the reason they have been somewhat overlooked in this capacity in the past. Readers interested in more information regarding cockroaches, both their biology and medical-economic significance, should consult Blatchley (1920), Cornwell (1968), Guthrie and Tindall (1968), Harwood and James (1979), Helfer (1963), McKittrick (1964), Ragge (1973), and Roth and Willis (1957, 1960).

The Blattaria share several features with the Mantodea (mantids) and are included with them in the order Dictyoptera by many authors. In that classification, the Blattaria and Mantodea are each given subordinal status. Another scheme of classification treats the Dictyoptera as a suborder of the order Orthoptera, with both the Blattoidea and Mantodea as superfamilies of the Dictyoptera. There is also disagreement with respect to the familial classification of the Blattaria but most authors recognize the following families.

Family Blaberidae. This is a large group of mainly tropical cockroaches that are characterized by smooth femora and folding of the anal area of the hindwing in a fanlike fashion when at rest. Some representatives are large (body length, over 50 mm), are capable of stridulation, and can secrete repellent odors.

Family Blattellidae (=Epilampridae). This is another large group of cockroaches although most are small in size (body length, 12 mm or less). An apical portion of the hindwing folds over when at rest, a subgenital plate is present and the fore-femora typically have 2 or 3 apical spines. This family includes the German cockroach, *Blattella germanica* (figure 12.20).

Family Blattidae. These cockroaches have spiny mid- and hind-femora, and fold their wings in a fanlike manner at rest. This is another large family and most represen-

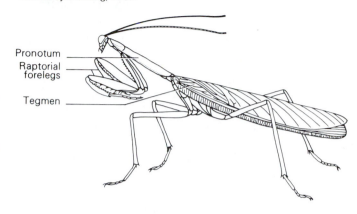

Figure 12.21   Mantodea. Chinese mantid, *Tenodera aridifolia.* (VL.)
Redrawn from Essig, 1942.

tatives are large (body length, over 25 mm in most species). The Oriental cockroach, *Blatta orientalis,* and numerous species of *Periplaneta* including the American cockroach, *P. americana,* are placed in this family.

Family Cryptoceridae. In this small family, abdominal sclerites on the venter of segments 6 and 7 are elongate and cover the distal portion of the abdomen. These cockroaches are wingless, lack a subgenital plate, and are usually found inside rotting logs.

Family Polyphagidae. Most of these cockroaches are small (body length, 5–25 mm). They characteristically have smooth femora and a hirsute (hairy) pronotum. In winged species, the anal area of the hindwing is held flat (without fanlike folding) at rest. Polyphagids are geographically widely distributed and several species occur in the southern United States. Some species are myrmecophilous, whereas some desert forms burrow in sand.

Order Mantodea

Mantid, soothsayer; mantids, praying mantids (figure 12.21).

Body characteristics small to very large (largest over 100 mm); body usually elongate, somewhat cylindrical, with elongate prothorax.

Mouthparts mandibulate; hypognathous.

Eyes and ocelli well-developed compound eyes; three ocelli or none.

Antennae filiform.

Wings usually present in males; may be reduced or absent in females; forewings are tegmina with small anal area; hindwings membranous with enlarged anal area.

Legs raptorial forelegs at anterior end of elongated prothorax; fore-coxae greatly lengthened; fore-femora and tibiae armed with spines; midlegs and hindlegs cursorial; 5-segmented tarsus.

Thorax prothorax elongate with a single tympanum on metasternum.

Abdomen comparatively short, multisegmented cerci; pair of styli usually associated with 9th sternum in males.

Comments head triangular with wide range of movement.

Praying mantids are solitary, exclusively carnivorous insects. They capture prey such as flies and grasshoppers with their well-developed raptorial forelegs. Some large South American species may even attack small birds and other small vertebrates. Mantids do not actively pursue prey, but wait motionless on vegetation for passing prey. They monitor the position of passing prey by turning their heads as prey passes and strike out with their forelegs when prey are in range. Few other insects are able to turn their heads so freely. Some species of mantids, especially in tropical regions, mimic flowers and leaves. This probably aids in prey capture and may also provide protection from predators. Mantids will run or fly and may make short jumps when disturbed. Some adopt a threat posture when disturbed, raising their sometimes brightly colored wings and striking out with their forelegs. If grasped by the mesothoracic or metathoracic legs, they may escape by autoamputating the imprisoned leg.

Mantids undergo hemimetabolous development. They encase their eggs in a frothy substance that solidifies into an egg case (ootheca). Egg cases (some containing more than 200 eggs) are typically glued to twigs or bark, and their shape and size tend to be species-specific. Some species of mantids are parthenogenetic.

As discussed for the previous order, mantids are sometimes included with cockroaches in the Dictyoptera. The Dictyoptera, in turn, is treated either as a separate order or as a suborder of the Orthoptera. Some authors include all mantids in one family, the Mantidae, but most recognize eight families, of which the following five are the largest.

Family Amorphoscelidae. These mantids have a comparatively short, quadrate prothorax, weakly spinose fore-tibiae, and long midlegs and hindlegs. A number of small species from Australia and the Old World tropics are included in this family.

Family Empusidae. The members of this family are large and elongate and often have bizarre body shapes. This is another Old World family and one genus, *Empusa*, occurs in southern Europe.

Family Eremiaphilidae. These are robust, brachypterous, desert-inhabiting mantids. Two genera are known, *Eremiaphila* and *Heteronutarsus*.

Family Hymenopodidae. These small- to medium-sized mantids often have a laterally expanded pronotum and banded or spirally-marked tegmina. Females of some species are brachypterous. This family is circumtropical in distribution, but most species occur in Africa or southern Asia.

Family Mantidae. This is the largest and most diverse mantid family. The European *Mantis religiosa* (which has been introduced to North America) belongs in this family. These mantids range in size from small to very large and the inner ventral spines on the fore-femora are alternately long and short. The tegmina are never bicolored.

Readers interested in more information concerning this order should consult Blatchley (1920); Gurney (1951); and Helfer (1963).

Order *Zoraptera*
Zor, pure; *ptera*, wings; zorapterans (figure 12.22).

Body characteristics minute; soft-bodied.

Mouthparts chewing.

Eyes and ocelli compound eyes and dorsal ocelli in winged forms but lacking in wingless forms.

Antennae 9-segmented; moniliform or filiform.

Wings 2 pairs when present; membranous; held flat over body at rest; forewings larger than hindwings; shed by breakage along a basal fracture line; reduced venation; both sexes of a given species may have winged and wingless forms; wingless forms more common.

Legs 2-segmented tarsi; thickened hind femora with heavily sclerotized species-specific spines.

Abdomen short, 10-segmented; 1-segmented cerci.

This is the second smallest order in the class Insecta. Only 30 species are known worldwide (Choe 1992); 2 of these species occur in North America (mainly in the southeastern United States) but they are uncommon. Zorapterans occur in all of the major zoogeographical regions of the world except the Palearctic. All described species are included in the family Zorotypidae.

Zorapterans occur primarily in tropical forests and live colonially under the bark of rotting logs. Members of this order have also been found in such places as humus, sawdust piles, termite and mammal nests, and rotting banana stems. These insects are principally scavengers, feeding on such materials as fungal spores and the remains of other arthropods, but they also specialize in capturing small mites, springtails, nematodes, etc. (Gurney 1938, Choe 1992).

Zorapterans are considered by some to provide an evolutionary link between the orthopteroid orders and the Psocoptera and hence the hemipteroid orders. However, Boudreaux (1979) disagrees and views them as highly specialized relatives of the cockroaches, termites, and mantids. They differ from the Psocoptera principally in the reduced wing venation and the beaded appearance of the antennae.

Figure 12.22 Zoraptera *Zorotypus hubbardi*. (*a*) Alate (winged) form of adult female. (*b*) Dealated form of adult female. (*c*) Adult female of apterous form. (Mi.)
Redrawn from Caudell, 1918.

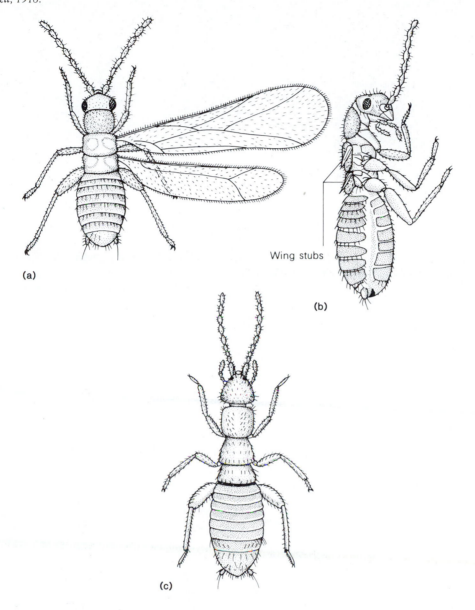

(a)

Wing stubs

(b)

(c)

Hemipteroid Orders

The orders grouped as hemipteroids are Psocoptera, Mallophaga, Anoplura, Thysanoptera, Hemiptera, and Homoptera (figure 12.7). Boudreaux (1979) applies the name Acercaridae to this group in reference to the complete absence of cerci. Members of these orders share certain features, including: (1) specialized mandibulate or haustellate mouthparts (lacinia slender, often stylets; labial palps absent or with no more than 2 segments); (2) enlarged clypeus associated with musculature of cibarial pump; (3) hindwings (if present) lacking a large anal lobe; (4) cerci lacking; (5) four or fewer Malpighian tubules; and (6) a tendency toward concentration of the ventral chain ganglia.

Development is hemimetabolous in this group of insects.

Order Psocoptera

Psoco, rub small; *ptera,* wing; Corrodentia; booklice, barklice, psocids (figure 12.23).

   Body characteristics minute; head capsule large compared to rest of body; enlarged clypeus.

   Mouthparts chewing; labial silk glands present.

Figure 12.23 Psocoptera. (*a*) Winged species, *Peripsocus californicus*. (*b*) Wingless species, *Troctes divinatorius*. (Mi.) *Redrawn from Essig, 1942.*

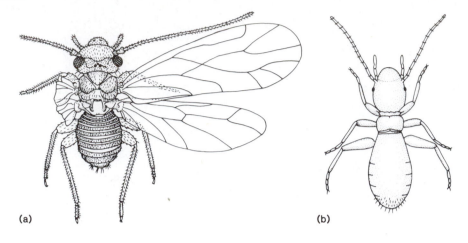

(a)   (b)

Eyes and ocelli compound eyes strongly or weakly developed; dorsal ocelli: 3 in winged forms, absent in wingless forms.

Antennae long; filiform.

Wings most with 2 pairs; membranous; reduced venation; held rooflike over body at rest; forewings larger than hindwings; vestigial in some species and absent in others.

Legs 2- or 3-segmented tarsi.

Abdomen ovipositor partly concealed and aedeagus completely concealed; cerci lacking.

Psocopterans live in a variety of terrestrial habitats: under bark (barklice), amid vegetation, and in bird and mammal nests. Many are gregarious. Several of the wingless species occur in human habitations, particularly around books and papers, and are commonly called booklice. The outdoor species feed on organic matter, including fungi, algae, lichens, pollen, and fragments of decaying organic material.

Most species are oviparous, although some are viviparous. Parthenogenesis occurs in some species.

Psocoptera may reach large enough populations among books, papers, stored cereal grains, insect collections, and other materials to constitute a pest situation, but they are generally of little economic importance.

Psocopterans are considered to be somewhat intermediate between the zorapterans (here treated as an orthopteroid group) and the remaining hemipteroids. The Psocoptera share certain morphological features with the lice (orders Mallophaga and Anoplura) and these three taxa (Psocoptera, Mallophaga, and Anoplura) are grouped together in the superorder Psocodea by some authors. The Mallophaga are believed to have evolved directly from the Psocoptera with which they share chewing mouthparts and several other traits. It would appear to

represent a relatively small evolutionary step for psocopterans inhabiting bird nests to transfer onto birds, and from that stage, to evolve additional traits characteristic of ectoparasites.

The order Psocoptera is most conveniently separated into three suborders: Psocomorpha, Troctomorpha, and Trogiomorpha. The Psocomorpha have 13 or fewer antennal segments, 2- or 3-segmented tarsi, 1- or 2-segmented labial palps and a thickened pterostigma in winged forms. This is a large suborder consisting of at least 22 families, the largest of which is the Psocidae with many bark-inhabiting species throughout the world. Members of the suborder Troctomorpha have 11–17 segmented antennae, 3-segmented tarsi, 1- or 2-segmented labial palps, and winged forms with an unthickened pterostigma. Four families are usually included in the Troctomorpha: Amphientomidae, Liposcelidae, Pachytroctidae, and Sphaeropsocidae. The Trogiomorpha have more than 20 antennal segments, and have 3-segmented tarsi, and 2-segmented labial palps; the pterostigma is unthickened in winged forms. Four families are typically included in the suborder Trogiomorpha: Lepidopsocidae, Psoquillidae, Psyllipsocidae, and Trogidae. For further information on psocids, the following references should be consulted: Chapman (1930), Gurney (1950), Mockford (1987); Mockford and Gurney (1956).

Order Mallophaga

Mallo, wool; *phaga,* eat; chewing lice (figure 12.24).

Body characteristics minute to small; dorsoventrally flattened; triangular head broader than thorax.

Mouthparts chewing.

Eyes and ocelli reduced compound eyes; dorsal ocelli lacking.

Antennae 3- to 5-segmented; usually capitate or filiform; when capitate, concealed beneath the head.

Figure 12.24 Mallophaga, chewing lice. (*a*) Dog-biting louse, *Trichodectes canis*. (*b*) Chicken body louse, *Menacanthus stramineus*. (*c*) African bush pig chewing louse, *Haematomyzus porci* (dorsal features to left of midline, ventral features to right). (Mi.)

(a) after U.S. Public Health Service; (b) redrawn from Boreal Labs key card; (c) after Emerson and Price, 1988.

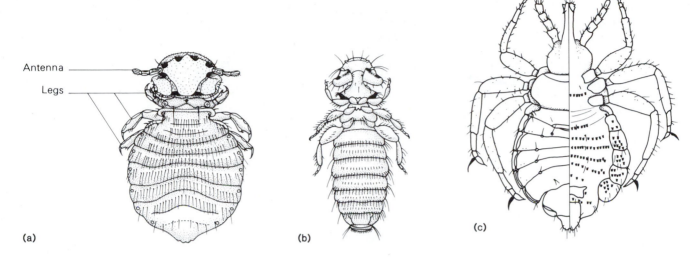

Wings lacking.

Legs tarsi modified for grasping hairs in some species with mammalian hosts; 1- or 2-segmented tarsi.

Abdomen cerci lacking.

Most Mallophaga are ectoparasites of birds, but close to 500 (ca. 15%) of the known species (members of the families Abrocomophagidae, Boopiidae, Gyropidae, Haematomyzidae, Trichophilopteridae, Trichodectidae and Trimenoponidae) have mammalian hosts. They feed on various organic fragments (e.g., from feathers and skin) and epidermal secretions, and certain species, for example, *Menacanthus stramineus* (figure 12.24*b*), are known to gnaw through skin or developing quills and obtain blood. Some obtain sebum and perhaps serum by attacking hair follicles. The mandibulate mouthparts of *Trochiloecetes* (parasites of hummingbirds) are significantly modified and can function as piercing and sucking organs. *Piagetiella* has an unusual life cycle in that it feeds on blood, tissue and mucus inside pelican's pouches, but must temporarily leave this habitat to oviposit on the feathers of its host.

The order Mallophaga is usually divided into three suborders: Amblycera, Ischnocera, and Rhyncophthirina. The Amblycera (figure 12.24*b*) possess short antennae that are usually concealed in grooves along the head; members of this group also possess maxillary palps and appear to be related to the Psocoptera. Representatives of the Ischnocera, Rhyncophthirina, and Anoplura all lack maxillary palps. Ischnocerans (figure 12.24*a*) have filiform antennae that are free from the head. The Rhyncophthirina (figure 12.24*c*) is a bizarre taxon consisting of just three known species (all in the genus *Hae-*

matomyzus in the family Haematomyzidae) that parasitize elephants (both African and Indian), warthogs, and African bush pigs, respectively. Although the mouthparts of the Rhyncophthirina are modified for chewing, the small mandibles are borne at the tip of an elongated rostrum.

Some authors, particularly in Europe but also in North America, prefer to include the Mallophaga with the Anoplura (sucking lice) in the order Phthiraptera. In fact, some recent cladistic analyses suggest that the Ischnocera and Rhyncophthirina may be more closely related to the Anoplura than to the Amblycera.

Chewing lice pass through three nymphal instars. The eggs are attached to host feathers or hairs, and generation after generation is spent on the same host. Individuals are typically transferred from one host to another by direct contact, and they soon die if separated from an acceptable host. Some species are transferred from host to host by means of phoresy (see "Interspecific Interactions" in chapter 9). Different species sometimes show different preferences as to the part of the host's body attacked so that one host species may be parasitized by five or more species of chewing lice.

Although most domestic animals are attacked by one or more species, chewing lice usually do not constitute major problems. However, if an infestation is large, some damage may occur, particularly to poultry, which may suffer restlessness, decrease in egg production, and loss of feathers. Three common chewing lice that attack poultry are *Menopon pallidum*, the common chicken louse; *Menacanthus stramineus* (figure 12.24*b*), the chicken body louse; and *Menopon gallinae*, the shaft louse. Cattle, horses, sheep, goats, dogs, cats, chickens,

Figure 12.25 Anoplura, sucking lice. (*a*) Crab louse, *Pthirus pubis*. (*b*) Human louse, *Pediculus humanus*. (*c*) Egg or nit of crab louse attached to host hair. (*d*) Egg of human head louse attached to host hair. (*e*) Foreleg of human louse showing tibiotarsal claw. (Mi.)

(*a*) and (*b*) after U.S. Public Health Service; (*c*) and (*d*) redrawn from Boreal Labs key card; (*e*) from Askew, 1973, after Keilin and Nuttall, 1930.

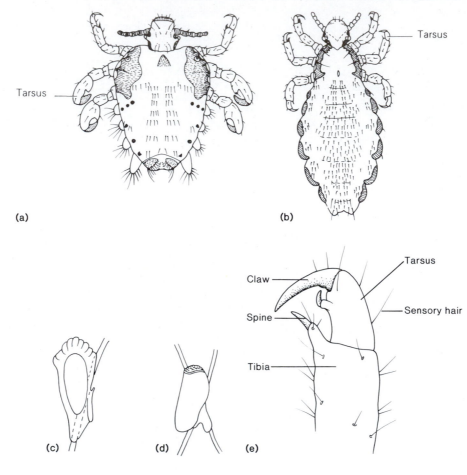

(a)   (b)

(c)   (d)   (e)

and turkeys are among the domestic animals attacked by species of chewing lice. Humans are not attacked by these lice, but may occasionally become infested by direct contact with infested animals; the lice do not survive long on a human host. Indirectly, however, the common dog louse, *Trichodectes canis* (figure 12.24*a*), does at times constitute a threat to human health because it serves as an intermediate host for the double-pored dog tapeworm, *Dipylidium caninum*. If an infested louse is inadvertently ingested by a human playing with an infested dog, the human can become infested with the tapeworm. Some chewing lice that are associated with aquatic birds are known to be vectors of parasitic nematodes, whereas other species associated with livestock animals can spread pathogenic skin fungi between their hosts. For further information on Mallophaga the following references are recommended: Askew (1973); Clay (1969); Emerson (1972); Emerson and Price (1981, 1985); Hopkins and Clay (1952); Marshall (1981).

Order Anoplura

Anopl, unarmed; *ura,* tail; Siphunculata, Pseudorhyncota, Lipognatha, Ellipoptera; sucking lice (figure 12.25).

Body characteristics minute to small; dorsoventrally flattened; head narrower than thorax.

Mouthparts piercing-sucking; retracted into head when not feeding.

Eyes and ocelli compound eyes weakly developed or lacking; dorsal ocelli lacking.

Antennae short; 3 to 5 segments.

Wings lacking.

Legs 1-segmented tarsi; tarsi, especially of midlegs and hindlegs, usually adapted for grasping hairs (figure 12.25*e*).

Abdomen cerci lacking.

Sucking lice are all blood feeders and, like the chewing lice, pass through three nymphal instars during their life cycle. They attack a wide variety of mammals, including some marine forms (seals, sea lions, walruses) but do not parasitize monotremes, marsupials, bats, pangolins, edentates, elephants, whales, or sea cows. In general, they are highly host specific. The space enclosed by the tibio-tarsal claws (figure 12.25e) is often characteristic for a given species of sucking louse and typically corresponds with the host hair diameter onto which the claws lock to aid in attachment.

The major pests of domestic animals in this order are in the families Haematopinidae (e.g., the hog louse, *Haematopinus suis,* the horse sucking louse, *H. asini,* and the cattle tail louse, *H. quadripertusus*) and Linognathidae (e.g., the long-nosed ox louse, *Linognathus vituli,* the sheep foot louse, *L. pedalis,* the dog sucking louse, *L. setosus,* and the little blue cattle louse, *Solenopotes capillatus*).

Two species of sucking lice attack man, *Pthirus pubis* (figure 12.25a) (family Pthiridae) and *Pediculus humanus* (figure 12.25b) (family Pediculidae). Only one other species of *Pthirus* is known and it parasitizes gorillas; all other species of *Pediculus* are associated with either Old or New World monkeys.

Pthirus pubis is the human pubic louse or crab louse and infests the pubic and armpit areas, although it may be found in other areas of the body with coarse hair (eyebrows, eyelashes, beards). This louse causes intense itching, and an infestation may result from contact with an already infested individual via toilet seats, blankets, and (principally) sexual intercourse. The life cycle ordinarily takes about 1 month to complete and the eggs, or *nits,* are attached to hairs on the host (figure 12.25c). Pubic lice are not known to transmit any human pathogens. In France, pubic lice are often referred to as "papillons d'amour" ("butterflies of love").

Presently available morphological evidence allows *Pediculus humanus* to be divided into two subspecies, *P. humanus capitis,* the head louse, and *P. humanus humanus,* the body louse. The first infests particularly the head region, but it has been found on other hairy areas of the body. The body louse infests regions that come into frequent or continuous contact with clothing (e.g., armpits, neck, and crotch). The eggs are usually attached to clothing, which provides an ideal vehicle for transmission to a new host. Head lice, on the other hand, attach their eggs to hairs (figure 12.25d), and a new infestation may result from contact with a stray hair bearing an egg or from infested hats, combs, or earphones. Both head lice and body lice are usually spread via direct contact. They develop quite rapidly, a complete generation occurring in about 3 weeks.

Infestation with lice is sometimes referred to as *pediculosis.* The attendant discomforts of large infestations range from intense itching to anemia and pathologic changes in the skin resulting from the chronic feeding of lice and the associated blood loss. In addition to the problems caused directly by their feeding, body lice are involved in the transmission of a number of human pathogens, the most important of which are epidemic relapsing fever, caused by a spirochete, *Borrelia recurrentis;* epidemic typhus, caused by a rickettsia, *Rickettsia prowazekii;* and particularly during episodes of war, the otherwise rare malady of trench fever, caused by the rickettsia *Rochalimaea quintana.* All three of these diseases have occurred in large epidemics and in the case of epidemic typhus have claimed thousands of lives, particularly during wartime when populations become highly concentrated and sanitary practices are lax. Cool climates in concert with crowding and poor sanitary conditions appear to promote outbreaks of epidemic typhus. Several other pathogenic microbes (e.g., *Salmonella*) may be transmitted by human lice. The etiologic agent of murine (endemic) typhus, *Rickettsia typhi,* may also be transmitted by human lice but the available evidence suggests that fleas typically are vectors of this microorganism to man. However, rodent sucking lice (various species of *Hoplopleura* and *Polyplax*) can transmit *R. typhi* from rat to rat and thus may serve to indirectly amplify an epidemic. Sylvatic epidemic typhus has recently been detected in North American flying squirrel populations and, on occasion, the rickettsial organism has been transmitted to humans from squirrels nesting in attics and eaves of houses. Because flying squirrel sucking lice (*Neohaematopinus sciuropteri*) and fleas (*Orchopeas howardi*) have tested positively for the microorganism, the consensus of opinion is that these ectoparasites are involved in transmission to people, although the exact route of infection remains unknown (because flying squirrel lice are not known to feed on humans, perhaps the *Rickettsia* is inhaled as minute airborne particles from dried louse and flea feces).

Sucking lice are also known to transmit pathogens to domestic and wild mammals. For example, *H. suis* is capable of transmitting Pox virus to hogs and various species of cattle-associated sucking lice can transmit *Anaplasma marginale,* the rickettsial agent of anaplasmosis. Similarly, certain species of wild rodent sucking lice have been shown to transmit to their hosts the microorganisms that cause murine eperythrozoonosis (*Eperythrozoon coccoides*), haemobartonellosis (*Haemobartonella muris*), rodent brucellosis (*Brucella brucei*), Tularemia (*Francisella tularensis*), Q fever (*Coxiella burnetii*), epidemic and endemic typhus.

A morphologically varied assemblage of sucking lice parasitizes wild mammals throughout the world, although the greatest proportion of described species (35%) occurs in Africa. Because most Anoplura are host specific, host-parasite coevolution and parallel evolution is widespread and louse phylogenies often mirror those of their hosts. The following references contain additional

Figure 12.26 Thysanoptera, thrips. (*a*) Member of Terebrantia. (*b*) Member of Tubulifera, female. (*a*) Mi; (*b*) Me.

(*a*) *after U.S. Public Health Service; (b) redrawn from Essig, 1942.*

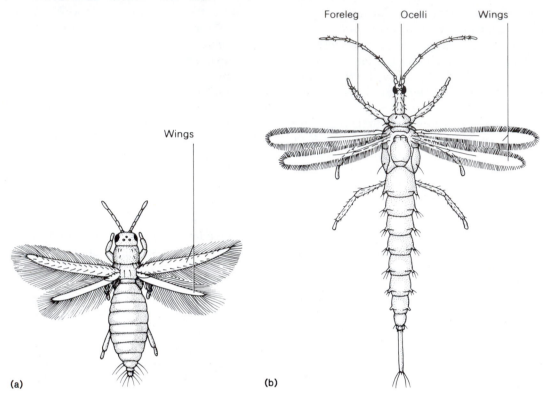

Foreleg   Ocelli   Wings

Wings

(a)   (b)

information about the sucking lice: Askew (1973); Clay (1973); Ferris (1951); Harwood and James (1979); Hopkins (1949); Keilin and Nuttall (1930); Kim (1985); Kim and Ludwig (1978); Kim et al. (1986).

Order Thysanoptera

Thysano, a fringe; *ptera,* wing; Physapoda; thrips (figure 12.26).

Body characteristics minute to small; slender; head cone-shaped; abdomen tapering; usually darkly colored.

Mouthparts rasping-sucking; asymmetrical, the right mandible lacking or reduced.

Eyes and ocelli small compound eyes; 3 dorsal ocelli in winged forms; dorsal ocelli lacking in wingless forms.

Antennae short; 6 to 10 segments.

Wings 2 pairs; narrow with fringe of hairs; coupled by basal hooks.

Legs 1- or 2-segmented tarsi are bladderlike at the tip.

Abdomen cerci lacking.

The order Thysanoptera is divided into two suborders, based mainly on the presence or absence of an ovipositor. Females in the suborder Terebrantia (figure 12.26*a*) have well-developed ovipositors used to deposit eggs in plant tissues; wings, if present, have veins. Females in the suborder Tubulifera (figure 12.26*b*) have either weakly developed ovipositors or none; wings, when present, are nearly or completely veinless. The posterior extremity of the abdomen is tubular, and eggs are deposited in crevices.

Although the order Thysanoptera is classified as hemimetabolous, the life cycle is strongly suggestive of holometabolism. Early instar nymphs are active and wingless but otherwise resemble adults. However, the last two or three instars are quiescent, sometimes in a cocoon or earthen cell during which time wings, if they are going to be present, develop. Generation time is relatively short, and there are usually several generations per year. Parthenogenesis is common in this group.

Some thrips are predaceous (all in Tubulifera), feeding on aphids and various other small insects or mites, but most are plant feeders, usually feeding on sap. They are widely distributed on all sorts of vegetation. Although they are generally weak fliers, some may be carried many miles by the wind.

Several species are pests, particularly on truck and floricultural crops, although some are pests of field crops (e.g., tobacco). They not only cause damage by feeding, but they also may transmit plant disease organisms (especially viruses). For example, the western flower thrips,

Figure 12.27   Hemiptera, Heteroptera, true bugs. Major external features of the boxelder bug, *Leptocoris trivittatus* (Rhopalidae), and lateral view showing mouthparts. (Me.)
Redrawn from Essig, 1942.

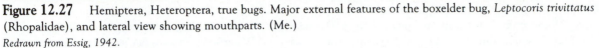

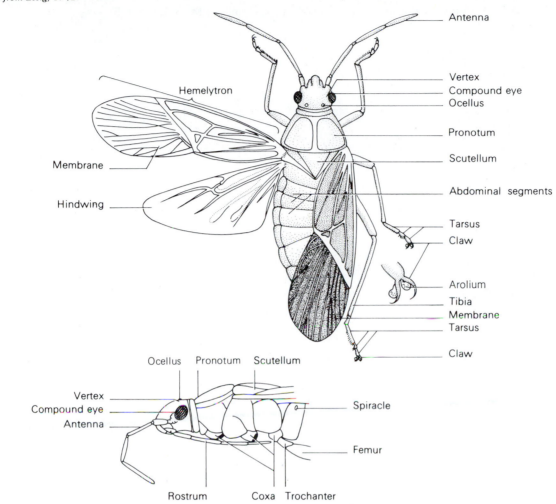

Frankliniella occidentalis, is a serious pest of numerous floricultural crops and serves as the vector of tomato spotted wilt virus, which attacks ornamentals and some vegetables.

Thrips can cause human discomfort by attempting to pierce the skin and may constitute a serious annoyance if present in large numbers.

Thrips are widespread in all of the world's major zoogeographical regions. Taxonomically, the suborder Terebrantia is more diverse than the suborder Tubulifera. Terebrantia is considered by most authors to include five families (Aeolothripidae, Heterothripidae, Merothripidae, Thripidae, and Uzelothripidae), whereas the Tubulifera includes just one (Phlaeothripidae). Two large families, Thripidae and Phlaeothripidae (one family from each suborder), include all of the economically important species. For further information on thrips, the reader should consult the following publications: Bailey (1957); Cott (1966); Lewis (1973); Palmer et al. (1989); Stannard (1957, 1968).

Order Hemiptera
Hemi, one-half; *ptera,* wing; true bugs (figures 12.27 through 12.29).

Body characteristics minute to very large.

Mouthparts piercing-sucking; degenerate palpi; labial sheath arises from the anterior part of the head.

Eyes well-developed in most.

Ocelli 2 or none.

Antennae 4- or 5-segmented; comparatively long; concealed in some.

Wings usually 2 pairs held horizontally over abdomen at rest; forewings hemelytra (basal portion leathery and apical portion membranous); many brachypterous and wingless forms; usually pronounced pronotum with a distinct triangular scutellum on mesothorax between forewing bases.

Figure 12.28 Representative Hydrocorisae and Amphibicorisae. (*a*) Water boatman (Corixidae). (*b*) Backswimmer, *Notonecta irrorata* (Notonectidae). (*c*) Water scorpion, *Nepa* sp. (Nepidae). (*d*) Giant water bug, *Lethocerus americanus* (Belostomatidae). (*e*) Creeping water bug, *Pelocoris femoratus* (Naucoridae). (*f*) Water strider, *Gerris buenoi* (Gerridae). (*g*) Toad bug, *Gelastocoris barberi* (Gelastocoridae). (*a*), S to Me; (*b*, *c*, *e*) and (*f*) Me; (*d*), L; (*g*), S. (*b*–*g*) *redrawn from Britton, 1923.*

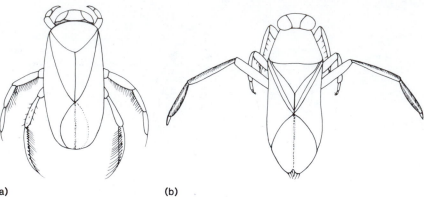

(a)   (b)

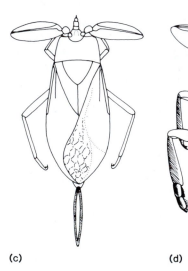

(c)   (d)   (e)

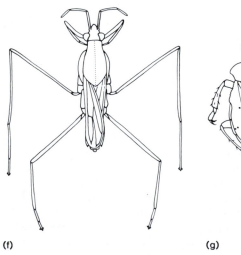   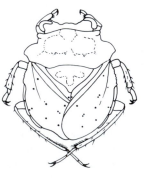

(f)   (g)

Figure 12.29   Representative Geocorisae. (*a*) Seed bug, *Ligyrocoris diffusus* (Lygaeidae). (*b*) Stink bug, *Apateticus cynicus* (Pentatomidae). (*c*) Tarnished plant bug, *Lygus lineolaris* (Miridae). (*d*) Squash bug, *Anasa tristis* (Coreidae). (*e*) Conenose bug (Reduviidae). (*f*) Common bed bug, *Cimex lectularius* (Cimicidae). (a, c, and f, s; b, S to Me; d, Me; e, Me to L.) (*a*)–(*d*) *redrawn from Britton, 1923, (e) and (f) after U. S. Public Health Service.*

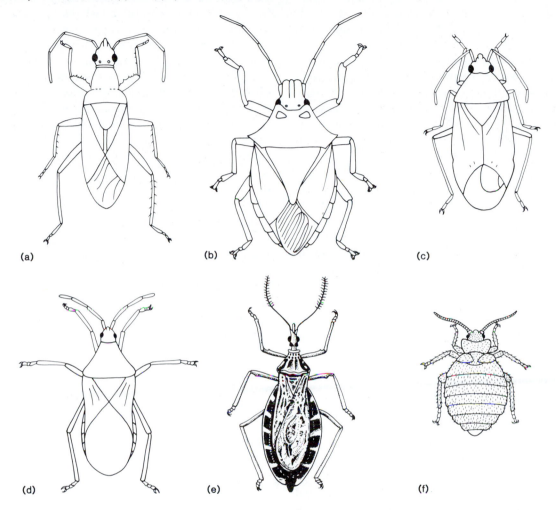

(a) (b) (c)

(d) (e) (f)

Legs variable; usually 2- or 3-segmented tarsi.

Abdomen cerci lacking.

This is the largest of the hemimetabolous orders. Some specialists lump the Hemiptera and Homoptera (aphids, leafhoppers, and relatives) into a single order, in which case the entire order is named Hemiptera and the true bugs are classified as the suborder Heteroptera and the homopterans as the suborder Homoptera (Henry and Froeschner 1988). Evidence cited in support of lumping includes characteristics of the family Peloridiidae that are found in Australia, New Zealand, and South America. Members of this family are transitional, displaying both heteropteran and homopteran traits (China 1962).

Many adult bugs possess repugnatorial or scent glands located in the region of the metathoracic coxae. In some nymphs, similar glands are located in the region of the metathoracic coxae, while in others they are located on the dorsum of the abdomen. Many species give off a very unpleasant odor when handled. Bed bugs (Cimicidae) have a characteristic odor, and a large infestation can be identified by smell alone.

Many species produce sounds (Richards and Davies 1977), and stridulatory organs have been identified in the prosternal region (Reduviidae and Phymatidae), abdominal sterna (some Pentatomidae), forelegs and clypeus (Corixidae), coxae (certain Nepidae), and dorsum of the abdomen in association with the wings (some Pentatomidae). The habitats in which hemipterans are found are diverse, with terrestrial, aquatic, semiaquatic, and ectoparasitic forms being known. Feeding habits are likewise diverse, some being herbivorous, others carnivorous, and some hematophagous.

The Hemiptera can be divided into three groups of families: "water" bugs (suborder Hydrocorisae), "surface-dwelling" bugs (suborder Amphibicorisae), and "terrestrial" bugs (suborder Geocorisae). These divisions are based largely on ecological groupings, and some taxonomists prefer to divide the Hemiptera into two suborders, the Gymnocerata (with free antennae) and Cryptocerata (with antennae concealed in grooves in the head).

The Hydrocorisae (series Cryptocerata) (figure 12.28a–e, g) have short antennae (shorter than the head) concealed in grooves on the venter of the head and are all aquatic or semiaquatic; the most common families are Corixidae, water boatmen; Notonectidae, backswimmers; Nepidae, water scorpions; Belostomatidae, giant water bugs; Naucoridae, creeping water bugs; and Gelastocoridae, toad bugs. The vast majority of these insects are predatory, and certain ones, such as backswimmers and giant water bugs, can inflict a painful "bite" if carelessly handled. Most are capable of flight and many appear in the vicinity of outside lights at night. These families show many interesting variations of mechanisms involved in ventilation in an aquatic habitat (see "Ventilation in Aquatic Insects" in chapter 4).

The Amphibicorisae (part of series Gymnocerata) are mostly aquatic, living on the surface of water or living along the shore. The antennae are not concealed and are longer than the head. This group includes the families Gerridae (figure 12.28f), Veliidae, Mesoveliidae, Hebridae (all commonly referred to as water striders), Hydrometridae (water measurers) and others. Members of most families are predatory, but water striders feed mainly on dead insects. The seagoing *Halobates* spp. (see chapter 9) belong to the family Gerridae.

The Geocorisae (part of series Gymnocerata) (figures 12.27 and 12.29) have unconcealed antennae longer than the head. There are terrestrial, aquatic, and parasitic forms. This group which contains a much larger number of families than the other two, usually is divided into several superfamilies.

Geocorisae contains many species of economic and medical significance. The chinch bug, *Blissus leucopterus* (Lygaeidae), attacks corn, sorghum, and small grains in the United States. In 1934 the chinch bug caused an estimated loss of $27.5 million to the corn crop and $28 million to wheat, barley, rye, and oats (Burkhardt 1978). Many species in the family Miridae (plant bugs) are serious pests of leguminous forage crops, particularly alfalfa. The families Pyrrhocoridae (red bugs or stainers), Coreidae (leaf-footed bugs), Pentatomidae (stink bugs or shield bugs), Tingidae (lace bugs), and others also contain important agricultural pests.

Cimicidae (bed bugs and relatives) and Reduviidae (assassin bugs and relatives) are the Geocorisae of medical importance. Cimicidae are wingless ectoparasites of birds and mammals (bats, various rodents, and others). Three species attack humans (Harwood and James 1979): *Cimex lectularius* (figure 12.29f), *Cimex hemipterus* (tropical America, Africa, Asia, East Indies, and certain Pacific islands), and *Leptocimex boueti* (tropical Africa).

Although experimentally bed bugs can transmit pathogens (including hepatitis B virus) that cause a number of human diseases, there is little evidence that they are the natural vectors of any pathogens. The reduviids (figure 12.29e) are mostly predaceous, feeding on other insects (many can inflict painful bites), but members of the subfamily Triatominae (kissing bugs and relatives) are exclusively ectoparasites of vertebrates. Several species (e.g., *Rhodnius prolixus* and *Triatoma* spp.) are vectors of *Trypanosoma cruzi,* the causative agent of Chagas' disease. On the positive side, *Rhodnius prolixus* has been the subject of much significant biological research, particularly that carried out by a well-known insect physiologist, Sir Vincent Wigglesworth. Another ectoparasitic family is the Polyctenidae whose members live on bats. Additional noteworthy families in the Geocorisae include the following: Phymatidae (ambush bugs) members of which usually occur on flower heads where they wait for potential prey (butterflies, flower flies, etc.) to alight; Rhopalidae (scentless plant bugs), which includes the boxelder bug, *Leptocoris trivittatus,* a common North American species; and Aradidae (flat bugs or fungus bugs), most of which live under tree bark or fungi.

The following sources should be consulted for additional information on the order Hemiptera: Blatchley (1926), Britton (1923), De Coursey (1971), Ghauri (1973), Henry and Froeschner (1988), Herring and Ashlock (1971), Hungerford (1959), Merritt and Cummins (1984), Polhemus (1978), Slater and Baranowski (1978), Usinger (1956).

Order Homoptera

Homo, uniform; *ptera,* wing; aphids, leafhoppers, and relatives (figures 12.30 through 12.33).

Body characteristics minute to very large; extremely variable.

Mouthparts piercing-sucking; degenerate palpi; labial sheath arises more posteriorly than in Hemiptera.

Eyes usually present.

Ocelli 2 or 3 ocelli or none.

Antennae filiform to bristlelike.

Wings 2 pairs, uniform in texture, held rooflike over body; absent in some.

Legs variable; 1- to 3-segmented tarsi.

Abdomen cerci lacking

The majority of these insects are small; exceptions include some of the fulgorid planthoppers and cicadas. The prothorax is usually inconspicuous, except in the Membracidae, where it takes on a variety of bizarre shapes. Members of several groups have wax glands, which may secrete a powdery material. The females of

Figure 12.30 Homoptera. Major external structures of a leafhopper, *Paraphlepsius irroratus* (Cicadellidae). (*a*) Dorsal view. (*b*) Anterior view. (S.)

Redrawn with modifications from Borror and DeLong, 1971.

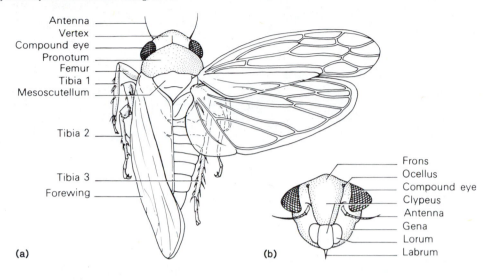

Antenna
Vertex
Compound eye
Pronotum
Femur
Tibia 1
Mesoscutellum
Tibia 2
Tibia 3
Forewing

Frons
Ocellus
Compound eye
Clypeus
Antenna
Gena
Lorum
Labrum

(a)   (b)

Figure 12.31 Homoptera, Auchenorrhyncha. (*a*) Treehopper, *Stictocephala bubalus* (Membracidae). (*b*) Same species as (*a*), dorsal view. (*c*) Spittlebug (Cercopidae). (*d*) Fulgorid planthopper (Fulgoridae). (*e–g*) Periodical cicada, *Magicicada septendecim*: (*e*) Last instar nymph; (*f*) Cast skin; (*g*) Adult female. ((*a–d*), S; (*e–g*), Me to L.)

(a) and (b) redrawn from Borror and DeLong, 1971; (c) and (d) redrawn from Britton, 1923; (e–g) redrawn from Boreal Labs key card.

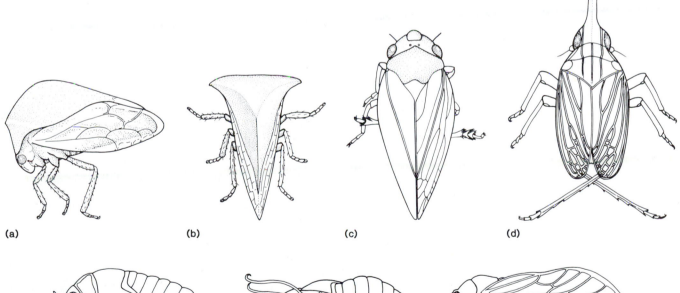

(a)   (b)   (c)   (d)

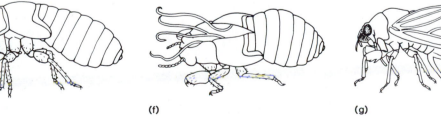

(e)   (f)   (g)

Figure 12.32 Homoptera, Sternorrhyncha. (*a*) Aphid, *Aphis pomi* (Aphididae), wingless viviparous female. (*b*) Same species as (*a*) winged viviparous female. (*c*–*f*) San Jose scale, *Aspidiotus perniciosus* (Diaspididae): (*c*) Scales on a twig; (*d*) Nymph; (*e*) Adult male: (*f*) Adult female with scale removed.
Redrawn from Boreal Labs key card.

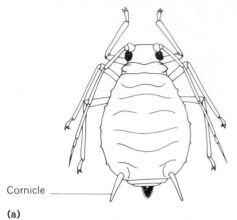

Cornicle

(a)

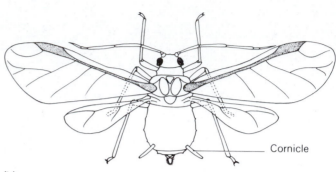

Cornicle

(b)

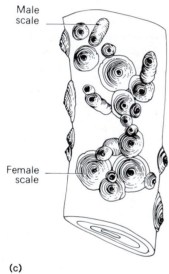

Male scale

Female scale

(c)

(d)

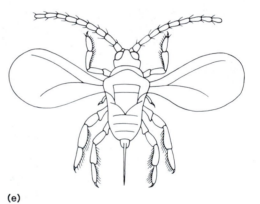

(e)

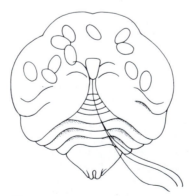

(f)

Figure 12.33 Life cycle of alternation of generations, and host plants of the bird cherry-oat aphid, *Rhopalosiphym padi* (Homoptera; Aphididae). (*A*) Fundatrix. (*B*) Apterous fundatrigenia. (*C*) Emigrant. (*D*) Apterous exule. (*E*) Alate exule. (*F*) Gynopara. (*G*) Male. (*H*) Ovipara. (*I*) Egg. Sexual reproduction occurs during autumn, parthenogenetic reproduction predominates during the rest of the year. *Redrawn from Dixon, 1973.*

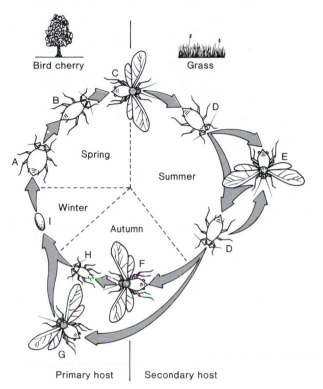

some groups have distinct ovipositors. Cicadas and certain leafhoppers produce sound by means of specialized structures on the dorsolateral portion of the base of the abdomen. Aphids, whiteflies, and others produce honeydew, a liquid material excreted from the anus.

The Homoptera differ from the Hemiptera in being exclusively terrestrial herbivores.

As with all the Hemiptera, most homopterans undergo typical hemimetabolous development. However, in certain scale insects, whiteflies, and phylloxerans the last nymphal instar is quiescent and resembles a pupa.

Homoptera may be divided into two groups: Auchenorrhyncha and Sternorrhyncha. In the Auchenorrhyncha (figure 12.31) the proboscis clearly arises from the posterior of the head and the antennae are short and bristlelike; members of this group are usually very active. In the Sternorrhyncha (figure 12.32) the proboscis appears to arise from the sternum between the forecoxae, and the antennae are generally long and filiform; many of these lead inactive or sedentary lives, in particular certain stages of the scale insects (superfamily Coccoidea).

There are no medically important Homoptera, but many are economically important pests. These insects are destructive to agricultural crops by feeding on plant juices,

causing oviposition damage, and serving as vectors of several plant disease organisms, particularly viruses. Especially significant families are Psyllidae (jumping plant lice), Aleyrodidae (whiteflies), Aphididae (aphids or plant lice), (figure 12.32*a,b*), Coccidae (wax and tortoise scales), Cicadellidae (leafhoppers) (figure 12.30), Cicadidae (cicadas) (figure 12.31*e–g*), Membracidae (treehoppers), Cercopidae (froghoppers and spittlebugs), Phylloxeridae (phylloxerans), Diaspididae (armored scales) (figure 12.32*c–f*), Pseudococcidae and Eriococcidae (mealybugs). Members of these families cause feeding damage and may also produce a toxic effect on their hosts. Aphids and leafhoppers are extensively involved in plant disease transmission. Aphids are well known for their complex life cycles, which involve various combinations of parthenogenesis, sexual generations, wingless and winged forms, and alternation of plant hosts (figure 12.33). They feed on plant juices extracted from nearly every part of a plant, roots included.

Although many homopterans are common, the cicadas (figure 12.31*e–g*) are particularly well known. These insects range in length from about 12 mm to approximately 10 cm, and certain tropical species are the largest homopterans. The nymphs are subterranean and feed on the roots of plants; the adults live above the ground in trees. The periodical cicadas, *Magicicada* spp., are a particularly well-known group; most spend either 13 or 17 years in the soil, depending on the species. At the end of their time in the soil, they emerge as adults, mate, and the females oviposit in slits in the twigs of trees, sometimes causing considerable damage. The emergence of a large brood may cause large patches of forest to appear brown owing to dead twigs and leaves. In a few weeks the newly emerged nymphs drop to the ground and burrow in, commencing another 13 or 17 years of feeding. There are several broods in the United States, and they emerge in different years. Another group of cicadas (e.g., *Tibicen* spp.) are referred to as "dog-day" cicadas and emerge from their subterranean home annually in May. Cicadas are chronic noisemakers, only the males being able to produce sound.

Another commonly encountered and interesting group of homopterous insects is the spittlebugs (Cercopidae; figure 12.31c), so called because the nymphs surround themselves in a frothy spittlelike substance. This substance apparently protects them from potential predators and parasites and prevents desiccation (Richards and Davies 1977). It is released through the anus. These are small insects, less than 12 mm in length, and the adults are active jumpers.

Representatives of several families of homopterans produce substances useful to humans and some of these insects have been cultured or "farmed" for this purpose. Examples include a crimson dye from cochineal insects, *Dactylopius coccus* (family Dactylopiidae) native to Mexico; chewing gum from *Cerococcus quercus* (family Asterolecaniidae) used by Native Americans in the southwestern United States; and *lac,* a substance used to manufacture shellac and varnishes, from *Laccifer lacca* (family Lacciferidae) in India.

The following sources should be consulted for additional information on the Homoptera: Britton (1923); Dixon (1973); Eastop and Hille Ris Lambers (1976); Kennedy and Stroyan (1959); Metcalf (1945, 1954–1963, 1962–1967); Nault and Rodriquez (1985).

Fossil Neopterous Exopterygote Orders

Carpenter (1977) recognizes four extinct neopterous exopterygote orders: Protorthoptera, Caloneurodea, Miomoptera, and Protelytroptera (see figure 11.14). The Protorthoptera were the largest, most diverse extinct order and displayed several orthopteroid traits, such as leathery forewings, hindwings with enlarged anal region, well-developed cerci, and mandibulate mouthparts. The Caloneurodea were also orthopteroid in appearance, but the forewings and hindwings were identical in appearance and the cerci were short. The Miomoptera were small insects with mandibulate mouthparts and short cerci. The forewings and hindwings were the same size and shape, and the venation was similar to the modern Psocoptera. The Protelytroptera had hardened forewings (elytra), beneath which the broad hindwings were folded at rest, and short cerci. The wing venation was similar to the Dermaptera.

CHAPTER 13

Survey of Class Insecta: II. Endopterygota (Holometabola)

D u r d e n / R o m o s e r

Members of all orders in the Endopterygota undergo complete metamorphosis. The immature stages typically are morphologically different and often have contrasting habitat and food requirements from adult forms. A pupal stage intervenes between the larval and adult instars. For information on the identification of larval forms, see Stehr (1987, 1991).

Endopterygota

Neopterous Endopterygotes

The orders in this category are generally viewed as being monophyletic. However, there is some disagreement as to the subgroups. Here we recognize division of the neopterous endopterygotes into coleopteroid, neuropteroid, panorpoid (mecopteroid), and hymenopteroid orders (figure 13.1). Kristensen (1975) and others include the coleopteroid orders as neuropteroids. Within the panorpoid orders, (a) the Mecoptera, Diptera, and Siphonaptera and (b) the Trichoptera and Lepidoptera are widely seen as two well-defined groups, the members of each displaying several similar features. The hymenopteroid category includes only one order, the Hymenoptera. These insects are sufficiently different that most agree to their separation from the rest of the neopterous endopterygotes.

Coleopteroids

The coleopteroid orders, Coleoptera and Strepsiptera, differ from the remaining endopterygote orders in the specialization of the metathorax for flight (Boudreaux 1979).

Order Coleoptera

Coleo, sheath; *ptera,* wing; beetles (figures 13.2, 13.3, and 13.4).

Body characteristics minute to very large; mostly hard-bodied; larvae: campodeiform, scarabeiform, or vermiform; pupae: exarate.

Mouthparts chewing, typically prognathous; some hypognathous; larvae: chewing.

Eyes and ocelli compound eyes present or absent; ocelli generally lacking, but 1 or 2 in certain groups; larvae: compound eyes lacking, but lateral ocelli usually present.

Antennae variable; small in larvae.

Wings forewings sclerotized elytra that protect the membranous hindwings at rest; some flightless with fused elytra; a few species are wingless.

Legs typically cursorial; some adapted for swimming, jumping, or digging; number of tarsal segments variable and taxonomically useful; larvae: commonly 6 legs, some legless.

Abdomen larvae: some with posterior pair of prolegs (no crochets); usually a pair of spiracles on each of the first 8 abdominal segments plus 1 or 2 pairs on the thorax.

Comments pronotum = single, conspicuous sclerite; vermiform and scarabeiform larvae are often called grubs.

This is the largest order of insects and therefore the largest order in the animal kingdom. Beetles comprise approximately 40% of all species of insects and 25% of all animal species.

As a group, beetles are found in nearly as wide a diversity of habitats as the entire class of insects. A similar statement can be made relative to their feeding habits.

Figure 13.1   Hypothetical phylogeny of endopterygote insects.
Based, with modifications, on Boudreaux, 1979.

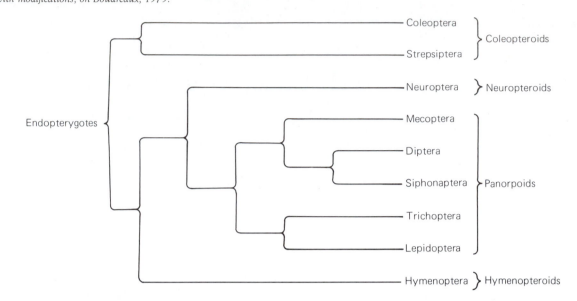

However, surprisingly few are parasitic. The largest numbers are herbivorous or predatory; while somewhat fewer are scavengers or feed on fungi. The predatory and scavenging species may be beneficial in that they either destroy other insects or take part in the breakdown of organic material.

Most beetles are oviparous, but a few are viviparous and some are parthenogenetic. Hypermetamorphosis occurs in some. The pupae are exarate. They often reside in earthen cells or within host plants. A number of beetles form pupal cocoons, which in some species are composed of a secretion from the Malpighian tubules or collapsed peritrophic membrane (see "Metamorphosis" in chapter 5), or they may be protected by the exuviae of the last larval instar.

The beetles rival moths in their economic importance. This order contains a large number of very destructive pests of agricultural crops; stored grains, seeds, and grain products; other stored products, such as tobacco, nuts, and chocolate; and shade trees and shrubs. Several species of beetles serve as vectors of plant disease.

Coleoptera may be divided into four suborders: Archostemata, Myxophaga, Adephaga (figure 13.3), and Polyphaga (figure 13.4). The first two groups are small and each is composed of two uncommon families. The suborder Archostemata includes the most primitive beetles and comprises the families Cupedidae (reticulated beetles), and Micromalthidae. The suborder Myxophaga includes the families Sphaeriidae (minute bog beetles), and Hydroscaphidae (skiff beetles). Members of the suborder Adephaga are characterized by having the first abdominal segment divided by the hind-coxae, whereas

members of the suborder Polyphaga lack this feature. Most beetles belong to one of these two suborders, although the Polyphaga is the larger of the two, in that 99 of the 111 families of beetles belong here (White 1983). Some large and/or economically important families of beetles are briefly discussed in the following paragraphs.

Suborder Adephaga   Family Cicindelidae (tiger beetles) (figure 13.3*a*). Most of the beetles in this family are blue or green and iridescent. The pronotum is narrower than the base of the elytra and the elytra are typically widest posteriorly. The eyes are large and bulging, and the legs are long and slender. Adults and larvae have sickle-shaped mandibles and are predaceous (these traits have given rise to the name "tiger beetles"), often pouncing on their prey. Larvae build vertical pits in hard soil in which they wait to seize passing prey. Adults can run and fly rapidly and usually frequent sandy or sunlit areas.

Family Carabidae (ground beetles) (figures 13.2, 13.3*b*). This is a very large family of beetles, most of which are entirely or partly black, with long legs and shiny, striated elytra. The antennae are filiform or moniliform and are inserted between the eye and the mandible. These beetles are very common especially on the ground where larvae and adults are predators, scavengers, or omnivores. Many species are nocturnally active. One North American species, *Stenolophus lecontei* (the seedcorn beetle), feeds on germinating seeds often resulting in significant crop losses.

Family Dytiscidae (predaceous diving beetles) (figure 13.3*d*). These beetles are aquatic and have characteristically flattened hindlegs that are fringed with setae. The body is elongate-oval, streamlined, and convex

Figure 13.2 Coleoptera, ground beetle, *Calosoma semilaeve* (Carabidae), with left elytron and hindwing spread. *Redrawn from Essig, 1942.*

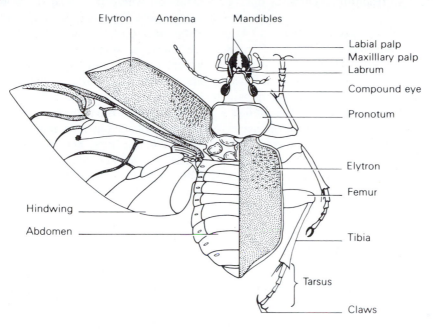

Figure 13.3 Representative members of the suborder Adephaga. (*a*) Tiger beetle (Cicindellidae). (*b*) Ground beetle (Carabidae). (*c*) Whirligig beetle (Gyrinidae). (*d*) Predaceous diving beetle (Dytiscidae). (S to L.) *Redrawn from Boreal Labs key card.*

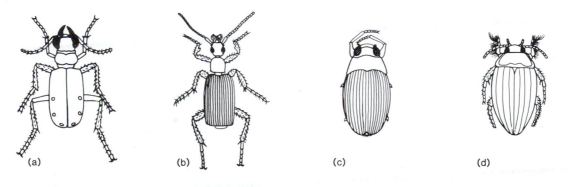

on both dorsal and ventral surfaces. The antennae are filiform and the body is generally black or brown with yellowish markings. These predators are often common in ponds, lakes, and slow-flowing rivers and streams where aquatic vegetation and small arthropods (potential prey) abound. Adults are usually active during the day and spend much of this time swimming beneath the water surface hunting; an air bubble trapped beneath the elytra is used for respiration.

Family Gyrinidae (whirligig beetles) (figure 13.3*c*). The eyes of these aquatic beetles are distinctly divided into two sections (permitting vision both above and below the water surface). The forelegs are long while the mid and hindlegs are short and flat. The body is flattened, elongate-oval in shape, and metallic black or green in color. These are the only beetles that regularly use the

water surface to support their bodies, but they are also capable of diving efficiently. Adults usually congregate in rivers, streams, ponds, lakes, etc., close to the bank or shore. Adults use a type of sonar and scavenge on animal and plant material, whereas larvae are mostly predaceous.

Suborder Polyphaga *Family Hydrophilidae (water scavenger beetles)* (figure 13.4*b*). These beetles have long maxillary palps and short, clubbed, 4-segmented antennae that are often concealed. The body is typically black or brown, smooth or shiny, oval or elliptical, flat ventrally, and convex dorsally. The hind-tarsi are often flattened and fringed with setae. Although most hydrophilids are aquatic, some live in fresh mammal dung, moist soil, or decaying vegetation. Adults are principally herbivorous, but all larvae are carnivorous or cannibalistic. Aquatic forms trap a layer of air along a pile of fine setae along the venter for respiratory purposes during diving.

Family Staphylinidae (rove beetles) (figure 13.4*f*). This is the largest beetle family in North America, Europe and most other regions. The body is slender and elongate but the elytra are short (usually exposing 3–6 abdominal segments). The abdomen is flexible and held aloft in some species. Adult staphylinids occur in a wide range of habitats from beneath stones and logs; along lake and ocean shores; on carrion, flowers and fungi; in manure; in ant, termite, bird, and mammal nests; under tree bark; in soil and leaf litter to caves. Adult rove beetles are efficient fliers. Many species are predaceous and are beneficial because they prey largely on potential insect pests.

Family Silphidae (carrion or burying beetles) (figure 13.4*e*). These minute- to large-sized beetles have clubbed antennae and are usually black or brown with red, yellow, or orange markings on the elytra and/or pronotum. The elytra are broadest posteriorly and either cover the abdomen loosely or are short (exposing 1–3 abdominal segments). Silphids are most common on carrion, but some occur in decaying vegetation. Some dig beneath carcasses and bury them before ovipositing, thus provisioning the larvae with a maggot-free food source (few flies can oviposit on buried carrion). Adult carrion beetles are largely predaceous on developing fly larvae prior to burying the carrion. Because of their association with dead animals, these beetles often have a strong characteristic odor. They may also carry *phoretic* mites, which use the beetle for transportation from one carrion source to another.

Family Cleridae (checkered beetles) (figure 13.4*g*). Adult clerids are elongate with a broad head and a narrow pronotum. The body is covered with bristly setae and typically marked with red, orange, yellow and/or blue. These insects are significant predators of wood-boring and bark beetles (families Cerambycidae and Scolytidae) and are known to have suppressed outbreaks of these pests in the past. Some clerids are scavengers, however, and one species, *Necrobia rufipes* (the red-legged ham beetle) is injurious to stored foodstuffs, especially meat products.

Family Lucanidae (stag beetles) (figure 13.4*c*). Most male stag beetles have spectacularly large or branched mandibles that they use in male-male competition. Both sexes have 10-segmented, elbowed antennae with a terminal club of 3–4 segments. The body is brown or black, elongate and robust with some species measuring greater than 60 mm in length. Larvae feed on dead or decaying wood, whereas adults apparently consume honeydew and tree sap.

Family Scarabaeidae (scarab beetles) (figure 13.4*a*). The terminal 3–4 antennal segments of scarab beetles are characteristically laminate. The body form is variable but most are stout and heavy. Some members of this large

and varied family are economically important. Larvae of most injurious species are called "white grubs" and inhabit soil where they feed on roots [e.g., *Popillia japonica* (the Japanese beetle), *Phyllophaga* spp., and *Polyphaga* spp.]. Larvae and adults of some other species feed on animal dung, rotten wood, carrion, humus, fungi, pollen, or plant sap. Some species inhabit termite or mammal nests. Adults of some *coprophagous* (dung-feeding) species prepare a dung ball in which the eggs are laid to provision for the larvae. June beetles belong in this family and include adults of various species of *Cotinus, Phyllophaga,* and *Polyphaga* (the name "June beetle" is applied to various species on a regional basis). Chafers (subfamily Melolonthinae) feed on plants and some are serious pests, [e.g., the Asiatic garden beetle (*Maladera castanea*) and the rose chafer (*Macrodactylus subspinosus*)]. The rhinoceros, hercules, and elephant beetles (subfamily Dynastinae) include the largest of all beetles and, with respect to mass, of all insects. For example, the North American hercules beetle *Dynastes tityus,* reaches a length of 62 mm and males of the South American *D. hercules* may reach a length of 180 mm.

Family Buprestidae (metallic wood-boring beetles) (figure 13.4*v*). Adult buprestids are typically metallic or bronzed with a bullet-shaped body. Elytral tips are pointed and have more or less parallel sides. The larvae of many species cause significant damage to the forestry industry through their tree-boring activities.

Family Elateridae (click beetles) (figure 13.4*d*). These beetles derive their vernacular name from the fact that a posterior elongation of the adult prosternum fits into a mesosternal depression and allows the beetle to rapidly click, often propelling it off the ground. Posterior pronotal angles are extended in most species and the body is long and narrow. Antennae are situated close to the eyes. Most larval elaterids are called wireworms and inhabit various soil types where they feed on roots, stems, and bulbs. Some wireworms cause heavy losses to agricultural crops (various larvae in the genera *Conoderus, Melanotus, Limonius,* and *Ctenicera*).

Family Dermestidae (carpet and museum beetles) (figure 13.4*s*). These minute- to small-sized beetles have elongate, robust bodies that are often covered with scales or setae arranged in characteristic patterns. The head is directed downwards and is partially concealed while the antennae are short and clubbed, often fitting into grooves beneath each side of the pronotum. Many economically important insects are included in this family. The genus *Anthrenus* includes species that consume household fabrics and dried animal products (including dried insects and other museum specimens). Various species of *Dermestes* are pests of dried foodstuffs or animal hides.

Family Cucujidae (flat bark and grain beetles) (figure 13.4*h*). These beetles are distinctly flattened dorsoventrally and have striate elytra and anteriorly-directed mandibles. Most species feed on dead animal or plant

Figure 13.4   Representative members of the suborder Polyphaga. (*a*) Scarab beetle (Scarabaeidae). (*b*) Water scavenger beetle (Hydrophilidae). (*c*) Stag beetle (Lucanidae). (*d*) Click beetle (Elateridae). (*e*) Carrion beetle (Silphidae). (*f*) Rove beetle (Staphylinidae). (*g*) Checkered beetle (Cleridae). (*h*) Cucujid beetle (Cucujidae). (*i*) Blister beetle (Meloidae). (*j*) Brentid beetle (Brentidae). (*k*) Darkling beetle (Tenebrionidae). (*l*) Languriid beetle (Languriidae). (*m*) Lady beetle (Coccinellidae). (*n*) Sap beetle (Nitidulidae). (*o*) Long-horned beetle (Cerambycidae). (*p*) Firefly (Lampyridae). (*q*) Tumbling flower beetle (Mordellidae). (*r*) Seed beetle (Bruchidae). (*s*) Carpet beetle (Dermestidae). (*t*) Snout beetle or weevil (Curculionidae). (*u*) Leaf beetle (Chrysomelidae). (*v*) Metallic wood borer (Buprestidae).

Redrawn from Boreal Labs key card.

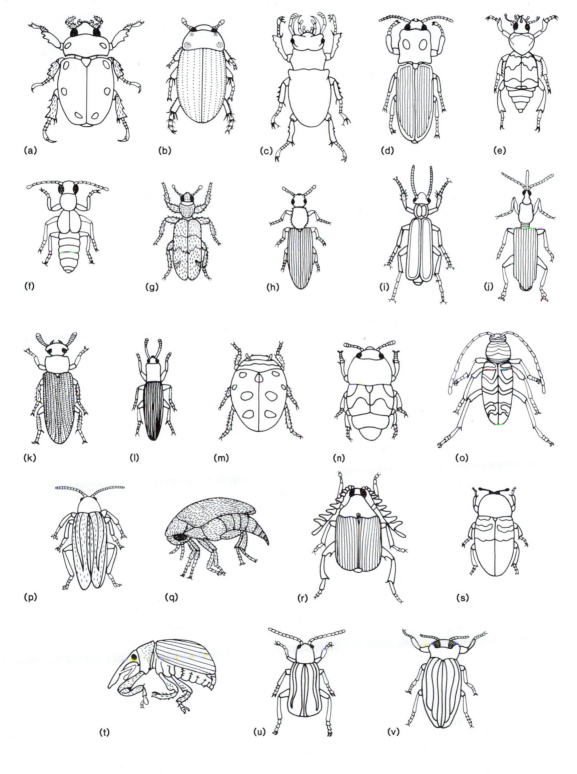

material, whereas others are predaceous. Some feed on stored food products and can reach pest proportions. For example, the saw-toothed grain beetle (*Oryzaephilus surinamensis*) is a pest on a worldwide basis and the square-necked grain beetle (*Cathartus quadricollis*) causes damage in the southern United States.

Family Coccinellidae (lady or ladybird beetles) (figure 13.4*m*). Members of this family are small with broadly oval or round, brightly-colored bodies that are strongly convex dorsally and flat ventrally. The antennae and pronotum are short and the head is partly or completely concealed from above. Most of these beetles are beneficial in that both larvae and adults prey on pest arthropods such as aphids, scale insects, and mites. Some coccinellids are herbivorous, however, and are considered to be pests. Examples of such pests include the Mexican bean beetle (*Epilachna varivestis*) and the squash beetle (*E. borealis*).

Family Lyctidae (powderpost beetles). These minute reddish-brown beetles (body length, 1–7 mm) have elongate, flattened bodies and a 2-segmented antennal club; the pronotum is widest anteriorly. Larvae and adults bore into dry, seasoned wood and some are pests (e.g., *Lyctus* spp.) because they form bore holes in furniture and other finished wood products.

Family Anobiidae (deathwatch and furniture beetles). These minute beetles (body length, 1–9 mm) have a hoodlike prothorax that conceals the head. The terminal 3 antennal segments are elongate and often clubbed. The body shape varies from almost spherical to elongate. Many of these beetles are pests. Both the drugstore beetle (*Stegobium paniceum*) and the cigarette beetle (*Lasioderma serricorne*) feed on stored organic materials including drugs, tobacco, seeds, cereals, and spices. Because of their boreholes, both the furniture beetle (*Anobium punctatum*) and the deathwatch beetle (*Xestobium rufovillosum*) can inflict serious damage to wood including furniture and structural supports of buildings.

Family Trogositidae (bark-gnawing beetles). These beetles are minute or small (body length, 5–15 mm) and have a terminal 3-segmented antennal club. Most adults are cylindrical or flattened, and the pronotum and elytra are separated by a *petiole* (a waistlike constriction). Adults and larvae of the cadelle (*Tenebroides mauritanicus*) feed on grain supplies and this species is a worldwide pest of stored products.

Family Tenebrionidae (darkling beetles) (figure 13.4*k*). Body shape in darkling beetles varies greatly but notched eyes, 11-segmented antennae, and a black or brown (rarely red) body are distinguishing features. Many of these scavenging beetles occur in arid environments, whereas others are found in such things as rotting wood, fungi, termite and ant nests, animal dung, stored food products, and decaying vegetation. Pests of stored foods

include mealworms (*Tenebrio molitor* and *T. obscurus*), the confused flour beetle (*Tribolium confusum*), and the red flour beetle (*Tribolium castaneum*).

Family Meloidae (blister beetles) (figure 13.4*i*). Blister beetles have a broad head and abdomen and a narrow thorax. The elytra loosely cover the abdomen and often are short (especially in females), exposing part of the abdomen. Most species are black or brown, but a few are brightly-colored. Meloid larvae are *hypermetamorphic* in that they hatch as a campodeiform *triungulin* stage but later transform, usually into a scarabaeiform larva. Many larvae are parasites inside bee nests or are predators of grasshopper eggs. Adults are herbivorous and some, such as various species of *Epicauta*, are crop pests in North America. When disturbed, blister beetles respond by reflex bleeding—hemolymph is secreted from various joints and can cause blisters in mammals including humans. The hemolymph contains *cantharidin* ("Spanish fly") and deters some predators; if accidentally eaten by grazing livestock or horses, some blister beetles can cause a toxic reaction.

Family Cerambycidae (long-horned beetles) (figure 13.4*o*). The antennae of these beetles are at least half as long as the body (and often are much longer than the body). Eyes typically are notched and the body is robust and broad at the base of the elytra. Some cerambycids are large (up to 60 mm in North America and much larger in Central America and South America). The woodboring habits of the larvae make many of them pests. Examples include larvae of the tile-horned prionus (*Prionus imbricornis*), the eastern larch borer (*Tetropium cinnamopterum*), the sugar maple borer (*Glycobius speciosus*), the southern pine sawyer (*Monochamus titillator*), and the round-headed apple tree borer (*Saperda candida*).

Family Chrysomelidae (leaf beetles) (figure 13.4*u*). Chrysomelid beetles are very diverse in shape and form but have 5-segmented tarsi with the fourth tarsal segment being short. Body size ranges from minute to small (1–16 mm). Larvae and adults are herbivorous and numerous species are pests. Examples of injurious forms include the Colorado potato beetle (*Leptinotarsa decemlineata*), corn rootworms (larvae of *Diabrotica* spp.), the cereal leaf beetle (*Oulema melanopus*), the strawberry rootworm (larvae of *Paria fragariae*), and the pale-striped flea beetle (*Systena blanda*).

Family Bruchidae (seed beetles) (figure 13.4*r*). These minute insects (body length, 1–8 mm) have egg-shaped bodies. The head is usually concealed from above and prolonged into a short beaklike rostrum anteriorly. The hind femora are enlarged, the eyes are notched and the antennae terminate in a 6–7 segmented club. The larvae of these beetles eat seeds, and some are serious pests. Despite the vernacular names of certain species, they are not true weevils. Examples of pestiferous species include the pea weevil (*Bruchus pisorum*) and the bean weevil (*Acanthoscelides obtectus*).

Family Curculionidae (weevils or snout beetles) (figure 13.4*t*). These beetles have a well developed snout that bears small mandibles at its tip and is often very elongate and narrow. The antennae have a terminal 3-segmented club and usually are elbowed. Weevils constitute the largest and possibly the most economically important family of beetles in the world. Herbivory is the rule for larvae and adults; some species feed externally on plants, whereas others burrow into stems, induce galls, or are leaf miners. A selection of the many pestiferous species follows: the alfalfa snout beetle (*Otiorhynchus ligustici*), the European snout beetle (*Phyllobius oblongus*), the alfalfa weevil (*Hypera postica*), the boll weevil (*Anthonomus grandis*), the rice weevil (*Sitophilus oryzae*), the granary weevil (*Sitophilus granarius*), and the pecan weevil (*Curculio caryae*).

Family Scolytidae (bark and ambrosia beetles). These beetles are minute (body length, 0.6–9 mm) with a brown or black elongate, cylindrical body. The head is usually concealed from above; the antennae are short, elbowed, and usually terminate in a 1–3 segmented club. The pronotum often bears anterior "teeth." Adults and larvae excavate galleries in tree bark and many species are pests. In mature North American timber forests, scolytid beetles are the greatest single cause of timber loss. One of the best known scolytids is the European bark beetle, *Scolytus multistriatus,* which transmits the fungus that causes Dutch elm disease in Europe and North America.

Additional information on beetles may be found in applied entomology texts such as Blackwelder and Arnett, Jr. (1977); Booth et al. (1990); Evans (1975); Jaques (1951); Leech and Chandler (1956); Leech and Sanderson (1959); Pfadt (1978); and Metcalf, Flint, and Metcalf (1962); and in books specializing in Coleoptera (Dillon and Dillon 1972, Arnett 1980, White 1983).

Order Strepsiptera

Strepsi, turning or twisting; *ptera,* wing; stylopids or twisted-wing parasites (figure 13.5).

Body characteristics minute.

Mouthparts reduced mandibulate.

Eyes and ocelli compound eyes present in males and free-living females.

Antennae males: flabellate, 4–7 segments, some segments with lateral processes.

Wings males: mesothoracic, clublike, similar to halteres of Diptera; metathoracic, membranous and fan-shaped with reduced venation; females: lacking.

Legs larvae and adults mostly legless; 2- to 5-segmented tarsi.

Abdomen 3 to 5 genital pores in adult females.

This is a very small order of mostly endoparasitic insects that mainly attack Hemiptera and Hymenoptera but also Blattaria, Diptera, Mantodea, Orthoptera, and Thysanura. Males are endoparasitic as larvae but emerge as free-living adults. Females of most parasitic species are parasitic throughout their lives and are grublike, lacking legs, antennae, and usually eyes also.

Strepsipteran larvae undergo hypermetamorphic development. The first instar (triungulin) is free-living and possesses legs. When this larva comes into contact with a host, it burrows through the host's cuticle and becomes endoparasitic. The second and third instars are vermiform. Full grown larvae pupate within the last larval skin, principally inside the host but with the head and thorax protruding slightly from an abdominal intersegmental membrane of the host. Adult females remain in this position and mate with the free-living males. Fertile eggs are retained inside the female strepsipteran, and hatchling triungulins disperse through the genital openings of their mother.

Parasitism by strepsipterans typically produces morphological and physiological changes in the host insect. Such hosts are stated to be *stylopized* and can be rendered reproductively sterile because of physical destruction of tissues or hormonal imbalances. Stylopized bees and wasps can often be recognized because of abdominal distortion. In theory, strepsipterans could play a role in the biological control of insect pests but in practice such control on a commercial scale has yet to be achieved. The following references are recommended for further information on strepsipterans: Askew (1973); Bohart (1941); Kathirithamby (1989); Kinzelbach (1991); Ulrich (1966).

Neuropteroids

Here the neuropteroid group is treated as a single order, Neuroptera, although some authorities view the neuropteran suborders—Megaloptera, Raphidiodea, and Plannipennia—as separate orders—Megaloptera, Raphidioptera, and Neuroptera, respectively. Among the ways neuropteroids differ from the other endopterygotes are details of ovipositor and wing structure (there are characteristically many cross veins in the costal space between the costa and subcosta in neuropteroids). Some authorities classify the coleopteroid orders as neuropteroids.

Order Neuroptera

Neuro, nerve; *ptera-,* wing; lacewings, dobsonflies, alderflies, antlions, owlflies, and related forms (figures 13.6, 13.7, and 13.8).

Body characteristics minute to very large; soft-bodied; larvae: mostly campodeiform; pupae: exarate in silken cocoon.

Mouthparts grasping-sucking or chewing.

Eyes and ocelli compound eyes present; dorsal ocelli present or absent.

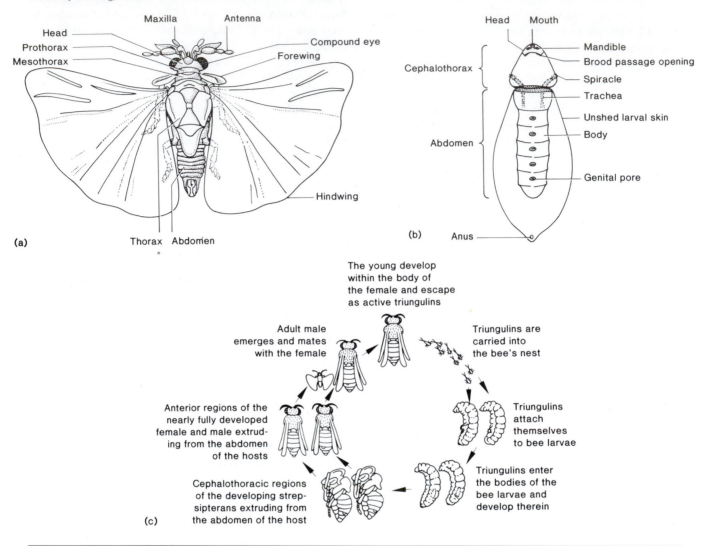

Figure 13.6 Neuroptera, Green lacewing (Chrysopidae), *Chrysopa perla*. (Me.)
Redrawn from Essig, 1942.

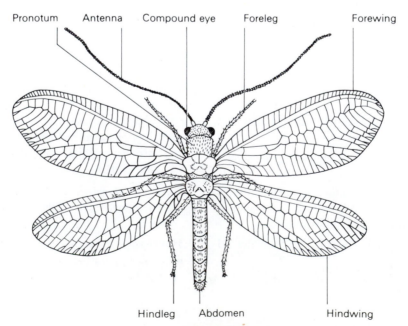

Figure 13.7 Neuroptera. (*a*) Snakefly, *Agulla bractea* (Raphidiodea; Raphidiidae). (*b*, *c*) and (*d*) Dobsonfly, *Corydalis cornutus*, female, male, and larva, respectively (Megaloptera; Corydalidae). (*e*) Alderfly, *Sialis mohri* (Megaloptera; Sialidae). (*a*), Me; (*b*–*d*), VL; (*e*), Me.

(a) redrawn from Wolglum and McGregor, 1958; (b–d) redrawn from Boreal Labs key card; (e) redrawn from Frison, 1937.

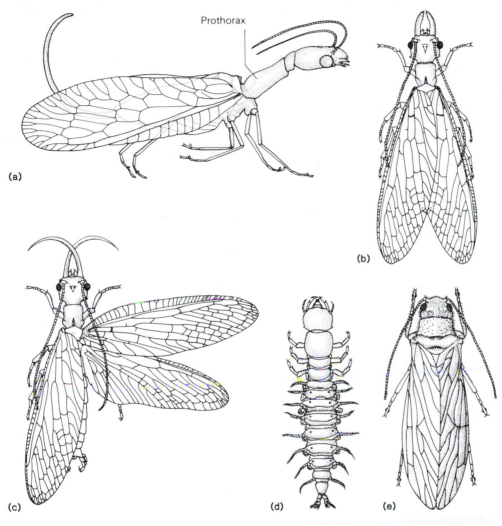

Antennae usually filiform.

Wings 2 pairs; similar in size and appearance; usually held rooflike over body at rest.

Legs 5-segmented tarsi.

Abdomen cerci lacking; variously modified ovipositor present.

Comments mostly weak fliers.

Neuropterans are found about vegetation and often near bodies of water, because many have aquatic larvae. Adults feed on fluid materials and various soft-bodied insects. Larvae are carnivorous, capturing and holding prey in their well-developed mandibles.

Some Neuroptera undergo hypermetamorphosis.

Neuroptera may be divided into three suborders: Megaloptera (figure 13.7*b*–*e*), Raphidioidea (figure 13.7*a*), and Planipennia (figures 13.6 and 13.8). Mega-

loptera have aquatic larvae with chewing mouthparts, hindwings basally slightly broader than forewings, with the anal region of the hindwings folding at rest. Members of the Raphidioidea (which has a single family, Raphidiidae), the snakeflies (figure 13.7*a*), are very distinct in appearance; the prothorax is elongate, with legs attached to the posterior end, giving the impression of a very long "neck." Snakefly larvae are terrestrial. Planipennia is the largest neuropteran suborder. Adults are characterized by having wings of similar shape and size with a nonfolding anal region on the hindwing. The larvae are usually terrestrial and have mandibulate sucking mouthparts.

Megaloptera contains two families, Corydalidae (dobsonflies and fishflies) and Sialidae (alderflies). Dobsonflies (figure 13.7*b*–*d*) are comparatively large and their aquatic larvae, or hellgrammites (figure 13.7*d*), are well-known to anglers as fish bait. Fishflies are smaller

Figure 13.8 Neuroptera, Planipennia, (*a*) Mantispid, *Mantispa brunnea* (Mantispidae). (*b*) Brown lacewing, *Sympherobius angustatus* (Hemerobiidae). (*c*) Spongillafly, *Climacia areolaris* (Sisyridae). (*d*) Antlion adult, *Hesperoleon abdominalis* (Myrmeleontidae). (*e*) Antlion larva, *Myrmeleon sp.* (*f*) Owlfly (Ascalaphidae). (*g*) Nemopteran adult (Nemopteridae). (*h*) Nemopteran larva. (*a*), (*e*), and (*h*), Me; (*b*) and (*c*), S; (*d*), (*f*), and (*g*), L. (*a*), (*b*), and (*f*) redrawn from Essig, 1942; (*c*) and (*d*) redrawn from Froeschner, 1947; (*e*) redrawn from Peterson, 1951; (*g*) after Smart, 1962; (*h*) after Riek, 1970.

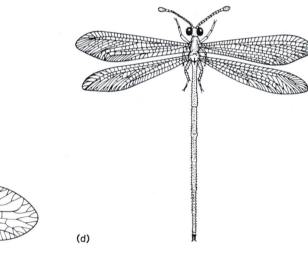

than dobsonflies. Alderflies (figure 13.7*e*) are also smaller than dobsonflies, also have aquatic larvae, and tend to be darkish in color.

The suborder Planipennia can be grouped into four superfamilies: Coniopterygoidea, Osmyloidea, Hemerobioidea and Myrmeleontoidea.

The Coniopterygoidea includes one family, the Coniopterygidae, the members of which are minute or small. The radial sector wing vein has two branches and none of the veins branch at the wing margins. The coniopterygids represent the smallest and most aberrant group of Neuroptera. Superficially, adults resemble aphids, but the mouthparts are mandibulate and larvae are campodeiform and very similar in form and habits to those of other Planipennia.

Members of the Osmyloidea have numerous costal veinlets and spotty markings on the wings. Three families are placed in this superfamily: Ithonidae, Osmylidae, and Neurorthidae. Osmylid and neurorthid larvae are aquatic or semi-aquatic.

Among the families in Hemerobioidea are Mantispidae (mantidflies; figure 13.8*a*), Chrysopidae (common or green lacewings; figure 13.6), Hemerobiidae (brown lacewings; figure 13.8*b*), and Sisyridae (spongillaflies; figure 13.8*c*). Mantidflies (figure 13.8*a*) bear a superficial resemblance to snakeflies but have raptorial forelegs attached anteriorly to the elongate prothorax. Their larvae are terrestrial and attack spiderlings inside nests of ground spiders. These larvae are campodeiform immediately after hatching but later undergo dramatic hypermetamorphosis, usually into an eruciform stage. Both brown and green lacewings are terrestrial groups. The larvae of both families live on vegetation and prey on aphids, mites, and other smaller soft-bodied insects. The common lacewings (figure 13.6) tend to be greenish in color and often have yellowish metallic-appearing eyes. They are often 2 or 3 cm long, and their eggs are deposited at the ends of stalks that are anchored to leaves. The larvae often carry a layer of debris about on their dorsum, which provides them with

Figure 13.8 (continued)

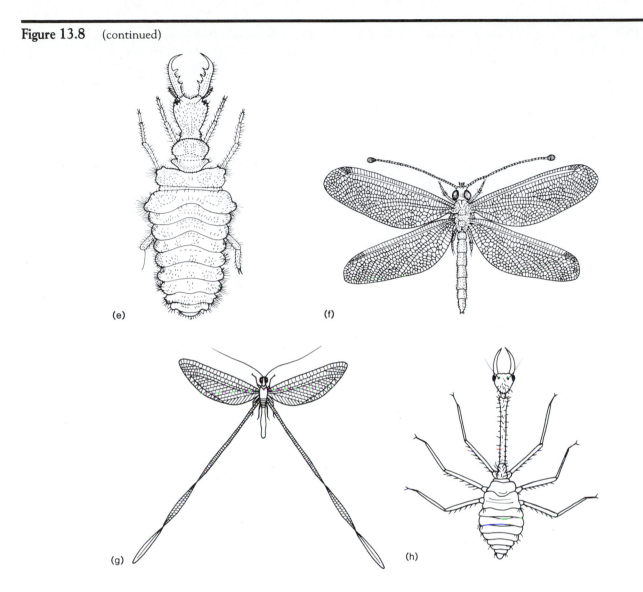

(e)

(f)

(g)

(h)

camouflage. The brown lacewings (figure 13.8b) are usually brownish in color and are smaller than the common lacewings. They do not deposit their eggs at the ends of stalks. Spongillaflies (figure 13.8c) are small insects and feed on freshwater sponges during the larval stage. Unlike the immature stages of other aquatic Planipennia, spongillafly larvae have ventral abdominal gills.

Myrmeleontoidea contains five families, the largest of which are Myrmeleontidae (antlions), Ascalaphidae (owlflies), and Nemopteridae. Antlion adults (figure 13.8d) superficially resemble damselflies and are similar in size. However, they are weaker fliers, have a very different pattern of wing venation, have longer distinctly clubbed antennae, and are soft-bodied. The larvae (figure 13.8e) are terrestrial and many live at the bottom of funnel-shaped pits in sand. When a potential prey insect or other small arthropod passes close to the pit, the antlion creates a miniature landslide, bringing the prey down within reach of its prehensile, sucking mandibles, and subsequently dines.

Owlflies (figure 13.8f) bear a resemblance to dragonflies but differ from them in the same features that the antlion adult differed from the damselflies. Their clubbed antennae are much longer than those of adult antlions. Their larvae are predatory and are often concealed by a layer of debris on their bodies.

Nemopterids are highly specialized in that the hindwings of adults and the prothorax of larvae are greatly elongated (figure 13.8g and h). These insects are confined to the Old World and Australia.

Members of this order have no medical significance. However, because of their predatory habits, many green and brown lacewings (both larval and adult stages) are highly beneficial in that they destroy garden and agricultural pests. Also, the larval stages of aquatic forms (especially Megaloptera) serve as important food sources for fish.

Additional information on the Neuroptera can be found in the following: Aspöck and Aspöck (1975); Chandler (1956); Evans (1978); Gurney and Parfin (1959); Riek (1970).

Figure 13.9 Mecoptera (*a*) Common scorpionfly (male), *Panorpa chelata* (Panorpidae). (*b*) Hanging scorpionfly, *Bittacus chlorostigma* (Bittacidae). (*c*) Snow scorpionfly, *Boreus californicus* (Boreidae). (*a*), Me; (*b*), Me to L; (*c*), Mi. (*a*) redrawn from Ross, 1962; (*b*) and (*c*) redrawn from Essig, 1942.

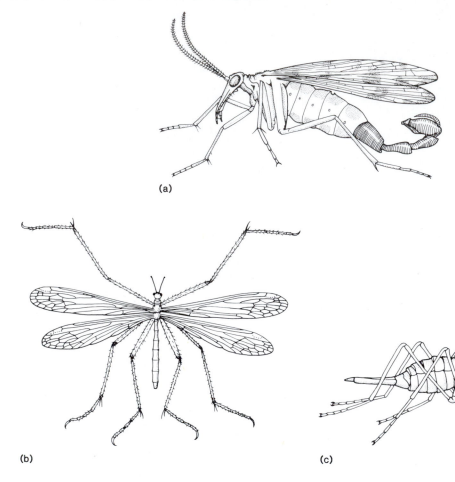

(a)

(b)

(c)

Panorpoids

The panorpoid orders (Mecoptera, Diptera, Siphonaptera, Trichoptera, and Lepidoptera; figure 13.1) are somewhat closely related to the neuropteroids. Few obvious features characterize this group (Boudreaux 1979). Included among the defining traits are extreme reduction of the externally apparent parts of the mesosternum and metasternum and the fact that the media and cubitus of the hindwing (when present) arise from a common stem. A fully developed appendicular ovipositor is not present in any member of this group, and specialized, telescoping terminal abdominal segments have come to serve as an ovipositor (see figure 2.44*e*). Each panorpoid order contains at least some members that spin a cocoon from silk secreted by larval labial glands.

Order Mecoptera

Meco, long; *ptera,* wing; scorpionflies (figure 13.9).

Body characteristics small to medium; fragile; larvae: usually eruciform; pupae: exarate.

Mouthparts chewing; commonly at apex of rostrum formed by elongation of head capsule; larvae: chewing.

Eyes and ocelli compound eyes present; 0 to 3 dorsal ocelli.

Antennae long and filiform.

Wings most species with 2 pairs; membranous; commonly with dark spots; some brachypterous and apterous (wingless) forms.

Legs long and slender; 5-segmented tarsi.

Abdomen long and slender in some and genitalia prominent; last segment in male carried upright suggestive of a scorpion; short cerci present.

Scorpionfly adults are terrestrial and are found in areas of decaying vegetation. The larvae are eruciform or scarabeiform with chewing mouthparts, and they live in or on the soil. Abdominal prolegs, when present, lack crochets and are 16 in number. The pupae are exarate.

Among the mecopteran families are Panorpidae, the common scorpionflies (figure 13.9*a*); Bittacidae, the

Figure 13.10 Diptera. A mosquito, *Anopheles maculipennis* (Culicidae) (female).
Redrawn from James and Harwood, 1969.

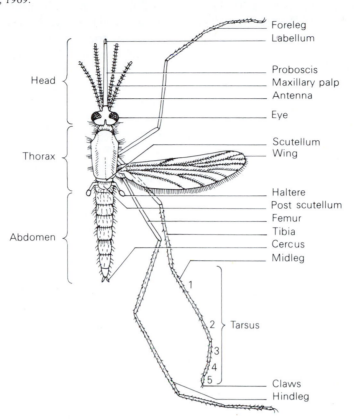

hanging scorpionflies (figure 13.9*b*); and Boreidae, the snow scorpionflies (figure 13.9*c*). Common scorpionflies feed mainly on dead insects and plant materials. Hanging scorpionflies superficially resemble crane flies and hang suspended from vegetation by their forelegs. They capture small flies with their raptorial tarsi. Snow scorpionflies inhabit moss and damp areas under stones and feed on plant material. They are sometimes found on the surface of snow.

For further information on the Mecoptera, see the following: Byers (1954, 1963, 1965, 1968); Carpenter (1931a, 1931b).

Order Diptera

Di, two: *ptera,* wing; true flies (figures 13.10 through 13.13).

Body characteristics minute to very large; larvae: apodous (legless), some with well-defined head and thorax and others wormlike (vermiform); pupae: obtect or coarctate.

Mouthparts adults: various modifications of sucking; larvae: various modifications of mandibulate mouthparts.

Eyes and ocelli compound eyes present; most species with 3 ocelli; ocelli lacking in some.

Antennae variable.

Wings 1 pair; membranous; metathoracic halteres; many secondarily apterous species.

Legs 5-segmented tarsi.

Abdomen cerci present or absent.

Diptera is the fourth largest order of insects, and its members are found in a wide diversity of habitats. Some live in biologically adverse environments; for example, the shore fly, *Psilopa petrolei* (Ephydridae) spends its larval stage in crude oil, and larvae of other ephydrid flies are found in the Great Salt Lake, Utah. A survey of the feeding habits of the order could well apply to the entire class Insecta. Members of several families are herbivorous as larvae. Some are mycetophagous (fungus-eaters). Many flies are detritivores and play a vital role in the breakdown of decaying organic material. A large number are predaceous and are to be counted among the beneficial forms. Many feed on the blood of vertebrates or are external or internal parasites. Such associations have resulted in their involvement with human and animal diseases.

Diptera is generally divided into three suborders: Nematocera, Brachycera, and Cyclorrhapha. Members of these suborders are separated on the basis of such features as antennal characteristics, the shape of the exit hole from the pupal case, the nature of the pupa, and larval characteristics.

The Nematocera (figures 13.10 and 13.11) are so called because, in the adult stage, they have comparatively long antennae (Nematocera = "thread horn"). The larvae (see figure 4.15b) have well-developed heads with mandibles that move in a horizontal plane. The pupae are obtect.

Flies in the suborder Brachycera (figure 13.12) have antennae that are shorter than the thorax (Brachycera = "short horn"). The pupae of these flies are exarate. Larvae have weakly developed heads, which are generally retractile, and the mandibles move in the vertical plane. As with Nematocera, Brachycera contains a large number of families. However, many of them are composed of uncommon species.

Members of the suborder Cyclorrhapha (figure 13.13) have coarctate pupae. Cyclorrhapha ("circular seam") refers to the circular shape of the opening through which the adult emerges from the puparium (see figure 5.24c,d). Near one end of the puparium there is a circular line of weakness along which the cuticle splits at adult emergence, much like a cap popping from a bottle. The split results from pressure being applied by means of an eversible bladderlike structure, the ptilinum, which protrudes from the head after emergence, usually leaving the frontal suture as a remnant. The larvae lack heads and are vermiform (maggots), with mouth hooks that operate in the vertical plane (see figure 2.32d). The adults generally have aristate antennae.

Diptera contains more insects of medical and veterinary importance than any other order. They cause serious problems for humans and domestic and wild animals in several ways: as vectors of the causative agents of many serious diseases, as blood feeders, as general nuisances, and by direct invasion of organs and tissues of humans and animals.

Flies associated with disease transmission typically, but not always, feed on the blood of vertebrates. There are several outstanding examples of fly-borne diseases. Mosquitoes (Nematocera, Culicidae; figures 13.10 and 13.11c) are probably responsible for the transmission of pathogens that cause more human diseases than any other group of insects. They are the vectors of the parasites that cause malaria (caused in humans by four species of Protozoa in the genus *Plasmodium*); filariasis (caused by nematodes such as *Wuchereria bancrofti*); and several viral diseases (yellow fever, dengue, and various types of encephalitis/encephalomyelitis). Two other nematoceran families also include disease-carrying members: Simuliidae (black flies; figure 13.11e) and Psychodidae (moth flies and sand flies; figure 13.11b and see also figure 8.4).

Black flies are vectors of onchocerciasis, a disease caused by a filarial worm, *Onchocerca volvulus,* which induces the development of painful nodules in the skin and often results in blindness. Psychodid flies in the genus *Phlebotomus* transmit protozoans in the genus *Leishmania* which cause various forms of leishmaniasis in humans. The tsetse flies, *Glossina* spp. (Cyclorrhapha, Glossinidae), serve as vectors of the trypanosomes that cause African sleeping sickness in humans (*Trypanosoma gambiae* and *T. rhodesiense*) and nagana (*T. brucei*), a serious disease of cattle in Africa.

Other biting flies capable of transmitting disease-causing organisms, but more often involved as nuisances through their blood-feeding activities, are the biting midges or "no-see-ums" (Ceratopogonidae; figure 13.11d); horse flies and deer flies (Tabanidae; figure 13.12a); certain snipe flies (Rhagionidae); and the stable fly, horn fly, and other flies in the family Muscidae. An interesting group of biting flies, which usually do not attack humans but are intermittent blood-feeding parasites of sheep, goats, and various wild birds and mammals, are the louse flies and bat flies. The larvae of these flies are retained in the female parent until nearly ready to pupate, at which time they are released (for this reason, these families of flies are often grouped together in a taxon termed "Diptera Pupipara"). The sheep ked (figure 13.13e), *Melophagus ovinus,* is a common wingless ectoparasite of sheep and goats.

Nonbiting flies that pass the larval stage in garbage, excrement, and other decaying organic matter may also be involved in the transmission of disease, generally through the contamination of food. Flies in this category are largely in the cyclorrhaphan families Muscidae (house fly, figure 13.13c, and relatives), Calliphoridae (blow flies; figure 13.13f), and Sarcophagidae (flesh flies). These flies may transmit pathogens that cause various gastrointestinal diseases, both of bacterial (dysentery, typhoid fever, and others), protozoan (*Entamoeba*), and related etiology (cholera, yaws, poliomyelitis, and eye infections). These flies also vector various organisms that cause disease in domestic animals, including several helminths (tapeworms; nematodes or roundworms). Chloropidae (frit flies and relatives) is another family of nonbiting flies, certain species of which transmit pathogens—those in the genus *Hippelates*. These are commonly called eye gnats and have been identified as probable vectors of the agents that cause conjunctivitis ("pinkeye"), yaws, and bovine mastitis. These flies are attracted to various body secretions and so tend to cluster around the mouth, eyes, nose, and genital and anal openings as well as sores exuding pus.

Larvae of many flies may invade the organs and tissues of humans and animals, sometimes causing severe injury and even death. Such invasions are referred to as myiasis. Invasion may be brought about accidentally (e.g., accidental ingestion of maggots in food) or may be due to adult female flies being attracted to an open wound or body opening and depositing eggs. Several flies (e.g., the

Figure 13.11 Representative nematocerous Diptera. (*a*) Crane fly (Tipulidae). (*b*) Sand fly, *Phlebotomus* (Psychodidae). (*c*) Mosquito, *Aedes aegypti* (Culicidae). (*d*) Biting midge, *Culicoides furens* (Ceratopogonidae). (*e*) Black fly, *Simulium venustum* (Simuliidae). (*f*) Gall midge (Cecidomyiidae). (*a*), Me to L; (*b*), (*c*), and (*e*), S; (*d*) and (*f*), Mi. *(a) and (f) redrawn from Boreal Labs key card; (b–e) after U.S. Public Health Service.*

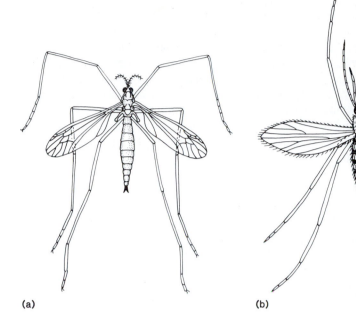

(a)

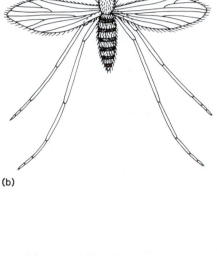

(b)

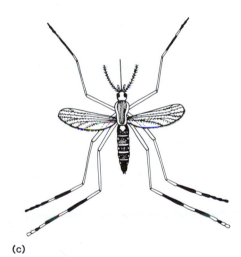

(c)

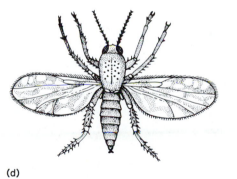

(d)

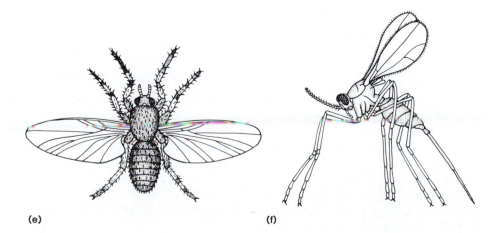

(e)

(f)

screwworm fly, *Cochliomyia hominivorax*) require an animal host in the larval stages. This situation is called *obligate myiasis*. Other examples of obligate myiasis occur in the bot flies in the families Gasterophilidae (horse bot flies) and Oestridae (warble flies and bot flies). Other species (especially in the family Calliphoridae) can survive either as parasitic or free-living larvae (*facultative myiasis*).

Surprisingly few true flies are pests of agricultural significance. Among those that are, the Hessian fly, *Phytophaga destructor* (family Cecidomyiidae), is outstanding. This fly is the most destructive of the insects that attack wheat in the United States. It was apparently introduced from Europe into the United States during the Revolutionary War in the straw bedding used by Hessian soldiers. Also, the Mediterranean fruit fly, and other related species discussed below are pestiferous.

A brief synopsis of the main morphological and biological features of some important dipteran families follows. Points covered above under different subject areas will generally not be repeated in the synopsis. For further information on flies, the following references should be consulted: Askew (1973); Cole (1969); Curran (1934); Greenberg (1971); Harwood and James (1979); McAlpine (1981, 1987, 1989); Merritt and Schlinger (1978); Oldroyd (1964); Smith (1973); Stone et al. (1965); Teskey (1978), Wirth (1956).

Suborder Nematocera   *Family Tipulidae (crane flies)* (figure 13.11*a*). Adult crane flies have very long legs and a V-shaped mesonotal suture. This is a very large cosmopolitan family. Larvae inhabit water, moist soil, or decaying vegetation and some species (leatherjackets) are pests of lawns and certain crops.

Family Psychodidae (moth and sand flies) (figure 13.11*b*). These minute or small flies have hirsute bodies with broad wings that are held either rooflike (moth flies) or together (sand flies) above the body when at rest. The radial wing vein has 5 branches. Moth flies (e.g., *Psychoda* spp.) are associated with drains, sewers, sewage treatment plants, and other places where decaying organic matter accumulates; these organic materials are consumed by the larvae. Sand flies (e.g., *Phlebotomus*) usually occur near water, and most larvae inhabit moist soil. Although moth flies do not bite, sand flies are blood feeders and certain species are known to transmit pathogens, including those that cause the various types of leishmaniasis and sand fly fever virus.

Family Culicidae (mosquitoes) (figures 13.10 and 13.11*c*). Adult mosquitoes are armed with an elongate proboscis and have long, narrow wings with scales along the veins and margins. The distal part of each wing has an unforked vein between two forked veins. Larvae are aquatic; most are detritivores but some are predators. Adult females of most species feed on vertebrate blood and their considerable significance with respect to disease transmission has been discussed already. Members of the genus *Toxorhynchites* are beneficial, however, because larvae in this taxon are voracious predators of other mosquito larvae and the adults do not feed on blood.

Family Ceratopogonidae (biting midges, no-see-ums, or punkies) (figure 13.11*d*). Most of these minute flies measure less than 3 mm. Wing venation is distinctive: radial branches are prominent, the medial vein has 2 branches, and the costal vein is thickened along 50% to 75% of its length. Larvae are snakelike and are either aquatic or occur in moist habitats. The majority of adult ceratopogonids feed on vertebrate blood, but a few imbibe fluids from invertebrates, and some are predators. The bites of these flies cause a significant nuisance to mammals including humans, and some are also capable of transmitting pathogens as discussed previously.

Family Chironomidae (midges or nonbiting midges). Adult chironomids have long, narrow wings; the medial vein is unbranched and a thickened section of the costal vein extends almost to the wing tip. Male antennae are plumose. Larvae of most species are aquatic and typically inhabit mud at the bottom of lakes and rivers. Both larvae and adults often occur in huge numbers. Larvae may account for the major component of the biomass of aquatic benthic habitats and feature prominently in the diets of many fish.

Family Simuliidae (black flies) (figure 13.11*e*). Black flies are generally 4 mm or less in length and have robust hump-backed bodies. The antennae are short, and the wings are broad at the base and have prominent anterior veins. Female flies are voracious biters and are a significant nuisance for livestock and humans; disease transmission by these flies has already been discussed. Larvae inhabit fast-flowing streams where they attach to solid objects and filter food particles from the current.

Family Cecidomyiidae (gall midges or gall gnats) (figure 13.11*f*). These minute flies (body length, 3 mm or less) are slender with long antennae and legs. Wing venation is reduced and 7 or fewer veins extend to the wing margin. This is a very large, widely distributed family. Larvae of most species induce galls in their foodplants. Some other phytophagous species are not associated with galls and still others are saprophagous, predaceous, or parasitic. Some cecidomyiid larvae are significant crop pests (e.g., the Hessian fly, discussed previously), the alfalfa gall midge (*Asphondylia websteri*), the sorghum midge (*Contarinia sorghicola*), and the wheat midge (*Sitodiplosis mosellana*).

Suborder Brachycera   *Family Tabanidae (horse and deer flies)* (figure 13.12*a*). These robust flies (body length, 6–35 mm) have an elongate third antennal segment and divergent fourth and fifth radial wing veins that encircle the wing tip. Eyes are iridescent and brightly-colored in many species. Tabanids are worldwide in distribution. Females feed on blood, whereas males usually feed on flower nectar. Larvae of most species are aquatic. Adults

Figure 13.12 Representative brachycerous Diptera. (*a*) Horse fly (Tabanidae). (*b*) Robber fly (Asilidae). (*c*) Long-legged fly (Dolichopodidae). (*d*) Bee fly (Bombyliidae). (*a*) and (*b*), Me to L; (*c*), S; (*d*), Me.
Redrawn from Boreal Labs key card.

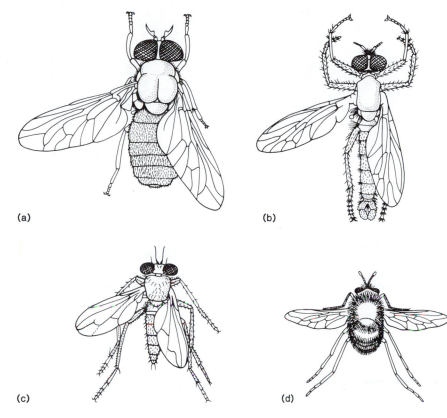

(a)

(b)

(c)

(d)

are often abundant close to swamps or ponds inhabited by the immature stages, despite the fact that they are strong fliers and are capable of traveling great distances from their breeding sites. Bites from female tabanids are a great nuisance mainly because they are painful; some species are also capable of transmitting pathogens (e.g., the viruses that cause equine infectious anemia and anthrax; *Trypanosoma evansi,* which causes "Surra" in certain livestock and domestic animals in tropical and subtropical regions; and *Loa loa,* the "eyeworm," an African filarial nematode that affects humans).

Family Asilidae (robber flies) (figure 13.12*b*). Robber flies are also robust (body length, 5–40 mm) with a relatively large thorax and legs and a long, frequently tapering abdomen. The head is hollowed between the eyes and the third antennal segment is elongate. Some species are hirsute. Adults are voracious predators and often perch on emergent vegetation to await flying insects that they may seize in mid-air. Larvae are found in soil or decaying logs and some are predators of other insect larvae.

Family Bombyliidae (bee flies) (figure 13.12*d*). These minute- to medium-sized flies are stout-bodied and hirsute, with some species mimicking bees. The first medial wing vein terminates posterior to the wing tip and 3 or 4 posterior wing cells plus a discal cell are present.

Adult bee flies feed on flower nectar, and the wings are held outstretched at rest. Most species are capable of efficient hovering during flight. Some species have patterned wings, and some possess an elongate proboscis. Most bee fly larvae are parasitoids of other insects.

Suborder Cyclorrhapha—Division Aschiza (frontal suture absent) *Family Phoridae (hump-backed flies).* These flies are minute and have a "hump-backed" thorax, thickened costal wing veins, flattened hind-femora and short antennae. This is a very large and diverse family and some species are very common. Most are associated with decaying vegetation or fungi, but some occur in ant or termite nests, and others are parasitoids of other insects.

Family Syrphidae (flower flies or hover flies) (figure 13.13*d*). These common and widespread flies have a "false" or spurious vein (a membrane thickening) between the radial and medial wing veins. Most species are brightly-colored and some are hirsute, often mimicking certain species of bees and wasps; syrphid flies do not bite or sting however. Adults are often observed hovering during flight or feeding on flowers. Larvae are predators (many of these species consume aphids and are beneficial), scavengers (living in dung, carrion, decaying organic matter or polluted aquatic habitats), herbivores

Figure 13.13   Representative cyclorrhaphous Diptera. (*a*) Fruit fly (Drosophilidae). (*b*) Picture-winged fly (Otitidae). (*c*) House fly (Muscidae). (*d*) Hover fly or flower fly (Syrphidae). (*e*) Sheep ked, *Melophagus ovinus* (Hippoboscidae). (*f*) Green bottle fly, *Phaenicia sericata* (Calliphoridae). (*a*), Mi; (*b–f*), S.
(*a–d*) *redrawn from Boreal Labs key card;* (*e*) *and* (*f*) *after U.S. Public Health Service.*

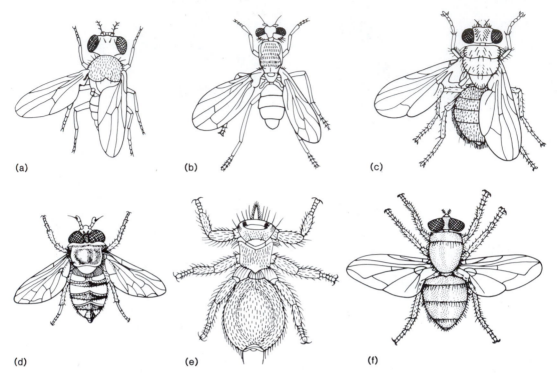

(a)   (b)   (c)

(d)   (e)   (f)

(some of these species are pests of crops or bulbs), or inhabit ant nests. Representatives of this family include the drone fly (*Eristalis tenax*), the lesser bulb fly (*Eumerus tuberculatus*), and the 'narcissus bulb fly' (*Lampetia equestris*).

Suborder Cyclorrhapha—Division Schizophora (frontal suture present)   *Family Tephritidae* (*= Trypetidae*) *(fruit flies).* These fruit flies are minute or small and often have brightly-colored, spotted or banded, wings. The apex of the subcostal wing vein curves sharply forward. Members of this large family are common and widespread. Adults usually are found on flowers or vegetation, but larvae are phytophagous. Larvae of numerous species are important pests of crops or fruit trees, for example, the Mediterranean fruit fly or Medfly (*Ceratitis capitata*), the Oriental fruit fly (*Dacus dorsalis*), the Mexican fruit fly (*Anastrepha ludens*), the cherry fruit fly (*Rhagoletis cingulata*), the black cherry fruit fly (*Rhagoletis fausta*), and the apple maggot (*Rhagoletis pomonella*).

Family Drosophilidae (fruit flies or pomace flies) (figure 13.13*a*). An interrupted costal wing vein and the presence of an anal cell characterize these common, minute, yellow/brown flies. Larvae of most species breed in decaying fungi or fruit and some of the latter may reach pest proportions. Species of *Drosophila* (especially *D. melanogaster*) have been used extensively in studies of genetics and heredity.

Family Chloropidae (frit flies). Frit flies are minute and have incomplete subcostal and interrupted costal wing veins; an anal cell is lacking. Many species occur in grassy areas, and larvae inhabit grass stems or decaying organic matter. Adults of some species (eye gnats) are attracted to the eyes (or to skin sores). Larvae of a few species are pests, for example, the wheat stem maggot (*Meromyza americana*).

Family Agromyzidae (leaf-miner flies). The flies belonging to this family are minute and usually black or yellow in color. The subcostal wing vein is incomplete or fused with the first radial vein, and the costal vein is interrupted. Adults are common on vegetation, whereas most larvae are leaf miners (other larvae feed inside stems or seeds). Some species are economically important, for example, the alfalfa blotch leaf miner (*Agromyza frontella*).

Families Hippoboscidae (louse flies or keds) (figure 13.13e), Streblidae (streblid bat flies), and Nycteribiidae (nycteribiid bat flies). These three families share several characteristics and collectively make up the 'Diptera Pupipara.' All are blood-feeding ectoparasites of birds and mammals (Hippoboscidae) or of bats (Streblidae and Nycteribiidae), and numerous species are wingless. Nycteribiids are all wingless and have long spidery legs. The sheep ked (*Melophagus ovinus*) (figure 13.13*e*) is an economically important species of hippoboscid.

Family Anthomyiidae (anthomyiid flies). The wings of these common and widespread flies have a parallel-sided fifth radial cell. Most adults resemble house flies. Larvae have varied habits; many are phytophagous, whereas others are scavengers or feed on decaying organic material. Larvae of some species are crop pests, for example, the seed-corn maggot (*Delia cilicrura*).

Families Gasterophilidae (horse bot flies), Oestridae (bot and warble flies), and Cuterebridae (robust bot flies). Larvae of flies belonging to these three families are endoparasites or subdermal parasites of mammals. The first two families have been discussed previously. Cuterebrids are typically parasites of New World rodents and lagomorphs, although one Central and South American species, the torsalo (*Dermatobia hominis*) is a subdermal parasite of livestock and humans. Interestingly, gravid female *D. hominis* attach their eggs to biting flies such as mosquitoes and the eggs hatch rapidly when the latter alight on a warm-blooded host. Penetration by the first instar larvae into the host dermis is a rapid process.

Family Muscidae (muscid or house flies) (figure 13.13*c*). These widespread, common, and often pestiferous flies have a short second anterior wing vein that fails to extend to the wing margin. The house fly (*Musca domestica*) and the little house fly (*Fannia canicularis*) breed in various kinds of organic matter and are capable of transmitting numerous pathogens. Larvae of the latrine fly (*Fannia scalaris*) and related muscids breed in animal waste. Adults of the stable fly (*Stomoxys calcitrans*) have a sharp, elongate proboscis with which they penetrate mammalian skin (inflicting a painful bite) to obtain blood. Adult horn flies (*Haematobia irritans*) feed on cattle blood, whereas the larval stages breed in cow pats. Face fly (*Musca autumnalis*) larvae also inhabit cow manure; adults imbibe body secretions from livestock and pose a significant nuisance to these animals.

Family Tachinidae (tachinid flies). Adult tachinids have the fifth radial wing cell narrowed or closed distally; they also possess a well-developed postscutellum. This is another large family; some of its members resemble house flies, whereas others are larger and may resemble bees. Larvae are parasitoids of other arthropods, especially insects, and numerous species are beneficial because they attack pests.

Family Calliphoridae (blow flies) (figure 13.13*f*). Calliphorid flies have wing characteristics similar to those of tachinids but the postscutellum is not developed, and the body is often a metallic blue or green. These flies are widely distributed and often abundant. Larvae of most species are scavengers living in carrion, dung, and other organic materials (e.g., *Calliphora vomitoria*). Larvae of other species feed on dead or living tissue of live animals including humans. This family includes examples of both obligate and facultative myiasis. Examples of obligate parasites include the larvae of the screwworm fly (*Cochliomyia hominivorax*) in the New World, and the tumbu fly (*Cordylobia anthropophaga*), which attacks various vertebrates, including humans, in Africa. Examples of facultative parasites are the larvae of *Phormia regina* and *Phaenicia sericata* (figure 13.13*f*), both of which may attack sheep, a condition termed *sheep strike*.

Family Sarcophagidae (flesh flies). Flesh flies are similar to blow flies but the body is never metallic and they often resemble house flies. Most sarcophagid larvae are scavengers but some are predators, parasitoids of other insects, or develop in vertebrate skin. Members of the genus *Wohlfahrtia* cause obligate myiasis in humans and other mammals.

Order Siphonaptera

Siphon, a tube; *aptera,* wingless; Aphaniptera; fleas (figures 13.14 and 13.15).

Body characteristics minute to small; adults: mostly hard-bodied; bilaterally compressed; many posteriorly directed spines; larvae: vermiform; pupae: exarate in silken cocoons.

Mouthparts adults: piercing-sucking; larvae: chewing.

Eyes and ocelli compound eyes usually lacking; 2 lateral ocelli or eyeless.

Antennae short; contained within antennal grooves.

Wings lacking; all fleas are secondarily apterous.

Legs coxae modified for jumping in most species; 5-segmented tarsi.

Abdomen male terminalia highly modified (male flea sexual organs morphologically constitute the most complex genitalia in the Animal Kingdom) with 1 pair of 2-segmented claspers; cerci lacking.

Fleas are ectoparasites of warm-blooded (endothermic) animals (mammals and birds) although under laboratory conditions some species can be induced to feed on reptiles. The bilaterally compressed body and posteriorly directed spines of most fleas facilitate movement between the hairs or feathers of the host. Although most fleas are excellent jumpers, the primary mode of locomotion is walking or running. Some 'nest' fleas, such as members of the genera *Conorhinopsylla*, *Megarthroglossus,* and *Wenzella,* have poorly developed or nonexistent jumping abilities. Fleas have retained their ancestral wing-hinge ligaments which have been displaced mid-laterally as a consequence of the bilateral compression of the body. These ligaments, in concert with direct flight muscles (subalars and basalars), leg muscles (tergotrochanteral depressors), and an elastic protein (resilin), situated in the pleural arch, all function to execute a jump. Essentially, compressed resilin pads are suddenly released at takeoff by muscle relaxation. Energy from the resilin pads is transferred to the hindlegs and the flea is catapulted into the air with an acceleration of up to 200 gravities so that 100 cm (over 3 feet) can be traveled in 0.002 second.

Figure 13.14 Siphonaptera, a flea. (Mi.)

After U.S. Public Health Service.

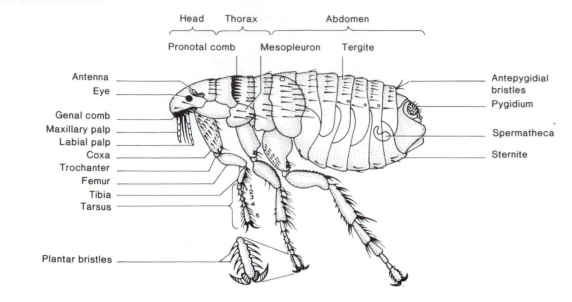

Figure 13.15 Life stages of a flea. (*a*) Egg. (*b*) Larva. (*c*) Pupa. (*d*) Adult.

After U.S. Public Health Service.

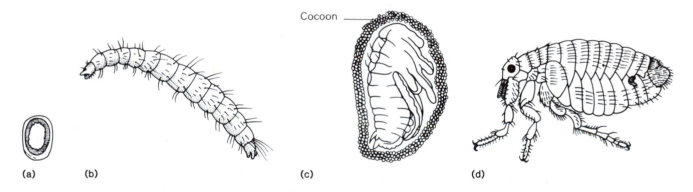

Female fleas generally require a blood meal before they are capable of developing eggs. The eggs (figure 13.15*a*) may or may not be deposited on the host, but ultimately they typically fall to the resting place of the host. The vermiform larvae (figure 13.15*b*) feed on organic detritus; some feed on partially digested blood voided in the feces of adults. Larvae of a few species have unusual feeding behaviors; those of *Uropsylla tasmanica* are subdermal parasites, whereas larvae of *Euhoplopsyllus* are ectoparasites, and those of *Tunga monositis* do not feed at all. Larvae are whitish in appearance, quite active, and pupate within silken cocoons (figure 13.15*c*). The pupae (figure 13.15*c*) are exarate. Adults are generally able to remain quiescent in the cocoon for long periods of time and exit in response to vibrations of the substrate and other stimuli associated with a potential host. It is common for persons with pets to return from vacation and be attacked by large numbers of fleas that have been quiescent within their pupal cocoons. This behavior decreases the probability of an adult flea becoming active when no potential hosts are available. In addition, most adult fleas can survive prolonged periods of starvation (months or, occasionally, years) under favorable environmental conditions. The life cycle varies from 3 to several weeks depending on the species.

Reproduction by the rabbit fleas *Spilopsyllus cuniculi* and *Cediopsylla simplex* is stimulated by the reproductive hormones present in the blood meal taken from the female host. In this way, these rabbit fleas are able to synchronize their reproduction with that of the host so that the emergence of a generation of adult fleas coincides with the birth of a litter of rabbits, thus ensuring maximal host availability.

Many fleas are morphologically well adapted to survival on their hosts. Host specific fleas, like lice, often exhibit coevolutionary or parallel evolutionary relationships with their hosts. These morphological adaptations are often seen in the positioning and thickness of various setae on the body of the fleas, which can reflect the vestiture of the host thereby serving to enhance attachment. Two sites of specialized thickened setae are arranged into combs (*ctenidia*) in many species of fleas; these are the genal and pronotal combs (figure 13.14). Additional cephalic and abdominal combs are present in some other fleas. These combs are thought to aid in attachment to the host (an alternative hypothesis suggests that they protect flexible joints), and in host-specific fleas the host hairs fit between adjacent ctenidial spines to "lock" into position.

Although fleas are highly specialized insects, they show definite affinities with members of the Diptera and in particular with the Mecoptera.

Fleas can pose problems from several standpoints. Their blood-feeding is irritating to humans as well as to many wild and domestic animals. Although fleas tend to show host preferences, most species will feed on a variety of available hosts. *Pulex irritans,* for example, is often referred to as the human flea, but it has many other hosts including pigs and carnivores. Other well-known fleas with wide geographical distributions include *Echidnophaga gallinacea,* the sticktight flea of poultry; *Ctenocephalides canis* and *C. felis,* the dog and cat fleas, respectively; *Xenopsylla cheopis,* the Oriental rat flea; *Nosopsyllus fasciatus,* the northern rat flea; and *Leptopsylla segnis,* the house mouse flea. Cat fleas are often a significant pest problem and are common not only on cats but also on dogs (dog fleas are currently rare in many regions), people, and many domestic and wild mammals.

Fleas are implicated in the transmission of the bacteria that cause plague. Plague is principally a disease of wild rodents and the causative bacterium, *Yersinia pestis,* is transmitted from rodent to rodent by fleas. Under certain conditions, humans may be bitten by infected fleas and develop the symptoms of plague. In regions of high population concentration and poor sanitary conditions, the potential may exist for epidemics of this disease. Throughout recorded human history there have been sporadic outbreaks of plague, or the Black Death as it has been called at times (Langer 1964). Fleas are also the primary vectors of murine (endemic) typhus caused by *Rickettsia typhi.* As with plague, wild rodents are the reservoir hosts for murine typhus. Both plague and murine typhus exist in nature as scattered foci around the world and, under certain conditions, these foci may change in intensity or shift geographically sometimes resulting in epidemics. Fleas also play a role in the spread of tularemia, listeriosis, and some other pathogenic diseases of humans, other mammals, and birds. North American flying squirrel fleas (*Orchopeas howardi*) may be involved in the transmission of the agent that causes sporadic epidemic typhus (see section on Anoplura in chapter 12). Some fleas, in addition to other arthropods, serve as intermediate hosts of the tapeworms *Dipylidium caninum* and *Hymenolepis diminuta,* both of which can par-

asitize humans if infested fleas are accidentally ingested. Dog-associated fleas are the principal intermediate hosts of a parasitic nematode (*Dipetalonema*) that parasitizes these domestic animals.

One species of flea, *Tunga penetrans,* the chigoe, is purported to burrow into the skin. These fleas do not burrow, however, but induce tissue changes that cause them to be partially enveloped by host tissues. Only the females "burrow," and they generally attack in the region of the toes, sometimes opening the way for secondary infection. In some instances, amputation of toes has been attributed to this flea. Chigoes are found in tropical and subtropical regions, particularly in Central and South America, the Caribbean islands, and Africa. Other species of *Tunga* are parasites of rodents. For further information on fleas, the following references are recommended: Askew (1973); Ewing and Fox (1943); Fox (1940); Harwood and James (1979); Holland (1964, 1985); Hopkins and Rothschild (1953–1971); Hubbard (1947); Lewis (1972–1975); Mardon (1981); Marshall (1981); Rothschild (1965, 1975); Smit (1973, 1987); Traub (1985); Traub and Starcke (1980); Traub et al. (1983).

Order Trichoptera

Tricho, hair; *ptera,* wing; caddisflies (figures 13.16, 13.17, and 13.18).

Body characteristics small to medium; generally somber colored, usually brownish; larvae: eruciform; pupae: exarate.

Mouthparts adults: some nonfeeding in adult stage; adapted for imbibing fluid; mandibles weakly developed or lacking; larvae: chewing.

Eyes and ocelli compound eyes present; 0 or 3 dorsal ocelli.

Antennae range from setaceous to filiform.

Wings 2 pairs; membranous; hindwings broader than forewings; covered with modified hairlike setae; held rooflike over body at rest.

Almost all caddisflies are aquatic in the egg, larval, and pupal stages (a few live in brackish water). This is the only order of endopterygote insects in which the immature stages are primarily aquatic. Adults are terrestrial and may be found some distance from water. Eggs are deposited in strings or masses near or in the water, on aquatic vegetation, beneath stones, and in similar locations. Most larvae (figures 13.17b and 13.18) construct cases with a silken foundation into which a variety of environmental materials are incorporated. Exactly which environmental materials, and the method by which they are incorporated into the case, is a family or sometimes a generic characteristic. The larvae that inhabit cases have hooked caudal appendages used to maintain a hold within the cases. Larval and pupal ventilation is by means of abdominal gills. The larvae feed on small aquatic plants and

Figure 13.16 Trichoptera, a caddisfly. (S to L.)
Redrawn from Essig, 1942.

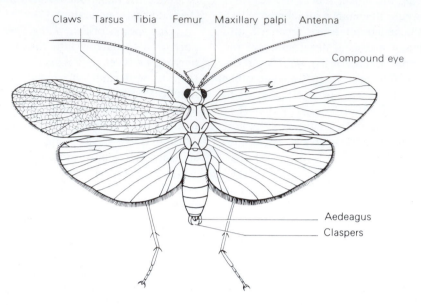

Figure 13.17 Caddisfly, *Rhyacophila fenestra* (Rhyacophilidae). (*a*) Adult in resting posture. (*b*) Larva.
Redrawn from Ross, 1938.

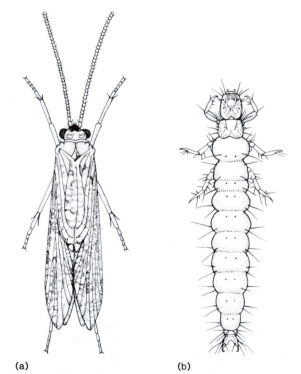

(a)   (b)

Figure 13.18   Examples of caddisfly larval cases. (*a*) *Leptocella*. (*b*) *Triaenodes*. (*c*) *Agrypnia*. (*d*) *Polycentropus*.
Redrawn from Cummins, Miller, Smith, and Fox, 1965.

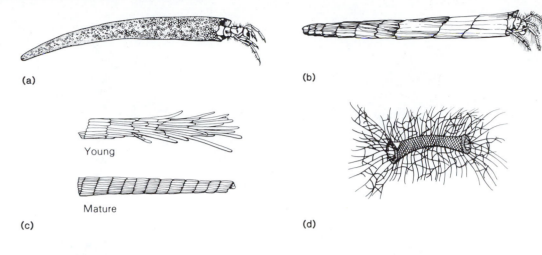

(a)

(b)

Young

Mature

(c)

(d)

animals, and some are important predators of black flies. Caddisflies are important as fish food. Larvae of many species are good bioindicators of aquatic pollution.

The order Trichoptera is usually divided into two suborders (Annulipalpia whose members have an annulated, elongate terminal segment of the maxillary palps, and Integripalpia, which possess a short, non-annulated terminal segment of these palps) and 23–34 families. Family classification is based principally on adult characters of the ocelli, maxillary palps, tibial spurs, and "thoracic warts" (swellings on the dorsal surface of the thorax). Among the larger families are the Hydroptilidae (microcaddisflies), Rhyacophilidae (primitive caddisflies), Phryganeidae (large caddisflies), Limnephilidae (northern caddisflies), Psychomyiidae (trumpet-net and tube-making caddisflies), Hydropsychidae (net-spinning caddisflies), Lepidostomatidae (lepidostomatids), and Leptoceridae (long-horned caddisflies). Further information on caddisflies can be found in the following references: Denning (1956); Pennak (1978); Ross (1944, 1959); Wiggins (1978).

Order Lepidoptera

Lepido, scale; *ptera,* wing; butterflies, moths, and skippers (figures 13.19, 13.20 and 13.21).

Body characteristics minute to very large; usually covered with scales; much diversity in color and color pattern; larvae: eruciform; pupae: usually obtect.

Mouthparts adults: usually long, sucking proboscis (the only exception to this is in one primitive family, whose members have mandibulate mouthparts); vestigial in some; larvae: strong mandibles, chewing.

Eyes and ocelli large compound eyes; 0 or 2 dorsal ocelli; larvae: clusters of lateral ocelli (stemmata).

Antennae prominent; vary in form.

Wings 2 pairs; membranous; covered with scales; large in proportion to body; few cross veins; a few wingless; forewings generally larger than hindwings.

Legs 5-segmented tarsi; larvae: 3 pairs of true legs.

Abdomen cerci lacking; larvae usually with 5 pairs of prolegs (false legs) on segments 3 through 6 and 10; apex of each proleg with tiny hooks (crochets).

Comments members in several families have organs of hearing (tympanal organs).

Lepidoptera is the second largest order of insects. It can be divided into suborders, Frenatae (or Heteroneura) and Jugatae (or Homoneura) on the basis of wing-coupling mechanisms. In the Frenatae the forewings and hindwings are coupled by a spine or a group of spines, the *frenulum* (see figure 2.40*b*), at the anterior base of the hindwing (which usually catches into a process, the *retinaculum* on the forewing) or by an enlarged area along the anterior base of the hindwing (humeral angle). The hindwings are smaller and have fewer veins than the forewings, hence the alternative name, Heteroneura ("mixed nerves"). The wings of the Jugatae are coupled by the jugum (see figure 2.40*c*), a projection from the base of the posterior edge of the forewing. The venation is similar in both forewings and hindwings, hence the alternative name Homoneura ("same nerves"). The vast majority of Lepidoptera are in the suborder Frenatae, the Jugatae being a small group of uncommon insects.

The suborder Frenatae may be further divided into two major groups: the Macrolepidoptera and the Microlepidoptera. The Macrolepidoptera are a diverse group of butterflies (figure 13.20*a*), skippers (figure 13.20*b*), and moths (figure 13.20*c*) of variable size, but usually with a wingspan of more than 2.5 cm. The Microlepidoptera are a large group of moths, most of which are much smaller

Figure 13.19 Stages in the life cycle of the monarch butterfly, *Danaus plexippus*. (*a*) Egg. (*b*) Larva. (*c*) Pupa (chrysalis). (*d*) Adult. (L.)
Redrawn from Boreal Labs key card.

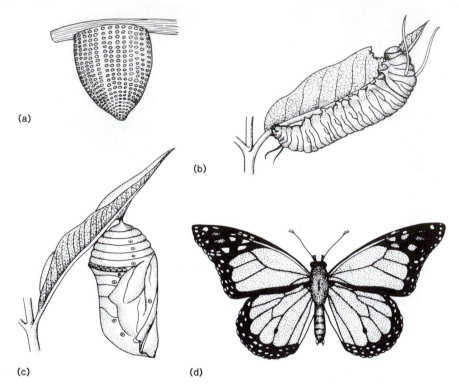

(a)

(b)

(c)

(d)

Figure 13.20 (*a*) Butterfly. (*b*) Skipper. (*c*) Moth.
Redrawn from U.S. Department of Agriculture, 1952.

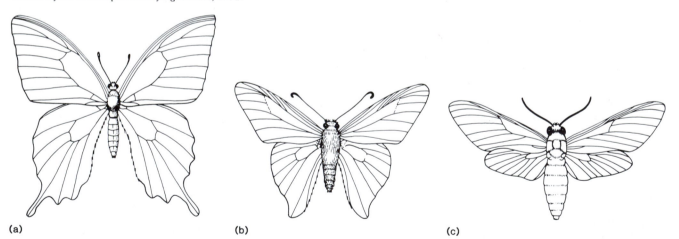

(a)

(b)

(c)

than Macrolepidoptera, usually with a wingspan of 20 mm or less. Wing venation and other characters are used in further subdivision of these two groups.

Some authors, however, divide the Lepidoptera into the suborders Rhopalocera (butterflies and skippers) and Heterocera (moths). The Rhopalocera are characterized by being typically diurnally active; holding the wings vertically at rest; having clubbed, or clublike, long unbranched antennae; and usually having a frenulum. The Heterocera, however, are typically nocturnal; hold the wings flat or rooflike over the body at rest; have variable antennae (often feathery but usually not clubbed); and most species have a frenulum, but some (the Jugatae) have a jugum to couple the wings. There are exceptions to most of these characteristics; for example, many skippers hold their wings horizontally when at rest, and there are numerous species of day-flying moths (notably, certain members of the families Zygaenidae, Arctiidae, Sesiidae and Uraniidae).

Many lepidopteran larvae (caterpillars) have repugnatorial glands of various sorts. The lobelike eversible *osmeteria,* located on the dorsum of the prothorax of papilionid (swallowtail butterflies) larvae, are examples of such glands. Caterpillars have the ability to produce silk. This substance is secreted by large labial silk glands, which open via a common duct into the highly modified labium.

Pupae (figure 13.19c) are generally obtect, although some are exarate, and are often encased in some sort of external protective structure constructed by the last instar larva. This protective structure may, for example, be a silken cocoon, with or without various kinds of detritus incorporated into it; detritus held together by a sticky secretion that hardens into a foundation matrix; a cell in the soil; or a tentlike structure made by "tying" edges of a leaf together. In many, no protective structure is constructed and they must depend upon various kinds of protective coloration and patterning. Such is the case with most butterflies that attach themselves to twigs or other objects by means of a silken thread and/or a silken patch at the tip of a projection from the posterior part of the body, the *cremaster.* These naked pupae are commonly referred to as *chrysalids.*

Although the vast majority of Lepidoptera are terrestrial, certain members of several families (Arctiidae, tiger and footman moths; Pyralidae, pyralid moths; and others) have been found in aquatic environments as larvae and sometimes as adults.

Most Lepidoptera feed on flowering plants and are particularly voracious in the larval stages. Some notable exceptions to phytophagy include certain Hawaiian geometrid moth larvae in the genus *Eupithecia* which are predaceous, and clothes-moth larvae (family Tineidae) which feed on woollen products and are capable of digesting keratin. Most adults are capable of obtaining only fluid meals from flowers and other sources and are unable to masticate plant material. Their plant-feeding habits make several species among the most important agricultural pests. Many attack stored grain and some attack fiber in clothing. Most pest species are moths that may attack various crops. A few lepidopteran species, particularly the silkworm moth, *Bombyx mori* (see chapter 14), are highly valued insects.

A few species are of medical or veterinary importance (Harwood and James 1979). For example, certain noctuid moths in Africa feed on the lachrymal secretions of cattle and may be the vectors of the etiological agent of infectious keratitis, a disease that may cause blindness. Some species have been observed to feed in the eyes of humans. Incredibly, a few species of moths in Malaysia, Thailand, and Laos are blood-feeders (Bänziger 1971, 1975). One of these moths, the noctuid *Calyptra eustrigata* is capable of piercing the intact skin of tapirs and wild buffalo to obtain blood. The caterpillars of certain members of at least 10 families possess urticating hairs, which may contain a toxin and can cause severe blistering and even blindness if they get into the eyes.

A brief synopsis of some of the major lepidopteran families follows. Although family classification is largely based on wing venation, veins are not always easy to decipher and distinctive wing coloration and markings are often easier to follow. For further information on this order, the following references should be consulted: Chu (1949); Common (1975); Covell (1984); Ehrlich and Ehrlich (1961); Emmel (1975); Holland (1931, 1968); Holloway et al. (1987); Howe (1975); Klots (1951, 1958); Mitchell and Zim (1964); Peterson (1948); Sargent (1976); Scott (1986); Dominick (1971 to present); Tietz (1973); Urquhart (1960); Vane-Wright and Ackery (1984); Watson and Whalley (1975).

Suborder Frenatae—Macrolepidoptera Rhopalocera (butterflies and skippers) *Family Papilionidae (swallowtails)* (figure 13.20a). The hindwings of these large butterflies typically are developed into tails. The radial vein on the forewing has 5 branches. The tiger swallowtail (*Pterourus glaucus*) is a common species in eastern North America.

Family Pieridae (whites, sulphurs, yellows, orange tips, etc.). These are small to large butterflies typically with white, yellow, orange, and occasionally black markings. The hindwing has 2 anal veins. Larvae of *Artogeia* (=*Pieris*) *rapae* are pests of cabbage and related crops. This species is known as the "small white" in Europe, but it has been introduced to North America where it is referred to as the cabbage butterfly or the imported cabbageworm. Larvae of *Colias eurytheme* (the alfalfa butterfly) and *Colias philodice* (the common sulphur) consume clovers and legumes and sometimes reach pest proportions in North America.

Figure 13.21 Representative lepidopteran families. (*a*) Sphingidae (*Deilephila elpenor*). (*b*) Nymphalidae (*Aglais urticae*). (*c*) Lycaenidae (*Polyommatus icarus*). (*d*) Saturniidae (*Saturnia pavonia*). (*e*) Geometridae (*Cleora rhomoidaria*). (*f*) Arctiidae (*Arctia caja*). (*g*) Noctuidae (*Noctua janthina*) (S to VL).
After Smart, 1962.

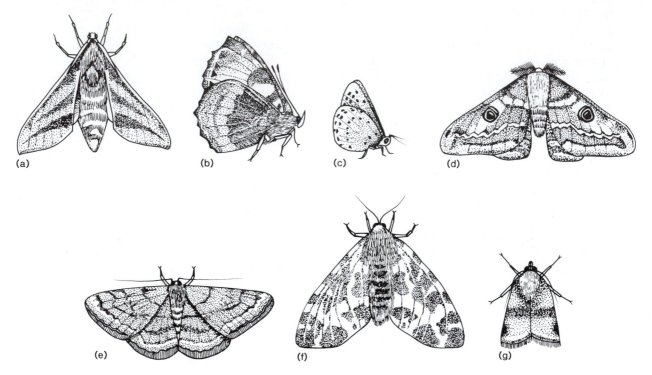

Family Lycaenidae (gossamer-winged butterflies) (figure 13.21*c*). This is a large family of generally small butterflies that includes the blues, coppers, harvesters, hairstreaks, and elfins. The hindwing lacks a humeral vein and the forelegs of males are typically reduced. Larvae of most species are sluglike and some complete their development inside ant nests.

Family Nymphalidae (brush-footed butterflies) (figure 13.21*b*). This is the largest family of butterflies and, as currently recognized, it includes the satyrines (formerly recognized as a distinct family, the Satyridae) (nymphs, satyrs, browns, heaths and arctics), heliconiines (long-wings), as well as the nymphalines (tortoiseshells, admirals, fritillaries, and relatives). The forelegs are greatly reduced.

Family Danaidae (milkweed butterflies) (figure 13.19*a–d*). These butterflies are brightly-colored. Larvae feed on milkweeds and accumulate toxic chemicals from these plants as a deterrent to potential predators. The monarch, *Danaus plexippus* (figure 13.19*a–d*) is a common North American insect but it also occurs in many other parts of the world, largely because of its renowned abilities of long-distance migration.

Family Hesperiidae (skippers) (figure 13.20*b*). Although these diverse, small- to medium-sized insects are frequently considered to be butterflies, many lepidopterists regard them as a separate group distinct from both moths and true butterflies. Skippers have large heads (at least as wide as the thorax) and typically possess 2 pairs of tibial spurs on the hindlegs.

Suborder Frenatae—Heterocera (moths) *Family Sphingidae (hawk or sphinx moths)* (figure 13.21*a*). These are medium to very large insects with robust bodies, relatively short wings, and thickened antennae. Adults are strong fliers and have a rapid wingbeat frequency; they often hover above flowers as they collect nectar. Most larvae possess a soft hornlike spine at the posterior end of the body and may rear up to adopt a sphinxlike pose when disturbed. Many larvae are called hornworms, and some of these can become pests, for example, the tobacco hornworm (*Manduca sexta*).

Family Saturniidae (giant silkmoths or silkworm moths) (figure 13.21*d*). With respect to total wing area, this family includes the largest known extant insects. Members of the Southeast Asian genus *Attacus* (the atlas moths) may have a wingspan of more than 25 cm, and *Coscinocera hercules* from New Guinea and Australia is even larger. Saturniids typically have broad wings that frequently are patterned with eyespots or transparent patches (figure 13.21*d*), atrophied mouthparts (adults do not feed), and only 1 anal vein on the hindwing. Pupation is inside a silken cocoon spun by the final instar larva. Examples include the luna moth (*Actias luna*) and the largest North American lepidopteran the cecropia moth (*Hyalophora cecropia*). The emperor moth (*Saturnia pavonia*) (figure 13.21*d*) is widespread in Europe.

Family Bombycidae (silkworm moths). These are stout-bodied, short-winged, white moths with poor flying abilities. The commercial silkworm (*Bombyx mori*) has been cultured for millenia, particularly in China, to obtain the high-quality silk spun by its larvae (see chapter 14).

Family Geometridae (geometer or inchworm moths) (figure 13.21*e*). Members of this large, widespread group range from minute to large in size. They have slender bodies and the subcostal vein of the hindwing has an abrupt basal angle. Larvae are slender and some mimic twigs; their looping movements have given rise to the names inchworms and measuring worms. Adults usually hold the wings horizontally at rest (figure 13.21*e*).

Family Lasiocampidae (lappet moths and tent caterpillars). These medium- to large-sized moths are hirsute and lack a frenulum (the humeral angle of the hindwing is expanded to articulate with the forewing). Antennae are plumose in many species. Both the eastern tent caterpillar (*Malacosoma americanum*) and the forest tent caterpillar (*M. disstria*) are gregarious and construct silken "tents"; in North America, they often completely defoliate fruit and other trees. Many lasiocampid larvae have urticating setae.

Family Arctiidae (tiger and footman moths) (figure 13.21*f*). These moths, many of which are diurnal, are small to medium in size and often are brightly banded or spotted. Many species are toxic to predators and "advertise" this through their colorful patterns (*aposematism*; see chapter 9). Larvae are usually hirsute and some are known as "woolly bears." Setae are urticatory in some larvae.

Family Noctuidae (noctuid moths) (figure 13.21*g*). This is the largest family in the order Lepidoptera; its members are diverse and widely distributed. Subcostal and radial veins on the hindwing are fused for a short distance, antennae are filiform, and the labial palps are elongate. Larvae of some noctuids are pests, for example, the corn earworm [*Helicoverpa* (formerly *Heliothis*) *zea*], the beet armyworm (*Spodoptera exiqua*), and the rose budworm (*Pyrrhia umbra*).

Suborder Frenatae—Microlepidoptera *Family Sesiidae (= Aegeridae) (clearwing moths).* These moths are wasplike in appearance with extensive wing areas devoid of scales. Forewings are long and narrow and adults of most species are diurnal. Larvae bore into roots, stems, trunks, etc. Many are pests, for example, the grape root borer (*Vitacea polistiformis*) and the strawberry moth (*Ramosia bibionipennis*).

Family Cossidae (leopard, carpenter, and goat moths). Although cossids are classified as microlepidopterans, some species are very large (certain tropical species have wingspans of more than 20 cm). These are stout-bodied moths with mottled or spotted wings with 2 complete anal veins and an accessory cell on the forewing. Larvae bore into wood and some species can damage trees [e.g., the carpenterworm (*Prionoxystus robiniae*)].

Family Limacodidae (slug caterpillar moths). As indicated by their vernacular name, the larvae of these small- to medium-sized moths are sluglike; many of them have stinging setae. Adults are stout-bodied with broad, rounded wings that are usually marked in green or silver. Two complete anal veins are present on the forewing.

Family Pyralidae (pyralid moths). Pyralids are minute or small with an elongate-triangular forewing and a broad, rounded hindwing. The labial palps are large and often protrude anteriorly. This is the largest microlepidopteran family, and its members are widespread and common. Larvae have diverse feeding habits and some are pests, for example, the European corn borer (*Ostrinia nubilalis*), the garden webworm (*Loxostege similalis*), and the alfalfa webworm (*L. commixtalis*).

Family Psychidae (bagworm moths). The females of these aberrant moths usually lack wings and may also be devoid of legs, antennae, and eyes. Mouthparts are vestigial. Males are winged and fly to the essentially immobile females to mate. Larvae construct portable "bags" consisting of small pieces of leaves and twigs, in which they eventually pupate. Females usually remain inside their bags after eclosion and also mate and oviposit on these structures. Some species damage trees, [e.g., the evergreen bagworm (*Thyridopteryx ephemeraeformis*)].

Family Olethreutidae (olethreutid moths). These are minute or small, brown or gray moths. The forewing is square-tipped and the upper surface of the cubital vein on the hindwing is typically fringed with long setae. Many larvae are pestiferous, for example, the Oriental fruit moth (*Grapholitha molesta*) and the cherry fruit moth (*G. packardi*).

Family Tortricidae (tortricid moths). Tortricids are similar to olethreutids but lack a fringe on the cubital hindwing vein. Most larvae roll or tie leaves together with silk and feed inside these structures. A significant North American pest species is the spruce budworm (*Choristoneura fumiferana*).

Family Tineidae (tineid and clothes moths). The forewings and hindwings of these minute or small moths are about equal in size. The head is bristly or has rough scales and maxillary palps are usually present but are folded at rest. Although larval habits are diverse, three species (the clothes moths) are exceptional in that they feed on woollen materials (*Tineola bisselliella, Tinea pellionella,* and *Trichophaga tapetzella*).

Figure 13.22 External features of Hymenoptera. An ichneumon wasp (Ichneumonidae). *Redrawn from Essig, 1958.*

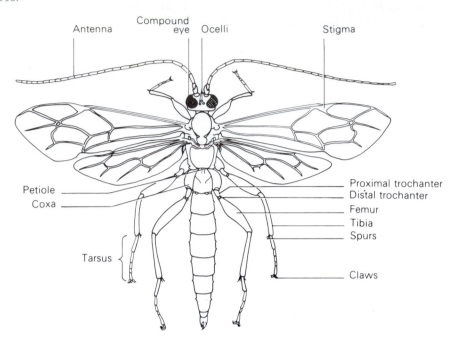

Suborder Jugatae *Family Micropterygidae (mandibulate moths).* These primitive insects are the only lepidopterans in which adult mouthparts are mandibulate. These mandibles are chiefly used to eat pollen. Larvae apparently feed on mosses or liverworts.

Family Hepialidae (ghost or swift moths). These moths range from small to very large in size. The maxillary palps are well developed and tibial spurs are lacking. Larvae bore into the roots of various species of trees.

Hymenopteroids

The single order, Hymenoptera (figure 13.22), is a highly specialized group, and there has been much disagreement regarding evolutionary relationships with other orders.

Order Hymenoptera

Hymen, membrane; *ptera,* wing; ants, bees, wasps, sawflies, horntails, and relatives (figures 13.22, 13.23, and 13.24).

Body characteristics minute to very large; larvae: usually legless with distinct head; some (suborder Symphyta) eruciform with legs and prolegs (which lack crochets); pupae: exarate, usually within a cocoon.

Mouthparts variable; chewing to lapping or sucking.

Eyes and ocelli compound eyes usually well developed; commonly 3 dorsal ocelli, but lacking in some.

Antennae commonly long and filiform or geniculate with elbowed scape and distal segments clubbed.

Wings most with 2 pairs of membranous wings; hindwings smaller than forewings; many wingless.

Legs usually 5-segmented tarsi.

Abdomen in the suborder Apocrita, the first abdominal segment (*propodeum*) is fused with the metathorax and is constricted posteriorly, forming the *petiole* between the thorax and abdomen; members of the suborder Symphyta lack a petiole; females with ovipositor or homolog modified for piercing plant tissues, sawing, or stinging.

On the basis of complexity and diversity of behavior the Hymenoptera are generally recognized as the most advanced group of insects.

Hymenoptera can be divided into two suborders, Symphyta (Chalastogastra) and Apocrita (Clistogastra). The most obvious characteristic separating these two suborders is the relationship between the thorax and abdomen. The Symphyta (figure 13.23) lack the petiole that is characteristic of the Apocrita (figure 13.24). Other differences include the following: the Symphyta are less behaviorally sophisticated than the Apocrita, parasitic forms are almost nonexistent, the ovipositor is usually fitted for sawing or piercing plant tissues, and the larvae are eruciform. Among the Apocrita are many parasitoid and predatory forms, and the ovipositor is specialized for piercing and/or stinging. The larvae are legless, and a distinct head is lacking in some of the parasitoids.

The habits of Hymenoptera are very diverse and represent nearly every mode of insectan life. The ants (Formicidae); paper wasps and relatives (Vespidae); one species of sphecid wasp, *Microstigmus comes* (Sphecidae; Matthews 1968); and many bees (superfamily Apoidea, e.g., Apidae) have elaborate patterns of social behavior (see chapter 8). Many of these insects, as well as many other Hymenoptera, are pollen-feeders and are important plant pollinators (see chapters 10 and 14). A large number of Hymenoptera are parasitoids (see chapters 8 and 9), for example, all members of the superfamilies Ichneumonoidea and Proctotrupoidea and most members of the Chalcidoidea. Members of the superfamilies Sphecoidea and Vespoidea are predatory. Most Symphyta are herbivorous, some being of major economic importance. Some members of the superfamily Cynipoidea induce plants to form galls (see chapter 10). A few Hymenoptera are found in aquatic habitats (Hagen 1978).

Hypermetamorphosis is common in certain families. Parthenogenesis is also very common, and parthenogenetic generations may alternate with sexual generations. In some groups parthenogenesis plays a role in sex determination; in the honey bee, *Apis mellifera*, and other social Hymenoptera, males (drones) develop from unfertilized eggs, whereas queens and workers (all females) develop from fertilized eggs (see chapter 5). Polyembryony, the development of more than one individual from one egg, has been reported in a few parasitoid species. Hymenopteran larvae are usually legless and generally have distinct heads (Apocrita). Some are eruciform and have thoracic legs and abdominal prolegs that lack the tiny hooks or crochets characteristic of lepidopteran larvae (Symphyta). The pupae are exarate and typically within a cocoon or waxen or earthen cell.

The members of this order are certainly more beneficial than harmful, since many are involved in plant pollination; are parasitoids or predatory on other, often harmful, insects; and produce useful products (e.g., honey and beeswax). However, there are several species that are destructive pests. Most of these are sawflies in the families Diprionidae (conifer sawflies), Tenthredinidae (common sawflies), and Cephidae (stem sawflies). The larvae of common sawflies feed on foliage of trees and can be very damaging. The larvae of stem sawflies bore into the stems of grasses and berries, and two species are important pests of wheat. Ants (Formicidae) may be a problem when they invade households and infest foodstuffs.

Although no hymenopteran insects serve as vectors of disease-causing microorganisms, many are capable of inflicting a painful sting that can lead to serious consequences if the person stung happens to be hypersensitive to the injected venom. Stinging Hymenoptera are typically social species and occur in the following seven superfamilies (Harwood and James 1979): (1) Chrysidoidea—cuckoo wasps and relatives; (2) Bethyloidea—bethyloid wasps; (3) Scolioidea—scoliid wasps, ants, velvet ants, and relatives; (4) Formicoldea—the ants; (5) Vespoidea—hornets, yellowjackets, spider wasps, and relatives; (6) Sphecoidea—sphecoid wasps; and (7) Apoidea—bees.

A synopsis of the major hymenopteran families follows. Principal characters used to separate families within this order are those of wing venation, antennae, legs, ovipositor, pronotum, and thoracic sutures. Some authors prefer to segregate the Hymenoptera by superfamilies rather than by families. For further information on this order of insects, the following references should be consulted: Andrews (1971); Askew (1973); Bohart and Menke (1976); Clausen (1940); Creighton (1950); Evans (1963); Evans and Eberhard (1970); von Frisch (1967); Hagen (1978); Hölldobler and Wilson (1990); Huber and Goulet (1992); Michener (1974); Mitchell (1960, 1962); Muesebeck et al. (1951–1967); Spradberry (1973); Wilson (1971, 1975); Yarrow (1973).

Suborder Symphyta *Families Tenthredinidae (common sawflies), Diprionidae (conifer sawflies), and Cephidae (stem sawflies)* (figure 13.23a). These families, in addition to a few others that are less well-known, constitute the superfamily Tenthredinoidea. In addition to the broad junction between the thorax and abdomen, these insects have 2 apical spurs on the fore-tibiae. As previously mentioned, larvae of some species defoliate crops [e.g., the gooseberry sawfly (*Nematus ribesi*)].

Families Siricidae (horntails), and Xiphydriidae (wood wasps) (figure 13.23b). These two families and some smaller ones, make up the superfamily Siricoidea. Adults are similar to tenthredinoids, but the fore-tibiae possess 1 apical spur, and females of most species have an extended ovipositor. Larvae typically bore into wood and some are pests.

Suborder Apocrita *Families Braconidae (braconids), Ichneumonidae (ichneumonids) (figures 13.22 and 13.24a), and Stephanidae (stephanids).* These three families make up the large and widespread superfamily Ichneumonoidea. The antennae are filiform with at least 16 segments, whereas hind trochanters are 2-segmented (figure 13.22) and the ovipositor arises anterior to the tip of the abdomen (it cannot be withdrawn and often is longer than the body). Larvae are parasitoids of insects and spiders and certain species are biological control agents of noxious insects [e.g., *Apanteles glomeratus* (Braconidae), which attacks larvae of cabbage butterflies].

Families Mymaridae (fairy flies), Trichogrammatidae (trichogrammatids), Eulophidae (eulophids), Encyrtidae (encyrtids), Chalcididae (chalcids) (figure 13.24b), Torymidae (torymids), Agaonidae (fig wasps),

Figure 13.23 Hymenoptera, Symphyta. (*a*) Sawfly (Tenthredinoidea). (*b*) Raspberry horntail, *Hartigia cressoni* (Siricoidea). (Me.)

(a) redrawn from U.S. Department of Agriculture, 1952; (b) redrawn from Essig, 1958.

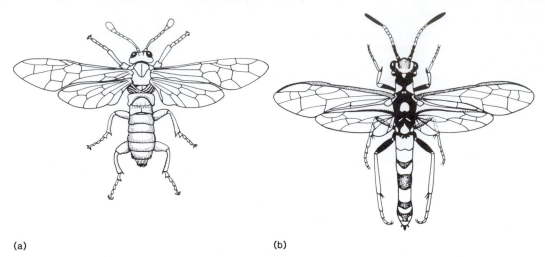

(a) (b)

Pteromalidae (pteromalids), and Eurytomidae (eurytomids or seed chalcids). These families are representatives of the large superfamily Chalcidoidea. These wasps are minute with a squarish pronotum and short, 5–13 segmented, elbowed antennae. At least one pair of trochanters is 2-segmented, and wing venation is significantly reduced. Larvae of most species are parasitoids or hyperparasites of other insects, but a few are phytophagous. Many of the former are useful in the biological control of pests. Fig wasps live symbiotically with fig trees, which they pollinate.

Family Cynipidae (gall wasps). Gall wasps (superfamily Cynipoidea) are "hump-backed" with filiform 13–16 segmented antennae. Trochanters are 1-segmented and the ovipositor arises anterior to the apex of the abdomen. Most larvae are associated with the formation of characteristic galls in various plants, particularly trees, but some are parasitoids of other insects.

Family Pelecinidae (pelecinids). This family is the sole representative of the superfamily Pelecinoidea. Although just one species, *Pelecinus polyturator,* occurs in North America, females are large (body length, ca. 5 cm) with short wings and a highly distinctive elongate, slender, abdomen. Pelecinid larvae are parasitoids of beetle larvae.

Families Scelionidae (scelionids) and Platygasteridae (platygasterids). These minute wasps are assigned to the superfamily Proctotrupoidea, the larval stages of which are all parasitoids of other insects. Antennae are elbowed, 7–12 segmented, and the abdomen is partially flattened dorso-ventrally. Some scelionids and platygasterids are highly important biocontrol agents of various crop pests.

Family Mutillidae (velvet ants) (figure 13.24*d*). Members of this family and the Formicidae are representatives of the superfamily Scolioidea. Adult velvet ants are hirsute and often brightly-colored. Males are winged, but females are wingless. Females are capable of inflicting very painful stings. Larvae are parasitoids of other insect larvae particularly those of ground-nesting bees and wasps.

Family Formicidae (ants) (figure 13.24*e*). Ants are common, almost ubiquitous insects. The first one or two abdominal segments are nodelike and have a dorsal protuberance. Antennae are elbowed and 6–13 segmented with an elongate first segment. All species are social with various forms of caste differentiation. Some species are carnivorous, whereas others are scavengers, herbivores, or omnivores. Most species can bite and/or sting and some are capable of emitting deterrent chemicals (e.g., formic acid). A wide spectrum of ecological niches is filled by various species of ants. A few species are considered to be pests, for example, the black carpenter ant (*Camponotus pennsylvanicus*), which can damage wooden structures, and the imported fire ant (*Solenopsis invicta*), an aggressive species that is capable of inflicting painful stings. For a wealth of information on ants, reference should be made to Hölldobler and Wilson (1990).

Family Vespidae (vespid wasps) (figure 13.24*f*). This family constitutes the major portion of the superfamily Vespoidea and includes the potter wasps, paper wasps, yellowjackets, and hornets. Stings of the last three are well-known and often necessitate prompt medical attention in cases of multiple stings or allergic reactions. Antennae are filiform, 12-segmented in females and

Figure 13.24 Hymenoptera, Apocrita. (*a*) Ichneumon wasp (Ichneumonoidea; Ichneumonidae). (*b*) Chalcid wasp (Chalcidoidea; Chalcididae). (*c*) California gallfly, *Andricus californicus* (Cynipoidea; Cynipidae). (*d*) Velvet ant (Scolioidea; Mutillidae). (*e*) Thief ant, *Solenopsis molesta* (Scolioidea; Formicidae). (*f*) Golden polistes, *Polistes aurifer* (Vespoidea; Vespidae). (*g*) Thread-waisted wasp (Sphecoidea; Sphecidae). (*h*) Honey bee, *Apis mellifera* (Apoidea; Apidae). (*a*), (*f*), (*g*), and (*h*), Me; (*b*), (*c*), and (*e*), Mi; (*d*), S to Me.

(*a*), (*b*), (*g*), and (*h*) redrawn from U.S. Department of Agriculture, 1952; (*c*) and (*f*) redrawn from Essig, 1958, (*d*) and (*e*) after U.S. Public Health Service.

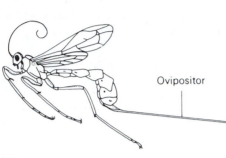
Ovipositor
(a)

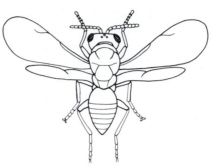
(b)

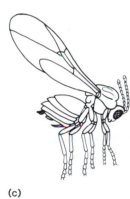
(c)

(d)

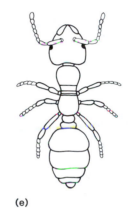
(e)

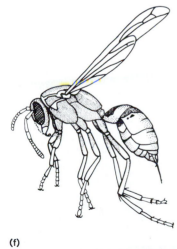
(f)

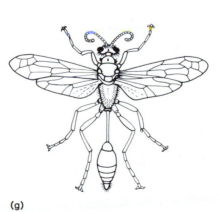
(g)

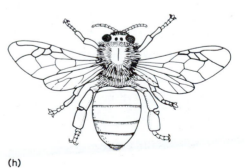
(h)

13-segmented in males. Trochanters are 1-segmented and the ovipositor, which also functions as a stinger, is apical. Many species are social and some construct large colonial paper nests.

Family Sphecidae (sphecid or thread-waisted wasps) (figure 13.24*g*). This is the principal family of the superfamily Sphecoidea. The short pronotum is collarlike and has a small lobe on each side. Sphecoids are solitary wasps and typically construct ground nests, provisioning these with characteristic invertebrate prey that is paralyzed (by stinging) prior to oviposition.

Family Megachilidae (leaf-cutter bees). This family and the next one are representatives of the large superfamily Apoidea. Megachilids are mostly stout-bodied and small to medium in size. The forewing has 2 submarginal cells that are approximately equal in size. These bees are common and usually nest in cavities or in the ground, provisioning their brood cells with severed leaves prior to oviposition.

Family Apidae (honey bees, bumble bees, carpenter bees, digger bees, etc.) (figure 13.24*h*). Apids have the first two segments of the labial palps flattened and elongate, whereas the glossa is long and slender. There are typically 3 submarginal cells on the hindwing. This group is extremely varied in size, morphology, and biology. Some species are social, and many can inflict painful stings. Large carpenter bees (*Xylocopa* spp.) excavate nests in wood, and sometimes are destructive to buildings. Carpenter bees are placed in a separate family, the Xylocopidae, by some authors. The honey bee (*Apis mellifera*) is renowned for its production of honey and wax as well as its importance in crop pollination.

Fossil Neopterous Endopterygotes

The Glosselytrodea (see figure 11.14) are the only extinct neopterous endopterygote insects that meet Carpenter's criteria (see chapter 11) for recognition as an order. The pattern of setae on the wings and appearance of head and thorax are similar to the Neuroptera, but the venation is more specialized and the forewings somewhat hardened.

PART FOUR

Applied Entomology

This section deals with the pragmatic aspects of the science of entomology. This is the point at which the results of both basic and applied research are brought to bear on specific problems. The approach will be first to discuss insects that are directly or indirectly beneficial and then to consider how insects are problems for human beings (chapter 14). In chapter 15, we discuss the many ways modern science has enabled us to fight back against the insects that, for a variety of reasons, we view as enemies. In chapter 16, we show, in the context of agroecosystems, how the various tools of insect control may be used in concert to achieve satisfactory pest insect population levels with economic efficiency and with minimal environmental damage.

CHAPTER 14

Beneficial and Harmful Insects

Beneficial Insects

Although many insects are pests and often cause serious problems, the vast majority are not, and in the overall scheme of nature must be considered to be very important animals. In this section we examine several ways in which insects are directly or indirectly beneficial.

Insect Products

Honey

Honey (figure 14.1) is a highly nutritive liquid material prepared from flower nectar by several species of bees. It is a sticky, viscous material that ranges from nearly colorless to dark amber. It is composed mainly of water and several sugars (levulose and dextrose in particular) with small amounts of various other substances, including fatty acids, proteins, vitamins, and minerals. The major producer of honey is the honey bee, *Apis mellifera,* which has been domesticated and is maintained in artificial hive containers throughout the world. Maintenance of honey bees for the purpose of harvesting honey, wax, and other products is called *apiculture,* or beekeeping.

Honey has been sought out and collected by humans for thousands of years and has found use mainly as a food material. Meads, beverages similar to wines, are prepared from honey and were possibly among the earliest alcoholic drinks (White 1963). It is interesting that the words "mead" and "medicine" share the same root. Mead was used as an "elixir" (Hogue 1987, Taylor 1975). Honey is also widely used in the preparation of baked goods, candies, and ice cream. World production of honey totaled 850,569 metric tons in 1972 (Atkins 1978).

Information on bees, beekeeping, and bee products may be found in Adams (1972); Butler (1971); Crane (1963, 1975); Gojmerac (1980); Gould and Gould (1988); Grout (1963); Jay (1986); More (1976); Moritz and Southwick (1992); Morse (1975); and Winston (1987). Townsend and Crane (1973) trace the history of apiculture. Morse (1978) and Bailey and Ball (1991) contain information on the pests, predators, and diseases of honey bees.

Beeswax

Beeswax is a yellowish white solid waxy material that is secreted by specialized epidermal glands between the abdominal sternites of honey bees. It is used to construct hexagonally cross-sectioned cells (honeycomb) in which bees store honey, pollen or rear larvae. Beeswax has long been used for a variety of purposes and was probably the major wax material of ancient times. It was used by people as early as the sixth century. In the past seventy years or so, several other wax or waxlike substances have come into prominence, causing the demand for beeswax to decrease somewhat. Nonetheless, it is still widely used in many cosmetics, nearly smokeless church candles, various pharmaceuticals, some polishes, dental wax, wax museum figures, and several other manufactured materials. One of the major uses to which beeswax is put is in the preparation of comb foundation, which is affixed to the frames of a commercial beehive. This foundation serves to induce the bees to construct honeycomb in the frames, which in turn makes the hive much easier to manage.

Silk

Although the ability to produce silk is widespread in the class Insecta, commercial silk is a product of some lepidopteran larvae, which use it in the construction of pupal

Figure 14.1 Commercial honey.

Figure 14.2 Silk neckties and a silk blazer from Thailand.

cocoon. It is composed of two proteins, fibroin and sericin. Approximately 4,000 years ago in China it was discovered that by boiling a cocoon, the filament of silk used to construct it became loose and could be unwound, fortunately in a single strand (Clausen 1954). The thread produced by winding several of these filaments together could be woven into a soft, lustrous, easily dyed fabric (figure 14.2). From that time on the silk industry (*sericulture*) developed and flourished in China, and its methods were closely guarded secrets. Indeed, it has been suggested that the silk trade was in large part responsible for the success of the Chinese Empire (Hogue 1987). Finally, in about A.D. 550 silkworm eggs and the secret techniques for preparing silk were smuggled to Constantinople, and the silk industry began in Europe (Clausen 1954). ·

Several species of silkworms have been cultured for the commercial production of silk, but *Bombyx mori* (Bombycidae) is and has been the most widely used. It has become, through generation after generation of careful selection, a totally domesticated insect. Larvae feed on mulberry leaves, and close, continuous attention and great care are required to rear them and harvest silk. Many of the steps in silk production require hours of tedious hand labor, a factor that has allowed the silk industry to flourish in Asia where such labor has been fairly cheap. However, silk has been produced in more than twenty countries throughout the world, including the United States (Yokoyama 1973). In addition to its use in producing fabric, the gummy material that is secreted by the silk glands and drawn out into the filaments by the silkworm is dissected directly from the glands and artificially drawn into thin threads that still find use in surgical stitching and have in the past been used as fishing leaders (Clausen 1954). In recent years cheaper and in many ways superior fibers have been prepared synthetically and have threatened to destroy the silk industry.

Yokoyama (1973) and Cherry (1987) trace the history of sericulture. Horie and Wantanabe (1980) consider recent advances in sericulture, including topics like nutrition and metabolism, silk protein synthesis, biochemical aspects of viral infections (see chapter 9), and defense mechanisms against disease.

Lac

Lac is the crude resinous material from which commercial shellac is prepared. It is a glandular secretion of the scale insect *Laccifer lacca* (Homoptera, Coccoidea), which inhabits trees in India and Burma. Tiny immature "crawlers" suck sap from branches of these trees and eventually cover themselves with lac, which serves as a protective shield. Thousands of crawlers feed in close proximity to one another and the lac comes to cover branches almost entirely. When they mature, the females remain wingless and sedentary, whereas winged males emerge and fertilize the females through tiny openings in the lac. When the fertilized eggs hatch, the young crawlers migrate to uninfested areas and commence to feed, beginning the cycle anew. Lac production is encouraged by removing twigs covered with mature females and tying them to uninfested trees. These twigs are referred to as brood lac, and crawlers emerge from them by the thousands and infest their new host tree (Clausen 1954). Harvesting of lac is done by removing branches covered with lac (stick lac) and grinding them. The resultant seed lac is washed, dried, and bleached in the sun, and, after drying, heated in cloth bags over open charcoal fires. As the lac melts, it is squeezed onto the floor and quickly pressed and stretched into thin sheets, which are then flaked. Shellac is prepared by dissolving this flake lac (Metcalf, Flint, and Metcalf 1962). Lac is a basic ingredient of stiffening agents in the toes and soles of shoes, and in felt, fur, and composition hats; shoe polishes; artificial fruits and flowers; lithographic ink; electrical insulation; protective coverings for wood, paper, fabric, wax emulsions, wood fillers, sealing wax and buttons; glazes on confections; coffee bean burnishing; paints; cements and adhesives, shellac varnishes and moldings, photographic products, phonographic records; playing card finishes; dental plates, pyrotechnics; foundry work and hair dyes—(Clausen 1954). Glover (1937) describes lac cultivation and production in India.

As with beeswax and silk, lac has been pretty much replaced by synthetic materials.

Cochineal

Cochineal is also a scale insect product, being secreted by *Coccus cacti,* which lives and feeds on the prickly pear. It is prepared from dried, ground bodies of the insects and in this form is a red pigment that has been used widely, particularly in the past, as a permanent dye, for example, in rouge, for cake decoration, and as a coloring agent for beverages and medicines (Metcalf, Flint, and Metcalf 1962). The major producers of cochineal are in Mexico, Honduras, and the Canary Islands. Approximately 70,000 insects are required to produce a pound of dye (Bishopp 1952). In the past, dyes have also been prepared from other species of scale insects.

Insect galls

Several kinds of insect galls have been the source of various pigments used for dyeing wool, skin, hair, leather, and so on, and for the production of permanent inks. Tannic acid, a substance widely used in tanning, dyeing, and preparation of inks, is also derived from insect galls. See Askew (1971) and Felt (1940) for information on gall-making insects.

Use of Insects in Medicine

Aside from the many purported medicinal uses recorded in ancient literature and folklore, insects or insect secretions have proved to be some value in the treatment of certain human ailments.

Maggot Therapy

During the last three or four centuries the observation was made on a number of occasions that wounded soldiers on the battlefield fared better if their wounds became infested with the larvae of certain flies (Leclercq 1969). These larvae appeared to devour necrotic tissue and inhibit infection. In the early 1930s the use of a strain of *Lucilia sericata* larvae, which attack only necrotic tissues, was advocated for the treatment of osteomyelitis and chronic open sores. It was subsequently discovered that allantoin, a nitrogenous waste excreted by the maggots, produced the same inhibition of infection as the whole insects. This substance was then produced commercially. In recent years the antibiotics have replaced maggot therapy and the use of allantoin.

Cantharidin

Cantharidin is derived from the bodies of blister beetles (Coleoptera, Meloidae). The best-known species is *Lytta vesicatoria,* the Spanish fly, found throughout Europe. When taken internally, cantharidin acts as a strong urogenital irritant. For this reason it has been used as an aphrodisiac (e.g., in cattle breeding), and for the treatment of certain urogenital diseases. It is a very dangerous substance and is no longer used in humans.

Miscellaneous

Several other insect substances have been used as medicaments or have promise for such use. Among these are bee venom, which has been used in the treatment of certain forms of arthritis, myalgias, neuralgias, and other ailments (Leclercq 1969; Metcalf, Flint, and Metcalf 1962). Cochineal is purported to relieve pain associated with whooping cough and neuralgia (Metcalf, Flint, and Metcalf 1962). The Aleppo gall or gallnut formed on several species of oaks in Asia and Europe has been used as a "tonic, astringent, and antidote for certain poisons" (Bishopp 1952). According to Leclercq (1969), there is evidence that antibiotic substances are present in the hemolymph of, or are secreted by, certain insect species. An ancient technique still applied by primitive peoples in various parts of the world is the use of insects such as ants and carabid beetles, which have well-developed mandibles, in the suturing of a wound (Leclercq 1969). The insect is induced to bite a wound so that the two edges are brought together and is then decapitated; the head remains in the clamped position and provides a suture.

According to Cherry (1991), the Australian aborigines have been known to use a variety of insects for medicinal purposes, for example, the bush cockroach for a local anesthetic; a drink prepared from green tree ants as an expectorant, antiseptic, cold remedy, and headache relief; and the silk bag constructed by gregarious Processionary caterpillars (Family Notodontidae) as a wound dressing.

Insects in Biological Research

Insects make ideal organisms for fundamental biological research. They are usually easy to collect and rear in large numbers, are small in size and can be easily manipulated within a small space, have comparatively short generation times, and as a group display a diversity in form and function nearly unapproached by other groups of animals. *Drosphila melanogaster* has been widely utilized as a research species in genetics. A large amount of experimental-developmental endocrine work has been carried out on the bug *Rhodnius prolixus* (Hemiptera, Reduviidae). Among the many other genera that appear often in biological literature are *Tribolium* (Coleoptera, Tenebrionidae), *Calliphora* and *Phormia* (Diptera, Calliphoridae), *Musca* (Diptera, Musicidae), *Periplaneta* (Blattaria), and *Apis* (Hymenoptera, Apidae), *Aedes* (Diptera, Culicidae). See chapter 1 for more information on the significance of insects in biological research.

No student should study entomology (or, for that matter, general biology) without reading Dethier's delightful, enlightening book *To Know a Fly* (1962).

Pollination by Insects

In addition to countless wild flowering plants (see chapter 10), domesticated plants pollinated by insects include vegetables such as tomatoes, peas, beans, and onions; most fruit crops; and field crops such as alfalfa, red and white clover, and tobacco. Insects involved in the pollination of flowering plants number in the thousands of species and are found mainly among the Hymenoptera, Diptera, Lepidoptera, and Coleoptera, although members of some smaller orders (e.g., Thysanoptera) may also be involved. Several Hymenoptera, in particular the bees (Apoidea), are the most important pollinators in commercial crops. In the past, wild, native pollinating insects were sufficient in number to accomplish the pollination of our food crops, but with the intensive land use, clean cultivation, and excessive application of insecticides characteristic of modern agriculture, the populations of many of these species have been so reduced that they can no longer maintain an adequate level of pollination. According to Bohart (1952), "An estimated 80 percent of the insect pollination of our commercial crops is performed by honey bees." Thus honey bees are no doubt our most valuable pollinators and have the advantage that they can be easily cultured and moved about at will. In fact, "It has been claimed that the value of bees in the pollination of crops is 10 to 20 times the value of the honey and wax which they produce" (Lovell 1963). Pimentel (1975) provides a graphic description of the pollinating abilities of honey bees.

A single honey bee may visit and pollinate 1,000 blossoms in a single day. In New York state, which has about 3 million domestic honeybee colonies, each with about 10,000 worker bees, honeybees could visit 30 trillion blossoms in a day. Wild bees pollinate a number at least equal to that. Thus on a bright, sunny day 60 trillion blossoms may be pollinated by the bees in New York, a task impossible for man to accomplish today.

Additional information on pollinating insects can be found in Bohart (1972), Free (1970), Jay (1986), Kevan and Baker (1983), Martin and McGregor (1973), McGregor (1976), Nørgaard Holm (1976), Richards (1978), and Real (1983).

Insects Consumed and as Consumers

As Food for Humans and Domestic Animals

Insects actually have a high nutritional value, being quite rich in protein and lipids, and may therefore be a very important supplement to the diets of otherwise vegetarian peoples (Leclercq 1969, Taylor 1975). In fact, at least some insects compare very favorably with other food

Figure 14.3   Examples of edible insects. The "Tequila" lollipop encases a "worm" (beetle larva).

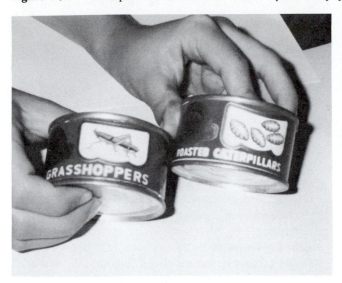 

items. For example, corn has 320–340 kcal/100 g in comparison to nine Mexican insect species that had 377–516 kcal/100 g (DeFoliart 1989). Insects that have been studied likewise rate favorably in regard to vitamins and minerals.

In Africa today, meals are sometimes supplemented with insects as a source of protein, which may partially stave off the protein-deficiency disease of children, kwashiorkor (LeClercq 1969). The Australian aborigines realize that grubs are an essential part of their diet, even though they do not realize that they constitute their sole source of protein (Clausen 1954, Cherry 1991). The "manna" referred to in the Bible that provided nourishment for the founders of the Jewish nations may have been secreted by scale insects feeding on the tamarisk plant (Hogue 1987). Other insects that have been or are still prepared in some way or eaten raw by various peoples in the world include termites, silkworm pupae, migratory locusts, ants, caterpillars, diving beetles, cicadas, and eggs of water boatmen. According to DeFoliart (1989):

> Current information suggests that 30 species or more are used by indigenous populations in many Third World countries, although the specific identity of relatively few of the species is known. . . . Where careful studies have been conducted, the volume of insects consumed as a percentage of the total animal protein has been shown to be appreciable.

In recognition of insect's potential high nutritional value, DeFoliart notes that there are strong advocates of increased use of insects as food sources:

> Insects offer a number of attributes, such as high food conversion efficiency compared with conventional meat animals, use of a wide array of organic substances not efficiently used in conventional agriculture, and producibility without the need for additional arable land, irrigation, fertilizers, herbicides, pesticides, or expensive equipment, that make their use highly compatible with the principles of low-input sustainable agriculture. Further, because many important food insects are also important pest insects, it may be possible in some cases to incorporate food insect harvest as part of pest management programs, thus helping to reduce the need for insecticides for crop protection.

In less acute food situations, various insects have for centuries been valued as delicacies and have even become articles of food commerce (figure 14.3). For example, "guasanos"—fried caterpillars, earthworms, and beetle grubs found in agave plants—are exported from Mexico and can be purchased in gourmet shops in the United States (Clausen 1954).

See Taylor (1975) and DeFoliart (1989) for discussions of insects as human food and Taylor and Carter (1976) for some recipes.

As Food for Wildlife

Aside from being eaten by a wide variety of arthropods, including their own kind, insects are major food organisms for many kinds of wildlife; in particular birds and fishes (see chapter 9). Insects probably comprise more than half the food consumed by the more than 1,400 species and subspecies of birds in North America (Swan 1964). Metcalf, Flint, and Metcalf (1962) cite the work of Forbes, who concluded that two fifths of the food of adult freshwater fishes is insects, the most important of which are bloodworms, mayfly larvae, and caddisfly larvae.

Insects as Consumers

The magnitude of the roles played by insects as plant eaters, scavengers, predators, and parasites in the total picture of nature is inestimable. They are a part of nearly every terrestrial and freshwater food web. Our understanding, all too incomplete at times, has enabled us in some instances to manipulate populations of insects to control other insects and, even, undesirable weeds (e.g., klamath weed and prickly pear). The food habits of insects are discussed in chapter 8.

Forensic Entomology

In recent years, insects have increasingly been recognized as valuable tools for use in certain criminal investigations, murder in particular. Thus the specialty of *forensic entomology* is coming into its own. The use of insects in murder investigations is based on the occurrence of a sequence of "colonizations" by various insects on a human corpse (i.e., faunal succession). The first wave of insects is made up mainly of flies, e.g., blow flies (Diptera, Calliphoridae), which deposit their eggs around natural body openings and open wounds. The larvae (maggots) from these eggs then proceed to devour the soft tissues right along with the various bacteria and fungi that are involved in the decomposition process. The second wave of insects are attracted to the corpse by the changes caused by the first wave and accompanying microbes. These insects [e.g., skin beetles (Coleoptera, Dermestidae)] feed on dried skin and cartilage. In addition to the insects that feed directly on the corpse are various predators and parasites of the scavenging insects, as well as a selection of omnivorous insects that feed on the corpse as well as other insects.

In a given region, characteristic insect species take part in such colonizations or "waves" in a consistent fashion, which is influenced to some extent by variations in environmental factors, especially temperature. This being the case, examination of the insects scavenging on a corpse at the time of discovery along with information on recent temperatures, etc., can be of value in establishing the time of death. Knowledge of the distributions of various insects which frequent decaying animal matter may also be useful in establishing whether or not a corpse has been transported from one place to another after death.

The following passage from a recent and fascinating article provides a good example of forensic entomology in action:

> The life cycle of blow flies were the key to our investigation of the body of a thirty-seven-year-old male, found by joggers in a swamp on the windward side of Oahu. The corpse was infested with maggots of two species of blow fly. Our analysis of the larvae, compared with both laboratory-rearing data and results of decomposition studies, suggested the victim had been dead for roughly 120 hours. That conclusion was corroborated by information obtained independently by the Honolulu police: the victim had last been seen alive 123 hours before the discovery of the corpse, and he had failed to report to work 121 hours before his body was found. Our estimate helped place the victim in the company of a suspect, who was later convicted of the murder (Goff 1991).

Keh (1985), the article by Goff (1991), and Smith (1986) provide a good introduction to forensic entomology.

Insects, Esthetics, Philosophy, and Blatant Anthropomorphism

Beyond the rigors of science, one cannot help but be in awe of insects as organisms worthy of admiration and respect. Along with pathogenic microbes, they are formidable adversaries. But, on the other hand, they often appeal to our emotions in a positive way. Many display great beauty and grace (see, for example, the marvelous photographs in Sandved and Brewer 1976, and in Dalton 1975). Their beauty has been celebrated in art forms ranging from oil paintings to jewelry (Akre et al. 1991) to postage stamps (figure 14.4). Some have even been elevated to gods (e.g., the dung-rolling scarab beetles in ancient Egypt; Kritsky 1991).

If not appealing to our esthetic senses, some insects entertain and astound us with the bizarre: devouring a mate during copulation (praying mantids), having strangely placed penises (dragonflies and damselflies), flitting around releasing disagreeable odors when nervous about being eaten, flying against burning light bulbs or into fires, rolling sundry pieces of fecal material into little balls, or using a "jet propulsion" system to escape from predators the way dragonfly larvae do.

Further, the organizational skills of insects are without equal. Intricately run termite, ant, bee, and wasp societies put our ever-present bureaucrats to shame.

Figure 14.4 Examples of postage stamps featuring insects. Note that both beneficial and harmful insects are portrayed.

Public awareness of the fascinating world of insects has increased in the past few years. Insects are becoming popular attractions at zoos and museums, for example, the Cincinnati Zoo and the National Museum of Natural History in Washington, D.C., both provide exhibits of living insects. *Time* Magazine, which usually reserves its cover for notable human beings of the moment, featured on the cover of its July 21, 1976 issue, a frontal view of a cicada killer wasp.

Insects have been "known" to be philosophers, like the cockroach "Archie" who befriended a journalist in the 1930s. This was fortunate, because that journalist, Don Marquis, conveyed Archie's outlook on life to us (see *Archie and Mehitabel,* 1960, and *Archie's Life of Mehitabel,* 1966):

> insects have their own point of view about civilization a man thinks he amounts to a great deal but to a flea or a mosquito a human being is merely something good to eat—Marquis (1960).

Archie typed his messages to Mr. Marquis by butting the keys, but was unable to reach the shift key and therefore typed everything in lower case.

In recent years ". . . the branch of investigation that addresses the influence of insects (and other terrestrial Arthropoda, including arachnids, myriapods, etc.) in literature, languages, music, the arts, interpretive history, religion, and recreation . . ." has become "cultural entomology" (Hogue 1987).

Harmful Insects

Fortunately, the vast majority of insects are directly or indirectly beneficial, but the relatively few that are pests will no doubt continue to tax our ingenuity to its fullest. It has been estimated that 150 to 200 insect species or species complexes frequently cause serious damage; an additional 400 to 500 species are occasional, but serious pests and may cause serious damage; and 6,000 or so are sometimes pests, but rarely cause severe damage (National Academy of Sciences 1969). These figures represent but a small fraction of the nearly 90,000 described species of insects in North America north of Mexico (Borror, Triplehorn, and Johnson 1989). A similar proportion probably prevails worldwide. But, however small the proportion, the magnitude of trouble caused by insect pests is very large:

> Estimates of the pest problem on a world scale suggest that, without insect pests, world food production could be increased by about a third. As this estimate represents the loss despite current control measures, it would clearly be catastrophic for mankind if control of insect pests were not attempted or should fail (van Emden 1989).

There is nothing, biologically speaking, that clearly defines an insect as a pest. Two similar insects may have almost identical biological patterns and even be in the same family, yet one will be considered a pest, and the other will not, simply because one attacks humans or something valued by humans. Rolston and McCoy (1966) illustrate this point with three chrysomelid beetles.

The Colorado potato beetle effectively defoliates the potato, and the dock beetle, *Gastrophysa cyanea* Melsheimer, defoliates its host with equal regularity. That the former is a pest and the latter is not reflects the value we place on the potato and our disinterest in dock. We may even encourage the destruction of some plants by insects. St. John's wort, or Klamath weed, has been cleared from thousands of acres of rangeland in western states by a group of insects, but principally by a leaf beetle, *Chrysolina quadrigemina* (Suffrian), introduced specifically to control this weed.

Many factors contribute to the development of pest situations. These may be grouped into three categories: modification of the environment, transportation, and human attitudes and demands.

Human activity has modified the environment, sometimes to advantage, often to disadvantage. One undesirable effect of human activity has been the creation of comparatively simple ecosystems. Generally, the simpler the system of interacting organisms, the less the system's inherent stability and the greater the likelihood of large fluctuations of populations of the component species. Thus it is not difficult to understand why there are a multitude of problems associated with the great monocultures of organisms, particularly plants, that humans have developed. These *agroecosystems* are much less complex than the "natural" ecosystems that preceded them and hence are prone to spawn pest situations.

A good example of a vast monoculture is the corn grown yearly in the United States. Literally millions of acres are involved and the value of this crop is in the billions of dollars. Insect damage to corn on an annual basis costs in the hundreds of millions of dollars (Burkhardt 1978).

Another concentration process we have brought about has been the storage of vast quantities of materials that are palatable to certain insects. Among these stored materials are millions of bushels of grains, fresh vegetables, various other food products, fiber, and so on. Such "habitats" furnish optimal conditions for certain species. Also, we have often concentrated ourselves and commonly under other than sanitary conditions. Such concentrations of human populations provide an ideal circumstance for the transmission of insect-borne pathogens (e.g., louse-borne typhus and plague).

A text by Service (1989) deals specifically with the relationships between human demography and vector-borne diseases.

With the many forms of modern transportation, insects are no longer limited in their dispersal capabilities by natural geographic barriers such as mountains and oceans, but are carried about as hitchhikers.

The introduction of an exotic insect into a new region away from its natural enemies may have disastrous results. For example, the European corn borer is a major pest of corn in the United States, but it is only a minor pest in its original home (Fronk 1978). Conversely, plants introduced into a new region may prove to be ideal food for indigenous insects, which then constitute a pest problem. For example, the cucumber originated in the East Indies, while one of its major pest insects, the striped cucumber beetle, originated in North America (Fronk 1978).

Human attitudes and demands may create a pest problem or greatly exaggerate what might, in fact, be a minor one. For instance, even slight evidence of insect damage may decrease or destroy the economic value of a vegetable or fruit, even though the food value may be unaffected, for the simple reason that the average consumer is very particular about such matters and demands blemish-free produce; often referred to as the *cosmetic effect*. This commonly forces producers to employ costly and ecologically undesirable means of insect control. Many insects are considered to be pests not because they attack us or destroy any of our goods, but simply because we find them distasteful.

The paper by Hahn and Ascerno (1991) focuses on public attitudes in an urban context and the bibliography provides entry into the literature of this fascinating area.

We will examine the conflict between humans and insects in three general categories on the basis of what is attacked by insects.

1. Growing plants,
2. Stored products, household goods, and structural materials,
3. Humans and wild and domestic animals valued by humans.

Examples of many pest insects are given in chapters 12 and 13. The applied entomology texts by Pfadt (1978), Davidson and Lyon (1987), and Pedigo (1989) provide good introductions to insect control and deal with pests of specific crops. Metcalf and Luckman (1975) and van Emden (1989) deal with fundamental ideas and methods in insect pest management. The following are interesting and informative essays on the impact of insects on the welfare of humans: Cloudsley-Thompson (1976), Cushing (1957), Pimentel (1975), McKelvey (1975), Ritchie (1979), and Southwood (1977a, b). Jones (1973), Philip

and Rozeboom (1973), and Schwerdtfeger (1973) consider various aspects of the history of agricultural, medical-veterinary, and forest entomology, respectively.

Growing Plants

Virtually every growing plant we value is shared, to a greater or lesser extent, with one or usually more species of insect. For example, it is estimated that approximately 400 species of insects infest the apple tree, and 25 of these are of economic importance (Johansen 1977). Almost all the damage these insects cause is associated with their feeding activities. Depending on the species and life stage, they remove chunks of tissue by chewing, sucking sap from, and/or boring through every part of a plant. Not only are all parts of a plant susceptible to insect attack but all life stages as well. The feeding of insects not only causes the removal of tissue and sap but also may poison a plant by the injection of toxic saliva. As a result of feeding, several species of insects are responsible for the introduction, transmission, and dissemination of plant diseases caused primarily by viruses but also by bacteria, fungi, and a few protozoans (see below and "Kinds of Intraspecific Interactions" in chapter 9).

Pfadt's (1978) description of the various kinds of injury to small grains (e.g., wheat, barley, and oats) and the guilty insects, illustrates the above points.

Insects injure small grains in the field from the time the seed is planted until the grain is harvested. Both wireworms and false wireworms feed on the planted seed. In dry soil they may even destroy a crop before rains stimulate germination and growth. After the grain germinates, these pests devour the tender sprouts just as they push out of the seeds.

Wireworms also kill seedlings by boring into and shredding the underground portion of stems. Wireworms and white grubs feed on roots and sever them from the plant. Cutworms, white grubs, and false wireworms cut off young plants near soil level. In addition, the wounds left by soil pests allow rot pathogens to enter the plant.

Grasshoppers, Mormon crickets, and armyworms may devour young plants completely; they may strip the leaves from older plants, feed on maturing heads, or cut through the stems below the heads.

By injecting toxic secretions while feeding, larvae of Hessian flies retard or kill seedlings and reduce the yields of older plants. Weakened stems of older plants are likely to cause the crop to lodge. Wheat stem sawfly, wheat jointworm, and wheat strawworm bore within culms and obstruct the flow of sap. This damage reduces the number and weight of kernels. Boring insects also cause grain to lodge.

Insects such as the chinch bug and various aphids impoverish plants by sucking juices from leaves or stems. Moreover, they produce fatal necroses by injecting toxic saliva. Small grain pests may transmit serious plant diseases, such as wheat streak mosaic by the wheat curl mite, striate by the painted leafhopper, barley yellow dwarf by several species of aphids, and aster yellows disease of barley by the aster leafhopper.

Some insects cause damage to growing plants through oviposition. Several kinds of insects have well-developed ovipositors that can penetrate rather hard surfaces like the bark of woody stems. The periodical cicada provides a good example of an insect that can cause considerable oviposition damage. The larvae of these insects spend 13 or 17 years feeding on plant roots and then emerge from the ground for a brief period of time to mate and lay eggs. Mated females deposit their eggs on young stems of shrubs and trees, causing severe damage and often necrosis of the affected area.

For information on insects and plant disease, see chapter 16 and the following references: Carter (1973), Maramorosch and Harris (1981), Nault and Ammar (1985), Pyenson and Barke (1977), and Sylvester (1980).

Stored Products, Household Goods, and Structural Materials

The wide variety of materials accumulated by man and stored in containers ranging from large bins to small boxes present ideal environmental conditions for certain insects to thrive, essentially free of any natural enemies. The presence of these insects may go undetected for long periods of time, and hence the potential extent of damage is great. Stored grains are particularly susceptible to attack and probably incur the most damage of any stored material. Stored-grain pests not only consume grain but also render large quantities useless by contaminating it with fecal material, webbing, odors, shed exoskeletons, and whole or fragmented dead individuals. The activities of these insects may also cause heating of grain. This heating causes moisture-laden warmed air to rise to the surface, where it is cooled, resulting in the condensation of the accumulated moisture on the surface of the grain. This in turn causes caking of the grain and affords an ideal situation for the growth of molds and encourages spoilage (Wilbur and Mills 1978).

Major stored-grain pests include the rice and granary weevils, the flour moths, and the Angoumois grain moth. The damage done by stored-grain insects is estimated to be 5% to 10% of the world's production, and much higher in certain areas.

Destruction of food by stored grain insects is a major factor responsible for the low levels of subsistence in many tropical countries. If these losses could be prevented, it would alleviate much of the food shortages in the famine areas of the world. According to one estimate, 130 million people could have lived for one year on the grain destroyed or contaminated by stored grain insects in 1968. Wilbur and Mills (1978).

Packaged food materials may be attacked by insects at any point from the processing plant to the consumer's home. What constitutes food for an insect is not necessarily human food. Certain insects, such as the cigarette beetle and drug-store beetle, both members of the Anobiidae (Coleoptera), eat a wide variety of nonhuman food items, including such delicacies as tobacco and several drugs.

Several species of insects have become common cohabitants with humans. These include cockroaches, silverfish and firebrats, and a large variety of ants, beetles, and moths. Stored-products insects also fall into this category. Even when these insects do little or no damage, their presence in a modern home is deemed highly undesirable. Several household items are susceptible to attack, including rugs, furniture, clothing, books, paper, and food.

Any structures made out of wood may be attacked by insects, and extensive and serious damage can result. Structural timbers in houses, buildings, and bridges, and wooden structures such as fences, railroad ties, and telephone poles are all susceptible to attack. Subterranean termites are the major pests involved, although insects such as powderpost beetles may also cause trouble.

Ebeling (1978) and Hill (1990) are useful texts in the area of stored-products pests.

People and Their Animals

The ways in which insects attack humans and the animals they value are essentially the same and hence will be treated together. The terms *medical entomology* and *veterinary entomology* are commonly used in reference to this area of applied entomology. As might be expected, this category also has strong economic overtones. Animals, wild or domestic, may be killed, weakened, and/or decreased in value as a result of being attacked by insects. Similarly, people may be killed or weakened and literally millions of hours of productive labor lost. Insect-borne diseases such as malaria and trypanosomiasis are particularly responsible for the latter. Vast areas of the world have been made uninhabitable or nearly so by the presence of insects and the disease-causing microorganisms they carry. For example, the tsetse flies, *Glossina* spp., because they transmit the trypanosomes that cause nagana in cattle and two types of African sleeping sickness in humans, have prevented the development of millions of square miles of tropical Africa (McKelvey 1973, Nash 1969).

Insects affect the health of humans and other animals in two ways: directly as the causative agents of disease and discomfort and indirectly as the transmitters (*vectors*) of disease-causing microorganisms (bacteria, viruses, protozoans, helminths, etc.).

Many of the diseases of humans and animals, which involve insect vectors, are mentioned in chapters 9 ("Interspecific Interactions") and in chapters 12 and 13 in discussions of the various insect groups.

Information on insects of medical and veterinary importance may be found in Benenson (1990); Busvine (1966, 1975); Gillett (1972); Greenberg (1971, 1973); Harwood and James (1979); Horsfall (1962); Kettle (1984); Lehane (1991); McKelvey et al. (1981); Monath (1988); Pfadt (1978); Reeves et al. (1990); Scott and Grumstrup-Scott (1988); Smith (1973); Snow (1974); Theiler and Downs (1973); and White (1987).

Insects as the Causative Agents of Disease and Discomfort

Insects cause disease and/or discomfort by various combinations of feeding activities, physical injury, secretions, invasion and infestation, and psychological disturbances.

The most important feeding activity that affects people and animals is the tendency of several species of insects, such as mosquitoes, black flies, other biting flies, bed bugs, and conenose bugs, to suck blood. These insects can make life miserable and may in some instances cause serious illness or even death. Illness due to blood loss in human beings is usually not significant since they can generally defend themselves or escape. However, the pain of being bitten and other complications make blood-feeding insects of major significance. Moreover, blood loss in wild and domestic animals may be considerable, and when the population of the offending insects is large, blood loss coupled with interference with feeding causes death.

An extreme instance is the oft-quoted census made in Rumania, Bulgaria, and Yugoslavia in the year 1923, where nearly 20,000 domestic animals, horses, cattle, sheep, and goats are said to have been killed, along with many wild animals, by the Golubatz fly, *Simulium columbaschense*—Oldroyd (1964).

Some nonblood-feeding insects are attracted to animals and humans and may feed on sebaceous and lachrymal secretions or perspiration and in this way cause considerable disturbance. Certain species of noctuid and other moths in Africa and Asia have been observed to feed on lachrymal secretions of cattle and sometimes of humans (Harwood and James 1979). Flies in the genus *Hippelates* (Diptera, Chloropidae) are attracted to lachrymal secretions as well as to mucous and sebaceous secretions, pus, and blood (Harwood and James 1979).

Physical injury may occur if an insect flies into the eye or ear, or as a result of a defensive reaction of an insect that is carelessly handled, or because of inadvertent contact with an insect. Many insects equipped with powerful jaws or legs may pinch, bite, or jab. For example, many of the predaceous Hemiptera, giant water bugs and backswimmers, may inflict a very painful bite. As with anything that causes a break in the skin of human or animal, secondary infection is always a possibility.

Harmful insect secretions may be divided into two groups: the inherently toxic *venoms* and the *allergens,* whose effect depends on the physiological response of the victim. Introduction of a venom usually follows a predictable course, causing pathological conditions; the response to the introduction of an allergen may vary from no reaction to anaphylactic shock, which may result in death. To become allergic or hypersensitive requires an initial introduction (injection or inhalation) of allergen, at which time there is no response except that antibodies are produced that will interact with the allergen when subsequently introduced and initiate the allergic response. A given secretion may act both as a venom and as an allergen (e.g., bee venom usually causes pain and a local reddening of the skin and if introduced into an allergic or hypersensitive individual may cause serious complications).

There are four ways by which venoms and allergens may be introduced: bite, sting, contact, and active projection (Harwood and James 1979).

Blood-feeding insects typically inject an amount of saliva into their host before they begin to remove blood. This saliva may act as an allergen, as has been demonstrated with mosquitoes, bed bugs, and others. Thrips, certain phytophagous Hemiptera, and certain predatory Hemiptera have also been observed to bite humans, and their salivary secretions may act as venoms or allergens or both. Several Hymenoptera (i.e., many ants, bees, and wasps) are equipped with stinging apparatuses and venom glands on the posterior part of the abdomen. These structures serve in defense and/or prey capture.

Venoms injected by stinging insects can act as allergens. Direct contact with secretions from the bodies of blister beetles (Coleoptera, Meloidae) can cause blistering of the skin. Some lepidopteran larvae in the families Saturniidae (e.g., the Io moth larva), Lymantriidae (e.g., the browntail moth larva), Megalopygidae (e.g., certain flannel moth larvae), and others possess urticating (stinging) hairs that contain ducts from one or more poison glands. These hairs can cause trouble if contact is made with either a live caterpillar or airborne fragments of dead, dried caterpillars. The effect of urticating hairs is much the same as stinging nettles. Butterfly scales and the dried exuviae and fragmented bodies of dried dead mayflies and other insects such as cockroaches may cause allergic reactions, sometimes very serious, when inhaled by sensitive individuals. Some insects, such as certain ants and predatory Hemiptera and others, can actively project or spray venom/allergen(s) from their bodies. This ability is associated with the defense mechanisms of insects (see chapters 8 and 9). Brenner et al. (1991), Frasier and Brown (1980), and Schmidt (1986) deal with allergic responses to insects. Akre and Davis (1978), Bucherl and Buckley (1972), Minton (1974), Piek (1986), and Tu (1977) contain information on insect venoms.

There has been much concern during the past two decades or so about the northward spread of the very aggressive Africanized honey bees from South America. These bees (*Apis mellifera adansonii*) were imported to Brazil from Africa because of their outstanding honey production. The hope was to cross them with local bees (previously imported from Europe) and produce a well-adapted, highly productive strain. Unfortunately, several African queens escaped in 1957, successfully interbred with wild populations, and began to spread. Since that time there have been problems with large numbers of these very aggressive bees attacking humans and various animals. They have continued to spread from South America through Central America and are finally reaching the United States. For more information on these bees, see Fletcher (1978), Michener (1975), Spivak et al. (1991), and Winston (1992).

Imported fire ants, *Solenopsis* spp., are another example of a major stinging insect problem. Two species, *S. richteri* and *S. invicta,* were imported accidentally into the United States many years ago, and they now constitute a serious problem in the southeastern states (see Lofgren, Banks, and Glancey 1975 and Rhoades 1977).

Several species of insects spend all or at least a portion of their existence on humans or other animals. Some of these are *facultative parasites;* that is, they can complete their life cycles without invading or infesting people or animals but can take advantage of such hosts should they become available. Others are *obligate parasites,* requiring a human or animal host in order to survive. Among the larvae of true flies there are several that will invade various cavities or open wounds of animals. This type of invasion is referred to as myiasis (Zumpt 1965). Other animal-infesting insects include the chewing (Mallophaga) and sucking (Anoplura) lice. Extensive information on arthropods which infest or invade the human body may be found in Parish et al. (1983).

Occasionally psychological disturbances are associated with the presence of insects, both in human and in wild and domestic animals. These disturbances may be totally unrelated to whether or not the offending insects are harmless. In humans, Pomeranz (1959) recognizes two categories of psychological disturbances caused by insects:

1. *Arthropod phobia* (*entomophobia* in specific reference to insects)— "the irrational, persistent fear of recurrent arthropod infestation."
2. *Hallucination of arthropod infestation*— "a condition in which the subject imagines he is being molested by small and difficult-to-locate forms which reach and localize on the body despite all sorts of extraordinary preventive measures."

In wild and domestic animals, the persistent buzzing and biting or oviposition attempts by flying insects may cause considerable behavioral disturbance and interfere severely with grazing. For example, horses become extremely nervous in response to the buzzing and oviposition attempts of adult female bot flies (Diptera, Gasterophilidae). For further information on psychological disturbances associated with insects, see Olkowski and Olkowski (1976).

CHAPTER 15

The Insect Control Arsenal

F e r r o / R o m o s e r

There are many tools for insect control found in today's arsenal, but no one method is without drawbacks. In this chapter we review the various major approaches that have been used in attempts to solve insect problems and some that show promise for the future.

The following are general treatments of various aspects of insect control: Barbosa and Schultz (1987), Burn et al. (1988), Curtis (1989), Drummond et al. (1987), Kogan (1986), Mallis (1990), Metcalf and Luckman (1982), Pedigo (1989), and Pimental (1990).

Biological Control

Under natural conditions, insect populations are kept in check by the influences of a multitude of environmental factors. Together these factors bring about the "natural control" of a given population. Among these natural control factors are parasites, predators, pathogenic microbes, and competing species (see chapter 9).

Biological control refers to the regulation of pest populations using predators, parasitoids, nematodes, and microbial agents. There are several advantages in using biological control agents. If a biological control agent is acclimated to the target area and pesticides are judiciously used, these agents should become a permanent fixture. Unlike pesticides, biological control agents are safe to use and do not pose any threat to the environment. The development of biological control agents is considerably less expensive than the development of an insecticide.

General treatments of biological control include Coppel and Mertins (1977), Croft (1990), DeBach (1974), van den Bosch and Messenger (1973), Hoy and Herzog (1985), Huffaker (1971), Huffaker and Messenger (1976), Huffaker, Luck, and Messenger (1977), Swan (1964), Sweetman (1936a, b), and Weiser (1991). Hagen and Franz (1973) trace the history of the development of biological control.

Parasitoids and Predators

Insect parasitoids and predators are discussed in chapter 9. The major parasitoids used in biological control are in the order Hymenoptera (including chalcid, braconid, and ichneumonid wasps) and Diptera (Tachinidae and several other families). More than two thirds of the cases of successful biological control have involved the use of Hymenoptera (DeBach 1974). Among the predatory insects that have been used in biological control are various Coleoptera (especially Coccinellidae, ladybird beetles), Neuroptera (Chrysopidae, green lacewings; Hemerobiidae, brown lacewings), Hymenoptera (e.g., certain ants), Diptera (e.g., Syrphidae and Asilidae), and certain Hemiptera.

Biological control involving the use of parasitoids and predators can be classified into four approaches: conservation, augmentation, inundation, and introduction. *Conservation* consists of conserving existing agents by creating refugia that are protected from pesticides and provide an alternative food source to its host. *Augmentation* is the process by which mass-reared agents are released to augment existing populations in the field or for reintroduction into habitats where natural enemies have been killed. *Inundation* is the process by which mass-reared agents are released to inundate the pest population, to control pest populations within the first generation of release, such as in greenhouses. Generally, this approach is too costly to use on a large scale and on a yearly basis.

The *introduction* of exotic species with hopes of controlling pest species has been by far the most successful of the four methods. There are two situations in which this method is appropriate: when there are "unoccupied niches in the life system of the pest, which could be filled by an introduced species," and when "a certain niche is occupied by an organism that is inherently inefficient as

Figure 15.1 Hypothetical relationship between a herbivore population and its predator, illustrating how the density of a predator increases as the density of the herbivore increases (numerical response).

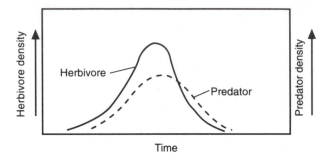

a regulator and that might be displaced by a more efficient exotic regulator" (National Academy of Science 1969). Both situations exist particularly when a given pest species has been accidentally introduced from another area. A high percentage of the pest insect species in the United States fit into this category. Introduced species find themselves in an environment in which they are free to multiply in the absence of their natural enemies, and if they happen to feed on something of value, they can quickly become problems.

The Chinese have been using predaceous ants in citrus groves since 1200 B.C. One of the best examples of introducing a predator is the vedalia bettle, *Rodolia cardinalis* (Coleoptera, Coccinellidae), for controlling the cottony cushion scale, *Icerya purchasi* (Homoptera, Coccoidea). The cottony cushion scale nearly destroyed the citrus industry in California in the late 1800s. This scale originated in New Zealand and Australia. Albert Koebele was sent to Australia in 1888 by C. V. Riley. He found larval and adult vedalia beetles feeding on the scale. A population of the vedalia beetle was brought to California, released in citrus groves, and within a few years it had reduced the cottony cushion scale populations to noneconomic levels. The total cost was $2,000.

There are certain biological and behavioral characteristics that describe a successful predator or parasitoid. Populations of a predator or parasitoid must be in synchrony with host populations. For example the ladybird beetle, *Coleomegilla maculata* (Coleoptera, Coccinellidae), is present in large numbers in the spring in woody areas surrounding potato fields. When the Colorado potato beetle (CPB) starts laying eggs in late spring, *C. maculata* adults move into the potato field to feed on CPB eggs. When the number of predators increases in response to an increase in prey density either through migration or reproduction, this is called a *numerical response* (figure 15.1).

The searching behavior of a natural enemy will dictate to a large extent the effectiveness of an agent at regulating the population dynamics of a pest insect. The convergent ladybird beetle, *Hippodamia convergens*, feeds on the walnut aphid, *Chromaphis juglandicola*, in California; however, if populations of aphids are too low, it leaves the tree in search of other aphid populations. Fortunately, another ladybird beetle, *Olla abdominalis*, is able to feed at much lower aphid densities and keeps the aphid populations below economically damaging levels. *O. abdominalis* moves along the leaf midvein and crisscrosses back and forth across the top and bottom of each leaf in search of aphids, whereas *H. convergens* goes down the midvein and along the edge of each leaf. The likelihood of *H. convergens* encountering an aphid is much less than that of *O. abdominalis*.

The searching and feeding behavior of natural enemies may change as pest population densities increase, so each individual agent kills more pests as pest densities increase. This behavior is referred to as a *functional response*. Holling has classified three types of functional responses. Type 1 represents a rather specialized situation and is not common in insects. The Type 2 response is probably the most common type in insects. There are four essential components in the Type 2 functional response: rate of successful search, time predator and prey are exposed, handling time and hunger. *C. maculata* adults feeding on CPB eggs alone or in the presence of alternative foods are representative of a Type 2 response (figure 15.2). The Type 3 response is a sigmoid response, and in addition to the above mentioned components of the Type 2 response, there is one more component that describes the Type 3 response, learning (figure 15.3). By understanding how a predator or parasite responds to changes in host density and having population estimates of natural enemies, the pest management practitioner can be in a better position to make management decisions.

Another factor that greatly influences the success of a natural enemy in regulating a pest population is its overwinter survivorship. If survivorship is low, an agent will generally be ineffective during its first generation. The number of generations predators and parasites go through can also affect their success as biological control agents. A multivoltine (more than one generation per year) agent is more likely to be in synchrony with its host and have a greater potential for increasing its numbers within a season than a univoltine agent (one generation per year).

There is considerable controversy over the merits of a generalist predator or parasite versus a specialist. A specialist will tend to inflict higher levels of mortality than a generalist because a specialist is dependent on a single host as a food source and/or egg-laying site. Because of this dependency, a specialist is more likely to reduce its

Figure 15.2 Comparison of functional response of *Coleomegilla maculata* adult females as predators of Colorado potato beetle eggs with and without the aphid *Myzus persicae* or corn pollen as alternate food.

Redrawn from Hazzard and Ferro (1991).

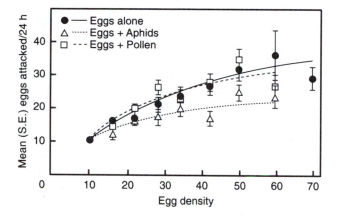

Figure 15.3 Postulated types of functional responses of parasites and predators to insect prey densities.

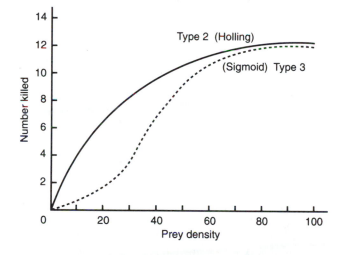

host to population levels below that which can support its own population. For this reason, the population density of a specialist is going to oscillate more than that of a generalist. The generalist is not as likely to inflict high levels of mortality on a pest species; however, a generalist is more likely to remain in high numbers in the host habitat by feeding on or parasitizing other hosts.

Eighty percent of all successful biological control programs have been accomplished by the introduction of a single species of natural enemy. In most cases, control has been achieved within three generations.

Potential biological control agents are collected from an area climatically and geographically (and photoperiodically) similar to the area where they are to be released. Efforts are made to study the biology of the agent in its home environment to identify any special needs that may prevent it from being a successful agent.

For example, the parasitic wasp *Edovum puttleri* (Eulophidae: Hymenoptera), which is endemic to Colombia, South America, was collected from eggs of *Leptinotarsa undecimlineata* (a close relative of the Colorado potato beetle, CPB). Laboratory and greenhouse studies showed it to cause high levels of parasitism of the CPB. However, even though it caused high levels of parasitism to second generation CPB, the wasp was not effective at parasitizing first generation CPB, nor could the wasp overwinter in the temperate areas of the Northeast. Because it was collected from close to the equator where temperatures vary little throughout the year, there was no selective pressure for it to be able to withstand cold winter temperatures. Further, a study carried out in Cali, Colombia, showed that the wasp feeds on honeydew produced by several homopteran species and that it is able to survive for only a few days without this honeydew or some other source of carbohydrate. No homopterans colonize potatoes in the northeastern United States until the second generation of CPB eggs. Thus, *E. puttleri* is unable to survive and parasitize first generation CPB eggs in the U.S. because of the lack of aphid honeydew. The better an agent is preadapted to the environment into which it is to be released, the greater the chance of successful biological control.

Because of geopolitical constraints, foreign exploration is often restricted to less than desirable areas. The Russian wheat aphid, *Diuraphis noxia,* within the past few years has been a major pest of cereal crops in Texas, through the central plain states, and northwest to Washington. Early efforts to collect natural enemies of this pest were centered in Turkey and Jordan, countries which are environmentally very different from most of the areas in North America where the Russian wheat aphid is a pest. Parasites, and to a lesser extent predators, tend to have very rigid behavior patterns and are only tolerant of a narrow range of conditions. Parasites collected from Turkey and Jordan were shipped to a USDA quarantine laboratory where they were examined for hyperparasites and disease organisms. Once clean cultures were available, parasites were released into an aphid colony under laboratory and greenhouse conditions in order to identify any nutritional, physiological or behavioral needs and to rear large numbers for field testing. Parasites were then made available for release in states where the Russian wheat aphid was established. Because the areas of collection were very different from the areas of release, survivorship and proliferation by the parasites depended on post-adaptation. More recently, in 1990 and 1991, natural enemies of the Russian wheat aphid have been collected from eastern Europe and western China. These agents should be better adapted for the hot summers and cold winters typical of the central plain states and the Pacific Northwest.

Many vertebrates (e.g., birds, reptiles and amphibians, fish, and some mammals) prey on insects, and many are active in the natural control of insect populations. Some vertebrates have been used in biological control, and many more have potential. A good example is the use of certain cyprinodont fishes, especially *Gambusia affinis*, against mosquito larvae. *Gambusia* has several characteristics that make it valuable in this regard: broad tolerance of salinity and organic pollution, viviparous reproduction, high fecundity, small adult size, surface-feeding habits, ability to penetrate regions where mosquito larvae breed, and ease of transport (Sweetman 1936a, Bay 1967). The giant toad *Bufo marinus* has also been used in insect control, for example, against sugarcane white grubs in Puerto Rico (Sweetman 1936a).

For more information on the use of parasitoids and predators in insect control, see van den Bosch (1975), Ridgway and Vinson (1977), Stehr (1975), and Waage and Greathead (1986).

Microbial Agents and Nematodes

Nematodes and microbial control agents are becoming more widely used in controlling insect pests. Although humans have long been aware of the "natural" control of insect populations by microbes, the first record of the idea of using them for insect control was in the eighteenth century. Insect microbiology/pathology did not become a serious topic of study until the early 1900s.

One of the major benefits of these agents is that they are generally very host specific. This means that these agents can be used in a pest management program where parasites and predators are being used without killing these natural enemies. Most commercially available agents have been isolated from insects collected in the field. However, biotechnology has provided new laboratory engineered organisms that in most cases are an improvement over field strains. Commercial development has progressed slowly until the past few years. Prior to the development of methods for protein "finger printing" different strains of the same organism, it was impossible for companies to patent the use of a particular organism. Now that such patents are being granted and companies have proprietary rights, they are willing to invest the capital in developing these organisms for insect control.

There is a wide variety of microorganisms that infect insects, including viruses, bacteria, fungi, and protozoans. These organisms can reach epidemic levels under natural conditions, causing high mortality to insect populations. Generally, these epidemics do not occur until the insect population has become very dense. For this reason it is generally not possible for a farmer to rely on natural epidemics to reduce pest populations to nondamaging levels. The future use of microbial agents is dependent on commercial development, that is, being able to mass produce an organism, formulate it in such a way that it can be applied with conventional spray equipment and can cause high enough mortality to reduce pest numbers to subeconomic levels. There has been virtually no commercial development of rickettsia or protozoans, and only limited development of fungi. Only recently has a virus been registered for use in controlling agricultural insect pests. Bacteria account for most of the commercial use of microbial agents.

Environmental conditions can greatly influence the effectiveness of microbial agents. Bacteria and fungi generally lose their virulence below 18°C, and many viruses do not replicate rapidly unless temperatures are between 21°C and 29°C. Many viruses and bacteria are quickly killed when exposed to sunlight (primarily ultraviolet light). Composition of the soil can be critical; for example, fungi seem to survive best in soils high in organic matter. Soil pH or the pH of the foliage is important. Acid conditions are unfavorable for *Bacillus popillae* spores (milky disease of the Japanese beetle), while alkaline conditions can destroy the polyhedral structure of polyhedrosis viruses. Environmental conditions in conjunction with the mode of entry by a microorganism into the insect's body will dictate to a large extent how the agent can be used.

Microorganisms invade insect hosts in several ways. All organisms can gain entry by being ingested or through damage to the insect's integument. Fungi and nematodes often enter the insect's body through the tracheae. Viruses, rickettsiae, and protozoans can be passed from adult females to their eggs (transovarial transmission). If an organism must be ingested, it is best to apply the organism at a time of day when the insect is most actively feeding.

Viruses

There are several types of viruses that infect insects. The most common types are the baculoviruses of which there are two occluded viruses of primary importance. The polyhedral and granulosis viruses are characterized by occlusion of virons in a paracrystalline protein matrix; the whole particle is known as a polyhedral inclusion body (PIB). This polyhedral structure is thought to protect the viron from adverse environmental conditions. The midgut is thought to be the only route of entry of these viruses. Passage of virus to gut cells begins with dissolution of the polyhedral protein in the alkaline gut contents and invasion of the gut epithelial cells (figure 15.4). There appears to be a virus enhancement factor (VEF) associated with the paracrystalline protein matrix that when released into the midgut lumen alters the structure of the peritrophic membrane, enabling the virus to gain access to the epithelial cells (figure 15.4).

Figure 15.4 Postulated mode of action of a polyhedral virus in the midgut of an insect larva. Polyhedral inclusion bodies (PIB) dissolve in the midgut lumen and release a virus enhancement factor (VEF) which immediately binds and degrades the peritrophic membrane. The virus particles then invade the epithelial cells.
Redrawn from Granados and Corsaro (1990).

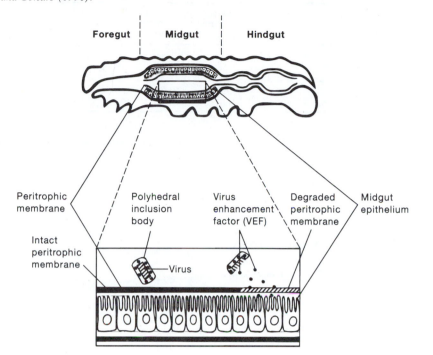

There are several factors in the field that affect the persistence of these viruses. Natural sunlight-UV (>290 nm) inactivates these viruses. Sunlight protectants, such as clays, titanium dioxide, polyflavanoids, and whitening agents have all increased the persistence of PIBs. Field temperatures in the 15°C to 45°C range have no effect on the stability of PIBs. Acid and alkaline conditions disrupt PIBs; however, viral activity is relatively unaffected at pH 4.0 to 9.0. When viruses are applied to plant foliage the PIBs may be exposed to repeated drying and wetting due to the formation of dew in the morning and subsequent drying; this action causes the PIBs to be dissolved, and a loss of viral activity results.

One of the limitations in using viruses is that the insects continue to feed until the infection has spread from the site of inoculation in the midgut to other cells. It may take several days from the time of ingestion to when the insect quits feeding. In some cases, the additional damage done during this time cannot be tolerated. For example, first instar codling moths ingest the granulosis virus while feeding on the surface of the apple; however, they do not die until they have penetrated the apple "skin." One of the advantages in using viruses is that the cadaver of a diseased insect acts as a primary inoculum source for infecting other larvae feeding on plant tissue associated with the cadaver. This process helps to maintain PIBs within

the pest habitat. In commercial settings, where it is not practical to allow a natural epizootic to occur, it may be necessary to make repeated applications of suspensions of PIBs.

Bacteria

There are several different genera of bacteria that infect insects; however, the only commercially available products are in the genus *Bacillus*. Members of this genus are spore-forming bacteria whose spores can remain viable in the soil for years. The two most important species are *B. popilliae* and *B. thuringiensis*.

B. popilliae is the causative agent of "milky disease" in soil inhabiting larvae of the Japanese beetle, *Popillia japonica*. After ingestion, the milky disease spores germinate within the gut of susceptible larvae. The vegetative bacteria induce localized infections in the midgut epithelium, followed by massive bacteremia. Only recently has it become possible to produce this bacterium *in vitro* so that viable spores can be used in the field. For this reason mass production of spores was done *in vivo* by injecting intrahemocoelically into larvae or adult *P. japonica*. Because this bacterium kills Japanese beetle larvae through massive infections, and it is only from infected larvae that spores are produced, it may take several years before the full benefits of applications of this bacterium are realized in a treated field.

Figure 15.5 Colorado potato beetle stage-specific larval mortality when fed foliage treated with *Bacillus thuringiensis* subsp. *san diego* (milligrams toxin per liter of water). CI = confidence interval.
From Ferro and Lyon (1991).

Larval stage	LC$_{50}$, mg/liter	Larval weight, mg	95% CI		
			Lower	Upper	df
Early 1st instar	2.03	1.0	1.46	2.60	14
Late 1st instar	3.92	2.3	2.02	6.27	36
Early 2nd instar	4.35	4.0	3.30	5.56	13
Late 2nd instar	14.45	7.8	10.75	19.50	33
Early 3rd instar	14.86	15.6	9.95	20.48	24

B. thuringiensis produces a parasporal crystal (protein delta-endotoxin) and a spore during sporulation. After the ingestion of spores and parasporal crystals by susceptible larvae, the crystals are solubilized and activated by alkaline (pH 10.5) gut proteases. The toxic subunits bind to receptor sites on the midgut epithelium within minutes of ingestion. This is quickly followed by lysis of these cells, causing a cessation of feeding. Although the spores penetrate into the hemocoel through lesions in the midgut to germinate, it is the starvation that kills the insect. This bacterium can be produced in large quantities using commercial fermentors. Formulations can be applied to foliage or other larval substrates in the same manner as most insecticides.

There are several subspecies (=varieties) of *B. thuringiensis* based on the serotype of flagellar antigens, and these subspecies tend to be host specific. For example, *B.t. israelensis* is effective only against Nematocera (Diptera), *B.t. kurstaki* against Lepidoptera, and *B.t. san diego* (=*tenebrionis*) against Chrysomelidae (Coleoptera). Because these bacteria are so host specific, they can be quickly incorporated into a pest management program in which biological control agents are an integral component. However, there are several operative factors affecting the effectiveness of these bacterial agents.

*B.t.*s are most effective against early instars (figure 15.5); their effectiveness is very dependent upon ambient temperatures (figure 15.6); the protein endotoxin is not very persistent; thorough coverage of foliage is necessary; and they are host specific. In many cropping systems, there is a complex of insect pests to control and often these need to be controlled at the same time. Because most *B.t.*s are so host specific, they cannot be effectively used when a complex of pest species is to be controlled. Because *B.t.*s are most effective against young larvae, the timing of applications is very critical, and generally this means sprays must begin earlier than when using synthetic insecticides. Because the protein endotoxin is not very persistent, and because the amount of toxin ingested is dependent on the amount of food consumed by the larvae, which in turn is dependent on temperature, the *B.t.*s are

Figure 15.6 Percentage mortality to early first instar Colorado potato beetle fed foliage treated with *Bacillus thuringiensis* subsp. *san diego* and held at different temperatures.
Redrawn from Ferro and Lyon (1991).

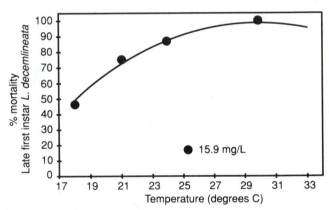

generally not very effective at low temperatures. New formulations and genetically engineered strains are being produced commercially that are more effective than commercial products in the past.

Field studies have shown the same amount of toxin formulated as an oil is more effective than when formulated as a wettable powder or aqueous flowable. The gene that controls the production of the delta endotoxin of *B.t. san diego* (kills the Colorado potato beetle) has been inserted into the bacterium *Pseudomonas fluorescens* and after the fermentation has been completed the broth is chemically treated and heated to kill the bacteria. During this process, the protein toxin becomes encapsulated by the cell wall. This encapsulation process appears to protect the protein against rapid degradation in the field, making it more persistent. In another case, the *B.t. san diego* (kills the Colorado potato beetle) toxin gene and the *B.t. kurstaki* (kills the European corn borer) toxin gene have been placed within the same bacterial cell to produce both toxins. This product then can be used for controlling both of these pests.

Figure 15.7   Adult Colorado potato beetle infected with *Beauveria bassiana*.
Courtesy of D. N. Ferro.

Fungi

The first microorganism to be recognized as an agent of disease was the fungus, *Beauveria bassiana* (figure 15.7), which Bassi demonstrated to be the causal agent of the white muscardine disease of the silkworm, *Bombyx mori*. Bassi de Lodi's "germ theory of disease" was developed from this host/pathogen life system. Although most microbial pathogens infect their host primarily via the digestive tract, fungi usually infect by direct penetration of the insect body wall. Phytophagous insects with sucking mouthparts are susceptible to few infectious diseases other than fungi. Fungi are known from almost all insect taxa. However, commercial production of fungi for use in agricultural systems has not proceeded as rapidly as has the development of bacteria and viruses. One of the major limitations in using fungal pathogens is that many species are easily killed by fungicides applied to control plant pathogens. Other environmental factors affect the initial levels of infection and secondary spread.

The Entomophthorales contain many species that frequently cause epizootics in aphid, beetle, fly and leafhopper populations. There are several environmental conditions that affect the spread of Entomophthorales and other fungi. Epizootics generally are positively correlated with leaf wetness either from rain, irrigation, or heavy dew. For example, spotted alfalfa aphid, *Therioaphis trifolii maculata* populations in the Bet Shan Valley, Israel, are decimated each year by epizootics of the fungus *Erynia radicans*. Although the Bet Shan Valley is a hot, arid area, moisture is provided by heavy dew and irrigation. Most fungal species can form and discharge conidia at 5° C and then germinate. However, it may take over 16 hours for these events to occur, whereas at an optimum of around 20° C, all of this may happen within a few hours.

Another important factor in the primary spread of fungal pathogens is the level of initial inoculum in the field at the beginning of the season. Secondary spread is by movement of infected larvae and adults coming into contact with noninfected hosts and by airborne dispersal of infective conidia.

There are several soil-borne fungi that commonly infect soil-dwelling insects. *Beauveria* spp. and *Metarhizium anisopliae* in particular infect Coleoptera. The soil habitat generally affords a relatively stable environment for these fungi. However, in many soils the upper few centimeters of soil reach temperatures well above 50° C, a temperature lethal to the vegetative stages of most insect pathogenic fungi. Applications of *B. bassiana* have been applied to potato foliage to control the Colorado potato beetle with little success. However, by using the same formulations and incorporating the material into the soil where larvae come in contact with the spores prior to pupation, there has been reasonable success.

Nematodes

Nematodes in the families Steinernematidae, Heterorhabditidae, and Mermithidae parasitize insects. The first two families are mutualistically associated with bacteria that kill the host quickly by causing septicemia. Two species of bacteria, *Xenorhabdus nematophilus* and *X. luminescens,* are symbionts of steinernematids and heterorhabditids, respectively. Also, most of the research on using nematodes as potential pest management tools have used these two families.

Parasitism by nematodes may result in sterility, reduced fecundity, delayed development, or death of the host. Most nematode species cause rapid host mortality which prevents or limits insect damage. However, host sterility may be just as effective as host death in regulating populations. *Deladenus siricidicola* sterilizes the wood wasp, *Sirex noctilio,* and has been successfully used to reduce wood wasp populations below economic threshold levels in *Pinus radiata* plantations in Australia.

Steinernematids and heterorhabditids have simple life cycles. They have a resistant stage called the "dauer," meaning durability or permanence. The dauer is the third stage nematode ensheathed in the second stage cuticle, and is the infective stage and contains cells of the symbiotic bacteria (*Xenorhabdus*) within the intestinal lumen. Host finding by infectives can be an active response to physical and chemical host cues. The dauers enter suitable host via natural body openings (mouth, anus or spiracles, exsheath, and penetrate mechanically into the hemocoel). The nematodes then release the *Xenorhabdus* cells causing septicemia, which kills the host within 24 to 48 hours. The nematodes feed upon the bacterial cells and host tissues, and as these resources become depleted, later generations develop into infective dauers that exit from the cadaver and seek new hosts. If no new host is found, the infectives can survive for a long time under humid conditions.

The effective host range in the field is limited by the nematodes' moisture requirement, exposure to radiation, and temperature extremes. These nematodes are very sensitive to dry conditions. Antidesiccants have been used to improve the survival of *Steinernema feltiae,* however not sufficiently to allow this species to be used against foliar feeding insects. Temperatures above 35° C are lethal, but they can recover from exposure to temperatures as low as −10° C. Because of these environmental limitations nematodes are most likely to be integrated into pest management programs to control soil-borne stages of insect pests or pests found in cryptic habitats, such as insect galleries in trees where humidity is high and the infective stages are protected from hostile environmental factors.

Microbial agents for controlling insect pest populations are likely to play a more important role in pest management programs in the future, especially as programs attempt to integrate biological control agents into programs.

For additional information on the use of microbes in insect control, see Briggs (1975); Bulla (1973); Burges and Hussey (1971); Burges (1981); Ferron (1978); Gaugler and Kaya (1990); Ignoffo (1988); Maddox (1975); Maramorosch (1977); N.Y. Academy of Sciences (1972); Nickle (1984); Poinar and Thomas (1978); Steinhaus (1963); Tinsley (1977); and Weiser (1969). Cameron (1973) outlines the history of insect pathology.

Genetic Control

Genetic control involves manipulation of the mechanisms of heredity.

An outstanding example of genetic control is the sterile-male method. Many years ago the idea was conceived that if a sufficient number of the matings in a given population in the field resulted in no offspring, then over a period of generations the population would decrease (Knipling 1955). Thus if sexually sterilized males are introduced into, or induced within a wild population each generation and if the matings of these sterilized individuals exceed normal matings, the population will decline. If the number of sterile individuals is kept constant (by additional releases) for each generation, the ratio between sterile and normal matings will increase rapidly and the rate of population decline will increase correspondingly.

The earliest application of the sterile-male technique was against the screwworm fly, *Cochliomyia hominivorax,* in the southeastern United States (Knipling 1959, Baumhover 1966). During 1958–1959, 3.7 billion sterilized (by gamma irradiation) screwworm pupae were reared and released throughout large portions of Florida and Georgia. This resulted in the successful eradication of the fly from this part of the country. Since that time there have been sporadic outbreaks traceable to the movement of infested animals into the territory, but the screwworm has not been a problem since 1959. Baumhover (1966) describes the cost and savings of the screwworm-eradication program:

> During the two-year campaign, 3.7 billion screwworm pupae were produced, and 6.3 million lb. of horsemeat and whale meat were used. Twenty light aircraft were used to release flies over a maximum area of 85,000 square miles, and peak employment, including plant personnel, fly distributors, field inspectors, and clerical and administrative help, totalled 500. However, for a research cost of only $250,000 and an eradication-program cost of $10 million, ranchers in the Southeast have experienced $140 million in savings since inception of the program in 1958.

The screwworm-eradication program has since been expanded and is now active in the southwestern United States and Mexico. A barrier zone of sterilized males is maintained to block the northward movement of flies from overwintering sites in Mexico. Although this program has faced difficulties (Novy 1978), it constitutes a notable success in insect control. As a result, losses of nearly $2 billion have been prevented (Scruggs 1978).

Because of the vast numbers of sterilized males that need to be released to ensure successful mating with wild females and effect eradication, the sterile-male method is practical only against insects that occur in relatively small populations as adults. Small populations may occur naturally or may be brought about by the employment of other methods of insect control, insecticides in particular. Further, it is very important that the sterilized flies have the vigor to compete with wild individuals for a mate. Finally, the species involved must be amenable to artificial rearing in huge numbers and at a reasonable cost. The sterile-male method is impractical against insects that are very prolific and widespread or against insects that appear in large numbers sporadically and unpredictably (e.g., floodwater mosquitoes) because large numbers of artificially reared individuals would have to be maintained at all times (National Academy of Sciences 1969).

Since the screwworm-eradication campaign, the use of the sterile-male method against other species has been studied and is currently being applied or in the pilot test stage for a number of pest species (Knipling 1972).

In addition to radiation, several chemicals have been discovered that induce sterility when ingested by insects. Out of approximately 6,000 compounds screened, more than 300 show promise as chemosterilants (National Academy of Sciences 1969). Chemicals produce sterility primarily by causing insects to fail to produce sperm or ova, causing the death of sperm or ova after they have been produced, or producing genetic defects in spermatozoa that prevent zygote development (National Academy of Sciences 1969). The last of these mechanisms for the production of sterility is the most desirable because such sterilized males are generally competitive with the unsterilized males in mating with available females.

Chemosterilants are useful in situations in which the sterile-release method would be inappropriate (e.g., against species that occur in very large populations and are difficult or impossible to rear in large numbers in the laboratory). Because the best chemosterilants so far known must be ingested to be effective, they are usually applied with baits. Chemosterilants that are effective by contact might be used in association with luring stimuli such as light and sex attractants. Another possibility is the application of a chemosterilant to breeding places. However, since all the promising chemosterilants are mutagenic agents, they present a serious hazard to other animals, including humans. According to Metcalf and Metcalf (1982), "Their use in pest management cannot be recommended."

Other genetic approaches to insect control include the use of such factors as (World Health Organization 1967):

1. Sex-ratio distorters.
2. Detrimental genes incorporated into chromosomes that have meiotic drive.
3. Chromosome translocations in heterozygous males.
4. Conditional lethal genes that allow the parents to survive in the laboratory but are lethal to their descendants under field conditions.

For more information on genetic control of insects, see Baumhover (1966); Börkovec (1975); Curtis (1986); Davidson (1974); Foster et al. (1972); Hoy and McKelvey (1979); Huffaker and Messenger (1976); Knipling (1972, 1979, 1982); Krafsur et al. (1987); Labrecque and Smith (1968); Lorimer (1981); Metcalf and Metcalf (1982); Pal and Whitten (1974); Richardson (1978); Smith and von Borstell (1972); and Whitten (1982).

Breeding Insect-Resistant Hosts

Under natural conditions, some individuals of a given species possess characteristics that make them more able than others to cope with various environmental stresses (e.g., insect attack). Individuals possessing such characteristics will tend to be more successful in reproducing and hence in passing the heredity determinants of these characteristics to their progeny. For this reason, over time a population exposed for many generations to a given stress will become composed predominately of members that are able to cope with this stress. This, in a very general way, is the process of natural selection. In a number of instances, artificial selection has been carried out in the laboratory to produce strains or varieties of organisms that are resistant, or more resistant than members of other strains or varieties, to an environmental stress. Another approach has been to encourage the proliferation of some naturally occurring resistant strain or variety.

Resistance may be defined as the heritable ability of an individual, strain, variety, race, and so on (plant or animal) to repel or withstand the effects of some environmental stress or stresses to a greater degree than other individuals or groups. In the context of this discussion insect pests are the "environmental stresses."

Much more emphasis has been placed on studies of insect resistance in plants than in animals. This is due at least in part to greater use and lower costs of plant breeding.

Host plant resistance to insect pests may act on the insect at any time in the sequence of events leading up to successful colonization and survivorship of offspring on a host plant (discussed in the following sections). It is important to understand these steps, as the rate of successful colonization is determined by the insect's ability to locate a suitable host.

Host-Habitat Location
The crop habitat is generally located through phototactic, anemotactic, and geotactic responses by the insect. This initial step is extremely important for migratory insect pests, especially those with a limited host range. For example, aphids that colonize cultivated plants are attracted to objects that have a peak reflectance of wavelengths of about 550 nm, which is the same for many cultivated plants, weeds, and herbaceous plants.

Host Location
Once within the crop habitat, the insect must find its host. It does so by using several sensory modalities. Color, shape, and odor are important cues for the cabbage maggot fly in orienting to its host. Once on the plant, the fly uses olfactory (kairomone) and tactile (pubescence, texture) cues to assess the quality of its host.

Host Recognition
This step is closely linked to host location and host suitability and is generally accomplished through chemical and tactile cues. Plant chemicals may be detected by olfaction, gustatory activity, or palpation of the plant surface. The Colorado potato beetle must palpate or chew the foliage before it can determine if the foliage is acceptable. The codling moth, *Cydia pomonella,* lays its eggs on waxy surfaces of leaves and apples and generally avoids the bark and stems.

Host Acceptance
Tactile cues, odor, and ingestion of plant material are all important for finding acceptable hosts. For example, aphids generally must probe tissue and ingest plant protoplasm before accepting the host. In this way, aphids can vector nonpersistent viruses in plants that are not acceptable hosts.

Host Suitability
Although all sensorial responses by the insect may indicate that it has colonized an acceptable host, the host still may not be suitable for survival and reproduction. Nutritionally, the plant may lack essential amino acids, it may have low concentrations of carbohydrates, or there may be an imbalance of these nutrients.

Plant resistance characteristics are under genetic control. However, some characters are very labile and fluctuate wildly under different environmental conditions. General types of resistance may be *genetic resistance* (primarily regulated by plant genotype) or *ecological resistance* (primarily regulated by environmental factors).

Cultural practices can take advantage of ecological resistance where there is phenological asynchrony between host and pest development. For example, the bean leaf beetle emerges in soybean fields in early September, after the crop is ready to harvest (short season cultivar).

Genetic resistance refers to nonpreference of host, host antibiosis, or plant tolerance (terms coined by Painter 1951). Nonpreference is based on resistance factors that influence the behavioral processes leading up to host acceptance. For example, the Asiatic rice borer, *Chilo suppressalis,* lays 15 times more eggs on susceptible plants than on resistant plants. This type of resistance is not nearly as dramatic as antibiosis, but when integrated with other pest management tactics such subtle differences may provide just enough reduction in pest populations to allow for good overall control.

Antibiosis effects may be due to morphological, nutritional or plant metabolites. There may be several adverse physiological effects when plant tissues are ingested by insects due to toxic plant metabolites (alkaloids, glucosides, quinones), lack of essential nutrients, or enzymes that inhibit normal digestion and assimilation of food:

1. First instars die.
2. Abnormal growth rates.
3. Abnormal metabolism of food.
4. Failure to pupate.
5. Failure of adult emergence.
6. Malformed adults.
7. Inability to store food for overwintering.
8. Decreased fecundity, reduced fertility.

Plant tolerance to insects refers to the ability of the plant to repair injury or to grow to produce adequate yields despite supporting pest numbers capable of causing losses in susceptible cultivars. These result from one or more of the following:

1. Vigorous plant growth.
2. Regrowth of damaged tissues.
3. Strength of stems and other plant structures.
4. Production of additional branches or foliage.
5. Compensation by adjacent plants.

Beck (1965); de Wilde and Schoonhoven (1969); Galun et al. (1975); Hanover (1975); Kogan (1975); Maxwell and Jennings (1979); Maxwell et al. (1972); Maxwell and Jennings (1980); Painter (1958); Pathak (1975); Singh (1986); Sondheimer and Simeone (1970); and van Emden (1966) should be consulted for more information on plant resistance to insects.

Ecological Control

Ecological control procedures involve the removal, destruction, modification, or isolation of materials that might favor the survival of an insect pest by affording food or making a site suitable for breeding and/or dormancy.

Such procedures may be applied in an agricultural context (*cultural control*) as well as in other pest insect contexts.

Cultural Control

Cultural practices refer to that broad set of management tactics or options that may be manipulated by farmers to achieve crop production goals, or the manipulation of the environment to improve crop production. Cultural control, on the other hand, is the deliberate manipulation of the cropping system or specific crop production practices to reduce pest populations or to avoid pest injury to crops. These tactics may include: impediments to pest colonization of the crop, creation of adverse biotic conditions that reduce survival of individuals or populations of the pest, or modifications of the crop in such a way that pest infestation results in reduced injury to the crop.

Destruction or Provision of Breeding or Overwintering Refugia

Many natural enemy species require food sources in the form of pollen, nectar, or innocuous arthropods that are not present in particular crop habitats. These food requirements may be provided to support natural enemy populations by encouraging deliberate development of certain wild vegetation habitats near plantings of the crop. For example, natural biological control of the grape leafhopper, *Erythroneura elegantula,* the most important pest of grapes in the San Joaquin Valley of California, can be achieved by an egg parasitoid, *Anagrus epos* Girault. However, *Anagrus* is only effective when vineyards are located within 3.5 miles of streams and rivers. *A. epos* does not successfully overwinter on the grape leafhopper, but does on populations of the blackberry leafhopper, *Dikrella californica* (Lawson), which survives on blackberry stands in stream and river bottoms. Vineyards planted near blackberry stands along rivers and streams have high levels of parasitism of *E. elegantula.*

Another example is aphids that vector viruses which infect horticultural crops in the temperate areas of North America. Most aphid species in this area overwinter as eggs on their primary woody host. The green peach aphid (GPA), *Myzus persicae,* overwinters on peach and wild cherry and a number of other *Prunus* spp. The GPA serves as a vector for the maize dwarf mosaic virus of sweet corn. Because the GPA does not migrate into horticultural crops until early July, it is primarily the later plantings of sweet corn that show symptoms and yield reductions. Thus growers plant early sweet corn near peach trees, so that the later plantings can escape heavy probing by the GPA. A similar situation exists with the GPA as a vector of cucumber mosaic virus of green pepper. Again, either the destruction of *Prunus* spp. or planting away from these primary hosts can reduce losses to these viruses.

Destruction or Provision of Alternate Hosts or Volunteer Plants

Many pest species feed on alternate hosts or volunteer plants that allow for populations to build up or act as trap crops. For example, the black cutworm, *Agrotis ipsilon,* is a major pest of corn seedlings in the corn belt states, especially in no-till cropping systems. Young larvae feed on weeds then move onto corn seedlings until they reach the fourth instar. They cause serious damage by cutting or drilling the plants. However, if the seedlings can reach the four-leaf stage before being infested, no significant yield reductions occur. However, this strategy is a two-edged sword. If the grower waits until the corn reaches the four-leaf stage before cultivating or using herbicides to control the weeds, yield reductions occur due to weed competition. So the best management tactic is to apply preplant herbicides at least 14 days before planting to reduce weed populations, hence minimizing the number of ovipositional sites and early instar food sources.

Crop Rotation or Maintenance of a Host-Free Habitat

By rotating crops or maintaining host-free habitats, the normal life cycle of an insect pest is interrupted by effectively placing the insect in a nonhost habitat. Rotation is generally most successful against arthropod pest species with long generation cycles and with limited dispersal capabilities. For example, in the white fringed weevil complex, *Graphognathus leucoloma* (Boheman) and *G. peregrinus* (Buchanan), adults lay more eggs when they feed on soybean, causing heavy damage to this crop. However, the grass crops, including corn, are in some way nutritionally deficient and will not support populations and do not suffer damage from this pest. So a soybean/corn rotation is effective and economical.

Another example is with the Colorado potato beetle. Overwintered Colorado potato beetle can disperse to colonize new fields; however, after they emerge from the soil, they need to regenerate their flight muscles and they will not oviposit until they have fed on a nutritionally acceptable host plant. So between the colonization process and physiological development, a rotated field of as little as 200 m from last year's field can be colonized 1 to 2 weeks later and generally at lower production densities than a nonrotated field. This saves 1 to 2 sprays for first generation larvae. In the Northeast, because the rotated field is colonized later, the first summer generation of adults do not emerge until after the first of August when the beetles are exposed to shorter days, which induces diapause. Those beetles emerging after August 1 do not produce eggs. Hence, there is only one larval generation to control (figure 15.8).

Figure 15.8 Relationship between date of emergence of first generation adult Colorado potato beetles from the soil and induction of diapause, as measured by number of eggs laid, So. Deerfield, Massachusetts, 1986. *From Voss, Ferro, and Logan (1988).*

Date of emergence	n^a	Days after emergence for first oviposition mean ± SEM	Total no. eggs laid	% ♀♀ ovipositing	n^b
25 July	11	7.8 ± 0.36	1,060	81.8	9[c]
1 Aug.	12	8.0 ± 0.00	128	8.3	1[c]
8 Aug.	12	—	0	0.0	0
15 Aug.	10	—	0	0.0	0
22 Aug.	11	—	0	0.0	0

[a] Number of females surviving.

[b] Number of females ovipositing.

[c] Significantly different ($P < 0.001$, Fisher's exact probability test).

Tillage

Tillage operations used to produce a crop include soil turning and residue-burying practices, seedbed preparation, and cultivation. Some forms of tillage can reduce pest populations indirectly by destroying wild vegetation (weeds) and volunteer crop plants in and around crop-production habitats. Overwintering populations of *Helicoverpa zea,* the corn earworm, may be greatly reduced by either fall or spring plowing operations. Overwintering survival of the soybean stem borer, *Dectes texanus,* is inversely related to depth of burial of soybean crop residue following harvest.

Timing of Planting or Harvest

Alterations in planting date and harvest date can frequently result in escape from damaging pest infestations. By delaying the planting date of potatoes (after June 10 emergence) in Massachusetts in nonrotated fields, fields suffer less damage from the Colorado potato beetle. Overwintered beetles remain in the field for about 5–7 days after emerging from the soil and if no host plants are present, leave the field by flight. Ninety percent of the overwintered population emerges by early June, and late planted fields are only colonized by a small proportion of this population. Because the summer generation beetles emerge after the onset of short days, the beetles enter diapause without laying any eggs; hence, there is only one larval generation to contend with.

The pink bollworm, *Pectinophora gossypiella,* overwinters as last instar larvae, and diapause is controlled by short days (<13 hours of light). By harvesting early, the number of overwintering larvae is reduced to low levels. The following practices are encouraged in the Southwest: (1) defoliate or desiccate the mature crop to cause all bolls to open at nearly the same time, expediting machine harvesting; (2) harvest the crop early, shred stalks, and plow under crop remnants immediately; (3) irrigate prior to planting in desert areas if water is available; (4) plant new crops during a designated planting period, which allows for maximum suicidal emergence of overwintering moths, that is, moths emerge and die before cotton fruit is available for oviposition. This tactic is only effective when short season cultivars are planted.

Trap Crops

Crop monocultures are often damaged more severely by pests than is the same crop located in an area with crop diversity. However, there are cases in which such diversity can aggravate pest problems. It is in these situations that trap crops can be important. *Lygus* bugs are a key pest of cotton in the San Joaquin Valley of California. A major habitat of the bugs is alfalfa fields, which are commonly interspersed with cotton fields. When the alfalfa is harvested, the bugs leave the field in large numbers and infest cotton fields. By "strip cutting" the alfalfa field (i.e., harvesting alternate strips of alfalfa so that there is always foliage available for the *Lygus* bugs), the number of bugs migrating to cotton can be minimized. A twist on this approach is to plant strips of alfalfa every 400 feet between cotton fields and then only harvest one half of the alfalfa strip at any one time. Stern (1969) demonstrated that *Lygus* bugs concentrate in these strips, and the need for insecticides was virtually eliminated. Furthermore, beneficial parasitoids and predators were extremely abundant in these strips and, as they moved to and from the adjacent cotton, an added benefit resulted.

Water or Nutrient Management

Water can be used directly for suffocating insects or indirectly by changing the overall health of the plant, whereas fertilizers can influence the injury to a crop primarily through alterations in crop growth or nutritional value to the pest. Some pest populations are enhanced by poor crop growth, whereas others are enhanced by succulent crop growth. Flooding of cranberry bogs is a valuable tool for controlling several insect pests of cranberry. Flood irrigation is frequently used to reduce populations of wireworms in vegetables and sugarcane crops. Likewise, flooding can be used to control white grubs in sugarcane, especially under conditions of high temperature. Furrow irrigated potato fields in many parts of California tend to crack upon drying, exposing potato tubers to ovipositing potato tuberworms. In areas where this is a problem, overhead sprinkler irrigation is recommended. Overhead sprinkler irrigation can enhance dissemination and infectivity of some entomopathogenic organisms, especially fungal pathogens. Such practices help to promote epizootics of *Nomuraea rileyi* in populations of the velvetbean caterpillar in soybean.

Enhancement of succulent cotton growth through fertilization renders the crop more attractive to populations of the cotton aphid (*Aphis gossypii*), cotton fleahopper (*Pseudatamoscelis seriatus*), and the cotton bollworm (*Heliothis zea*). Undernourished plants are often more attractive to colonizing aphids because they are more yellow and reflect more light in the 540 nm range.

Cultural control and using cultural practices are generally necessary components of a pest management program that uses biological control agents and other noninsecticidal tactics, as it is by using these practices that pest populations can be reduced to levels in which biological control agents can have an impact.

Ecological Control of Nonagricultural Insect Pests

One of the most effective ways of controlling many insect pests is the maintenance of high standards of sanitation. Accumulations of trash, garbage, and untreated sewage provide food and breeding sites for numerous annoying and disease-carrying insects (not to mention rats). Thus these materials must be removed from their source of accumulation and be properly treated and disposed of. Garbage disposals, trash and garbage collection services, sewage collection systems and treatment plants, sanitary landfills, and incineration are among the methods used.

Naturally or artificially produced aquatic or semiaquatic habitats can produce severe insect problems, for example, outbreaks of mosquitoes, biting midges, and horse and deer flies. Examples of naturally occurring aquatic or semiaquatic habitats that provide suitable conditions for the breeding of these insects are saltwater and freshwater marshes, swamps, and various depressions in which water can collect. Artificial habitat creation includes the construction of water impoundments for

a variety of reasons (e.g., flood control, hydroelectric power, water storage, drinking water for animals, recreational purposes, borrow pits) and the development of water distribution systems, particularly for irrigation.

To effect insect control, marshes, swamps, and depressions containing water may be either flooded or drained, depending on a variety of factors. For example, along the east coast of Florida, many of the salt marshes, which produce large numbers of salt-marsh mosquitoes, *Aedes taeniorhynchus* and *A. sollicitans,* and biting midges, particularly *Culicoides furens,* have been impounded, a measure that has provided successful control.

Careful management of artificially impounded water and water distribution systems is necessary to avoid insect problems. Practices include (1) proper construction so that water does not stand in any one place longer than necessary (e.g., proper land grading for irrigation purposes, appropriate drain construction); (2) deepening or filling very shallow areas; (3) periodical removal of vegetation within the water-level fluctuations boundaries; and (4) periodic removal of vegetation and accumulated debris from drains to avoid clogging, seepage, and overflow.

The use of properly designed and constructed storage boxes, bins, rooms, and so on, in which cool, dry conditions prevail are usually ideal for the inhibition of pest species of insects that attack stored materials.

Chemical Control

This section deals with those natural or synthetic chemicals that cause directly the death, repulsion, or attraction of insects.

Insecticides

An *insecticide* may be defined as a chemical or a mixture of chemicals employed to kill insects and related arthropods. The term *pesticide* is more inclusive, referring to insecticides as well as herbicides, rodenticides, and other substances.

Insecticides were apparently used before recorded history. The writings of Greeks, Romans, Persians, and Chinese all allude to the use of various substances, such as sulfur, hellebore (a poisonous herb), and arsenic. Prior to the 1940s the insecticidal value of a number of inorganic chemicals (e.g., arsenic, mercuric chloride, and carbon disulfide) and organic chemicals of botanical origin (e.g., pyrethrum, cube, and nicotine) was known and put to extensive use. The discovery of DDT by Paul Müller in Europe in 1939 revolutionized insect control and marked the beginning of the development and application of synthetic organic insecticides.

Since that time hundreds of compounds of varying insecticidal value have been discovered, and thousands of new potential toxicants are being evaluated each year by

Figure 15.9 Pyrethrum compounds consist of four esters, which are the combinations of two different alcohols with two different acids. The proportions may vary with the strain of flowers, growing conditions and method of extracting and concentrating these toxins.

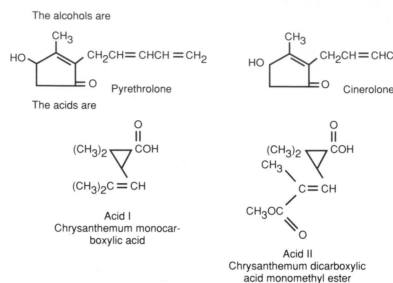

a detailed screening process. The vast majority (about 90%) of currently used pesticides are of the synthetic organic variety. General treatments of chemical control include Bohmont (1990); Haskell (1985); Hassall (1990); and Worthing and Hance (1991).

Insecticides may be classified in different ways. Some insecticides are more useful than others for a given stage in the insect life cycle. Thus there are ovicides, larvicides, and adulticides. Another useful classification is based on the primary route of entry: *stomach poisons* act via ingestion and absorption from the alimentary canal; *contact poisons* are absorbed through the cuticle; and *fumigants* enter through the spiracles and tracheal system or body wall in the gaseous state. These categories are by no means completely exclusive (e.g., many contact insecticides also act quite effectively via the oral route).

Insecticides can also be classified based on the chemical nature of the insecticide with the primary division being inorganic versus organic.

Inorganic

There are few inorganic insecticides used today. In the past, lead arsenate and sulfur were used. The mineral sodium aluminum fluoride (kryocide) is mined in Greenland and is a by-product of the aluminum industry. This insecticide is still used to control foliage feeding pests of potatoes and grapes. Because it is applied at 10 lbs/acre per application, there is concern about long-term residues.

Organic

There are a wide range of organic insecticides (including petroleum and vegetable oils, botanicals and synthetic chemicals) used to control insect pests.

Plant products have several uses in insect control. Some act as attractants (geraniol and methyl eugenol), some as repellents (citronella and oil of cedar), some as solvents (cottonseed oil), and some as carriers of insecticides (pulverized walnut-shell). Yet, the primary use of plant derivatives is as insect toxicants. Nicotine, extracted from *Nicotiana tabacum* and *N. rustica,* has been used for several hundred years for insect control. It is highly toxic to a great number of insects and appears to directly affect the central nervous system. The insecticidal activity of pyrethrum (figure 15.9) was discovered by the Persians around 1800. "Persian Powder" was sold at extravagant prices for louse and flea control in human habitats. Pyrethrum is extracted from the flowers of *Chrysanthemum coccineum* and *C. carneum.* The primary source of natural pyrethrum today is Kenya. This chemical attacks the insects peripheral nervous system and for this reason has a rapid knockdown. Rotenone (figure 15.10) is found in 68 species of leguminous plants in the genus *Derris,* grown principally in the Far East, and in the genus *Lonchocarpus,* found mostly in the Amazon Basin of South America. This chemical is found in the roots of these plants. Indians in South America mash the roots and allow the exudate to flow into streams to kill fish for food. Rotenone is a metabolic inhibitor (i.e.,

Figure 15.10 Rotenone is one of six rotenoids that occur naturally and are extracted from plants in the genera *Lonchocarpus* (from South America) and *Derris* (from Asia).

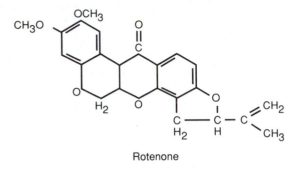

Rotenone

Figure 15.11 Structure of DDT molecule.

DDT

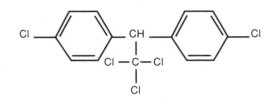

Figure 15.12 Parathion is oxidized within the insect to paraoxon, a more potent anticholinesterase.

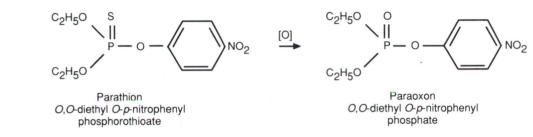

Parathion
O,O-diethyl *O-p*-nitrophenyl
phosphorothioate

Paraoxon
O,O-diethyl *O-p*-nitrophenyl
phosphate

it inhibits the respiratory chain, the oxidation of NADH-linked substrate). Although these botanicals are still in use, they are extremely expensive to produce and for this reason they are rarely used in a commercial setting.

Oils in their natural state are highly phytotoxic, but when used in an emulsion, they may be safely applied to plants. Mineral oil is a heterogeneous mixture of saturated and unsaturated chains and cyclic hydrocarbons. Certain fractions of this mixture are much more useful as insecticides than other fractions. The quality of the oil is based on the viscosity, boiling or distillation range, and sulfonation rating (purity or degree of refinement). Generally, the lower the viscosity the safer it is to use with respect to phytotoxicity. Phytotoxicity increases with increase in distillation range, because the greater the distillation range the less volatile the oil. However, the heavier spray oil fractions are more effective at killing insects than the lighter oils. Oils are composed of both saturated and unsaturated hydrocarbons. The unsaturated hydrocarbons are unstable and readily form compounds which are toxic to plants. To test the degree of purity, a sulfonation test is conducted. The oil to be tested is reacted with strong sulfuric acid and the unsaturated hydrocarbons that react with the acid precipitates. The unreacted part, the *unsulfonated residue* (U.R.), is measured in % and is used as the measure of purity. Dormant oils have a U.R. rating of 50% to 90%, and the more highly refined summer oils have a rating of 90% to 96%.

One of the first synthetic organic insecticides was DDT, which is representative of the organochlorine chemicals (figure 15.11). DDT was developed in 1938 by Dr. Paul Müller of the Geigy Company of Switzerland. Organochlorine molecules tend to be very stable because of the placement of the chlorine ions in the molecule. Most of these chemicals have been banned from use because of their persistence in the environment and toxicity to nontarget organisms. Soon after DDT was released into the market in the early 1940s, primarily to control lice and fleas that were vectoring disease organisms to humans in war torn Europe, the organophosphorous insecticides were developed.

In 1934 Dr. Schrader working for I. G. Farben of Germany discovered that certain organophosphorous compounds were active insecticides. But because of the political scene in Germany at the time the military pushed the mammalian toxicity element—first group of chemicals now described as nerve gases. These chemicals kill insects and vertebrates by binding with acetyl cholinesterase in the synaptic junctions of the nervous system. This results in a continuous flow of electrical-chemical signals along the nerve, which results in repeated muscle contraction. Parathion is considered to be the first modern organophosphorous insecticide (figure 15.12). It was synthesized by Schrader in 1944. These chemicals are still widely used to control insect pests in agriculture.

Figure 15.13 Carbaryl is the most widely marketed carbamate insecticide and is registered to control over 100 different insect species.

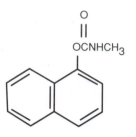

The carbamate insecticides (figure 15.13) were developed in the early 1950s and their mode of action is very similar to that of the organophosphates. Like the organophosphates, the carbamates tend to break down rapidly once applied, leaving no harmful residues. However, if these chemicals are incorporated into the soil where they are not exposed to light and the soil pH is low, they may persist for 1–2 years. The systemic insecticide Temik (aldicarb) was used in the northeastern United States to control insect pests of potato. However, soon after it was introduced in the late 1970s, residues of aldicarb or its metabolites were found in well water. This chemical is very water soluble when compared to other insecticides. If heavy rains occurred before the potato plants developed a root system, which absorbed the chemical, aldicarb was quickly moved through the top soil and into the water saturated zone. Carbamates still are widely used worldwide.

The pyrethroid insecticides (figure 15.14) are synthetic analogues of the natural pyrethrins and tend to be more stable because of the lower reactivity of the side-chains. Although the pyrethroids affect the peripheral nervous system, causing a quick knockdown, the primary target seems to be the ganglia of the central nervous system. These chemicals are now widely used throughout the world.

Nomenclature

Insecticides have three names. The trade name is the name listed on the label and is determined by the company marketing the material (e.g., Ambush). The common name is the name accepted internationally by most of the industry to use instead of the chemical name [e.g., permethrin (Ambush)]. The chemical name provides a chemical description [e.g., (3-phenoxyphenyl)-methyl (+ or −) cis-trans-3-(2,2-dichloroethenyl)-2, 2-dimethylcyclopropane-carboxylate (Ambush)]; however, it is extremely cumbersome to use.

Toxicity

The toxicity of an insecticide is established by exposing test animals (insects and vertebrates) to a range of doses and determining the number killed at each dose. By plotting the number killed against the range of doses using log-probit paper, it is possible to extrapolate the dose that kills 50% of the test animals (figure 15.15). When the exact amount of insecticide being applied per body weight (μg of toxicant per mg body weight of the insect) is known, the lethal dose which kills 50% of the population can be determined (LD_{50}). If the insects have been dipped in different concentrations or fed foliage dipped in different concentrations, then the lethal concentration is established (e.g., LC_{50}). Once these values are calculated for different insects and for vertebrates, it is possible to compare these values to other insecticides. For additional information on insecticide toxicity, see Matsumura (1985).

Mode of Action

This refers to the way in which the chemical acts upon the insect to kill it (Corbett et al. 1984, Narahashi and Chambers 1989). Much research needs to be done in this area of toxicology, but some generalities can be made. There are five main groups: physical poisons, protoplasmic poisons, metabolic inhibitors, nerve poisons, and molting inhibitors/insect growth regulators (IGRs).

Physical poisons kill by some physical action such as excluding air (mineral oil) or by abrasion or sorptive actions resulting in loss of water (e.g., silica aerogel or borax). *Protoplasmic poisons,* such as arsenicals, kill by precipitating intercellular proteins. *Metabolic inhibitors* affect the metabolic pathways. For example, the respiratory poisons such as hydrogen cyanide, rotenone, and dinitrophenols deactivate respiratory enzymes. Materials like piperonyl butoxide inhibit the mixed function oxidase system. The foramidine compounds like chlordimeform (Fundal) inhibit amine metabolism. *Nerve poisons* directly affect the peripheral or central nervous system of insects and vertebrates. The organophosphorous and carbamate insecticides tie up cholinesterase in the synaptic junction so that there is a build up of acetylcholine, which results in continuous nerve transmission and contraction of muscles. *Molting inhibitors and growth regulators* kill insects by interfering with the molting process, preventing insects from molting from one instar to the next.

Synergists

Some chemicals have the property of greatly increasing toxicity of certain insecticides. When the increased toxicity is markedly greater than the sum of the two used

Figure 15.14   Permethrin is one of several recently introduced pyrethroid insecticides and is used to control agricultural and medically important insect pests.

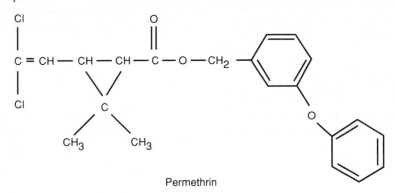

Permethrin

Figure 15.15   Hypothetical probit regressions for a resistant insect population (RR) that is 100× more resistant than the susceptible population (SS) based on comparison of the LD_{50} values.

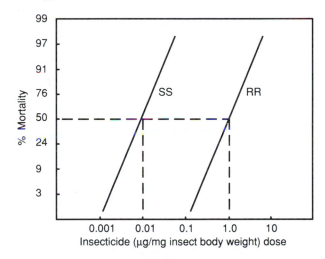

Figure 15.16   (a) Sesamin is one of the active principals of sesame oil, which has long been known to synergize the action of pyrethrin. (b) Sesamolin is another active principal of sesame oil.

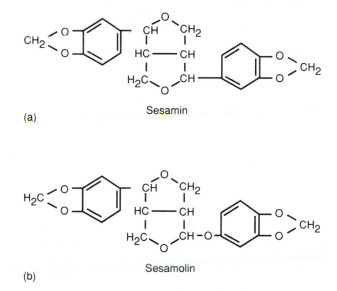

(a)

Sesamin

(b)

Sesamolin

separately, it is called a synergistic action. Most of these synergists have been used with pyrethrum or pyrethroid insecticides. They act by preventing the hydroxylation of these insecticides by the mixed function oxidase system. The most common synergist is piperonyl butoxide. Sesamin (figure 15.16a) and sesamolin (figure 15.16b) are also common synergists and are active principals of sesame oil. These materials have been used to synergize the action of pyrethrins for over 50 years.

Formulations

The residual activity of an insecticide and its utility as a management tool can be altered according to how it is formulated (Cross and Scher 1988). Most insecticides today are applied at 1/10th to 1 lb/acre. Try to imagine spreading a pound of sugar evenly over an acre of crop foliage. This would be very difficult. However, if the sugar is first dissolved in 50 gallons of water, then it can be sprayed uniformly over the crop. Insecticides are formulated so that it is possible to obtain uniform coverage by forming an emulsion or suspension in water or by diluting with some inert material such as clay. An insecticide, as it appears on the market, is composed of a toxicant or active ingredient (poison) and one or more inert materials (nonactive, nonpoisonous). These inert materials may function to dissolve the poison; act as a carrier; dilute the toxicant; or act as an emulsifier, dispersant, or spreader-sticker.

Emulsifiable Concentrate

Most insecticides are only slightly soluble in water and must be dissolved in an organic solvent (such as xylene or methylnaphthalene). The insecticide/solvent is not miscible in water (figure 15.17a); however, by adding an emulsifying agent (similar to household detergent) an emulsion of insecticide/solvent in water can be made and sprayed onto the crop (figure 15.17b). An emulsifying agent is generally a long-chained hydrocarbon in which one end of the chain is lipophilic and the other end is hydrophilic (figure 15.18). These are referred to as emulsifiable concentrates, and each gallon of formulated material usually contains from 2 to 4 lbs of active ingredient.

Wettable Powder

The insecticide/solvent can also be applied to organic or mineral particles 1–10 microns (e.g., talc) and allowed to dry. Then by adding a wetting agent that breaks the surface tension of the water, the powder (impregnated with the insecticide) can be suspended in water and applied as a water-based spray. This is referred to as a wettable powder. Generally, the finished product is 50% to 75% poison.

Granule

The same technique used for making wettable powders can be used, except instead of applying the insecticide to a small particle like talc, a larger particle (100–500 microns) is used. About 5% to 20% of the finished product is poison. These insecticide-laced granules can be placed in the soil and when the granules become wet, the insecticide is slowly released into the soil where soil-inhabiting insects come in contact with it, or the insecticide is absorbed by the plant's roots and translocated to the foliar parts of the plant where it is consumed by insect pests. Insecticides that are absorbed by the plant and translocated to other parts are called *systemic insecticides*.

Flowable

Some insecticides in their raw form cannot be easily dissolved in an organic solvent or in water. These insecticides are finely ground in oil (oil-based flowable), water (water-based flowable), or with no lubricant (dry flowable). The particles are ground to about 4 microns in either water or oil, then a suspending agent is added, and, in the case of the water-based material, an anti-freeze agent is added. The final product has the consistency and drying properties of a latex paint. Insecticides formulated in this manner tend to provide better residual activity than other spray formulations.

There are other formulations, such as aerosols and dusts; however, the previously mentioned formulations account for most of the materials currently being used.

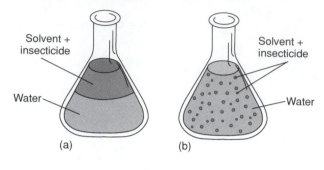

Figure 15.17 Beaker with layer of solvent plus insecticide floating on water. (a) Beaker in which an emulsifying agent has been added and contents stirred to produce an emulsion. (b) Emulsion of solvent plus insecticide in water.

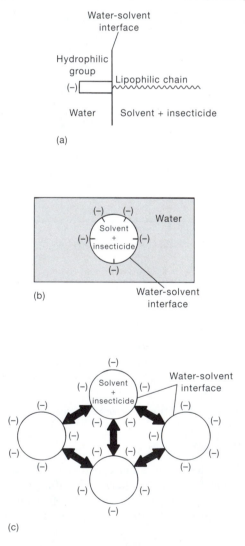

Figure 15.18 (a) Lipophilic chain of an emulsifying agent within organic solvent/insecticide and hydrophilic group in water. (b) Emulsifying agent in droplet of solvent/insecticide. (c) Stabilized emulsion due to repulsion of charged droplets.

Repellents

Repellents are substances that are mildly toxic or nontoxic to pests, but prevent damage by causing the pests to make oriented movements away from the source. In very early times, smoke from wood fires was used to keep away biting and annoying insects. As early as 1897, oil of citronella was used as a mosquito repellent (Shepard 1951). Most of the earlier repellent substances were quite odorous and perhaps somewhat repellent to humans as well as insects. However, many of the more modern, synthetic repellents have little, if any, disagreeable odor.

The ideal insect repellent would be more than merely repellent to insects. It would be nontoxic, nonirritating, and nonallergenic to humans and domestic animals; inoffensive in odor; harmless to fabrics; persistent (not easily removed by perspiration, rubbing, laundering, and so on); and effective against a broad spectrum of pest species. In addition, the ideal repellent would be cheap and nondamaging to plastics, painted surfaces, and the like.

The use of good repellents has a number of advantages: they afford individual protection from insects without the necessity for expensive and time-consuming population eradication; they do not damage or kill beneficial animals or plants; and the ones that are available for use are nontoxic to humans. On the other hand, disadvantages may include the following: repellents are at best a temporary measure (a few hours at most) and tend to evaporate from or rub off skin or clothing due to perspiration and the like; they commonly have an oily feel and may have a somewhat disagreeable odor; they must be applied in comparatively large doses (in the range of 20–40 mg/cm^2 of skin; (Smith 1966); and they may damage certain plastics or painted surfaces.

So far, repellents have been primarily used for the protection of man and animals against attacks from bloodsucking or otherwise annoying insects. The armed forces have over the years had a particular interest in repellents, realizing that protection of troops from disease vectors would greatly increase their military advantage. However, interest has also developed in the use of repellents in plant protection. Metcalf, Flint, and Metcalf (1962) describe three general groups of repellents: those used against crawling insects, the egg laying of insects, and the feeding of insects. Repellents used against crawling insects usually consist of a repellent barrier interposed between an insect and whatever material happens to be attractive to it. For example, creosote has been used as a barrier against the migration of chinch bugs, *Blissus leucopterus,* and trichlorobenzene and other repellent insecticidal chemicals to protect buildings from termite invasion. Creosotes derived from coal and wood tar have been used extensively for the protection of wood against termites, powderpost beetles, and rot organisms (Shepard 1951). Creosote and other oils apparently ". . . smother the natural attraction of the insect to its food or oviposition site" (Shepard 1951). Several chemicals have been found that are reasonably effective in repelling insects from feeding. Washes containing bordeaux mixture, lime, and other materials are used to repel leafhoppers and some chewing insects, and inert dusts have been useful on cucurbits to repel cucumber beetles (Metcalf, Flint, and Metcalf 1962). An ideal repellent for plant protection would be one that would somehow block the natural attractants (kairomones; see chapter 8) to which pest species respond. Diethyltoluamide, considered to be one of the best general repellents yet discovered, dimethyl phthalate, ethyl hexanediol, dimethyl carbate, and powdered sulfur are examples of repellent chemicals that have been applied to human skin and/or clothing.

When a broad spectrum of effectiveness is desired, a mixture of repellents may be appropriate. For example according to Rolston and McCoy (1966):

> The Armed Forces used a repellent containing dimethyl phthalate, ethyl hexanediol, and dimethyl carbate for application to skin, and a repellent containing benzyl benzoate, *n*-butylacetanilide, and 2-butyl-2-ethyl-1,3-propanediol for application to clothing. Both mixtures provided protection against medically important mosquitoes, fleas, ticks, and chiggers.

Smokes, smudges, and burning of pyrethrum (insecticidal and repellent) are useful repellent measures for outdoors. Chemicals that have been used as repellents against pests of livestock include low concentrations of pyrethrums, butoxypolypropylene glycol, and dibutyl succinate.

In recent years there has been interest in the development of a systemic insect repellent (i.e., one that would be taken orally and following ingestion would appear in the skin or skin secretions). Ideally such a systemic would be repellent to a wide variety of arthropods, of low toxicity, and effective for a long period of time (12–24 hours or more after ingestion; Sherman 1966). There have been repeated failures in the search for a workable systemic repellent, but research is still being conducted along these lines (National Academy of Sciences 1969).

For more information on insect repellents, see Metcalf and Metcalf (1982) and Shorey and McKelvey (1977).

Attractants

Chemical cues that elicit a behavioral response by an insect pest or its natural enemies can be used in a number of different ways in insect control. Chemical attractants are one of the most widely used tools in pest management programs today. The sensory mechanisms involved in many phases of insect behavior, including searching for

Figure 15.19 A visual and odor trap used for monitoring cabbage maggot fly populations.

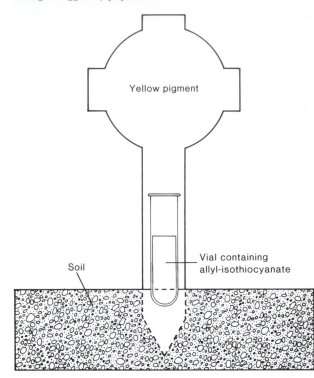

Yellow pigment

Soil

Vial containing allyl-isothiocyanate

Figure 15.20 Sentry® trap used for monitoring the European corn borer.

food, oviposition sites, and mates are stimulated and controlled by chemicals. Chemicals that deliver behavioral messages are termed *semiochemicals*. Interspecific semiochemicals that favor the producer/emitter are called allomones, whereas those that favor the receiver are called kairomones, and semiochemicals used for intraspecific communication between individuals of the same species are called pheromones (see chapter 8).

Food or ovipositional lures are natural chemicals (kairomones) present in many plant hosts or produced by microorganisms associated with the plant that directs the insect pest toward suitable sites for feeding or ovipositing. Ovipositing seedcorn maggots, *Delia platura,* oviposit near germinating seeds because of stimulatory substances produced by certain microorganisms growing in exudates from germinating seeds.

Many insect species use vision to find their hosts. Plant architecture, contrast between the plant and its background, and plant color all are important visual cues. By mimicking these cues, visual traps can be used to monitor for insect pests. For example, the cabbage maggot fly, *Delia radicum,* uses both visual and olfactory cues for finding host plants. A cross-shaped trap (20 by 20 cm) painted yellow mimics the plant color and architecture. When a mustard oil (allyl-isothiocyanate), an important plant volatile that the flies orient to, are combined, they make a very effective trap for monitoring flight activity of this pest (figure 15.19).

There are several different types of pheromones; however, it is the sex pheromones used by insects to locate a mate that have been most widely used in pest management programs. Sex pheromones have been identified for a wide range of insect pests. The chemical composition of the pheromone, release rate of pheromones, trap design, and trap placement within the field are important parameters that determine the effectiveness of the traps. For example, there are three known strains of the European corn borer, *Ostrinia nubilalis,* in North America. The 97:3 (Z)- and (E)-11-tetradecenyl acetates and 3:97 (Z)- and (E)-11-tetradecenyl acetates populations occur sympatrically in many areas. Where this occurs, separate traps baited with each of the blends must be placed in the field at the same time in order to detect the abundance of each strain. The Sentry™ trap (figure 15.20) has been shown to be much more efficient at trapping than the wing trap. Because moths mate in the weedy areas adjacent to corn fields where the humidity within the plant canopy is high, the traps are placed in these weedy areas and not in the field.

For an attractant to be useful, the insect behavior with which it is associated must be understood. Further, the attractant must be extracted and chemically identified, and if possible it should be synthesized. Ideally the synthesized attractant or an analogue would be even more attractive to an insect than the original. Another avenue of approach is to test various chemicals for attractiveness. The various sex pheromones that are commercially available in the United States are listed in Caswell (1977). Beroza (1976); Mayer and McLaughlin (1990); Metcalf and Metcalf (1982); Shorey and Mc-Kelvey (1977). Ridgway et al. (1991); Rockstein (1978); and Roelofs (1975, 1978), contain information on the use of chemicals to control insect behavior.

Other Chemical Controls

Additional substances with promise as chemical control agents include *antimetabolites, feeding deterrents,* and *hormones.*

Antimetabolites chemically resemble essential nutrients and interfere with metabolism. They are low in mammalian toxicity and are thus safe to use (e.g., for insect-proofing fabrics). They may be effective against insects that have access only to treated food; however, they have limited value against polyphagous insects (National Academy of Sciences 1969).

Feeding deterrents or antifeeding compounds may be defined as compounds "which will prevent the feeding of pests on a treated material, without necessarily killing or repelling them" (Wright 1963). Antifeeding compounds have been used for several years in the mothproofing of fabric. However, the use of these compounds in the protection of crops is a fairly new idea. The use of feeding deterrents is still in the experimental stage. They offer considerable specificity because they would affect only the insects that feed on treated plants and would spare the parasites and predators of these pest species. They are also low in mammalian toxicity. Further information on feeding deterrents may be found in Shorey and Mc-Kelvey (1977).

Because growth, development, and sexual maturation are largely regulated by hormones (see "Control of Growth and Metamorphosis" in chapter 5), these substances have potential value in insect control. Ecdysone, juvenile hormone, and various analogues of these compounds have been shown to disrupt the development of an insect if applied at appropriate times and in appropriate doses (Williams 1970). For example, cyasterone, a substance related to ecdysone, when injected into a diapausing *Cynthia* (moth) pupa in a very low dose (0.2 microgram), stimulates termination of diapause and formation of a normal moth. However, if a high dose (10 micrograms) is injected, the developmental events are accelerated and their sequence disrupted. This results in the premature deposition of cuticle, which literally "locks in" the epidermal tissues before they have completed the developmental changes necessary for survival into the adult stage. Compounds related to ecdysone (i.e., the phytoecdysones) have been found in many different kinds of plants, particularly ferns. In addition to lethal effects, there is evidence that certain of the phytoecdysones may function as feeding deterrents. As with ecdysone and its relatives, juvenile hormone and its mimics (e.g., methoprene) can be applied with lethal effects, either by preventing the transformation of the pupa into an adult or by inhibiting the development of eggs. Again, as with ecdysone, several species of plants have been shown to produce compounds that mimic juvenile hormone activity. Understanding the chemistry and physiological effects of ecdysone and juvenile hormone and their analogues (i.e., of insect growth regulators or IGRs) may ultimately provide the key to the synthesis of insecticides with extreme specificity.

Due to the expense of production, inability to penetrate the cuticle, and wide range of activity, it is unlikely that ecdysones will be used commercially in the future. However, the juvenile hormone analogues methoprene (figure 15.21) and kinoprene are used to control stored-product pests and pests found in food processing plants, and for controlling insect pests of ornamental plants and vegetable seed crops, respectively. They are active against a wide range of insect pests, including beetles, flies (including mosquitoes), ants, aphids, leafhoppers, scales, and caterpillars. These chemicals interfere with metamorphosis.

The benzoylphenyl urea chemicals (figure 15.22) are a new class of IGRs that are starting to be used in controlling agricultural and forestry insect pests. In general, they tend to be more selectively toxic than other synthetic insecticides. Chemicals in this group inhibit chitin synthesis, and for this reason it takes longer to kill the insect than it would take by using most other synthetic insecticides. It usually takes 2–5 days before the complete effects are noticed.

Williams (1967) traces the early developments in the area of the applied use of insect hormones. Menn and Pallos (1975) review the use of IGRs in insect control. See also Hedin (1991), Menn and Beroza (1972), Morgan and Mandava (1987), Riddiford (1972), and Schneiderman (1972).

The Positives and Negatives of Chemical Control

Among the advantages afforded by chemicals are the following: they are usually very effective, they generally act within a short period of time, they are effective when applied against large pest populations, and they are readily available for use when needed. The use of insecticides has played an important role in the development of agriculture as we know it today. Both increased yield per unit land area or animal and improved quality of agricultural

Figure 15.21 Kinoprene is a juvenile hormone analogue used for controlling aphids, scales, and whiteflies in greenhouses, whereas methoprene, another juvenile hormone analogue, is used to control fleas, flies, ants, lice, mosquitoes, and stored products' insect pests.

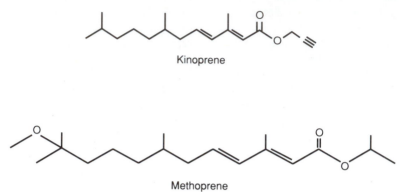

Kinoprene

Methoprene

Figure 15.22 Teflubenzuron is a novel benzoylphenyl urea that acts as a chitin synthesis inhibitor.

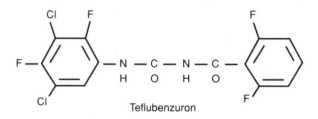

Teflubenzuron

products are in large part attributable to the successful use of insecticides. Agricultural yield showed a 54% increase in the 20 years following the general use of synthetic organic insecticides (National Academy of Sciences 1969). Undoubtedly other factors played a role in this impressive increase, but application of insecticides is certainly among the most important. Insecticides have also enabled people to bring several vector-borne diseases under control or at least reduce their extent.

On the negative side, application of an insecticide can be very hazardous, and direct contact with a highly toxic insecticide can cause severe illness and death. This is why the careful reading and following of instructions on the labels on pesticide containers and constant concern with safety during their application are so important. Other important considerations are the storage of pesticides in well-labeled containers out of reach of children; taking extreme care in application, including avoidance of spillage on clothing or skin, or inhalation of sprays or dusts; not smoking or eating when working with toxicants; and the proper disposal of empty insecticide containers.

Also of concern is the presence of residues in food products, plant and animal, that have been treated with pesticides at some point in their production. Every pesticide that is shipped in interstate commerce in the United States must be registered under the Federal Insecticide,

Fungicide, and Rodenticide Act. The objective of registration is the protection of the consumer, not only in terms of safety, but also with regard to the value of the products. When an insecticide is registered, the amount of residue (tolerance) to be allowed in various food products is established. Some insecticides have such low mammalian toxicities that they are tolerance exempt, whereas others, which are considered dangerous following an extensive series of tests, are subject to a rigid tolerance. If these tolerance levels are exceeded, various degrees of legal action against the violators result. In addition to the strict controls applied by the federal and state governments, the development of a new insecticide is a rigorous, time-consuming process that increases the probability that an insecticide that reaches the market will be effective and reasonably safe if directions are followed carefully.

Despite the hazards involved, the safety record of pesticide usage in the United States has not been as bad as one might suspect:

In recent years there has been so much publicity about pesticide hazards that much of the general public has developed a distorted impression of the magnitude of the problem. Vital statistics show that pesticides account for only 1 out of every 700 accidental deaths and only 5% of all poison deaths in the United States. As a cause of death, pesticides are far outranked by common drugs and by household agents such as cleaners, polishes, and solvents. For many years, the ubiquitous aspirin tablet has caused about the same number of accidental deaths annually in the United States as all the pesticides combined, and since 1957 aspirin has caused slightly more accidental deaths. As a matter of record, the death rate for all pesticides has remained constant at slightly over one per million of population for the past 25 years.—(National Academy of Sciences 1969)

Environmental pollution with insecticides has become a matter of great concern. Highly residual insecticides can pass well beyond their intended targets and may reduce populations of beneficial insects and wildlife. DDT in particular has been attacked in this regard. Two major factors have contributed to its being a problem: it is highly residual, and it becomes concentrated in a stepwise fashion along a food chain. Woodwell et al. (1969) provide a graphic description of this process of "biological concentration":

> . . . in the food web which the herring gull is a scavenger in Lake Michigan, DDT (DDE and DDD) concentrations in the bottom muds at 33–96 feet averaged 0.014 part per million. In a shrimp (*Pontoporeia affinis*) they were 0.44 ppm, more than ten times higher. Levels increased in fish to the range of a few ppm (3.3 alewife; 4.5 chub; 5.6 whitefish), another tenfold increase, and jumped in the scavenging, omnivorous herring gull to 98.8 ppm, twenty times higher still, and 7000 times as high as in the mud.

Unfortunately, such accumulations of insecticides kill or inhibit reproduction of certain animals, birds in particular. Further, the long-term indirect effects pesticide residues may have by disrupting the intricate balance of ecosystems is of great concern.

Rachel Carson (1962) in her book *Silent Spring* painted a particularly vivid and frightening picture of what could happen. Brown (1978); Haque and Freed (1975); Jepson (1990); Khan (1977); McEwen and Stephenson (1979); Miller and Berg (1969); Perring and Mellanby (1977); Rudd (1964); van den Bosch (1978); White-Stevens (1977); and Woodwell et al. (1971) discuss various aspects of insecticides in the environment. Johansen (1977) considers the effects of pesticides on pollinating insects.

Another major problem associated with the use of insecticides has been the development of insecticide resistance in strains of several pest species. The term *resistance* as commonly used by entomologists is relative and as such implies different levels of susceptibility by a pest population to a pesticide. To the field entomologist, if it takes 2 or 3 times the recommended rate to control a pest, it is categorized as being resistant. Whereas to the toxicologist, resistance is not conferred until it takes up to 100 times the dosage to kill 50% of the test population. However, it is through the highly quantitative approaches of the toxicologist that levels of resistance are determined and theories of how to manage for resistance have been developed.

There are now over 450 species of mites and insects that have developed resistance to insecticides (figure 15.23). There is a genetic basis to insecticide resistance (i.e., there are genes in an insect population allowing some individuals to survive exposure to pesticide doses that are

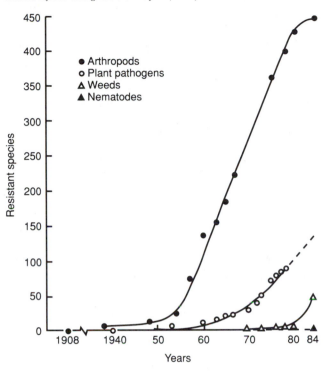

Figure 15.23   Chronological increase in number of cases of resistant species.
Redrawn from Georghiou and Taylor (1986).

lethal to the rest of the population). It has become accepted that the selection of resistant genotypes is an inevitable consequence of pesticide use, so that a new pesticide has a limited useful life, and the course of action to deal with this situation is to use an alternative pesticide. The gradual depletion of substitute pesticides, as resistance to these alternatives develops, shows the limitations of this practice.

In the most common situation the resistant insect possesses a detoxification mechanism that enables it to render a particular insecticide or group of insecticides nontoxic. For example, resistant house flies, *Musca domestica* (Diptera, Muscidae), produce an enzyme, DDT dehydrochlorinase, that catalyzes the breakdown of DDT to HCl and nontoxic DDE (National Academy of Sciences 1969). *Culex tarsalis,* a mosquito (Diptera, Culicidae), is able to detoxify the organophosphate insecticide malathion by means of a carboxyesterase enzyme. Strains of some species have been observed to possess a "behavioral" resistance to certain pesticides. According to Hoskins (1963):

> The chief example is a hyperirritability which causes insects to leave a treated surface before picking up a lethal dose. Thus female *Anopheles albimanus* mosquitoes were observed to survive DDT house spraying in certain districts of Panama

Figure 15.24 Proportion of Colorado potato beetle resistant (RR) or susceptible (SS) to transgenic plants expressing the *Bacillus thuringiensis* delta-endotoxin using a seed mixture of 90% transgenic plants and 10% susceptible, and survivorship of RS individuals is varied.
From Ferro (in press).

Prop. Surv. RR	F_4		F_6		F_8		F_{10}		F_{12}	
	RR	SS	RR	SS	RR	SS	RR	SS	RR	SS
50%	0	.83	.23	.14	.71	.01	.91	0	.97	0
10%	0	1.0	0	.98	0	.93	.04	.74	.46	.19
0%	0	1.0	0	1.0	0	1.0	0	1.0	0	1.0

in much greater numbers than in former years. When confined over treated paper they quickly left the paper and showed signs of restlessness to a greater degree than did mosquitoes from other districts.

Other resistant individuals have been discovered to have cuticles or gut walls that allow much slower penetration of toxicant than the nonresistant individuals. For example, certain strains of German cockroaches (*Blattella germanica,* Blattaria) are resistant to the carbamate Sevin as a result of the slow penetration of the insecticide through the integument. This has been called structural resistance (Hoskins 1963).

The first known case of insecticide resistance in the United States was in 1908, when it was found that strains of San Jose scale, *Aspidiotus perniciosus* (Homoptera, Coccoidea), were resistant to lime-sulfur sprays (National Academy of Sciences 1969). In 1946, DDT-resistant house flies were first discovered in Sweden. Prior to that time only 9 species of insects and ticks were known to have developed insecticide resistance (Brown 1968). Resistance in these cases was, of course, to inorganic and botanical insecticides.

Most resistance appears to be controlled by a single gene; however, resistance may be polygenic or controlled by more than one gene. Dominance has an important influence on the eventual fate of the resistant (R) gene. When the R allele is rare, it occurs almost exclusively in the heterozygous state (RS). Therefore, the susceptibility of RS during the early generations is the main determinant of the rate of resistance evolution (figure 15.24).

The presence of R genes at low frequencies suggests that they exist because of recurrent mutations and are maintained at this low frequency by a balance between mutation and selection, or there is no selective advantage for individuals carrying the R gene (i.e., they are less "fit"). It appears, based on research by population geneticists, that a realistic value for mutation rates at equilibrium would be 10^{-4} to 10^{-6} for R alleles.

If it is assumed or known that biochemical or genetic factors conferring resistance to a pesticide exist within a population, then the rate of evolution of resistance development is going to be related to a number of parameters. Georghiou and Taylor (1986) classified these factors as follows:

Genetic
　frequency of resistance alleles
　number of resistant alleles
　dominance of resistant alleles
　penetrance; expressivity; interactions of resistant alleles
　past selection by other chemicals
　extent of integration of resistant genome with fitness factors
Biological/Ecological
　Biotic
　　number of generations per season
　　offspring per generation
　　monogamous vs. polygamous; parthenogenesis
　Behavioral/Ecological
　　isolation; mobility; migration
　　monophagous, oligophagous, polyphagous
　　refugia within treated area
Operational
　Chemical characteristics
　　chemical nature of pesticide
　　relationship of chemical to previously used chemicals
　　persistence of residues; formulations
　Application characteristics
　　application threshold (ETL or action threshold)
　　selection threshold
　　life stage(s) selected
　　mode of application
　　alternating selection

Figure 15.25 Patterns of resistance development based on a simulation model. (*a*) Resistance steadily increases, no immigration, dose = 0.01. (*b*) Stable cycle, 100 immigrants per day, dose = 0.03. (*c*) Increasing cycles, 50 immigrants per day, dose = 0.01.

Redrawn from Tabashniky and Croft (1982).

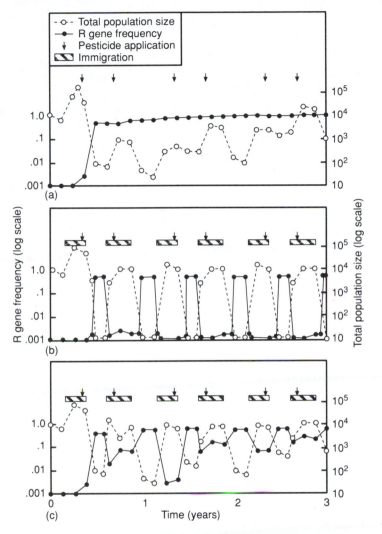

Those factors in the first two categories cannot easily be controlled, and the importance of some may not be determined until resistance is expressed. The operational factors affecting resistance development are human related and can be managed. Resistance management of pest populations is becoming an extremely important component of any pest management program, especially because fewer new pesticides are being produced.

Resistance steady-state occurs when various factors affecting resistance (e.g., application and selection, or alternating chemicals) are manipulated so that populations initially may show a limited increase in resistance to an insecticide but then revert back rapidly or frequently enough after selection that they never enter the log phase of resistance development (figure 15.24).

In these cases resistance never develops or develops very slowly. A number of ecological factors have been implicated in greatly affecting the stability of resistance steady-state and rate of resistance development. Immi-

gration of susceptible individuals into treated areas and presence of untreated refugia within treated areas appear to be two of the more important factors governing the rate of resistance development (figure 15.25).

Unfortunately, natural enemies develop resistance slower than pests. There are two well accepted theories about this subject. Pests are preadapted to detoxify pesticides because they detoxify plant toxins, whereas natural enemies are not preadapted. Natural enemies starve or emigrate following sprays that eliminate hosts, hence reducing natural enemy populations under selective pressure. Much work has been done in California by Dr. Marjorie Hoy in developing resistant strains of the predatory mite, *Metaseiulus occidentalis*. She has developed strains resistant to carbaryl (carbamate), azinphosmethyl (organophosphate), and permethrin (pyrethroid). This means that these predators will survive when the above materials are used to control nonmite pests.

For more information on the development of resistance to insecticides, see Georghiou and Saito (1983), Georghiou and Taylor (1986), National Research Council (1986), and Rousch and Tabashnik (1990). Brewer (1991) discusses a very modern approach to detecting the occurrence of insecticide resistance (i.e., the use of techniques from molecular biology).

Physical Control

Physical control involves the use of devices that destroy insects directly or act indirectly as barriers, excluders, or collectors. Physical controls include the use of heat, light, electricity, X rays, and so on, to kill insects directly, reduce their reproductive capacity, or to attract them to something that will kill them.

Physical methods include the use of simple manual techniques such as handpicking, swatting and crushing, jarring and shaking, and the use of various kinds of barriers, excluders (e.g., screens), and traps. Handpicking is effective only when comparatively small numbers of insects are to be dealt with (e.g., the handpicking of bagworms from an ornamental shrub or the handpicking of tomato and tobacco hornworms, extensively used in the past, but now limited to small-garden situations). The fly swatter, the bare hand, or any of a wide variety of implements can be useful against a few insects within a dwelling or on one's person. Jarring and shaking or hand beating of shrubs or trees, especially fruit trees, is sometimes used to remove insects, particularly beetles. Sheets, or buckets of kerosene or other material, can be used beneath the plant being shaken to trap the insects that fall into them. Various sorts of collecting devices have been used against insects. These have ranged from a bucket and paddle to horse-drawn "hopper-dozers" and similar machines that were used for grasshopper control in the western United States in the late 1800s (National Academy of Sciences 1969).

Several mechanical means are employed to act as barriers to insect movement. Sticky materials in which insects become hopelessly entangled have been used, for example, in the form of flypaper that traps numerous flying insects. Sticky materials have also been applied in bands about the trunks of trees to protect them from oviposition damage caused by the periodical cicada. Red spheres that mimic ripe apples can be coated with Stickum™ and placed in apple trees to entrap apple maggot flies, *Rhagoletis pomonella,* in search of egg laying sites. Metal collars around tree trunks have also been used for the same purpose and are effective against any nonflying insects that would otherwise attack the branches and foliage. Metal is also used in the construction of shields

around the foundation of houses and buildings to prevent attack by subterranean termites. Screens of metal, cloth, fiberglass, or plastic have been used to cover various openings (doors, windows, vents, etc.) to containers and dwellings to allow the passage of air but exclude most insects. Cloth netting (e.g., cheesecloth) is useful in excluding the periodical cicada from young fruit trees and has been used extensively to protect sleeping persons from mosquitoes and other biting insects.

In high value vegetable and fruit crops, row covers made of artificial webs of polymer fibers can exclude insect pests from the plants. Row covers are used to prevent cabbage maggot flies, *Delia radicum,* from ovipositing on crucifer crops and to exclude bacteria-laden striped cucumber beetles, *Acalymma vittatum,* from feeding on cucumber seedlings in the spring, which prevents the spread of bacterial wilt. Other barrier techniques include the protective packaging of food products, low sheet-metal fences against the flightless Mormon and coulee crickets, plastic sheeting and bags as liners and containers, and the digging of deep furrows around fields being threatened by chinch bug and armyworm attacks (Metcalf, Flint, and Metcalf 1962).

Traps are used for control, survey, and surveillance purposes. Control traps are usually used in conjunction with some attractive stimulus (e.g., light, food, or sex pheromone) and with some means of killing the insects that enter (e.g., a pesticide or an electrically charged grid). Survey and surveillance traps are used to detect the presence of potential pest species, to evaluate the effectiveness of any control procedures that may have been carried out in a given area, and to monitor levels of various economically or medically important species.

Both high and low temperatures have been used to destroy pest insects in a variety of situations. Most insects become inactive at temperatures of about 4° C or below, and many stored products maintained at such temperatures are not damaged, although the insects present would not likely be killed. However, cold can be used to kill dry-wood termites in furniture in vaults for 4 days at −9° C (National Academy of Sciences 1969). High temperatures are more effective. Few, if any insects can survive prolonged exposure to temperatures much above 60° − 66° C. Submersion of mangoes in a hot water bath (46° C) for 60 min killed 100% of the West Indian fruit fly, *Anastrepha obliqua,* without damaging the fruit. High temperatures have also been used against insects that infest stored grain, coffee bean, various seeds, citrus fruits, clothing, bedding, furniture, baled fibrous materials, bulbs, soil, and logs. Recent studies using controlled-atmosphere cold storage at 0° C, >95% RH, and atmospheric components of 1.0%–1.5% O_2, < 1% CO_2, and a balance of N_2 showed that no nondiapausing codling moth, *Cydia pomonella,* survived after apples had been stored for 13 weeks. So for early harvested apple cultivars, standard controlled-atmosphere cold storage could be used as a quarantine treatment. Whether low or high temperatures are used depends in part on the nature of the product to be protected or disinfested.

Insects such as clothes moths and carpet beetles prefer soiled wool garments. Hence laundering or drycleaning not only kills any insects that may be present, but also makes clothing less susceptible to attack.

The burning of crop stubble is highly effective against all life stages of any insects. In the Pacific Northwest, the stubble remaining after grass seed has been harvested is burned, effectively killing any insects, plant pathogens, or weed seeds on the soil surface. Some potato growers use propane burners to kill the potato vines in preparation for harvesting. If this is done before the Colorado potato beetle enters diapause, all life stages remaining in the field are killed, effectively reducing the overwintering population.

Electrically charged grids near windows and doors or in combination with an attracting light source have been used to kill insects on contact.

Several other possible physical means for insect control may eventually come into their own. These include high-frequency electric fields, particularly against stored-grain species; ionizing radiation such as X rays, and high-energy beta particles against insects attacking materials that would not themselves be harmed by the radiation. For example, gamma radiation (75–100 Gy) has been used to irradiate mangoes being shipped abroad from Australia; this procedure was used to kill eggs and larvae of the fruit fly, *Bactrocera tryoni.*; laser beams, short-duration light used, for example, to induce overwintering species to break diapause and succumb to unfavorable environmental conditions; light reflection from aluminum foil against aphids; and sounds that destroy, repel, attract, or confuse insects.

Regulatory Control

Regulatory control involves the enactment and enforcement of quarantines. Quarantines are designed to prevent the entry of potential pest species, to confine them to as small an area as practicable once introduced, or to prevent them from being exported to other countries. Even if quarantine measures only retard the spread of a given species, the money saved may very well justify the cost.

In the definition of regulatory control the phrase "potential pest" is especially significant. It should be borne in mind that because a given species is not a major pest in its native land does not mean that it will not be one in another region where few or none of its natural enemies exist.

Prior to the advent of human means of transport, insect dispersal consisted entirely of natural means (migration, etc.). However, when people developed methods of covering long distances, a new means of dispersal became available. The environmental changes that people have brought about have also profoundly influenced the dispersal and distribution of insects. For example, the boll weevil and harlequin bug extended their ranges from Mexico into the United States because of the extensive irrigation of large areas of the desert that was once an effective natural barrier (Swain 1952).

Examples of the introduction of insect pests directly attributable to human transport are numerous. A particularly interesting example is the way in which the gypsy moth, an important forest pest, was introduced into the United States. In 1869 an amateur entomologist who was studying silkworms acquired living specimens of the gypsy moth and brought them to Massachusetts (Swan 1964). Since that time the gypsy moth has spread to several other states in the northeastern United States and poses a constant threat to other areas of the country where susceptible trees are present.

Apparently the first regulatory control legislation was passed in Germany in 1873. It was intended to prohibit the entry into that country of any materials that might harbor the grape phylloxera, *Phylloxera vitifoliae,* from America (National Academy of Sciences 1969). In the United States, although there was earlier regulatory legislation both at the state and national levels, the first major and effective legislation was passed in 1905. This was the Federal Insect Pest Act, which provided for the regulation of importation and interstate movement of potentially injurious insects. The Plant Quarantine Act of 1912 supplemented and extended earlier legislation and gave the Secretary of Agriculture the authority to enforce laws designed to protect the agriculture of the United States from insect pests and plant diseases by the regulation of importation and interstate movement of potential carrier materials. Additional pertinent legislation has included the Postal Terminal Inspection Act of 1915 and subsequent amendments and several more recent legislative actions, both at state and national levels. Under the authority provided by these acts, trained inspectors at key locations are able to examine materials as they cross the international boundaries into the United States or cross boundaries of regions within the United States that are under quarantine for one or more plant diseases or insect pests. These inspectors have the power to prevent movement of infested or infected materials across these boundaries or to render these materials "safe" by appropriate treatment whether it be fumigation, application of liquid insecticides, dry heat, or otherwise. Also the regulatory control legislations of other countries are respected, and potentially pestiferous materials that are to be moved from the United States require inspection and export certification.

There are currently several quarantines being enforced in the United States and in the past there have been many effective as well as not-so-effective quarantines. The Japanese beetle has been under quarantine regulation since 1919, and as a result many areas of this country are uninfested although they provide favorable environments for this insect (National Academy of Sciences 1969). Current and past control and quarantine programs have retarded movement of the gypsy moth to regions of the northeastern United States, but unfortunately have failed to completely contain this major forest pest. Past domestic quarantines, in combination with other kinds of control, have been in part responsible for the virtual elimination of several pest species of insects and ticks—the cattle tick, *Boophilus annulatus;* red tick, *Rhipicephalus evertsi,* from Florida; parlatoria date scale, *Parlatoria blanchardii,* from Arizona and California; Mediterranean fruit fly, *Ceratitis capitata,* from Florida and Texas; and many others (National Academy of Sciences 1969). Quarantines against the European corn borer, the satin moth, the Asiatic garden beetle, and others were terminated because these insects became widespread in spite of the quarantine and other control measures. In recent years, the Asian tiger mosquito, *Aedes albopictus,* the Asian Gypsy moth, and the sweet potato whitefly have found their way into the United States.

CHAPTER 16

Integrated Pest Management in Agroecosystems

Ferro

Integrated pest management (IPM) is the integration of insect, other animal, or plant management tactics, such as biological control, cultural control, chemical control and plant resistance, to maintain populations below damaging levels using the most economically and environmentally compatible tactics. Although there is little "natural" about agricultural cropping systems, the crop environment and those habitats surrounding the crop environment constitute an agricultural ecosystem, or agroecosystem. In most agricultural systems, soil, water, weather, plant and animal interactions are rarely constant enough to provide the ecological stability or equilibrium characteristic of nonagricultural ecosystems. However, if these interactions and the economics of a cropping system are understood, it is possible to manage most insect pests in an economically efficient and environmentally safe manner. This chapter focuses on the ecological and economic basis of IPM and discusses the major management tactics used in IPM programs for insect control. This chapter is dedicated to insect pest management rather than the broader issue of all pests—such as weeds, plant pathogens, nematodes, and vertebrates—because insects are the focus of this book. However, it is important to recognize that insect pest management tactics must be integrated with other pest management strategies and cultural practices.

Many citations of literature pertinent to IPM can be found in chapter 15.

Chemical Control and IPM

Chemicals have historically been used to control insect pests. The Greeks used sulfur to control plant pests, Marco Polo used mineral oil to control mange of camels, and the Persians used the botanical insecticide pyrethrum for controlling human pests. However, it only has been over the past 50 years that synthetic insecticides have played a prominent role in controlling insect pests of crops.

After World War II, the benefits to farmers from using DDT and other organochlorine insecticides was so dramatic that in many cases growers substituted heavy reliance on insecticides for sound pest management practices. Growers soon stepped onto the "pesticide treadmill." The more dependent they became on pesticides the more difficult it was to step off, and the use of pesticides has steadily increased (figure 16.1). Pests in cotton and high value vegetable crops became resistant to one insecticide after another. There were outbreaks of secondary pests, either because they developed resistance to these insecticides or because their natural enemies were killed by the insecticides. Yet, as long as chemical companies could introduce new products growers did not worry. The situation has drastically changed.

It has become extremely expensive for companies to develop and register new insecticides, and for this reason they are being very selective in marketing new products. In many cases, pests have become resistant to all federally registered materials, or materials have been pulled from the market because of toxicity (acute or chronic) to humans, wildlife, and nontarget insects; residue on food products; or environmental pollution. Where these situations have occurred, growers have quickly embraced IPM programs. Insecticides presently play a vital role in most IPM programs, and it is unlikely their role will change in the near future. For this reason, it is extremely important to have an understanding of how insecticides are formulated (see chapter 15), their mode of action (how they kill insects), and how they can best be integrated into a pest management program. However, it must be remembered that it was the recognition of misuse of pesticides or environmental contamination that was the impetus for federally funded pilot IPM programs in the early 1970s. The environmental concerns of the 1970s have changed with respect to the specific chemicals but not with respect to the broader issues. Contamination of

Figure 16.1 The amounts of synthetic pesticides (insecticides, herbicides, and fungicides) produced in the United States (Pimentel and Hanson, eds., *CRC Handbook of Pest Management in Agriculture*, Boston: CRC Press, 1991. The decline in total amount produced is in large part due to the increase in toxicity and effectiveness of pesticides brought into the market since 1975.

Redrawn from Pimentel (1991).

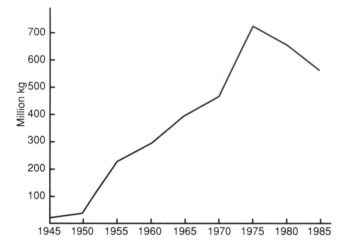

groundwater and other nontarget sites by pesticides, adverse effects of pesticides on nontarget organisms, and insecticide resistance are still the driving forces behind the implementation of IPM programs.

IPM Program Development

IPM programs tend to go through three phases of development. How quickly the program goes from one phase to the next depends on the existing knowledge of the agroecosystem and the level of sophistication desired. The first phase and the one that is usually the easiest to implement is the *pesticide management phase.* To implement this phase, there must be an established relationship between pest densities and resulting damage so that pesticides are not being applied unnecessarily. The efficacy of the pesticide should be well established, and the susceptibility of the different plant growth stages to insect feeding should be known. This approach alone can reduce the number of sprays by as much as 50%.

The second phase can be referred to as the *cultural management phase,* and this phase usually requires a thorough understanding of the pest's biology and its relationship to the cropping system. This would include tactics such as delayed planting dates, crop rotation, altering harvest dates, and planting resistance cultivars. The primary components of an agroecosystem are the individual organisms, and it is on the individual that natural selection works. For this reason, it is important to study pest biology, behavior, and physiology in response to other

species and to abiotic factors (temperature, humidity, wind, etc.) in the environment. By understanding those factors that regulate population growth, it may be possible to use this information to minimize the impact of insect pests on a crop. For example, the pink bollworm, *Pectinophora gossypiella,* enters diapause when the day light hours drop below 13 hours, and if all cotton is harvested before the pink bollworm completes development and enters diapause, there will be few individuals that successfully overwinter. It was only through detailed physiological studies that this management tactic was developed.

The last and most difficult phase to implement is the *biological control phase.* In addition to having a thorough understanding of the pest's biology, we need to have a thorough understanding of the pest's natural enemies, and a quantitative assessment of the effectiveness of these agents in controlling the pest. Generally it is not possible to rely solely on biological control agents. The key element for a biological control agent to successfully control a pest is to have sufficient numbers of the agent present at the right time. If the presence of a biological control agent and the pest are not in synchrony, it may be possible to alter the planting date, so that the time of colonization by the pest is either moved up or delayed. This could help to make sure the vulnerable pest stage is present when the biological control agent is present.

Most often within a cropping system there is more than one insect to control and it may be necessary to use an insecticide. This can be very hazardous to natural enemies and requires either proper timing of pesticide application to minimize exposure of biological control agents to pesticides or the use of pesticide resistant agents or "soft" pesticides (those pesticides not toxic to nontarget organisms). It is extremely difficult to integrate biological control agents into IPM programs, and this can be accomplished only with a thorough understanding of the entire agroecosystem.

IPM and Weather Factors

Climate (average weather conditions over long periods of time) in many cases defines the geographical range and distribution of an insect pest and its natural enemies. When the number of generations of the European corn borer, *Ostrinia nubilalis,* based on climatic zones is compared to the actual distribution of European corn borer, the two distributions are quite similar (figure 16.2). Weather (temperature, humidity, wind, radiation, and so on, at any point in time) directly influences the physiology and behavior of an insect. Weather and climate, then, regulate and influence insect populations by directly affecting the physiology and behavior of insect pests, and by indirectly affecting their food supply, habitat and natural enemies.

Figure 16.2 (*a*) Number of generations of European corn borer to be expected based on climatic zones. (*b*) Approximate distribution of European corn borer by generations per year in North America. *Redrawn from Showers (1979).*

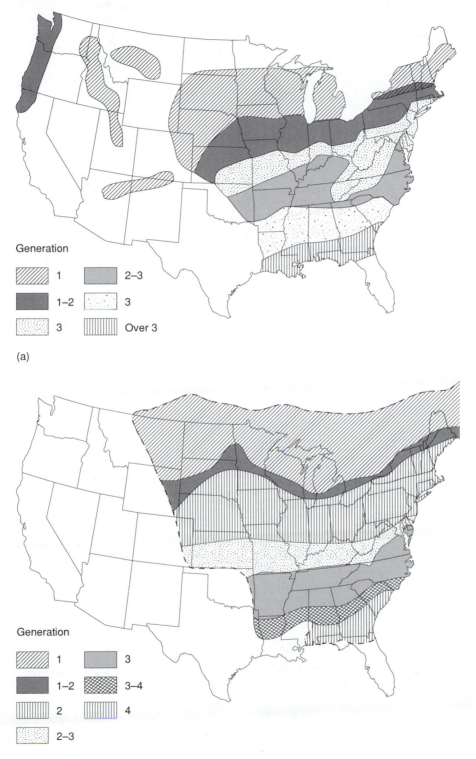

Figure 16.3   Relationship of insect development and exposure temperature of insect.

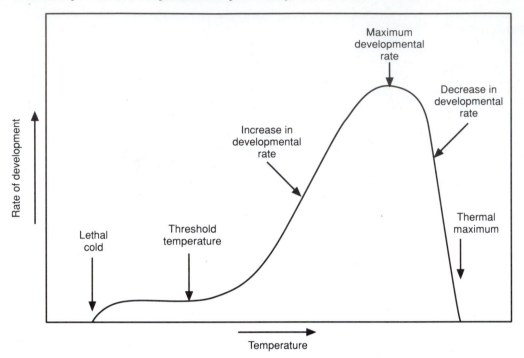

Temperature

Insects are poikilothermic; that is, chemical reactions that take place within insects are directly regulated by temperature. And, for this reason, developmental rates of insects are regulated by temperature. Developmental rate is the most important variable influencing the intrinsic rate of increase of colonizing species. Although insects may have a wide geographic distribution, it is generally in those areas where optimal temperatures occur that insects become pests.

The developmental rate of an insect is determined by the accumulation of heat units, or degree-days, above a threshold temperature (figure 16.3). The threshold temperature is that temperature below which development ceases. Above the threshold temperature there is a geometric increase in rate of development with increasing temperature, until the maximum rate of development occurs. At temperatures above the maximum rate of increase, the developmental rate rapidly decreases, until the thermal maximum or death is reached. Some insects can develop and survive over a wide range of temperatures, whereas others tolerate only a narrow temperature range. By monitoring daily temperatures and relating these temperatures to the specific temperature-dependent growth curves for an insect, it is possible to predict peak activity or first occurrence of pests.

There are two types of temperature driven models used to predict insect development. The degree-day models are the simplest and are linear models that are based on the linear portion of the developmental curve. The more complex models are based on enzyme kinetics

and enzyme denaturization models (Logan et al. 1976). These later models are extremely accurate and are often used for simulations modeling pest population dynamics.

The simplest approach for developing a degree-day model is to find the thermal constant, $k = y(x - a)$, for a given pest using the accumulation of mean daily temperatures above the lower developmental threshold.

Where: k = thermal constant expressed in degree-days
y = time required to complete development
$$x = \frac{\text{maximum temp.} + \text{minimum temp.}}{2}$$
a = threshold temperature

The most accurate way of establishing thermal constants is to conduct rearing experiments under controlled laboratory conditions using constant temperatures. However, it is possible to generate degree-days (d.d.) by monitoring maximum and minimum field temperatures and accumulating the degrees above threshold for each day until the insect completes its development. For example, the Colorado potato beetle, *Leptinotarsa decemlineata* (Say), overwinters in the soil and by monitoring soil temperatures and knowing the post-diapause developmental threshold temperature, it is possible to determine how many degree-days must accumulate before the beetles emerge from the soil. For Colorado potato beetle, $a = 10°$ C and by subtracting this value from the average soil temperature at 10 cm for each day and accumulating these degrees until the beetle emerges from the soil, it is possible to determine the degree-days for emergence of overwintered beetles (figure 16.4). By accumulating the degree-days for soil temperatures at 10 cm

Figure 16.4 Mean daily soil temperature at 10 cm in South Deerfield, Massachusetts, and mean average daily temperature above threshold temperature (10° C) for post-diapause Colorado potato beetle. Overwintered adult CPB emergence after about 180-degree days have accumulated.

	Soil temperature °C at 10 cm	Temperature above threshold		Soil temperature °C at 10 cm	Temperature above threshold
April 9	16.4	6.4	May 1	14.4	4.4
April 10	13.8	3.8	May 2	14.0	4.0
April 11	12.7	2.6	May 3	11.1	1.1
April 12	11.7	1.6	May 4	14.0	4.0
April 13	8.6	—	May 5	16.6	6.6
April 14	11.5	1.5	May 6	12.8	2.8
April 15	9.4	—	May 7	13.3	3.3
April 16	12.1	2.1	May 8	14.0	4.0
April 17	16.3	6.3	May 9	13.4	3.4
April 18	8.8	—	May 10	15.9	5.9
April 19	10.9	0.9	May 11	17.4	7.4
April 20	10.7	0.6	May 12	19.7	9.7
April 21	7.6	—	May 13	21.6	11.6
April 22	7.7	—	May 14	19.0	9.0
April 23	9.4	—	May 15	21.3	11.3
April 24	10.6	0.6	May 16	22.1	12.0
April 25	12.3	2.3	May 17	22.9	12.9
April 26	14.7	4.6	May 18	18.8	8.8
April 27	10.9	0.9	May 19	18.3	8.3
April 28	10.7	0.6			
April 29	16.3	6.3	Total		
April 30	16.5	6.5	Degree-days		181.0

Adult CPB emergence

in 1991, the predicted date of emergence would be May 19. However, beetles were first observed on May 17. For most management purposes, this discrepancy of two days would be of little consequence.

The degree-day models can be used to predict pest flight activity, to predict initiation of egg hatch, to monitor when insect pests should begin, or to better time insecticide applications.

Humidity

The survival of an insect pest is influenced by its ability to tolerate fluctuations in body water. Insects are able to maintain a balance between the water taken in orally or through the integument and the water lost by excretion and transpiration through the integument or through respiration. If an insect loses water faster than it takes it in, it will die, and the rate of transpiration is closely linked to humidity. For this reason, the average humidity for any particular environment can influence the distribution of an insect pest.

Humidity can also influence insect behavior. The calling behavior of European corn borer (i.e., the release of pheromones by females to attract males) is affected by humidity. At low humidities the females will not call. This may explain why calling females are generally found in the canopy of thickly growing vegetation around fields where the humidity is higher than within fields. For this reason, pheromone traps should be placed in the dense vegetation around fields rather than in the middle of a corn field.

Wind

Wind is the primary mode of transportation by which many insects disperse and colonize crops. For example, many insect pests of crops grown in temperate areas are colonized each year by insects that are blown in from southern states. The corn leaf aphid, *Rhopalosiphum maidis* (Fitch), after taking to flight is carried north to infest corn as far north as Minnesota, where it vectors maize dwarf mosaic virus (MDMV). The potato leafhopper, *Empoasca fabae* (Harris), overwinters in Louisiana and is carried by wind to the north central states, and then secondary populations are transported east and infest alfalfa, potato, apples and many other plants. Because wind and storms vary greatly from one year to the next, it is difficult to predict when wind-borne insects will arrive. Often insect pests are carried by wind independent of their natural enemies, and because of this their populations increase rapidly without being kept in check by their natural enemies.

Sampling and Monitoring

The relationship between pest densities and pest damage is the basis of IPM. This process begins with proper identification of pests to be sampled. Estimates of insect pest densities are determined by sampling for the pest. Measurements taken to estimate population densities in agricultural crops can be categorized into three groups: absolute methods, relative methods, and population indices. Absolute sampling methods provide estimates of

Figure 16.5 (*a*) Person using a sweep net for sampling insects in alfalfa. (*b*) Multipher® pheromone trap for monitoring Fall armyworm moth flight activity.

(a)

(b)

density per unit area (number of larvae per m²), whereas relative methods estimate density per some other unit than area (e.g., number of weevils per 20 sweeps, or number of moths caught per pheromone trap, figures 16.5*a* and *b*). Population indices do not count insects at all, but rather they measure insect products (e.g., exuviae or frass) or effects (e.g., plant damage). The trade-off between the level of precision necessary for making management decisions and the amount of time (in terms of labor cost) that can be devoted to sampling will dictate the type of sampling program.

Ideally we like to have absolute estimates of pest densities for making management decisions; however, it is generally too costly an investment to obtain these estimates on a routine basis. Once absolute estimates have been established and correlated with pest damage, it is possible to then relate relative estimates to absolute estimates using regression analysis. Then the relative estimates can be used for making management decisions.

Sampling Design

First, the area to be sampled must be determined. If there is considerable variation between manageable units within the same field, it may be necessary to divide the field into separate sections for sampling so that each section can be managed separately. Once the area to be sampled has been determined, the sample unit must be established before sampling begins. Ideally, it should be a biologically meaningful unit (e.g., an apple, a leaf), yet not so large

as to be unwieldy. For example, when sampling a cabbage plant it may not be necessary to examine the entire head, especially when the variability in the data is greatest between plants and not within plants. In this case, the sample unit may be a quarter of the head. Or, if the pest complex is restricted to the wrapper leaves, then it is only necessary to sample these leaves.

It is not possible to examine every plant within a field and for this reason a sample of the population is taken. A sample consists of a small number of individuals taken from the larger population. The number of observations (*n*) in the sample is referred to as the *sample size*. The number of sample units per observation can vary. For example, when sampling for potato aphids the sample unit is a compound leaf and generally 3–5 leaves are sampled for each plant to be sampled. However, when sampling for the Colorado potato beetle, the sample unit is a single stalk for each plant and only one stalk is sampled per plant. In each case it is possible to determine the number of sample units per plant, and by knowing the number of plants per m², it is possible to obtain an absolute estimate of the population density of each pest.

Because there can be considerable variability between sample units, it is necessary to determine a common measure of spread or variation among the observations, i.e., the standard deviation(s). If the standard deviation is very large when compared to the mean, it might be necessary to take more samples so that the level of precision is such that you are comfortable with the decision to treat or not to treat.

Most sampling is based on taking random samples throughout the field, and for statistical reasons, this is preferred. Sometimes this is not the most efficient way to sample a field, especially if the pest is not randomly distributed. In this case it may be more efficient to stratify the area to be sampled. For example, the common stalk borer overwinters in the grassy areas around the edge of corn fields, and the larvae move in from the grassy areas to feed on the rows of corn closest to the edge of the field. In this case, the strata to be randomly sampled should be the 10 rows closest to the edge of the field, and not the entire field. This type of sampling is called *stratified random*. It is efficient, increases precision, and reduces variance between sample units.

Sequential Sampling

One of the most efficient sampling schemes used for making treatment recommendations in IPM programs is *sequential sampling*. Sequential sampling is based on a thorough knowledge of variance and distribution of a population. The number of samples is variable, and sampling stops once it is known that a pest population is at a certain density. For example, if we know that 15 or more Colorado potato beetle egg masses/plant will cause economic losses, and less than 10 egg masses will not, then we can establish a series of rules to determine whether to treat or not treat based on the accumulated number of egg masses per plant. A minimum number of plants are sampled and if the accumulated number of egg masses is significant, then a recommendation to treat is based on only a few plants. If the accumulated number falls into an area of uncertainty, then more samples will need to be taken. To determine the relationship between the accumulated number of egg masses per plant and the number of samples to be taken, it is necessary to determine the level of error that is acceptable (e.g., are you willing to be wrong 5% of the time?).

Relative Sampling

Many IPM programs are based on relative sampling methods, because such methods are much more efficient than obtaining absolute estimates of population densities. These methods often involve using different types of traps, such as pheromone traps, host mimic traps, pan traps, and emergence traps. Many insects use chemicals (pheromones) for attracting mates, and these pheromones have been synthesized and placed inside traps. The baited traps are then placed in areas where the pest insects commonly mate. The insects are attracted to these pheromone traps and become entrapped once inside. Aphids are attracted to yellow objects that act as "super foliage mimics." Cake pans painted yellow on the inside and filled with soapy water are used for monitoring winged aphids and allow

for early detection. The apple maggot fly uses visual and olfactory cues for finding egg-laying sites (apples). Red spheres coated with a sticky material are used for monitoring for this pest. It is possible to relate pest damage to the number of insects trapped; in this way it is possible to quickly assess the presence of pests and to make management decisions based on these relative sampling methods.

Economics of IPM

From a growers perspective, the bottom line of an integrated pest management program is the financial rewards for using the program. Without understanding the economics of the system, it is not possible to make the decisions necessary to determine when a particular control tactic should or should not be employed. It is necessary to weigh the benefits against the costs and to consider the environmental effects.

Cost/Benefit

To understand the benefits of the application of a particular control tactic, the impact of the pest on reducing yields must be known. Studies devoted to this type of research have been labeled crop loss assessment. There are few studies to date that have truly established this relationship. The three key elements are plant yield (dependent on plant cultivar, phenology and partitioning of photosynthates, soil factors, and pest damage), economics (cost of production, market and cost of control tactic), and pest complex (biology and population dynamics of pests and soil factors). Crop loss assessment is easiest to develop for a single pest and becomes considerably more complex as the number of pests increases. The environmental risks, human risks, and risks to biological control agents must also be considered.

Plant Response to Damage
In general, crops can withstand some pest damage; this is particularly true of defoliators or pests that feed on the nonmarketable parts of the plant. For example, if a grower is only concerned with selling the beet root, the plant can tolerate rather high population densities of leafminers. However, if the root and the "greens" are to be marketed, damage to the foliage cannot be tolerated.

This brings us to the concept of *economic injury level* (EIL) and *economic threshold level* (ETL). An insect is, in economic terms, not a pest until the insect reaches the density where it must be controlled. The EIL is the population density in which the financial gains are equal to or greater than the cost of the control. The ETL is the population density in which a control tactic must be implemented to prevent the population from reaching the EIL. Some times these levels have not been established,

Figure 16.6 Economic injury level (EIL), economic threshold level (ETL) and general equilibrium. (*a*) Banks grass mite. (*b*) Potato leafhopper. (*c*) Green peach aphid. (*d*) Corn earworm. Note: Pesticide applied when pest population density reaches the ETL, so that density does not overshoot EIL by very much.

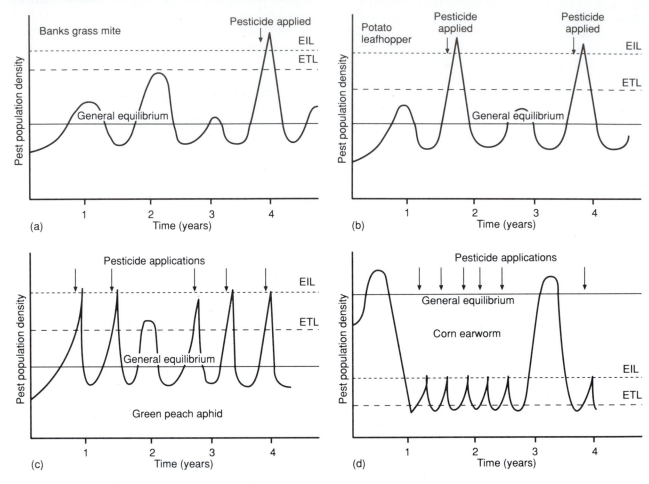

especially when a relative estimate, such as one based on pheromone trap counts, is used for measuring pest population densities. In this case, an *action threshold* is used. The action threshold is generally based on empirical results from working with growers and often the level becomes less conservative as the knowledge base increases.

The establishment of an EIL or ETL is a complex endeavor based on intense studies of pest and plant ecology as affected by weather, natural enemies, host plant resistance, plant growth, and environmental consequences of the control measure. Pest population densities tend to oscillate over a long period of time and reach a general equilibrium, and the magnitude of these oscillations is dependent on the above-mentioned factors that affect plant and insect ecology. If the general equilibrium is usually below the EIL, then control measures are rarely needed. This is the situation with Banks grass mite (not an insect, but a useful example nonetheless) in field corn

(figure 16.6*a*). However, outbreaks of the Banks grass mite can result from changes in weather or the application of an insecticide for another pest that kills the natural enemies of the Banks grass mite. The next situation is one in which the general equilibrium level is normally below the EIL but periodically exceeds the EIL. This is exemplified by the potato leafhopper, which periodically becomes a pest in the Northeast (figure 16.6*b*). This insect overwinters in the southern United States and migrates to the north-central states in the spring. In early summer, it may migrate eastward to infest potato. If it arrives early enough in the growing season, it will likely produce populations that exceed the EIL. So in this case the prevailing winds determine its potential as a pest. Another situation is one in which the general equilibrium is below the EIL, however, pest densities routinely oscillate above the EIL (figure 16.6*c*). The green peach aphid, *Myzus persicae,* normally reaches pest densities each year and must be controlled with insecticides. Finally, there is the situation in which the general equilibrium of a pest is always above the EIL (figure 16.6*d*). This is typical of most insects that are pests of the commodity to be sold, and the only way to keep the population below the EIL is to make routine applications of insecticides. Each female corn earworm, *Helicoverpa zea,* produces about 1,500 eggs and these eggs are laid on the silk protruding from an ear of corn. For this reason, it only takes a few fertile females to produce enough offspring to cause economic damage. When this pest is present during the hottest part of the summer, it is necessary to apply insecticides every three days from silking to harvest based on the number of moths captured in pheromone traps and plant phenology.

Reality of Pest Control

The cost/benefit analysis has assumed that the farmer knows with certainty factors such as the EIL, the effect of alternate management tactics, the effects of weather, the soil fertility, and the pH on yield. It also assumes each pest control decision is independent of every other decision as well as other inputs, such as cultural practices. Many grower practices, in particular pesticide applications, are a form of insurance—to minimize risk. Generally growers select plant cultivars and cultural practices independent of pest competition. Hence, farmers are interested in expenditures that reduce uncertainty or risk. However, any insurance should be linked to some probability function rather than some formulated schedule. It is the aim of an IPM program to reduce the risks and needs for scheduled insecticide applications by providing accurate information on factors such as crop growth, weather, and pest densities so that insurance sprays are not necessary.

Insect Vectors of Plant Pathogens and IPM

Insects that vector plant pathogens often cause direct feeding damage to plants. However, this damage is usually quite minor when compared to the damage caused by the plant pathogen introduced by the insect. IPM programs that deal with an insect vector and a plant pathogen require special consideration. For example, when potatoes are being grown for seed, they must be disease free. Thus, the economic threshold level for the green peach aphid, which vectors potato leafroll virus, is much lower for seed potatoes than for potatoes grown as table stock. There are several orders of insects that are involved in the transmission of plant pathogens: Homoptera, Coleoptera, Thysanoptera, and Diptera. However, aphids, leafhoppers, and whiteflies represent the most important groups.

Figure 16.7 Life cycle of the green peach aphid in northern temperate areas.
Redrawn from University of California Press Publication No. 3316 (1986).

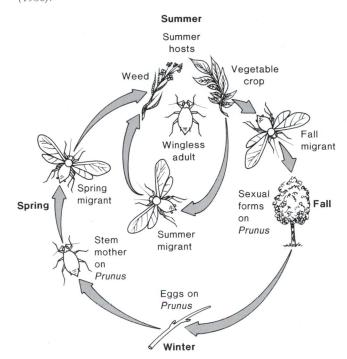

Aphids

The biology, feeding behavior, and worldwide distribution make aphids well-suited for transmitting plant viruses. Of the 3,700 described species of aphids, only about 300 have been tested as vectors of plant viruses, and of these only 180 are demonstrated vectors. Most of these aphid species are relatively host specific.

In order to appreciate the transmission process, spread of plant pathogens, and control tactics, it is necessary to understand the biology of the insect and the pathogens it vectors. The green peach aphid (GPA), *Myzus persicae,* is a known vector of over 60 plant viruses. The following is a description of its life cycle in temperate areas of the world (figure 16.7). The GPA overwinters in the egg stage on its primary host (*Prunus* spp.). In the spring the eggs hatch to produce only wingless female aphids that go through one or more generations on the tree, reproducing parthenogenetically. Winged females migrate from the primary host to colonize crops and weeds, again reproducing parthenogenetically. They go through 3–12 generations on these plants before returning to the primary host, where the winged female produces males and females that mate on the primary host; fertile females then lay eggs on primary hosts.

In nontemperate areas, the GPA can reproduce parthenogenetically year round. In fact, in New Zealand the egg stage has never been found.

Aphid host selection behavior, when coupled with the mechanism of feeding (inserting the maxillary tips into the cells and after ingesting sap and closing the feeding sites with a salivary plug), make aphids ideal vectors of viruses. Aphid-borne viruses can be classified as nonpersistent (noncirculative) and persistent (circulative) based on how the aphid transmits the virus.

Nonpersistent Viruses

If a virus neither multiplies nor circulates, exhibits no discernible latent period between acquisition and transmission, and is lost during multiple probing behavior on plants it is defined as a nonpersistent virus. Nonpersistent viruses can be acquired during brief probes (as little as 5 seconds), retained for several days, and transmitted within minutes. These viruses can be transmitted by both colonizing and noncolonizing species. For example, the GPA does not colonize sweet corn; however, it will alight on sweet corn and insert its mouthparts and ingest sap to determine if the corn is an acceptable host. During these short probes, it transmits the maize dwarf mosaic virus to the corn plant. Because there is a constant influx of individual aphids into crops and because individuals remain in the field for brief periods, insecticides are not effective at preventing the transmission of nonpersistent viruses.

Persistent Viruses

These viruses persist in the aphid for long periods (days) and this appears to be related to the ability of the virus particles to permeate the salivary gland. There is evidence that the virus/vector specificity of some persistent transmissions are mediated by vector salivary gland/virus coat protein interactions. These viruses tend to be species specific and are only vectored by colonizing species of aphids.

Control Tactics

Many possible control tactics may be used in concert, in order to prevent or minimize virus transmission by the GPA. These tactics are as follows:

I. Breed resistant plant cultivars
II. Insecticides
    A. Insecticides are effective in preventing the spread of persistent viruses because of the longer acquisition and inoculation feedings required for transmission, and because there is a latent period between the acquisition and inoculation phases.

    B. Insecticides are ineffective against the spread of nonpersistent viruses due to the brevity of probes in the epidermis, the absence of a detectable latent period, relatively slow acting insecticides, and the behavior of alate aphids making brief probes on several different plants before settling on a host plant for prolonged feeding.
III. Control aphids on primary hosts before they migrate in the spring.
IV. Remove primary hosts (e.g., *Prunus* spp., overwintering host of the GPA).
V. Remove intermediate hosts (generally weeds) between primary hosts and crops.
VI. Mineral oils—mineral and vegetable oils and milk lipids have been shown to inhibit normal transmission of nonpersistent viruses.
VII. Reflective mulches have been shown to repel alate aphids, reducing the number landing and probing.
VIII. Use only virus-free seed. For example, this is the key to minimizing the effects of potato leaf roll virus of potato. Lettuce mosaic virus inoculum levels of 1 lettuce seed in 25,000 is enough to cause economic losses.

Biological Control and Virus Spread

Biological control agents in general tend to aggravate the problem. Aphids when disturbed by biological control agents often avoid these enemies by dispersing or dropping to the ground, and then feed on a different plant. This encourages the spread of the virus.

Leafhoppers

There are 130 known vector species, vectoring 38 viruses, 31 mycoplasm-like organisms (MLOs) and 4 spiroplasms. Most leafhopper-borne viruses are transmitted in a persistent manner and most multiply in their vectors. Some viruses are vectored in a nonpersistent manner.

Whiteflies

Whiteflies transmit plant pathogens mainly in tropical or semitropical areas. They can be a problem in temperate areas, especially where populations have been allowed to build up in a greenhouse or have been shipped north with plants (often already infected). This is a common occurrence in the vegetable industry.

Beetles

Beetles most commonly associated with the transmission of plant pathogens are in the family Chrysomelidae and generally have a limited host range. Both the larvae and adults are known to be vectors, however, there is no evidence that virus acquired by larvae can be carried through pupation to the adult stage. Beetles are most important as vectors of bacteria.

There are about 180 bacterial plant pathogens. Most species do not form spores, and thus are not resistant to environmental stress. Structurally, the bacterial cell is thin walled, fragile, and easily damaged by desiccation, heat, and irradiation. Bacteria overwinter in the gut of beetles where they are protected from environmental stress; when the beetles feed the following spring, they transmit the bacteria into the damaged tissue. Beetles are also capable of dispersing pathogens over long distances.

Case Studies

One of the more instructive ways to appreciate how pest management tactics are integrated is to examine case studies. The following cases have been chosen because they represent two very different cropping systems from two different regions of North America.

Cotton

There are 12 relatively well-defined cotton production zones in the United States, each utilizing different cultivars and production practices, and each with different pest problems. Most of the world's cotton, including that grown in the United States, belongs to various strains selected from *Gossypium hirsutum*. One of the best examples of a cotton IPM program is the Texas short-season cotton program for the central Texas blacklands, the coastal plains, the Wintergarden area and the lower Rio Grande Valley. The key insect pests are the boll weevil, *Anthonomus grandis* Boheman; the bollworm, *Helicoverpa zea* (Boddie); and the tobacco budworm, *H. virescens* (F.). Basic to this program are the multi-adversity resistant cotton cultivars (TAMCOTs), possessing rapidly fruiting, short-season characteristics. These cultivars were originally bred for disease resistance (seedling disease, bacterial blight, verticillium wilt, fusarium wilt/root-knot nematode, and cold and seed deterioration). However, it was the rapid fruiting characteristic of these cultivars that made them so useful as a major component in insect management. The precocity of fruiting not only allows the majority of cotton fruit to escape late season insect populations, but also lengthens the host-free period following harvest, thus promoting greater overwintering mortality of key insect pests, such as the boll weevil.

The cotton fleahopper, *Pseudatomoscelis seriatus* (Reuter), a pest unique to Texas cotton production, is an early season pest that causes damage to primordial fruit. This pest largely resides on wild host plants outside of the cotton field and may, depending on moisture conditions, migrate into cotton and cause significant damage. Successful management of the cotton fleahopper is necessary not only to prevent the loss of valuable fruit set early in the growing season, but also to prevent a delay in overall fruiting that may nullify the early-season effect so essential to the IPM program. Careful field monitoring of cotton fleahopper adults and nymphs is critical early in the season. Control of this insect and control of spring boll weevil populations frequently overlap. The goal with both cotton fleahopper and boll weevil in the early season is to stop insecticide applications as early as possible, 14–21 days before bloom, to allow natural enemies to build up as biological control agents for successive populations of bollworm and tobacco budworm.

The boll weevil is the pivotal insect around which cotton IPM programs in Texas are structured. The majority of the management tactics for boll weevil control are based upon the cultural manipulation of the crop. The two critical areas of crop management are time of planting and post-harvest phytosanitation practices. In the central and southern parts of Texas, early and uniform planting is recommended in order to encourage early fruit initiation. Field sampling and boll weevil pheromone traps are used to provide information that directs early-season insecticide applications toward weevils emerging from diapause and entering cotton in the spring. The use of short-season cultivars allows fruiting to take place when weevil densities are low. The goal is to encourage the maturation of bolls to at least 10–12 days of age. Bolls that are 10–12 days old have a significantly lower probability of being damaged than younger fruit. Prompt and area-wide post-harvest cotton stalk destruction is practiced in order to force a long, host-free period. A uniform delay in cotton planting is practiced in the more northern production regions of the state where high overwintering boll weevil mortality is frequently encountered due to much lower winter temperatures. A delay in planting allows weevils to emerge in the spring without suitable cotton fruit sites in which to feed and reproduce, thus producing a high level of weevil mortality. Control of pre-diapausing adults through the application of insecticides in the fall is practiced in west Texas cotton production regions. Diapause control programs conducted for 28 years in the western Rolling Plains region of the state have successfully stopped the westward movement of the boll weevil into the High Plains region where 3.2 million acres of cotton are grown.

The bollworm/tobacco budworm complex was not a pest problem prior to the boll weevil becoming established in this region of Texas. Early-season sprays for the boll weevil destroyed the suite of natural enemies that kept the bollworm/budworm complex in check. Once this complex gained pest status, insecticide sprays were needed to control them. This in turn resulted in the bollworm and budworm becoming resistant to all registered insecticides, effectively making it uneconomical to grow cotton in this area. The introduction of the short-season cultivars eliminated the need to apply season-long insecticide applications for these pests, and thus avoided destroying the beneficial arthropods that keep these pests in check.

In 1970, 104,000 acres of cotton were harvested in San Patricio and Nueces counties (Coastal Bend area of Texas) and by 1975 it was down to 50,400 acres. In the mid-1970s, the short-season cultivars were introduced in conjunction with an area-wide IPM program, and by 1979 236,500 acres of cotton were harvested. The tactics involved in these programs included (1) early and uniform planting of a short-season cotton; (2) reduced application of nitrogen and irrigation where appropriate; (3) intensive field sampling (scouting for early season pests such as overwintering boll weevils so they could be controlled if damaging levels were predicted), pheromone trapping and bollworm/budworm computer forecasting; (4) continued intensive field scouting of key insects and application of insecticide based on appropriate action levels; (5) use of harvest-aid chemicals for desiccation or defoliation; (6) early harvest; and (7) complete and area-wide stalk destruction as soon after harvest as possible (Frisbie et al. 1989). This is an excellent example of how an IPM program can positively impact a farming community.

Biointensive Potato IPM Program

White potato, *Solanum tuberosum,* is grown throughout the world, and in the northeastern United States there are about 160,000 acres harvested annually at a value of about $285 million. Although the insect pest complex is the same for the Northeast, there are regional differences in the pest status of these insects. The key insect pest is the Colorado potato beetle, which can completely defoliate a field by mid-season if left uncontrolled. The beetle has become resistant to every currently registered synthetic insecticide, and because of these high levels of resistance, growers make repeated applications of insecticides. In 1991, potato farmers on Long Island, New York, on average made 12 applications costing about $388/acre to control this pest. The potato aphid, *Macrosiphum euphorbiae,* and the green peach aphid, *Myzus persicae,* cause direct damage to potato plants and vector several plant viruses. The potato leafhopper, *Empoasca fabae,* overwinters in the southern states and moves into the Northeast in late June. Even very low population densities (5 leafhoppers/leaf) of this pest can cause severe damage to plants. While feeding, the leafhopper releases a toxin into the plant that causes the foliage to turn brown and die.

The key to controlling the Colorado potato beetle is to delay colonization of potato by this pest and to reduce the number of colonizing adults. The beetles overwinter as adults in the previous year's field and in the undisturbed habitats around the field. If beetles emerge into a field planted to potato, they will feed and lay eggs for about two weeks then migrate to other habitats. If they emerge into fallow lands or into uncultivated areas adjacent to fields, they begin walking and if they do not find any hosts within 5 days they take to flight. These are the early colonizers of first-year potato fields (rotated fields). Rotated fields are colonized one to three weeks later than nonrotated fields, depending on the distance these fields are from overwintering sites. Summer generation beetles that emerge after the "diapause switch" (about August 1 in the Northeast) do not produce any eggs. Even a one week delay in colonization can cause most of the summer adults to emerge after the switch. Effectively, this means that growers only have to contend with one generation of larvae.

Because the rotated fields are colonized by fewer egg laying beetles, egg densities are lower and predatory insects, such as the twelve-spotted ladybird beetle, *Coleomegilla maculata,* can cause high levels of mortality (50% or more). *C. maculata* populations reach high densities in field corn where the larvae feed on pollen and aphids. These ladybird beetles overwinter at the base of trees in the woody areas surrounding corn fields. When potatoes are planted to these fields the following spring, large numbers of ladybird beetles move into the fields to feed on Colorado potato beetle eggs. For this reason, a potato/corn rotation is very effective. There are several other biological control agents that are useful against the beetle, including a prepupal tachinid parasitoid (*Myiopharus* spp.) and the fungal pathogen *Beauvaria bassiana.*

Myiopharus can cause 80% or more parasitism of the second generation larval population, whereas the soil-borne fungal pathogen, *B. bassiana,* can cause 75% mortality to summer generation beetles. Prophylactic use of the fungicides chlorothalonil and mancozeb that are commonly used to control the spread of late blight, *Phytophora infestans,* also kill *B. bassiana.* However, if growers use the fungicide metalaxyl, which is innocuous to *B. bassiana,* the week prior to fourth instars dropping to the soil to pupate, *B. bassiana* will cause high levels of mortality. In general, the biological control agents are not capable of maintaining beetle populations below damaging levels, even on rotated fields, and an insecticide must be used. However, the broad-spectrum insecticides registered for controlling the beetle are very toxic to the beetles' natural enemies.

There are several new strains of *Bacillus thuringiensis* that are toxic to Colorado potato beetle larvae, and nontoxic to its natural enemies. These biorational insecticides are a key component to the biointensive IPM program for potatoes in the Northeast. These formulations of *B. thuringiensis* are not toxic to aphids or leafhoppers.

Synthetic insecticides are still necessary for controlling aphids and leafhoppers. Aphids colonize potatoes over a very short two week period beginning in late June, and this is about the same time that leafhoppers first appear. Growers have two options for controlling these pests. The most preferable approach is to use a systemic insecticide at planting or first hilling. Because systemic insecticides are taken up by the plant versus being applied to the foliage, natural enemies of Colorado potato beetle are not affected. The second approach is to apply a foliar spray in early July before the second flush of Colorado potato beetle eggs, and at a time when its predators are virtually absent from the field.

This IPM program integrates the cultural practice of crop rotation with conservation of existing native natural enemies of Colorado potato beetle, selective applications of *B. thuringiensis* based on weekly scouting of fields to control the beetle, and the use of systemic insecticides for controlling aphids and leafhoppers. This means that only a single application of a broad-spectrum, synthetic insecticide is needed, and the cost of insecticides is about half that of conventional growers. Where growers have used this program for several years, we have seen a reversion in levels of resistance by the Colorado potato beetle to the pyrethroid insecticides.

Comments on Case Studies

Although the management tactics used in both the cotton and potato IPM programs appear to be rather simple and straightforward, the development of these tactics and the integration of them into a complete program took many years and only came to fruition after considerable research was completed on the biology, physiology, and ecology of the pests and their natural enemies. These tactics are based on the scientific process of identifying a problem, designing experiments to address the problem, and ultimately finding a solution to the problem. However, when IPM programs fail it is generally not due to an incomplete database but to sociological problems. Growers are often resistant to change, and it is only through the efforts of creative and innovative IPM practitioners that growers embrace IPM tactics. It is also important to acknowledge that IPM programs are not static, and recommendations and tactics are constantly changing as new information becomes available.

A

a- without; lacking.

ab- away from; off.

abiotic nonliving, both in characteristics and origin.

accessory cells cells involved with or secreting the cuticular structures that make up a sensory receptor; for example, trichogen and tormogen cells (figure 6.10).

accessory glands secretory cells/tissues associated with the internal reproductive organs (figures 5.1 and 5.6).

acetylcholine a kind of neurotransmitter.

acetylcholine esterase an enzyme that breaks down acetylcholine.

acinar resembling a cluster.

acinus resembling a single grape in a cluster.

acoustical communication communication by sound and hearing.

acquired active immunity immunity induced by injection of toxoids or killed or attenuated microorganisms.

acron anterior, preoral body cap of a segmented animal, also called the *prostomium* (figure 11.12).

acrosternite small sclerite between the antecostal suture and intersegmental membrane on the abdominal sternum; compare with acrotergite.

acrotergite small sclerite between the antecostal suture and intersegmental membrane on the abdominal tergum (figure 2.11b).

action threshold the population density at which pest control measures are applied to avoid economic loss.

activation center a region in a fertilized egg where cleavage cells are induced to undergo further growth and development.

active ventilation ventilation aided by movements such as muscular pumping.

ad- next to; at.

adecticous without functional mandibles (movable mouthparts) used for escape from a cocoon or pupal cell; *compare with* decticous.

adipocytes cells composing fat body (figure 4.11b).

adjuvants substances (e.g., dusts, solvents, emulsifiers, and adhesives) that are combined with insecticides to increase their effectiveness.

aedeagus *see* penis.

aeropyles channels in insect eggs by which the gas-filled layer communicates with the outside of the egg.

aerosols sprays made of extremely small droplets, 0.1–50 micrometers in diameter.

afferent neuron *see* sensory neuron.

aging all changes in structure and function that occur from the beginning to the termination of life.

agroecosystem a relatively simple ecosystem created by humans for agricultural purposes, in which the major producer—the crop—is usually a single plant species.

alarm pheromones chemicals released by one insect that elicit dispersal or aggressive behavior by conspecifics.

alary muscles segmented, fibromuscular structures associated with the heart and having a role in the functioning of that organ (figure 4.6).

alate winged.

alinotum wing-bearing portion of the tergum (figures 2.18 and 2.19).

all-, allo- other; different.

allelochemics defensive chemicals. *See also* secondary plant substances.

allergens insect secretions, usually proteins, whose effect depends on the physiological response of the victim.

allometric growth *see* heterogonic growth.

allomone chemical emitted by an individual of one species and received by one of another species that induces a response favorable to the emitter.

altruistic behavior self-sacrifice.

ametabolous development without change of form.

amnion inner membrane around the developing embryo (figures 5.14, 5.18, and 5.19).

an- without.

anal margin the posterior edge of an insect wing (figure 2.23).

anamorphosis the addition of abdominal segments at the time of molting.

anapleurite one of three theoretical, primitive subcoxal sclerites (figure 2.20).

anautogenous requiring an exogenous source of protein for vitellogenesis.

androconia specialized scales and associated sex-pheromone-producing glands among the wing scales of some male Lepidoptera.

anemotaxis directed response relative to an air current.

anlage, anlagen precursor; the first recognizable sign of a developing organ in an embryo.

Anoplura, order (*anopl*, unarmed; *ura*, tail) sucking lice (figure 12.25).

ante- before; in front of.

antecosta an internal ridge, associated with the tergum and sternum of an abdominal segment (figure 2.11).

antennae elongate sensory appendages found on the insect head and usually bearing olfactory and tactile receptors (figure 2.12).

antennal lobes lobes in the insect brain that receive both sensory and motor axons from the antennae (figure 3.7).

antennal suture a suture that surrounds the antennal base and delineates the antennal sclerite (figure 2.14).

antennifer articular point of antenna located in the antennal socket (figure 2.14).

anterior imaginal ring ring of embryonic cells located at the posterior end of the larval foregut. *See also* posterior imaginal ring.

anterior intestine the portion of the hindgut between the Malpighian tubules and the rectum (figure 4.1).

anterior notal process anterior wing articulation point on the notum (figure 2.18a and b).

anthoxanthins whitish or yellow pigments found in insects.

anthraquinones red and orange pigments common in scale insects.

anthropocentric human-centered point of view.

anthropomorphism interpreting behavior patterns observed in animals in terms of human characteristics.

anti- against.

antibiosis substance produced by one organism has an adverse physiological effect on another organism.

antibiotic substance that prevents, inhibits, or destroys life.

antimetabolites chemicals resembling essential nutrients that interfere with metabolism.

anus posterior opening of the alimentary system (figures 4.1 and 4.2).

aorta anterior subdivision of the dorsal vessel, lacking ostia and alary muscles (figures 4.6 and 4.7).

aphins red, orange, and yellow pigments found only in aphids.

apical distal.

apical margin the outer edge of an insect wing (figure 2.23).

apiculture beekeeping

apneustic the closed type of ventilatory system, without functional spiracles.

apodeme an internal inflection of the integument to which muscles attach (figure 2.10).

apolysis the separation of the old cuticle from the epidermis during molting.

apophysis spinelike inflection (apodeme) of the insect integument to which muscles attach (figure 2.10).

aporous without a pore.

aposematic coloration warning coloration such as that of dangerous or poisonous animals.

appetitive behavior behavior that increases the probability of exposure to a releasing stimulus.

apposition image image formed when the light reaches the rhabdom of an individual eye unit only via the dioptric apparatus of the same eye unit (figure 6.13).

apterous wingless.

Apterygota primitively wingless insects that do not undergo metamorphosis.

arboviruses arthropod-borne viruses.

arch- primitive; original.

archedictyon irregular network of veins found between the principal wing veins in many fossil insect groups; similar networks in the wings of contemporary forms (e.g., dragonflies) are probably remnants of the archedictyon.

Archeognatha, order (*archeo*, ancient; *gnath*, jaw) jumping bristletails (figure 12.6a).

arolium bulbous, lobelike structure found between the tarsal claws, as in Orthoptera (figure 2.38b).

arthropodins a group of soluble proteins found in the insect cuticle.

arthropod phobia the persistent fear of infestation by arthropods.

assembling scents chemicals (pheromones) that, when released, result in the formation of a group of individuals of the same species.

association neurons *see* interneurons.

associative learning learning based on the presence of reward or punishment; includes classical conditioning, instrumental learning, and shock-avoidance conditioning.

asynchronous muscles *see* resonating muscles.

atrial aperture the outside opening of a chambered or atriate spiracle (figure 4.14b and c).

atrium spiracular chamber housing the opening to a trachea (figure 4.14b and c).

augmentation a biological control technique that involves the release of massed reared parasitoids or predators that are released to augment existing parasitoid or predator populations or to reintroduce parasitoids or predators.

auto- self; same.

autoamputation the spontaneous release of a leg that is being grasped by a predator; characteristic of nymphal walking sticks.

autogenous able to carry out vitellogenesis without an exogenous source of protein; *compare with* anautogenous.

autotomy *see* autoamputation.

axon a fiber of a nerve cell that transmits impulses from the cell body (figure 3.1a, b, and c).

B

balanced polymorphism the occurrence of several different forms resulting from variation in selection pressures acting on the same species.

basal proximal.

basalar sclerite small sclerite anterior to the pleural wing process.

basal lamina *See* basement membrane.

basement membrane the noncellular innermost layer of the insect integument (figure 2.1a) or the noncellular layer on the hemocoel-side of the gut epithelium (figure 4.3) or any such layer associated with an epithelium.

basisternum portion of the sternum between the presternum and sternellum (figure 2.18b and c).

Batesian mimic a palatable and otherwise harmless species that gains protection from predators by virtue of its close resemblance to an unpalatable or harmful species.

bi- two.

binomen the two parts, genus and epithet, that make up an organism's scientific name.

bio- life; living.

biological species a group of interbreeding organisms reproductively isolated from other closely similar groups.

bioluminescence light produced by living organisms.

biosphere the land, water, and atmospheric portions of our planet in which life exists.

biotic potential the maximum level of increase that a population is capable of under optimal conditions.

bipolar neuron neuron with a cell body bearing an axon and a single, branched or unbranched dendrite (figure 3.1*b*).

biramous appendages two-branched appendage; for example, characteristic of Crustacea.

bivoltine producing two generations per year.

blasto- embryo.

blastocoel an internal cavity surrounded by blastoderm.

blastoderm the first distinct layer of cells to form in morphogenesis.

blastokinesis all displacements, rotations, or revolutions of the embryo in the egg.

blastomeres cells composing the blastoderm.

Blattaria, order (*blatta*, cockroach) cockroaches (figure 12.20).

bombykol female sex pheromone of the silkworm moth.

botanical insecticides secondary plant substances having a toxic, repellent, deterrent, or hormonal effect on insects.

brachypterous having very short wings.

brain the anterior composite ganglion located dorsal to the foregut in the insect head.

brain hormone the hormone that stimulates secretory activity of the prothoracic gland; also called *prothoracicotropic hormone* or *ecdysiotropin*.

brood care parental care of offspring.

bursa copulatrix a saclike vagina.

bursicon neurosecretory hormone that stimulates sclerotization of the cuticle following ecdysis; also called *tanning hormone*.

C

calyx cavity joined with the lateral oviduct and into which the pedicels of the ovarioles open (figure 5.6).

campaniform sensilla (sensilla campaniformia) type of mechanoreceptor consisting of a pit covered with a cuticular dome (figures 6.1 and 6.2).

carapace dorsal plate that covers the cephalothorax in many crustaceans (figure 11.6).

carbon-based defenses allelochemicals that make food plants less nutritious (table 10.1).

cardenolides secondary plant substances in milkweed plants that may render milkweed-feeding insects unpalatable to vertebrate predators.

cardia a structure composed of both foregut and midgut tissue and in which the foregut-midgut junction is located; *see also* proventriculus and stomodael valve (figures 4.1 and 4.2).

cardiac glycosides chemicals from milkweed plants that render insects that feed on these plants unpalatable to predators.

cardial epithelium the most anterior tissue of the midgut.

cardo proximal segment of the maxilla (figure 2.15).

carnivores zoophagous or animal-eating animal.

carotenoids insect pigments of red and yellow, derived from plant tissues.

carrying capacity the maximum density of a population of a given species that an ecosystem can sustain (figure 9.2).

caryo- nucleus.

castes morphologically and functionally different types within a colony of social insects as in Isoptera (figures 12.18 and 12.19) and some Hymenoptera.

caudal ganglion the posteriormost abdominal ganglion; innervates the genitalia (figure 3.3).

caudal visceral system the part of the visceral nervous system associated with the posterior segments of the abdomen.

cauterization destruction of tissue by burning, usually with an electrical current.

cavernicolous found in caves.

cell a functional unit of living matter bounded by a membrane and containing one or more nuclei or nuclear materials; a space on an insect

wing delineated by various combinations of longitudinal and cross veins and the wing margins.

cement layer outermost layer of the epicuticle.

central body neuropilar mass connecting the two lobes of the protocerebrum (figure 3.7).

central nervous system the chain of paired segmental ganglia connected by paired cords; the location of the integrative activities of the insect nervous system (figures 3.3 and 3.11).

central place foragers insects like ants and bees that return to their nests or hives after foraging.

cephalopharyngeal skeleton reduced, invaginated mouthparts that provide articular points for the mouthparts of higher dipteran larvae (figure 2.32*d*).

cephalothorax fused head and thorax (figure 11.4).

cerci appendages of the eleventh abdominal segment of insects (figures 2.12, 2.24, and 2.42).

cervical sclerites lateral plates found on the membranous region connecting the head with the thorax (figures 2.12 and 2.13*c*).

cervix membranous region connecting the head with the thorax (figures 2.12 and 2.13*c*).

chaetotaxy the arrangement of setae in constant patterns, useful in systematics.

character a trait.

chelicerae the anteriormost pair of preoral appendages found in spiders, ticks, scorpions, and relatives (figure 11.4).

chemoreception the perception of atoms and molecules as stimuli.

chitin a high-molecular-weight polymer of *N*-acetyl-D-glucosamine that resembles cellulose and constitutes one-fourth to one-half of the dry weight of the exo- and endocuticle (figure 2.3).

chordotonal organ *see* scolopophorus organ.

chorion eggshell secreted by the ovarian follicle cells (figure 5.10).

chrom-, -chrome color; pigment.

chromosomal puffs bulges at various points along a chromosome associated with localized uncoiling of the DNA molecule; taken to indicate mRNA transcription associated with the expression of a specific gene or genes.

chrysalid a butterfly pupa that lacks a protective cover (figure 13.19c).

cibarium the portion of the preoral cavity between the hypopharynx and the labrum (figures 2.16 and 4.1).

circadian rhythms behavioral or physiological events that occur at approximately 24-hour intervals.

circumesophageal connectives two large nerves, one on each side of the esophagous, that run between the brain and the subesophageal ganglion (figure 3.3).

cladistics a school of systematic thought grounded in the belief that classification can and should reflect phylogeny.

cladogram a diagram based on cladistic analysis of a group of organisms.

classical conditioning *see* associative learning.

cleavage the divisions of the zygote and subsequent daughter cells that lead to the formation of the blastoderm (figure 5.12*b* and *c*).

cleavage center location where cleavage and migration of cleavage nuclei begin.

cleptoparasitism a process in which members of one species take over the nest of another species for their own use.

clypeus lobelike structure located on the anterior part of the insect head just beneath the epistomal suture; hinged with the labrum in many insects (figures 2.13*a* and *c*, 2.15, and 2.16).

co- with; together.

coadapted *see* coevolved.

coarctate pupae a type of pupa typical of cyclorrhaphous Diptera, in which the next-to-last larval skin forms a hard barrel-shaped case, the puparium. The puparium splits at emergence, and the adult fly pushes away the top using an eversible head sac, the ptilinum (figure 5.24*c* and *d*).

cocoonase an enzyme that digests the silken cocoon in saturniid and bombyliid moths.

coel- cavity; chamber.

coelomic sacs small cavities in the mesoderm of a developing insect (figure 5.18).

coevolved closely associated species that have acted as agents of natural selection on one another.

cohesion mutual attraction of atoms or molecules; for example, the cohesion of water molecules.

Coleoptera, order (*coleo-*, sheath; *ptera*, wing) beetles (figures 13.2, 13.3, and 13.4).

collateral axon branch (figure 3.1*b*).

Collembola, order (*coll*, glue; *embol*, a wedge) springtails (figure 12.5).

colleterial glands glands that secrete adhesive materials.

collophore a nongenital structure found on the venter of the first abdominal segment in Collembola (figure 12.5).

colon posterior region of the anterior intestine of the hindgut.

com- together.

combination cleavage total cleavage of the egg followed by peripheral cleavage in blastoderm formation, for example, in Collembola.

commensalism a symbiotic association in which one individual benefits and the other is neither helped nor harmed.

common oviduct the duct that joins the lateral oviducts to the vagina (figure 5.6).

common salivary duct the duct that joins the lateral salivary ducts to the salivarium (figure 4.1).

communal behavior cooperation in nest construction but not brood rearing.

communication the influence of signals from one organism on the behavior and/or physiology of another organism.

competitive displacement one species, having a competitive advantage, eventually eliminates another species that occupies the same or a closely similar niche.

competitive exclusion *see* competitive displacement.

compound eyes located on either side of the insect head, compound eyes are composed of a number of individual units, the ommatidia (figures 2.12, 2.13, 6.12, and 6.13).

concave veins longitudinal veins in troughs formed by wings folded in a pleated fashion.

conjunctiva soft membranous area separating sclerites in the insect skeleton.

constitutive defenses plant defenses present at all times, for example, physical defenses.

consummatory act behavior that follows appetitive behavior when an appropriate releasing stimulus is encountered.

contact poisons toxic substances that are absorbed through the cuticle.

convergent evolution the process by which nonhomologous structures become similar in appearance due to similar selection pressure.

convex veins longitudinal veins located on crests resulting from wings folded in a pleated fashion.

corbiculum specialized structure on the legs of honey bees, used to store pollen for transport to the hive; also called a *pollen basket* (figure 2.37*a*).

cornea the cuticular part of an ommatidium; *see* corneal lens.

corneagen cells a layer of cells that secretes the corneal lens of a dorsal ocellus.

corneal lens the outermost part of an ommatidium (figure 6.12) or dorsal ocellus (figure 6.17), derived from cuticle.

corneal pigment cells cells of an ommatidium, which contain pigments responsible for blocking the entry of light from an adjacent ommatidium (figures 6.12 and 6.13).

cornicles a pair of lobelike projections on the posterior dorsum of the abdomen in aphids (figures 12.32, 12.33*a* and *b*); secrete alarm pheromones.

coronal suture suture along dorsal midline of head (figure 2.13*a* and *b*).

corpora allata endocrine glands that secrete juvenile hormone and store and release some brain hormones (figures 3.4, 3.6*b*, 3.9, 3.16, and 5.27).

corpora cardiaca neurohaemal organs where some brain hormones are stored and released (figures 3.4, 3.9, 3.15*a*, 3.16, and 5.27).

corpora pedunculata cell masses in the insect brain that contain association neurons and are considered to be the location of the "higher centers" controlling the most complex behavior; also called *mushroom bodies* (figure 3.7).

corporotentorium body of tentorium formed by the fusion of the anterior and posterior tentorial arms (figure 2.13*d*).

costal margin anterior edge of the wing.

coxa the basal segment of the leg (figures 2.21, 2.36, and 2.37).

coxopleurite one of three primitive, subcoxal sclerites (figure 2.20).

coxopodite basal segment of a genital appendage.

cranium head capsule.

cremaster a structure, on the end of the abdomen, that bears small hooks by which some lepidopterous pupae attach to a substrate.

crochets minute hooks found on the prolegs of lepidopterous larvae (figure 2.42c).

crop enlarged posterior region of the esophagus that functions as a site for temporary food storage (figures 4.1 and 4.2).

cross veins transverse supporting structures in the insect wing that connect the longitudinal veins.

crypsis combination of form, color, and pattern that facilitates "hiding" from predators and/or parasites.

cryptobiosis showing no signs of metabolic activity.

cryptonephridia distal ends of Malpighian tubules encased by rectal tissues (figure 4.21d).

crystalline cone a translucent structure just beneath the corneal lens of an ommatidium (figure 6.12).

ctenidia specialized, thickened setae arranged into combs (figure 13.14) characteristic of many species of fleas.

cultural control manipulation of cropping system or specific crop production practices to reduce or avoid pest damage.

cultural management phase second phase of an IPM program; it involves the application of various cultural control tactics.

cursorial legs adapted for running and walking (figure 2.36a).

cutaneous ventilation gaseous exchange directly across the cuticle of the exoskeleton.

cuticle the noncellular outermost layer of the insect integument, secreted by the epidermis (figure 2.1).

cuticular gills see spiracular gills.

cuticular phase the period of time between the first appearance of new cuticle beneath the old and ecdysis.

cuticulin layer a layer of epicuticle, probably lipoprotein, that covers the entire integumental surface, including tracheoles and gland ducts.

-cyte, cyto- cell.

D

decticous having functional mandibles used for escape from a cocoon or pupal case.

degree-days number of days above a particular threshold temperature; used in prediction of insect development (figure 16.4).

dendrites tiny branching processes associated with or near the cell body of a neuron (figure 3.1a, b, and c).

dendrogram see phenogram.

derm- skin; covering.

dermal glands modified epidermal cells on the outer surface of the insect body that secrete irritants, poisons, wax, scents, silk, and so on (figures 2.1a and 3.12).

Dermaptera, order (derm, skin; ptera, wings) earwigs (figure 12.16).

deutocerebrum the lobe of the insect brain that receives both sensory and motor fibers from the antennae (figure 3.6).

deutoplasm see yolk.

di- two.

diapause state of arrested development in which an insect is able to survive unfavorable conditions.

diastasis a period of rest between successive heartbeats (figure 4.8).

diastole relaxation phase of the heartbeat cycle (figure 4.8).

dichotomous double-branching, as in a taxonomic key (figure 11.1).

dicondylic joint a joint consisting of two condyles and restricting movement to a single plane (figure 7.1b and c).

differentiation center a location in the insect egg that induces differentiation of embryonic cells.

digestion reducers see carbon-based defenses.

dioptric apparatus the parts of an ommatidium (cornea, crystalline cone, and pigment cells) responsible for the collection of light (figure 6.12).

Diplura, order (dipl, two; ura, tail) campodeids and japygids (figure 12.4b).

Diptera, order (di, two; ptera, wings) true flies (figures 13.10–13.13).

direct wing muscles muscles that attach to the wing bases (figure 7.9).

distal distant from the point of attachment.

diverticula blind sacs that form as evaginations from the alimentary canal, for example, the esophageal diverticula of mosquitoes (figure 4.2).

dormancy suppression of activity associated directly with changes in the abiotic environment.

dorsal diaphragm fibromuscular septum that encloses the alary muscles associated with the heart and divides the hemocoel into the pericardial and perivisceral sinuses (figure 4.12); also called pericardial septum.

dorsal light reaction a process in which longitudinal and transverse axes of the body are kept perpendicular to a directed source of light.

dorsal longitudinal trunks major tracheae located on either side of the dorsal midline of the body (figure 4.12).

dorsal ocelli simple eyes usually accompanying compound eyes and found dorsally on the insect head (figures 2.12, 2.13a and c, and 6.17b).

dorsal vessel principal conducting tube of the circulatory system, extending the length of the dorsal midline (figures 4.6, 4.7, and 4.12).

ductule cell a duct forming cell associated with a simple exocrine gland (figure 3.12b).

Dyar's rule the ratio of a given linear dimension in one instar to the same dimension in the next instar is constant throughout the larval instars and for a given species.

E

ecdysial cleavage lines preformed lines of weakness along which the shed cuticle of the cranium splits during ecdysis; cuticle in the region of these lines lacks the exocuticle (figure 2.8c).

ecdysial glands see prothoracic glands.

ecdysiotropin see brain hormone.

ecdysis shedding of the exuvia.

ecdysone a hormone that initiates the growth and molting activity of epidermal cells.

eclosion the emergence of an immature stage from the egg; sometimes used to refer to pupal-adult ecdysis.

eclosion hormone a hormone, synthesized by the median neurosecretory cells, which influences behavior associated with pupal-adult ecdysis.

ecological niche the total of needed resources (food, shelter, mates, and so on) of a species.

economic injury level the lowest density of pest population that causes economically significant damage.

economic threshold the density of pest population at which control measures should be applied to prevent the pest from reaching the economic injury level.

ectad outward relative to the midline of the body.

ecto- outside; external.

ectoderm the outermost germ layer (figures 5.15 and 5.16).

ectognathous having mouthparts not pulled into the head.

ectotherm organism that relies on heat from the external environment for thermoregulation.

effective environment all environmental factors that directly influence the fate of a population (figure 9.1).

efferent neurons *see* motor neurons.

egg development neurosecretory hormone secretion of median neurosecretory cells of mosquitoes that induces ovaries to release ecdysone (figure 5.11).

ejaculatory duct a duct formed by the joining of vas deferens from each testis and involved in the propulsion of semen (figure 5.1a).

elytra hardened forewings that protect membranous hindwings; characteristic of beetles (figures 2.39c and 13.2).

Embioptera, order (*embio*, lively; *ptera*, wing) webspinners (figure 12.2).

embryogenesis developmental events between zygote formation and exit from the egg.

empodium spinelike or lobelike structure arising between the tarsal claws (figure 2.38a).

encapsulation a protective mechanism in which large numbers of hemocytes become layered around a foreign entity, such as a parasitic worm, that has invaded the hemocoel (figure 4.9).

endo- within.

endocrine glands ductless glands whose products are released directly into the insect hemolymph.

endocuticle the inner, unstabilized layer of the chitinous cuticle—which may or may not be pigmented—that is digested during molting.

endoderm innermost germ layer.

endogenous arising from within.

endopterygote refers to insects in which the wings develop internally during the larval stage, i.e., insects that undergo complete metamorphosis; immature instars dissimilar to adults' and usually adapted to different environmental conditions; *syn.* holometabolous.

endotherm organism that metabolically generates heat required for thermoregulation.

entad inward relative to the midline of the body.

ento- within.

entognathous having mouthparts somewhat pulled into the head.

entomophagous insect-eating.

entomophilous insect-loving.

entomophobia irrational and persistent fear of insect infestation or attack.

environmental resistance the collective action of environmental factors that prevents manifestation of the biotic potential of a given species.

Ephemeroptera, order (*ephemero*, for a day; *ptera*, wings) mayflies (figure 12.10).

epicranial suture coronal plus frontal sutures (figure 2.13a).

epicuticle unpigmented outer layer of the insect cuticle, 0.05–4.0 micrometers thick (figure 2.1).

epidermis the cellular layer of the insect integument (figure 2.1).

epididymis coiled vas deferens in which sperm are stored.

epimeron posterior portion of the thoracic pleuron (figure 2.18b).

epipharynx a lobe on the epipharyngeal wall.

epiproct dorsal plate of the eleventh abdominal segment (figure 2.24).

episternum anterior portion of a thoracic pleuron (figure 2.18b).

epistomal suture suture above the clypeus and connecting the subgenal sutures (figure 2.13a).

esophagous conducting tube of the foregut immediately posterior to the pharynx (figures 4.1 and 4.2).

eusocial truly social.

eusternum large plate of the thoracic sternum (figure 2.18c).

exarate pupae pupa with free appendages, usually not covered by a cocoon (figure 5.24a).

exocrine glands glands that discharge their products through apertures or ducts into the external world or into lumens of various viscera (figure 3.12).

exocuticle stabilized layer of cuticle between the endo- and epicuticular layers; resistant to molting fluid and often pigmented (figure 2.1).

exogenous arising from without.

exopterygotes describes insects having wings that become externally apparent in the immature instars (immatures resemble adults), i.e., undergo simple metamorphosis; *syn.* hemimetabolous.

exoskeleton a skeleton located outside the body, as in arthropods.

extant presently existing form.

extrinsic muscles muscles that originate on the inner surface of the body wall of the head, thorax, or abdomen and insert into an appendage (figure 3.20).

exuvia the undigested exocuticle and epicuticle shed during ecdysis.

exuvial fluid *see* molting fluid.

exuvial membrane a thin, homogeneous membrane that may appear in the exuvial space during molting (figure 2.7b and c).

exuvial phase period of time between apolysis and the beginning of new cuticle secretion.

exuvial space the space between the epidermis (and later the new cuticle) and the old cuticle that forms as the old cuticle is digested during molting (figure 2.7b and c).

F

facultative diapause more than one generation between the termination of one diapause and the initiation of the next diapause.

facultative mutualism capable of, but not obliged to, form a mutualistic relationship.

facultative parasite a parasite that can complete its life cycle without invading a host but can also take advantage of a host if one is available.

feeding deterrent a chemical that prevents feeding of pests on treated material.

felt chamber atrium of a spiracle lined with fine hairs.

femur long leg segment between the trochanter and tibia (figure 2.36).

fenestrae openings in the dorsal diaphragm through which hemolymph can pass.

fiber tracts groups of axons within a ganglion that run parallel to one another.

filter chambers close association between anterior and posterior regions of the alimentary canal, found in insects like spittlebugs, which ingest food with high water content; facilitate passage of water in such a way that the midgut is bypassed.

flagellum long, apical part of the antenna, commonly composed of several subsegments (figure 2.14).

flavines greenish yellow pigments found in insects.

follicle structure composed of oocyte, nurse cell, and follicular epithelium (figures 5.7, 5.8, and 5.9).

follicular epithelium layer of cells that invests a developing oocyte and that secretes the chorion (figures 5.7, 5.8, and 5.9).

fontanelle depression on dorsum of head in many species of termites.

foramen magnum posterior opening of the head capsule through which internal structures (alimentary canal, dorsal nerve cord, and so on) pass to the thorax (figure 2.13b).

foregut anterior region of the insect alimentary canal, also called the stomodaeum (figure 4.1).

forensic entomology use of entomological information in criminal investigations.

fossorial legs bearing sclerotized claws for digging (figure 2.36b).

frenulum spinelike wing-coupling mechanism characteristic of several Lepidoptera (figure 2.40b).

frons frontal region delimited by the frontal sutures (figure 2.13a and c).

frontal ganglion ganglion of the stomatogastric system, located on the dorsal midline of the foregut just anterior to the brain (figures 3.3, 3.4, and 3.6a and b).

frontal sutures two sutures that diverge ventrally across the anterior portion of the head (figure 2.13a and c).

frontoclypeal suture *see* epistomal suture.

fumigants insecticides that enter through the spiracles and tracheal system in gaseous form.

furca sternal apophysis in the form of a V (figure 2.17b).

furcal pits external manifestations of the sternal apophyses (figure 2.18c).

furcula fork-shaped structure on the abdomen of the springtail that engages the tenaculum and is used to propel the insect through the air (figure 12.5).

G

galea tonguelike, distal lobe attached to the stipes of the maxilla (figure 2.15).

galls abnormal growths on plants caused by insect larvae feeding on the plant tissue (figure 10.8).

ganglion a group of neurons (figures 3.1d and 3.2).

gaster collective term for the hymenopteran abdominal segments posterior to the petiole.

gastric caeca a group of diverticula commonly found in the midgut (figure 4.1).

gastric nerves nerves connecting the hypocerebral ganglion and ventricular ganglion (figure 3.6b).

gastrulation formation of the mesoderm and endoderm (figure 5.15).

-gen, -geny producing; generation.

gena lateral "cheek" region of the head delimited by the compound eye, subgenal suture, and occipital suture (figure 2.13a and c).

genetic resistance genetically controlled plant resistance to insect attack.

genital ridge embryonic tissue of mesodermal origin that, together with the germ cells, are the precursors of the gonads (figure 5.18c).

germarium the apical tissue of a testicular follicle or ovariole from which the germ cells originate (figures 5.2 and 5.6).

germ band thickened area of columnar cells of the bastoderm that develops into the embryo (figure 5.12e and f).

germ cells cells that differentiate into gametes—eggs and sperm (figure 5.12d and e).

giant chromosomes chromosomes that are much larger than ordinary chromosomes, consisting of a very large number of identical DNA molecules; polytene chromosomes.

glial cells cells that invest neurons and aid in the functioning of these neurons (figure 3.2).

glomeruli spheroidal collections of neuropile in each antennal lobe of the brain.

glossae inner, distal lobes found on the prementum of the labium (figure 2.15).

gnathal segments the posterior three segments of the developing insect head that give rise to the mouthparts (figure 11.13, segments 4, 5, and 6).

gnathobasic mandibles appendage with basal process that functions as a jaw.

gonopods appendages composing the ovipositor.

gonopore external reproductive opening.

Grylloblattodea, order (*gryll*, cricket; *blatta*, cockroach) rock crawlers or ice bugs (figure 12.17).

gula structure formed from the fusion of the postgenae, characteristic of many beetles and neuropterous insects (figure 2.34e).

gulamentum fusion of the gula and mentum.

gustation contact chemoreception; "taste."

gynandromorphs insects displaying some typically female body parts and some typically male body parts.

H

habituation a form of associative learning in which a given stimulus no longer elicits a response.

hair-pencils eversible structures covered with fine hairs, providing extensive surface area that facilitates dispersal of pheromones.

hair plate cluster of tiny trichoid sensilla, located in appressed or overlapping areas of the body, that have a proprioceptive function (figure 6.4).

hallucination of arthropod infestation a condition in which a person imagines he or she is being molested by small, difficult-to-locate forms, which localize on the body despite extraordinary preventive measures.

halteres hindwings modified to club-shaped structures important in flight stability.

hamuli tiny hooks on the costal margin of the hindwings that attach to the anal margin of the forewings, affecting wing coupling; characteristic of the Hymenoptera (figure 2.40a).

haplodiploidy a reproductive process in which females develop from fertilized (diploid) eggs, whereas males develop from unfertilized (haploid) eggs, as in the social Hymenoptera.

harmonic growth a growth pattern in which all parts of the insect body, and the body as a whole, increase by the same ratio during each molt; also called *isogonic growth*.

haustellate having mouthparts adapted for sucking activities (figures 2.30 and 2.31).

heart posterior division of the dorsal vessel; associated with alary muscles and contains ostia (figures 4.6 and 4.7).

hematophagous blood-feeding.

hemelytra partly hardened forewings with membranous distal portions, as in Hemiptera (figures 2.39*d* and 12.27).

hemimetabolous *see* exopterygote.

hemipneustic number of spiracles reduced in comparison with most generalized pattern.

Hemiptera, order (*hemi*, one-half; *ptera*, wing) true bugs (figures 12.27, 12.28, and 12.29).

hemocoel body cavity of the insect in which the viscera are located and the hemolymph circulates.

hemocoelic insemination *see* traumatic insemination.

hemocytes insect blood cells.

hemocytopoietic organs organs that produce blood cells.

hemogram a count of the total number of hemocytes and an estimate of the numbers of different types of hemocytes in an insect at a given time.

hemolymph insect blood that, as implied by its name, carries out the functions of both blood and lymph, which are different fluids in vertebrates.

hemostasis inhibition of hemolymph loss; coagulation.

hermaphroditic an individual that possesses both male and female sex organs.

hetero- other; different.

heterogametic possessing the XO or XY chromosome configuration.

heterogonic growth growth patterns in which some parts of the insect body develop at different rates than others.

heteromorphous describes occurrence of a regenerated appendage different in form from the appendage it has replaced.

hindgut posterior region of the insect alimentary canal (figure 4.1); also called *proctodaeum*.

hist- tissue.

holoblastic cleavage method of blastoderm formation characteristic of Collembola and certain parasitic wasps and associated with eggs that contain little yolk; the entire zygote divides (compare with meroblastic cleavage).

holocrine secretion cell products released by complete breakdown of a cell.

holometabola the group of insects that undergoes complete metamorphosis.

holometabolous undergoing complete metamorphosis; *see also* endopterygote.

holopneustic pertaining to open ventilatory system that consists of two lateral rows of 10 spiracles each.

homeo-, homo- like; similar.

homogametic sex possessing two X chromosomes.

homologous characters in two or more species that are fundamentally similar and attributable to common ancestry.

Homoptera, order (*Homo*, uniform; *ptera*, wing) aphids, leafhoppers, and relatives (figures 12.30 through 12.33).

hooked trichomes minute, sharp, hooked hairs found on certain plants that provide defense against insect attack (figure 10.11).

humeral blood-borne.

hyaline clear; transparent.

hydro- water; fluid; hydrogen.

hydrophilic having water-attractant characteristics; wettable.

hydrophobic having water-repellent characteristics; nonwettable.

hydrostatic balance maintenance of body position in the water by aquatic insects.

hygroreception perception of moisture in the air.

Hymenoptera, order (*hymen*, membrane; *ptera*, wings) ants, bees, wasps, and relatives (figures 13.22, 13.23, and 13.24).

hyper- over; above; excessive.

hypermetamorphosis a form of holometabolous development in which one or more larval instars is distinctly different from the others (figure 5.22*e*).

hyperparasitism a parasite serving as a host for another species of parasite.

hyperparasitoidism a parasitoid serving as a host for another species of parasitoid.

hypo- under; lower.

hypocerebral ganglion ganglion located just posterior to the brain and dorsal to the foregut (figures 3.3 and 3.9).

hypognathous mouthparts hung vertically from the head capsule (figure 2.33*a*).

hypopharynx in generalized forms, the tonguelike structure that is surrounded by the externally visible mouthparts and that separates the cibarium from the salivarium (figure 2.16); may be in the form of a stylet and channel in some sucking insects.

hypostomal bridge closure of the occipital foramen by fusion of the posterior region of the subgenae (figure 2.34).

I

ileum differentiated region of the anterior intestine of the hindgut in some insects.

imaginal buds *see* imaginal disc.

imaginal disc mass of cells that persist in an undifferentiated state throughout the larval instars and become adult tissues (figure 5.25).

imago adult insect.

indirect wing muscles muscles that contribute to insect flight but are not directly attached to the insect wings (figures 3.21 and 7.9).

inducible defenses defense mechanism brought into play in response to damage as, for example, in herbivore feeding on a plant.

innate behavior genetically more-or-less fixed behavior.

inquilines insects that are residents in the shelters or nests of other insects.

Insecta the taxon made up of insects; means literally "in-cut," describing the segmented appearance of insects.

insecticide a chemical employed to kill insects.

insectoverdins green pigments found in insects.

insect pest management a modern approach to dealing with pest insects; control measures chosen from all available methods and applied on the basis of carefully evaluated ecological/economic parameters.

insemination transfer of semen from the male reproductive system to the female reproductive system.

insertion movable attachment area of skeletal muscles.

in situ in a natural or original position.

instar each developmental form of an insect's life cycle; the insect between successive ecdyses or between successive apolyses.

instrumental conditioning trial and error learning.

integrated pest management application of several different insect control methods to maintain a pest population below economically damaging levels.

integument general body covering.

inter- between.

interneurons neurons, found within ganglia, that may synapse with one another or with sensory or motor neurons; also called *association neurons* (figure 3.1d).

internuncial neurons *see* interneurons.

intersegmental muscles the principal longitudinal muscles running between successive antecostae in insects with secondary segmentation (figure 2.11b and c).

interspecific among different species.

intima chitinous lining of the foregut and hindgut (figure 4.1) and tracheae (figure 4.13a), which is continuous with the cuticle of the integument.

intra- within.

intraspecific among different species.

intrinsic muscles both origins and insertions entirely within a given appendage (figure 3.20a).

inundation very large number of mass-reared biological control agents released to overcome a pest population.

iso- equal; uniform.

isogonic growth *see* harmonic growth.

Isoptera, order (*iso*, equal; *ptera*, wings) termites (figures 12.18 and 12.19).

J

Johnston's organ sensory structure located in the antennal pedicel (figure 6.6).

jugum lobelike wing-coupling mechanism characteristic of some Lepidoptera (figure 2.40c).

juvabione juvenile hormone mimic from balsam fir.

juvenile hormone a hormone that promotes larval development and inhibits development of adult characteristics.

K

kairomone a chemical that is adaptively favorable to the receiving organism.

kin-, kino- motion; activity.

kineses undirected locomotor responses to stimuli.

klinotactic orientation a taxis characterized by moving the body back and forth across a stimulus field and moving toward or away from the region of maximum stimulation.

L

labella apical spongelike structure on the proboscis of nonbiting muscoid flies (figure 2.31b).

labial glands glands associated with the labial segment, commonly having salivary functions.

labial palpi a pair of segmented appendages associated with the labium (figure 2.15).

labial suture boundary between the proximal mentum, and the distal prementum, of the labium (figure 2.15).

labium the posterior mouthpart or "lower lip" (figure 2.15).

labrum the anterior mouthpart or "upper lip" (figure 2.15).

lac a natural body secretion of the scale insect *Laccifer lacca* (Homoptera) that is used to make shellac, shoe polish, paint, adhesives, and many other products.

lacinia distal, mesal lobe of the stipes of a maxilla; bears teeth on its inner edge (figure 2.15).

larvae immature instars of insects; usage of the term is commonly limited to immatures of insects with complete metamorphosis.

larviparity a form of viviparity in which the female gives "birth" to larvae instead of depositing eggs.

latent learning learning without apparent reward or punishment.

lateral ocelli *see* stemmata.

lateral oviduct ducts connecting each ovary to the medial, common oviduct (figure 5.6).

lateral salivary duct duct that carries secretion products from a salivary gland to the common salivary duct (figure 4.1).

learned behavior behavior acquired as a result of interaction with the environment.

Lepidoptera, order (*lepido-*, scale; *ptera*, wings) butterflies, moths, skippers (figures 13.19, 13.20, and 13.21).

longitudinal veins wing supporting structures with a basal to apical orientation (figure 2.23).

M

macro- large.

macroevolution major phylogenetic patterns that develop over wide spans of geological time.

macrotrichia *see* setae.

Mallophaga, order (*mallo*, wool; *phaga*, eat) chewing lice (figure 12.24).

Malpighian tubules elongate excretory structures of the digestive tract in most insects; located at the anterior end of the hindgut (figures 4.1 and 4.2).

mandibles a pair of highly sclerotized, unsegmented "jaws" located between the labrum and maxillae (figures 2.12, 2.13, 2.15, 2.28, 2.32, and 2.34); stylets in many insects with piercing-sucking mouthparts (figure 2.30).

mandibulate having primitive, usually chewing, mouthparts (figures 2.28a and 2.29).

Mantodea, order (*mantid*, soothsayer) praying mantids (figure 12.21).

maxillae paired, segmented, secondary jaws that aid in holding and chewing foods; located between the mandibles and the labium (figures 2.12, 2.13, and 2.15); stylets in many insects with piercing-sucking mouthparts (figure 2.30); variously modified in other insects (figures 2.31, 2.32, and 2.34).

maxillary palpus appendage arising from maxilla.

mechanoreception the perception of stimuli as various mechanical changes.

mechanosensilla sensory structures that respond to stimuli such as stretching, bending, or compression (figures 6.4, 6.5, 6.6, 6.7, 6.8, and 6.9).

Mecoptera, order (*meco*, long; *ptera*, wing) scorpion flies (figure 13.9).

median neurosecretory cells cells in the pars intercerebralis region of the brain that secrete hormones, including brain and eclosion hormones (figures 3.7 and 3.9).

mega- large.

melanins insect pigments of yellow, black, or brown.

melanization darkening.

mentum distal portion of the postlabium (figure 2.15).

meroblastic cleavage method of blastoderm formation characteristic of most insects and associated with eggs containing a large amount of yolk; the fusion nucleus and associated cytoplasm proliferate mitotically, and the daughter nuclei migrate to the periphery of the egg and form the blastoderm (figure 5.12).

meroistic ovarioles polytrophic and telotrophic ovarioles taken together.

meron the posterior portion of the coxa.

mesenteron *see* midgut.

mesentoderm a longitudinal band of tissue in the insect embryo that is surrounded by the ectoderm.

meso- middle.

mesocuticle a procuticular layer lying between the exocuticle and endocuticle.

mesoderm the middle germ layer (figure 5.16).

mesothorax middle segment of the thorax (figure 2.12).

meta- later; latter; changing.

metameres essentially identical primary body segments found in embryos and thought to be characteristic of the primitive arthropod; also called somites (figures 11.12 and 11.13).

metamorphosis change in form.

metathorax the posterior segment of the thorax (figure 2.12).

micro- small.

microevolution evolutionary changes that occur in populations, e.g., industrial melanism and development of resistance to pesticides.

micropyle channel through which spermatozoa enter an egg (figure 5.10).

microtrichia minute fixed hairs that lack the basal articulation characteristic of setae.

microvilli numerous, fingerlike processes that serve to increase absorptive surface area, for example, in midgut epithelial cells (figure 4.3).

midgut middle region of the insect alimentary canal; site of most digestion and absorption of food; also called mesenteron (figure 4.1).

mimicry close resemblance in form, color, pattern, and so on, between one species and another that is interpreted as conferring a benefit on one or both of these species.

molting the process of digesting the old cuticle, secreting the new cuticle, and shedding the old cuticle (figure 2.7).

molting fluid a substance secreted by the epidermal cells; contains chitinase and protease and digests old endocuticle (figure 2.7).

molting hormone ecdysone; hormone that stimulates molting.

monocondylic (articulation) joint containing a single condyle, allowing considerable movement (figure 7.1).

monolectic describes pollinating insects that visit a single plant taxon; compare with polylectic.

monophyletic evolving from a single taxon.

morph-, morpho- form.

morphogenesis all events occurring between the formation of the zygote and the emergence of a sexually mature adult.

mosaic theory model of insect vision in which each ommatidium of the compound eye "sees" only a small portion of the external environment, the perception of the total environment appearing as a mosaic of the contribution from each ommatidium.

motor neuron a unipolar neuron that stimulates muscles; also called efferent neuron (figure 3.1*d*).

mouthparts organs of ingestion.

multipolar neurons neurons with an axon and several branched dendrites.

multiterminal innervation pattern of insect motor axons that branch out and form terminals at several regularly spaced points along a muscle fiber.

muscle fiber basic structural unit of a muscle (figure 3.23).

muscle unit a group of 10 to 20 muscle fibers organized as a discrete morphological entity and enveloped by a tracheolated membrane (figure 3.25).

mushroom bodies *see* corpora pedunculata.

mutualism intimate relationships between two or more kinds of animals or between plants and animals that are advantageous to both.

mycetocytes cells specialized to house microbes essential to the normal growth and development of certain insects.

mycetomes tissues composed of mycetocytes and associated with the gut, fat body, or gonads.

mycetophagous fungus-eating.

myiasis a condition caused by the invasion of open wounds or cavities by certain fly larvae (maggots).

myo- muscle.

myofibril unit of muscle fiber composed of myofilaments (figure 3.23).

myrmecophilous "ant-loving"; refers to organisms, other than ants, that interact with ants symbiotically.

N

naiad immature instar of an aquatic hemimetabolous insect.

natatorial (legs) legs adapted for swimming (figure 2.36*e*).

neo- new.

neopterous possessing a wing-flexion mechanism.

nephr- kidney.

nerve a bundle of axons that provides connections among ganglia and other parts of the nervous system.

neural lamella outer connective tissue layer of a ganglion (figure 3.2).

neuroblasts embryonic cells that differentiate into neural tissues (figure 5.18).

neurohaemal organs structures that serve as storage and release sites for neurosecretory material such as corpora cardiaca.

neuron a basic functional unit of the nervous system; an elongated, excitable cell that carries information in the form of electrical charges (figure 3.1).

neuropile central region of a ganglion, consisting of intermingling, synapsing axons encapsulated by glial cell processes (figures 3.2 and 3.7).

Neuroptera, order (*neuro*, nerve; *ptera*, wing) lacewings, dobsonflies, and their relatives (figures 13.6, 13.7, and 13.8).

neurosecretory cells modified neurons that secrete hormones (figures 3.7, 3.9, 3.13, and 3.14).

neurotransmitters molecules, such as acetylcholine, that are secreted at the terminal arborizations of one neuron and that initiate an action potential via the dendrites or axon of another neuron.

nicotine secondary plant substance in the common tobacco plant that is toxic to insects.

nidi groups of regenerative cells located at the bases of midgut epithelial cells.

nonresonating muscles muscles that require a nervous impulse for each contraction (compare with resonating muscles); also called synchronous muscles.

notum dorsal region of a thoracic segment (figures 2.17*a*, 2.18*a* and *b*, and 2.19).

nuclear cytoplasm cytoplasm around the nucleus of an insect egg (figure 5.10*a*).

nurse cells nutritive cells in polytrophic and telotrophic ovarioles; also called trophocytes (figures 5.7, 5.8, and 5.9*b* and *c*).

nymphs immature instars of ametabolous and hemimetabolous insects (figure 5.22*a*–*c*).

O

obligate parasites organisms that require a host to survive.

oblique sternals lateral abdominal muscles in other than longitudinal or dorsal-ventral orientation (figure 3.22).

obtect pupae pupa with wings, antennae, mouthparts, and legs glued close to the body; commonly covered by a cocoon (figure 5.24*b*).

occiput the posterior head region located between the occipital and postoccipital sutures (figure 2.13*c*).

occipital suture a line that runs from the posterior end of the coronal suture to just above the mandibles on either side of the head (figure 2.13*b* and *c*).

ocular sutures sutures found around the periphery of the compound eyes of some insects (figure 2.13*c*).

olfaction distance chemoreception.

olfactory lobes *see* antennal lobes.

oligophagous feeding on a few different plant species.

ommatidium individual functional unit of the insect compound eye (figure 6.12).

ommochromes red, yellow, and brown pigments commonly found as insect eye pigments.

omnivores insects or other animals that fall into two or more of the following categories: detritivores, herbivores, mycetophagous, or carnivores.

oo- egg.

oocytes germ cells in females (figures 5.7 and 5.9).

oogonia primary female germ cells located in the germarium (figure 5.9) of an ovariole.

oosome specialized region of the egg associated with differentiation of germ cells in the embryo (figure 5.12*a* and *b*).

ootheca "egg purse"; several eggs contained in a common capsule.

opisthognathous mouthparts mouthparts directed ventroposteriorly relative to the head capsule (figure 2.33*c*).

optic lobes bilateral portions of the brain that receive sensory input from the compound eyes (figures 3.6 and 3.7).

optomotor responses compensating adjustments based on visual input during flight.

origin stationary attachment area of skeletal muscle.

Orthoptera, order (*ortho-*, straight; *ptera*, wing) grasshoppers, locusts, crickets, katydids, and related forms. (figures 12.14 and 12.15).

osmeterium a usually Y-shaped eversible glandular structure located behind the head in swallowtail butterfly larvae; secretes a repugnatorial substance.

ostium valvular openings in the heart (figures 4.6 and 4.7).

ov-, ovi- egg.

ovaries paired gonads in the female reproductive system (figure 5.6*a*).

ovarioles egg-producing units of the ovaries (figures 5.6 and 5.9).

ovariole sheath cellular layer that surrounds an ovariole (figure 5.8*b*).

oviparous deposition of chorionated egg.

ovipositor egg-laying apparatus (figures 2.44*a*–*d* and 6.5).

oviscapt the abdomen when used as an "ovipositor" (figure 2.44*e*).

ovitubus *see* oviscapt.

ovoviviparity a condition in which a fully developed egg with a chorion hatches within the parent and a larva is "born" instead of an egg deposited.

ovulation the movement of an oocyte from the ovariole.

P

pacemaker a region from which waves of contraction in the heartbeat cycle are propagated.

paleopterous wing flexion mechanism absent.

palpifer lateral sclerite located on the stipes of a maxilla and bearing the maxillary palpus (figure 2.15).

palpigers lateral sclerites located bilaterally on the prelabium that bear the labial palps (figure 2.15).

panoistic type of ovariole that lacks trophocytes (figure 5.9*a*).

para- alongside of.

parabiosis joining the hemolymph circulation of one individual with that of another.

paraglossae two outer lobes found on the prementum of the labium (figure 2.15).

paraprocts plates of the eleventh abdominal segment located on either side of the anus (figure 2.24).

parasitism an intimate relationship in which one individual (the parasite) benefits at the expense of another (the host).

parasitoids insects that resemble both parasite and predator; they begin life as small parasites but may outgrow and eventually devour the host.

pars intercerebralis cell mass, located in the middorsal region of the brain above the protocerebral bridge, that contains the median neurosecretory cells (figures 3.7 and 3.9).

parthenogenesis production of individuals from unfertilized eggs; asexual reproduction.

passive suction ventilation sucking in of air through nearly closed or fluttering spiracles as a result of negative pressure in the tracheae.

passive ventilation simple diffusion of gases into the tracheae and tracheoles without the aid of pumping or other movements.

paurometabolous undergoing a gradual metamorphosis, passing through immature and adult instars in essentially the same environment.

pedicel second antennal segment, between the scape and the flagellum; the constricted region between the thorax and abdomen of Hymenoptera, Apocrita, which is also called a petiole (figure 2.14).

pediculosis infestation with lice.

pedogenesis reproduction by immature insects, for example, the beetle *Micromalthus debilis*.

penis male copulatory organ; also called aedeagus (figure 5.1).

peri- surrounding.

pericardial surrounding the heart.

pericardial cells large, multinucleate athrocytes found in association with the heart, alary muscles, and surrounding connective tissue (figure 4.11*a*).

pericardial septum *see* dorsal diaphragm.

pericardial sinus body cavity region above the dorsal diaphragm (figure 4.6).

perikaryon cell body of a neuron; also called the soma or cell body.

perineural surrounding nervous tissue.

perineurium cellular layer located beneath the neural lamella of a ganglion that probably secretes the neural lamella (figure 3.2).

periodicity recurrence of particular behavior(s) or physiological events in a regular pattern.

peripheral nervous system all the nerves emanating from the ganglia of the central and visceral nervous systems.

periplasm cytoplasm located around the periphery of the yolk of an insect egg (figure 5.12*a*).

periproct posterior somite of a segmented animal, in which the anus is located; also called the telson (figure 11.12*a*).

peristaltic contractions rhythmic contractions of longitudinal and circular muscles surrounding or composing a tubular organ such as the alimentary canal.

peritrophic membrane a chitinous, noncellular membrane that surrounds the food bolus in most insects (figure 4.1).

perivisceral sinus body cavity region below the dorsal diaphragm and above the ventral diaphragm (if present) (figure 4.6).

pesticide chemical employed to kill insects, weeds, rodents, and other pests.

petiole the constriction following the first abdominal segment of Hymenoptera, Apocrita; also called a pedicel (figure 2.41*a*).

phagostimulant a stimulus that induces feeding.

pharate period of time between apolysis and ecdysis, while the insect is still within the confines of the old cuticle.

pharynx foregut structure into which the true mouth opens; often, a well-developed pump.

Phasmida, order (*phasm*, phantom) stick and leaf insects (figure 12.13).

phenetics a school of systematic thought based on the assumption that phylogeny cannot be deduced with certainty and therefore classification should be based on as many different characteristics as possible (compare with cladistics).

pheromone chemical produced by one individual that influences the behavior of another individual of the same species.

-phore carrier.

phoresy a commensalistic relationship in which one kind of animal attaches to another and thereby gains a mode of transportation.

photo- light.

photoreception reception of stimuli in the form of light.

phototaxis directed response relative to light.

phragma platelike inflections of the terga (nota) that provide surface area for the attachment of the dorsal longitudinal wing muscles (figures 2.18*a* and *b*, 2.19, and 7.9*a*).

-phyll leaf.

physical poisons kill by physical action, e.g., exclusion of air or abrasion.

-phyte, phyto- plant.

phytoecdysones compounds found in certain plants that mimic the action of ecdysone.

phytophagous plant-eating.

placoid sensilla platelike cuticular sensory structure (figures 6.1*e* and 6.2).

plasm-, plasmo-, -plasm something molded or modeled.

plasma fluid portion of insect blood.

plastron fine hydrofuge hairs or other fine cuticular networks which hold a thin layer of gas permanently in place; acts as a permanent "physical gill."

Plecoptera, order (*pleco*, pleat; *ptera*, wing) stone flies (figure 12.11).

pleural apophysis armlike projection of the pleural ridge (figure 2.17*a*).

pleural membrane the unsclerotized cuticle in the pleural region of the abdomen that separates the terga and sterna (figure 2.24).

pleural wing process ventral articulation of a wing (figures 2.18*b* and 7.9).

pleuron lateral region of the thorax (figures 2.17*a* and 2.18*b*).

pleurosternal sutures sutures dividing the sternum from the laterosternites (figure 2.18*c*).

poikilothermic animals that remain at ambient temperature, that is, "cold-blooded."

pollen basket *see* corbiculum.

poly- many.

polyembryony mitotic division of an egg resulting in the development of several embryos from a single egg; found among several groups of parasitic Hymenoptera.

polylectic describes pollinating insects that visit plants of many different taxa; compare with monolectic.

polymorphism many forms, for example, one or both sexes of a single species may occur in two or more clearly distinct forms.

polyneuronal innervation the situation in which insect skeletal muscle units receive more than one motor axon (figure 3.25).

polyphagous eating a wide variety of plants.

polyphyletic evolving from more than one taxon.

polytrophic (ovarioles) nurse cells within each ovarian follicle (figure 5.9*c*).

pore canals tiny tubes connecting the cellular and cuticular layers of the insect integument (figure 2.1*a* and *c*).

postalar bridge exoskeletal region of a wing-bearing segment that unites the postnotum with the epimeron (figure 2.18*b*).

postcoxal bridges exoskeletal region of a wing-bearing segment that unites the epimeron with the sternum (figure 2.18*b*).

posterior imaginal ring ring of embryonic cells located at the anterior end of the larval hindgut. *See also* anterior imaginal ring.

posterior intestine *see* rectum.

posterior notal process articular point for a wing (figure 2.18*a*).

postgena lateral region of the head above each subgenal suture and between the occipital and postoccipital sutures; the "cheek" (figure 2.13*c*).

postgenal bridge fusion of lobes of the postgenae found in some insects (figure 2.34*d*).

postmentum basal portion of the labium and composed of the proximal submentum and distal mentum (figure 2.15).

postnotum portion of a thoracic notum that bears the internally inflected phragma (figures 2.18a and b, 2.19, and 7.9a).

postoccipital suture a line surrounding the foramen magnum and indicative of an inflected ridge to which muscles are attached; located posterior to the occipital suture (figure 2.13b and c).

postocciput head sclerite posterior to the postoccipital suture; surrounds most of the foramen magnum (figure 2.13c).

precoxal bridge exoskeletal region of a wing-bearing segment that unites the episternum with the sternum (figure 2.18b).

predator a free-living organism that feeds on other, commonly smaller, living organisms (prey); the prey is killed and eaten.

prementum apical portion of the labium that bears the labial palpi (figure 2.15).

preoral cavity cavity formed by the mouthparts (figure 2.16).

prescutal suture line that delineates the prescutum (figure 2.18a and b).

prescutum anterior sclerite divided off from the alinotum of a wing-bearing segment (figure 2.18a and b).

presocial behavior that includes one or two of the three characteristics that define social behavior.

presternal suture the line that separates the presternum from the rest of the sternum (figure 2.18c).

presternum a sclerite on the anterior part of the sternum divided off by the presternal suture (figure 2.18c).

pretarsus distal leg segment, usually bearing claws (figure 2.21).

prey organisms eaten by predators.

primary oocyte oocyte that is located proximally in an ovariole and that is ready to undergo vitellogenesis.

primary orientation assumption and maintenance of the body in space.

primary segmentation the arrangement of body units such that intersegmental grooves form the attachment points for longitudinal muscles (figure 2.11a and b).

primary spermatocytes diploid cells that will undergo meiotic cell division and eventually form into spermatozoans.

pro- before.

proctodael feeding ingestion by one individual of material from the anus of another individual.

proctodael invagination ectodermal invagination that develops into the hindgut.

proctodaeum *see* hindgut.

procuticle endocuticular and exocuticular layers (figure 2.1a and b) together.

prognathous anteriorly directed mouthparts (figure 2.33b).

prolegs bilateral abdominal legs characteristic of lepidopteran and many hymenopteran larvae (figure 2.42c).

propodeum first abdominal segment of Hymenoptera-Apocrita; it is completely associated with the thorax and anterior to the petiole (figure 2.41a).

prostomium *see* acron.

prot-, proto- first; primary.

protein epicuticle the innermost layer of the epicuticle (figure 2.1c, inner epicuticle).

prothoracic gland gland, also called ecdysial gland, that secretes ecdysone (figures 3.15b, 3.16 and 5.27).

prothoracic gland hormone *see* ecdysone.

prothoracicotropic hormone *see* brain hormone.

prothorax anterior segment of the thorax (figure 2.12).

protocephalon bilaterally expanded anterior region of the germ band in an insect embryo (figure 5.17a).

protocerebrum dorsal lobe of the insect brain (figure 3.6); contains optic lobes and other major cell masses and regions of neuropile (figure 3.7).

protoplasmic poisons substances that kill by precipitating intercellular proteins.

Protura, order (*prot*, first; *ura*, tail) proturans (figure 12.4a).

proventriculus a structure posterior to the crop that bears sclerotized teeth or spines (figure 4.1); the stomodael valve in Diptera.

proximal adjacent to or nearest the point of attachment.

pseudo- false; temporary.

Psocoptera, order (*psoco*, rub small; *ptera*, wings) book lice and relatives (figure 12.23).

pterins red, yellow, and white pigments commonly found as insect eye pigments.

ptero- wing.

pterothorax collective term describing the wing-bearing segments.

Pterygota taxon composed of winged or secondarily wingless insects.

pterygote with wings.

ptilinum an eversible bladder in the head of coarctate pupae, used to force open the puparial case.

pubescent covered with fine hairs.

pulvillus padlike structure on the lower surface of tarsi in some Orthoptera or in association with each pretarsal claw in Diptera (figure 2.38a).

pupa transitional, metamorphic stage between larva and adult in holometabolous insects (figures 5.22d and e and 5.24).

puparium hardened cuticle of the next-to-last larval instar that encloses a pupa (figure 5.24c and d).

pygidial glands exocrine glands that secrete a repugnatorial substance and that are located at the posterior end of the abdomen in some species of beetles.

pygopodia eversible accessory locomotor appendages that arise from the terminal abdominal segment in some hexapodous larval forms.

pyloric valve valvular structure at the anterior extremity of the insect hindgut (figure 4.1).

pyrethrum botanical insecticide extracted from *Chrysanthemum* flowers (figure 15.9).

Q

quadrivoltine four life cycles per year.

quiescence inactivity in response to adverse environmental conditions that reverses when favorable environmental conditions return.

R

raptorial (legs) forelegs modified for grabbing and holding prey (figure 2.36c).

receptor apparatus portion of an ommatidium of a compound eye in which light is converted to a nervous impulse (figure 6.12c and d).

receptor cells portion of a sense organ in which stimuli are converted to a nervous impulse (figure 6.1, sense cell; figures 6.3, 6.7, 6.8*b*; 6.9, sense cells; figure 6.10).

rectal papillae pads projecting into the lumen of the rectum that play an important role in the excretory system.

rectum muscular, enlarged, posteriormost region of the intestine (figures 4.1 and 4.2).

recurrent nerve neural connection between the frontal ganglion and the hypocerebral ganglion (figures 3.6*b* and 3.9).

reflex bleeding release of hemolymph in response to a threat.

regenerative cells replacement cells at the bases of the midgut epithelial cells.

replacement cells *see* regenerative cells.

resilin a cuticular protein with rubberlike properties; sometimes found in pure form in skeletal articulations.

resonating muscles muscles capable of undergoing several successive contractions when stimulated by a single nerve impulse; usually associated with insect flight and sound production; also called asynchronous muscles.

respiration intracellular oxidative breakdown of carbohydrate to carbon dioxide and water.

retina portion of a dorsal ocellus in which light is converted to a nervous impulse (figure 6.17, retinular cells).

retinular cells cells in photosensitive organs in which light is converted to a nervous impulse (figures 6.12 and 6.17).

rhabdom centrally located rod structures formed by contributions from each retinular cell (figures 6.12 and 6.13).

rhabdomere portion of each retinular cell contributed to the formation of the rhabdom.

rheotaxis directed response in relation to water current.

rhodopsins light-absorbing pigments involved with photosynthesis.

ring gland small tissue mass found in cyclorrhaphous Diptera that is thought to be homologous with the corpora allata, the corpora cardiaca, and the thoracic glands.

S

salivarium the cavity between the hypopharynx and the labium into which the common salivary duct opens (figures 2.16 and 4.1).

salivary glands exocrine glands typically associated with the labial segment; highly variable in size, structure, and function (figure 4.1).

saltatorial (legs) legs adapted for jumping (figure 2.36*d*).

saprophagous feeding on decaying organic matter.

sarcolemma the outer layer of a muscle fiber (figures 3.18*a* and 3.23).

sarcoplasm cytoplasm of a muscle fiber (figure 3.23).

scape basal segment of an insect antenna (figure 2.14).

sclerites the hardened plates of the insect skeleton.

sclerotins stabilized proteins responsible for the hard, horny character of cuticle.

sclerotization hardening.

scolopidia *see* scolopophorous organ.

scolopophorous organ typically, a bundle of sensilla consisting of one or more specialized, bipolar neurons (scolopidia; also called scolopophores) stretched between two internal integumental surfaces; often has a proprioceptive function; also called chordotonal organ (figure 6.3).

scotopic (eye) *see* superposition (eye).

scutellum posterior plate of the alinotum (figure 2.18 *a* and *b*).

scutoscutellar suture line dividing the alinotum into the anterior scutum and posterior scutellum (figure 2.18*a* and *b*).

scutum relatively large plate on the alinotum anterior to the scutoscutellar suture (figure 2.18*a* and *b*).

secondary compound *see* allelochemics and secondary plant substances.

secondary orientation positioning of the insect body relative to various external or internal stimuli; superimposed on primary orientation.

secondary plant substances chemicals, found in plants, that have no apparent metabolic role and that often influence the behavior and/or physiology of herbivorous insects.

secondary segmentation characterized by membranous areas between adjacent segments slightly anterior to the attachment points of the longitudinal muscles; typical of adult and many larval insects.

secondary spermatocytes developing male gametes that form from the first meiotic divisions of the primary spermatocytes; they form into spermatids (figure 5.2) with second meiotic division.

segmental muscles the principal longitudinal muscles running between the intersegmental constrictions associated with primary segmentation.

sematectonic communication stimulation of behavior or physiological change in one animal in response to stimuli associated with the result of evidence of work by another animal.

semen spermatozoa plus various glandular secretions.

seminal vesicle dilated region of the vas deferens in which spermatozoa are stored (figure 5.1*a*).

semiochemicals chemicals that deliver behavioral messages.

sensescence changes in structure and function that decrease an individual's capacity for survival and lead to death.

sensilla specialized structures that collect information from the external and internal environments and transmit their information to the central nervous system (figure 6.1).

sensory neuron usually bipolar neuron with cell body located peripherally in the insect and dendrites associated with sensory structures; also called afferent neuron (figure 3.1*d*).

sericulture the silk industry.

serosa outer membrane around the yolk, amnion, and embryo (figures 5.12*e* and *f*, 5.14, and 5.15*c*, *e*, and *f*, and 5.16).

sesamin a mixed-function oxidase inhibitor produced by the *Chrysanthemum* plant.

setae macrotrichia; unicellular, cuticular processes (figure 2.9).

sex ratio proportion of males to females.

sexual dimorphism differences between the sexes.

sieve plates porous covers associated with spiracles that are thought to retard water loss and prevent entry of airborne particles.

sign stimuli *see* token stimuli.

Siphonaptera, order (*siphon*, a tube; *aptera*, wingless) fleas (figures 13.14 and 13.15).

skeletal muscles muscles attached at both ends to the integument (figures 3.17 through 3.22).

skeleton supportive shell.

sociotomy method of nest foundation among the termites in which immatures and secondary reproductives separate from the parent colony or a migrating colony divides into two daughter colonies.

-soma, somat-, -some body.

soma *see* perikaryon.

somites *see* metameres.

specific epithet second part of a species name, indicating a particular species.

spermatheca outpocketing of the vagina in which spermatozoa are stored prior to fertilization (figure 5.6a).

spermatids developing male gametes that form from division (second meiotic) of the secondary spermatocytes (figure 5.2).

spermatocytes diploid cells that undergo meiotic division and become spermatozoa.

spermatodesms bundles of spermatozoa.

spermatogenesis development of male reproductive cells (spermatozoa) in the testicular follicles (figures 5.2 and 5.3).

spermatogonia male germ cells that proliferate mitotically and that are capable of developing into spermatocytes (figure 5.2).

spermatophore spermatozoa encased in a gelatinous packet that is introduced into the female reproductive tract or deposited on the substrate and taken up by the female (figure 5.5).

spermatozoa male reproductive cells (figures 5.2, 5.3, 5.4, and 5.10b).

spina internally inflected median process on the spinasternum (figure 2.18b and c) of the thorax.

spinal pit the externally evident indication of the spina on the spinasternum (figure 2.18c).

spinasternum small posterior sclerite on the venter of a thoracic segment found in many generalized pterygote insects (figure 2.18b and c).

spinneret silk-spinning apparatus of lepidopteran larvae (figure 2.32e) and others.

spiracles small openings through tracheae that communicate with the external environment (figures 2.12, 2.18b, 2.24, 4.12, 4.13d, 4.14, 4.15).

spiracular gills in aquatic insects with closed ventilatory systems, filamentous outgrowths of thin cuticle that open directly into the tracheae; also called cuticular gills (figure 4.18c).

stadium interval of time passed in an instar.

stemmata simple eyes found on the sides of the insect head when compound eyes are lacking (figure 2.27); also called lateral ocelli.

sternacosta apodeme associated with the sternacostal suture (figure 2.17).

sternacostal suture line on the sternum that connects the furcal pits and that is associated with the sternacosta (figure 2.18c).

sternal apophyses internal projections arising from the eusternum (figure 2.17a).

sternellum one of the three parts of the eusternum in many insects (figure 2.18c).

sternites sclerites that form subdivisions of the abdominal sternum (figures 2.12 and 2.24).

sternopleurite one of three theoretical subcoxal sclerites that have come to form the insect pleuron (figure 2.20).

sternum ventral region of a thoracic or abdominal segment (figures 2.18c and 2.24).

stigma pigmented cell immediately posterior to the costal vein and near the apex of the wings of dragonflies and damselflies (figures 12.8 and 12.9); thickened portion of the wing membrane near the apex of the costal edge of the wing (figure 13.22).

stimulus filtering the process by which the multitude of stimuli that impinge on the insect nervous system are reduced to those which are pertinent to the behavior of the insect.

stipes comparatively large part of a maxilla, borne on the cardo, that bears the galea and lacinia (figure 2.15).

stomach poisons toxicants that must be ingested to have a lethal effect.

stomatogastric system the part of the visceral nervous system associated with the brain, aorta, and foregut; also called the stomodael system (figure 3.6).

stomodael invagination the embryonic ectodermal invagination that forms into the foregut (figure 5.19).

stomodael valve a structure formed by an intussusception of foregut and midgut tissues (figure 4.1).

stomodaeum *see* foregut.

Strepsiptera, order (*strepsi*, turning or twisting; *ptera*, wing) stylopids or twisted-wing parasites (figure 13.5).

stretch receptors mechanoreceptors composed of multipolar neurons associated with muscles, the alimentary canal, and other viscera; these receptors produce a nervous impulse when associated tissue undergoes a change in length.

stridulation production of sound by means of a frictional mechanism.

stylets swordlike or needlelike modifications of the generalized mouthpart structures (figure 2.30).

styli simple, bilateral appendages borne on the abdominal segments (figure 2.42a).

stylopized a condition in which an insect, such as a bee, is parasitized by a Strepsipteran (figure 13.5c).

subalar sclerite small sclerite posterior to the pleural wing process (figure 2.18b).

subcuticle newly secreted endocuticle in which the microfibers are not yet oriented.

subesophageal ganglion the ganglion, ventral to the foregut, that innervates sense organs and muscles associated with mouthparts, salivary glands, and the neck region (figures 3.3, 3.4, 3.6, and 3.8b).

subgenual organ group of chordotonal sensilla located in the basal portion of the tibial leg segment (figure 6.9).

subgenual sutures lines on the insect head located immediately above the bases of the mandibles and maxillae.

subimago winged "subadult" that undergoes one molt and becomes the adult; unique to mayflies (Ephemeroptera).

submentum proximal portion of the postlabium (figure 2.15).

subocular sutures vertical lines that may run beneath the compound eyes (figure 2.13a and c).

sulcus(i) an alternate name for a suture.

superposition (eye) compound eye type in which a superposition image is formed (figures 6.12d and 6.13b and c).

superposition image image formed when the light reaching the rhabdom of an individual eye unit, ommatidium, enters via the diopteric apparatuses of more than one adjacent eye unit (figure 6.13b and c).

sutures external grooves in the insect skeleton; also called sulci (figure 2.10).

sym-, syn- together.

synapse region of close association between terminal arborizations and dendrites (figure 3.1*d*).

synaptic cleft space between the terminal arborizations of one neuron and the dendrites of the next.

synchronous muscles *see* nonresonating muscles.

synergist chemical that, when mixed with an insecticide, renders the mixture more toxic than the simple sum of the toxicities of the individual components.

systemic insecticides a synthetic organic insecticide that, when absorbed by plants or ingested and absorbed by animals, makes them toxic to insects.

systole contraction phase of the heartbeat cycle (figure 4.8).

T

tactile referring to the sense of touch.

taenidia helical fold in the cuticular lining of tracheae and tracheoles (figure 4.13*a* and *b*).

tagmata the three major regions of the insect body; head, thorax, and abdomen (figure 2.12).

tanning hormone *see* bursicon.

tarsomeres subdivisions of the tarsus.

tarsus distal, segmented part of the insect leg attached to the tibia (figure 2.21).

taxes directed response relative to a stimulus source.

tegmen (tegmina) parchmentlike forewing characteristic of the **Orthoptera, order** (*ortho*, straight; *ptera*, wing) grasshoppers, locusts, crickets, katydids, and related forms. (figures 12.14 and 12.15).

teleology ascribing purposiveness to natural phenomena.

telotrophic (ovarioles) trophocytes connected to oocytes by nutritive cords (figure 5.9*b*).

telson *see* periproct.

tenaculum abdominal structure found on the venter of Collembola; serves as a "catch" for the furcula prior to "springing" (figures 2.42*b* and 12.5*a*).

tendinous epidermal cells specialized cells that produce the microtubules that make up the tonofibrillae (figures 3.18*b* and 3.19*a*).

tenent hairs tiny hairs on the pulvilli and tarsal pads that allow some insects to cling to smooth surfaces (figure 7.3*c*).

teneral newly emerged, pale, soft-bodied individuals prior to the completion of melanization and sclerotization.

tentorial bridge *see* corporotentorium.

tentorial pits small, externally apparent mouths of the invaginations that form the anterior and posterior tentorial arms (figure 2.13).

tentorium an internal framework in the insect head that provides rigidity and areas for muscle attachment (figure 2.13*d*).

tergites sclerites that form the tergum (figure 2.24).

tergosternals abdominal muscles in a dorsoventral orientation (figure 3.22).

tergum dorsal region of the thorax and abdomen (figure 2.24); usually called notum in reference to the thorax.

terminal arborizations branching processes found at the end of an axon (figure 3.1*a, b,* and *c*).

terminal filament a thread made up of a contribution from the apex of each ovariole, which anchors the cephalad end of each ovary to the dorsal diaphragm (figure 5.6).

testes in the male reproductive system, the bilateral, paired structures housing the germ cells (figures 5.1*a* and 5.3).

testicular follicles one (figure 5.3) to several (figure 5.1*b* and *c*) sperm tubes composing the testes (figure 5.1).

thermoreception the reception of stimuli as heat.

thermoregulation adjustment of body temperature.

thigmotaxis a taxis based on lateral contact.

thoracic ganglia three ganglia located posterior to the subesophageal ganglion and innervating the three thoracic segments.

thoracic glands masses of tissue, usually closely associated with tracheae, that secrete the molting hormone, ecdysone; also called prothoracic glands (figures 3.15*b*, 3.16, and 5.27) or ecdysial glands.

Thysanoptera, order (*thysano,* a fringe; *ptera,* wing) thrips (figure 12.26).

Thysanura, order (*thysan,* fringe; *ura,* tail) silverfish, firebrats, and relatives (figure 12.6*b* and *c*).

tibia long leg segment located between the femur and the tarsus (figure 2.21).

token stimuli stimuli that by themselves may not have relevance to survival (e.g., substances that lack nutritive value but that act as olfactory attractants to a food source).

tonofibrillae fibrillar structures (bundles of microtubules) of skeletal muscle that anchor muscle to the integument (figures 3.18*b* and 3.19), microtubules.

tormogen cell cell that forms the socket and membrane around a seta (figures 2.9, 6.1, and 6.10).

tracheae a system of branching tubules through which a gaseous exchange is accomplished in insects (figures 4.12 and 4.13 *a–c*).

tracheal end cell end of a trachea; typically branches into several tracheoles (figure 4.13*c*).

tracheal gills integumental evaginations, covered by thin cuticle and supplied with tracheae and tracheoles; found in many aquatic (figure 4.18*a* and *b*) and some endoparasitic insects.

transovarial via the egg, for example, transovarial transmission of symbiotic microbes from one generation to the next.

transpiration loss of water through the integument or tracheae by evaporation.

transverse orientation body oriented at a fixed angle relative to stimulus direction.

transverse tracheal commissures connections between longitudinal tracheal trunks (figure 4.12).

traumatic insemination ejaculation of semen directly into the hemocoel; characteristic of bed bugs and close relatives; also called hemocoelic insemination.

trichogen cell cell that forms a seta (figures 2.9, 6.1, and 6.10).

trichoid sensillum sensory hair.

Trichoptera, order (*tricho,* hair; *ptera,* wing) caddisflies (figures 13.16, 13.17, and 13.18).

tritocerebrum ventral lobe of the insect brain (figure 3.6).

triungulin active, first instar larva characteristic of the Strepsiptera (figure 13.5) and certain parasitoid beetles.

trivoltine three generations per year.

trochanter typically small segment of the insect leg that articulates with the coxa and usually forms an immovable attachment with the femur (figure 2.21).

trochantin a small sclerite on the pleuron of many generalized pterygote insects; provides one of the coxal articulation points (figure 2.18*b*).

trophallaxis nutrient exchange between members of a social insect colony; may be from the mouth (stomodael) or from the anus (proctodael).

trophocytes *see* nurse cells.

true mouth the opening at the base of the hypopharynx, within the preoral (cibarial) cavity formed by the mouthparts (figure 2.16).

tubular body array of microtubules in the distal process of a receptor cell; deformation of this structure initiates a nervous impulse.

tympanic organs auditory organs composed of a thin integumental area (tympanic membrane) and a group of chordotonal sensilla attached to the entad surfaces (figures 6.7 and 6.8).

Tyndall blue a blue color produced by colloidal-sized granules [for example, in the epidermal cells of some dragonflies and in neurosecretory cells (figure 3.13)].

U

ungue pretarsal claw (figure 2.38).

unipolar neuron neuron with a single stalk from the cell body connecting with the axon and a collateral (figure 3.1*a*).

uniporous with a single pore.

univoltine one generation per year.

urate cells cells in fat body; may store nitrogenous waste.

V

vagina external opening of the female reproductive system (figure 5.6*a*).

valvifers proximal structures of the ovipositor that bear the valvulae (figure 2.24).

valvulae distal structures of the ovipositor borne on the valvifers (figure 2.24).

vas deferens a duct, into which the vas efferentia empty, that connects each testis to a seminal vesicle (figure 5.1).

vas efferens tiny ducts leading from each testicular follicle to a common lateral duct, the vas deferens (figure 5.1*b*).

vector carrier; for example, *Anopheles* mosquitoes serve as vectors for malaria parasites, carrying the parasites from one human host to another.

venom inherently toxic insect (or other animal) secretion, harmful to humans and animals.

venter the ventral region.

ventilation transport of oxygen to and carbon dioxide from the insect tissues.

ventral diaphragm a fibromuscular septum that is located ventral to the alimentary canal and dorsal to the ventral nerve cord (figure 4.6).

ventral visceral system subsystem of the visceral nervous system associated with the ventral nerve cord (figure 3.11).

ventricular ganglion small ganglionic masses found on either side of the posterior region of the foregut and connected via nerves with the hypocerebral ganglion.

ventricular nerves nerves that connect the hypocerebral ganglion with the ventricular ganglion (figure 3.6*b*).

ventriculus the usually somewhat enlarged region of the midgut that serves as the insect's stomach (figure 4.1).

vermiform resembling a worm.

vertex the dorsal portion of the cranium bisected by the coronal suture (figure 2.13*a*).

visceral muscles muscles responsible for movements of internal organs; usually attached to other muscles.

visceral nervous system collective term for the stomatogastric, ventral visceral, and caudal visceral systems (figure 3.11).

vitellarium the region of an ovariole in which oocytes undergo vitellogenesis (figures 5.6 and 5.9).

vitellin yolk protein.

vitelline membrane cell membrane encasing the egg (figure 5.10*a*).

vitellogenesis the process of deposition of nutrients in oocytes; yolk deposition.

vitellogenin protein produced in the fat body that is transferred to oocytes during vitellogenesis.

vitellophage yolk cells considered responsible for the initial digestion of yolk (figure 5.12*c*, *d*, and *e*).

viviparous insects female insects that give "birth" to larvae.

W

wax filaments tiny canals, continuous with pore canals, that penetrate the protein epicuticle and cuticulin layers and are thought to be involved in transporting wax molecules to the epicuticle; also called wax canals.

wax layer an epicuticular lipid layer between the cuticulin layer that contributes to the permeability characteristics of the cuticle (figure 2.1*c*).

Weisman's ring *see* ring gland.

wing flexion the folding of the wings posteriorly over the abdomen.

wing pads externally apparent developing wings of hemimetabolous insects.

wing veins longitudinal and transverse supportive framework of the wings (figure 2.23).

Y

yolk nutrients (carbohydrates, proteins, and lipids) found in mature insect eggs (figures 5.10*a* and 5.12*a–e*); also called deutoplasm.

Z

zoophagous the habit of eating animals.

Zoraptera, order (*zor*, pure; *ptera*, wing) zorapterans (figure 12.22).

zygote fertilized egg (figure 5.12*a*).

References

CHAPTER 1

Arnett, R. H., Jr. American insects. New York: Van Nostrand Reinhold Company; 1985.

Atkins, M. D. Introduction to insect behavior. New York: Macmillan; 1980.

Blackwelder, R. E. Taxonomy: a text and reference book. New York: Wiley; 1967.

Blum, M. S., editor. Fundamentals of insect physiology. New York: Wiley; 1985.

Borror, D. J. Dictionary of word roots and combining forms. Palo Alto, CA: Mayfield Publishing Company; 1960.

Borror, D. J.; Triplehorn, C. A.; Johnson, N. F. An introduction to the study of insects. 6th ed. Philadelphia: Saunders College Publishing; 1989.

Calisher, C. H.; Thompson, W. H., editors. California serogroup viruses. New York: Alan R. Liss; 1983.

Carson, R. L. Silent spring. Boston: Houghton Mifflin; 1962.

Chamberlin, W. J. Entomological nomenclature and literature. 3rd ed. Dubuque, IA: Wm. C. Brown; 1952.

Chapman, R. F. The insects—structure and function. 3rd ed. Cambridge, MA: Harvard University Press; 1982.

Cloudsley-Thompson, J. L. Insects and history. London: Weidenfeld & Nicolson; 1976.

C.S.I.R.O. (Commonwealth Scientific and Industrial Research Organization). The insects of Australia. 2nd ed. Ithaca, NY: Cornell University Press; 1991. 2 Vol.

Cushing, E. C. History of entomology in World War II. Washington, DC: Smithsonian Institution Publications; 1957.

Davidson, R. H.; Lyon, W. F. Insect pests of farm, garden, and orchard. 8th ed. New York: Wiley; 1987.

Dethier, V. G. The hungry fly. A physiological study of the behavior associated with feeding. Cambridge, MA: Harvard University Press; 1976.

Erwin, T. L. Tropical forests: Their richness in Coleoptera and other arthropod species. Coleopterists Bulletin 36:74–76; 1982.

Erwin, T. L. The tropical forest canopy: The heart of biotic diversity. In: Wilson, E. O., ed. Biodiversity. Washington, DC: National Academy Press; 1988.

Essig, E. O. A history of entomology. New York: Macmillan; 1931.

Evans, H. E. Insect biology—a textbook of entomology. Reading, MA: Addison-Wesley Publishing Company; 1984.

Foote, R. H. Thesaurus of entomology. College Park, MD: The Entomological Society of America; 1977.

Gaston, K. J. The magnitude of global insect species richness. Conservation Biology 5:283–296; 1991.

Gilbert, P.; Hamilton, C. J. Entomology—a guide to information sources. 2nd ed. London: Mansell Publishing Limited; 1990.

Hammack, G. M. The serial literature of entomology—a descriptive study. College Park, MD: The Entomological Society of America; 1970.

Harwood, R. F.; James, M. T. Entomology in human and animal health. 7th ed. New York: Macmillan; 1979.

Hölldobler, B.; Wilson, E. O. The ants. Cambridge, MA: The Belknap Press of Harvard University Press; 1990.

Howard, L. O. A history of applied entomology. Washington, DC: Smithsonian Institution, Miscellaneous Publications; 1930.

Kettle, D. S. Medical and veterinary entomology. London: Croom-Helm; 1984.

Kerkut, G. A.; Gilbert, L. I., editors. Comprehensive insect physiology, biochemistry, and pharmacology. Oxford and New York: Pergamon; 1985. 13 Vol.

Leftwich, A. W. A dictionary of entomology. New York: Crane Russack; 1976.

Mallis, A. American entomologists. New Brunswick, NJ: Rutgers University Press; 1971.

Matthews, R. W.; Matthews, J. R. Insect behavior. New York: Wiley; 1978.

Nichols, S. W., compiler. The Torre-Bueno glossary of entomology. New York: The New York Entomological Society in cooperation with the American Museum of Natural History; 1989.

Osborn, H. Fragments of entomological history. Columbus, OH: H. Osborn; 1937.

Price, P. W. Insect ecology. 2nd ed. New York: Wiley; 1984.

Richards, O. W.; Davies, R. G. Imm's general textbook of entomology. 10th ed. Vol. 1. Structure, physiology, and development. Vol. 2. Classification and biology. New York: Wiley (Halsted Press); 1977.

Richards, O. W.; Davies, R. G. Imm's outlines of entomology. 6th ed. New York: Wiley; 1978.

Sharov, A. G. Basic arthropodan stock with special reference to insects. New York: Pergamon; 1966.

Smith, R. C. Guide to the literature of the zoological sciences. 5th ed. Minneapolis: Burgess; 1958.

Smith, R. F.; Mittler, T. E.; Smith, C. N., editors. History of entomology. Palo Alto, CA: Annual Reviews Inc.; 1973.

Smith, R. C.; Reid, W. M.; Luchsinger, A. E. Smith's guide to the literature of the life sciences. Minneapolis: Burgess Publishing Company; 1980.

Southwood, T. R. E. Entomology and mankind. American Scientist 65:30–39; 1977.

Wigglesworth, V. B. The principles of insect physiology. 7th ed. London: Chapman & Hall; 1972.

Wigglesworth, V. B. Insects and the life of man. New York: Wiley; 1976.

CHAPTER 2

Anderson, S. O. Cuticular enzymes and sclerotization in insects. In: Hepburn, H. P., ed. The insect integument. New York: Elsevier Publishing Co.; 1976:p. 121–144.

Anderson, S. O. Biochemistry of insect cuticle. Annual Review of Entomology 24:29–61; 1979.

Anderson, S. O. Sclerotization. In: Binnington, K.; Retnakaran, A., eds. Physiology of the insect epidermis. East Melbourne, Victoria, Australia: C.S.I.R.O. Publications; 1991:p. 123–140.

Barrett, F. M. Phenoloxidases and the integument. In: Binnington, K.; Retnakaran, A., eds. Physiology of the insect epidermis. East Melbourne, Victoria, Australia: C.S.I.R.O. Publications 1991:p. 123–140.

Beament, J. W. L. The effect of temperature on the waterproofing mechanism of an insect. Journal of Experimental Biology 35:494–519; 1958.

Bennet-Clark, H. C. Active control of the mechanical properties of insect endocuticle. Journal of Insect Physiology 8:627–633; 1962.

Bennet-Clark, H. C. Energy storage in jumping insects. In: Hepburn, H. R., ed. The insect integument. New York: Elsevier Scientific; 1976:p. 421–443.

Binnington, K.; Retnakaran, A., editors. Physiology of the insect epidermis. East Melbourne, Victoria, Australia: C.S.I.R.O. Publications; 1991.

Borror, D. J.; D. M. DeLong. An introduction to the study of insects, revised edition. New York: Holt, Rinehart and Winston, Inc.; 1964.

Borror, D. J.; Triplehorn, C. A.; Johnson, N. F. An introduction to the study of insects. 6th ed. Philadelphia: Saunders College Publishing; 1989.

Chapman, R. F. The insects—structure and function. 3d ed. Cambridge, MA: Harvard University Press; 1982.

Cohen, E. Chitin biochemistry. Annual Review of Entomology 32:71–94; 1987.

Cohen, E. Chitin biochemistry. In: Binnington, K.; Retnakaran A., eds. Physiology of the insect epidermis. East Melbourne, Victoria, Australia: C.S.I.R.O. Publications; 1991:p. 94–112.

Comstock, J. H. The wings of insects. Ithaca, NY: Cornell University Press; 1918.

Comstock, J. H. An introduction to entomology. 9th ed. Ithaca, NY: Cornell University Press; 1940.

C.S.I.R.O. (Commonwealth Scientific and Industrial Research Organization). The insects of Australia, 2nd ed. Ithaca, NY: Cornell University Press; 1991. 2 Vol.

de Renobales, M.; Nelson, D. R.; Blomquist, G. J. Cuticular lipids. In: Binnington, K.; Retnakaran, A., eds. Physiology of the insect epidermis. East Melbourne, Victoria, Australia: C.S.I.R.O. Publications; 1991: 240–251

DuPorte, E. M. Manual of insect morphology. New York: Van Nostrand Reinhold; 1961.

Ebeling, W. Permeability of insect cuticle. In: M. Rockstein, ed. The physiology of insecta. 2nd ed. New York: Academic Press; 1974:p. 271–343, vol. VI.

Ebeling, W. Insect integument: a vulnerable organ system. In: Hepburn, H. R., ed. The insect integument. New York: Elsevier Scientific; 1976:p. 383–400.

Essig, E. O. College entomology. New York: Macmillan; 1942.

Essig, E. O. Insect and mites of western North America. New York: Macmillan; 1958.

Folsom, J. W.; Wardle, R. A. Entomology, with special reference to its ecological aspects. 4th ed. Philadelphia: Blakiston; 1934.

Fuzeau-Braesch, S. Pigments and colour changes. Annual Review of Entomology 17:403–424; 1972.

Hackman, R. H. Chemistry of the insect cuticle. In: Rockstein, M., ed. The physiology of insecta. 2nd ed. New York: Academic Press; 1974:p. 215–270, vol. VI.

Hackman, R. H. The interactions of cuticular proteins and some comments on their adaptation to function. In: Hepburn, H. R., ed. New York: Elsevier Scientific; 1976:p. 107–120.

Hamilton, K. G. A. The insect wing. Part II. Vein homology and the archetypal insect wing. Journal of the Kansas Entomological Society 45:54–58; 1972.

Hebard, M. The Dermaptera and Orthoptera of Illinois. Illinois Natural History Survey, Bulletin 20(3); 1934.

Herms, W.; James, M. T. Medical entomology. 5th ed. New York: Macmillan Publishing Co., Inc.; 1961.

Hepburn, H. R., editor. The insect integument. New York: Elsevier Scientific; 1976.

Hepburn, H. R. The integument. In: Blum, M. S., ed. Fundamentals of insect physiology. New York: Wiley & Sons; 1985.

Hepburn, H. R.; Joffe, I. On the material properties of insect exoskeletons. In: Hepburn, H. R., ed. The insect integument. New York: Elsevier Scientific; 1976:p. 207–235.

Hinton, H. E. Recent work on physical colours of insect cuticle. In: Hepburn, H. R., ed. The insect integument. New York: Elsevier Scientific; 1976:p. 475–496.

Hopkins, T. L.; Kramer, K. J. Catecholamine metabolism and the integument. In: Binnington, K.; Retnakaran, A., eds. Physiology of the insect epidermis. East Melbourne, Victoria: Australia; 1991:p. 213–239.

James, M. T.; Harwood, R. F. Herms' medical entomology. New York: Macmillan Publishing Co., Inc.; 1969.

Jenkin, P. M.; Hinton, H. W. Apolysis in arthropod molting cycles. Nature 211(5051):871; 1966.

King, R. C.; Akai, H., editors. Insect ultrastructure. Vol. 1. New York: Plenum; 1982.

Locke, M. Permeability of the insect cuticle to water and lipids. Science 147:295–298; 1965.

Locke, M. Body wall. In: Tipton. V. J., ed. Syllabus: introductory entomology. Provo, UT: Brigham Young University Press; 1973:p. 165–182.

Locke, M. The structure and formation of the integument in insects. In: M. Rockstein, ed. The physiology of insecta. 2nd ed. New York: Academic Press; 1974:p. 124–213.

Manton, S. M. The evolution of arthropodan locomotory mechanisms. Part 10: Locomotory habits, morphology, and evolution of the hexapod classes. Zoology—Journal of the Linnean Society 51:203–400; 1972.

Matsuda, R. Morphology and evolution of the insect head. Memoirs of the American Entomological Institute 4; 1965.

Matsuda, R. Morphology and evolution of the insect thorax. Entomological Society of Canada, Memoir 76; 1970.

Matsuda, R. Morphology and evolution of the insect abdomen. Elmsford, NY: Pergamon; 1976.

Matsuda, R. Morphologie du thorax et des appendices thoraciques des insectes. In: P.-P. Grasse, ed. Traité de Zoologie: Anatomie, Systematique, Biologie. Tome VIII. Fase. II. Insectes, Thorax, Abdomen. Paris: Masson; 1979: 1–289.

Matsuda, R. The origin of insect wings (Arthropoda: Insecta). International Journal of Insect Morphology and Embryology 10:387–398; 1981.

Metcalf, C. L.; Flint, W. P.; Metcalf, R. L. Destructive and useful insects. New York: McGraw-Hill; 1962.

Miller, T. A. Cuticle techniques in arthropods. New York: Springer-Verlag; 1980.

Needham, A. E. Insect biochromes: their chemistry and role. In: Rockstein, M., ed. Biochemistry of insects. New York: Academic Press; 1978:p. 233–305.

Neville, A. C. Cuticle ultrastructure in relation to the whole insect. In: Neville, A. C., ed. Insect ultrastructure. London: Royal Entomological Society, Symposium 5; 1970.

Neville, A. C. Biology of the arthropod cuticle. New York: Springer-Verlag; 1975.

Noble-Nesbitt, J. Cuticular permability and its control. In: Binnington, K.; Retnakaran, A., eds. Physiology of the insect epidermis. East Melbourne, Victoria, Australia: C.S.I.R.O. Publications; 1991: 252–283.

Packard, A. S. A textbook of entomology. New York: Macmillan; 1898.

Peterson, A. Larvae of insects. Part I. Lepidoptera and plant infesting Hymenoptera. Columbus, OH: A. Peterson; 1948.

Peterson, A. Larvae of insects. Part II. Coleoptera, Diptera, Neuroptera, Siphonaptera, Mecoptera, Trichoptera. Columbus, OH: A. Peterson; 1951.

Rempel, J. G. The evolution of the insect head: the endless dispute. Quaestiones Entomologicae 11:7–25; 1975.

Richards, A. G. The integument of arthropods. Minneapolis, MN: University of Minnesota Press; 1951.

Richards, A. G. Anatomy and morphology. In: Smith, R. F.; Mittler, T. E.; Smith, C. N., eds. History of entomology. Palo Alto, CA: Annual Reviews, Inc.; 1973:p. 185–202.

Richards, A. G. The chemistry of insect cuticle. In: Rockstein, M., ed. Biochemistry of insects. New York: Academic Press; 1978:p. 205–232.

Richards, O. W.; Davies, R. G. Imm's general textbook of entomology. 10th ed. Vol. 1. Structure, physiology, and development. Vol. 2. Classification and biology. New York: Wiley (Halsted Press); 1977.

Riley, C. V. The habits of *Thalessa* and *Tremex*. U.S. Division of Entomology Ins. Life 1:168–179; 1888.

Ross, H. H. How to collect and preserve insects. Illinois Natural History Survey, Circular 39; 1962.

Schneider, D. Insect antennae. Annual Review of Entomology 9:103–122; 1964.

Scudder, G. G. E. Comparative morphology of insect genitalia. Annual Review of Entomology 16:379–406; 1971.

Skaife, S. H. Dwellers in darkness. New York: Doubleday; 1961.

Smart, J. Notes on the mesothoracic musculature of Diptera. In: Studies in Invertebrate Morphology. Smithsonian Institution, Miscellaneous Collections 157:331–364; 1959.

Smith, E. L. Evolutionary morphology of external insect genitalia. I. Origin and relationships to other appendages. Annals of the Entomological Society of America 62:1051–1079; 1969.

Snodgrass, R. E. Principles of insect morphology. New York: McGraw-Hill; 1935.

Snodgrass, R. E. The dragonfly larva. Smithsonian Institution, Miscellaneous Collections 123(2); 1954.

Snodgrass, R. E. A revised interpretation of the external reproductive organs of male insects. Smithsonian Miscellaneous Collections 135(6):1–60; 1957.

Snodgrass, R. E. The anatomical life of the mosquito. Smithsonian Institution, Miscellaneous Collections 138(8); 1959.

Snodgrass, R. E. Facts and theories concerning the insect head. Smithsonian Institution, Miscellaneous Collections 142(1):1–61; 1960.

Snodgrass, R. E. The caterpillar and the butterfly. Smithsonian Institution, Miscellaneous Collections 143(6); 1961.

Snodgrass, R. E. A contribution toward an encyclopedia of insect anatomy. Smithsonian Institution, Miscellaneous Collections 146(2):1–48; 1963.

Sugumaran, M. Molecular mechanisms of sclerotization. In: Binnington, K.; Retnakaran, A., eds. Physiology of the insect epidermis. East Melbourne, Victoria, Australia: C.S.I.R.O. Publications; 1991:p. 141–168.

Tuxen, S. L. Taxonomist's glossary of genitalia in insects. 2nd ed. New York: Stechert-Hafner Service Agency, Inc.; 1970.

U.S. Public Health Service. Pictorial keys to arthropods, reptiles, birds and mammals of public health significance. Atlanta, GA: National Communicable Disease Center; 1969.

Vogel, R. Zur Kenntnis des Baues und der Funktion des Stachels und des Vorderdarmes der Kleiderlaus. Zoologische Jahrbuecher, Anatomie 42:p. 229–258; 1921.

West, L. S. The housefly: its natural history, medical importance, and control. Ithaca, NY: Comstock Publishing Company, Inc.; 1951.

Wharton, G. W.; Richards, A. G. Water vapor exchange kinetics in insects and acarines. Annual Review of Entomology 23:309–328; 1978.

Wigglesworth, V. B. Transpiration through the cuticle of insects. Journal of Experimental Biology 21:p. 97–114; 1945.

Wigglesworth, V. B. The control of growth and form: a study of the epidermal cell in an insect. Ithaca, NY: Cornell University Press; 1959.

Wigglesworth, V. B. The principles of insect physiology. 7th ed. London: Chapman & Hall; 1972.

Wigglesworth, V. B. The distribution of lipid in the cuticle of *Rhodnius*. In: Hepburn, H. R., ed. The insect integument. New York: Elsevier Scientific; 1976:p. 89–106.

Wolbert, P.; Schafer, F. G. Macromolecular changes during metamorphosis of the integument. In: Binnington, K.; Retnakaran, A., eds. Physiology of the insect epidermis. East Melbourne, Victoria, Australia: C.S.I.R.O. Publications; 1991:p. 169–184.

Zacharuk, R. Y. Structural changes associated with moulting. In: Hepburn, H. R., ed. The insect integument. New York: Elsevier Scientific; 1976:p. 299–321.

CHAPTER 3

Aidley, D. J. Muscular contraction. In: Kerkut, G. A.; Gilbert, L. I., eds. Comprehensive insect physiology, biochemistry, and pharmacology. New York: Pergamon Press; 1985:p. 407–437, vol. 5.

Bailey, E. Biochemistry of insect flight. Part 2—fuel supply. In: Candy, D. J.; Kilby, B. A., eds. Insect biochemistry and function. New York: Wiley; 1975.

Birch, M. C.; Poppy, G. M.; Baker, T. C. Scents and eversible scent structures of male moths. Annual Review of Entomology 35:25–58; 1990.

Blum, M. S. Biosynthesis of arthropod exocrine compounds. Annual Review of Entomology 32:381–413; 1987.

Börkovec, A. B.; Kelly, T. J., eds. Insect neurochemistry and neurophysiology. New York: Plenum Press; 1984.

Breer, H. Neurochemical aspects of cholinergic synapses in the insect brain. In: Gupta, A. P., ed. Arthropod brain. New York: John Wiley & Sons; 1987:p. 415–437.

Bullock, T. H.; Horridge, G. A. Structure and function in the nervous systems of invertebrates. Vols. I & II. San Francisco: W. H. Freeman and Co.; 1965.

Carlson, S. D.; Saint Marie, R. L. Structure and function of insect glia. Annual Review of Entomology 35:597–621; 1990.

Caveney, S. Muscle attachment related to cuticle architecture in apterygota. Journal of Cell Science 4:541–559; 1969.

Chapman, R. F. The insects—structure and function. 2nd ed. New York: American Elsevier; 1971.

Christensen, T. A.; Hildebrand, J. G. Male-specific, sex pheromone-selective projection neurons in the antennal lobes of the moth *Manduca sexta*. Journal of Comparative Physiology A, 160:553–569; 1987.

Cook, B. J.; Holman, G. M. Peptides and kinins. In: Kerkut, G. A.; Gilbert, L. I., eds. Comprehensive insect physiology, biochemistry, and pharmacology. New York: Pergamon Press; 1985:p. 531–593, vol. 11.

Cornwell, P. B. The cockroach. Vol. 1. London: Hutchinson; 1968.

Crabtree, B.; Newsholme, E. A. Comparative aspects of fuel utilization and metabolism by insects. In: Usherwood, P. N. R., ed. Insect muscle. New York: Academic Press; 1975:p. 405–500.

Dethier, V. G. The physiology of insect senses. New York: John Wiley & Sons, Inc.; 1963.

Dethier, V. G. The hungry fly. Cambridge, MA: Harvard Univ. Press; 1976.

Downer, R. G. H., editor. Energy metabolism in insects. New York: Plenum Press; 1981.

DuPorte, E. M. Manual of insect morphology. New York: Van Nostrand Reinhold; 1961.

Engelmann, F. The physiology of insect reproduction. New York: Pergamon; 1970.

Finlayson, L. H. Development and degeneration. In: Usherwood, P. N. R., ed. Insect muscle. New York: Academic Press; 1975:p. 75–149.

French, S. A. Transduction mechanisms of mechanosensilla. Annual Review of Entomology 33:39–58; 1988.

Gilmour, D. The metabolism of insects. San Francisco: Freeman; 1965.

Goldsworthy, G. J. The endocrine control of flight metabolism in locusts. Advances in Insect Physiology 17:149–204; 1983.

Gupta, A. P., editor. Arthropod brain: its evolution, development, structure, and functions. New York: John Wiley & Sons; 1987.

Gupta, A. P., editor. Neurohemal organs of arthropods. Springfield, IL: Charles C. Thomas, Pub.; 1983.

Herried, C. F., II. Biology. New York: Macmillan; 1977:p. 884.

Highnam, K. C.; Hill, L. The comparative endocrinology of the invertebrates. 2nd ed. New York: Elsevier Scientific; 1980.

Holman, G. M.; Nachman, R. J.; Wright, M. S. Insect neuropeptides. Annual Review of Entomology 35:201–217; 1990.

Homberg, W.; Christensen, T. A.; Hildebrand, J. G. Structure and function of the deutocerebrum in insects. Annual Review of Entomology 34:477–501; 1989.

Howard, R. W.; Blomquist, G. J. Chemical ecology and biochemistry of insect hydrocarbons. Annual Review of Entomology 27:149–172; 1982.

Howes, P. F. Brain structure and behavior in insects. Annual Review of Entomology 20:359–379; 1975.

Hoyle, G. Neural control of skeletal muscle. In: Rockstein, M., ed. The physiology of insecta. New York: Academic Press; 1965:p. 407–449, vol. 2.

Hoyle, G. Neural control of skeletal muscle. In: Rockstein, M., ed. The physiology of insecta, 2nd ed. New York: Academic Press; 1974. Vol. IV, p. 176–269.

Hoyle, G. Muscles and their neural control. New York: John Wiley & Sons; 1983.

Huber, F. Neural integration (central nervous system). In: Rockstein, M., ed. The physiology of insecta. 2nd ed. New York: Academic Press; 1974:p. 4–100, vol. IV.

Huddart, H. Visceral muscle. In: Kerkut, G. A.; Gilbert, L. I., eds. Comprehensive insect physiology, biochemistry, and pharmacology. New York: Pergamon Press; 1985:p. 131–194, vol. 11.

Huxley, H. E. The contraction of muscle. In: The living cell, readings from Scientific American. San Francisco, CA: Freeman; 1965:p. 279–289.

Johnson, C. G. Migration and dispersal of insects by flight. London: Methuen; 1969.

Kerkut, G. A.; Gilbert, L. I., editors. Comprehensive insect physiology, biochemistry, and pharmacology. New York: Pergamon Press; 1985. 13 vol.

Kramer, K. J. Vertebrate hormones in insects. In: Kerkut, G. A.; Gilbert, L. I., eds. Comprehensive insect physiology, biochemistry, and pharmacology. New York: Pergamon Press; 1985:p. 511–536, vol. 7.

Lane, N. J. Structure of components of the nervous system. In: Kerkut, G. A.; Gilbert, L. I., eds. Comprehensive insect physiology, biochemistry, and pharmacology. New York: Pergamon Press; 1985:p. 1–47, vol. 5.

Lunt, G. G.; Olsen, R. W. Comparative invertebrate neurochemistry. New York: Cornell Univ. Press.; 1988.

Maeda, S. Expression of foreign genes in insects using baculovirus vectors. Annual Review of Entomology 34:351–372; 1989.

Margolis, F. L.; Getchell, T. V., editors. Molecular neurobiology of the olfactory system. New York: Plenum Press; 1988.

Maruyama, K. Biochemistry of muscle contraction. In: Kerkut, G. A.; Gilbert, L. I., eds. Comprehensive insect physiology, biochemistry, and pharmacology. New York: Pergamon Press; 1985:p. 487–498, vol. 10.

Mayer, M. S.; Mankin, R. W. Neurobiology of pheromone perception. In: Kerkut, G. A.; Gilbert, L. I., eds. Comprehensive insect physiology, biochemistry, and pharmacology. New York: Pergamon Press; 1985:p. 95–144, vol. 9.

Menn, J. J.; Börkovec, A. B. Insect neuropeptides: potential new insect control agents. Journal of Agriculture and Food Chemistry 37:271–278; 1989.

Miller, T. A. Insect visceral muscle. In: Usherwood, P. N. R., ed. Insect Muscle. New York: Academic Press; 1975:p. 545–606.

Mobbs, P. G. Brain structure. In: Kerkut, G. A.; Gilbert, L. I., eds. Comprehensive insect physiology, biochemistry, and pharmacology. New York: Pergamon Press; 1985:p. 299–370, vol. 5.

Nässel, D. R. Strategies for neuronal marking in arthropod brains. In: Gupta, A. P., ed. Arthropod brain. New York: John Wiley & Sons; 1987:p. 549–570.

Noirot, C.; Quennedy, A. Fine structure of insect epidermal glands. Annual Review of Entomology 19:61–80; 1974.

Packard, A. S. A textbook of entomology. New York: Macmillan; 1898.

Parnas, I.; Dagan, D. Functional organization of giant axons in the central nervous systems of insects: new aspects. Advances in Insect Physiology 8:95–144; 1971.

Penzlin, H. Stomatogastric nervous system. In: Kerkut, G. A.; Gilbert, L. I., eds. Comprehensive insect physiology, biochemistry, and pharmacology. New York: Pergamon Press; 1985:p. 371–406, vol. 5.

Pichon, Y.; Manaranche, R. Biochemistry of the nervous system. In: Kerkut, G. A.; Gilbert, L. I., eds. Comprehensive insect physiology, biochemistry, and pharmacology. New York: Pergamon Press; 1985:p. 417–450, vol. 10.

Piek, T. Neurotransmission and neuromodulation of skeletal muscles. In: Kerkut, G. A; Gilbert, L. I., eds. Comprehensive insect physiology, biochemistry, and pharmacology. New York: Pergamon Press; 1985:p. 55–118, vol. 11.

Raabe, M. Insect neurohormones. New York: Plenum Press; 1982.
Raabe, M. Recent developments in insect neurohormones. New York: Plenum Pub. Corp.; 1989.
Raabe, M. The neurosecretory-neurohaemal system of insects; anatomical, structural and physiological data. Advances in Insect Physiology 17:205–303; 1983.
Rees, H. H. Insect biochemistry. New York: Wiley (Halsted Press); 1977.
Richards, A. G. The integument of arthropods. Minneapolis, MN: University of Minnesota Press; 1951.
Richards, O. W.; Davies, R. G. Imm's general textbook of entomology. 10th ed., Vol. 1, Structure, physiology, and development; Vol. 2, Classification and biology. New York: Wiley (Halsted Press); 1977.
Rudall, K. M.; Kenchington, W. Arthropod silks: the problem of fibrous proteins in animal tissues. Annual Review of Entomology 16:73–96; 1971.

Scharrer, B. Insects as models in neuroendocrine research. Annual Review of Entomology 32:1–16; 1987.
Schwartz, L. M.; Truman, J. W. Hormonal control of muscle atrophy and degeneration in the moth *Antheraea polyphemus*. Journal of Experimental Biology 111:13–30; 1984.
Singh, R. N.; Strausfeld, N. J. Neurobiology of sensory systems. New York: Plenum Pub. Corp.; 1989.
Smith, D. S.; Treherne, J. E. Functional aspects of the organization of the insects nervous system. Advances in Insect Physiology 1:401–484; 1963.
Smith, D. S. Insect cells: their structure and function. Edinburgh, Scotland: Oliver & Boyd; 1968.
Smith, D. S. The structure of insect muscles. In: King, R. C.; Akai, H., eds. Insect ultrastructure. New York: Plenum Press; 1984:p. 111–150, vol. 2.
Snodgrass, R. E. Principles of insect morphology. New York: McGraw-Hill; 1935.
Snodgrass, R. E. A textbook of arthropod anatomy. Ithaca, NY: Comstock; 1952.
Snodgrass, R. E. The anatomy of the honey bee. In: Grout, R. A., ed. The hive and the honey bee. Hamilton, IL: Dadant and Sons, Inc.; 1963:p. 141–190.
Staddon, B. W. The scent glands of Heteroptera. Advances in Insect Physiology 14:351–418; 1979.
Steele, J. E. Hormonal control of metabolism in insects. Advances in Insect Physiology 12:239–323; 1976.
Strausfeld, N. J. Atlas of an insect brain, New York: Springer-Verlag; 1976.
Strausfeld, N. J.; Miller, T. A., editors. Neuroanatomical techniques-insect nervous system. New York: Springer-Verlag; 1980.
Sturm, H. Die Paarung beim Silberfischen, *Lepisma saccharina* Z. Tier-psychol. 13:1–2; 1956.

Treherne, J. E. Insect neurobiology. New York: American Elsevier; 1974

Usherwood, P. N. R., editor. Insect muscle. New York: Academic Press; 1975.

Wigglesworth. V. V. The principles of insect physiology, 7th ed. London: Chapman & Hall; 1972.

CHAPTER 4

Abelson, P. H. Support for plant biology. Science 246(4932):865; 1989.
Angioy, A. M. Reflex cardiac response to a feeding stimulus in the blowfly *Calliphora vomitoria* L. Journal of Insect Physiology 34:21–27; 1988.
Applebaum, S. W. Biochemistry of digestion. In: Kerkut, G. A.; Gilbert, L. I., eds. Comprehensive insect physiology, biochemistry, and pharmacology. New York: Pergamon Press; 1985:p. 279–311, vol. 4.
Arnold, J. W. The hemocytes of insects. In: Rockstein, M., ed. The physiology of insecta. 2nd ed. New York: Academic Press; 1974:p. 202–254, vol. V.

Barbosa, P.; Letourneau, D. K. Novel aspects of insect-plant interactions. New York: John Wiley & Sons; 1988.
Bernays, E. A. Regulation of feeding behaviour. In: Kerkut, G. A.; Gilbert, L. I., eds. Comprehensive insect physiology, biochemistry, and pharmacology. New York: Pergamon Press; 1985:p. 1–32, vol. 4.
Bernays, E. A.; Simpson, S. J. Control of food intake. Advances in Insect Physiology 16:59–118; 1982.
Berridge, M. J. A structural analysis of intestinal absorption. In: Neville, A. C., ed. Insect ultrastructure. London: Royal Entomological Society, Symposium 5; 1970:p. 135–151.
Billingsley, P. F. The midgut ultrastructure of hematophagous insects. Annual Review of Entomology 35:219–248; 1990.
Boman, H. G.; Hultmark, D. Cell-free immunity in insects. Annual Review of Microbiology 41:103–126; 1987.
Boudreaux, H. B. Arthropod phylogeny with special reference to insects. New York: John Wiley & Sons; 1979.
Bradley, T. J. The excretory system: structure and physiology. In: Kerkut, G. A.; Gilbert, L. I., eds. Comprehensive insect physiology, biochemistry, and pharmacology. New York: Pergamon Press; 1985:p. 421–465, vol. 4.
Bradley, T. J. Physiology of osmoregulation in mosquitoes. Annual Review of Entomology 32:439–462; 1987.
Bradley, T. J.; Miller, T. A., editors. Measurement of ion transport and metabolic rate in insects. New York: Springer-Verlag; 1984.
Brammer, J. D.; White, R. H. Vitamin A deficiency: effect on mosquito eye ultrastructure. Science 163:821–823; 1969.
Brattsten, L. B. Biochemical defense mechanisms in herbivores against plant allelochemicals. In: Rosenthal, G. A.; Janzen, D. H., eds. Herbivores: their interaction with secondary plant metabolites. New York: Academic Press; 1979:p. 199–270.
Brattsten, L. B.; Ahmad, S., editors. Molecular aspects of insect-plant associations. New York: Plenum Press; 1986.

Brehélin, M., editor. Immunity in invertebrates. Cells, molecules, and defense reactions. New York: Springer-Verlag; 1986.

Brooks, M. A. Symbiosis and aposymbiosis in arthropods. Symposium of the Society for General Microbiology 13:200–231; 1963.

Brower, L. P.; Nelson, C. J.; Seiber, J. N.; Fink, L. S.; Bond, C. Exaptation as an alternative to coevolution in the cardenolide-based chemical defense of monarch butterflies. In: Spencer, F., ed. Chemical mediation of coevolution. New York: Academic Press; 1988:p. 447–475.

Brown, M. R.; Grim, J. W.; Lea, A. O. FMRFamide- and pancreatic polypeptide-like immunoreactivity of endocrine cells in the midgut of a mosquito. Tissue Cell 18:419–428; 1986.

Buck, J. B. Physical properties and chemical composition of insect blood. In: Roeder, K., ed. Insect physiology. New York: John Wiley & Sons; 1953:p. 147–190.

Buchner, P. Endosymbiosis of animals with plant micro-organisms. New York: Interscience; 1965.

Campbell, B. C. On the role of microbial symbiotes in herbivorous insects. In: Bernays, E. A., ed. Insect-plant interactions. Boca Raton, FL: CRC Press, Inc.; 1989:p. 1–44, vol. I.

Chapman, R. F. The insects—structure and function. New York: American Elsevier Publishing Company, Inc.; 1982.

Chapman, R. F. Structure of the digestive system. In: Kerkut, G. A.; Gilbert, L. I., eds. Comprehensive insect physiology, biochemistry, and pharmacology. New York: Pergamon Press; 1985:p. 165–211, vol. 4.

Chapman, R. F.; Joern, A., editors. Biology of grasshoppers. New York: John Wiley & Sons; 1990.

Chen, P. S. Amino acid and protein metabolism. In: Kerkut, G. A.; Gilbert, L. I., eds. Comprehensive insect physiology, biochemistry, and pharmacology. New York: Pergamon Press; 1985:p. 177–217, vol. 10.

Chino, H. Lipid transport: biochemistry of hemolymph lipophorin. In: Kerkut, G. A.; Gilbert, L. I., eds. Comprehensive insect physiology, biochemistry, and pharmacology. New York: Pergamon Press; 1985:p. 115–135, vol. 10.

Chippendale, G. M. The functions of carbohydrates in insect life processes. In: Rockstein, M., ed. Biochemistry of insects. New York: Academic Press; 1978:p. 1–55.

Cochran, D. G. Nitrogenous excretion. In: Kerkut, G. A.; Gilbert, L. I., eds. Comprehensive insect physiology, biochemistry, and pharmacology. New York: Pergamon Press; 1985a:p. 467–506, vol. 4.

Cochran, D. G. Nitrogen excretion in cockroaches. Annual Review of Entomology 30:29–49; 1985b.

Cook, B. J.; Holman, G. M. Peptides and kinins. In: Kerkut, G. A.; Gilbert, L. I., eds. Comprehensive insect physiology, biochemistry, and pharmacology. New York: Pergamon Press; 1985:p. 531–593, vol. 11.

Cornwell, P. B. The cockroach, a laboratory insect and industrial pest. Vol. 1. London: Hutchinson Press; 1968.

Crossley, A. C. The cytophysiology of insect blood. Advances in Insect Physiology 11:117–221; 1975.

Crossley, A. C. Nephrocytes and pericardial cells. In: Kerkut, G. A.; Gilbert, L. I., eds. Comprehensive insect physiology, biochemistry, and pharmacology. New York: Pergamon Press; 1985:p. 487–515, vol. 3.

Cruden, D. L.; Markovetz, A. J. Microbial ecology of the cockroach gut. Annual Review of Microbiology 41:617–643; 1987.

Dadd, R. H. Digestion in insects. In: Florkin, M.; Scheer, B. T., eds. Chemical zoology. New York: Academic Press; 1970:p. 117–145, vol. 5. Arthropoda.

Dadd, R. Insect nutrition: current developments and metabolic implications. Annual Review of Entomology 18:381–420; 1973.

Dadd, R. H. Nutrition: organisms. In: Kerkut, G. A.; Gilbert, L. I., eds. Comprehensive insect physiology, biochemistry, and pharmacology. New York: Pergamon Press; 1985:p. 313–390, vol. 4.

Davey, K. G.; Treherne, J. E. Studies on crop function in the cockroach (Periplaneta americana L.) I–III. Journal of Experimental Biology 40:763–773, 775–780; 41:513–524; 1963.

Dean, R. L.; Locke, M.; Collins, J. V. Structure of the fat body. In: Kerkut, G. A.; Gilbert, L. I., eds. Comprehensive insect physiology, biochemistry, and pharmacology. New York: Pergamon Press; 1985:p. 155–210, vol. 3.

DelValle, J.; Yamada, T. The gut as an endocrine organ. Annual Review of Medicine 41:447–455; 1990.

Dethier, V. G. The hungry fly. A physiological study of the behavior associated with feeding. Cambridge, MA: Harvard University Press; 1976.

Dow, J. A. T. Insect midgut function. Advances in Insect Physiology 19:187–328; 1986.

Downer, R. G. H., editor. Energy metabolism in insects. New York: Plenum Press; 1981.

Dunn, P. E. Biochemical aspects of insect immunology. Annual Review of Entomology 31:321–333; 1986.

Eisner, T.; Johnessee, J. S.; Carrel, J.; Hendry, L. B.; Meinwald, J. Defensive use by an insect of a plant resin. Science 184:996–999; 1974.

Essig, E. O. College entomology. New York: Macmillan; 1942.

Florkin, M.; Jeuniaux, C. Hemolymph: composition. In: Rockstein, M., ed. The physiology of insecta. 2nd ed. New York: Academic Press; 1974:p. 255–307, vol. V.

Fraenkel, G. S. Untersuchungen über die Koordination von Reflexen und automatisch-nervösen Rhythmen bei Insekten III. Zeitschrift fuer VerGleichende Physiologie 16:394–460; 1932.

Friedman, S. Carbohydrate metabolism. In: Kerkut, G. A.; Gilbert, L. I., eds. Comprehensive insect physiology, biochemistry, and pharmacology. New York: Pergamon Press; 1985:p. 44–76, vol. 10.

Gelperin, A. Control of crop emptying in the blowfly. Journal of Insect Physiology 12:331–345; 1966.

Gelperin, A. Regulation of feeding. Annual Review of Entomology 16:365–378; 1971.

Gillott, C. Entomology. New York: Plenum Press; 1980.

Goodchild, A. J. P. Evolution of the alimentary canal in Hemiptera. Biological Reviews 41:97–140; 1966.

Götz, P.; Boman, H. G. Insect immunity. In: Kerkut, G. A.; Gilbert, L. I., eds. Comprehensive insect physiology, biochemistry, and pharmacology. New York: Pergamon Press; 1985:p. 453–485, vol. 3.

Greene, E. A diet-induced developmental polymorphism in a caterpillar. Science 243:643–646; 1989.

Gupta, A. P., editor. Insect hemocytes. New York: Cambridge Univ. Press; 1979.

Gupta, A. P., editor. Arthropod phylogeny. New York: Van Nostrand Reinhold; 1979:p. 762.

Gupta, A. P., editor. Hemocytic and humoral immunity in arthropods. New York: John Wiley & Sons; 1986:p. 535.

Guthrie, D. M.; Tindall, A. R. The biology of the cockroach. London: Edward Arnold, Ltd.; 1968.

Heinrich, B., editor. Insect thermoregulation. New York: John Wiley & Sons; 1981.

Hinton, H. E. Spiracular gills. Advances in Insect Physiology 5:65–162; 1968.

Holman, G. M.; Nachman, R. J.; Wright, M. S. Insect neuropeptides. Annual Review of Entomology 35:201–217; 1990.

House, C. R. Physiology of invertebrate salivary glands. Biological Reviews 55:417–473; 1980.

House, C. R.; Ginsborg, B. L. Salivary gland. In: Kerkut, G. A.; Gilbert, L. I., eds. Comprehensive insect physiology, biochemistry, and pharmacology. New York: Pergamon Press; 1985:p. 195–224, vol. 11.

House, H. L. Nutrition. In: Rockstein, M., ed. The physiology of insecta. 2nd ed. New York: Academic Press; 1974a:p. 1–62, vol. V.

House, H. L. Digestion. In: Rockstein, M., ed. The physiology of insecta. 2nd ed. New York: Academic Press; 1974b:p. 63–117, vol. V.

Huddart, H. Visceral muscle. In: Kerkut, G. A.; Gilbert, L. I., eds. Comprehensive insect physiology, biochemistry, and pharmacology. New York: Pergamon Press; 1985:p. 131–194, vol. 11.

Ishaaya, I. Nutritional and allelochemic insect-plant interactions relating to digestion and food intake. In: Miller, J. R. and Miller, T. A., eds. Insect-plant interactions. New York: Springer-Verlag; 1986:p. 191–223.

Jones, J. C. The circulatory system of insects. In: Rockstein, M., ed. The physiology of insecta. New York: Academic Press; 1964:p. 1–107, Vol. III.

Jones, J. C. The circulatory system of insects. Springfield, IL: Charles C. Thomas, Pub.; 1977.

Jones, J. C. The anatomy of the grasshopper (*Romalea microptera*). Springfield, IL: Charles C. Thomas, Pub.; 1981.

Kayser, H. Pigments. In: Kerkut, G. A.; Gilbert, L. I., eds. Comprehensive insect physiology, biochemistry, and pharmacology. New York: Pergamon Press; 1985:p. 367–415, vol. 10.

Keeley, L. L. Physiology and biochemistry of the fat body. In: Kerkut, G. A.; Gilbert, L. I., eds. Comprehensive insect physiology, biochemistry, and pharmacology. New York: Pergamon Press; 1985:p. 211–248, vol. 3.

Kenchington, W. Adaptations of insect peritrophic membranes to form cocoon fabrics. In: Hepburn, H. R., ed. The insect integument. New York: Elsevier Scientific; 1976:p. 497–513.

Kerkut, G. A.; Gilbert, L. I., editors. Comprehensive insect physiology, biochemistry, and pharmacology. New York: Pergamon Press; 1985.

Koch, A. Insects and their endosymbionts. In: Henry, S. M., ed. Symbiosis. New York: Academic Press; 1967:p. 1–106, vol. 2.

Lackie, A. M., editor. Immune mechanisms in invertebrate vectors. Oxford, England: Clarendon Press; 1986.

Lackie, A. M. Haemocyte behaviour. Advances in Insect Physiology 21:83–178; 1988.

Levenbook, L. Insect storage proteins. In: Kerkut, G. A.; Gilbert, L. I., eds. Comprehensive insect physiology, biochemistry, and pharmacology. New York: Pergamon Press; 1985:p. 307–346, vol. 10.

Locke, M. The structure and development of the vacuolar system in the fat body of insects. In: King, R. C.; Akai, H., eds. Insect ultrastructure. New York: Plenum Press; 1984:p. 151–197, vol. 2.

Maddrell, S. H. P. The mechanisms of insect excretory systems. Advances in Insect Physiology 8:200–324; 1971.

Martoja, R.; Ballan-Dufrancais, C. The ultrastructure of the digestive and excretory organs. In: King, R. C.; Akai, H., eds. Insect ultrastructure. New York: Plenum Press; 1984:p. 199–268, chapt. 6, vol. 2.

Matheson, R. Handbook of the mosquitos of North America. 2nd ed. Ithaca, NY: Cornell University Press; 1944.

Meeusen, R. L.; Warren, G. Insect control with genetically engineered crops. Annual Review of Entomology 34:373–381; 1989.

Menn, J. J.; Börkovec, A. B. Insect neuropeptides: potential new insect control agents. Journal of Agricultural and Food Chemistry 37:271–278; 1989.

Miles, P. W. The saliva of Hemiptera. Advances in Insect Physiology 9:183–255; 1972.

Mill, P. J. Respiration: aquatic insects. In: Rockstein, M., ed. The physiology of insecta. 2nd ed. New York: Academic Press; 1974:p. 403–467, vol. VI.

Mill, P. J. Structure and physiology of the respiratory system. In: Kerkut, G. A.; Gilbert, L. I., eds. Comprehensive insect physiology, biochemistry, and pharmacology. New York: Pergamon Press; 1985:p. 517–593, vol. 3.

Miller, P. L. Respiration—aerial gas transport. In: Rockstein, M., ed. The physiology of insecta. New York: Academic Press, Inc.; 1964:p. 557–615, vol. 3.

Miller, P. L. Respiration—aerial gas transport. In: Rockstein, M., ed. The physiology of insecta. 2nd ed. New York: Academic Press; 1974a:p. 345–402, vol. VI.

Miller, T. A. Electrophysiology of the insect heart. In: Rockstein, M., ed. The physiology of insecta. 2nd ed. New York: Academic Press; 1974:p. 169–200, vol. V.

Miller, T. A. Neurosecretion and the control of visceral organs in insects. Annual Review of Entomology 20:133–149; 1975.

Miller, T. A. Heart and diaphragms. In: Kerkut, G. A.; Gilbert, L. I., eds. Comprehensive insect physiology, biochemistry, and pharmacology. New York: Pergamon Press; 1985a:p. 119–130, vol. 11.

Miller, T. A. Structure and physiology of the circulatory system. In: Kerkut, G. A.; Gilbert, L. I., eds. Comprehensive insect physiology, biochemistry, and pharmacology. New York: Pergamon Press; 1985b:p. 289–353, vol. 3.

Mullins, D. E. Chemistry and physiology of the hemolymph. In: Kerkut, G. A.; Gilbert, L. I., eds. Comprehensive insect physiology, biochemistry, and pharmacology. New York: Pergamon Press; 1985:p. 356–400, vol. 3.

Mullins, D. E.; Cochran, D. G. Nutritional ecology of cockroaches. In: Slansky, F., Jr.; Rodriques, J. G., eds. Nutritional ecology of insects, mites, spiders, and related invertebrates. New York: John Wiley & Sons; 1987:p. 885–902.

Nappi, A. J.; Carton, Y. Cellular immune responses and their genetic aspects in Drosophila. In: Brehélin, M., ed. Immunity in invertebrates. New York: Springer-Verlag; 1986:p. 171–187.

Nicolas, G.; Sillans, D. Immediate and latent effects of carbon dioxide on insects. Annual Review of Entomology 34:97–116; 1989.

Noirot, C.; Noirot-Timothée, C. The structure and development of the tracheal system. In: King, R. C.; Akai, H., eds. Insect ultrastructure. New York: Plenum Press; 1982:p. 351–381, vol. 1.

O'Shea, M.; Adams. M. Proctolin: from "gut factor" to model neuropeptide. Advances in Insect Physiology 19:1–28; 1986.

Packard, A. S. A textbook of entomology. New York: Macmillan; 1898.

Patton, R. L. Introductory insect physiology. Philadelphia: W. B. Saunders Co.; 1963.

Peters, W. Investigations on the peritrophic membranes of Diptera. In: Hepburn, H. R., ed. The insect integument. New York: Elsevier Scientific; 1976:p. 515–543.

Peters, W. Peritrophic membranes New York: Springer-Verlag; 1992.

Piek, T. Insect venoms and toxins. In: Kerkut, G. A.; Gilbert, L. I., eds. Comprehensive insect physiology, biochemistry, and pharmacology. New York: Pergamon Press; 1985:p. 595–633, vol. 11.

Price, G. M. Protein and nucleic acid metabolism in insect fat body. Biological Reviews 48:333–375; 1972.

Ratcliffe, N. A.; Rowley, A. F., editors. Invertebrate blood cells. Vol. 2. New York: Academic Press; 1981.

Reinecke, J. P. Nutrition: artificial diets. In: Kerkut, G. A.; Gilbert, L. I., eds. Comprehensive insect physiology, biochemistry, and pharmacology. New York: Pergamon Press; 1985:p. 391–419, vol. 4.

Ribeiro, J. M. C. Role of saliva in blood-feeding by arthropods. Annual Review of Entomology 32:463–478; 1987.

Richards, A. G.; Richards, P. A. The peritropic membranes of insects. Annual Review of Entomology 22:219–240; 1977.

Rizki, T. M.; Rizki, R. M. The cellular defense system of Drosophila melanogaster. In: King, R. C.; Akai, H., eds. Insect ultrastructure. New York: Plenum Press; 1984:p. 579–604.

Rodriquez, J. G., editor. Insect and mite nutrition: significance and implication in ecology and pest management. New York: American Elsevier Pub. Co., Inc.; 1972.

Rosenthal, G. A.; Berenbaum, M. R., editors. Herbivores: their interaction with secondary plant metabolites, Vol. II. New York: Academic Press; 1992.

Rosenthal, G. A.; Janzen, D. H., editors. Herbivores: their interaction with secondary plant metabolites. New York: Academic Press; 1979.

Ryan, C. A. Proteinase inhibitors. In: Rosenthal, G. A.; Janzen, D. H., eds. Herbivores: their interaction with secondary plant metabolites. New York: Academic Press; 1979: chapt. 17, pp. 599–618.

Ryan, C. A. Proteinase inhibitor gene families: strategies for transformation to improve plant defenses against herbivores. Bioassays 10:20; 1989.

Schmidt, J. O. Biochemistry of insect venoms. Annual Review of Entomology 27:339–368; 1982.

Schneiderman, M. A.; Williams, C. M. An experimental analysis of the discontinuous respiration of the cecropia. Biological Bulletin 109:123–143; 1955.

Shapiro, J. P.; Law, J. H.; Wells, M. A. Lipid transport in insects. Annual Review of Entomology 33:297–318; 1988.

Singh, P.; Moore, R. F., editors. Handbook of insect rearing. Vols. I and II. New York: Elsevier Scientific; 1985.

Slansky, F., Jr.; Panizzi, A. R. Nutritional ecology of seed-sucking insects. In: Slanksy, F., Jr.; Rodriques, J. G., eds. Nutritional ecology of insects, mites, spiders, and related invertebrates. New York: John Wiley & Sons; 1987:p. 283–320.

Slansky, F., Jr.; Rodriquez, J. G., editors. Nutritional ecology of insects, mites, spiders, and related invertebrates. New York: John Wiley & Sons; 1987.

Snodgrass, R. E. Principles of insect morphology. New York: McGraw-Hill; 1935.

Snodgrass, R. E. The anatomy of the honey bee. In: Grout, R. A., ed. The hive and the honey bee. Hamilton, IL: Dadant and Sons, Inc.; 1963:p. 141–190.

Spence, K. D. 1991. Structure and physiology of the peritrophic membrane. In: Binnington, K.; Retnakaran, A., eds. Physiology of the insect epidermis. East Melbourne, Victoria, Australia: C.S.I.R.O. Publications; 1991.

Stobbart, R. H.; Shaw, J. Salt and water balance; excretion. In: Rockstein, M., ed. The physiology of insecta. 2nd ed. New York: Academic Press; 1974:p. 362–446, vol. V.

Taylor, C. W. Calcium regulation in insects. Advances in Insect Physiology 19:155–186; 1986.

Terra, W. R. Evolution of digestive systems of insects. Annual Review of Entomology 35:181–200; 1990.

Terriere, L. C. Induction of detoxification enzymes in insects. Annual Review of Entomology 29:71–88; 1984.

Thompson, S. N. Nutrition and *in vitro* culture of insect parasitoids. Annual Review of Entomology 31:197–219; 1986.

Truman, J. W. Hormonal release of stereotyped motorprogrammes from the isolated nervous system of the *Cecropia* silkmoth. Journal of Experimental Biology 74:151–174; 1978.

Tu, A. T., editor. Handbook of natural toxins. Vol. 2. Insect poisons, allergens, and other invertebrate venoms. New York: Marcel Dekker, Inc.; 1984.

Turunen, S. Absorption. In: Kerkut, G. A.; Gilbert, L. I., eds. Comprehensive insect physiology, biochemistry, and pharmacology. New York: Pergamon Press; 1985:p. 241–277, vol. 4.

Vinson, S. B. How parasitoids deal with the immune system of their host: An overview. Archives of Insect Biochemistry and Physiology 23:3–27; 1990.

Wasserthal, L. T. Antagonism between haemolymph transport and tracheal ventilation in an insect wing (*Attacus atlas* L.). Journal of Comparative Physiology 147:27–40; 1982.

West, L. S. The housefly: its natural history, medical importance, and control. Ithaca, NY: Comstock Publishing Company, Inc.; 1951.

Whitten, J. M. Comparative anatomy of the tracheal system. Annual Review of Entomology 17:373–402; 1972.

Wigglesworth, V. B. The formation of the peritrophic membrane in insects with special reference to the larvae of mosquitoes. Quarterly Journal of Microscopic Science 73:593–616; 1930.

Wigglesworth, V. B. The distribution of lipid in the cuticle of *Rhodnius*. In: Hepburn, H. R., ed. The insect integument. New York: Elsevier Scientific; 1976:p. 89–106.

Wigglesworth, V. B. The physiology of insect tracheoles. Advances in Insect Physiology 17:85–148; 1983.

Yeager, J. F. Mechanographic method of recording insect cardiac activity, with reference to effect of nicotine on isolated heart preparation of *Periplaneta americana*. Journal of Agricultural Research 56:267–306; 1938.

Zlotkin, E. Toxins derived from arthropod venoms specifically affecting insects. In: Kerkut, G. A.; Gilbert, L. I., eds. Comprehensive insect physiology, biochemistry, and pharmacology. New York: Pergamon Press; 1985:p. 499–546, vol. 10.

CHAPTER 5

Adiyodi, K. G.; Adiyodi, R. G., editors. Reproductive biology of invertebrates. Vol. I. Oogenesis, oviposition, and oosorption. Vol. II. Spermatogenesis and sperm function. Vol. III. Accessory sex glands. New York: John Wiley & Sons; 1983.

Agrell, I. P. S.; Lundquist, A. M. Physiological and biochemical changes during insect development. In: Rockstein, M., ed. The physiology of insecta. 2nd ed. New York: Academic Press; 1973:p. 159–247, vol. I.

Alcock, J. Animal behavior: an evolutionary approach. Sunderland, MA: Sinauer Associates, Inc. Pub.; 1975.

Anderson, D. T. Embryology and phylogeny in annelids and arthropods. New York: Pergamon; 1973.

Anderson, D. T. Embryos, fate maps, and the phylogeny of arthropods. In: Gupta, A. P., ed. Arthropod phylogeny. New York: Van Nostrand Reinhold; 1979:p. 59–105.

Ashburner, M. Function and structure of polytene chromosomes during insect development. Advances in Insect Physiology 7:1–95; 1970.

Ashburner, M. Puffing patterns in *Drosophila melanogaster* and related species. In: Beerman, N., ed. Developmental studies on giant chromosomes. New York: Springer-Verlag; 1972:p. 101–151.

Baccetti, B. Insect sperm cells. Advances in Insect Physiology 9:316–384; 1972.

Beerman, W., editor. Developmental studies on giant chromosomes. Berlin: Springer-Verlag; 1977.

Bergerard, J. Environmental and physiological control of sex determination and differentiation. Annual Review of Entomology 17:57–74; 1972.

Berry, S. J. Maternal direction of oogenesis and early embryogenesis in insects. Annual Review of Entomology 27:205–227; 1982.

Blum, M. S.; Blum, N. A., editors. Sexual selection and reproductive competition in insects. New York: Academic Press; 1979.

Bodenstein, D., editor. Milestones in developmental physiology of insects. New York: Appleton-Century-Crofts; 1971.

Bonhag, P. F. Ovarian structure and vitellogenesis in insects. Annual Review of Entomology 3:136–160; 1958.

Bownes, M. Expression of the genes coding for vitellogenin (yolk protein). Annual Review of Entomology 31:507–531; 1986.

Breland, O. P.; Eddleman, C. D.; Biesele, J. J. Studies of insect spermatozoa I. Entomological News 78:197–216; 1968.

Browder, L. W., editor. Developmental biology, a comprehensive synthesis. Vol. 1. Oogenesis. New York: Plenum Press; 1985.

Chen, P. S. Biochemical aspects of insect development. Vol. 3. Monographs in developmental biology. New York: S. Karger; 1971.

Chen, P. S. The functional morphology and biochemistry of insect male accessory glands and their secretions. Annual Review of Entomology 29:233–255; 1984.

Cheng, K. M.; Siegel, P. B. Quantitative genetics of multiple matings. Animal Behavior 40:406–407; 1990.

Clark, A. M.; Rockstein, M., editors. Aging in insects. In: Rockstein, M., ed. The physiology of insecta. New York: Academic Press; 1964:p. 227–281, vol. 1.

Collatz, K.-G.; Sohal, R. S., editors. Insect aging: strategies and mechanisms. New York: Springer-Verlag; 1986.

Cottrell, C. B. Insect ecdysis with particular emphasis on cuticular hardening and darkening. Advances in Insect Physiology 2:175–218; 1964.

Counce, S. J. The causal analysis of insect embryogenesis. In: Counce, S. J.; Waddington, C. H., eds. Developmental systems: insects. New York: Academic Press; 1973: vol. 2.

Counce, S. J.; Waddington, C. H., editors. Developmental systems: insects. New York: Academic Press; 1972, 1973, vols. 1 & 2.

Danks, H. V. Insect dormancy: an ecological perspective. Ottawa, Canada: Biological Survey of Canada; 1987.

Davey, K. G. Reproduction in the insects. Edinburgh, Scotland: Oliver & Boyd; 1965.

Davey, K. G. The male reproductive tract. In: Kerkut, G. A.; Gilbert, L. I., eds. Comprehensive insect physiology, biochemistry, and pharmacology. New York: Pergamon Press; 1985:p. 1–14, vol. 1.

Denlinger, D. L. Dormancy in tropical insects. Annual Review of Entomology 31:239–264; 1986.

de Wilde, J.; de Loof, A. Reproduction. In: Rockstein, M., ed. The physiology of insecta. 2nd ed. New York: Academic Press; 1973a:p. 12–95, vol. I.

de Wilde, J.; de Loof, A. Reproduction—endocrine control. In: Rockstein, M., ed. The physiology of insecta. 2nd ed. New York: Academic Press; 1973b:p. 97–157, vol. I.

deWilde, J.; Beetsma, J. The physiology of caste development in social insects. Advances in Insect Physiology 16:167–246; 1982.

Doane, W. W. Role of hormones in insect development. In: Counce, S. J.; Waddington, C. H., eds. Developmental systems: insects. New York: Academic Press; 1973, vol. 2.

Downer, G. H.; Laufer, H. Endocrinology of insects. New York: Alan R. Liss, Inc.; 1984.

Dyar, H. G. The number of moults of lepidopterous larvae. Psyche 5:420–422; 1890.

Engelmann, F. Endocrine control of reproduction in insects. Annual Review of Entomology 13:1–26; 1968.

Engelmann, F. The physiology of insect reproduction. New York: Pergamon; 1970.

Essig, E. O. Insects and mites of western North America. New York: Macmillan; 1958:p. 1050.

Etkin, W.; Gilbert, L. I., editors. Metamorphosis: a problem in developmental biology. Amsterdam: North-Holland; 1968.

Flanagan, R. T.; Hagedorn, H. H. Vitellogenin synthesis in the mosquito: the role of juvenile hormone in the development of responsiveness to ecdysone. Physiological Entomology 2:173–178; 1977.

Fox, R. M.; Fox, J. W. Introduction to comparative entomology. New York: Van Nostrand Reinhold; 1964.

Fraenkel, G. S.; Hsiao, C. Bursicon, a hormone which mediates tanning of the cuticle in the adult fly and other insects. Journal of Insect Physiology 11:513–526; 1965.

Furneaux, P. J. S.; Mackay, A. L. The composition, structure, and formation of the chorion and the vitelline membrane of the insect egg-shell. In: Hepburn, H. R., ed. The insect integument. New York: Elsevier Scientific; 1976:p. 157–176.

Gehring, W. J.; Nöthiger, R. The imaginal discs of Drosophila. In: Counce, S. J.; Waddington, C. H., eds. Developmental systems: insects. New York: Academic Press; 1973:p. 212–290, vol. 2.

Gerber, G. H. Evolution of the methods of spermatophore formation in pterygote insects. Canadian Entomologist 102:358–362; 1970.

Gilbert, L. I. Physiology of growth and development: endocrine aspects. In: Rockstein, M., ed. The physiology of insecta. New York: Academic Press, Inc.; 1964: vol. 1.

Gilbert, L. I., editor. The juvenile hormones. New York: Plenum Press; 1976.

Gilbert, L. I.; Frieden, E., editors. Metamorphosis—a problem in developmental biology. New York: Plenum Press; 1981.

Gilbert, L. I.; King, D. S. Physiology of growth and development: endocrine aspects. In: Rockstein, M., ed. The physiology of insecta. 2nd ed. New York: Academic Press; 1973:p. 249–370, vol. I.

Gupta, A. P., editor. Morphogenetic hormones of arthropods—discoveries, syntheses, metabolism, evolution, modes of action, and techniques. New Brunswick, NJ: Rutgers Univ. Press; 1990.

Hagedorn, H. H. The control of vitellogenesis in the mosquito, Aedes aegypti. American Zoologist 14:1207–1217; 1974.

Hagedorn, H. H.; Kunkel, J. G. Vitellogenin and vitellin in insects. Annual Review of Entomology 24:475–505; 1979.

Hagen, R. R. Embryology of the viviparous insects. New York: Ronald Press; 1951.

Happ, G. M. Structure and development of male accessory glands. In: King, R. C.; Akai, H., eds. Insect ultrastructure. New York: Plenum Press; 1984:p. 365–396, vol. 2.

Highnam, K. C.; Hill, L. The comparative endocrinology of the invertebrates. 2nd ed. New York: Elsevier Scientific; 1980.

Hinton, H. E. Concealed phases in the metamorphosis of insects. Nature (London) 157:552–553; 1946.

Hinton, H. E. On the origin and function of the pupal stage. Transactions of the Royal Entomological Society of London (London) 99:395–409; 1948.

Hinton, H. E. Concealed phases in the metamorphosis of insects. Scientific Progress (London) 46:260–275; 1958.

Hinton, H. E. The origin and function of the pupal stage. London: Proceedings of the Royal Entomological Society of London [A] 40:96–113; 1963.

Hinton, H. E. Sperm transfer in insects and the evolution of haemocoelic insemination. In: Highnam, K. C., ed. Insect reproduction. London: Royal Entomological Society [A], Symposium 3; 1964:p. 95–107.

Hinton, H. E. Respiratory systems of insect egg shells. Annual Review of Entomology 14:343–368; 1969.

Hinton, H. E. Some neglected phases in metamorphosis. Proceedings of the Royal Entomological Society (London) [C], 35:55–64; 1971.

Hinton, H. E. Neglected phases in metamorphosis: A reply to V. B. Wigglesworth. Journal of Entomology [A] 48:57–68; 1973.

Hinton, H. E. Accessory functions of seminal fluid. Journal of Medical Entomology 11:19–25; 1974.

Hinton, H. E. Notes on neglected phases in metamorphosis, and a reply to J. M. Whitten. Annals of the Entomological Society of America 69(3):560–566; 1976.

Hinton, H. E. Biology of insect eggs. New York: Pergamon; 1979. 3 vol.

Ivanova-Kasas, O. M. Polyembryony in insects. In: Counce, S. J.; Waddington, C. H., eds. Developmental systems: insects. New York: Academic Press; 1972:p. 243–271, vol. 1.

Jamieson, B. G. M. The ultrastructure and phylogeny of insect spermatozoa. New York: Cambridge Univ. Press; 1987.

Jenkin, P. M.; Hinton, H. E. Apolysis in arthropod molting cycles. Nature 211(5051):871; 1966.

Johannsen, O. A.; Butt, F. H. Embryology of insects and myriapods. New York: McGraw-Hill; 1941.

Kafatos, F. C. The cocoonase zymogen cells of silk moths: A model of terminal cell differentiation for specific protein synthesis. Current Topics in Developmental Biology 7:125–191; 1972.

Kafatos, F. C.; Tartakoff, A. M.; Law, J. H. Cocoonase. I. Preliminary characterization of a proteolytic enzyme from silkmoths. Journal of Biological Chemistry 242:1477–1487; 1967.

Karlson, P.; Sekeris, C. E. Control of tyrosine metabolism and cuticle sclerotization by ecdysone. In: Hepburn, H. R., ed. The insect integument. New York: Elsevier Scientific; 1976:p. 145–156.

Kaulenas, M. S. Insect accessory reproductive structures—function, structure, and development. New York: Springer-Verlag; 1992.

Kennedy, J. S., editor. Insect polymorphism. Royal Entomological Society (London), Symposium 1; 1961.

Kerkut, G. A.; Gilbert, L. I., editors. Comprehensive insect physiology, biochemistry, and pharmacology. New York: Pergamon Press; 1985. 13 vol.

King, R. C. Symposium on reproduction of arthropods of medical and veterinary importance. I. Insect gametogenesis. Journal of Medical Entomology 11:1–7; 1974.

King, R. C.; Akai, H. Insect ultrastructure. Vol. 1. New York: Plenum Press. 1982.

Kroeger, H. Gene activities during insect metamorphosis and their control by hormones. In: Etkin, W.; Gilbert, L. I., eds. Metamorphosis: a problem in developmental biology. Amsterdam, North-Holland; 1968: p. 185–219.

Larsen-Rapport, E. W. Imaginal disc determination: molecular and cellular correlates. Annual Review of Entomology 31:145–175; 1986.

Law, J. H. Molecular entomology. New York: Alan R. Liss, Inc.; 1987.

Lawrence, P. A. Polarity and patterns in the post-embryonic development of insects. Advances in Insect Physiology 7:197–266; 1970.

Lawrence, P. A. The development of spatial patterns in the integument of insects. In: Counce, S. J.; Waddington, C. H., eds. Developmental systems: insects, vol. 2. New York: Academic Press; 1973:p. 157–209.

Lawrence, P. A. The making of a fly—the genetics of animal design. Cambridge, MA: Blockwell Scientific Publications: 1992.

L'Helias, C. Chemical aspects of growth and development in insects. In: Florikin, M.; Sheer, B. J., eds. Chemical Zoology, Vol. V, Part A, Arthropoda. New York: Academic Press; 1970:343–400.

Leopold, R. A. The role of male accessory glands in insects. Annual Review of Entomology 21:199–221; 1976.

Lipke, H.; Sugumaran, M.; Henzel, W. Mechanisms of sclerotization in dipterans. Advances in Insect Physiology 17:1–84; 1983.

Lüscher, M. Social control of polymorphism in termites. In: Kennedy, J. S., ed. Insect polymorphism. London: Royal Entomological Society, Symposium 1; 1961:p. 57–67.

Lüscher, M. Phase and caste determination in insects. Elmsford, New York: Pergamon; 1976.

Lüscher, M.; Springhetti, A. Functions of the corpora allata in the development of termites. Washington: Proceedings XVI International Congress of Zoology 4:244–250; 1963.

Mahowald, A. P. Oogenesis. In: Counce, S. J.; Waddington, C. H., eds. Developmental systems: insects. New York: Academic Press; 1972:p. 1–47, vol. I.

Matsuda, R. Morphology and evolution of the insect abdomen. Elmsford, N. Y.: Pergamon; 1976.

Menn, J. J.; Beroza, M., editors. Insect juvenile hormones—chemistry and action. New York: Academic Press; 1972.

Michiels, N. K.; Dhondt A. A. . Direct and indirect estimates of sperm precedence and displacement in the dragonfly *Sympetrum danae* (Odonata: Libellulidae). Behavioral Ecology and Sociobiology 23:257–263; 1988.

Michod, R. E.; Levin, B., editors. The evolution of sex. Sunderland, MA: Sinauer; 1988.

Miller, P. L. Respiration—aerial gas transport. In: Rockstein, M., ed. The physiology of insecta. 2nd ed. New York: Academic Press; 1974:p. 345–402, vol. VI.

Morgan, E. D.; Poole, C. F. The extraction and determination of ecdysones in arthropods. Advances in Insect Physiology 12:17–62; 1976.

Naisse, J. Controle endocrine de la différenciation sexuelle chez l'insecte *Lampyris noctiluca*. I–III, Arch. Biol., Liege, 77:139–201. Gen. Comp. Endocrinol., 7:85–104, 105–110; 1966.

Naisse, J. Role des neurohormones dans la différenciation sexuelle de *Lampyris noctiluca*. Journal of Insect Physiology 15:877–892; 1969.

Novák, V. J. A. Insect hormones. New York: John Wiley & Sons (Halsted Press); 1975.

Oberlander, H. The imaginal discs. In: Kerkut, G. A.; Gilbert, L. I., eds. Comprehensive insect physiology, biochemistry, and pharmacology. New York: Pergamon Press; 1985:p. 151–182, vol. 3.

Packard, A. S. A textbook of entomology. New York: Macmillan; 1898.

Papaj, D. R.; Prokopy, R. J. Ecological and evolutionary aspects of learning in phytophagous insects. Annual Review of Entomology 34:315–350; 1989.

Parker, G. A. Sperm competition and its evolutionary consequences in the insects. Biological Reviews 45:525–567; 1970.

Phillips, D. M. Insect sperm: their structure and morphogenesis. Journal of Cellular Biology 44:243–277; 1970.

Raabe, M. Insect reproduction: regulation of successive steps. Advances in Insect Physiology 19:29–154; 1986.

Rees, H. H. Insect biochemistry. New York: John Wiley & Sons (Halsted Press); 1977.

Rempel, J. G. The evolution of the insect head: the endless dispute. Quaestiones Entomologicae 11:7–25; 1975.

Reynolds, S. E. Integration of behaviour and physiology of ecdysis. Advances in Insect Physiology 14:475–595; 1980.

Richards, O. W. An introduction to the study of polymorphism in insects. In: Kennedy, J. S., ed. Insect polymorphism. London: Royal Entomological Society, Symposium 2; 1961:p. 2–10.

Richards, O. W.; Davies, R. G. Imm's general textbook of entomology. 10th ed. Vol. 1, Structure, physiology, and development. p. 418. Vol. 2, Classification and biology, p. 1354. New York: John Wiley & Sons (Halsted Press); 1977.

Riddiford, L. M. Hormone action at the cellular level. In: Kerkut, G. A.; Gilbert, L. I., eds. Comprehensive insect physiology, biochemistry, and pharmacology. New York: Pergamon Press; 1985:p. 37–84, vol. 8.

Riddiford, L. M.; Truman, J. W. Biochemistry of insect hormones and insect growth regulators. In: Rockstein, M., ed. Biochemistry of insects. New York: Academic Press; 1978:p. 307–357.

Ridley, M. Mating frequency and fecundity in insects. Biological Reviews 63:509–549; 1988.

Riek, E. F. Four-winged Diptera from the upper Permian of Australia. Proceedings of the Linnean Society of New South Wales 101:250–255; 1977.

Rockstein, M.; Miquel, J. Aging in insects. In: Rockstein, M., ed. The physiology of insecta. 2nd ed. New York: Academic Press; 1973:p. 371–478, vol. I.

Sander, K. Specification of the basic body pattern in insect embryogenesis. Advances in Insect Physiology 12:125–238; 1976.

Sander, K.; Gutzeit, H. O.; Jackle, J. Insect embryogenesis: morphology, physiology, genetical, and molecular aspects. In: Kerkut, G. A.; Gilbert, L. I., eds. Comprehensive insect physiology, biochemistry, and pharmacology. New York: Pergamon Press; 1985:p. 319–385, vol. 1.

Schal, C.; Bell, W. J. Ecological correlates of paternal investment of urates in a tropical cockroach. U.S.A.: Science 218:170–172; 1982.

Schaller, F. Indirect sperm transfer by soil arthropods. Annual Review of Entomology 16:407–446; 1971.

Schneiderman, H. A.; Gilbert, L. I. Control of growth and form in insects. Science, 143:325–333; 1964.

Schwalm, F. E. Insect morphogenesis. Monographs in developmental biology. New York: Karger Publications; 1988.

Sheppard, P. M. Recent genetical work on polymorphic mimetic Papilios. In: Kennedy, J. S., ed. Insect polymorphism. London: Royal Entomological Society, Symposium 1; 1961:p. 20–29.

Sharov, A. G. Basic arthropodan stock with special reference to insects. New York: Pergamon; 1966.

Sláma, K.; Romanůk, M.; Sörm, F. Insect hormones and bioanalogues. New York: Springer-Verlag; 1974.

Smith, R. L., editor. Sperm competition and the evolution of animal mating systems. New York: Academic Press; 1984.

Snodgrass, R. E. Principles of insect morphology. New York: McGraw-Hill; 1935.

Staal, G. B. Anti-juvenile hormone agents. Annual Review of Entomology 31:391–429; 1986.

Stearns, S. C., editor. The evolution of sex and its consequences. Experientia/Supplementum. Basel, Switzerland: Birkhauser; 1987.

Sturm, H. Die Paarung beim Silberfischen, *Lepisma saccharina* Z. Tier-psychol. 13:1–2; 1956.

Telfer, W. H. The mechanism and control of yolk formation. Annual Review of Entomology 10:161–184; 1965.

Telfer, W. H. Development and physiology of the oocyte-nurse cell syncytium. Advances in Insect Physiology 11:223–319; 1975.

Thomson, J. A. Major patterns of gene activity during development in holometabolous insect. Advances in Insect Physiology 11:321–398; 1976.

Thornhill, R.; Alcock, J. The evolution of insect mating systems. Cambridge, MA: Harvard University Press; 1983.

Tobe, S. S.; Stay, B. Structure and regulation of the corpus allatum. Advances in Insect Physiology 18:305–432; 1985.

Treherne, J. E.; Smith, P. J. S.; Howes, E. A. Neural repair and regeneration in insects. Advances in Insect Physiology 21:35–84; 1988.

Truman, J. W. Physiology of insect ecdysis I–III. Journal of Experimental Biology 54:804–814; Biological Bulletin of Marine Biology Laboratory, Woods Hole 144:200–211; Journal of Experimental Biology 48:821–829; 1971–1973.

Ursprung, H.; Nöthiger, R. The biology of imaginal discs. New York: Springer-Verlag; 1972.

Waddington, C. H. The morphogenesis of patterns in *Drosophila*. In: Counce, S. J.; Waddington, C. H., eds. Developmental systems: insects. New York: Academic Press; 1973:p. 499–535, vol. 2.

Weaver, R. F.; Hedrick, P. W. Genetics. Dubuque, IA: Wm. C. Brown Pubs.; 1989.

Weygoldt, P. Significance of later embryonic stages and head development in arthropod phylogeny. In: Gupta, A. P., ed. Arthropod phylogeny. New York: Van Nostrand Reinhold; 1979:p. 107–135.

White, M. J. D. Cytogenetic mechanisms in insect reproduction. In: Highnam, K. C., ed. London: Royal Entomological Society, Symposium 2; 1964:p. 1–12.

Whitten, J. M. Metamorphic changes in insects. In: Etkin, W.; Gilbert, L. I., eds. Metamorphosis—a problem in developmental biology. Amsterdam, North-Holland; 1968:p. 43–105.

Whitten, J. M. Definition of insect instars in terms of "apolysis" or "ecdysis." Annals of the Entomological Society of America 69(3):556–559; 1976.

Wigglesworth, V. B. The physiology of insect metamorphosis. Cambridge, England: Cambridge University Press; 1954.

Wigglesworth, V. B. The control of growth and form: a study of the epidermal cell in an insect. Ithaca, NY: Cornell University Press; 1959.

Wigglesworth, V. B. Insect hormones. Edinburgh, Scotland: Oliver & Boyd; 1970.

Wigglesworth, V. B. The hormonal regulation of growth and reproduction in insects. Advances in Insect Physiology 2:247–336; 1964.

Wigglesworth, V. B. The principles of insect physiology. 7th ed. London: Chapman & Hall; 1972.

Wigglesworth, V. B. The significance of "apolysis" in the moulting of insects. Journal of Entomology [A], 47:141–149; 1973.

Williams, C. M. Hormonal interactions between plants and insects. In: Sondheimer, E.; Simeone, J. B., eds. Chemical ecology. New York: Academic Press; 1970:p. 103–132.

Willis, J. H. Morphogenetic action of insect hormones. Annual Review of Entomology 19:97–115; 1974.

Wyatt, G. R. Biochemistry of insect metamorphosis. In: Etkin, W.; Gilbert, L. I., eds. Metamorphosis—a problem in developmental biology. Amsterdam: North-Holland; 1968:p. 143–184.

Zeh, D. W.; Zeh, J. A.; Smith, R. L. Ovipositors, amnions, and eggshell architecture in the diversification of terrestrial arthropods. Quarterly Review of Biology 64:147–168; 1989.

CHAPTER 6

Allan, S. A.; Day, J. F.; Edman, J. D. Visual ecology of biting flies. Annual Review of Entomology 32:297–316; 1987.

Altner, H.; Loftus, R. Ultrastructure and function of insect thermo- and hygroreceptors. Annual Review of Entomology 30:273–296; 1985.

Baldus, K. Experimentelle Untersuchungen über die Entfernungslokalisation der Libellen (Aeschna cyanea). Z. vergl. Physiol. 3:375–505; 1926.

Barth, F. G. Insects and flowers. The biology of partnership. Princeton, NJ: Princeton University Press; 1985.

Bell, W. J.; Cardé, R. T., editors. Chemical ecology of insects. New York: Chapman and Hall; 1984.

Bennet-Clark, H. C. The mechanism and efficiency of sound production in mole crickets. Journal of Experimental Biology 52:619–652; 1970.

Bernhard, C. G., editor. The functional organization of the compound eye. Oxford: Pergamon; 1966.

Blest, A. D. The turnover of phototransductive membrane in compound eyes and ocelli. Advances in Insect Physiology 20:1–54; 1988.

Burns, M. D. Structure and physiology of the locust femoral chordotonal organ. Journal of Insect Physiology 20:1319–1339; 1974.

Butler, C. G. The honey bee colony—life history. In: Groud, R. A., ed. The hive and the honey bee. Hamilton, IL: Dadant & Sons; 1963: chapter 3.

Cade, W. Acoustically orienting parasitoids: fly phonotaxis to cricket song. Science 190:1312–1313; 1975.

Carlson, S. D.; Chi, C. The functional morphology of the insect photoreceptor. Annual Review of Entomology 24:379–416; 1979.

Carlson, S. D.; Marie, R. L. S.; Chi, C. The photoreceptor cells. In: King, R. C.; Akai, H., eds. Insect ultrastructure. New York: Plenum Press; 1984:p. 397–434, vol. 2.

Chapman, R. F. Chemoreception: the significance of receptor number. Advances in Insect Physiology 16:247–346; 1982a.

Chapman, R. F. The insects—structure and function. 3rd ed. Cambridge, MA: Harvard University Press; 1982b.

Claridge, M. F. Acoustic signals in the Homoptera: Behavior, taxonomy, and evolution. Annual Review of Entomology 30:297–318; 1985.

Comstock, J. H. An introduction to entomology. 9th ed. Ithaca, NY: Cornell University Press; 1940.

Dethier, V. G. The physiology of insect senses. New York: John Wiley & Sons; 1963.

Dethier, V. G. The physiology and histology of the contact chemoreceptors of the blowfly. Quarterly Review of Biology 30:348–371; 1955.

Dethier, V. G. Chemoreception. In: Roeder, K., ed. Insect physiology. New York: John Wiley & Sons; 1953.

Dethier, V. G. The hungry fly. A physiological study of the behavior associated with feeding. Cambridge, MA: Harvard University Press; 1976.

Dethier, V. G. Discriminative taste inhibitors affecting insects. Chemical Senses 12:251–263; 1988.

DeVries, P. J. Enhancement of symbioses between butterfly caterpillars and ants by vibrational communication. Science 248:1104–1106; 1990.

Eaton, J. L. Insect photoreceptor: An internal ocellus is present in sphinx moths. Science 173:822–823; 1971.

Essig, E. O. College entomology. New York: Macmillan; 1942.

Exner, S. Die Physiologie der facettierten Augen von Krebsen und Insekten. Leipzig; 1891.

Frisch, K. von. Bees, their vision, chemical senses, and language. Ithaca, NY: Cornell University Press; 1950.

Ganesalingam, W. K. Mechanisms of discrimination between parasitized and unparasitized hosts by Venturia canescens (Hymenoptera: Ichneumonidae). Entomologia Experimentalis et Applicata 17:36:44; 1974.

Goldsmith, T. H.; Bernard, G. D. The visual system of insects. In: Rockstein, M., ed. The physiology of insecta. 2nd ed. New York: Academic Press; 1974:p. 165–272, vol. II.

Goldsworthy, G. J.; Wheeler, C. H., editors. Insect flight. Boca Raton, FL: CRC Press, Inc.; 1989.

Goodman, L. J. The structure and function of the insect dorsal ocellus. Advances in Insect Physiology 7:97–195; 1970.

Gould, J. The case for magnetic field sensitivity in birds and bees. American Scientist 68:256–267; 1980.

Gould, J. Ethology—the mechanisms and evolution of behavior. New York: Norton & Company; 1982.

Haskell, P. T. Insect sounds. Chicago: Quadrangle; 1961.

Haskell, P. T. Sound production. In: Rockstein, M., ed. The physiology of insecta. New York: Academic Press; 1974:p. 354–410.

Hertz, M. Die Organization des optischen Feldes bei der Biene I. Z. vergl. Physiol. 8:693–748; 1929.

Hess, W. N. The chordotonal organs and pleural discs of cerambycid larvae. Annals of the Entomological Society of America 10:63–78; 1917.

Hinkle, N. C.; Koehler, P. G.; Patterson, R. S. Egg production, larval development, and adult longevity of cat fleas (Siphonaptera: Pulicidae) exposed to ultrasound. Journal of Economic Entomology 83:2306–2309; 1990.

Hodgson, E. S. Chemoreception in arthropods. Annual Review of Entomology 3:19–36; 1958.

Hodgson, E. S. The chemical senses and changing viewpoints in sensory physiology. Viewpoints in Biology 4:83–124; 1965.

Hodgson, E. S. Chemoreception. In: Rockstein, M., ed. The physiology of insecta. New York: Academic Press; 1974:p. 127–164, vol. II.

Hoeglund, G.; Hamdorf, K.; Langer, H.; Paulsen, R.; Schwemer, J. The photopigments in an insect retina. In: Langer, H., ed. Biochemistry and physiology of visual pigments. New York: Springer-Verlag; 1973.

Horn, E. Gravity In: Kerkut, GA; Gilbert, L. I., editors. Comprehensive insect physiology, biochemistry, and pharmacology. New York: Pergamon Press; 1985:p. 557–576.

Horridge, G. A. The compound eye and vision of insects. Oxford: Clarendon Press; 1975.

Horridge, G. A. The compound eye and vision of insects. Scientific American 237:108–120; 1977.

Huber, F.; Moore, T. E.; Loher, W., editors. Cricket behavior and neurobiology. Ithaca, NY: Cornell University Press; 1989.

Järvilehto, M. The eye: vision and perception. In: Kerkut, G. A.; Gilbert, L. I., eds. Comprehensive insect physiology, biochemistry, and pharmacology. New York: Pergamon Press;1985:p. 356–429, vol. 6.

Kaissling, K.-E. Chemo-electrical transduction in insect olfactory receptors. Annual Review of Neuroscience 9:121–145; 1986.

Kalmring, K.; Elsner, N., editors. Acoustic and vibrational communication in insects. New York: Paul Parey Scientific Publications; 1985.

Kellogg, F. E. Water vapour and carbon dioxide receptors in Aedes aegypti. Journal of Insect Physiology 16:99–108; 1970.

Kerkut, G. A.; Gilbert, L. I., eds. Comprehensive insect physiology, biochemistry, and pharmacology. Vols. 6 and 9. New York: Pergamon Press; 1985.

Lewis, C. T. Structure and function in some external receptors. In: Neville, A. C., ed. Insect ultrastructure. Royal Entomological Society (London), Symposium 5; 1970: p. 59–76.

Lloyd, J. E. Aggressive mimicry in Photuris fireflies: signal repertoires in femmes fatales. Science 187:452–453; 1975.

Lloyd, J. E. Bioluminescence and communication in insects. Annual Review of Entomology 28:131–160; 1983.

Long, M. E. Secrets of animal navigation. National Geographic 179:70–99; 1991.

McElroy, W. D.; Seliger, H. H. Biological luminescence. Scientific American 207:76–89; 1962.

McElroy, W. D.; Seliger, H. H.; DeLuca, M. Insect bioluminescence. In: Rockstein, M., ed. The physiology of insecta. 2nd ed. New York: Academic Press; 1974:p. 411–460, vol. II.

McElroy, W. D.; DeLuca, M. Biochemistry of insect luminescence. In: Kerkut, G. A.; Gilbert, L. I., eds. Comprehensive insect physiology, biochemistry, and pharmacology. New York: Pergamon Press; 1985:p. 553–564, vol. 4.

McIver, S. B. Structure of cuticular mechanoreceptors of arthropods. Annual Review of Entomology 20:381–398; 1975.

McIver, S. B. Mechanoreception. In: Kerkut, G. A.; Gilbert, L. I., eds. Comprehensive insect physiology, biochemistry, and pharmacology. New York: Pergamon Press; 1985:p. 71–132, vol. 6.

Markl, H. Schweresinnesorgane bei Ameisen und anderen Hymenopteren. Z. vergl. Physiol. 44:475–569; 1962.

Matthews, R. W.; Matthews, J. R. Insect behavior. New York: John Wiley & Sons, 1978: 507.

Mazokhin-Porshnyakov, G. A. Insect vision. New York: Plenum; 1969

Michelsen, A.; Nocke, H. Biophysical aspects of sound communication in insects. Advances in Insect Physiology 10:247–296; 1973.

Mill, P. J., editor. Structure and function of proprioceptors in the invertebrates. New York: John Wiley & Sons (Halsted Press); 1976.

Miller, L. A. Structure of the green lacewing tympanal organ (Chrysopa carnea, Neuroptera). Journal of Morphology 131:359–382; 1970.

Miller, T. A. Electrophysiology of the insect heart. In: Rockstein, M., ed. The physiology of insecta. 2nd ed. New York: Academic Press; 1974:p. 169–200, vol. V.

Miller W. H.; Bernard, C. D.; Allen, J. L. The optics of insect compound eyes. Science 162:760–767; 1968.

Muller, H. J. Formen der Dormanz bei Insekten. Nova Acta Leopoldina 35:1–27; 1826.

Nasci, R. S.; Harris, C. W.; Porter, C. K. Failure of an electrocuting device to reduce mosquito biting. Mosquito News 43:180–184; 1983.

Payne, T. L.; Birch, M. C.; Kennedy, C. E. J., editors. Mechanisms in insect olfaction. Oxford, England: Oxford University Press; 1986.

Prokopy, R. J.; Owens, E. D. Visual detection of plants by herbivorous insects. Annual Review of Entomology 28:337–364; 1983.

Richards, O. W.; Davies, R. G. Imm's general textbook of entomology. 19th ed. Vol. 1, Structure, physiology, and development. Vol. 2, Classification and biology. New York: John Wiley & Sons (Halsted Press); 1977.

Roeder, K. D. A physiological approach to the relation between prey and predator. In: Studies in invertebrate morphology. Smithsonian Institution, Miscellaneous Collections 157; 1959b:p. 287–306.

Roeder, K. D. Moths and ultrasound. Scientific American 212:94–102; 1965.

Roeder, K. D. Auditory system of noctuid moths. Science 154:1515–1521; 1966.

Roeder, K. D. Nerve cells and insect behavior, revised edition. Cambridge, MA: Harvard University Press; 1967.

Roeder, K. D. Acoustic and mechanical sensitivity of the distal lobe of the pilifer in choerocampine hawkmoths. Journal of Insect Physiology 18:1249–1964; 1972.

Ruck, P. Retinal structures and photoreception. Annual Review of Entomology 9:83–102; 1964.

Rust, M. K.; Parker, R. W. Lack of behavioral responses of the cat flea to a broad spectrum of ultrasound. Journal of Medical Entomology 25:144–146; 1988.

Schneider, D. Insect olfaction: deciphering system for chemical messages. Science 163:1031–1037; 1969.

Schneider, D. The sex-attractant receptor of moths. Scientific American 231:28–35; 1974.

Schön, A. Bau and Entwicklung des tibialen Chordotonalorgane bei der Honigbiene und bei Ameisen. Zoologische Jahrbuecher, Anatomie 31:439–472; 1911.

Schoonhoven, L. M. Insect chemosensory responses to plant and animal hosts. In: Shorey, H. H.; McKelvey, J. J., Jr., eds. Chemical control of insect behavior. New York: John Wiley & Sons; 1977:p. 7–14.

Schoonhoven, L. M.; Blom, F. Chemoreception and feeding behaviour in a caterpillar: towards a model of brain functioning in insects. Entomologia Experimentalis et Applicata 49:123–129; 1988.

Schwabe, J. Beiträge zur Morphologie und Histologie der tympanalen Sinnesapparate der Orthopteren. Stuttgart: Zoologica; 1906.

Schwartzkopff, J. Mechanoreception. In: Rockstein, M., editor. The physiology of insecta, 2nd ed. New York: Academic Press; 1974: p. 273–352, vol. II.

Seabrook, W. D. Insect chemosensory responses to other insects. In: Shorey, H. H.; McKelvey, J. J., Jr., eds. Chemical control of insect behavior. New York: John Wiley & Sons; 1977:p. 15–43.

Slifer, E. H. The structure of arthropod chemoreceptors. Annual Review of Entomology 15:121–142; 1970.

Soper, R. S.; Shewell, G. E.; Tyrrell, D. *Colcondamyia auditrix* nov. sp. (Diptera: Sarcophagidae), a parasite which is attracted by the mating song of its host, *Okanagana rimosa* (Homoptera: Culicidae). Canadian Entomologist 108:61–68; 1976.

Spangler, H. G. Moth hearing, defense, and communication. Annual Review of Entomology 33:59–82; 1988.

Stoffolano, J. G., Jr.; Yin, L. R. S. Structure and function of the ovipositor and associated sensilla of the apple maggot, *Rhagoletis pomonella* (Walsh) (Diptera: Tephritidae). International Journal of Insect Morphology & Embryology 16:41–69; 1987.

Summers, K. M.; Howells, A. J.; Phyliotis, N. A. Biology of eye pigmentation in insects. Advances in Insect Physiology 16:119–166; 1982.

Truman, J. W. Extraretinal photoreception in insects. Photochemical Photobiology 23:215–225; 1976.

Visser, J. H. Host odor perception in phytophagous insects. Annual Review of Entomology 31:121–144; 1986.

Wehner, J. W. Astronavigation in insects. Annual Review of Entomology 29:277–298; 1984.

Wehner, R. Information processing in the visual systems of arthropods. New York: Springer-Verlag: 1971.

Wehner, R. Polarized-light navigation by insects. Scientific American 235:106–115; 1976

Weis-Fogh, T. Flying insects and gravity. In: Gordon, S. A.; Cohen, M. J., eds. Gravity and the organism. Chicago, IL: University of Chicago Press; 1971:p. 177–184.

Wendler, G. Gravity orientation in insects: the role of different mechanoreceptors. In: Gordon, S. A.; Cohen, M. J., eds. Gravity and the organism. Chicago, IL: University of Chicago Press; 1971:p. 195–201; 1971.

Wigglesworth, V. B. The sensory physiology of the human louse, *Pediculus humanus corporis* DeGeer (Anoplura). Parasitology 33:67–109; 1941.

Wigglesworth, V. B. The life of insects. London: Weidenfeld & Nicolson; 1964.

Wigglesworth, V. B. The principles of insect physiology. 7th ed. London: Chapman & Hall; 1972.

Wilson, E. O. Chemical communication within animal species. In: Sondheimer, E.; Simeone, J. B., eds. Chemical ecology. New York: Academic Press; 1970:p. 133–155.

Yamaoka, K.; Hoshino, M.; Hirao, T. Role of sensory hairs on the anal papillae in oviposition behavior of *Bombyx mori*. Journal of Insect Physiology 17:897–911; 1971.

Zacharuk, R. Y.; Shields, V. D. Sensilla of immature insects. Annual Review of Entomology 36:331–354; 1991.

CHAPTER 7

Aidley, D. J. Structure and function in flight muscle. In: Goldsworthy, G. J.; Wheeler, C. H., eds. Insect flight. Boca Raton, FL: CRC Press, Inc.; 1989:

Barth, R. Muskulatur und Bewegungsart der Raupen. Zool. Jahrb. 62:507–566; 1937.

Bennet-Clark, H. C. Energy storage in jumping insects. In: Hepburn, H. R., ed. The insect integument. New York: Elsevier Scientific; 1976:p. 421–443.

Blum, M. S. Biochemical defenses of insects. In: Rockstein, M., ed. Biochemistry of insects. New York: Academic Press; 1978:p. 466–513.

Dalton, S. Borne on the wing. The extraordinary world of insects in flight. New York: Dutton; 1975.

Delcomyn, F. Walking and running. In: Kerkut, G. A.; Gilbert, L. I., eds. Comprehensive insect physiology, biochemistry, and pharmacology. New York: Pergamon Press; 1985: chap. 11, vol. 5.

Gillett, J. D.; Wigglesworth, V. B. The climbing organ of an insect, *Rhodnius prolixus* (Hemiptera: Reduviidae). Proceedings of the Royal Society (London), Series B 111:364–376; 1932.

Goldsworthy, G. J.; Wheeler, C. H., editors. Insect flight. Boca Raton, FL: CRC Press, Inc.; 1989.

Hughes, G. M.; Mill, P. J. Locomotion: terrestrial. In: Rockstein, M., ed. The physiology of insecta. 2nd ed. New York: Academic Press; 1974:p. 335–379, vol. III.

Kammer, A. E. Flying. In: Kerkut, G. A.; Gilbert, L. I., eds. Comprehensive insect physiology, biochemistry, and pharmacology. New York: Pergamon Press; 1985:p. 491–552, vol. 5.

Kerkut, G. A.; Gilbert, L. I., editors. Nervous system: structure and motor function. In: Kerkut, G. A.; Gilbert, L. I., eds. Comprehensive insect physiology, biochemistry, and pharmacology. New York: Pergamon Press; 1985: vol. 5.

Kevan, D. K. McE. Soil animals. New York: Philosophical Library; 1962.

Mangan, A. La Locomotion chez les Animaux, I: le Volume des Insectes. Paris: Hermann & Cie; 1934.

McConnell, E.; Richards, A. G. How fast can a cockroach run? Bulletin of the Brooklyn Entomological Society 50:36–43; 1955.

Möhl, B. Sense organs and the control of flight. In: Goldsworthy, G. J.; Wheeler, C. H., eds. Insect flight. Boca Raton, FL: CRC Press, Inc.; 1989:p. 75–97.

Nachtigall, W. A. Uber Kinematik, Dynamik und Energetik des Schwimmens einheimischen Dytisciden. Z. vergl. Physiol. 43:48–118; 1960.

Nachtigall, W. A. Zur Lokomotionsmechanik schwimmender Dipterenlarven. Z. vergl. Physiol. 46:449–466; 1963.

Nachtigall, W. A. Insects in flight. New York: McGraw-Hill; 1974.

Nachtigall, W. Swimming in aquatic insects. In: Kerkut, G. A.; Gilbert, L. I., eds. Comprehensive insect physiology, biochemistry, and pharmacology. New York: Pergamon Press; 1985: vol. 5.

Nachtigall, W. *Calliphora* as a model system for analysing insect flight. In: Kerkut, G. A.; Gilbert, L. I., eds. Comprehensive insect physiology, biochemistry, and pharmacology. New York: Pergamon Press; 1985: vol. 5.

Nachtigall, W. Mechanics and aerodynamics of flight. In: Goldsworthy, G. J.; Wheeler, C. H., eds. Insect flight. Boca Raton, FL: CRC Press, Inc.; 1989:p. 1–29.

Pringle, J. W. S. Insect flight. Cambridge, England: Cambridge University Press; 1957.

Romoser, W. J.; Nasci, R. S. Functions of the ventral air space and first abdominal spiracles in *Aedes aegypti* (Diptera: Culicidae). Journal of Medical Entomology 15(2): 109–114; 1979.

Rothschild, M.; Schlein, Y.; Parker, K.; Neville, A. C.; Sternberg, S. The flying leap of the flea. Scientific American 229:92–100; 1973.

Snodgrass, R. E. The caterpillar and the butterfly. Smithsonian Institution, Miscellaneous Collections 143(6); 1961.

Snodgrass, R. E. Principles of insect morphology. New York: McGraw-Hill; 1935.

Stellwaag, F. Wie steuern die Insekten wahrend des Gluges. Biol. Zentrabl. 36:30–44; 1916.

Wigglesworth, V. B. The principles of insect physiology. 6th ed. London: Methuen; 1965.

Wigglesworth, V. B. The principles of insect physiology. 7th ed. London: Chapman & Hall; 1972.

Wooton, R. J. The mechanical design of insect wings. Scientific American 263(5):114–120; 1990.

CHAPTER 8

Ahmad, S., editor. Herbivorous insects. Host-seeking behavior and mechanisms. New York: Academic Press; 1983.

Alexander, R. D. Acoustical communication in arthropods. Annual Review of Entomology 12:495–596; 1967.

Alexander, R.D.; Moore, T. E. The evolutionary relationships of 17-year and 13-year cicadas, and three new species (Homoptera, Cicadidae, Magicicada). University of Michigan, Misc. Publ. Mus. Zool. No. 121; 1962.

Alexander, R. D. Natural selection and specialized chorusing behavior. In: Pimentel, D., ed. Insects, science and society. New York: Academic Press; 1975:p. 35–77.

Allan, S. A.; Day, J. F.; Edman, J. D. Visual ecology of biting flies. Annual Review of Entomology 32:297–316; 1987.

Alloway, T. M. Learning and memory in insects. Annual Review of Entomology 17:43–56; 1972.

Alloway, T. M. Learning in insects except *Apoidea*. In: Corning, W. C.; Dyal, J. A.; Willows, A. O. D., eds. Invertebrate learning. New York: Plenum; 1973:p. 131–171, vol. 2.

Andersson, M. The evolution of eusociality. Annual Review of Ecology and Systematics 15:165–189; 1984.

Andrewartha, H. G.; Miethke, P. M.; Wells, A. Induction of diapause in the pupa of *Phalaenoides glycinae* by a hormone from the subesophageal ganglion. Journal of Insect Physiology 20:679–701; 1974.

Andrews, C. The lives of wasps and bees. New York: American Elsevier; 1971.

Askew, R. R. Parasitic insects. New York: American Elsevier; 1971.

Atkins, M. D. Introduction to insect behavior. New York: Macmillan; 1980.

Attygalle, A. B.; Morgan, E. D. Ant trail pheromones. Advances in Insect Physiology 18:1–29; 1985.

Baker, R. R. Insect territoriality. Annual Review of Entomology 28:65–90; 1983.

Barash, D. P. Sociobiology and behavior. New York: Elsevier North-Holland; 1977.

Barbosa, P.; Letourneau, D. K., editors. Novel aspects of insect-plant interactions. New York: John Wiley & Sons, Inc.; 1988.

Bartell, R. J. Behavioral responses of Lepidoptera to pheromones. In: Shorey, H. H.; McKelvey, J. J., Jr., eds. Chemical control of insect behavior. New York: John Wiley & Sons; 1977:p. 201–213.

Barth, R. H.; Lester, L. J. Neuro-hormonal control of sexual behavior in insects. Annual Review of Entomology 18:455–472; 1973.

Barton Browne, L. The experimental analysis of insect behavior. New York: Springer-Verlag; 1974.

Barton Browne, L. Regulatory mechanisms in insect feeding. Advances in Insect Physiology 11:1–116; 1975.

Bastock, M. A gene mutation which changes a behavior pattern. Evolution 10:421–439; 1956.

Batra, S. W. T.; Batra, L. R. The fungus gardens of insects. Scientific American 217:112–120; 1967.

Beach, R. Mosquitoes: biting behavior inhibited by ecdysone. Science 205:829–831; 1979.

Beament, J. W. L.; Treherne, J. E., editors. Insects and physiology. London: Oliver & Boyd; 1967.

Beck, S. D. Insect photoperiodism. New York: Academic Press; 1968.

Beck, S. D. Insect thermoperiodism. Annual Review of Entomology 28:91–108; 1983.

Bell, W. J. Searching behavior patterns in insects. Annual Review of Entomology 35:447–468; 1990.

Bell, W. J. Searching behaviour—The behavioural ecology of finding resources. New York: Chapman Hall; 1991.

Bell, W. J.; Cardé, R. T., editors. Chemical ecology of insects. New York: Chapman Hall; 1984.

Bentley, D. R. Single gene cricket mutations: effects on behavior, sensilla, sensory neurons, and identified interneurons. Science 187:760–764; 1975.

Bentley, D. R.; Hoy, R. Genetic control of the neuronal network generating cricket (*Telogryllus gryllus*) song patterns. Animal Behavior 20:478–492; 1972.

Bentley, D. R.; Hoy, R. The neurobiology of cricket song. Scientific American 231:34–44; 1975.

Bentley, M. D.; Day, J. F. Chemical ecology and behavioral aspects of mosquito oviposition. Annual Review of Entomology 34:401–422; 1989.

Benzer, S. Genetic dissection of behavior. Scientific American 229:24–37; 1973.

Bernays, E. A.; Simpson, S. J. Control of food intake. Advances in Insect Physiology 16:59–118; 1982.

Beroza, M., editor. Chemicals controlling insect behavior. New York: Academic Press; 1970.

Birch, M. C., editor. Pheromones. Amsterdam: North-Holland; 1974.

Birukow, G. Orientation behavior in insects and factors which influence it. In: Haskell, P. T., ed. Insect behavior. London: Royal Entomological Society of London, Symposium 3; 1966:p. 2–12.

Blum, M. S. Alarm pheromones. Annual Review of Entomology 14:47–80; 1969.

Blum, M. S; Brand, J. M. Social insect pheromones: their chemistry and function. American Zoologist 12:5553–5576; 1972.

Blum, M. S. Biochemical defenses of insects. In: Rockstein, M., ed. Biochemistry of insects. New York: Academic Press; 1978:p. 466–513.

Blum, M. S.; Blum, N. A. Sexual selection and reproductive competition in insects. New York: Academic Press; 1979.

Borden, J. H. Behavioral responses of Coleoptera to pheromones, allomones, and kairomones. In: Shorey, H. H.; McKelvey, J. J., Jr., eds. Chemical control of insect behavior. New York: John Wiley & Sons; 1977:p. 169–198.

Bornemissza, G. F. Sex attractant of male scorpionflies. Nature 203:786–787; 1964.

Bowen, M. F. The sensory physiology of host-seeking behavior in mosquitoes. Annual Review of Entomology 36:139–158; 1991.

Brady, J. The physiology of insect circadian rhythms. Advances in Insect Physiology 10:1–115; 1974.

Brower, L. P. Monarch migration. Natural History 86:40–53; 1977.

Brown, W. L.; Eisner, T.; Whittaker, R. H. Allomones and kairomones: transpecific chemical messages. BioScience 20:21–22; 1970.

Brues, C. T. Insect dietary. Cambridge, MA: Harvard University Press; 1946.

Buck, J. B.; Buck, E. Mechanisms of rhythmic synchronous flashing of fireflies. Science 159:1319–1328; 1968.

Buck, J. B.; Buck, E. Synchronous fireflies. Scientific American 234:74–85; 1976.

Bulla, L. A., editor. Regulation of insect populations by microorganisms. Annals of the New York Academy of Science, vol. 217, 1973.

Bünning, E. The physiological clock. 2nd ed. rev. New York: Springer-Verlag; 1967.

Bursell, E. Environmental aspects—temperature. In: Rockstein, M., ed. The physiology of insecta. New York: Academic Press; 1974a:p. 2–41.

Bursell, E. Environmental aspects—humidity. In: Rockstein, M., ed. The physiology of insecta. 2nd ed. New York: Academic Press; 1974b:p. 44–84, vol. 1.

Butenandt, A.; Beckmann, R.; Stamm, D; Hecker, E. Über den Sexual-Lockstoff des Seidenspinners *Bombyx mori*. Reindarstellung und Konstitution. Zeitschrift fuer Naturforschung 14:283–284; 1959.

Carew, T. J.; Sahley, C. L. Invertebrate learning and memory: from behavior to molecules. Annual Review of Neuroscience 9:435–487; 1986.

Carthy, J. D. The behavior of arthropods. San Francisco: Freeman; 1965.

Claridge, M. F. Acoustic signals in the Homoptera: behavior, taxonomy, and evolution. Annual Review of Entomology 30:297–318; 1985.

Clausen, C. P. Phoresy among entomophagous insects. Annual Review of Entomology 21:343–368; 1976.

Corbet, P. S. The role of rhythms in insect behavior. In: Haskell, P. T., ed. Insect behavior. London: Royal Entomological Society of London, Symposium 3; 1966:p. 13–28.

Danilevskii, A. S. Photoperiodism and seasonal development of insects. Edinburgh, Scotland: Oliver & Boyd; 1965.

Danks, H. V. Insect dormancy: An ecological perspective. Ottawa, Canada: Biological Survey of Canada Monograph. series no. 1. 1987.

Dawkins, R. The selfish gene. Oxford: Oxford University Press; 1976.

Denlinger, D. L. Dormancy in tropical insects. Annual Review of Entomology 31:239–264; 1986.

Dethier, V. G. Feeding behavior. In: Haskell, P. T., ed. Insect behavior. London: Royal Entomological Society of London, Symposium 3; 1966:p. 46–58.

Dethier, V. G. Chemical interaction between plants and insects. In: Sondheimer, E.; Simeone, J. B., eds. Chemical ecology. New York: Academic Press; 1970:p. 83–102.

Dethier, V. G. The hungry fly. A physiological study of the behavior associated with feeding. Cambridge, MA: Harvard University Press; 1976.

Dingle, H. Migration strategies of insects. Science 175:1327–1335; 1972.

Dingle, H., editor. Evolution of insect migration and diapause. New York: Springer-Verlag; 1978a.

Dingle, H. Migration and diapause in tropical, temperate, and island milkweed bugs. In: Dingle, H., ed. Evolution of insect migration and diapause. New York: Springer-Verlag; 1978b:p. 254–276.

Drake, V. A.; Farrow, R. A. The influence of atmospheric structure and motions on insect migration. Annual Review of Entomology 33:183–210; 1988.

Dudai, Y. Neurogenetic dissection of learning and short-term memory in *Drosophila*. Annual Review of Neuroscience 11:537–563; 1988.

Duffey, S. S. Arthropod allomones: chemical effronteries and antagonists. Proceedings of the XV International Congress of Entomology; 1977:p. 323–394.

Eberhard, W. G. Sexual selection and animal genitalia. Cambridge, MA: Harvard University Press; 1985.

Ebling, J.; Highnam, K. C. Chemical communication. Studies in biology, No. 19. New York: Crane Russak; 1970.

Ehrman, L.; Parsons, P. A. The genetics of behavior. Sunderland, MA: Sinauer Associates; 1976.

Eisenstein, E. M.; Cohen, M. J. Learning in an isolated insect ganglion. Animal Behavior 13:104–108; 1966.

Eisenstein, E. M.; Reep, R. L. Behavioral and cellular studies of learning and memory in insects. In: Kerkut, G. A.; Gilbert, L. I., eds. Comprehensive insect physiology, biochemistry, and pharmacology. New York: Pergamon Press; 1985:p. 513–549, vol. 9.

Eisner, T.; Meinwald, Y. C. Defensive secretion of a caterpillar. Science 150:1733–1735; 1966.

Eisner, T. Chemical defense against predation in arthropods. In: Sondheimer, E.; Simeone, J. B., eds. Chemical ecology. New York: Academic Press; 1970:p. 157–217.

Eisner, T.; Hicks, K.; Eisner, M.; Robson, D. S. "Wolf-in-sheep's-clothing" strategy of a predaceous insect larva. Science 199:790–793; 1978.

Eisner, T.; Meinwald, Y. C. Defensive secretion of a caterpillar. Science 150:1733–1735; 1965.

Eisner, T.; van Tassell, E.; Carrel, J. E. Defensive use of a "fecal shield" by a beetle larva. Science 158:1471–1473; 1967.

Elsner, N.; Popov, A. V. Neuroethology of acoustic communication. Advances in Insect Physiology 13:229–356; 1978.

Erber, J. The dynamics of learning in the honeybee. Journal of Comparative Physiology 99:231–255; 1975.

Evans, D. L.; Schmidt, J. O., editors. Insect defenses: adaptive mechanisms and strategies of prey and predators. Ithaca, NY: State University of New York Press; 1991.

Evans, H. E. Studies on the comparative ethology of digger wasps of the genus *Bembix*. Ithaca, NY: Comstock; 1957.

Evans, H. E.; Eberhard, M. T. W. The wasps. Ann Arbor, MI: University of Michigan Press; 1970.

Evans, P. D. Octopamine. In: Kerkut, G. A.; Gilbert, L. I., eds. Comprehensive insect physiology, biochemistry, and pharmacology. New York: Pergamon Press; 1985:p. 499–530, vol 11.

Ewing, A. W. Communication in Diptera. In: Sebeok, T. A., ed. How animals communicate. Bloomington, IN: Indiana University Press; 1977:p. 403–417.

Ewing, A. W.; Manning, A. The evolution and genetics of insect behavior. Annual Review of Entomology 12:471–494; 1967.

Ewing, A. W. Arthropod bioacoustics: neurobiology and behavior. Ithaca, NY: Cornell University Press; 1990.

Fabre, J. H. Souvenirs entomologiques. 10 Volumes. Paris, France: Delagrave; 1879–1908. (Translations by A. T. de Mattos later published by Garden City Publishing Co., New York).

Fletcher, B. S. Behavioral responses of Diptera to pheromones, allomones, and kairomones. In: Shorey, H. H.; McKelvey, J. J., Jr., eds. Chemical control of insect behavior. New York: John Wiley & Sons; 1977:p. 129–148.

Fraenkel, G. S.; Gunn, D. L. The orientation of animals. New York: Dover (reprint of 1940 Oxford edition); 1961.

Free, J. B. The social organization of honey bees. London: Edward Arnold; 1977.

Friend, W. G.; Smith, J. J. B. Factors affecting feeding by bloodsucking insects. Annual Review of Entomology 22:309–331; 1977.

Frings, H.; Frings, M. Animal communication. 2nd ed. rev. Norman, OK: University of Oklahoma Press; 1977.

Fukuda, S.; Takeuchi, S. Diapause factor-producing cells in the subesophageal ganglion of the silkworm, *Bombyx mori* L. Proceedings of the Japanese Academy of Science 43:41–56; 1967.

Fuller, J. L.; Thompson, W. R. Behavior genetics. New York: John Wiley & Sons; 1960.

Galun, R. The physiology of hematophagous insect/animal host relationships. Proceedings of the XV International Congress of Entomology; 1977a:p. 257–265.

Galun, R. Responses of bloodsucking arthropods to vertebrate hosts. In: Shorey, H. H.; McKelvey, J. J., eds. Chemical control of insect behavior—theory and application. New York: John Wiley & Sons: 1977b:p. 103–115.

Gamboa, G. J.; Reeve, H. K; Pfennig, D. W. The evolution and ontogeny of nestmate recognition in social wasps. Annual Review of Entomology 31:431–454; 1986.

Gelperin, A. Regulation of feeding. Annual Review of Entomology 16:365–378; 1971.

Goetsch, W. The ants. Ann Arbor, MI: University of Michigan Press; 1957.

Goldsworthy, G. J.; Wheeler, C. H. Insect flight. Boca Raton, FL: CRC Press, Inc.; 1989.

Gould, J. L. Ethology—the mechanisms and evolution of behavior. New York: W. W. Norton & Co.; 1982.

Guthrie, D. M., editor. Aims and methods in neurobiology. Manchester, England: Manchester University Press; 1987.

Hailman, J. P. Communication by reflected light. In: Sebeok, T. A., ed. How animals communicate. Bloomington, IN: Indiana University Press; 1977:p. 184–210.

Hall, J. C. Genetics of circadian rhythms. Annual Review of Genetics 24:659–697; 1990.

Hall, J. C.; Rosbash, M. Mutations and molecules influencing biological rhythms. Annual Review of Neuroscience 11:343–393; 1988.

Hamilton, W. D. The genetical theory of social behavior, I and II. Journal of Theoretical Biology 7:1–52; 1964.

Hansell, M. Ethology. In: Kerkut, G. A.; Gilbert, L. I., eds. Comprehensive insect physiology, biochemistry, and pharmacology. New York: Pergamon Press; 1985:p. 1–93, chapter 9.

Haskell, P. T. Stridulation and associated behavior in certain Orthoptera. 3. The influence of the gonads. Animal Behavior 8:76–81; 1960.

Haskell, P. T. Sound production. In: Rockstein, M., ed. The physiology of insecta. 2nd ed. New York: Academic Press; 1974:p. 354–410.

Heinrich, B. Thermoregulation in endothermic insects. Science 185:747–756; 1974.

Heinrich, B.; Bartholomew, G. A. Temperature control in flying moths. Scientific American 226:70–77; 1972.

Hermann, H. R. Social insects. Vol. 1. New York: Academic Press; 1979.

Hirsch, J., editor. Behavior—genetic analysis. New York: McGraw-Hill; 1967.

Hirsch, J.; Erlenmeyer-Kimling, L. Sign of taxis as a property of the genotype. Science 134:835–836; 1961.

Hirsch, J.; Erlenmeyer-Kimling, L. Studies in experimental behavioral genetics. IV. Chromosome analyses for geotaxis. Journal of Comparative and Physiological Psychology 55:732–739; 1962.

Hocking, B. Blood-sucking behavior of terrestrial arthropods. Annual Review of Entomology 16:1–26; 1971.

Hölldobler, B. Communication in social Hymenoptera. In: Sebeok, A., ed. How animals communicate. Bloomington, IN: Indiana University Press; 1977:p. 418–471.

Hölldobler, B.; Wilson, E. O. The ants. Cambridge, MA: Harvard University Press; 1990.

Horridge, G. A. Learning of leg position by headless insects. London: Nature 193:697–698; 1962b.

Horridge, G. A. Learning of leg position by the ventral nerve cord of headless insects. London: Proceedings of the Royal Society (B) 157:33–52; 1962a.

Hotta, Y.; Benzer, S. Mapping of behavior in *Drosophila* mosaics. London: Nature 240:527–535; 1972.

Howard, R. W.; Blomquist, G. J. Chemical ecology and biochemistry of insect hydrocarbons. Annual Review of Entomology 27:149–172; 1982.

Howse, P. E. Brain structure and behavior in insects. Annual Review of Entomology 20:359–379; 1975.

Hoy, R. R. Genetic control of acoustic behavior in crickets. American Zoologist 14:1067–1080; 1974.

Hoy, R. R.; Paul, R. C. Genetic control of song specificity in crickets. Science 180:82–83; 1972.

Hoyle, G. Approaches to understanding the neurophysiological bases of behavior. In: Fentriss, J. C., ed. Simpler networks and behavior. Sunderland, MA: Sinauer Assoc., Inc.; 1976:p. 21–38.

Hoyle, G. Cellular mechanisms underlying behavior—neuroethology. Advances in Insect Physiology 7:349–444; 1970.

Huber, F. Neural integration (central nervous system). In: Rockstein, M., ed. The physiology of insecta. 2nd ed. New York: Academic Press; 1974:p. 4–100, vol. IV.

Huber, F. Invertebrate neuroethology: guiding principles. In: Camhi, J. M., ed. Invertebrate neuroethology. Experientia (Basel) 44:428–431; 1988.

Huber, F.; Moore, T. E.; Loher, W., editors. Cricket behavior and neurobiology. Ithaca, NY: Cornell University Press; 1989.

Huber, I.; Masler, E. P.; Rao, B. R., editors. Cockroaches as models for neurobiology: applications in biomedical research. Vols. I & II. Boca Raton, FL: CRC Press, Inc.; 1990.

Huber, F.; Thorson, J. Cricket auditory communication. Scientific American 253:60–68; 1985.

Huettel, M., editor. Evolutionary genetics of invertebrate behavior. New York: Plenum Press; 1986.

Huheey, J. E. Warning coloration and mimicry. In: Bell, W. J.; Cardé, R. T., eds. Chemical ecology of insects. New York: Chapman Hall; 1984:p. 257–300.

Ishay, J. Acoustical communication in wasp colonies (Vespinae). Proceedings of the XV International Congress of Entomology; 1977:p. 406–435.

Jacobson, M. Insect sex attractants. New York: Wiley-Interscience; 1965.

Jacobson, M. Insect sex pheromones. New York: Academic Press; 1972.

Jacobson, M. Insect pheromones. In: Rockstein, M., ed. The physiology of insecta. 2nd ed. New York: Academic Press; 1974:p. 229–276, vol. III.

Jander, R. Insect orientation. Annual Review of Entomology 8:94–114; 1963.

Jander, R. Ecological aspects of spatial orientation. Annual Review of Ecology and Systematics 6:171–188; 1975.

Jeanne, R. L. Evolution of social behavior in the Vespidae. Annual Review of Entomology 25:371–396; 1980.

Johnson, C. G. A functional system of adaptive dispersal by flight. Annual Review of Entomology 11:233–260; 1966.

Johnson, C. G. Insect migration. Carolina Biology Reader, No. 84. Burlington, NC: Carolina Biological Supply Co.; 1976.

Johnson, C. G. The aerial migration of insects. Scientific American 209:132–138; 1963.

Johnson, C. G. Migration and dispersal of insects by flight. London: Methuen; 1969.

Johnson, C. G. Insect migration: aspects of its physiology. In: Rockstein, M., ed. The physiology of insecta. 2nd ed. New York: Academic Press; 1974:p. 279–334, vol III.

Jones, J. C. The feeding behavior of mosquitoes. Scientific American 238:138–148; 1978.

Kalmring, K.; Elsner, N., editors. Acoustic and vibrational communication in insects. New York: Paul Parey Scientific Publications; 1985.

Kandel, E. R. Nerve cells and behavior. Scientific American 223:57–70; 1970.

Karlson, P.; Butenandt, A. Pheromones (ectohormones) in insects. Annual Review of Entomology 4:39–58; 1959.

Kennedy, J. S. Insect dispersal. In: Pimentel, D., ed. Insects, science and society. New York: Academic Press; 1975:p. 103–119.

Kennedy, J. S. Olfactory responses to distant plants and other odor sources. In: Shorey, H. H.; McKelvey, J. J., Jr., eds. Chemical control of insect behavior—theory and application. New York: John Wiley & Sons; 1977:p. 67–91.

Kennedy, J. S. Animal motivation: the beginning of the end? In: Chapman, R. F.; Bernays, E. A.; Stoffolano, J. G., Jr., eds. Perspectives in chemoreception and behavior. New York: Springer-Verlag, 1987:p. 17–31.

Kerkut, G. A.; Gilbert, L. I., editors. Comprehensive insect physiology, biochemistry, and pharmacology. New York: Pergamon Press; 1985.

Kogan, M. The role of chemical factors in insect/plant relationships. Proceedings of the XV International Congress of Entomology; 1977:p. 103–146.

Kring, J. B. Flight behavior of aphids. Annual Review of Entomology 17:461–492; 1972.

Krishna, K.; Weesner, F. M., editors. Biology of termites. New York: Academic Press; 1969:p. 598, vol. 1.

Krishna, K.; Weesner, F. M., editors. Biology of termites. New York: Academic Press; 1970:p. 643, vol. 2.

Law, J. H.; Regnier, F. E. Pheromones. Annual Review of Biochemistry 40:533–548; 1971.

Lees, A. D. The physiology of diapause in arthropods. Cambridge, England: Cambridge University Press; 1955.

Lewis, T., editor. Insect communication. New York: Academic Press; 1984.

Lindauer, M. Social behavior and mutual communication. In: Rockstein, M., ed. The physiology of insecta. 2nd ed. New York: Academic Press; 1974:p. 149–228, vol. III.

Lloyd, J. E. Studies on the flash communication system in *Photinus* fireflies. University of Michigan: Miscellaneous Publications Museum of Zoology 1966:p. 95, No. 130.

Lloyd, J. E. Bioluminescent communication in insects. Annual Review of Entomology 16:97–122; 1971.

Lloyd, J. E. Bioluminescence and communication. In: Sebeok, T. A., ed. How animals communicate. Bloomington, IN: Indiana University Press; 1977:p. 164–183.

Lloyd, J. E. Bioluminescence and communication in insects. Annual Review of Entomology 28:131–160; 1983.

Lorenz, K. Z. King Solomon's ring: new light on animal ways. New York: T. Y. Crowell; 1952.

Lüscher, M. Air-conditioned termite nests. Scientific American 205:138–145; 1961.

Lynch, G.; McGaugh, J. L., Weinberger, N. M., editors. Neurobiology of learning and memory. New York: Guilford; 1984.

Manning, A. Sexual behavior. In: Haskell, P. T., ed. Insect behavior. London: Royal Entomological Society, Symposium 3; 1966:p. 59–68.

Mansingh, A. Physiological classification of dormancies in insects. Canadian Entomologist 103:983–1009; 1971.

Markl, H. Insect behavior: functions and mechanisms. In: Rockstein, M., ed. The physiology of insecta. 2nd ed. New York: Academic Press; 1974:p. 3–148, vol. III.

Markl, H.; Lindauer, M. Physiology of insect behavior. In: Rockstein, M., ed. The physiology of insecta. New York: Academic Press; 1965:p. 3–122, vol. II.

Masaki, S. Summer diapause. Annual Review of Entomology 25:1–26; 1980.

Matthews, J. R.; Matthews, R. W., editors. Insect behavior—a sourcebook of laboratory and field exercises. Boulder, CO: Westview Press; 1982.

Matthews, R. W.; Matthews, J. R. Insect behavior. New York: John Wiley & Sons; 1978:p. 507.

May, M. L. Insect thermoregulation. Annual Review of Entomology 24:313–349; 1979.

McClearn, G. E.; DeFries, J. C. Introduction to behavior genetics. San Francisco: Freeman; 1973.

McElroy, W. D.; Seliger, H. H. Biological luminescence. Scientific American 207:76–89; 1962.

McElroy, W. D.; Seliger, H. H.; DeLuca, M. Insect bioluminescence. In: Rockstein, M., ed. The physiology of insecta. 2nd ed. New York: Academic Press; 1974:p. 411–460, vol. II.

McGuire, T. R. Learning in three species of Diptera: the blow fly, *Phormia regina,* the fruit fly, *Drosophila melanogaster,* and the house fly, *Musca domestica.* Behavioral Genetics 14:479–526; 1984.

McNeil, J. N. Behavioral ecology of pheromone-mediated communication in moths and its importance in the use of pheromone traps. Annual Review of Entomology 36:407–430; 1991.

Menzel, R.; Erber, J. Learning and memory in bees. Scientific American 239:102–110; 1978.

Menzel, R.; Mercer, A., editors. Neurobiology and behavior of honeybees. New York: Springer-Verlag; 1987.

Michelsen, A.; Fink, F.; Gogala, M.; Trane, D. Plants as transmission channels for insect vibrational songs. Behavioral Ecology and Sociobiology 11:267–281; 1982.

Michener, C. D. The social behavior of bees. A comparative study. Cambridge, MA: Belknap Press/Harvard University Press; 1974:p. 404.

Milburn, N. S.; Roeder, K. D. Control of efferent activity in the cockroach terminal abdominal ganglion by extracts of the corpora cardiaca. General and Comparative Endocrinology 2:70–76; 1962.

Milburn, N. S.; Weiant, E. A.; Roeder, K. D. The release of efferent nerve activity in the roach, *Periplaneta americana*, by extracts of the corpus cardiacum. Woods Hole: Biological Bulletin of Marine Biology Laboratory 118:111–119; 1960.

Miller, P. L. The neural basis of behavior. In: Treherne, J. E., ed. Insect neurobiology. New York: American Elsevier; 1974:p. 359–430.

Mitchell, R. Insect behavior, resource exploitation, and fitness. Annual Review of Entomology 26:373–396; 1981.

Montagner, H. Learning and communication. Proceedings of the XV International Congress of Entomology; 1977:p. 397–399.

Müller, H. J. Formen der Dormanz bei Insecten. Nova Acta Leopoldina 34:1–27; 1970.

Nelson, W. A.; Keirans, J. E.; Bell, J. F.; Clifford, C. M. Host-ectoparasite relationships. Journal of Medical Entomology 12:143–166; 1975.

Oster, G. F.; Wilson, E. O. Caste ecology in the social insects. Princeton, NJ: Princeton University Press; 1978.

Otte, D. Effects and functions in the evolution of signalling systems. Annual Review of Ecology and Systematics 5:385–417; 1974.

Otte, D. Communication in orthoptera. In: Sebeok, A., ed. How animals communicate. Bloomington, IN: Indiana University Press; 1977:p. 334–361.

Page, T. L. Clocks and circadian rhythms. In: Kerkut, G. A.; Gilbert, L. I., eds. Comprehensive insect physiology, biochemistry, and pharmacology. New York: Pergamon Press; 1985:p. 577–652, vol. 6.

Papaj, D. R.; Prokopy, R. J. Ecological and evolutionary aspects of learning in phytophagous insects. Annual Review of Entomology 34:315–350; 1989.

Pasteels, J. M. Evolutionary aspects in chemical ecology and chemical communication. Proceedings of the XV International Congress of Entomology; 1977:p. 281–293.

Pasteels, J. M.; Grégoire, J.-C.; Rowell-Rathier, M. The chemical ecology of defense in arthropods. Annual Review of Entomology 28:263–290; 1983.

Plowright, R. C.; Laverty, T. M. The ecology and sociobiology of bumble bees. Annual Review of Entomology 29:175–200; 1984.

Prestwich, G. D. Defense mechanisms of termites. Annual Review of Entomology 29:201–232; 1984.

Price, P. W. Insect ecology. New York: John Wiley & Sons; 1984.

Prokopy, R. J.; Owens, E. D. Visual detection of plants by herbivorous insects. Annual Review of Entomology 28:131–160; 1983.

Quinn, W. G.; Greenspan, R. J. Learning and courtship in *Drosophila*: two stories with mutants. Annual Review of Neuroscience 7:67–93; 1984.

Rabe, W. Beitrage zum Orientierungsproblem der Wasserwanzen. Zeitschrift fuer Vergleichende Physiologie. 35:300–325; 1953.

Rainey, R. C., editor. Insect flight. Royal Entomological Society (London), Symposium 7. London: Blackwell Scientific; 1976.

Reichstein, T.; von Euw, J.; Parsons, J. A.; Rothschild, M. Heart poisons in the monarch butterfly. Science 161:861–866; 1968.

Richard, G. The historical development of nineteenth and twentieth century studies on the behavior of insects. In: Smith, R. F.; Mittler, T. E.; Smith, C. N., eds. History of entomology. Palo Alto, CA: Annual Reviews; 1973: p. 477–502.

Richards, O. W. An introduction to the study of polymorphism in insects. In: Kennedy, J. S., ed. Insect polymorphism. London: Royal Entomological Society, Symposium 1; 1953a:p. 2–10.

Richards, O. W. The social insects. New York: Harper & Row; 1953b:p. 219.

Riddiford, L. M.; Truman, J. W. Hormones and insect behavior. In: Barton Browne, L., ed. Experimental analysis of insect behavior. New York: Springer-Verlag; 1974: p. 286–296.

Roeder, K. D. Moths and ultrasound. Scientific American 212:94–102; 1965.

Roeder, K. S. Auditory system of noctuid moths. Science 154:1515–1521; 1966.

Roeder, K. D. Nerve cells and insect behavior. rev. ed. Cambridge, MA: Harvard University Press; 1967:p. 238.

Roeder, K. D. Episodes in insect brains. American Scientist 58:378–389; 1970.

Roeder, K. D. Some neuronal mechanisms of simple behavior. Advances in the Study of Behavior 4:2–46; 1974.

Roelofs, W. L. Insect communication—chemical. In: Pimentel, D., ed. Insects, science, and society. New York: Academic Press; 1975.

Roelofs, W. L.; Cardé, R. T. Responses of Lepidoptera to synthetic sex pheromone chemicals and their analogues. Annual Review of Entomology 22:377–405; 1977.

Rosenthal, G. A.; Janzen, D. H., editors. Herbivores—their interaction with secondary plant metabolites. New York: Academic Press; 1979.

Roth, L. M.; Eisner, T. Chemical defenses of arthropods. Annual Review of Entomology 7:107–136; 1962.

Rothenbuhler, W. C. Behavior genetics of nest cleaning in honey bees IV. Response of F_1 and backcross generations to disease-killed brood. American Zoologist 4:111–123; 1964.

Rothenbuhler, W. C. Genetic and evolutionary considerations of social behavior of honey bees and some related insects. In: Hirsch, J., ed. Behavior—genetic analysis. New York: McGraw-Hill; 1967:p. 61–106.

Rothschild, M. Secondary plant substances and warning coloration in insects. In: van Emden, H. F., ed. Insect/plant relationships. Royal Entomological Society (London), Symposium 6. London: Blackwell Scientific; 1972:p. 59–83.

Rusak, B.; Bina, K. G. Neurotransmitters in the mammalian circadian system. Annual Review of Neuroscience 13:387–402; 1990.

Ruse, M. Sociobiology: sense or nonsense? Dordrecht: Netherlands: Reidel; 1979.

Saunders, D. S. Circadian rhythms and photoperiodism in insects. In: Rockstein, M., ed. The physiology of insecta. 2nd ed. New York: Academic Press; 1974:p. 461–533, vol. II.

Saunders, D. S. Insect clocks. New York: Pergamon; 1976a.

Saunders, D. S. The biological clock of insects. Scientific American 234:114–121; 1976b.

Schaefer, G. W. Radar observations of insect flight. In: Rainey, R. C., ed. Insect flight. Royal Entomological Society (London), Symposium 7. London: Blackwell Scientific; 1976:p. 157–197.

Schildknecht, H.; Holoubek. Die Bombardier Käfer und ihre Explosionschemie. V. Mitteilung über insekten Abwehrstoffe. Angewandte Chemie 73(1):1–6; 1961.

Schmidt, J. M.; Smith. J. J. B. Host examination walk and oviposition site selection of Trichogramma minutum: studies on spherical hosts. Journal of Insect Physiology 2:143–171; 1989.

Schneider, D. The sex-attractant receptor of moths. Scientific American 231:28–35; 1974.

Schneider, D. Pheromone communication in moths and butterflies. In: Galun, R.; Hillman, P.; Pranas, I.; Werman, R., eds. Sensory physiology and behavior. New York: Plenum; 1975:p. 173–193.

Schneider, F. Dispersal and migration. Annual Review of Entomology 7:223–242; 1962.

Schneirla, T. C. Army ants—a study in social organization. San Francisco: Freeman; 1971.

Schoonhoven, L. M. Plant recognition by lepidopterous larvae. In: van Emden, H. F., ed. Insect/plant relationships. Royal Entomological Society (London), Symposium 6. London: Blackwell Scientific; 1972.

Schwartzkopff, J. Mechanoreception. In: Rockstein, M., ed. The physiology of insecta. 2nd ed. New York: Academic Press; 1974:p. 273–352, vol. II.

Seabrook, W. D. Neurobiological contributions to understanding insect pheromone systems. Annual Review of Entomology 23:471–485; 1978.

Sebeok, T. A., editor. Animal communication. Techniques of study and results of research. Bloomington, IN: Indiana University Press; 1968:p. 686.

Sebeok, T. A., editor. How animals communicate. Bloomington, IN: Indiana University Press; 1977:p. 1128.

Shorey, H. H. Behavioral responses to insect pheromones. Annual Review of Entomology 18:349–380; 1973.

Shorey, H. H. Animal communication by pheromones. New York: Academic Press; 1976.

Shorey, H. H. Pheromones. In: Sebeok, T. A., ed. How animals communicate. Bloomington, IN: Indiana University Press; 1977a:p. 137–163.

Shorey, H. H. The adaptiveness of pheromone communication. Proceedings of the XV International Congress of Entomology; 1977b:p. 294–307.

Shorey, H. H.; McKelvey, J. J., Jr., editors. Chemical control of insect behavior—theory and applications. New York: John Wiley & Sons; 1977.

Silberglied, R. E. Communication in the Lepidoptera. In: Sebeok, T. A., ed. Bloomington, IN: Indiana University Press; 1977:p. 362–402.

Skaife, S. H. Dwellers in darkness. New York: Doubleday; 1961.

Slansky, F., Jr.; Rodriguez, J. G., editors. Nutritional ecology of insects, mites, spiders, and related invertebrates. New York: John Wiley & Sons; 1987.

Smith, K. G. V., editor. Insects and other arthropods of medical importance. London: British Museum (Natural History); 1973.

Smith, R. L., editor. Sperm competition and the evolution of animal mating systems. New York: Academic Press; 1984.

Smith, W. J. The behavior of communicating—an ethological approach. Cambridge, MA: Harvard University Press; 1977.

Spangler, H. G. Moth hearing, defense, and communication. Annual Review of Entomology 33:59–81; 1988.

Spradbery, J. P. Wasps. An account of the biology and natural history of social and solitary wasps. Seattle, WA: University of Washington Press; 1973:p. 416.

Staedler, E. Sensory aspects of insect plant interactions. Proceedings of the XV International Congress of Entomology; 1977:p. 228–248.

Stinner, R. E.; Barfield, C. S.; Stimac, J. L; Dohse, L. Dispersal and movement of insect pests. Annual Review of Entomology 23:319–335; 1983.

Stoffolano, J. G., Jr. Control of feeding and drinking in diapausing insects. In: Barton Browne, L., ed. Experimental analysis of insect behavior. New York: Springer-Verlag; 1974:p. 32–47.

Stuart, A. M. Some aspects of communication in termites. Proceedings of the XV International Congress of Entomology; 1977:p. 400–405.

Sturm, H. Die Paarung beim Silberfischen, Lepisma saccharina Zeitschrift fuer Tierphysiologie, Tierernaehrung, und Futtermittelkunde 13:1–2; 1956.

Sudd, J. H. An introduction to the behavior of ants. New York: St. Martin's Press; 1967:p. 200.

Sullivan, D. J. Insect hyperparasitism. Annual Review of Entomology 32:49–70; 1987.

Suzuki, Y.; Tsuji, H.; Sasakawa, M. Sex allocation and effects of superparasitism on secondary sex ratios in the gregarious parasitoid, Trichogramma chilonis (Hymenoptera: Trichogrammatidae). Animal Behavior 32:478–484; 1984.

Tallamy, D. W.; Wood, T. K. Convergence patterns in subsocial insects. Annual Review of Entomology 31:369–390; 1986.

Tauber, M. J; Tauber, C. A. Insect seasonality: diapause maintenance, termination, and postdiapause development. Annual Review of Entomology 21:81–107; 1976.

Thiele, H. U. Remarks about Mansingh's and Müller's classification of dormancies in insects. Canadian Entomologist 105:925–928; 1973.

Thompson, J. N.; Pellmyr, O. Evolution of oviposition behavior and host preference in Lepidoptera. Annual Review of Entomology 36:65–90; 1991.

Thornhill, R.; Alcock, J., editors. The evolution of insect mating systems. Cambridge, MA: Harvard University Press; 1983.

Thorpe, W. H. A note on detour experiments with *Ammophila pubescens*. Curt. Behavior 12:257–263; 1950.

Thorpe, W. H. Learning and instinct in animals. 2nd ed. London: Methuen; 1963.

Tinbergen, N. The study of instinct. London: Clarendon Press of Oxford University Press; 1951.

Tinbergen, N. The animal in its world. Explorations of an ethologist, 1932–1972. Vol. 1. Field studies. Cambridge, MA: Harvard University Press; 1972.

Traniello, J. F. A. Foraging strategies of ants. Annual Review of Entomology 34:191–210; 1989.

Trivers, R. L.; Hare, H. Haplodiploidy and the evolution of social insects. Science 191:249–263; 1976.

Truman, J. How moths "turn on:" a study of the action of hormones on the nervous system. American Naturalist 61:700–706; 1973.

Truman, J. W.; Riddiford, L. M. Hormonal mechanisms underlying insect behavior. Advances in Insect Physiology 10:297–352; 1974.

Tully, T. *Drosophila* learning: behavior and biochemistry. Behavioral Genetics 14:527–557; 1984.

Tully, T.; Hirsch, J. Behavior-genetic analysis of *Phormia regina*. I. Isolation of pure-breeding lines for high and low levels of the central excitatory state (CES) from an unselected population. Behavioral Genetics 12:395–415; 1982.

Urquhart, F. A. Found at last: the monarch's winter home. National Geographic 150:161–173; 1976.

van Alphen, J. J. M.; Visser, M. E. Superparasitism as an adaptive strategy for insect parasitoids. Annual Review of Entomology 35:59–80; 1990.

Vinson, S. B. Host selection by insect parasitoids. Annual Review of Entomology 21:109–133; 1976.

Visser, J. H. Host odor perception in phytophagous insects. Annual Review of Entomology 31:121–144; 1986.

von Frisch, K. Bees—their vision, chemical senses, and language. Ithaca, NY: Cornell University Press; 1950.

von Frisch, K. The dance, language, and orientation of bees. Cambridge, MA: Harvard University Press; 1967. Translated by L. E. Chadwick.

von Frisch, K. Bees—their vision, chemical senses, and language. Ithaca, NY: Cornell University Press; 1971.

Waage, J. K. Dual function of the damselfly penis: sperm removal and transfer. Science 203:916–918; 1979.

Waldbauer, G. P.; Friedman, S. Self-selection of optimal diets by insects. Annual Review of Entomology 36:43–64; 1991.

Ward, J. V.; Stanford, J. A. Thermal responses in the evolutionary ecology of aquatic insects. Annual Review of Entomology 27:97–117; 1982.

Watson, J. D. The double helix. New York: New American Library; 1968.

Weaver, N. Chemical control of behavior—intraspecific. In: Rockstein, M., ed. Biochemistry of Insects. New York: Academic Press; 1978a:p. 360–389.

Weaver, N. Chemical control of behavior—interspecifc. In: Rockstein, M., ed. Biochemistry of Insects. New York: Academic Press; 1978b:p. 392–418.

Weber, N. A. Fungus-growing ants. Science 153:587–609; 1966.

Weber, N. A. Gardening ants—the attines. Philadelphia: American Philosophical Society; 1972:p. 146.

Wehner, R. Polarized-light navigation by insects. Scientific American 235:106–115; 1976.

Wehner, R. Astronavigation in insects. Annual Review of Entomology 29:277–298; 1984.

Wehner, R.; Menzel, R. Do insects have cognitive maps? Annual Review of Neuroscience 13:403–414; 1990.

Wells, P. H. Honey bees. In: Corning, W. C.; Dyal, J. A.; Willows, O.D., eds. Invertebrate learning. New York: Plenum; 1973:p. 173–185.

Wenner, A. M.; Wells, P. H. Anatomy of a controversy: the question of a "language" among bees. New York: Columbia Univ. Press; 1990.

Wheeler, D. A.; Kyriacou, C. P.; Greenacre, M. L.; Yu, Q.; Rutila, J. E.; Rosbash, M.; Hall, J. C. Molecular transfer of a species-specific behavior from *Drosophila simulans* to *Drosophila melanogaster*. USA: Science 251:1082–1085; 1991.

Wheeler, W. M. Ants. New York: Columbia University Press; 1910:p. 663.

Whittaker, R. H.; Feeney, P. P. Allelochemics: chemical interactions between species. Science 171:757–770; 1971.

Williams, C. M. Physiology of insect diapause. I–IV. Biological Bulletin 90:234–243; 93:90–98; 94:60–65; 103:120–138; 1946–1953.

Wilson, D. M. Genetic and sensory mechanisms for locomotion and orientation in animals. American Scientist 60:358–365; 1972.

Wilson, E. O. Pheromones. Scientific American 208:100–114; 1963.

Wilson, E. O. Chemical communication in the social insects. Science 149:1064–1071; 1965.

Wilson, E. O. Chemical communication within animal species. In: Sondheimer, E.; Simeone, J. B., eds. Chemical ecology. New York: Academic Press; 1970:p. 133–155.

Wilson, E. O. The insect societies. Cambridge, MA: Belknap Press/Harvard University Press; 1971.

Wilson, E. O. Sociobiology—the new synthesis. Cambridge, MA: Belknap Press/Harvard University Press; 1975.

Winston, M. The biology of the honey bee. Cambridge, MA: Harvard University Press; 1987.

Zacharuk, R. Y.; Shields, V. D. Sensilla of immature insects. Annual Review of Entomology 36:331–354; 1991.

Zeh, D. W.; Zeh, J. A.; Smith, R. L. Ovipositors, amnions, and eggshell architecture in the diversification of terrestrial arthropods. Quarterly Review of Biology 64:147–168; 1989.

Zumpt, F. Myiasis in man and animals in the Old World. London: Butterworth; 1965.

CHAPTER 9

Adam, G. Plant virus studies in insect vector cell cultures. In: Mayo, M. A.; Harrap, K. A., eds. Vectors in virus biology. New York: Academic Press; 1984:p. 37–62.

Ahmadjian, V.; Paracer, S. Symbiosis—an introduction to biological associations. Hanover, NH: University Press of New England; 1986.

Alstad, D. N.; Edmunds, G. F., Jr. Effects of air pollutants on insect populations. Annual Review of Entomology 27:369–384; 1982.

Andrewartha, H. G.; Birch, L. C. The distribution and abundance of animals. Chicago: University of Chicago Press; 1954.

Andrewartha, H. G.; Birch, L. C. The history of insect ecology. In: Smith, R. F.; Mittler, T. E.; Smith, C. N., eds. History of entomology. Palo Alto, CA: Annual Reviews; 1973:p. 229–266.

Asahina, E. Freezing and frost resistance in insects. In: Meryman, H. T., ed. Cryobiology. London: Academic Press; 1966.

Baker, E. W.; Evans, T. M.; Gould, D. J.; Hull, W. B.; Keegan, H. L. A manual of parasitic mites of medical or economic importance. New York: National Pest Control Association; 1956.

Bates, H. W. Contributions to the insect fauna of Amazon Valley. Transactions of the Linnean Society (London) 23:495–566; 1862.

Baust, J. G.; Morrissey, R. E. Strategies of low temperature adaptation. Proceedings of the XV International Congress of Entomology; 1977:p. 173–184.

Begon, M.; Harper, J. L.; Townsend, C. R. Ecology—individuals, populations, and communities. London: Blackwell Scientific Publications; 1990.

Bishop, J. A.; Cook, L. M. Moths, melanism and clean air. Scientific American 232:90–99; 1975.

Blest, A. D. The function of eyespot patterns in the Lepidoptera. Behaviour 11:209–256; 1957.

Boucher, D. H. The ecology of mutualism. Annual Review of Entomology 13:315–417; 1982.

Boucher, D. H., editor. The biology of mutualism—ecology and evolution. New York: Oxford University Press; 1985.

Bowman, H. G.; Hultmark, D. Cell-free immunity in insects. Annual Review of Microbiology 41:103–126; 1987.

Brooks, W. M. Protozoan infections. In: Cantwell, G. E., ed. Insect diseases. New York: Marcel Dekker; 1974:p. 237–300.

Brower, L. P. Ecological chemistry. Scientific American 220(2):22–29; 1969.

Brower, L. P.; Brower, J. V. Z. Birds, butterflies, and plant poisons: A study in ecological chemistry. Zoologica [New York] 49:137–159; 1964.

Brower, L. P.; Glazier, S. C. Localization of heart poisons in the monarch butterfly. Science 188:19–25; 1975.

Brower, L. P.; Ryerson, W. N.; Coppinger, L. L.; Glazier, S. C. Ecological chemistry and the palatability spectrum. Science 161:1349–1351; 1968.

Bursell, E. Environmental aspects—temperature. In: Rockstein, M., ed. The physiology of insecta. 2nd ed. New York: Academic Press; 1974a:p .2–41, vol. I.

Bursell, E. Environmental aspects—humidity. In: Rockstein, M., ed. The physiology of insecta. 2nd ed. New York: Academic Press; 1974b:p. 44–84, vol. I.

Calvert, W. H.; Hedrick, L. E.; Brower, L. P. Mortality of the monarch butterfly (Danaus plexippus L.): avian predation at five overwintering sites in Mexico. Science 204:847–851; 1979.

Cantwell, G. E., editor. Insect diseases. New York: Marcel Dekker; 1974. 2 Vol.

Carruthers, R. I.; Soper, R. S. Fungal diseases. In: Fuxa, J. R.; Tanada, Y., eds. Epizootiology of insect diseases. New York: John Wiley & Sons; 1987:p. 357–416.

Chauvin, R. The world of an insect. New York: McGraw-Hill Book Company; 1967. Translated from the French by H. Oldroyd.

Cheng, L., editor. Marine insects. Amsterdam: North-Holland; 1976.

Cherrett, J. M.; Powell, R. J.; Stradling, D. J. The mutualism between leaf-cutting ants and their fungus. In: Wilding, N., et al., eds. Insect-fungus interactions. New York: Academic Press; 1987:p. 93–120.

Clark, L. R.; Geier, P. W.; Hughes, R. D.; Morris, R. F. The ecology of insect populations in theory and practice. London: Methuen; 1967.

Clausen, C. P. Phoresy among entomophagous insects. Annual Review of Entomology 21:343–368; 1976.

Clements, A. N. The physiology of mosquitoes. New York: Macmillan; 1963.

Cleveland, L. R.; Hall, S. R.; Saunders, E. P.; Collier, J. The wood-feeding roach, Cryptocerus, its protozoa, and the symbiosis between protozoa and roach. Memoirs of the American Academy of Science 17:185–342; 1934.

Cloudsley-Thompson, J. L. Adaptations of arthropods to arid environments. Annual Review of Entomology 20:261–283; 1975.

Cott, H. B. Adaptive coloration in animals. London: Methuen; 1957.

Dasch, G. A.; Weiss, E.; Kwang-Poo Chang, B. Endosymbionts of insects. In: Krieg, N. R.; Holt, J. G., eds. Bergey's manual of systematic bacteriology. Baltimore, MD: Williams & Wilkins; 1984.

David, W. A. L. The status of viruses pathogenic for insects and mites. Annual Review of Entomology 20:97–117; 1975.

Dempster, J. P. The control of Pieris rapae with DDT. I. The natural mortality of the young stages of Pieris. Journal of Applied Ecology 4:485–500; 1967.

Dempster, J. P. Animal population ecology. New York: Academic Press; 1975.

Downes, J. A. Adaptations of insects in the Arctic. Annual Review of Entomology 10:257–274; 1965.

Drake, V. A.; Farrow, R. A. The influence of atmospheric structure and motions on insect migrations. Annual Review of Entomology 33:183–210; 1988.

Duffey, S. S. Arthropod allomones: Chemical effronteries and antagonists. Proceedings of the XV International Congress of Entomology; 1977:p. 323–394.

Duffey, S. S. Sequestration of plant natural products by insects. Annual Review of Entomology 25:447–477; 1980.

Dunn, J. A. The natural enemies of the lettuce root aphid, *Pemphigus bursarius* (L.). Entomological Research Bulletin 51:271–278; 1960.

Dunn, P. E. Biochemical aspects of insect immunology. Annual Review of Entomology 31:321–339; 1986.

Dunning, D. C.; Roeder, K. D. Moth sounds and the insect-catching behavior of bats. Science 147:173–174; 1965.

Edmunds, M. The evolution of cryptic coloration. In: Evans, D. L.; Schmidt, J. O., eds. Insect defenses—adaptive mechanisms and strategies of prey and predators. Albany, NY: State University of New York Press; 1990.

Evans, D. L.; Schmidt, J. O., editors. Insect defenses—adaptive mechanisms and strategies of prey and predators. Albany, NY: State University of New York Press; 1990.

Evans, H. F.; Entwistle, P. F. Viral diseases. In: Fuxa, J. R.; Tanada, Y., eds. Epizootiology of insect diseases. New York: John Wiley & Sons; 1987.

Faust, R. M. Bacterial diseases. In: Cantwell, G. E., ed. Insect diseases. New York: Marcel Dekker; 1974:p. 87–183, vol. 1.

Folsom, J. W.; Wardle, R. A. Entomology, with special reference to its ecological aspects. 4th ed. Philadelphia: Blakiston; 1934.

Frost, S. W. Insect life and insect natural history. 2nd ed. New York: Dover; 1959.

Fullard, J. H. The sensory ecology of moths and bats: global lessons in staying alive. In: Evans, D. L.; Schmidt, J. O., eds. Insect defenses—adaptive mechanisms and strategies of prey and predators. Albany, NY: State University of New York Press; 1990.

Fulton, J. P.; Gergerich, R. C.; Scott, H. A. Beetle transmission of plant viruses. Annual Review of Phytopathology 25:111–123; 1987.

Futuyma, D. J. Evolutionary biology. 2nd ed. Sunderland, MA: Sinauer Associates, Inc.; 1986.

Fuxa, J. R.; Tanada, Y., editors. Epizootiology of insect diseases. New York: John Wiley & Sons; 1987.

Gaugler, R., Kaya, H. K. Entomopathogenic nematodes in biological control. Boca Raton, FL: C.R.C. Press; 1990.

Gibbs, A. J., editor. Viruses and invertebrates. New York: Elsevier; 1973.

Gibo, D. L.; Pallett, M. J. Soaring flight of monarch butterflies, *Danaus plexippus* (Lepidoptera: Danaidae), during the late summer migration in southern Ontario. Canadian Journal of Zoology 57:1393–1401; 1979.

Gilbert, L. E.; Raven, P. H. Coevolution of animals and plants. Austin, TX: University of Texas Press; 1975.

Gochnauer, T. A. Diseases and enemies of the honey bee. In: Grout, R. A., ed. The hive and the honey bee. Hamilton, IL: Dadant & Sons; 1963:p. 477–511.

Goff, L. J. Symbiosis and parasitism: another viewpoint. BioScience 32(4):257–259; 1985.

Gordon, H. T. Growth and development of insects. In: Huffaker, C. B.; Rabb, R. L., eds. Ecological entomology. New York: John Wiley & Sons; 1984.

Gossard, T. W.; Jones, R. E. Journal of Applied Ecology 14:65–71; 1977.

Graham, S. A.; Knight, F. B. Principles of forest entomology. 4th ed. New York: McGraw-Hill; 1965.

Griffin, D. R. Echoes of bats and man. Garden City, NY: Doubleday (Anchor Books); 1959.

Grosch, D. S. Environmental aspects: radiation. In: Rockstein, M., ed. The physiology of insecta. 2nd ed. New York: Academic Press; 1974:p. 85–126, vol. II.

Guilford, T. The evolution of aposematism. In: Evans, D. L.; Schmidt, J. O., eds. Insect defenses—adaptive mechanisms and strategies of prey and predators. Albany, NY: State University of New York Press; 1990.

Harcourt, D. G. The development and use of life tables in the study of natural insect populations. Annual Review of Entomology 14:175–196; 1969.

Harcourt, D. G.; Leroux, E. J. Population regulation in insects and man. American Scientist 55(4):400–415; 1967.

Harrison, B. D.; Murant, A. F. Involvement of virus-coded proteins in transmission of plant viruses by vectors. In: Mayo, M. A.; Harrap, K. A., eds. Vectors in virus biology. New York: Academic Press; 1984.

Harwood, R. F.; James, M. T. Entomology in human and animal health. 7th ed. New York: Macmillan; 1979.

Heinrich, B. Insect thermoregulation. New York: John Wiley & Sons; 1981.

Heslop-Harrison, Y. Carnivorous plants. Scientific American 238:104–115; 1978.

Hinton, H. E. A new Chironomid from Africa, the larva of which can be dehydrated without injury. Proceedings of the Zoological Society (London) 121:371–380; 1951.

Hinton, H. E. Cryptobiosis in the larva of *Polypedilum vanderplanki* Hint. (Chironomidae). Journal of Insect Physiology 5:286–300; 1960.

Horsfall, F. L.; Tamm, I., editors. Viral and rickettsial infections of man. 4th ed. Philadelphia: Lippincott; 1965.

Houk, E. J.; Griffiths, G. W. Intracellular symbiotes of the Homoptera. Annual Review of Entomology 25:161–187; 1980.

Huffaker, C. B.; Barryman, A. A.: Laing, J. E. Natural control of insect populations. In: Huffaker, C. B.; and Rabb, R. L., eds. Ecological entomology. New York: John Wiley & Sons; 1984.

Huffaker, C. B.; Rabb, R. L., editors. Ecological entomology. New York: John Wiley & Sons; 1984.

Hughes, R. D.; Jones, R. E.; Gutierrez, A. P. Short-term patterns of population change: the life system approach to their study. In: Huffaker, C. B.; Rabb, R. L., eds. Ecological entomology. New York: John Wiley & Sons; 1984.

Janzen, D. H. Why are there so many species of insects? Proceedings of the XV International Congress of Entomology; 1977:p. 84–94.

Jones, J. C. The circulatory system of insects. Springfield, IL: Charles C. Thomas; 1977.

Keller, S.; Zimmermann, G. Mycopathogens of soil insects. In: Wilding, N. et al., eds. Insect-fungus interactions. New York: Academic Press; 1989:p. 239–270.

Kettle, D. S. Medical and veterinary entomology. New York: Wiley-Interscience; 1987.

Kettlewell, H. B. D. Darwin's missing evidence. Scientific American 200:48–53; 1959.

Kettlewell, H. B. D. The phenomenon of industrial melanism in Lepidoptera. Annual Review of Entomology 6:245–22; 1961.

Kettlewell, H. B. D. The evolution of melanism. The study of a recurring necessity. With special reference to industrial melanism in the Lepidoptera. New York: Clarendon; 1973.

Krieg, A. Diseases caused by bacteria and other prokaryotes. In: Fuxa, J. R.; Tanada, Y., eds. Epizootiology of insect diseases. New York: John Wiley & Sons; 1987:p. 323–355.

Kuno, E. Sampling and analysis of insect populations. Annual Review of Entomology 36:285–304; 1991.

Lawton, J. H.; Hassell, M. P. Interspecific competition in insects. In: Huffaker, C. B.; Rabb, R. L., eds. Ecological entomology. New York: John Wiley & Sons; 1984:p. 451–495.

Lee, R. E., Jr.; Denlinger, D. L., editors. Insects at low temperature. New York: Chapman & Hall; 1991.

Lee, K. E.; Wood, T. G. Termites and soils. New York: Academic Press; 1971.

Maramorosch, K.; Harris, K. F., editors. Leafhopper vectors and disease agents. New York: Academic Press; 1979.

Margulis, L. Symbiosis in cell evolution—life and its environment on the early earth. San Francisco: W. H. Freeman and Company; 1981.

Martin, M. M. The biochemical basis of the fungus-attine ant symbiosis. Science 169:16–20; 1970.

Masek Fialla, K. Die Korpertemperatur Poikilothermes Thiere im Abhangigkeit vom Kleinklima. Zeitschrift fuer Wissenschaftliche Zoologie 154:170–247; 1941.

May, M. Aerial defense tactics of flying insects. American Scientist 79:316–328; 1991.

Mayo, M. A.; Harrap, K. A., editors. Vectors in virus biology. New York: Academic Press; 1984.

McDonald, L. L.; Manly, B. F. J.; Lockwood, J. A.; Logan, J. A., editors. Estimation and analysis of insect populations. Berlin: Springer-Verlag; 1989.

Monath, T. P. The arboviruses. Boca Raton, FL: CRC Press; 1988. 5 Vol.

Müller, J. F. T. *Ituna* and *Thyridia:* a remarkable case of mimicry in butterflies. Transactions of the Entomological Society (Proceedings) 1879:20–28; 1879.

N. A. S. Insect-pest management and control. Subcommittee on Insect Pests. Washington, DC: National Academy of Sciences; 1969.

Needham, J. G.; Traver, J. R.; Hsu, Y. C. The biology of mayflies, with a systematic account of North American species. Ithaca, NY: Comstock; 1935.

Nickle, W. R., editor. Plant and insect nematodes. New York: Marcel Dekker, Inc.; 1984.

Odum, E. P.; Odum, H. T. Fundamentals of ecology. 2nd ed. Philadelphia: Saunders; 1959.

Park, T. Beetles, competition and populations. Science 138:1369–1379; 1962.

Pfadt, R. E., editor. Fundamentals of applied entomology. 3rd ed. New York: Macmillan; 1985.

Pielou, E. C. Population and community ecology—principles and methods. New York: Gordon and Breach; 1974.

Platt, R. B.; Griffiths, J. F. Environmental management and interpretation. New York: Reinhold; 1964.

Poinar, G. O., Jr. Nematodes as facultative parasites of insects. Annual Review of Entomology 17:103–122; 1972.

Poinar, G. O., Jr. Entomogenous nematodes. Leiden, Netherlands: Brill; 1975.

Portmann, A. Animal camouflage. Ann Arbor, MI: Ann Arbor Science Library, University of Michigan Press; 1950.

Price, P. Insect ecology. New York: John Wiley & Sons; 1984.

Reeves, W. C. Epidemiology and control of mosquito-borne arboviruses in California, 1943–1987; Sacramento, CA: California Mosquito and Vector Control Association, Inc.; 1990.

Reichstein, T.; von Euw, J.; Parsons, J. A.; Rothschild, M. Heart poisons in the monarch butterfly. Science 161:861–866; 1968.

Rettenmeyer, C. W. Insect mimicry. Annual Review of Entomology 15:43–74; 1970.

Ricklefs, R. F. Ecology. 3rd ed. San Francisco: W. H. Freeman and Company; 1990.

Riemer, J.; Whittaker, J. B. Air pollution and insect herbivores: observed interactions and possible mechanisms. In: Bernays, E. A., ed. Insect-plant interactions. Boca Raton, FL: CRC Press; 1989:p. 73–105.

Ritland, D. B.; Brower, L. P. The viceroy butterfly is not a batesian mimic. Nature 350:497; 1991.

Roeder, K. D. Acoustic sensory responses and possible bat evasion tactics of certain moths. In: Burd, M. D. B., ed. Proceedings of the Canadian Society of Zoologists Annual Meeting. Ottawa: National Research Council of Canada; 1974:71–78.

Rolston, L. H.; McCoy, C. E. Introduction to applied entomology. New York: Ronald Press; 1966.

Rothschild, M. Fleas. Scientific American 213:44–50; 1965a.

Rothschild, M. The rabbit flea and hormones. Endeavor 24:162–168; 1965b.

Rothschild, M. Secondary plant substances and warning colouration in insects In: van Emden, H. F., ed. Insect/plant relationships. Royal Entomological Society (London), Symposium 6. London: Blackwell Scientific; 1972:p. 59–83.

Rothschild, M.; Ford, B. Hormones of the vertebrate host controlling ovarian regression and copulation of the rabbit flea. Nature 211:261–266; 1966.

Rothschild, M.; Ford, B. Hormones of the vertebrate host controlling ovarian regression and copulation of the rabbit flea. Nature 211:261–266; 1972.

Salt, R. W. Role of glycerol in the cold-hardening of *Bracon cephi* (Gahan). Canadian Journal of Zoology 37:59–69; 1959.

Salt, R. W. Principles of insect cold-hardiness. Annual Review of Entomology 6:55–74; 1961.

Sargent, T. D. Startle as an anti-predator mechanism, with special reference to the underwing moths (*Catocala*). In: Evans, D. L.; Schmidt, J. O., eds. Insect defenses—adaptive mechanisms and strategies of prey and predators. Albany, NY: State University of New York Press; 1990.

Schaller, F. Soil animals. Ann Arbor, MI: The University of Michigan Press; 1968.

Schmidt, G. D.; Roberts, L. S. Foundations of parasitology. 3rd ed. St. Louis, MO: Times Mirror/Mosby College Publishing; 1985.

Schwemmler, W.; Gassner, G., editors. Insect endocytobiosis: morphology, physiology, genetics, evolution. Boca Raton, FL: CRC Press, Inc.: 1989.

Seastedt, T. R. The role of microarthropods in decomposition and mineralization processes. Annual Review of Entomology 29:25–46; 1984.

Seber, G. A. F. The estimation of animal abundance and related parameters. 2nd ed. London: Griffin; 1982.

Smith, K. M. Insect virology. New York: Academic Press; 1967.

Smith, K. M. Virus-insect relationships. New York: Longmans; 1977.

Smith, D. C.; Douglas, A. E. The biology of symbiosis. Baltimore, MD: Edward Arnold; 1987.

Smith, R. L. Ecology and field biology. 4th ed. New York: Harper and Row, Publishers; 1990.

Solomon, M. E. Population dynamics. The Institute of Biology's Studies in Biology, no. 18. London: Edward Arnold; 1969.

Southwood, T. R. E. Ecological methods with particular reference to the study of insect populations. London: Methuen; 1966.

Southwood, T. R. E. The dynamics of insect populations. In: Pimentel, D., ed. Insects, science and society. New York: Academic Press; 1975:p. 151–199.

Starr, M. P. A generalized scheme for classifying organismic associations. In: Jennings, D. H.; Lee; D. L., eds. Symbiosis: Symposia of the Society for Experimental Biology, No. 29. Cambridge, England: Cambridge University Press; 1975:p. 1–20.

Steinhaus, E. A. Principles of insect pathology. New York: McGraw-Hill; 1949.

Steinhaus, E. A. Insect pathology, an advanced treatise. New York: Academic Press; 1963. 2 vol.

Sweetman, H. L. The biological control of insects. Ithaca, NY.: Comstock; 1936a.

Sweetman, H. L. The principles of biological control. Dubuque, IA: Wm. C. Brown; 1936b.

Tauber, M. J.; Tauber, C. A.; Masaki, S. Adaptations to hazardous seasonal conditions: dormancy, migration, and polyphenism. In: Huffaker, C. B.; Rabb, R. L., eds. Ecological entomology. New York: John Wiley & Sons; 1984:p. 19–50.

Taylor, F. Ecology and evolution of physiological time in insects. American Naturalist 117:1–23; 1981.

Theiler, M.; Downs, W. G. Arthropod borne viruses of vertebrates. New Haven, CT: Yale University Press; 1973.

Tiivel, T. Endocytobiosis of leafhoppers with prokaryotic microorganisms. In: Schwemmler, W.; Gassner, G., eds. Insect endocytobiosis: morphology, physiology, genetics, evolution. Boca Raton, FL: CRC Press, Inc.; 1989.

Usinger, R. L., editor. Aquatic insects of California, with keys to North American genera and California species. Berkeley: University of California Press; 1956.

Varley, G. C.; Gradwell, G. R. Recent advances in insect population dynamics. Annual Review of Entomology 15:1–24; 1970.

Varley, G. C.; Gradwell, G. R.; Hassell, M. P. Insect population ecology: an analytical approach. Oxford, England: Blackwell Scientific Publications; 1973.

Vulinec, K. Collective security: aggregation by insects as a defense. In: Evans, D. L.; Schmidt, J. O., eds. Insect defenses—adaptive mechanisms and strategies of prey and predators. Albany, NY: State University of New York Press; 1990.

Vaughn, J. L. Viruses and rickettsial diseases. In: Cantwell, G. E., ed. Insect diseases. New York: Marcel Dekker; 1974:p. 49–85, vol. 1.

Ward, J. V.; Stanford, J. A. Thermal responses in the evolutionary ecology of aquatic insects. Annual Review of Entomology 27:97–117; 1982.

Way, M. J. Mutualism between ants and honeydew-producing Homoptera. Annual Review of Entomology 8:307–344; 1963.

Webber, J. F.; Gibbs, J. N. Insect dissemination of fungal pathogens of trees. In: Wilding, N. et al., eds. Insect-fungus interactions. New York: Academic Press; 1989:p. 161–193.

Wellington, W. G.; Trimble, R. M. Weather. In: Huffaker, C. B.; Rabb, R. L., eds. Ecological entomology. New York: John Wiley & Sons; 1984:p. 19–50.

Welch, H. E. Entomophilic nematodes. Annual Review of Entomology 10:275–302; 1965.

Whitcomb, R. F.; Shapiro, M.; Granados, R. R. Insect defense mechanism against microorganisms and parasitoids. In: Rockstein, M., ed. The physiology of insecta. 2nd ed. New York: Academic Press; 1974:p. 448–536, vol. V.

Wickler, W. Mimicry in plants and animals. New York: McGraw-Hill; 1968.

Wigglesworth, V. B. The principles of insect physiology. 7th ed. London: Chapman & Hall; 1972.

Wilding, N.; Collins, N. M.; Hammond, P. M.; Webber, J. F., editors. Insect-fungus interactions, 14th Symposium of the Royal Entomological Society of London in Collaboration with the British Mycological Society. New York: Academic Press; 1987.

Wilson, E. O.; Bossert, W. H. A primer of population biology. Stamford, CT: Sinauer Associates; 1971.

Wilson, E. O. Sociobiology—the new synthesis. Cambridge, MA: Belknap Press/Harvard University Press; 1975.

Wood, T. G.; Thomas, R. J. The mutualistic association between Macrotermitinae and *Termitomyces*. In: Wilding, N., et al., eds. Insect-fungus interactions, 14th Symposium of the Royal Entomological Society of London in Collaboration with the British Mycological Society. New York: Academic Press; 1987.

Yen, J. H.; Barr, A. R. The etiological agent of cytoplasmic incompatibility in *Culex pipiens*. Journal of Invertebrate Pathology 22:242–250; 1973.

CHAPTER 10

Abrahamson, W. G. Plant-animal interaction. New York: McGraw-Hill; 1989.

Ananthakrishnan, T. N., editor. Biology of gall insects. London: Edward Arnold; 1984.

Bach, C. E. Effects of plant density and diversity on the population dynamics of a specialist herbivore, the striped cucumber beetle, *Acalymma vittata* (Fab.) Ecology 61: 1515–1530; 1980.

Bach, C. E. Host plant growth form and diversity: effects on abundance and feeding preference of a specialist herbivore, *Acalymma vittata* (Coleoptera: Chrysomelidae). Oecologia 50:370–375; 1981.

Baker, H. G.; Baker, I. Amino-acids in nectar and their evolutionary significance. Nature 241:543–545; 1973.

Banerjee, B. An analysis of the effect of latitude, age, and area on the number of pest species of tea. Journal of Applied Ecology 18:339–342; 1981.

Barbosa, P.; Schultz, J. C., editors. Insect outbreaks. San Diego, CA: Academic Press; 1987.

Barth, F. Insects and flowers. Princeton, N. J.: Princeton University Press; 1985.

Bazzaz, F. A. The response of natural ecosystems to the rising global CO_2 levels. Annual Review of Ecology and Systematics 21:167–196; 1990.

Bernays, E. A. The insect on the plant—a closer look. Proceedings of the 5th International Symposium on Insect-Plant Relationships, Wageningen, The Netherlands. Wageningen, Holland: Centre Agricultural Publication Documentation; 1982.

Borror, D. J.; DeLong, D. M.; Triplehorn, C. A. An introduction to the study of insects, 4th ed. New York: Holt, Rinehart and Winston; 1976.

Clements, F. Plant succession: an analysis of the development of vegetation. Carnegie Institution of Washington Publication No. 242. 1916. 512 pp.

Coley, P. D. Herbivory and defensive characteristics of tree species in a lowland tropical forest. Ecological Monographs 53:209–233; 1983.

Edwards, J. S. Arthropods of alpine aeolian ecosystems. Annual Review of Entomology 32:163–169; 1987.

Erwin, T. L. Tropical forests: their richness in Coleoptera and other arthropod species. Coleopterists Bulletin 36:74–75; 1982.

Feeny, P. Seasonal changes in oak leaf tannins and nutrients as a cause of spring feeding by winter moth caterpillars. Ecology 51:565–581; 1970.

Fox, L. R.; Macauley, B. J. Insect grazing on *Eucalyptus* in response to variation in leaf tannins and nitrogen. Oecologia 29:145–162; 1977.

Fox, L. R.; Morrow, P. A. Estimates of grazing damage to *Eucalyptus* trees. Australian Journal of Ecology 8:139–144; 1983.

Fox, L. R.; Morrow, P. A. On comparing herbivore damage in Australian and north temperate systems. Australian Journal of Ecology 11:387–393; 1986.

Frankel, G. S. The *raison d'etre* of secondary plant substances. Science 129:1466–1470; 1959.

Furniss, R. L.; Carolin, V. M. Western forest insects. Miscellaneous Publications—United States Department of Agriculture 1339; 1977.

Gaston, K. The magnitude of global species richness. Conservation Biology 5:283–296; 1991.

Gilbert, L. E. Butterfly-plant coevolution: has *Passiflora adenopoda* won the selectional race with heliconiine butterflies? Science 172:585–586; 1971.

Gleason, H. A. The individualistic concept of the plant association. Bulletin of the Torrey Botanical Club 53:331–368; 1926.

Graebe, J. E. Gibberellin biosynthesis and control. Annual Review of Physiology 38:419–465; 1987.

Grant, V.; Grant, K. Flower pollination in the Phlox family. New York: Columbia University Press; 1965.

Hairston, N. G.; Smith, F. E.; Slobodkin, L. B. Community structure, population control, and competition. American Naturalist 94:421–425; 1960.

Handel, S. N.; Beattie, A. J. Seed dispersal by ants. Scientific American 263:76–83; 1990.

Harborne, J. B. Introduction to ecological biochemistry. 3d ed. London; San Diego, CA: Academic Press; 1988.

Haukioja, H. Induction of defenses in trees. Annual Review of Entomology 36:25–42; 1991.

Heatwole, H.; Lowman, M. D. Dieback—death of an Australian landscape. Sydney, Australia: Reed Publishing; 1986.

Heinrich, B. Bumblebee economics. Cambridge, MA: Harvard University Press; 1979.

Hespenheide, H. A. Bionomics of leaf-mining insects. Annual Review of Entomology 26:535–560; 1991.

Hodson, A. C. Minnesota springs; a fifty-year record. Minnesota Horticulturist 119:15; 1991.

Howe, H. F.; Westley, L. C. Ecological relationships of plants and animals. New York: Oxford University Press; 1988.

Huxley, C. R. The ant-plants *Myrmecodia* and *Hydnophytum* (Rubiaceae) and the relationships between their morphology, ant occupants, physiology, and ecology. New Phytologist 80:231–268; 1978.

Huxley, C. R.; Cutler, D. F., editors. Ant-plant interactions. New York: Oxford University Press; 1991.

Janzen, D. H. Coevolution of mutualism between ants and acacias in Central America. Evolution 20:249–275; 1966.

Janzen, D. H. Seed predation by animals. Annual Review of Ecology and Systematics 2:465–492; 1971a.

Janzen, D. H. Euglossine bees as long-distance pollinators of tropical plants. Science 171:203–205; 1971b.

Johnson, R. J.; Lincoln, D. F. Sagebrush carbon allocation patterns and grasshopper nutrition: The influence of CO_2 enrichment and soil mineral limitation. Oecologia 87:127–134; 1991.

Jones, D. A. Coevolution and cyanogenesis. In: Heywood, V. H., ed. Taxonomy and Ecology. London: Academic Press; 1973:p. 213–242.

Karowe, D. N.; Martin, M. M. The effects of quantity and quality of diet nitrogen on the growth, efficiency of food utilization, nitrogen budget, and metabolic rate of fifth-instar *Spodoptera eridania* larvae (Lepidoptera: Noctuidae). Journal of Insect Physiology 35:699–708; 1989.

Lawton, J. H. Plant architecture and the diversity of phytophagous insects. Annual Review of Entomology 28:23–29; 1983.

Levin, D. A. The role of trichomes in plant defense. Quarterly Review of Biology 48:3–15; 1973.

Lewis, T. Thrips: their biology, ecology, and economic importance. London: Academic Press; 1973.

Liss, W. J.; Gut, L. J.; Westigard, P. H.; Warren, C. E. Perspectives on arthropod community structure, organization, and development in agricultural crops. Annual Review of Entomology 31:455–478; 1986.

Lowman, M. D. Seasonal variation in insect abundance among several Australian rain forests, with particular reference to phytophagous types. Australian Journal of Ecology 7:353–361; 1982.

Lowman, M. D. Spatial and temporal variability in herbivory of five canopy trees in Australia. Australian Journal of Ecology 10:7–24; 1985.

Lowman, M. D.; Fox J. Variation in leaf toughness and phenolic content among 5 species of Australian rain forest trees. Australian Journal of Ecology 8:17–25; 1983.

McNeill, S.; Southwood, T. R. E. The role of nitrogen in the development of insect/plant relationships. In: Harborne, J. B., ed. Biochemical aspects of plant and animal coevolution. London: Academic Press; 1978:p. 77–98.

Mattson, W. J. Herbivory in relation to plant nitrogen content. Annual Review of Ecology and Systematics 11:119–161; 1980.

Meyer, J. Plant galls and gall inducers. Berlin: Gebruder Borntraeger; 1987.

Mitter, C. B.; Farrell, B.; Wiegmann, B. The phylogenetic study of adaptive zones: has phytophagy promoted insect diversification? American Naturalist 132:107–128; 1988.

Mitton, J. B.; Sturgeon, K. B. Bark beetles in North American conifers: a system for the study of evolutionary biology. Austin, TX: University of Texas Press; 1982.

Moran, N. A.; Whitham, T. G. Population fluctuations in complex life cycles: an example from *Pemphigus* aphids. Ecology 69:1214–1218; 1988.

Moran, V. C.; T.R.E. Southwood. The guild composition of arthropod communities in trees. Journal of Animal Ecology 51:289–306; 1982.

Morrow, P. A. The role of sclerophyllous leaves in determining insect grazing damage. Ecological Studies 43:509–524; 1983. Kruger, F. J.; Mitchell, D. T.; Jarvis, J. U. M. eds. Mediterranean-type ecosystems. Berlin: Springer-Verlag.

Morrow, P. A.; Fox, L. R. Effects of variation in *Eucalyptus* essential oil yield on insect growth and herbivore load. Oecologia 45:109–129; 1980.

Morrow, P. A.; Fox, L. R. Estimates of pre-settlement insect damage in Australian and North American forests. Ecology 70:1055–1060; 1989.

Pate, J. S. Uptake, assimilation, and transport of nitrogen compounds by plants. Soil Biology and Biochemistry 5:109–119; 1973.

Price, P. W. Colonization of crops by arthropods: non-equilibrium communities in soybean fields. Environmental Entomology 5:605–611; 1976.

Price, P. W. General concepts on the evolutionary biology of parasites. Evolution 31:405–420; 1977.

Price, P. W. Evolutionary biology of parasites. Monographs in Population Biology 15. Princeton, NJ: Princeton University Press; 1980.

Price, P. W. The role of natural enemies in insect populations. In: Barbosa, P.; Schultz, J. C., eds. Insect outbreaks. San Diego, CA: Academic Press; 1987:p. 287–312.

Price, P. W.; Cobb, N.; Craig, T. P.; Fernandes, G. W.; Itami, J. K.; Mopper, S.; Preszler, R. W. Insect herbivore population dynamics on trees and shrubs: new approaches relevant to latent and eruptive species and life table development. In: Bernays, E. A., ed. Insect-plant interactions. Boca Raton, FL: CRC Press; 1990:p. 1–38, vol. 2.

Price, P. W.; Lewinsohn, T. M.; Fernandes, G. W.; Benson, W. W., editors. Plant-animal interactions-evolutionary ecology in tropical and temperate regions. Somerset, NJ: John Wiley & Sons, Inc.; 1991.

Richards, O. W.; Davies, R. G. Imm's general textbook of entomology. London: Chapman and Hall; 1977, vol. 3.

Riemer, J.; Whittaker, J. B. Air pollution and insect herbivores: observed interactions and possible mechanisms. In: Bernays, E. A., ed. Insect-plant interactions. Boca Raton, FL: CRC Press; 1989:p. 73–106, vol. 1.

Root, R. B. Organization of a plant-arthropod association in simple and diverse habitats: the fauna of collards (*Brassica oleracea*). Ecological Monographs 43:95–124; 1973.

Roseman, R.; Howe, H.; Westley. Ecological relationships of plants and animals. New York: Oxford University Press; 1988.

Rosenthal, G. A.; Janzen, D. H., editors. Herbivores: their interaction with secondary plant metabolites. New York: Academic Press; 1979.

Ryan, C. A. Proteinase inhibitors. In: Rosenthal, G. A.; Janzen, D. H., eds. Herbivores: their interaction with secondary plant metabolites. New York: Academic Press; 1979:p. 599–618.

Ryan, C. A. Insect-induced chemical signals regulating natural plant protection responses. In: Denno, R. F.; McClure, M., eds. Variable plants and herbivores in natural and managed systems. New York: Academic Press; 1983:p. 43–60.

Schultz, J. C. Impact of variable plant defensive chemistry on susceptibility of insects to natural enemies. ACS Symposium Series 208:37–54; 1983.

Scriber, J. M.; Slansky, F., Jr. The nutritional ecology of immature insects. Annual Review of Entomology 26:183–212; 1981.

Sherman, I. W.; Sherman, V. G. Biology—a human approach. New York: Oxford University Press; 1983.

Singh, B. B.; Hadley, H. H.; Bernard, R. L. Morphology of pubescence in soy beans and its relationship to plant vigour. Crop Science 11:13–16; 1971.

Smart, J.; Hughes, N. F. The insect and the plant: progressive paleoecological integration. In: van Emden, H. F., ed. Insect/plant relationships. Symposia of the Royal Entomological Society of London 6: 1972:p. 142–155.

Strong, D. R.; Lawton, J. H.; Southwood, T.R.E. Insects on plants. Cambridge, MA : Harvard University Press; 1984.

Southwood, T. R. E. The number of species of insect associated with various trees. Journal of Animal Ecology 30:1–8; 1961.

Southwood, T. R. E. The insect/plant relationship—an evolutionary perspective. In: van Emden, H. F., ed. Symposia of the Royal Entomological Society of London 6: 1973:p. 3–30.

Southwood, T. R. E. The component of diversity. Symposium of the Royal Entomological Society of London 9:19–40; 1978.

Southwood, T. R. E.; Moran, V. C.; Kennedy, C. E. J. The assessment of arboreal insect fauna: comparisons of knockdown sampling and faunal lists. Ecological Entomology 7:331–40; 1982a.

Southwood, T. R. E.; Moran V. C.; Kennedy, C. E. J. The richness, abundance, and biomass of the arthropod communities on trees. Journal of Animal Ecology 51:635–649; 1982b.

Tallamy, D. W.; Raupp, M. J., editors. Phytochemical induction by herbivores. Somerset, NJ.: John Wiley & Sons, Inc.; 1991.

Thompson, J. N. Reversed animal-plant interactions: the evolution of insectivorous and ant-fed plants. Biological Journal of the Linnean Society 16:147–155; 1981.

Thompson, J. N. Interaction and coevolution. New York: John Wiley & Sons; 1982.

Thompson, J. N. Concepts of coevolution. Trends in Ecology and Evolution 61:179–183; 1989.

Tilman, G. D. Cherries, ants, and tent caterpillars: timing of nectar production in relation to susceptibility of caterpillars to ant predation. Ecology 59:686–692; 1978.

Turner, J. R. G. Adaptation and evolution in *Heliconius:* a defense of neoDarwinism. Annual Review of Ecology and Systematics 12:99–121; 1981.

Van Cleve, K.; Oliver, L.; Schlentnee, R.; Viereck, L. A.; Dyrness, C. T. Production and nutrient cycling in taiga forest ecosystems. Canadian Journal of Forest Research 13:747–766; 1983.

Wainhouse, D.; Cross, D. J.; Howell, R. S. The role of lignin as a defense against the spruce bark beetle *Dendroctonus micans:* effect on larvae and adults. Oecologia 85:257–265; 1990.

Waldbauer, G. P. The consumption and utilization of food by insects. Advances in Insect Physiology 5:229–289; 1968.

Waterman, P. G.; Mole, S. Extrinsic factors influencing production of secondary metabolites in plants. In: Bernays, E. A., ed. Insect-plant interactions. Boca Raton, FL: CRC Press; 1989:p. 107–134, vol. 1.

Waterhouse, D. F. The insects of Australia. Melbourne, Australia: University Press; 1970.

Wedin, D.; Tilman, G. D. Nitrogen cycling, plant competition, and the stability of tallgrass prairie. In: Smith, D. D.; Jacobs, C. A., eds. Proceeding of the Twelfth North American Prairie Conference: Recapturing a Vanishing Heritage. Cedar Falls, IA: University of Northern Iowa Press; 1992.

Weins, J. A.; Cates, R. G.; Rotenberrry, J. T.; Cobb, N.; Van Horne, B.; Redak, R. A. Arthropod dynamics on sagebrush (*Artemisia tridentata*): effects of plant chemistry and avian predation. Ecological Monographs 6:299–321; 1991.

Weis, A. E.; Berenbaum, M. R. Herbivorous insects and green plants. In: Abrahamson, W. G., ed. 1989. Plant-animal interactions. New York: McGraw-Hill; 1989:p. 123–162.

Whitham, T. G. Habitat selection by *Pemphigus* aphids in response to resource limitation and competition. Ecology 59:1164–1176; 1978.

Whitham, T. G. Host manipulation of parasites: within-plant variation as a defense against rapidly evolving pests. In: Denno, R. F.; McClure, M. S., eds. Variable plants and herbivores in natural and managed systems. New York: Academic Press; 1983:p. 15–39.

Whitham, T. G. Plant hybrid zones as sinks for pets. Science 244:149–151; 1989.

Wilson, E. O. The arboreal ant fauna of Peruvian Amazon forests: a first assessment. Biotropica 2:245–251; 1987.

Wilson, E. O., editor. Biodiversity. Washington, DC: National Academy of Science Press; 1988.

Zwolfer, H. Evolutionary and ecological relationships of the insect fauna of thistles. Annual Review of Entomology 33: 103–122; 1988.

CHAPTER 11

Alexander, R. D.; Brown, W. L., Jr. Mating behavior and the origin of insect wings. Ann Arbor, MI: University of Michigan, Occasional Papers of the Museum of Zoology No. 628; 1963.

Anderson, D. T. Embryology and phylogeny in annelids and arthropods. New York: Pergamon; 1973.

Anderson, D. T. Embryos, fate maps, and the phylogeny of arthropods. In: Gupta, A. P., ed. Arthropod phylogeny. New York: Van Nostrand Reinhold; 1979:p. 59–105.

Ballard, J. W. O.; Olsen, G. J.; Faith, D. P.; Odgers, W. A.; Rowell, D. M.; Atkinson, P. W. Evidence from 12S ribosomal RNA sequences that Onychophorans are modified Arthropods. Science 258:1345–1348; 1992.

Bergström, J. Morphology of fossil arthropods as a guide to phylogenetic relationships. In: Gupta, A. P., ed. Arthropod phylogeny. New York: Van Nostrand Reinhold; 1979:p. 3–56.

Berlocher, S. H. Insect molecular systematics. Annual Review of Entomology 29:403–433; 1984.

Blackwelder, R. E. Taxonomy: a text and reference book. New York: John Wiley & Sons; 1967.

Bland, R. C.; Jaques, H. E. How to know the insects. 3d ed. Dubuque, IA: Wm. C. Brown; 1978.

Borror, D. J.; Triplehorn, C. A.; Johnson, N. F. An introduction to the study of insects. 6th ed. Philadelphia: Saunders College Publishing; 1989.

Borror, D. J.: White, R. E. A field guide to the insects of America north of Mexico. Boston: Houghton Mifflin; 1970.

Boudreaux, H. B. Arthropod phylogeny with special reference to insects. New York: John Wiley & Sons; 1979.

Callahan, P. S. The evolution of insects. New York: Holiday House; 1972.

Carpenter, F. M. The geological history and evolution of insects. American Scientist 41:256–270; 1953.

Carpenter, F. M. Geological history and evolution of the insects. In: Tipton, V. J., ed. Syllabus—introductory entomology. Provo, UT: Brigham Young University Press; 1977:p. 77–88.

Cisne, J. L. Trilobites and the origin of arthropods. Science 186:13–18; 1974.

Clarke, K. U. The biology of the arthropods. New York: American Elsevier; 1973.

Cummins, K. W.; Miller, L. D.; Smith, N. A.; Fox, R. M. Experimental entomology. New York: Van Nostrand Reinhold; 1965.

Danks, H. V. Systematics in support of entomology. Annual Review of Entomology 33:271–296; 1988.

DeSalle, R.; Giddings, L. V.; Kaneshiro, K. Y. Mitochondrial DNA variability in natural populations of Hawaiian Drosophila II. Genetic and phylogenetic relationships of natural populations of D. silvestris and D. heteroneura. Heredity 56:87–96; 1986.

Douglas, M. M. Thermoregulatory significance of thoracic lobes in the evolution of insect wings. Science 211:84–86; 1981.

Dowling, T. E.; Moritz, C.; Palmer, J. D. Nucleic acids II: Restriction site analysis. In: Hillis, D. M.; Moritz, C., eds. Molecular systematics. Sunderland, MA: Sinauer Press; 1990:p. 250–317.

Entomological Society of America. Common names of insects and related organisms. Lanham, MD: Entomological Society of America; 1989.

Felsenstein, J. Numerical methods for inferring evolutionary trees. Quarterly Review of Biology 57:379–404; 1982.

Flower, J. W. On the origin of flight in insects. Journal of Insect Physiology 10:81–88; 1964.

Futuyma, D. J. Evolutionary biology. 2nd ed. Sunderland, MA: Sinauer Press; 1986.

Golenberg, E. M.; Giannasi, M. T.; Clegg, C. J.; Durbin, M.; Henderson, D.; Zurawski, G. Chloroplast DNA sequence from a Miocene Magnolia species. Nature 344:656–658; 1990.

Gupta, A. P., editor. Arthropod phylogeny. New York: Van Nostrand Reinhold; 1979.

Hackman, R. H.; Goldenberg, M. Peripatus: its affinities and its cuticle. Science 190:582–583; 1975.

Hamilton, K. G. A. The insect wing. Part I. Origin and development of wings from notal lobes. Journal of the Kansas Entomological Society 44:421–433; 1971.

Hamilton, K. G. A. The insect wing. Part II. Vein homology and the archetypal insect wing. Journal of the Kansas Entomological Society 45:54–58; 1972.

Handlirsch, A. Die fossilen Insekten und die Phylogenie der rezenten Formen. Ein Handbuch für Paleontologen und Zoologen (2 vols). Leipzig: Engelman; 1908.

Hendriks, L.; Huysmans, E.; Vandenberghe, A.; De Wachter, R. Primary structures of the 5S ribosomal RNAs of 11 arthropods and applicability of 5S RNA to the study of metazoan evolution. Journal of Molecular Evolution 24:103–109; 1986.

Hennig, W. Phylogenetic systematics. Annual Review of Entomology 10:97–116; 1965.

Hennig, W. Phylogenetic systematics. Chicago: University of Chicago Press; 1966.

Hennig, W. Die Stammesgeschichte der Insekten. Frankfurt-am-Main, Dramer; 1969.

Hillis, D. M.; Larson, A.; Davis, S. K.; Zimmer, E. A. Nucleic acids III: Sequencing. In: Hillis, D. M.; Morita, C., eds. Molecular systematics. Sunderland, MA: Sinauer Press; 1990:p. 318–370.

Hillis, D. M.; Moritz, C. An overview of applications of molecular systematics. In: Hillis, D. M.; Morita, C., eds. Molecular systematics; Sunderland, MA: Sinauer Press; 1990:p. 502–515.

Hinton, H. E. The origin of flight in insects. London: Proceedings of the Royal Entomological Society [C] 28; 1963:p. 24–25.

Hinton, H. E. Enabling mechanisms. Proceeding of the XVth International Congress of Entomology; 1977:p. 71–83.

Janzen, D. H. Why are there so many species of insects? Proceedings of the XVth International Congress of Entomology; 1977:p. 84–94.

Jeannel, R. Classification et phylogénie des insectes. In: Grassé, P. P., ed. Traité de Zoologie, 9:1–110. Paris, Masson; 1949.

Kamp, J. W. Numerical classification of the orthopteroids, with special reference to the Grylloblattodea. Canadian Entomologist 105:1235–1249; 1973.

Kim, K. C. Systematics resources and the Entomological Society of America. Bulletin of the Entomological Society of America 35(3):170–176; 1989.

Kim, K. E.; Ludwig, H. W. The family classification of the Anoplura. Systematic Entomology 3(3):249–284; 1978.

King, E. G; Leppla, N. C. Advances and challenges in insect rearing. ARS-USDA, Washington, DC: U.S. Government Printing Office; 1984.

Kingsolver, J. G.; Koehl, M. A. R. Aerodynamics, thermoregulation, and the evolution of insect wings: differential scaling and evolutionary change. Evolution 39:488–504; 1985.

Knudsen, J. W. Biological techniques. New York: Harper & Row; 1966.

Kristensen, N. P. The phylogeny of hexapod "orders." A critical review of recent accounts. Zeitscrift für Zoologische Systematik und Evolutions forschung. 13:1–44; 1975.

Kristensen, N. P. Phylogeny of insect orders. Annual Review of Entomology 26:135–157; 1981.

Kristensen, N. P. Insect phylogeny based on morphological evidence. In: Fernholm, B.; Bremer, K.; Jornvall, H., eds. The hierarchy of life—molecules and morphology in phylogenetic analysis: Proceedings from the Nobel Symposia 70. New York: Elsevier Scientific; 1989.

Kukalová-Peck, J. Pteralia of the paleozoic insect orders Palaeodictyopotera, Megasecoptera, and Diaphanopterodea (Paleoptera). Psyche 81:416–430; 1974.

Kukalová-Peck, J. Origin of the insect wing and wing articulation from the arthropodan leg. Canadian Journal of Zoology 61:1618–1669; 1983.

Kukalová-Peck, J. New carboniferous Diplura, Monura, and Thysanura, the hexapod ground plan, and the role of thoracic side lobes in the origin of wings (Insecta). Canadian Journal of Zoology 65:2327–2345; 1987.

Kukalová-Peck, J. The "Uniramia" do not exist: the ground plan of the Pterygota as revealed by Permian Diaphanopterodea from Russia (Insecta: Paleodictyopteroidea) Canadian Journal of Zoology 70: 236–255; 1992.

Labandeira, C. C.; Beall, B. S.; Hueber, F. M. Early insect diversification: evidence from a Lower Devonian bristletail from Quebec. Science 242:913–916; 1988.

Lindroth, C. H. Systematics specializes between Fabricius and Darwin: 1800–1859. In: Smith, R. F.; Mittler, T. E.; Smith, C. N., eds. History of entomology. Palo Alto, CA: Annual Reviews; 1973:p. 119–154.

Linsley, E. G; Usinger, R. L. Taxonomy. In: Gray, P., ed. The encyclopedia of the biological sciences. New York: Van Nostrand Reinhold; 1961.

Mackerras, I. M. Evolution and classification of insects. In: The insects of Australia. Canberra: C.S.I.R.O.; 1970:p. 152–167.

Manton, S. M. Mandibular mechanisms and the evolution of arthropods. Philosophical Transactions of the Royal Society (London), B. Biological Sciences 247:1–183; 1964.

Manton, S. M. Arthropoda: introduction. In: Florkin, M.; Sheer, B. J., eds. Chemical zoology, vol. 5. New York: Academic Press; 1970:1–34.

Manton, S. M. The evolution of arthropodan locomotory mechanisms. Part 10: Locomotory habits, morphology, and evolution of the hexapod classes. Zoological Journal. Linnean Society 51:203–400; 1972.

Manton, S. M. The Arthropoda, habits, functional morphology, and evolution. London: Oxford University Press; 1977.

Manton, S. M. Functional morphology and the evolution of the hexapod classes. In: Gupta, A. P., ed. Arthropod phylogeny. New York: Van Nostrand Reinhold; 1979:387–465.

Mayr, E. Animal species and evolution. Cambridge, MA: Harvard University Press; 1963.

Mayr, E. The growth of biological thought. Cambridge, MA: Belknap Press; 1982.

Mayr, E.; Linsley, E. G.; Usinger, R. L. Methods and principles of systematic zoology. New York: McGraw-Hill; 1953.

Meglitsch, R. A. Invertebrate zoology. London: Oxford University Press; 1967.

Menken, S. B. J.; Ulenberg, S. A. Biochemical characters in agricultural entomology. Agricultural Zoology Reviews 2:305–360; 1987.

Michener, C. D.; Sokal, R. R. A quantitative approach to a problem in classification. Evolution 11:130–162; 1957.

Mikulic, D. G.; Briggs, D. E. G.; Kluessendorf, J. A Silurian soft-bodied biota. Science 228:715–717; 1985.

Murphy, R. W.; Sites, J. W., Jr.; Buth, D. G.; Haufler, C. H. Proteins I: Isozyme electrophoresis. In: Hillis, D. M.; Moritz, C., eds. Molecular systematics. Sunderland, MA: Sinauer Press; 1990:p. 45–126.

Needham, J. G.; Galtsoff, P. S.; Lutz, F. E.; Welch, P. S. Culture methods for invertebrate animals. New York: Dover; 1937.

Pashley, D. P. Host-associated genetic differentiation in fall armyworm (Lepidoptera: Noctuidae): a sibling species complex? Annals of the Entomological Society of America 79:898–904; 1986.

Patterson, C. Introduction. In: Patterson, C., ed. Molecules and morphology in evolution: conflict or compromise? Cambridge, England: Cambridge University Press; 1987:p. 1–22.

Peterson, A. Entomological techniques. Ann Arbor, MI: Edwards Brothers; 1959.

Rempel, J. G. The evolution of the insect head: the endless dispute. Quaestiones Entomologicae 11:7–25; 1975.

Riek, E. F. Fossil history. In: C.S.I.R.O. The insects of Australia. Melbourne: Melbourne University Press; 1970:p. 168–186.

Robison, R. A. Earliest-known uniramous arthropod. Nature 343:163–164; 1990.

Rohdendorf, B. B. Palaontologie. In: Helmcke, J. G.; Starck, D.; Wermuth, H., eds. Handbuch der Zoologie, 4(2), Lfg. 9. Berlin: Walter de Gruyter; 1969:p. 1–27.

Rohdendorf, B. B. The history of paleoentomology. In: Smith, R. F.; Mittler, T. E.; Smith, C. N., eds. History of entomology. Palo Alto, CA: Annual Reviews; 1973:p. 155–170.

Ross, H. H. The evolution of the insect orders. Entomological News 66(8):197–208; 1955.

Ross, H. H. How to collect and preserve insects. Illinois Natural History Survey, Circular 39; 1962.

Ross, H. H. A textbook of entomology. 3d ed. New York: John Wiley & Sons; 1965.

Ross, H. H. Evolution and phylogeny. In: Smith, R. F.; Miller, T. E.; Smith, C. N., eds. History of entomology. Palo Alto, CA: Annual Reviews; 1973:p. 171–184.

Schlüter, T. Fossil insect localities in Gondwanaland. Entomologia Generalis 15:61–76; 1990.

Scudder, G. G. E. Recent advances in the higher systematics and phylogenetic concepts in entomology. Canadian Entomologist 105:1251–1263; 1973.

Sharov, A. G. Basic arthropodan stock with special reference to insects. New York: Pergamon; 1966.

Shear, W. A. End of the 'Uniramia' taxon. Nature 359: 477–478; 1992.

Shear, W. A.; Bonamo, P. M.; Grierson, J. D.; Rolfe, W. D. I.; Smith, E. L.; Norton, R. A. Early land animals in North America: evidence from Devonian age arthropods from Gilboa. New York. Science 224:492–494; 1984.

Simpson, G. G. Principles of animal taxonomy. New York: Columbia University Press; 1961.

Siverly, R. E. Rearing insects in schools. Dubuque, IA: Wm. C. Brown; 1962.

Smart, J.; Hughes, N. F. The insect and the plant: progressive palaeoecological integration. In: van Emden, H. F., ed. Insect/plant relationships, Symposium 6. London: Blackwell Scientific, 1972:p. 143–155.

Snodgrass, R. E. Principles of insect morphology. New York: McGraw-Hill; 1935.

Snodgrass, R. E. A textbook of arthropod anatomy. Ithaca, NY: Comstock; 1952.

Sokal, R. R.; Sneath, P. A. A. Principles of numerical taxonomy. San Francisco, CA: Freeman; 1963.

Stehr, F. W. Techniques for collecting, rearing, preserving, and studying immature insects. In: Stehr, F. W., ed. Immature insects. Vol. 1. Dubuque, IA: Kendall/Hunt Publishing; 1987:p. 7–18.

Stiles, K. A.; Hegner, R. W.; Boolootian, R. College zoology. 8th ed. New York: Macmillan; 1969.

Swofford, D. L.; Olsen, G. J. Phylogeny reconstruction. In: Hillis, D. M.; Moritz, C., eds. Molecular Systematics; Sunderland, MA: Sinauer Press; 1990:p. 411–501.

Thomas, R. H.; Schaffner, W.; Wilson, A. C.; Paabo, S. DNA phylogeny of the extinct marsupial wolf. Nature 340:465–467; 1989.

Tiegs, O. W.; Manton, S. M. The evolution of the Arthropoda. Biological Reviews 33:255–337; 1958.

Trueman, J. W. H. Comment-evolution of insect wings: a limb exite plus endite model. Canadian Journal of Zoology 68:1333–1335; 1990.

Tuxen, S. L. Entomology systematizes and describes: 1700–1815. In: Smith, R. F.; Mittler, T. E.; Smith, C. N., eds. History of entomology. Palo Alto, CA: Annual Reviews; 1973:p. 95–118.

Villee, C. A.; Dethier, V. G. Biological principles and processes. Philadelphia: Saunders; 1976.

Vossbrinck, C. R.; Friedman, S. A 28s ribosomal RNA phylogeny of certain cyclorrhaphous Diptera based upon a hypervariable region. Systematic Entomology 14:417–431; 1989.

Warburton, F. E. The purposes of classification. Systematic Zoology 16(3):241–245; 1967.

Weygoldt, P. Arthropoda interrelationships—the phylogenetic-systematic approach. Zeitschrift für Zoologische Systematik und Evolutionsforschung. 24:19–35; 1986.

Whalley, P.; Jarzembowski, E. A. A new assessment of Rhyniella, the earliest known insect, from the Devonian of Rhynie, Scotland. Nature 291:317; 1981.

Wheeler, W. C. The systematics of insect ribosomal DNA. In: Ferholm, B.; Bremer, K.; Jornvall, H., eds. The hierarchy of life. New York: Elsevier; 1989:p. 307–321.

Wigglesworth, V. B. The evolution of insect flight. In: Rainey, R. C., ed. Insect flight. London: Royal Entomological Society, Symposium 7; Blackwell Scientific; 1976:p. 255–269.

Wille, A. The phylogeny and relationshps between the insect orders. Revista de Biologia Tropical. 8:93–123; 1960.

Willmer, P. Invertebrate relationships. Patterns in animal evolution. Cambridge, England: Cambridge University Press; 1990.

Wilson, E. O. Time to revive systematics. Science 230:1227; 1985.

Wilson, H. F.; Doner, M. H. The historical development of insect classification. St. Louis, MO: J. S. Swift; 1937.

Wooton, R. J. Palaeozoic insects. Annual Review of Entomology 26:319–344; 1981.

Zeh, D. W.; Zeh, J. A. Ovipositors, amnions and eggshell architecture in the diversification of terrestrial arthropods. The Quarterly Review of Biology 64(2):147–168; 1989.

CHAPTER 12

Alba-Tercedor, J.; Sanchez-Ortega, O., editors. Overview and strategies of Ephemeroptera and Plecoptera. Gainesville, FL: Sandhill Crane Press; 1991.

Alexander, R. D.; Borror, D. J. The songs of insects [Phonograph record]. Ithaca, NY: Cornell University Press; 1956.

Arnett, R. H., Jr. American insects. A handbook of the insects of America north of Mexico. New York: Van Nostrand Reinhold; 1985.

Askew, R. R. Parasitic insects. London: Heinemann Educational Books; 1973.

Bailey, S. F. The thrips of California. Part I. Suborder Terebrantia. Bulletin of the California Insect Surveyz 4:143–220; 1957.

Behnke, F. L. A natural history of termites. New York: Scribner; 1977.

Bland, R. C.; Jacques, H. E. How to know the insects. 3d ed. Dubuque, IA: Wm. C. Brown; 1978.

Blatchley, W. S. Orthoptera of northeastern America. Indianapolis, IN: Nature Publishing Co.; 1920.

Blatchley, W. S. Heteroptera or true bugs of eastern North America, with especial reference to the faunas of Indiana and Florida. Indianapolis, IN: Nature Publishing Co.; 1926.

Borror, D. J.; DeLong, D. M. An introduction to the study of insects. 3d ed. New York: Holt, Rinehart and Winston, Inc.; 1971.

Borror, D. J.; DeLong, D. M.; Triplehorn, C. A. An introduction to the study of insects. 4th ed. New York: Holt, Rinehart & Winston; 1976.

Borror, D. J.; Triplehorn, C. A.; Johnson, N. F. An introduction to the study of insects. 6th ed. Philadelphia: Saunders; 1989.

Borror, D. J.; White, R. E. A field guide to the insects of America north of Mexico. Boston: Houghton Mifflin; 1970.

Boudreaux, H. B. Arthropod phylogeny with special reference to insects. New York: John Wiley & Sons; 1979.

Britton, W. E., editor. The Hemiptera or sucking insects of Connecticut. Connecticut State Geology and Natural History Survey Bulletin 34; 1923.

Burkhardt, C. C. Insect pests of corn. In: Pfadt, R. E., ed. Fundamentals of applied entomology. 3d ed. New York: Macmillan; 1978:p. 303–334.

Burks, B. D. The mayflies, or Ephemeroptera, of Illinois. Illinois Natural History Survey Bulletin 26:1–216; 1953.

Butcher, J. W.; Snider, R.; Snider, R. J. Biology of edaphic Collembola and Acarina. Annual Review of Entomology 6:249–288; 1971.

Carpenter, F. M. Geological history and evolution of the insects. Proceedings of the XV International Congress of Entomology; 1977:p. 63–70.

Caudell, A. N. Zoraptera not an apterous order. Proceedings of the Entomological Society of Washington 22:84–97; 1918.

Chapman, P. J. Corrodentia of the United States of America. I. Suborder Isotecnomera. Journal of the New York Entomological Society 38:219–290, 319–403; 1930.

Chapman, R. F.; Joern, J. A., editors. Biology of grasshoppers. New York: John Wiley & Sons; 1990.

Cheng, L., editor. Marine insects. Amsterdam: North-Holland; 1976.

China, W. E. South American Peloridiidae (Hemiptera-Homoptera: Coleorhyncha). Transactions of the Royal Entomological Society of London 114:131–161; 1962.

Choe, J. C. Zoraptera of Panama with a review of the morphology, systematics, and biology of the order. In: Quintero, D.; Aiello, A., eds. Insects of Panama and Mesoamerica: selected studies. Oxford: Oxford University Press; 1992.

Christiansen, K. Bionomics of Collembola. Annual Review of Entomology 9:147–178; 1964.

Christiansen, K. Aquatic Collembola. In: Merritt, R. W.; Cummins, K. W., eds. An introduction to the aquatic insects of North America. Dubuque, IA: Kendall/Hunt; 1978:p. 51–55.

Christiansen, K.; Bellinger, P. The Collembola of North America north of the Rio Grande. Los Angeles: Entomological Reprint Specialists; 1978.

Chu, H. F. How to know the immature insects. Dubuque, IA: Wm. C. Brown; 1949.

Claassen, P. W. Plecoptera nymphs of North America (north of Mexico). Thomas Say Foundation Publication No. 3. College Park, MD: Entomological Society of America; 1931.

Clay, T. A key to the genera of the Menoponidae (Amblycera: Mallophaga: Insecta). Bulletin of the British Museum (Natural History), Entomology Series 24:1–26; 1969.

Clay, T. Phthiraptera (lice). In: Smith, K. G. V., ed. Insects and other arthropods of medical importance. London: British Museum (Natural History); 1973:p. 395–397.

Comstock, J. H. An introduction to entomology. 9th ed. Ithaca, NY: Cornell University Press; 1940.

Corbet, P. S. A biology of dragonflies. Chicago: Quadrangle Books; 1963.

Cornwell, P. B. The cockroach. Vol. 1. London: Hutchinson; 1968.

Cott, H. E. Systematics of the suborder Tubulifera (Thysanoptera) of California. Berkeley: University of California Press; 1966.

C.S.I.R.O. The insects of Australia, a textbook for students and research workers. 2nd ed. Canberra: C.S.I.R.O.; 1991.

Davies, R. G. Outlines of entomology. New York: Chapman and Hall; 1988.

Day, W. C. Ephemeroptera. In: Usinger, R. L., ed. Aquatic insects of California. Berkeley: University of California Press; 1956:p. 79–105.

DeCoursey, R. M. Keys to the families and subfamilies of the nymphs of North American Hemiptera-Heteroptera. Proceedings of the Entomological Society of Washington 73:413–428; 1971.

Dirsh, V. M. Classification of the acridomorphoid insects. Faringdon, UK: E. W. Classey; 1975.

Dixon, A. F. G. Biology of aphids. The Institute of Biology's Studies in Biology No. 44. London: Edward Arnold; 1973.

Eastop, V. F.; Hille Ris Lambers, D. Survey of the world's aphids. The Hague: Dr. W. Junk; 1976.

Edmondson, W. T., editor. Freshwater biology. New York: John Wiley & Sons; 1959.

Edmunds, G. F., Jr. Ephemeroptera. In: Edmondson, W. T., ed. Freshwater biology. New York: John Wiley & Sons; 1959a:p. 908–916.

Edmunds, G. F., Jr. Biogeography and evolution of the Ephemeroptera. Annual Review of Entomology 17:21–42; 1959b.

Edmunds, G. F., Jr. Ephemeroptera. In: Merritt, R. W.; Cummins, K. W., eds. An introduction to the aquatic insects of North America. Dubuque, IA: Kendall/Hunt; 1978:p. 57–80.

Edmunds, G. F., Jr.; Jensen, S. L.; Berner, L. The mayflies of North and Central America. Minneapolis: University of Minnesota Press; 1976.

Eisenbeis, G.; Wichard, W. Atlas of the biology of soil arthropods. New York: Springer-Verlag; 1987.

Eisner, T. Defensive spray of a phasmid insect. Science 148:966–968; 1965.

Emerson, K. C. Checklist of the Mallophaga of North America (north of Mexico). Part 1. Suborder Ischnocera. Part 2. Suborder Amblycera. Part 3. Mammal host list. Part 4. Bird host list. Dugway, UT: Desert Test Center, Proving Ground; 1972.

Emerson, K. C.; Price, R. D. A host-parasite list of the Mallophaga on mammals. Miscellaneous Publications of the Entomological Society of America 12:1–72; 1981.

Emerson, K. C.; Price, R.D. Evolution of Mallophaga on mammals. In: Kim, K. C., ed. Coevolution of parasitic arthropods and mammals. New York: John Wiley & Sons; 1985:p. 233–255.

Emerson, K. C.; Price, R. D. A new species of *Haematomyzus* (Mallophaga: Haematomyzidae) of the bush pig, *Pomatochoerus porcus,* from Ethiopia, with comments on lice found on pigs. Proceedings of the Entomological Society of Washington 90:338–342; 1988.

Essig, E. O. College entomology. New York: Macmillan; 1942.

Ewing, H. E. The Protura of North America. Annals of the Entomological Society of America 33:495–551; 1940.

Ferris, G. F. The sucking lice. Memoirs of the Pacific Coast Entomological Society 1:1–320; 1951.

Frison, T. H. The stoneflies, or Plecoptera, of Illinois. Illinois Natural History Survey Bulletin 20:281–471; 1935.

Garman, P. A. The Odonata of Connecticut. Connecticut State Geological and Natural History Survey, Bulletin 39; 1927.

Gaston, K. J. The magnitude of global insect species richness. Conservation Biology 5:283–296; 1991.

Ghauri, M. S. K. Hemiptera (bugs). In: Smith, K. G. V., ed. Insects and other arthropods of medical importance. London: British Museum (Natural History); 1973:p. 373–393.

Gloyd, L. K.; Wright, M., Jr. Odonata. In: Edmondson, W. T., ed. Freshwater biology. New York: John Wiley & Sons; 1959:p. 917–940.

Gurney, A. B. A synopsis of the Order Zoraptera, with notes on the biology of *Zorotypus hubbardi* Caudell. Proceedings of the Entomological Society of Washington 40:57–87; 1938.

Gurney, A. B. The taxonomy and distribution of the Grylloblattidae. Proceedings of the Entomological Society of Washington 50:86–102; 1948.

Gurney, A. B. Corrodentia. In: Pest control technology, entomological section. New York: National Pest Control Association; 1950:p. 129–163.

Gurney, A. B. Praying mantids of the United States: native and introduced. Smithsonian Institution Report #1950; 1951:p. 339–362.

Guthrie, D. M.; Tindall, R. The biology of the cockroach. London: Edward Arnold; 1968.

Harper, P. P. Plecoptera. In: Merritt, R. W.; Cummins, K. W., eds. An introduction to the aquatic insects of North America. Dubuque, IA: Kendall/Hunt; 1978:p. 105–118.

Harwood, R. F.; James, M. T. Entomology in human and animal health. 7th ed. New York: Macmillan: 1979.

Hebard, M. The Dermaptera and Orthoptera of Illinois. Illinois Natural History Survey Bulletin 20(3); 1934.

Helfer, J. R. How to know the grasshoppers, cockroaches, and their allies. Dubuque, IA: Wm. C. Brown; 1963.

Hennig, W. Die Stammesgeschichte der Insekten. Frankfurt-am-Main: Dramer; 1969.

Henry, T. J.; Froeschner, R. C. Catalog of the Heteroptera, or true bugs, of Canada and the continental United States. Gainesville, FL: Sandhill Crane Press; 1988.

Herring, J. L.; Ashlock, P. D. A key to the nymphs of the families of Hemiptera (Heteroptera) of North America north of Mexico. Florida Entomologist 54:207–212; 1971.

Hitchcock, S. W. The Plecoptera or stoneflies of Connecticut. Guide to the insects of Connecticut VII, Bulletin 107; 1974.

Hopkins, G. H. E. The host associations of the lice of mammals. Proceedings of the Zoological Society of London 119:387–604; 1949.

Hopkins, G. H. E.; Clay, T. A checklist of the genera and species of Mallophaga. London: British Museum (Natural History); 1952.

Hubbard, M. D. Mayflies of the world: a catalog of the family and genus group taxa (Insecta: Ephemeroptera). [Flora and fauna handbook No. 8] Gainesville, FL: Sandhill Crane Press; 1990.

Hungerford, H. B. Hemiptera. In: Edmondson, W. T., ed. Freshwater biology. New York: John Wiley & Sons; 1959:p. 959–972.

Hynes, H. B. N. Biology of Plecoptera. Annual Review of Entomology 21: 135–153; 1976.

Hynes, H. B. N. A key to the adults and nymphs of stoneflies (Plecoptera). Scientific Publications of the Freshwater Biological Association, No. 14; 1977.

Jewett, S. G., Jr. Plecoptera. In: Usinger, R. L., ed. Aquatic insects of California. Berkeley: University of California Press; 1956.

Keilin, D.; Nuttall, G. H. F. Iconographic studies of *Pediculus humanus.* Parasitology 22:1–10; 1930.

Kennedy, J. S.; Stroyan, H. L. G. Biology of aphids. Annual Review of Entomology 4:139–160; 1959.

Kevan, D. K. McE. Soil animals. New York: Philosophical Library; 1962.

Kim, K. C. Evolution and host associations of Anoplura. In: Kim, K. C., ed. Coevolution of parasitic arthropods and mammals. New York: John Wiley & Sons; 1985:p. 197–231.

Kim, K. C.; Ludwig, H. W. The family classification of the Anoplura. Systematic Entomology 3:249–284; 1978.

Kim, K. C.; Pratt, H. D.; Stojanovich, C. J. The sucking lice of North America. An illustrated manual for identification. University Park, PA: Pennsylvania State University Press; 1986.

Krishna, K.; Weesner, F. M., editors. Biology of termites. Vol. 1. New York: Academic Press; 1969.

Krishna, K.; Weesner, F. M. Biology of termites. Vol. 2. New York: Academic Press; 1970.

Kristensen, N. P. The phylogeny of hexapod "orders." A critical review of recent accounts. Zeitscrift für Zoologische Systematik und Evolutionsforschung 13:1–44; 1975.

Kühnelt, W. Soil biology with special reference to the animal kingdom. London: Faber and Faber; 1961.

Lee, K. E.; Wood, T. G. Termites and soils. New York: Academic Press; 1971.

Lewis, T. Thrips—their biology, ecology, and economic importance. New York: Academic Press; 1973.

Lindauer, M. Social behavior and mutual communication. In: Rockstein, M., ed. The physiology of Insecta. Vol. II. New York: Academic Press; 1965:p. 124–187.

Manton, S. M. Arthropoda: introduction. In: Florkin, M.; Sheer, B. J., eds. Chemical Zoology. Vol. 5. New York: Academic Press; 1970:p. 1–34.

Marshall, A. G. The ecology of ectoparasitic insects. London: Academic Press; 1981.

Matsuda, R. Morphology and evolution of the insect thorax. Memoirs of the Entomological Society of Canada 76; 1970.

Maynard, E. A. A monograph of the Collembola or springtail insects of New York State. Ithaca, NY: Comstock; 1951.

McCafferty, W. P. The burrowing mayflies (Ephemeroptera: Ephemeroidea) of the United States. Transactions of the American Entomological Society 101:447–504; 1975.

McCafferty, W. P. Toward a phylogenetic classification of the Ephemeroptera (Insecta)—a commentary on systematics. Annals of the Entomological Society of America 84:343–360; 1991.

McKittrick, F. A. Evolutionary studies of cockroaches. Memoirs of the Cornell University Experiment Station 389; 1964.

Merritt, R. W.; Cummins, K. W., editors. An introduction to the aquatic insects of North America. Dubuque, IA: Kendall/Hunt; 1984.

Metcalf, Z. P. A bibliography of the Homoptera (Auchenorrhyncha). Vols. 1 and 2. Raleigh, NC: North Carolina State College, Dept. of Zoology and Entomology; 1945.

Metcalf, Z. P. General catalog of the Homoptera. Raleigh, NC: University of North Carolina at Raleigh; 1954–1963.

Metcalf, Z. P. General catalog of the Homoptera. USDA Agricultural Research Service; 1962–1967.

Mills, H. B. A monograph of the Collembola of Iowa. Ames, IA: Iowa State College Press; 1934.

Mockford, E. L. Order Psocoptera. In: F. W. Stehr, ed. Immature Insects. Dubuque, IA: Kendall/Hunt; 1987:p. 196–214.

Mockford, E. L.; Gurney, A. B. A review of the psocids, or book-lice and bark-lice of Texas (Psocoptera). Journal of the Washington Academy of Science 46:353–368; 1956.

Nault, L. R.; Rodriguez, J. G., editors. Leafhoppers and planthoppers. New York: John Wiley & Sons; 1985.

Needham, J. G.; Claassen, P. W. A monograph of the Plecoptera or stoneflies of America north of Mexico. Thomas Say Foundation Publication No. 2. College Park, MD: Entomological Society of America; 1925.

Needham, J. G.; Traver, J. R.; Hsu, Y. C. The biology of mayflies with a systematic account of North American species. Ithaca, NY: Comstock; 1935.

Needham, J. G.; Westfall, M. J. A manual of the dragonflies of North America (Anisoptera). Berkeley: University of California Press; 1955.

Otte, D. The North American grasshoppers. Vol. 1. Acrididae, Gomphocerinae, and Acridinae. Cambridge, MA: Harvard University Press; 1981.

Otte, D. The North American grasshoppers. Vol. 2. Acrididae, Oedipodinae. Cambridge, MA: Harvard University Press; 1984.

Palmer, J. M.; Mound, L. A.; du Heaume, G. J. Guides to insects of importance to man. No. 2. Thysanoptera. London: C.A.B. International; 1989.

Pennak, R. W. Fresh-water invertebrates of the United States. New York: Ronald Press; 1978.

Polhemus, J. T. Aquatic and semiaquatic Hemiptera. In: Merritt, R. W.; Cummins, K. W., eds. An introduction to the aquatic insects of North America. Dubuque, IA: Kendall/Hunt; 1978:p. 119–131.

Popham, E. S. A key to Dermaptera subfamilies. The Entomologist 98:126–136; 1965.

Ragge, D. R. Dictyoptera (cockroaches and mantises). In: Smith, K. G. V., ed. Insects and other arthropods of medical importance. London: British Museum (Natural History); 1973:399–403.

Remington, C. L. The "Apterygota." In: Kessel, E. L., ed. A century of progress in the natural sciences, 1853–1953. San Francisco: California Academy of Sciences; 1954:p. 495–505.

Remington, C. L. The suprageneric classification of the order Thysanura (Insecta). Annals of the Entomological Society of America 47:277–286; 1956.

Richards, O. W.; Davies, R. G. Imm's general textbook of entomology. 10th ed. Vol 1. Structure, physiology, and development. Vol 2. Classification and biology. New York: John Wiley & Sons (Halsted Press); 1977.

Ricker, W. E. Plecoptera. In: Edmondson, W. T., ed. Fresh-water biology. New York: John Wiley & Sons; 1959.

Ross, E. S. A revision of the Embioptera of North America. Annals of the Entomological Society of America 33:629–676; 1940.

Ross, E. S. A revision of the Embioptera, or webspinners, of the New World. Proceedings of the U.S. National Museum 94:401–504; 1944.

Ross, E. S. Biosystematics of the Embioptera. Annual Review of Entomology 15:157–172; 1970.

Ross, E. S. A synopsis of the Embiidina of the United States. Proceedings of the Entomological Society of Washington 86:82–93; 1984.

Roth, L. M.; Willis, E. R. The medical and veterinary importance of cockroaches. Smithsonian Institution Miscellaneous Collections 134(10); 1957.

Roth, L. M.; Willis, E. R. The biotic associations of cockroaches. Smithsonian Institution Miscellaneous Collections 141; 1960.

Salmon, J. T. An index to the Collembola. Bulletin of the Royal Society of New Zealand No. 7. Vols. 1–3; 1964–1965.

Schaller, F. Soil animals. Ann Arbor, MI: University of Michigan Press; 1968.

Scott, H. G. Collembola: pictorial keys to the Nearctic genera. Annals of the Entomological Society of America 54:104–113; 1961.

Sharov, A. G. Basic arthropodan stock with special reference to insects. New York: Pergamon; 1966.

Slabaugh, R. E. A new thysanuran, and a key to the domestic species of Lepismatidae (Thysanura) found in the United States. Entomological News 5:95–98; 1940.

Slater, J. A.; Baranowski, R. M. How to know the true bugs (Hemiptera-Heteroptera). Dubuque, IA: Wm. C. Brown; 1978.

Smart, J. Instructions for collectors No. 4A. Insects. 4th ed. London: British Museum (Natural History); 1962.

Smith, E. L. Evolutionary morphology of external insect genitalia. I. Origin and relationships to other appendages. Annals of the Entomological Society of America 62:1051–1079; 1969.

Smith, E. L. Biology and structure of some California bristletails and silverfish (Apterygota: Microcoryphia, Thysanura). Pan-Pacific Entomologist 46:212–225; 1970.

Smith, R. F.; Pritchard, A. E. Odonata. In: Usinger, R. L., ed. Aquatic insects of California. Berkeley: University of California Press; 1956.

Stannard, L., Jr. The phylogeny and classification of the suborder Tubulifera (Thysanoptera). Illinois Biological Monographs No. 25; 1957.

Stannard, L., Jr. The thrips, or Thysanoptera, of Illinois. Illinois Natural History Survey Bulletin 29:215–552; 1968.

Stehr, F. W., editor. Immature insects. Vols. 1 and 2. Dubuque, IA: Kendall/Hunt; 1987 & 1991.

Steinman, H. World catalogue of Dermaptera. Budapest: Kluwer Academic Publishers; 1989.

Tuxen, S. L. The Protura. A revision of the species of the world with keys for determination. Paris: Hermann; 1964.

Tuxen, S. L. Taxonomist's glossary of genitalia in insects. 2nd ed. New York: Stechert-Hafner Service Agency, Inc.; 1970.

Usinger, R. L., editor. Aquatic insects of California, with keys to North American genera and California species. Berkeley, CA: University of California Press; 1956.

Uvarov, B. Grasshoppers and locusts. A handbook of general acridology. Vol. 1. Anatomy, physiology, development, phase polymorphism, introduction to taxonomy. Cambridge, England: Cambridge University Press; 1966.

Uvarov, B. A handbook of general acridology. Vol. 2. Behavior, ecology, biogeography, population dynamics. London: Centre for Overseas Pest Research; 1977.

Walker, E. M. The Odonata of Canada and Alaska. Vol. 1. General, the Zygoptera—damselflies. Toronto: University of Toronto Press; 1953.

Walker, E. M. The Odonata of Canada and Alaska. Vol. 2. The Anisoptera—four families. Toronto: University of Toronto Press; 1958.

Walker, E. M.; Corbet, P. S. The Odonata of Canada and Alaska. Vol. 3. The Anisoptera—three families. Toronto: University of Toronto Press; 1975.

Wallwork, J. A. Ecology of soil animals. New York: McGraw-Hill; 1970.

Weesner, F. M. The termites of the United States—a handbook. Elizabeth, NJ: National Pest Control Association; 1965.

Westfall, M. J. Odonata. In: Merritt, R. W.; Cummins, K. W., eds. An introduction to the aquatic insects of North America. Dubuque, IA: Kendall/Hunt; 1978:p. 81–98.

Wilson, E. O. The insect societies. Cambridge, MA: Belknap Press/Harvard University Press; 1971.

Wilson, E. O. Sociobiology—the new synthesis. Cambridge, MA: Belknap Press/Harvard University Press; 1975.

Wygodzinsky, P. A revision of the silverfish (Lepismatidae, Thysanura) of the United States and the Caribbean area. American Museum Novitates 2481:1–26; 1972.

CHAPTER 13

Andrews, C. The lives of wasps and bees. New York: American Elsevier; 1971.

Arnett, R. H. The beetles of the United States. 2nd ed. Marlton, NJ: World Natural History; 1980.

Askew, R. R. Parasitic insects. London: Heinemann Educational Books; 1973.

Aspöck, H.; Aspöck, A. The present state of knowledge on the Raphidioptera of America (Insecta, Neuroptoidea). Polskie Pismo Entomologiczne 45:537–546; 1975.

Bänziger, H. Bloodsucking moths of Malaya. Fauna 1:5–16; 1971.

Bänziger, H. Skin-piercing bloodsucking moths I: Ecological and ethological studies on Calpe eustrigata (Lepidoptera, Noctuidae). Acta Tropica 32:125–144; 1975.

Blackwelder, R. E.; Arnett, R. H., Jr. Checklist of the beetles of Canada, United States, Mexico, Central America, and the West Indies. Marlton, NJ: World Natural History; 1977.

Bohart, R. M. A revision of the Strepsiptera with special reference to the species of North America. University of California Publications in Entomology 7:91–160; 1941.

Bohart, R. M.; Menke, A. S. Sphecid wasps of the world. Berkeley: University of California Press; 1976.

Booth, R. G.; Cox, M. L.; Madge, R. B. Guides to insects of importance to man. No. 3. Coleoptera. London: C.A.B. International; 1990.

Boudreaux, H. B. Arthropod phylogeny with special reference to insects. New York: John Wiley & Sons; 1979.

Byers, G. W. Notes on North American Mecoptera. Annals of the Entomological Society of America 47:484–510; 1954.

Byers, G. W. The life history of *Panorpa nuptialis* (Mecoptera: Panorpidae). Annals of the Entomological Society of America 56:142–149; 1963.

Byers, G. W. Families and genera of Mecoptera. Proceedings of the XII International Congress of Entomology; 1965:123.

Byers, G. W. Ecological distribution and structural adaptation in the classification of Mecoptera. Proceedings of the XIII International Conference of Entomology 1:486; 1968.

Carpenter, F. M. The biology of the Mecoptera. Psyche 38:41–55; 1931a.

Carpenter, F. M. Revision of Nearctic Mecoptera. Bulletin of the Museum of Comparative Zoology, Harvard University 72:205–277; 1931b.

Chandler, H. P. Megaloptera. In: Usinger, R. L., ed. Aquatic insects of California. Berkeley: University of California Press; 1956:p. 229–233.

Chu, H. F. How to know the immature insects. Dubuque, IA: Wm. C. Brown; 1949.

Clausen, C. P. Entomophagous insects. New York: McGraw-Hill; 1940.

Cole, F. R. The flies of western North America. Berkeley: University of California Press; 1969.

Common, I. F. B. Evolution and classification of the Lepidoptera. In: The insects of Australia. Canberra: C.S.I.R.O.; 1975:p. 765–866.

Covell, C. V., Jr. A field guide to the moths of eastern North America. Boston: Houghton Mifflin; 1984.

Creighton, W. S. The ants of North America. Harvard University Museum of Comparative Zoology Bulletin 104; 1950.

Cummins, K. W.; Miller, L. D.; Smith, N. A.; Fox, R. M. Experimental entomology. New York: Van Nostrand Reinhold Company; 1965.

Curran, C. H. The families and genera of North American Diptera. Woodhaven, NY: H. Tripp; 1934.

Denning, D. G. Trichoptera. In: Usinger, R. L., ed. Aquatic insects of California. Berkeley: University of California Press; 1956:p. 237–270.

Dillon, E. S.; Dillon, L. S. A manual of common beetles of eastern North America. New York: Dover; 1972. 2 vol.

Dominick, R. B., editor. The moths of America north of Mexico. Faringdon, UK: E. W. Classey; 1971–present.

Ehrlich, P. R.; Ehrlich, A. H. How to know the butterflies. Dubuque, IA: Wm. C. Brown; 1961.

Emmel, T. C. Butterflies: their world, their life cycle, their behavior. New York: Knopf; 1975.

Essig, E. O. College entomology. New York: MacMillan; 1942.

Essig, E. O. Insects and mites of Western/North America. New York: Macmillan Publishing Co., Inc.; 1958.

Evans, E. D. Megaloptera and aquatic Neuroptera. In: Merritt, R. W.; Cummins, K. W., eds. An introduction to the aquatic insects of North America. Dubuque, IA: Kendall/Hunt; 1978:p. 133–145.

Evans, G. The life of beetles. London: Allen and Unwin; 1975.

Evans, H. E. Wasp farm. Garden City, NY: Natural History Press; 1963.

Evans, H. E.; Eberhard, M. T. W. The wasps. Ann Arbor, MI: University of Michigan Press; 1970.

Ewing, H. E.; Fox, I. The fleas of North America. USDA Miscellaneous Publications No. 500; 1943.

Fox, I. Fleas of eastern United States. New York: Hafner; 1940.

Frison, T. H. Descriptions of Plecoptera. Illinois Natural History Survey, Bulletin 21(3):78–99; 1937.

Froeschner, R. C. Notes and keys to the Neuroptera of Missouri, Annals of the Entomological Society of America 40(1):123–36.

Greenberg, B. Flies and disease. Vol 1. Ecology, classification, and biotic associations. Princeton, NJ: Princeton University Press; 1971.

Gurney, A. B.; Parfin, S. Neuroptera. In: Edmondson, W. T., ed. Freshwater biology. New York: John Wiley & Sons; 1959:p. 973–980.

Hagen, K. S. Aquatic Hymenoptera. In: Merritt, R. W.; Cummins, K. W., eds. An introduction to the aquatic insects of North America. Dubuque, IA: Kendall/Hunt; 1978: p. 233–243.

Harwood, R. F.; James, M. T. Entomology in human and animal health. 7th ed. New York: Macmillan; 1979.

Holland, G. P. Evolution, classification, and host-relationships of Siphonaptera. Annual Review of Entomology 9:123–146; 1964.

Holland, G. P. The fleas of Canada, Alaska, and Greenland (Siphonaptera). Memoirs of the Entomological Society of Canada 130:1–631; 1985.

Holland, W. J. The moth book. New York: Dover; 1968. (reprint of 1903 Doubleday edition).

Holland, W. J. The butterfly book. Revised ed. Garden City, NY: Doubleday; 1931.

Hölldobler, B.; Wilson, E. O. The ants. London: Springer-Verlag; 1990.

Holloway, J. D.; Bradley, J. D.; Carter, D. J. Guides to insects of importance to man. No. 1. Lepidoptera. London: C.A.B. International; 1987.

Hopkins, G. H. E.; Rothschild, M. An illustrated catalogue of the Rothschild collection of fleas (Siphonaptera) in the British Museum (Natural History) (5 vols.). London: British Museum (Natural History); 1953–1971.

Howe, W. H. The butterflies of North America. Garden City, NY: Doubleday; 1975.

Hubbard, C. A. Fleas of western North America. New York: Hafner; 1947.

Huber, J.; Goulet, H., editors. Manual of Nearctic Hymenoptera. Ottawa: Agriculture Canada; 1992.

James, M. T.; Harwood, R. F. Herm's medical entomology. New York: Macmillan Publishing Co., Inc.; 1969.

Jaques, H. E. How to know the beetles. Dubuque, IA: Wm. C. Brown; 1951.

Kathirithamby, J. Review of the order Strepsiptera. Systematic Entomology 14:41–92; 1989.

Kinzelbach, R. The systematic position of the Strepsiptera (Insecta). American Entomologist 36:292–303; 1991.

Klots, A. B. A field guide to the butterflies. Boston: Houghton Mifflin; 1951.

Klots, A. B. The world of butterflies and moths. New York: McGraw-Hill; 1958.

Kristensen, N. P. The phylogeny of the hexapod "orders." A critical review of recent accounts. Zeitschrift für Zoologische Systematik und Evolutionforschung 5:144–297; 1975.

Langer, W. L. The black death. Scientific American 210: 114–121; 1964.

Leech, H. B.; Chandler, H. P. Aquatic Coleoptera. In: Usinger, R. L., ed. Aquatic insects of California. Berkeley: University of California Press; 1956:p. 293–371.

Leech, H. B.; Sanderson, M. W. Coleoptera. In: Edmondson, W. T., ed. Fresh-water biology. New York: John Wiley & Sons; 1959:p. 981–1023.

Lewis, R. E. Notes on the geographic distribution and host preferences in the order Siphonaptera. Parts 1–6. Journal of Medical Entomology 9:511–520, 10:255–260, 11:147–167, 403–413, 525–540, 658–676; 1972–1975.

Mardon, D. K. An illustrated catalogue of the Rothschild collection of fleas (Siphonaptera) in the British Museum (Natural History). Volume VI. Pygiopsyllidae. London: British Museum (Natural History); 1981.

Marshall, A. G. The ecology of ectoparasitic insects. London: Academic Press; 1981.

Matthews, R. W. *Microstigmus comes:* sociality in a sphecid wasp. Science 160:787–788; 1968.

McAlpine, F., editor. Manual of Nearctic Diptera. Vols. 1–3. Ottawa: Agriculture Canada; 1981, 1987, 1989.

Merritt, R. W.; Schlinger, E. I. Aquatic Diptera. Part 2. Adults of aquatic Diptera. In: Merritt, R. W.; Cummins, K. W., eds. An introduction to the aquatic insects of North America. Dubuque, IA: Kendall/Hunt; 1978:p. 259–283.

Metcalf, C. L.; Flint, W. P.; Metcalf, R. L. Destructive and useful insects. New York: McGraw-Hill; 1962.

Michener, C. D. The social behavior of bees. A comparative study. Cambridge, MA: Belknap Press/Harvard University Press; 1974.

Mitchell, R. T.; Zim, H. S. Butterflies and moths. New York: Golden Press; 1964.

Mitchell, T. B. Bees of the eastern United States. Vols. 1 and 2. Technical Bulletin, North Carolina Agricultural Experiment Station Nos. 141 & 152; 1960 and 1962.

Muesebeck, C. F. W.; Krombein, K. V.; Townes, H. K.; et al. Hymenoptera of America north of Mexico; synoptic catalog. USDA Agricultural Monograph 2, first & second supplements to monograph 2; 1951, 1958, 1967.

Oldroyd, H. The natural history of flies. New York: Norton; 1964.

Pennak, R. W. Fresh-water invertebrates of the United States. New York: Ronald Press; 1978.

Peterson, A. Larvae of insects. Part 1. Lepidoptera and plant infesting Hymenoptera. Columbus, OH: A. Peterson; 1948.

Peterson, A. Larvae of insects, Part II, Coleoptera, Diptera, Neuroptera, Siphonaptera, Mecoptera, Trichoptera. Columbus, OH: A. Peterson; 1951.

Pfadt, R. E. Fundamentals of applied entomology. 3d ed. New York: Macmillan; 1978.

Riek, E. F. Neuroptera (lacewings). In: The insects of Australia. A textbook for students and research workers. Melbourne, Australia: Melbourne University Press; 1970:p. 472–494.

Ross, H. H. Descriptions of nearctic caddisflies. Illinois Natural History Survey Bulletin 21(4); 1938.

Ross, H. H. The caddisflies or Trichoptera of Illinois. Illinois Natural History Survey Bulletin 23:1–326; 1944.

Ross, H. H. Trichoptera. In: Edmondson, W. T., ed. Fresh-water biology. New York: John Wiley & Sons; 1959:p. 1024–1049.

Ross, H. H. How to collect and preserve insects. Illinois Natural History Survey, Circular 39; 1962.

Rothschild, M. Fleas. Scientific American 213:44–50; 1965.

Rothschild, M. Recent advances in our knowledge of the order Siphonaptera. Annual Review of Entomology 20:241–259; 1975.

Sargent, T. D. Legion of the night: The underwing moths. Amherst, MA: University of Massachusetts Press; 1976.

Scott, J. A. Butterflies of North America. Stanford, CA: Stanford University Press; 1986.

Smart, J. Instructions for collectors No. 4A. Insects. 4th ed. London: British Museum (Natural History); 1962.

Smit, F. G. A. M. Siphonaptera (fleas). In: Smith, K. G. V., ed. Insects and other arthropods of medical importance. London: British Museum (Natural History); 1973.

Smit, F. G. A. M. An illustrated catalogue of the Rothschild collection of fleas (Siphonaptera) in the British Museum (Natural History). Vol. VII. Malacopsylloidea. Oxford: Oxford University Press and London: British Museum (Natural History); 1987.

Smith, K. G. V., editor. Insects and other arthropods of medical importance. London: British Museum (Natural History); 1973.

Spradbery, J. P. Wasps. An account of the biology and natural history of social and solitary wasps. Seattle, WA: University of Washington Press; 1973.

Stehr, F. W., editor. Immature insects. Dubuque: IA: Kendall/Hunt; 1987, vol. 1.

Stehr, F. W., editor. Immature insects. Dubuque, IA: Kendall/Hunt; 1991, vol. 2.

Stone, A.; Sabrosky, C. W.; Wirth, W. W.; Foote, R. H.; Coulson, J. R., editors. A catalog of the Diptera of America north of Mexico. USDA Agricultural Handbook 276; 1965.

Teskey, H. J. Aquatic Diptera. Part 1. Larvae of aquatic Diptera. In: Merritt, R. W.; Cummins, K. W., eds. An introduction to the aquatic insects of North America. Dubuque, IA: Kendall/Hunt; 1978:p. 245–257.

Tietz, H. M. An index to the described life histories, early stages, and hosts of the Macrolepidoptera of the continental United States and Canada. Hampton, U.K.: E. W. Classey; 1973.

Traub, R. Coevolution of fleas and mammals. In: Kim., K. C., ed. Coevolution of parasitic arthropods and mammals. New York: John Wiley & Sons; 1985:p. 295–437.

Traub, R.; Rothschild, M.; Haddow, J. In: Rothschild, M.; Traub, R. eds. The Rothschild collection of fleas. The Ceratophyllidae: key to the genera and host relationships with notes on their evolution, zoogeography, and medical importance. London: Academic Press; 1983.

Traub, R.; Starcke, H., editors. Fleas. Rotterdam: A. A. Balkema; 1980.

Ulrich, W. Evolution and classification of the Strepsiptera. Proceedings of the 1st International Congress of Parasitology 1:609–611; 1966.

Urquhart, F. W. The monarch butterfly. Toronto: University of Toronto Press; 1960.

U.S. Department of Agriculture. Insects—the yearbook of agriculture. Washington, DC: U.S. Department of Agriculture; 1952.

Vane-Wright, R. I.; Ackery, P. R. Biology of butterflies. London: Academic Press; 1984.

von Frisch, K. The dance, language, and orientation of bees. Cambridge, MA: Harvard University Press; 1967. Translated from the German by L. E. Chadwick.

Watson, A.; Whalley, P. E. S. The dictionary of butterflies and moths in colour. London: Michael Joseph; 1975.

White, R. E. A field guide to the beetles of North America. Boston: Houghton Mifflin; 1983.

Wiggins, G. B. Larvae of the North American caddisfly genera (Trichoptera). Toronto: University of Toronto Press; 1978.

Wilson, E. O. The insect societies. Cambridge, MA: Belknap Press/Harvard University Press; 1971.

Wilson, E. O. Sociobiology—the new synthesis. Cambridge, MA: Belknap Press/Harvard University Press; 1975.

Wirth, W. W.; Stone, A. Aquatic Diptera. In: Usinger, R. L., ed. Aquatic insects of California. Berkeley: University of California Press; 1956:p. 372–482.

Wolglum, R. S.; McGregor, F. A. Observations on the life history and morphology of *Agulla Bractea* Carpenter (Neuroptera: Raphidiodea: Raphidiidae, Annals of the Entomological Society of America 51:129–141.

Yarrow, I. H. H. Hymenoptera (bees, wasps, ants, etc.). In: Smith, K. G. V., ed. Insects and other arthropods of medical importance. London: British Museum (Natural History); 1973:p. 409–411.

CHAPTER 14

Adams, J. F. Beekeeping—the gentle craft. New York: Avon Books; 1972.

Akre, R. D.; Davis, H. G. Biology and pest status of venomous wasps. Annual Review of Entomology 23:215–238; 1978.

Akre, R. D.; Hansen, L. D.; Zack, R. S. Insect jewelry. American Entomologist 37(2):90–95; 1991.

Askew, R. R. Parasitic insects. New York: American Elsevier; 1971.

Atkins, M. D. Insects in perspective. New York: Macmillan; 1978.

Bailey, L.; Ball, B. Honey bee pathology. 2d ed. San Diego: Academic Press; 1991.

Benenson, A. S., editor. Control of communicable diseases in man. 15th ed. New York: The American Public Health Association; 1990.

Bishopp, F. C. Insects as helpers. In: Yearbook of agriculture. Washington, DC: U.S. Department of Agriculture; 1952:p. 79–87.

Bohart, G. E. Pollination by native insects. In: Yearbook of agriculture. Washington, DC: U.S. Department of Agriculture; 1952:p. 107–121.

Bohart, G. E. Management of wild bees for the pollination of crops. Annual Review of Entomology 17:287–312; 1972.

Borror, D. J.; Triplehorn, C. A.; Johnson, N. F. An introduction to the study of insects. 6th ed. Philadelphia: Saunders College Publishing; 1989.

Brenner, R. J.; Barnes, K. C.; Helm, R. M.; Williams, L. W. Modernized society and allergies to arthropods. American Entomologist 37(3):143–155; 1991.

Bucherl, W.; Buckley, E. E., editors. Venomous animals and their venoms. Vol. 3. Venomous invertebrates. New York: Academic Press; 1972.

Burkhardt, C. C. Insect pests of corn. In: Pfadt, R. E., ed. Fundamentals of applied entomology. 3d ed. New York: Macmillan; 1978:p. 303–334.

Busvine, J. R. Insects and hygiene. London: Methuen; 1966.

Busvine, J. R. Arthropod vectors of disease. London: Edward Arnold; 1975.

Butler, C. G. The world of the honeybee. London: Collins; 1971.

Carter, R. L. Insects in relation to plant disease. 2d ed. New York: John Wiley & Sons; 1973.

Cherry, R. H. History of sericulture. Bulletin of the Entomological Society of America 33(2):83–84; 1987.

Cherry, R. H. Use of insects by Australian aborigines. American Entomologist 37(1):9–13; 1991.

Clausen, L. W. Insect fact and folklore. New York: Macmillan (Collier Books); 1954.

Cloudsley-Thompson, J. L. Insects and history. London: Weidenfeld and Nicolson; 1976.

Crane, E. The world's beekeeping—past and present. In: Grout, R. A., ed. The hive and the honey bee. Rev. ed. Hamilton, IL: Dadant & Sons; 1963:p. 1–18.

Crane, E., editor. Honey, a comprehensive survey. London, Heineman; 1975.

Cushing, E. C. History of entomology in World War II. Smithsonian Institute Publication 4294; 1957.

Dalton, S. Borne on the wind. The extraordinary world of insects in flight. New York: Dutton; 1975.

Davidson, R. H.; Lyon, W. F. Insect pests of farm, garden, and orchard. 8th ed. New York: John Wiley & Sons; 1987.

DeFoliart, G. The human use of insects as food and as animal feed. Bulletin of the Entomological Society of America 35:27–35; 1989.

Dethier, V. G. To know a fly. San Francisco: Holden-Day; 1962.

Ebeling, W. Urban entomology. Berkeley: University of California Press; 1978.

Felt, E. P. Plant galls and gall makers. Ithaca, NY: Comstock; 1940.

Fletcher, D. J. C. The African bee, *Apis mellifera adansonii,* in Africa. Annual Review of Entomology 23:151–171; 1978.

Frazier, C. A.; Brown, F. K. Insects and allergy and what to do about them. Norman, OK: University of Oklahoma Press; 1980.

Free, J. B. Insect pollination of crops. New York: Academic Press; 1970.

Fronk, W. D. Chemical control. In: Pfadt, R. E., ed. Fundamentals of applied entomology. 3d ed. New York: Macmillan; 1978:p. 209–240.

Gillett, J. D. The mosquito: its life, activities, and impact on human affairs. Garden City, NY: Doubleday & Co.; 1972.

Glover, P. M. Lac cultivation in India. San Francisco: Freeman; 1937.

Goff, M. L. Feast of clues. The Sciences 31(4):30–35. The New York Academy of Sciences; 1981.

Gojmerac, W. L. Bees, beekeeping, honey, and pollination. Westport, CN: AVI Publishing Co; 1980.

Gould, J. L.; Gould, C. G. The honey bee. New York: Scientific American Books; 1988.

Greenberg, B. Flies and disease. Vol. I. Ecology, classification, and abiotic associations. Princeton, NJ: Princeton University Press; 1971.

Greenberg, B. Flies and disease. Vol. II. Biology and disease transmission. Princeton, NJ: Princeton University Press; 1973.

Grout, R. A. The production and uses of beeswax. In: Grout, R. A., ed. The hive and the honey bee. Hamilton, IL: Dadant & Sons; 1963:p. 425–436.

Hahn, J. D.; Ascerno, M. E. Public attitudes toward urban arthropods in Minnesota. American Entomologist 37(3):179–184; 1991.

Harwood, R. F.; James, M. T. Entomology in human and animal health. 7th ed. New York: Macmillan Publishing Co., Inc.; 1979.

Hill, D. S. Pests of stored products and their control. Boca Raton, FL: CRC Press; 1990.

Hogue, C. Cultural entomology. Annual Review of Entomology 32:181–199; 1987.

Horie, Y.; Wantanabe, H. Recent advances in sericulture. Annual Review of Entomology 25:49–71; 1980.

Horsfall, W. R. Medical entomology—arthropods and human disease. New York: The Ronald Press Co.; 1962.

Jay, S. C. Spatial management of honey bees on crops. Annual Review of Entomology 31:49–65; 1986.

Johansen, C. A. Pesticides and pollinators. Annual Review of Entomology 22:177–192; 1977.

Jones, D. P. Agricultural entomology. In: Smith, R. F.; Mittler, T. E.; Smith, C. N., eds. History of entomology. Palo Alto, CA: Annual Reviews; 1973:p. 307–332.

Keh, B. Scope and application of forensic entomology. Annual Review of Entomology 30:134–154; 1985.

Kettle, D. S. Medical and veterinary entomology. New York: John Wiley & Sons; 1984.

Kevan, P. G.; Baker, H. G. Insects as flower visitors and pollinators. Annual Review of Entomology 28:407–453; 1983.

Kritsky, G. Beetle gods of ancient Egypt. American Entomologist 37(2):85–89; 1991.

Leclercq, M. L. Entomological parasitology. New York: Pergamon Press, Inc.; 1969.

Lehane, M. J. Biology of blood-sucking insects. London: Harper Collins Academic; 1991.

Lofgren, C. S.; Banks, W. A.; Glancey, B. M. Biology and control of imported fire ants. Annual Review of Entomology 20:1–30; 1975.

Lovell, H. B. The honey bee as a pollinating agent. In: Grout, R. A., ed. The hive and the honey bee. Hamilton, IL: Dadant & Sons; 1963:p. 463–476.

Maramorosch, K.; Harris, K. F., editors. Plant diseases and vectors: ecology and epidemiology. New York: Academic Press; 1981.

Marquis, D. Archy and Mehitabel. Garden City, NY: Dolphin Books, Doubleday & Company, Inc.; 1960.

Marquis, D. Archy's life of Mehitabel. Garden City, NY: Dolphin Books, Doubleday & Company, Inc.; 1966.

Martin, E. C.; McGregor, S. E. Changing trends in insect pollination of commercial crops. Annual Review of Entomology 18:207–226; 1973.

McGregor, S. E. Insect pollination of cultivated crop plants. U.S. Department of Agriculture, Agriculture Handbook No. 496; 1976.

McKelvey, J. J., Jr. Man against tsetse—struggle for Africa. Ithaca, NY: Cornell University Press; 1973.

McKelvey, J. J., Jr. Insects and human welfare. In: Pimentel, D., ed. Insects, science and society. New York: Academic Press; 1975:p. 13–24.

McKelvey, J. J., Jr.; Eldridge, B. F.; Maramorosch, K. Vectors of disease agents—interactions with plants, animals, and man. New York: Praeger; 1981.

Metcalf, C. L.; Flint, W. P.; Metcalf, R. L. Destructive and useful insects. New York: McGraw-Hill Book Company; 1962.

Metcalf, R. L.; Luckman, W. H. Introduction to insect pest management. New York: John Wiley & Sons; 1975.

Michener, C. D. The Brazilian bee problem. Annual Review of Entomology 20:399–416; 1975.

Minton, S. A. Venom diseases. Springfield, IL: C. C. Thomas; 1974.

Monath, T. P., editor. The arboviruses: epidemiology and ecology. Vol. 1–5. Boca Raton, FL: CRC Press, Inc.; 1988.

More, D. The bee book. New York: Universe Books; 1976.

Moritz, R. F. A.; Southwick, E. E. Bees as super organisms. New York: Springer-Verlag, Inc.; 1992.

Morse, R. A. Bees and beekeeping. Ithaca, NY: Cornell University Press; 1975.

Morse, R. A., ed. Honey bee pests, predators, and diseases. Ithaca, NY: Cornell University Press; 1978.

Nash, T. A. M. Africa's bane—the tsetse fly. St. Jame's Place, London: Collins; 1969.

National Academy of Sciences. Insect-pest management and control. Washington, DC: Subcommittee on Insect Pests, N.A.S.; 1969.

Nault, L. R.; Ammar, E. D. Leafhopper and planthopper transmission of plant viruses. Annual Review of Entomology 34:504–529; 1985.

Nørgaard Holm, S. The utilization and management of bumble bees for red clover and alfalfa seed production. Annual Review of Entomology 11:155–182; 1976.

Oldroyd, H. The natural history of flies. New York: W. W. Norton Company, Inc.; 1964.

Olkowski, H.; Olkowski, W. Entomophobia in the urban ecosystem; some observations and suggestions. Bulletin of the Entomological Society of America 22(3):313–317; 1976.

Parish, L. C.; Nutting, W. B.; Schwartzman, R. M., editors. Cutaneous infestations of man and animal. New York: Praeger; 1983.

Pedigo, L. P. Entomology and pest managment. New York: Macmillan Publishing Co.; 1989.

Pfadt, R. E., editor. Fundamentals of applied entomology. 3d ed. New York: Macmillan; 1978.

Philip, C. B.; Rozeboom, L. E. Medico-veterinary entomology: a generation of progress. In: Smith, R. F.; Mittler, T. E.; Smith, C. N., eds. History of entomology. Palo Alto, CA: Annual Reviews, Inc.; 1973.

Piek, T., editor. Venoms of the hymenoptera—biochemical, pharmacological, and behavioral aspects. New York: Academic Press; 1986.

Pimentel, D. Introduction. In: Pimentel, D., ed. Insects, science, and society. New York: Academic Press; 1975.

Pomeranz, C. Arthropods and psychic disturbances. Bulletin of the Entomological Society of America 5:65–67; 1959.

Pyenson, L. L.; Barke, H. E. Fundamentals of entomology and plant pathology. Westport, CT: AVI Publishing Co., Inc.; 1977.

Real, L. Pollination biology. New York, NY: Academic Press, Inc.; 1983.

Reeves, W. C.; Asman, S. M.; Hardy, J. L.; Milby, M. M.; Reisen, W. K. Epidemiology and control of mosquito-borne arboviruses in California, 1943–1987. Sacramento, CA: California Mosquito and Vector Control Association, Inc.; 1990.

Rhoades, R. B. Medical aspects of the imported fire ant. Gainesville, FL: University Presses of Florida; 1977.

Richards, A. J., editor. The pollination of flowers by insects. New York, NY: Academic Press, Inc.; 1978.

Ritchie, C. I. A. Insects, the creeping conquerors. New York: Elsevier/Nelson Books; 1979.

Rolston, L. H.; McCoy, C. E. Introduction to applied entomology. New York: The Ronald Press Company; 1966.

Sandved, K. B.; Brewer, J. Butterflies. New York: Harry N. Abrams, Inc., Publishers; 1976.

Schmidt, J. O. Allergy to Hymenoptera venoms. In: Piek, T., ed. Venoms of the Hymenoptera—Biochemical, pharmacological, and behavioral aspects. New York: Academic Press; 1986:p. 509–546.

Scott, T. W.; Grumstrup-Scott, J. The role of vector-host interactions in disease transmission. Proceedings of a Symposium, Entomological Society of America, Miscellaneous Publication No. 68; 1988.

Schwerdtfeger, F. Forest entomology. In: Smith, R. F.; Mittler, T. E.; Smith, C. N., eds. History of entomology. Palo Alto, CA: Annual Reviews, Inc.; 1973.

Service, M. W., editor. Demography and vector-borne diseases. Boca Raton, FL: CRC Press, Inc.; 1989.

Smith, K. G. V. A manual of forensic entomology. Ithaca, NY: Cornell University Press; 1986.

Smith, K. G. V., editor. Insects and other arthropods of medical importance. London: British Museum (Natural History); 1973.

Snow, K. R. Insects and disease. New York: John Wiley & Sons; 1974.

Southwood, T. R. E. Entomology and mankind. Proceedings of the XV International Congress of Entomology; 1977a:36–51.

Southwood, T. R. E. Entomology and mankind. American Scientist 65:30–39; 1977b.

Spivak, M.; Fletcher, D. J. C.; Breed, M. D., editors. The "African" honey bee. Boulder, CO: Westview; 1991.

Swan, L. A. Beneficial insects. New York: Harper & Row; 1964.

Sylvester, E. S. Circulative and propagative virus transmission by aphids. Annual Review of Entomology 25:257–286; 1980.

Taylor, R. L. Butterflies in my stomach. Santa Barbara, CA: Woodbridge; 1975.

Taylor, R. L.; Carter; B. J. Entertaining with insects. Santa Barbara, CA: Woodbridge Press; 1976.

Theiler, M.; Downs, W. G. Arthropod borne viruses of vertebrates. New Haven, CT: Yale University Press; 1973.

Townsend, G. F.; Crane, E. History of apiculture. In: Smith, R. F.; Mittler, T. E.; Smith, C. N., eds. History of entomology. Palo Alto, CA: Annual Reviews, Inc.; 1973.

Tu, A. T. Venoms: chemistry and molecular biology. New York: John Wiley & Sons; 1977.

van Emden, H. F. Pest control. 2nd ed. London and New York: Edward Arnold; 1989.

White, G. B., editor. Medical entomology. In: Manson-Bahr, P. E. C.; Bell, D. R., eds. Manson's tropical diseases. 19th ed. London & Philadelphia: Bailliere Tindall; 1987:p. 1381–1488.

White, J. W. Honey. In: Grout, R. A., ed. The hive and the honey bee. Hamilton, IL: Dadant & Sons; 1963:p. 369–406.

Wilbur, D. A.; Mills, R. B. Stored grain insects. In: Pfadt, R. E., ed. Fundamentals of applied entomology. 3d ed. New York: Macmillan; 1978:p. 573–603.

Winston, M. L. The biology of the honeybee. Cambridge, MA: Harvard Univ. Press; 1987.

Winston, M. L. Killer bees—The Africanized honey bee in the Americas. Cambridge, MA: Harvard University Press; 1992.

Yokoyama, T. The history of sericultural science in relation to industry. In: Smith, R. F.; Mittler, T. E.; Smith, C. N., eds. History of entomology. Palo Alto, CA: Annual Reviews, Inc.; 1973.

Zumpt, F. Myiasis in man and animals in the Old World. London: Butterworth; 1965.

CHAPTER 15

Barbosa, P.; Schultz, J. C., editors. Insect outbreaks. New York: Academic Press; 1987.

Baumhover, A. H. Eradication of the screwworm fly. Journal of the American Medical Association 196(3):150–158; 1966.

Bay, E. C. Mosquito control by fish: a present day appraisal. WHO chronicle 21(10):415–423; 1967.

Beck, S. D. Insect photoperiodism. New York: Academic Press, Inc.; 1965.

Beroza, M., editor. Pest management with insect sex attractants. Washington, DC: American Chemical Society;1976.

Bohmont, B. C. Standard pesticide user's guide: revised and enlarged. Englewood Cliffs, NJ: Prentice-Hall; 1990.

Börkovec, A. B. Control of insects by sexual sterilization. In: Jacobson, M., ed. Insecticides of the future. New York: Marcel Dekker, Inc.; 1975.

Brewer, K. K. Application of molecular techniques to improved detection of insecticide resistance. American Entomologist 37(2):96–103; 1991.

Briggs, J. D. Biological regulation of vectors—the saprophytic and aerobic bacteria and fungi. DHEW Publication No.(NIH); 1975:77–1180.

Brown, A. W. A. Insecticide resistance comes of age. Bulletin of the Entomological Society of America 14(1):3–9; 1968.

Brown, A. W. A. The ecology of pesticides. New York: Wiley-Interscience; 1978.

Bulla, L. A., editor. Regulation of insect populations by microorganisms. Annals of the New York Academy of Science, Vol. 217; 1973.

Burges, H. D. Microbial control of pests and plant diseases, 1970–1980. New York: Academic Press; 1981.

Burges, H. D.; Hussey, N. W., editors. Microbial control of insects and mites. New York: Academic Press; 1971.

Burn, A. J.; Coaker, T. H.; Jepson, P. C., editors. Integrated pest management. New York: Academic Press; 1988.

Cameron, J. W. M. Insect pathology. In: Smith, R. F.; Mittler, T. E.; Smith, C. N., eds. History of entomology. Palo Alto, CA: Annual Reviews, Inc.; 1973:p. 267–284.

Carson, R. L. Silent Spring. Boston, MA: Houghton Mifflin; 1962.

Caswell, R. L., editor. Pesticide handbook—entoma. College Park, MD: The Entomological Society of America; 1977.

Coppel, H. C.; Mertins, J. W. Biological insect pest suppression. New York: Springer-Verlag; 1977.

Corbett, J. R.; Wright, K.; Baillie, A. C. Biochemical mode of action of pesticides. 2nd ed. New York: Academic Press; 1984.

Croft, B. A. Arthropod biological control agents and pesticides. New York: John Wiley & Sons; 1990.

Cross, B.; Scher, H. B., editors. Pesticide formulations: innovations and developments. Washington, DC: American Chemical Society; 1988.

Curtis, C. F. Genetic control of insect pests—growth industry or lead balloon. Biological Journal of the Linnean Society 26:359–374; 1986.

Curtis, C. F. Appropriate technology in vector control. Boca Raton, FL: CRC Press, Inc.; 1989.

Davidson, G. Genetic control of insect pests. New York: Academic Press; 1974.

DeBach, P. Biological control by natural enemies. London: Cambridge University Press; 1974.

deWilde, J.; Schoonhoven, L. M. editors. Insect and host plant. Amsterdam: North-Holland; 1969.

Drummond, R. O.; George, J. E.; Kuntz, S. E. Control of arthropod pests of livestock: a review of technology. Boca Raton, FL: CRC Press, Inc.; 1987.

Ferro, D. N. Potential for resistance to *Bacillus thuringiensis*: Colorado potato beetle, a model system. American Entomologist, in press.

Ferro, D. N.; Lyon, S. M. Colorado potato beetle larval mortality: operative effects of *Bacillus thuringiensis subspecies* San Diego. Journal of Economic Entomology 84(3):806–809.

Ferron, P. Biological control of insect pests by entomogenous fungi. Annual Review of Entomology 23:409–442; 1978.

Foster, G. G.; Whitten, M. J.; Prout, T.; Gill, R. Chromosome rearrangements for the control of insect pests. Science 176:875–880; 1972.

Galun, R. L.; Starks, K. J.; Guthrie, W. D. Plant resistance to insects attacking cereals. Annual Review of Entomology 20:337–357; 1975.

Gaugler, R.; Kaya, H. K. Entomopathogenic nematodes in biological control. Boca Raton, FL: CRC Press, Inc.; 1990.

Georghiou, G. P.; Saito, T. Pest resistance to pesticides. New York: Plenum; 1983.

Georghiou, G. P.; Taylor, C. E. Factors influencing the evolution of resistance. In: Pesticide resistance strategies and tactics for management, National Research Council. Washington, DC: National Academy Press; 1986.

Granados, R. R.; Corsaro, B. G. Baculovirus enhancing proteins and their implications for insect control. Vth International Colloquium on Invertebrate Pathology and Microbial Control; 1990.

Hagen, K. S.; Franz, J. M. A history of biological control. In: Smith, R. F.; Mittler, T. E.; Smith, C. N., eds. History of entomology. Palo Alto, CA: Annual Reviews, Inc.; 1973.

Hanover, J. W. Physiology of tree resistance to insects. Annual Review of Entomology 20:75–95; 1975.

Hague, R.; Freed, V. H. Environmental dynamics of pesticides. New York: Plenum Publishing Company; 1975.

Haskell, P. T., editor. Pesticide application. New York: Oxford University Press; 1985.

Hassall, K. A. The biochemistry and uses of pesticides. New York: VCH Publishers, Inc.; 1990.

Hazzard, R. V.; Ferro, D. N. Feeding responses of adult *Coleomegilla maculata* to eggs of the Colorado potato beetle and the green peach aphid. Environmental Entomology 20(2):644–651; 1991.

Hedin, P. A., editor. Naturally occurring pest bioregulators. Washington, DC: American Chemical Society; 1991.

Hoskins, W. M. Resistance to insecticides. International Review of Tropical Medicine 2:119–174; 1963.

Hoy, M. A.; Herzog, D.C., editors. Biological control in agricultural IPM systems. New York: Academic Press; 1985.

Hoy, M. A.; McKelvey, J. J., Jr. Genetics in relation to insect management. New York: The Rockefeller Foundation; 1979.

Huffaker, C. B. Biological control. New York: Plenum Publishing Corporation; 1971.

Huffaker, C. B.; Messenger, P. S. Theory and practice of biological control. New York: Academic Press; 1976.

Huffaker, C. B.; Luck, R. F.; Messenger, P. S. The ecological basis of biological control. In: Proceedings of the XV International Congress of Entomology. College Park, MD: The Entomological Society of America; 1977:560–586.

Ignoffo, C., editor. Microbial insecticides. Boca Raton, FL: CRC Press; 1988.

Jepson, P. C., editor. Pesticides and non-target invertebrates. New York: VCH Publishers, Inc.; 1990.

Johansen, C. A. Pesticides and pollinators. Annual Review of Entomology 22:177–192; 1977.

Khan, M. A. Q., editor. Pesticides in aquatic environments. New York: Plenum; 1977.

Knipling, E. F. Possibilities of insect control or eradication through the use of sexually sterile males. Journal of Economic Entomology 48:459–462; 1955.

Knipling, E. F. Screwworm eradication: concepts and research leading to the sterile male method. Smithsonian Institution Publication 4365; 1959.

Knipling, E. F. Sterilization and other genetic techniques. In: Pest control strategies of the future. Washington, DC: National Academy of Science; 1972:p. 272–287.

Knipling, E. F. The basic principles of insect population suppression and management. U.S. Dept. of Agriculture Handbook 512, Chapter 10; 1979.

Knipling, E. F. Present status and future trends of the SIT approach to the control of arthropod pests. In: Sterile insect techniques and radiation in insect control. Proceedings of the International Symposium on the Sterile Insect Technique and Use of Radiation in Genetic Insect Control, June 29 to July 3, 1981. Vienna, Austria: International Atomic Energy Agency; 1982:p. 3–23.

Kogan, M. The role of chemical factors in insect/plant relationships. In Proceedings of the XV International Congress of Entomology. College Park, MD: The Entomological Society of America; 1975:p. 211–227.

Kogan, M., editor. Ecological theory and integrated pest management practice. New York: John Wiley & Sons; 1986.

Krafsur, E. S.; Whitten, C. J.; Novy, J. E. Screwworm eradication in North and Central America. Parasitology Today 3:131–137; 1987.

Labrecque, G. C.; Smith, C. N., editors. Principles of insect chemosterilization. New York: Appleton-Century-Crofts; 1968.

Lorimer, N. Genetic means for controlling agricultural insects. In: Pimental, D., ed. CRC handbook of pest management in agriculture, 2. Boca Raton, FL: CRC Press, Inc.; 1981:p. 299–305.

Maddox, J. V. Use of diseases in insect pest management. In: Metcalf, R. L.; Luckmann, W. H., eds. Introduction to insect pest management. New York: John Wiley & Sons; 1975:p. 189–233.

Mallis Handbook of Pest Control. Cleveland, OH: PCT Books Department; 1990.

Maramorosch, K., editor. The atlas of insect and plant viruses. New York: Academic Press; 1977.

Matsumura, F. Toxicology of insecticides. 2nd ed. New York: Plenum; 1985.

Maxwell, F. G; Jenkins, J. N.; Parrott, W. S. Resistance of plants to insects. Advances in Agronomy 24:187–265; 1972.

Maxwell, F. G.; Jennings, P. R., editors. Breeding plants resistant to insects. New York: Wiley-Interscience; 1979.

Maxwell, F. G.; Jennings, P. R., editors. Breeding plants resistant to insects. New York; John Wiley & Sons; 1980.

Mayer, M. S.; McLaughlin, J. R. CRC handbook of insect pheromones and sex attractants. Boca Raton, FL: CRC Press, Inc.; 1990.

McEwen, F. L.; Stephenson, G. R. The use and significance of pesticides in the environment. New York: Wiley-Interscience; 1979.

Menn, J. J.; Beroza, M., editors. Insect juvenile hormones—chemistry and action. New York: Academic Press; 1972.

Menn, J. J.; Pallos, F. M. Development of morphogenetic agents in insect control. In: Jacobson, M., ed. Insecticides of the future. New York: Marcel Dekker, Inc.; 1975:p. 71–88.

Metcalf, G. L.; Flint, W. P.; Metcalf, R. L. Destructive and useful insects. New York:McGraw Book Company; 1962.

Metcalf, R. L.; Luckmann, W. H., editors. Introduction to insect pest management. New York: John Wiley & Sons; 1982.

Metcalf, R. L.; Metcalf, R. A. Attractants, repellents, and genetic control in pest management. In: Metcalf, R. L.; Luckmann, W. H., eds. Introduction to insect pest management. 2nd ed. New York: John Wiley & Sons; 1982:p. 299–314.

Miller, M. W.; Berg, G. C., editors. Chemical fallout. Springfield, IL: Charles C. Thomas; 1969.

Morgan, E. D.; Mandava, N. B. Insect growth regulators. Boca Raton, FL: CRC Press, Inc.; 1987.

Narahashi, T.; Chambers, J. E., editors. Insecticide action: from molecule to organism. New York: Plenum; 1989.

National Academy of Science. Insect-pest management and control. Washington, DC: Subcommittee on Insect Pests; 1969.

National Research Council. Pesticide resistance strategies and tactics for management. Washington, DC: National Academy Press; 1986.

New York Academy of Sciences. Regulation of insect populations by microorganisms. Conference on Regulation of Insect Populations by Microorganisms; 1972.

Nickle, W. R., editor. Plant and insect nematodes. New York: Marcel Dekker, Inc.; 1984.

Novy, J. E. Operation of a screwworm eradication program. In: Richardson, R. H., ed. The screwworm problem—evolution of resistance to biological control. Austin, TX: University of Texas Press; 1978.

Painter, R. H. Insect resistance in crop plants. New York: Macmillan; 1951.

Painter, R. H. Resistance of plants to insects. Annual Review of Entomology 3:267–290; 1958.

Pal, R.; Whitten, M. J., editors. The use of genetics in insect control. New York: Elsevier Scientific; 1974.

Pathak, M. D. Utilization of insect-plant interactions in pest control. In: Pimentel, D., ed. Insects, science and society. New York: Academic Press; 1975:p. 121–148.

Pedigo, L. P. Entomology and pest management. New York: Macmillan Publ. Co.; 1989.

Perring, F. H.; Mellanby, K., editors. Ecological effects of pesticides, No. 5 in the Linnean Society Symposium Series. New York: Academic Press; 1977.

Pimental, D., editor. CRC handbook of pest management in agriculture. 2nd ed. Boca Raton, FL: CRC Press, Inc.; 1990. 3 Vol.

Poinar, G. O.; Thomas, G. M. Diagnostic manual for the identification of insect pathogens. New York: Plenum Publishing Corporation; 1978.

Richardson, R. H. The screwworm problem—evolution of resistance to biological control. Austin, TX: University of Texas Press; 1978.

Riddiford, L. M. Juvenile hormone and insect embryonic development: its potential role as an ovicide. In: Menn, J. J.; Beroza, M., eds. Insect juvenile hormones—chemistry and action. New York: Academic Press; 1972:p. 95–111.

Ridgway, R. L.; Silverstein, R. M.; Inscoe, M. N. Behavior-modifying chemicals for insect management: applications of pheromones and other attractants. New York: Marcel Dekker, Inc.; 1991.

Ridgway, R. L.; Vinson, S. B., editors. Biological control by augmentation of natural enemies. New York: Plenum Publishing Corporation; 1977.

Rockstein, M., editor. Biochemistry of insects. New York: Academic Press; 1978.

Roelofs, W. L. Manipulating sex pheromones for insect suppression. In: Jacobson, M., ed. Insecticides of the future. New York: Marcel Dekker; 1975:p. 41–59.

Roelofs, W. L. Chemical control of insects by pheromones. In: Rockstein, M., ed. Biochemistry of insects. New York: Academic Press; 1978:p. 419–464.

Rolston, L. H.; McCoy, C. E. Introduction to applied entomology. New York: The Ronald Press Company; 1966.

Rousch, R. T.; Tabashnik, B. E. Pesticide resistance in arthropods. London: Chapman & Hall; 1990.

Rudd, R. L. Pesticides and the living landscape. Madison, WI: University of Wisconsin Press; 1964.

Schneiderman, H. A. Insect hormones and insect control. In: Menn, J. J.; Beroza, M., eds. Insect juvenile hormones—chemistry and action. New York: Academic Press; 1972:p. 3–27.

Scruggs, C. G. The origin of the screwworm control program. In: Richardson, R. H., ed. The screwworm problem—evolution of resistance to biological control. Austin, TX: University of Texas Press; 1978:p. 11–18.

Shepard, H. H. The chemistry and action of insecticides. New York: McGraw-Hill Book Company; 1951.

Sherman, J. L. Development of a systemic insect repellent. Journal of the American Medical Association 196(3):166–168; 1966.

Shorey, H. H.; McKelvey, J. J., Jr. Chemical control of insect behavior—theory and application. New York: John Wiley & Sons; 1977.

Singh, D. P. Breeding for resistance to diseases and insect pests. New York: Springer-Verlag; 1986.

Smith, C. N. Personal protection from blood sucking arthropods. Journal of the American Medical Association 196(3):146–149; 1966.

Smith, R. H.; von Borstell, R. C. Genetic control of insect populations. Science 178:1164–1174; 1972.

Sondheimer, E.; Simeone, J. B., editors. Chemical ecology. New York: Academic Press; 1970.

Stehr, F. W. Parasitoids and predators in pest management. In: Metcalf, R. L.; Luckmann, W. H., eds. Introduction to insect pest management. New York: John Wiley & Sons; 1975:p. 147–188.

Steinhaus, E. A., editor. Insect pathology, an advanced treatise. New York: Academic Press; 1963. 2 vol.

Stern, V. M. Interplanting alfalfa in cotton to control lygus bugs and other insect pests. Proceedings of the Tall Timbers Conference on Ecology, Animal Control, and Habitat Management 1:55–69; 1969.

Swan, L. A. Beneficial insects. New York: Harper & Row, Inc.; 1964.

Swain, R. B. How insects gain entry. In: Insects—the yearbook of agriculture. Washington, DC: U.S. Department of Agriculture; 1952:p. 350–355.

Sweetman, H. L. The biological control of insects. Ithaca, NY: Comstock Publishing Company, Inc.; 1936a.

Sweetman, H. L. The principles of biological control. Dubuque, IA: William C. Brown Company; 1936b.

Tabashniky, B. E; Croft, B.A. Managing pesticide resistance in crop-arthropod complexes: interactions between biological and operational factors. Environmental Entomology 11:1137–1144.

Tinsley, T. W. Viruses and the biological control of insect pests. BioScience 27(10):659–661; 1977.

van den Bosch, R. Biological control of insects by predators and parasites. In: Jacobson, M., ed. Insecticides of the future. New York: Marcel Dekker: 1975:p. 5–21.

van den Bosch, R. The pesticide conspiracy. Garden City, NY: Doubleday; 1978.

van den Bosch, R.; Messenger, P. S. Biological control. New York: Intext Educational Publishers; 1973.

van Emden, H. F. Plant insect relationships and pest control. World Reviews of Pest Control 5:115–123; 1966.

Voss, R. H.; Ferro, D. N.; Logan, J. A. Role of reproductive diapause in the population dynamics of the Colorado potato beetle in Western Massachusetts. Environmental Entomology 17(5):863–871; 1988.

Waage, J.; Greathead, D., editors. Insect parasitoids. New York: Academic Press; 1986.

Weiser, J. An atlas of insect diseases. Shannon, Ireland: Irish University Press; 1969.

Weiser, J. Biological control of vectors. New York: John Wiley & Sons; 1991.

White-Stevens, R., editor. Pesticides in the environment. Vol. 3. New York: Marcel Dekker; 1977.

Whitten, C. J. The sterile insect technique in the control of the screwworm. In: Sterile insect techniques and radiation in insect control. Proceedings of the International Symposium on the Sterile Insect Technique and Use of Radiation in Genetic Insect Control; 1981. June 29–July 3; Vienna, Austria. Internat. A.E.C. 1982:79–84.

Williams, C. M. Third-generation pesticides. Scientific American 217:13–17; 1967.

Williams, C. M. Hormonal interactions between plants and insects. In: Sondheimer, E.; Simeone, J. B., eds. Chemical ecology. New York: Academic Press; 1970.

Woodwell, G. M.; Malcolm, W. M.; Whittaker, R. H. A-bombs, bug bombs, and us. In: Shepard, P.; McKinley, D., eds. The subversive science—essays toward an ecology of man. Boston, MA: Houghton Mifflin; 1969:p. 230–241.

Woodwell, G. M.; Craig, P. P.; Johnson, H. A. DDT in the biosphere: where does it go? Science 174:1101–1107; 1971.

World Health Organization. The genetic control of insects. WHO Chronicle 21(12):517–524; 1967.

Worthing, C. R.; Hance, R. J., editors. The pesticide manual—a world compendium. 9th ed. Oxford: Blackwell Scientific Publications; 1991.

Wright, D. P., Jr. Anti-feeding compounds for insect control. In: New approaches to pest control and eradication, advances in chemistry series 41. Washington, DC: American Chemical Society; 1963.

CHAPTER 16

Frisbie, R. E.; Crawford, J. L.; Bonner, C. M.; Zalcom, F. G., editors. Implementing IPM in cotton. In: Frisbie, R. E.; Crawford, J. L.; Bonner, C. M.; Zalcom, F. G., eds. Integrated pest management systems and cotton production. New York: John Wiley & Sons; 1989.

Granados, R. R.; Corsaro, B. G. Baculovirus enhancing proteins and their implications for insect control. Vth International Colloquium on Invertebrate Pathology and Microbial Controls; 1990.

Logan, J. A.; Wollking, D. J.; Hoyt, S. L.; Tanigoshi, L. K. An analytical model for description of temperature dependent rate phenomena in arthropods. Environmental Entomology 5:1130–1140; 1976.

Pimentel, D., editor. CRC handbook of pest management in agriculture. 2nd ed. Boca Raton, FL: CRC Press, Inc.; 19.

Showers, W. B. Effect of diapause on the migration of the European corn borer into the southeastern United States. In: Rabb, R. L.; Kennedy, G. G., eds. Movement of highly mobile insects: concepts and methodology in research. Raleigh: North Carolina State University; 1979:p.420–430.

University of California, Integrated pest management for potatoes in the western United States. Oakland, CA: University of California Press; 1986: 3316.

Credits

LINE ART AND TEXT

Chapter 2

Fig. 2.5*a:* From V. B. Wigglesworth, "Transpiration Through the Cuticle of Insects" in *Journal of Experimental Biology,* 21:97–114, 1945. Copyright © 1945 Cambridge University Press. Reprinted by permission from The Company of Biologists, Ltd., Cambridge, England. Fig. 2.5*b:* From J. W. L. Beament, "The Effect of Temperature on the Waterproofing Mechanism of an Insect" in *Journal of Experimental Biology,* 35:494–519, 1958. Copyright © 1958 Cambridge University Press. Reprinted by permission from The Company of Biologists, Ltd., Cambridge, England. Fig. 2.7: Reprinted from V. B. Wigglesworth: *The Control of Growth and Form: A Study of the Epidermal Cell in an Insect.* Copyright © 1959 by Cornell University. Used by permission of the publisher, Cornell University Press. Figs. 2.8, 2.33, 2.34*a–d:* From R. E. Snodgrass, *Facts and Theories Concerning the Insect Head,* Misc. Coll. 142(1), publ. 4427. Copyright © 1960 Smithsonian Institution Press, Washington, DC. Reprinted by permission. Figs. 2.11, 2.13, 2.18*a,b,* 2.19, 2.20, 2.27*b,* 2.30*c,* 2.34*f,* 2.38, 2.44*a,b,d:* Reprinted from Robert E. Snodgrass, *Principles of Insect Morphology.* Copyright © 1993 Ellen Burden and Ruth Roach. Used by permission of the publisher, Cornell University Press. (in press). Figs. 2.15, 2.30*d,* 2.31*b:* Reprinted with the permission of Macmillan Publishing Company from *Entomology in Human and Animal Health,* Seventh Edition by Maurice T. James and Robert F. Harwood. Copyright © 1979 by Macmillan Publishing Company. Figs. 2.22, 2.23, 2.40: Source: J. H. Comstock, *The Wings of Insects,* 1918, Comstock/Cornell University Press, Ithaca, NY. Fig. 2.24: From E. M. DuPorte, *Manual of Insect Morphology.* Copyright © 1961 Van Nostrand Reinhold Company, New York, NY. Reprinted by permission. Fig. 2.26*a:* From R. E. Snodgrass, *The Dragonfly Larva,* Misc. Coll. 123(1), publ. 4175. Copyright © 1954 Smithsonian Institution Press, Washington, DC. Reprinted by permission. Figs. 2.26*b,* 2.29*a,* 2.39*a,c,d,* 2.41*b:* Reprinted with the permission of Macmillan Publishing Company from *Insects and Mites of Western North America* by E. O. Essig. Copyright © 1958 by Macmillan Publishing Company. Figs. 2.26*c,* 2.27*c,* 2.36*a–d,* 2.39*b,* 2.42*d,e:* Reprinted by permission from General Biological, Inc., Chicago, IL. Fig. 2.26*d:* Reprinted from J. H. Comstock: *An Introduction to Entomology,* Ninth Edition, Revised. Copyright 1933, 1936, 1940 by Comstock Publishing Company, Inc. Used by permission of the publisher, Cornell University Press. Figs. 2.26*f,g,* 2.36*e,* 2.37*b:* Source: J. W. Folsom and R. A. Wardle, *Entomology, with Special Reference to Its Ecological Effects,* 4th edition, 1934, McGraw-Hill Book Company, New York, NY. Figs. 2.27*a,* 2.31*a,* 2.42*c:* From R. E. Snodgrass, *The Caterpillar and the Butterfly,* Misc. Coll. 143(6), publ. 4472. Copyright © 1961 Smithsonian Institution Press, Washington, DC. Reprinted by permission. Fig. 2.29*c:* From Alvah Peterson, *Larvae of Insects, Part II, Coleoptera, Diptera, Neuroptera, Siphonaptera, Mecoptera, Trichoptera.* Copyright © 1951 Alvah Peterson. Reprinted by permission of the author. Figs. 2.30*b,* 2.32*c,* 2.34*e:* From R. E. Snodgrass, *The Anatomical Life of the Mosquito,* Misc. Coll. 138(8). Copyright © 1959 Smithsonian Institution Press, Washington, DC. Reprinted by permission. Fig 2.31*c:* Reprinted with the permission of Macmillan Publishing Company from *Medical Entomology,* Fifth Edition by William M. Herms, Herbert P. Herms, George W. Herms and Maurice T. James. Copyright © 1961 by Macmillan Publishing Company. Figs. 2.32*a,* 2.43*c:* From R. E. Snodgrass, *The Dragonfly Larva,* Misc. Coll. 123(2), publ. 4175. Copyright © 1954 Smithsonian Institution Press, Washington, DC. Reprinted by permission. Fig. 2.32*b,e:* From Alvah Peterson, *Larvae of Insects, Part I, Lepidoptera and Plant Infesting Hymenoptera.* Copyright © 1948 Alvah Peterson. Reprinted by permission of the author. Fig. 2.32*d:* From C. L. Metcalf, et al., *Destructive and Useful Insects.* Copyright © 1962 McGraw-Hill Book Company, New York, NY. Figs. 2.35*a,* 2.43*b:* Source: H. H. Ross, *How to Collect and Preserve Insects,* Ill. Nat. Hist. Surv. Circ. 39, 1962. Fig. 2.35*c:* From *An Introduction to the Study of Insects,* third edition, by Donald J. Borror and Dwight M. DeLong. Copyright © 1964, 1971 by Holt, Rinehart and Winston, Inc. Copyright 1954 by Donald J. Borror and Dwight M. DeLong.

Fig. 2.35*d:* From E. O. Essig, *College Entomology.* Copyright © Macmillan Publishing Company, New York, NY. Reprinted by permission. Fig. 2.37*a:* Reprinted from R. E. Snodgrass: *Anatomy of the Honeybee.* Copyright © 1956 by Cornell University. Used by permission of the publisher, Cornell University Press. Fig. 2.39*e:* From J. Smart, "Notes on the Mesothoracic Musculature of Diptera" in *Studies in Invertebrate Morphology,* Misc. Coll. 157, pages 331–364, 1959. Copyright © 1959 Smithsonian Institution Press, Washington, DC. Reprinted by permission. Fig. 2.41*c:* From S. H. Skaiffe, *Dwellers in Darkness.* Copyright © 1961 Longman Group, Ltd., London, England. Reprinted by permission. Fig. 2.43*a:* Source: M. Hebard, *The Dermaptera and Orthoptera of Illinois,* 1934, Ill. Nat. Survey, Bull. 20(3). Fig. 2.44*e:* Reprinted from L. S. West, *The Housefly: Its Natural History, Medical Importance, and Control.* Copyright 1951 by Comstock Publishing Company, Inc. Used by permission of the publisher, Cornell University Press.

Chapter 3

Fig. 3.2: From D. S. Smith and J. E. Treherne, "Functional Aspects of the Organization of the Insect Nervous System" in *Advances in Insect Physiology,* I:401–485, 1963. Copyright © 1963 Academic Press Inc., London, England. Reprinted by permission. Figs. 3.6, 3.20: Reprinted from Robert E. Snodgrass, *Principles of Insect Morphology.* Copyright © 1993 Ellen Burden and Ruth Roach. Used by permission of the publisher, Cornell University Press. (in press). Fig. 3.7: From R. F. Chapman, *The Insects—Structure and Function,* 2d ed. Copyright © 1971 Elsevier North Holland, Inc., New York, NY. Reprinted by permission. Fig. 3.8*a,b:* From T. A. Christensen and J. G. Hildebrand, "Male-Specific, Sex Pheromone-Selective Projection Neurons in the Antennal Lobes of the Moth Manduca Sexta" in *Journal of Comparative Physiology A,* 160:553–569, 1987. Copyright © 1987 Springer-Verlag, Berlin, Germany. Fig. 3.9*a:* "Reprinted from F. Engelman, *The Physiology of Insect Reproduction,* Copyright 1970, with permission from Pergamon Press Ltd., Headington Hill Hall, Oxford OX3 Ø BW, UK." Fig. 3.10: Reprinted by permission of the publishers from *The Hungry Fly: A Physiological Study of the Behavior Associated with Feeding* by V. G. Dethier, Cambridge, Mass.: Harvard University Press, Copyright © 1976 by the President and Fellows of Harvard College. Reprinted by permission. Fig. 3.11: From P. B. Cornwell, *The Cockroach,* Volume I, 1968. Copyright © 1968 Rentokil Limited, West Sussex, England. Reprinted by permission. Table 3.1: From H. Breer, "Neurochemical Aspects of Cholinergic Synapses in the Insect Brain" in *Arthropod Brain,* edited by A. P. Gupta. Copyright © 1987 John Wiley & Sons, Inc., New York, NY. Reprinted by permission of John Wiley & Sons, Inc. Fig. 3.16: "Reprinted from *Comprehensive Insect Physiology, and Pharmacology,* Volume II, B. J. Cook and G. M. Holman, "Peptides and Kinins," Pages 531–593, Copyright 1985, with permission from Pergamon Press Ltd., Headington Hill Hall, Oxford OX3 ØBW, UK." Fig. 3.18: From A. Glenn Richards, *The Integument of Arthropods.* Copyright © 1951 The University of Minnesota Press, Minneapolis, MN. Reprinted by permission. Fig. 3.19: From S. Caveney, "Muscle Attachment Related to Cuticle Architecture in Apterygota" in *J. Cell Science,* 4:541–59, 1969. Copyright © 1969 Cambridge University Press, New York, NY. Reprinted by permission. Fig. 3.21: From R. E. Snodgrass, "The Anatomy of the Honey Bee" in *The Hive and the Honey Bee,* R. A. Grout, editor. Copyright © 1963 Dadant and Sons, Inc., Hamilton, IL. Reprinted by permission. Fig. 3.23*a:* Reprinted with the permission of Macmillan Publishing Company from *Biology,* Second Edition by C. F. Herreid. Copyright © 1977 by Macmillan Publishing Company. Fig. 3.25: From G. Hoyle, "Neural Control of Skeletal Muscles" in *The Physiology of Insecta,* 2:407–49, 1965. Copyright © 1965 Academic Press, Orlando, FL. Reprinted by permission.

Chapter 4

Fig. 4.3: From M. J. Berridge, *A Structural Analysis of Intestinal Absorption* in *Insect Ultrastructure,* A. C. Neville, Editor, Symposia of the Royal Entomological Society of London, No. 5. Copyright © 1970 The Royal Entomological Society. Reprinted by permission of Blackwell Scientific Publications Limited. Fig. 4.4: From W. R. Terra, "Evolution of Digestive Systems of Insects" in *Annual Review of Entomology,* 35:181–200, 1990. Copyright © 1990 Annual Reviews, Inc., Palo Alto, CA. Reprinted by permission of the publisher and author. Fig. 4.5: From J. A. T. Dow, "Insect Midgut Function" in *Advances in Insect Physiology,* 19:187–328, 1986. Copyright © 1986 Academic Press Ltd., London, England. With permission of Academic Press Ltd., London. Fig. 4.7*b:* Reprinted from L. S. West, *The Housefly: Its Natural History, Medical Importance, and Control.* Copyright 1951 by Comstock Publishing Company, Inc. Used by permission of the publisher, Cornell University Press. Table 4.1: From S. B. Vinson, "How Parasitoids Deal with the Immune System of Their Hosts: An Overview" in *Archives of Insect Biochemistry and Physiology,* 13:3–27, 1990. Copyright © 1990 John Wiley & Sons, Inc., New York, NY. Reprinted by permission of John Wiley & Sons, Inc. Fig. 4.10*a:* "Reprinted from *Comprehensive Insect Physiology, Biochemistry and Pharmacology,* Volume 3, A. C. Crossley, "Nephrocytes and Pericardial Cells," Pages 487–515, Copyright 1985, with permission from Pergamon Press Ltd., Headington Hill Hall, Oxford OX3 ØBW, UK." Figs. 4.12. 4.15*a:* From E. O. Essig, *College Entomology.* Copyright © Macmillan Publishing Company, New York, NY. Reprinted by permission. Fig. 4.13*a,d:* From R. E. Snodgrass, "The Anatomy of the Honey Bee" in *The Hive and the Honey Bee,* R. A. Grout, editor. Copyright © 1963 Dadant and Sons, Inc., Hamilton, IL. Reprinted by permission. Fig. 4.13*b:* From P. L. Miller, "Respiration—Aerial Gas Transport" in *The Physiology of Insecta,* 3:557–615, 1964. Copyright © 1964 Academic Press, Orlando, FL. Reprinted by permission. Fig. 4.13*c:* From V. B. Wigglesworth, "A Theory of Tracheal Respiration in Insects" in *Proc. Roy. Soc. Lond. B.,* 106:229–250, 1930. Reprinted by permission of the author. Figs. 4.14, 4.15*b:* Reprinted from Robert E. Snodgrass, *Principles of Insect Morphology.* Copyright © 1993 Ellen Burden and Ruth Roach. Used by permission of the publisher, Cornell University Press. (in press). Fig. 4.17: From H. A. Schneiderman and C. M. Williams, "An Experimental Analysis of the Discontinuous Respiration of the Cecropia Silkworm" in *The Biological Bulletin,* 109:123–143, 1955. Copyright © 1955 Marien Biological Laboratory, Woods

Hole, MA. Reprinted by permission. Fig. 4.18*a,b:* Reprinted by permission from General Biological, Inc., Chicago, IL. Fig. 4.19: Reprinted from Robert Matheson: *Handbook of the Mosquitos of North America,* Second Edition. Copyright © 1944 by Cornell University. Used by permission of the publisher, Cornell University Press. Fig. 4.21: From R. L. Patton, *Introductory Insect Physiology.* Copyright © 1963 W. B. Saunders Company, Orlando, FL. Reprinted by permission. Fig. 4.22, Table 4.5: From C. Gillott, *Entomology.* Copyright © 1980 Plenum Publishing Corporation, New York, NY. Reprinted by permission of the publisher and author.

Chapter 5

Figs. 5.1, 5.2, 5.6, 5.15, 5.16: Reprinted from Robert E. Snodgrass, *Principles of Insect Morphology.* Copyright © 1993 Ellen Burden and Ruth Roach. Used by permission of the publisher, Cornell University Press. (in press). Fig. 5.4: From O. P. Breland, et al., "Studies of Insect Spermatozoa I" in *Entomol. News,* 79(8):197–216, 1968. Copyright © 1968 The American Entomological Society, Philadelphia, PA. Reprinted by permission. Fig. 5.5: From von H. Sturm, "Die Paarung beim Silberfischchen, *Lepisma saccharina,*" in *Zeitschrift fur Tierpsychologie,* 13:1–12, 1956. Copyright © 1956 Paul Parey Publisher, Lindenstrasse, Germany. Reprinted by permission. Fig. 5.10*a:* From H. R. Hagen, *Embryology of the Viviparous Insects.* Copyright © 1951 The Ronald Press Company, New York, NY. Reprinted by permission. Figs. 5.12, 5.18, 5.19: Source: O. A. Johannsen and F. H. Butt, *Embryology of Insects and Myriapods,* 1941, McGraw Hill Book Company, New York, NY. Fig. 5.13: From M. Raabe, "Insect Reproduction: Regulation of Successive Steps" in *Advances in Insect Physiology,* 19:29–154, 1986. Copyright © 1986 Academic Press, Ltd., London, England. With permission of Academic Press Ltd., London. Figs. 5.22*c,* 5.23*c,e,f:* Reprinted with the permission of Macmillan Publishing Company from *Insects and Mites of Western North America* by E. O. Essig. Copyright © 1958 by Macmillan Publishing Company. Fig. 5.24*a:* Reprinted by permission from General Biological, Inc., Chicago, IL. Fig. 5.24*c,d:* From V. B. Wigglesworth, *Insect Hormones.* Copyright © 1970 Oliver and Boyd, Edinburgh, Scotland. Reprinted by permission. Figs. 5.25, 5.26: From H. Urspring and R. Nothiger, *The Biology of Imaginal Discs.* Copyright © 1972 Springer-Verlag Publishing Company, New York, NY. Reprinted by permission. Fig. 5.27: From L. I. Gilbert, "Physiology of Growth and Development: Endocrine Aspects" in *The Physiology of Insecta,* 1:149–225, 1964. Copyright © 1964 Academic Press, Orlando, FL. Reprinted by permission. Table 5.1: From A. M. Clark and M. Rockstein, "Aging in Insects" in *The Physiology of Insecta,* Volume I, pages 227–281, edited by M. Rockstein. Copyright © 1964 Academic Press, Inc., Orlando, FL. Reprinted by permission of the publisher and authors.

Chapter 6

Fig. 6.3: From W. N. Hess, "The Chordotonal Organs and Pleural Discs of Cerambycid Larvae" in *Ann. Entomol. Soc. Amer.,* 10:63–78, 1917. Copyright © 1917 Entomological Society of America, College Park, MD. Reprinted by

permission. Fig. 6.4: From H. Markl, "Schweresinnersorgane bei Ameisen und anderen Hymenopteren" in *Z. verfl. Physiol.,* 44:475–596, 1962. Copyright © 1962 Springer-Verlag, Berlin, Germany. Fig. 6.5*a:* "Reprinted from *International Journal of Insect Morphology and Embryology,* Volume 16, J. G. Stoffolano, Jr., and L. R. S. Yin, "Structure and Function of the Ovipositor and Associated Sensilla of the Apple Maggot, Rhagoletis Pomenella (Walsh) (Diptera:) (Tephritiade)," Pages 41–69, Copyright 1987, with permission from Pergamon Press Ltd., Headington Hill Hall, Oxford OX3 ØBW UK." Fig. 6.7: From K. D. Roeder, "A Physiological Approach to the Relation Between Prey and Predator" in *Studies in Invertebrate Morphology,* Misc. Coll. 137, pages 287–306. Copyright © 1959 Smithsonian Institution Press, Washington, DC. Reprinted by permission. Fig. 6.14: From M. Hertz, "Die Organization des Optischen Feldes bei der Biene, I" in *Z. vergl. Physiology,* 8:693–748, 1929. Copyright © Springer-Verlag, Berlin, Germany. Fig. 6.15: From K. Baldus, "Experimentelle Untersuchungen uber die Entfernungslokalisation der Libellen (Aeschna cyanea)" in *Z. vergl. Physiol.,* 3:375–505, 1926. Copyright © Springer-Verlag, Berlin, Germany. Fig. 6.16: Reprinted from Karl von Frisch: *Bees: Their Vision, Chemical Senses, and Language.* Copyright © 1950 by Cornell University. Used by permission of the publisher, Cornell University Press. Fig. 6.18*a:* From E. O. Essig, *College Entomology.* Copyright © Macmillan Publishing Company, New York, NY. Reprinted by permission. Fig. 6.18*b:* Reprinted from J. H. Comstock: *An Introduction to Entomology,* Ninth Edition, Revised. Copyright 1933, 1936, 1940 by Comstock Publishing Company, Inc. Used by permission of the publisher, Cornell University Press.

Chapter 7

Figs. 7.1, 7.9, 7.10: From R. E. Snodgrass, *Principles of Insect Morphology.* Copyright © 1935 McGraw-Hill, Inc., New York, NY. Fig. 7.3: From J. D. Gillett and V. B. Wigglesworth, "The Climbing Organ of an Insect, *Rhodnius Prolixus* (Hemiptera: Reduviidae)," in *Proc. Roy. Soc. Lond. B.,* 111:364–376, 1932. Copyright © 1932 The Royal Society, London, England. Reprinted by permission. Fig. 7.5: From R. E. Snodgrass, *The Caterpillar and the Butterfly,* Misc. Coll. 143(6), publ. 4472. Copyright © 1961 Smithsonian Institution Press, Washington, DC. Reprinted by permission. Fig. 7.7: From Werner Nachtigall, "Uber Kinematik, Dynamik und Energetik des Schwimmens Einheimischer Dytisciden" in *Z. vergl. Physiol.,* 43:48–118, 1960. Copyright © 1960 Springer-Verlag, Berlin, Germany. Reprinted by permission of the author. Fig. 7.8: From Werner Nachtigall, "Zur Lokomotionsmechanik Schwimmender Dipterenlarven" in *Z. vergl. Physiol.,* 46:449–466, 1963. Copyright © 1963 Springer-Verlag, Berlin, Germany. Reprinted by permission of the author.

Chapter 8

Fig. 8.1: From R. Jander, "Insect Orientation" in *Ann. Rev. Entomol.,* 8:94–114, 1963. Copyright © 1963 Annual Reviews, Inc., Palo Alto, CA. Reprinted by permission. Fig. 8.2: From J. Truman, "How Moths 'turn on': A Study of the

Action of Hormones on the Nervous System" in *American Naturalist,* 61:700–706, 1973. Copyright © 1973 University of Chicago Press, Chicago, IL. Reprinted by permission. Fig. 8.3: From W. C. Rothenbuhler, "Behavior Genetics of Next Cleaning in Honey Bees IV. Response of F1 and Backcross Generations to Disease-Killed Brood" in *Amer. Zool.,* 4:111–23, 1964. Copyright © 1964 American Society of Zoologists, Thousand Oaks, CA. Reprinted by permission. Fig. 8.5: "Reproduced, with permission, from the *Annual Review of Entomology* Volume 16. © 1971 by Annual Reviews Inc." Fig. 8.6: From H. Schildknecht and K. Holoubek, "Die Bombardierkafer und ihre Expolsionschemis V. Mitteilung uber insekten. Abwehrstoffe" in *Angew. Chem.,* 73:(1):1–6, 1961. Copyright © 1961 Verlag Chemie GmbH. Reprinted by permission. Fig. 8.7: From von H. Sturm, "Die Paarung beim Silberfischehen, Lepisma saccharina," in *Zeitschrift fur Tierpsychologie,* 13:1–12, 1956. Copyright © 1956 Paul Parey Publisher, Lindenstrasse, Germany. Reprinted by permission. Table 8.2: Reprinted by permission of the publishers from *The Insect Societies* by Edward O. Wilson, Cambridge, Mass.: The Belknap Press of Harvard University Press, Copyright © 1971 by the President and Fellows of Harvard College. Reprinted by permission.

Chapter 9

Fig. 9.1: From L. R. Clark, et al., *Ecology of Insect Populations in Theory and Practice.* Copyright © 1967 Associated Book Publishers Ltd., London, England. Reprinted by permission. Fig. 9.3: From L. H. Rolston and L. E. McCoy, *Introduction to Applied Entomology.* Copyright 1966 The Ronald Press Company, New York, NY. Reprinted by permission. Fig. 9.5: From R. Chauvin, *The World of an Insect.* Copyright © 1967 Weidenfeld & Nicholson Publishing Co., Ltd., London, England. Reprinted by permission of Weidenfeld & Nicholoson Publishing Co., Ltd., and McGraw-Hill Book Company. Fig. 9.6: From H. G. Andrewartha and L. C. Birch, *The Distribution and Abundance of Animals.* Copyright © 1954 University of Chicago Press, Chicago, IL. Reprinted by permission. Fig. 9.7: From S. A. Graham and F. B. Knight, *Principles of Forest Entomology,* 4th ed. Copyright © 1965 McGraw-Hill Book Company, New York, NY. Reprinted by permission of McGraw-Hill, Inc.

Chapter 10

Fig. 10.2: Data from Southwood 1978 and Price 1977 as appeared in D. R. Strong, et al., *Insects on Plants-Community Patterns and Mechanisms.* Copyright © 1984 Harvard University Press, Cambridge, MA. Reprinted by permission. Fig. 10.3: From "Effects of Plant Density and Diversity on the Population Dynamics of a Specialist Herbivore, the Striped Cucumber Beetle, Acalymma Vittata (Fab.)" by C. E. Bach, *Ecology,* 1980, Ecological Society of America. Reprinted by permission. Fig. 10.5: Data from Moran and Southwood 1982 as appeared in D. R. Strong, et al., *Insects on Plants.* Copyright © 1984 Harvard University Press, Cambridge, MA. Reprinted by permission. Fig. 10.6: From J. R. G. Turner, "Adaptation and Evolution in Heliconius: A Defense of neoDarwinism" in *Annual Review*

of *Ecology and Systematics,* 12:99–121, 1981. Copyright © 1981 Annual Reviews, Inc., Palo Alto, CA. Figs. 10.7, 10.8*d:* From W. G. Abrahamson, *Plant-Animal Interactions.* Copyright © 1989 McGraw-Hill, Inc., New York, NY. Reprinted by permission of McGraw-Hill, Inc. Fig. 10.8*a:* From E. O. Essig, *Insects and Mites of Western North America.* Copyright © 1958 Macmillan Publishing Company, New York, NY. Fig. 10.8*b:* Reprinted by permission from General Biological, Inc., Chicago, IL. Fig. 10.8*c:* From "Habitat Selection by Pemphigus Aphids in Response to Resource Limitation and Competition" by T. G. Whitman, *Ecology,* 1978, 59, 1164–1176. Copyright © 1978 by the Ecological Society of America. Reprinted by permission. Fig. 10.10*a:* Source: W. Mattson, "Herbivory in Relation to Plant Nitrogen Content" in *Annual Review of Ecology and Systematics,* 11:119–162, 1980, Annual Reviews Inc., Palo Alto, CA. Fig. 10.10*b:* From P. A. Morrow and L. A. Fox, "Effects of Variation in *Eucalyptus* Essential Oil Yield on Insect Growth and Herbivore Load" in *Oecologia,* 45:209–219, 1980. Copyright © 1980 Springer-Verlag New York, Inc., New York, NY. Reprinted by permission. Fig. 10.12: From J. E. Graebe, "Gibberellin Biosynthesis and Control" in *Annual Review of Plant Physiology,* page 425, 1987. Copyright © 1987 Annual Reviews, Inc., Palo Alto, CA. Reprinted by permission.

Chapter 11

Fig. 11.3: Reprinted with the permission of Macmillan Publishing Company from *College Zoology,* Eighth Edition by R. A. Stiles, R. W. Hegner and R. Boolootian. Copyright © 1969 by Macmillan Publishing Company. Figs. 11.7, 11.12: From H. H. Ross, *A Textbook of Entomology,* 3d ed. Copyright © 1965 John Wiley & Sons, Inc., New York, NY. Reprinted by permission of John Wiley & Sons, Inc. Fig. 11.8: Reprinted by permission from General Biological, Inc., Chicago, IL. Fig. 11.13: From J. G. Rempel, "The Evolution of the Insect Head: The Endless Dispute" in *Quaestiones Entomologicae,* 11:7–25, 1975. Copyright © 1975 Quaestiones Entomologicae, Edmonton, Alberta, Canada. Reprinted by permission.

Chapter 12

Fig. 12.4*a:* From H. E. Ewing, "The Protura of North America" in *Ann. Entomol. Soc. Amer.,* 33:495–551, 1940. Copyright © 1940 The Entomological Society of America, College Park, MD. Reprinted by permission. Figs. 12.4*b,* 12.11*a,b,* 12.12, 12.14, 12.15, 12.16*a,b,* 12.17, 12.18*a–e,* 12.21, 12.23, 12.26*b,* 12.27: From E. O. Essig, *College Entomology.* Copyright © Macmillan Publishing Company, New York, NY. Reprinted by permission. From E. O. Essig, *College Entomology.* Copyright © Macmillan Publishing Company, New York, NY. Reprinted by permission. Fig. 12.5: Reprinted from Elliot A. Maynard: *A Monograph of the Collembola or Springtail Insects of New York State.* Copyright © 1951 by Comstock Publishing Company, Inc. Used by permission of the publisher, Cornell University Press. Figs. 12.8, 12.9: From P. A. Garman, *The Odonata of Connecticut.* Connecticut State Geological and Natural History Survey, Bulletin 39, 1927. Reprinted by permission. Fig. 12.10: Source: B. D. Burks, *The Mayflies, or*

ILLUSTRATIONS

Rolin Graphics

3.8*a,b,* 3.9*b,* 3.10, 3.16, 4.4, 4.10*a,* 4.22, 5.9*c,* 5.13, 5.25, 5.26, 6.5*a,* 10.1, 10.3, 10.7, 10.8*a–d,* 10.9, 10.14, 10.15, 10.16, 12.16*c,d,* 12.24*c,* 12.25*e,* 12.33, 13.8*g,h,* 13.21, 15.4, 15.18*a–c,* 15.19, 16.7

Index

S

Sacrifice, 241
Salicylic acid, 280
Salivarium, 23, 91
Salivary glands, 68, 91–92
Salix alba, 272
Salt, and water balance, 120–22
Saltortorial legs, 41, 42
Sample, and description, 295
Sample size, and IPM, 434
Sampling, 247
    and IPM, 433
Sand flies, 369
    survey of, 370
San Jose scale, 424
Sap beetle, survey of, 359
Saprophagous insects, 220
Sarchophaga, 154
Sarcolemma, 74
Sarcophagidae, 368
    survey of, 373
Sarcoplasm, 77–78
Sarcoplasmic reticula, 77–78
Sassafras, 270
Saturnia pavonia, survey of, 380, 381
Saturniidae, survey of, 380–81
Satyridae, 13
Sawflies, 275
    survey of, 382–86
Scale insects, 67
Scape, 22
Scarabaeidae, survey of, 359, 360
Scarabaeiform larval type, 154
Scarab beetles, survey of, 359, 360
Scatophaga stercoraria, 236
Scelionidae, survey of, 384
Scent glands, 67
Schistocerca, 171, 202
Schistocerca gregaria, 118, 162, 220, 224, 229
Schizophora, survey of, 372
Science citation index, 5
Scientific names, 299
Sclerites, 18–19
Sclerotins, 12, 277
Sclerotization, 12
Sclerotized, 9
Scolioidea, survey of, 385
Scolopendra heros, 303
Scolophores, 169
Scolopidia, 169
Scolopophores, 172
Scolopophorous organ, 169
Scolytidae, survey of, 361
Scolytus multistraitus, 261
Scolytus scolytus, 275
Scorpion, 301, 302
Scorpion flies, 234
    survey of, 366–67
Scotopic eye, 181, 182
Scotoscutellar suture, 25
Screwworm flies, 236, 370
    survey of, 373
Scudderia, 50

Scudderia furcata, 331
Scutellum, 25
Scutigera coleoptrata, 303
Secondary compounds, 280
Secondary dorsal organ, 145
Secondary ecological events, 248–49
Secondary orientation, 206–7
Secondary plant substances, 223
Secondary segmentation, 19
Secondary spermatocytes, 128
Secretogogue mechanisms, 91
Seed beetles, survey of, 359, 360–61
Seed bugs, 349
Seed chalcids, survey of, 384
Seed-eaters, 277
Segmental muscles, 19
Segmentation, 18–19, 141–43, 300
Segmentation genes, 146
Segmented tarsus, 28
Segmented worms, 300
Self-pollination, 287
Sematectonic communication, 218
Semen, 135
Seminal transfer, 135–36
Seminal vesicle, 126
Semiochemicals, 420
Semisocial behavior, 239
Senescence, 163–65
Sense organs
    methods used to study, 169–70
    morphology of, 167
Sensilla, 62, 63, 167–69, 203, 233
    transformation of stimuli in, 63–65
Sensilla ampullacea, 168, 169
Sensilla basiconica, 167, 168
Sensilla campaniformia, 168, 169
Sensilla chaetica, 167, 168
Sensilla coeloconica, 168, 169
Sensilla placoedea, 169
Sensilla squamiformia, 167, 168
Sensilla trichoidea, 167
Sensory adaptation, 65
Sensory coding, 178
Sensory fields, 170
Sensory mechanisms, 167–70
Sensory neurons, 55–56
Sentry trap, 420
Septicemia, 408
Sequential sampling, and IPM, 435
Sericulture, 390
Series, and description, 295
Series on entomology, 3–4
Serosa, 140, 142
Serrate antennal type, 31
Serratia, 259
Serritermitidae, survey of, 338
Sesamin, 417
Sesamolin, 417
Sesidae, survey of, 381
Sesiidae, 379

Setaceous antennal type, 31
Setae, 17–18
Sex chromosomes, 137–38
Sex determination, 135, 137–39
Sex pheromones, 232
Sex ratio, 247
Sexual dimorphism, 30, 161
Sheep ked, survey of, 368, 372
Shock-avoidance conditioning, 210
Shock-avoidance learning, 210
Short-horned grasshoppers, survey of, 332
Short-term memory, 211
Shrimp, 302
Sialidae, 154, 363
Sialis, 154
Sialis mohri, 363
Sibling species, 295
Sieve plates, 110
Sign stimuli, 223
Silent Spring, 3, 423
Silk, 92, 389–91
    and building, 67
Silkworm moths, 81, 82, 159, 170, 177, 206, 211, 212, 230, 232, 233, 379, 407
    survey of, 381
Silphidae, survey of, 359, 360
Silurian, 309
Silverfish, 48, 119, 128, 152, 234, 235
Simple eyes, 20
Simple groups, 238
Simuliidae, 368, 369
    survey of, 370
Simulium, 32
Simulium columbaschense, 399
Simulium venustum, 369
Sinuses, of heart, 95–96
Siphlonuridae, survey of, 325
Siphonaptera, 67, 131
    relative size within Insecta, 314
    survey of, 366, 373–75
Siphoning, 34
Siricidae, survey of, 383
Siricoidea, survey of, 384
Sisyridae, 364
Skeletal muscles, 73–81
    groups of, 74–76
    structure of, 77–78
Skeleton, 9, 18–19
    hydrostatic skeleton, 98–99
Skeletons, 195
Skippers, survey of, 377–82
Slug caterpillar moth, survey of, 381
Smell, 59
Sminthurus viridis, 317
Snakeflies, 363
Snipe flies, 368
Snout beetles, survey of, 359
Snow scorpionflies, survey of, 366
Social behavior, 238–41
Social biology, 238–41

Social parasitism, 223
Sociobiology, 241
Sociotomy, 337
Sodium, 120–22
    and hemolymph, 99
Soil
    pH, 404
    temperature, 433
Solanum tuberosum, 440
Soldiers, 239, 335
Solenopsis, 400
Solenopsis molesta, survey of, 385
Solidago, 282
Soma, 53
Somatic layer, 143–44
Somatogastric nervous system, 60
Somites, 18, 307
Sorbitol, 251
Sound, 218
    and sound production, 188–90
    and sound sense, 170, 173–75
Souvenirs Entomologiques, 233
Sowbugs, 302, 303
Spanish fly, 391
Spatial summation, 65–66
Speciation, 293, 300
Specidae, survey of, 386
Specid wasps, survey of, 386
Species
    defined, 298
    host-specific, 272–73
    and intraspecies communication, 67
    intraspecific and interspecific, 218–19
    and morphogeographical species, 298
    placed in hierarchy, 298
    and scientific names, 299
    and sibling species, 295
    type concept of, 298
Specific epithet, 299
Specimens, and type specimens, 294
Spectra, humans and honey bees, 185
Spectrometry, 68
Spermatheca, 129, 135
Spermathecal reproductive glands, 68
Spermatids, 128
Spermatodesms, 128
Spermatogenesis, 127–28
Spermatogonia, 127
Spermatophores, 128, 135, 136
Spermatozoa, 125–28, 135
Spermatozoon, 138
Sperm competition, 136
Sperm precedence, 136
Sphecidae, survey of, 385
Sphecoidea, survey of, 385
Sphingidae, survey of, 380
Sphinx moths, 101, 203, 231
    survey of, 380
Spiders, 302